D1163676

THIRD EDITION

FIBER-REINFORCED COMPOSITES

Materials, Manufacturing, and Design

THIRD EDITION

FIBER-REINFORCED COMPOSITES

Materials, Manufacturing, and Design

P.K. Mallick
Department of Mechanical Engineering
University of Michigan-Dearborn
Dearborn, Michigan

CRC Press
Taylor & Francis Group
Boca Raton London New York

CRC Press is an imprint of the
Taylor & Francis Group, an **informa** business

CRC Press
Taylor & Francis Group
6000 Broken Sound Parkway NW, Suite 300
Boca Raton, FL 33487-2742

Printed in the United States of America on acid-free paper
10 9 8 7 6 5 4 3

International Standard Book Number-13: 978-0-8493-4205-9 (Hardcover)

Library of Congress Cataloging-in-Publication Data

Mallick, P.K., 1946-
Fiber-reinforced composites : materials, manufacturing, and design / P.K. Mallick.
-- 3rd ed.
p. cm.
Includes bibliographical references and index.
ISBN-13: 978-0-8493-4205-9 (alk. paper)
ISBN-10: 0-8493-4205-8 (alk. paper)
1. Fibrous composites. I. Title.

TA418.9.C6M28 2007
620.1'18--dc22 2007019619

Visit the Taylor & Francis Web site at
http://www.taylorandfrancis.com

and the CRC Press Web site at
http://www.crcpress.com

To
my parents

Contents

Appendixes

Preface to the Third Edition

Almost a decade has gone by since the second edition of this book was published. The fundamental understanding of fiber reinforcement has not changed, but many new advancements have taken place in the materials area, especially after the discovery of carbon nanotubes in 1991. There has also been increasing applications of composite materials, which started mainly in the aerospace industry in the 1950s, but now can be seen in many nonaerospace industries, including consumer goods, automotive, power transmission, and biomedical. It is now becoming a part of the "regular" materials vocabulary.

The third edition is written to update the book with recent advancements and applications.

Almost all the chapters in the book have been extended with new information, example problems and chapter-end problems. Chapter 1 has been rewritten to show the increasing range of applications of fiber-reinforced polymers and emphasize the material selection process. Chapter 2 has two new sections, one on natural fibers and the other on fiber architecture. Chapter 7 has a new section on carbon–carbon composites. Chapter 8 has been added to introduce polymer-based nanocomposites, which are the most recent addition to the composite family and are receiving great attention from both researchers as well as potential users.

As before, I have tried to maintain a balance between materials, mechanics, processing and design of fiber-reinforced composites. This book is best-suited for senior-level undergraduate or first-level graduate students, who I believe will be able to acquire a broad knowledge on composite materials from this book. Numerous example problems and chapter-end problems will help them better understand and apply the concepts to practical solutions. Numerous references cited in the book will help them find additional research information and go deeper into topics that are of interest to them.

I would like to thank the students, faculty and others who have used the earlier editions of this book in the past. I have received suggestions and encouragement from several faculty on writing the third edition—thanks to them. Finally, the editorial and production staff of the CRC Press needs to be acknowledged for their work and patience—thanks to them also.

P.K. Mallick

Author

P.K. Mallick is a professor in the Department of Mechanical Engineering and the director of Interdisciplinary Programs at the University of Michigan-Dearborn. He is also the director of the Center for Lightweighting Automotive Materials and Processing at the University. His areas of research interest are mechanical properties, design considerations, and manufacturing process development of polymers, polymer matrix composites, and lightweight alloys. He has published more than 100 technical articles on these topics, and also authored or coauthored several books on composite materials, including *Composite Materials Handbook and Composite Materials Technology*. He is a fellow of the American Society of Mechanical Engineers. Dr Mallick received his BE degree (1966) in mechanical engineering from Calcutta University, India, and the MS (1970) and PhD (1973) degrees in mechanical engineering from the Illinois Institute of Technology.

1 Introduction

1.1 DEFINITION

Fiber-reinforced composite materials consist of fibers of high strength and modulus embedded in or bonded to a matrix with distinct interfaces (boundaries) between them. In this form, both fibers and matrix retain their physical and chemical identities, yet they produce a combination of properties that cannot be achieved with either of the constituents acting alone. In general, fibers are the principal load-carrying members, while the surrounding matrix keeps them in the desired location and orientation, acts as a load transfer medium between them, and protects them from environmental damages due to elevated temperatures and humidity, for example. Thus, even though the fibers provide reinforcement for the matrix, the latter also serves a number of useful functions in a fiber-reinforced composite material.

The principal fibers in commercial use are various types of glass and carbon as well as Kevlar 49. Other fibers, such as boron, silicon carbide, and aluminum oxide, are used in limited quantities. All these fibers can be incorporated into a matrix either in continuous lengths or in discontinuous (short) lengths. The matrix material may be a polymer, a metal, or a ceramic. Various chemical compositions and microstructural arrangements are possible in each matrix category.

The most common form in which fiber-reinforced composites are used in structural applications is called a laminate, which is made by stacking a number of thin layers of fibers and matrix and consolidating them into the desired thickness. Fiber orientation in each layer as well as the stacking sequence of various layers in a composite laminate can be controlled to generate a wide range of physical and mechanical properties for the composite laminate.

In this book, we focus our attention on the mechanics, performance, manufacturing, and design of fiber-reinforced polymers. Most of the data presented in this book are related to continuous fiber-reinforced epoxy laminates, although other polymeric matrices, including thermoplastic matrices, are also considered. Metal and ceramic matrix composites are comparatively new, but significant developments of these composites have also occurred. They are included in a separate chapter in this book. Injection-molded or reaction injection-molded (RIM) discontinuous fiber-reinforced polymers are not discussed; however, some of the mechanics and design principles included in this book are applicable to these composites as well. Another material of great commercial interest is classified as particulate composites. The major constituents in these composites are particles of mica, silica, glass spheres, calcium carbonate, and others. In general, these particles do not contribute to the load-carrying capacity of the material and act more like a filler than a reinforcement for the matrix. Particulate composites, by themselves, deserve a special attention and are not addressed in this book.

Another type of composites that have the potential of becoming an important material in the future is the nanocomposites. Even though nanocomposites are in the early stages of development, they are now receiving a high degree of attention from academia as well as a large number of industries, including aerospace, automotive, and biomedical industries. The reinforcement in nanocomposites is either nanoparticles, nanofibers, or carbon nanotubes. The effective diameter of these reinforcements is of the order of 10^{-9} m, whereas the effective diameter of the reinforcements used in traditional fiber-reinforced composites is of the order of 10^{-6} m. The nanocomposites are introduced in Chapter 8.

1.2 GENERAL CHARACTERISTICS

Many fiber-reinforced polymers offer a combination of strength and modulus that are either comparable to or better than many traditional metallic materials. Because of their low density, the strength–weight ratios and modulus–weight ratios of these composite materials are markedly superior to those of metallic materials (Table 1.1). In addition, fatigue strength as well as fatigue damage tolerance of many composite laminates are excellent. For these reasons, fiber-reinforced polymers have emerged as a major class of structural materials and are either used or being considered for use as substitution for metals in many weight-critical components in aerospace, automotive, and other industries.

Traditional structural metals, such as steel and aluminum alloys, are considered isotropic, since they exhibit equal or nearly equal properties irrespective of the direction of measurement. In general, the properties of a fiber-reinforced composite depend strongly on the direction of measurement, and therefore, they are not isotropic materials. For example, the tensile strength and modulus of a unidirectionally oriented fiber-reinforced polymer are maximum when these properties are measured in the longitudinal direction of fibers. At any other angle of measurement, these properties are lower. The minimum value is observed when they are measured in the transverse direction of fibers, that is, at 90° to the longitudinal direction. Similar angular dependence is observed for other mechanical and thermal properties, such as impact strength, coefficient of thermal expansion (CTE), and thermal conductivity. Bi- or multidirectional reinforcement yields a more balanced set of properties. Although these properties are lower than the longitudinal properties of a unidirectional composite, they still represent a considerable advantage over common structural metals on a unit weight basis.

The design of a fiber-reinforced composite structure is considerably more difficult than that of a metal structure, principally due to the difference in its

TABLE 1.1
Tensile Properties of Some Metallic and Structural Composite Materials

Material[a]	Density, g/cm^3	Modulus, GPa (Msi)	Tensile Strength, MPa (ksi)	Yield Strength, MPa (ksi)	Ratio of Modulus to Weight,[b] 10^6 m	Ratio of Tensile Strength to Weight,[b] 10^3 m
SAE 1010 steel (cold-worked)	7.87	207 (30)	365 (53)	303 (44)	2.68	4.72
AISI 4340 steel (quenched and tempered)	7.87	207 (30)	1722 (250)	1515 (220)	2.68	22.3
6061-T6 aluminum alloy	2.70	68.9 (10)	310 (45)	275 (40)	2.60	11.7
7178-T6 aluminum alloy	2.70	68.9 (10)	606 (88)	537 (78)	2.60	22.9
Ti-6A1-4V titanium alloy (aged)	4.43	110 (16)	1171 (170)	1068 (155)	2.53	26.9
17-7 PH stainless steel (aged)	7.87	196 (28.5)	1619 (235)	1515 (220)	2.54	21.0
INCO 718 nickel alloy (aged)	8.2	207 (30)	1399 (203)	1247 (181)	2.57	17.4
High-strength carbon fiber–epoxy matrix (unidirectional)[a]	1.55	137.8 (20)	1550 (225)	—	9.06	101.9
High-modulus carbon fiber–epoxy matrix (unidirectional)	1.63	215 (31.2)	1240 (180)	—	13.44	77.5
E-glass fiber–epoxy matrix (unidirectional)	1.85	39.3 (5.7)	965 (140)	—	2.16	53.2
Kevlar 49 fiber–epoxy matrix (unidirectional)	1.38	75.8 (11)	1378 (200)	—	5.60	101.8
Boron fiber-6061 A1 alloy matrix (annealed)	2.35	220 (32)	1109 (161)	—	9.54	48.1
Carbon fiber–epoxy matrix (quasi-isotropic)	1.55	45.5 (6.6)	579 (84)	—	2.99	38
Sheet-molding compound (SMC) composite (isotropic)	1.87	15.8 (2.3)	164 (23.8)		0.86	8.9

[a] For unidirectional composites, the fibers are unidirectional and the reported modulus and tensile strength values are measured in the direction of fibers, that is, the longitudinal direction of the composite.

[b] The modulus–weight ratio and the strength–weight ratios are obtained by dividing the absolute values with the specific weight of the respective material. Specific weight is defined as weight per unit volume. It is obtained by multiplying density with the acceleration due to gravity.

properties in different directions. However, the nonisotropic nature of a fiber-reinforced composite material creates a unique opportunity of tailoring its properties according to the design requirements. This design flexibility can be used to selectively reinforce a structure in the directions of major stresses, increase its stiffness in a preferred direction, fabricate curved panels without any secondary forming operation, or produce structures with zero coefficients of thermal expansion.

The use of fiber-reinforced polymer as the skin material and a lightweight core, such as aluminum honeycomb, plastic foam, metal foam, and balsa wood, to build a sandwich beam, plate, or shell provides another degree of design flexibility that is not easily achievable with metals. Such sandwich construction can produce high stiffness with very little, if any, increase in weight. Another sandwich construction in which the skin material is an aluminum alloy and the core material is a fiber-reinforced polymer has found widespread use in aircrafts and other applications, primarily due to their higher fatigue performance and damage tolerance than aluminum alloys.

In addition to the directional dependence of properties, there are a number of other differences between structural metals and fiber-reinforced composites. For example, metals in general exhibit yielding and plastic deformation. Most fiber-reinforced composites are elastic in their tensile stress–strain characteristics. However, the heterogeneous nature of these materials provides mechanisms for energy absorption on a microscopic scale, which is comparable to the yielding process. Depending on the type and severity of external loads, a composite laminate may exhibit gradual deterioration in properties but usually would not fail in a catastrophic manner. Mechanisms of damage development and growth in metal and composite structures are also quite different and must be carefully considered during the design process when the metal is substituted with a fiber-reinforced polymer.

Coefficient of thermal expansion (CTE) for many fiber-reinforced composites is much lower than that for metals (Table 1.2). As a result, composite structures may exhibit a better dimensional stability over a wide temperature range. However, the differences in thermal expansion between metals and composite materials may create undue thermal stresses when they are used in conjunction, for example, near an attachment. In some applications, such as electronic packaging, where quick and effective heat dissipation is needed to prevent component failure or malfunctioning due to overheating and undesirable temperature rise, thermal conductivity is an important material property to consider. In these applications, some fiber-reinforced composites may excel over metals because of the combination of their high thermal conductivity–weight ratio (Table 1.2) and low CTE. On the other hand, electrical conductivity of fiber-reinforced polymers is, in general, lower than that of metals. The electric charge build up within the material because of low electrical conductivity can lead to problems such as radio frequency interference (RFI) and damage due to lightning strike.

TABLE 1.2
Thermal Properties of a Few Selected Metals and Composite Materials

Material	Density (g/cm^3)	Coefficient of Thermal Expansion (10^{-6}/°C)	Thermal Conductivity (W/m°K)	Ratio of Thermal Conductivity to Weight (10^{-3} m^4/s^3 °K)
Plain carbon steels	7.87	11.7	52	6.6
Copper	8.9	17	388	43.6
Aluminum alloys	2.7	23.5	130–220	48.1–81.5
Ti-6Al-4V titanium alloy	4.43	8.6	6.7	1.51
Invar	8.05	1.6	10	1.24
K1100 carbon fiber–epoxy matrix	1.8	−1.1	300	166.7
Glass fiber–epoxy matrix	2.1	11–20	0.16–0.26	0.08–0.12
SiC particle-reinforced aluminum	3	6.2–7.3	170–220	56.7–73.3

Another unique characteristic of many fiber-reinforced composites is their high internal damping. This leads to better vibrational energy absorption within the material and results in reduced transmission of noise and vibrations to neighboring structures. High damping capacity of composite materials can be beneficial in many automotive applications in which noise, vibration, and harshness (NVH) are critical issues for passenger comfort. High damping capacity is also useful in many sporting goods applications.

An advantage attributed to fiber-reinforced polymers is their noncorroding behavior. However, many fiber-reinforced polymers are capable of absorbing moisture or chemicals from the surrounding environment, which may create dimensional changes or adverse internal stresses in the material. If such behavior is undesirable in an application, the composite surface must be protected from moisture or chemicals by an appropriate paint or coating. Among other environmental factors that may cause degradation in the mechanical properties of some polymer matrix composites are elevated temperatures, corrosive fluids, and ultraviolet rays. In metal matrix composites, oxidation of the matrix as well as adverse chemical reaction between fibers and the matrix are of great concern in high-temperature applications.

The manufacturing processes used with fiber-reinforced polymers are different from the traditional manufacturing processes used for metals, such as casting, forging, and so on. In general, they require significantly less energy and lower pressure or force than the manufacturing processes used for metals. Parts integration and net-shape or near net-shape manufacturing processes are also great advantages of using fiber-reinforced polymers. Parts integration reduces the number of parts, the number of manufacturing operations, and also, the number of assembly operations. Net-shape or near net-shape manufacturing processes, such as filament winding and pultrusion, used for making many fiber-reinforced polymer parts, either reduce or eliminate the finishing operations such as machining and grinding, which are commonly required as finishing operations for cast or forged metallic parts.

1.3 APPLICATIONS

Commercial and industrial applications of fiber-reinforced polymer composites are so varied that it is impossible to list them all. In this section, we highlight only the major structural application areas, which include aircraft, space, automotive, sporting goods, marine, and infrastructure. Fiber-reinforced polymer composites are also used in electronics (e.g., printed circuit boards), building construction (e.g., floor beams), furniture (e.g., chair springs), power industry (e.g., transformer housing), oil industry (e.g., offshore oil platforms and oil sucker rods used in lifting underground oil), medical industry (e.g., bone plates for fracture fixation, implants, and prosthetics), and in many industrial products, such as step ladders, oxygen tanks, and power transmission shafts. Potential use of fiber-reinforced composites exists in many engineering fields. Putting them to actual use requires careful design practice and appropriate process development based on the understanding of their unique mechanical, physical, and thermal characteristics.

1.3.1 Aircraft and Military Applications

The major structural applications for fiber-reinforced composites are in the field of military and commercial aircrafts, for which weight reduction is critical for higher speeds and increased payloads. Ever since the production application of boron fiber-reinforced epoxy skins for F-14 horizontal stabilizers in 1969, the use of fiber-reinforced polymers has experienced a steady growth in the aircraft industry. With the introduction of carbon fibers in the 1970s, carbon fiber-reinforced epoxy has become the primary material in many wing, fuselage, and empennage components (Table 1.3). The structural integrity and durability of these early components have built up confidence in their performance and prompted developments of other structural aircraft components, resulting in an increasing amount of composites being used in military aircrafts. For example, the airframe of AV-8B, a vertical and short take-off and landing (VSTOL) aircraft introduced in 1982, contains nearly 25% by weight of carbon fiber-reinforced epoxy. The F-22 fighter aircraft also contains ~25% by weight of carbon fiber-reinforced polymers; the other major materials are titanium (39%) and aluminum (16%). The outer skin of B-2 (Figure 1.1) and other stealth aircrafts is almost all made of carbon fiber-reinforced polymers. The stealth characteristics of these aircrafts are due to the use of carbon fibers, special coatings, and other design features that reduce radar reflection and heat radiation.

TABLE 1.3
Early Applications of Fiber-Reinforced Polymers in Military Aircrafts

Aircraft	Component	Material	Overall Weight Saving Over Metal Component (%)
F-14 (1969)	Skin on the horizontal stabilizer box	Boron fiber–epoxy	19
F-11	Under the wing fairings	Carbon fiber–epoxy	
F-15 (1975)	Fin, rudder, and stabilizer skins	Boron fiber–epoxy	25
F-16 (1977)	Skins on vertical fin box, fin leading edge	Carbon fiber–epoxy	23
F/A-18 (1978)	Wing skins, horizontal and vertical tail boxes; wing and tail control surfaces, etc.	Carbon fiber–epoxy	35
AV-8B (1982)	Wing skins and substructures; forward fuselage; horizontal stabilizer; flaps; ailerons	Carbon fiber–epoxy	25

Source: Adapted from Riggs, J.P., *Mater. Soc.*, 8, 351, 1984.

The composite applications on commercial aircrafts began with a few selective secondary structural components, all of which were made of a high-strength carbon fiber-reinforced epoxy (Table 1.4). They were designed and produced under the NASA Aircraft Energy Efficiency (ACEE) program and were installed in various airplanes during 1972–1986 [1]. By 1987, 350 composite components were placed in service in various commercial aircrafts, and over the next few years, they accumulated millions of flight hours. Periodic inspection and evaluation of these components showed some damages caused by ground handling accidents, foreign object impacts, and lightning strikes.

FIGURE 1.1 Stealth aircraft (note that the carbon fibers in the construction of the aircraft contributes to its stealth characteristics).

TABLE 1.4
Early Applications of Fiber-Reinforced Polymers in Commercial Aircrafts

Aircraft	Component	Weight (lb)	Weight Reduction (%)	Comments
Boeing				
727	Elevator face sheets	98	25	10 units installed in 1980
737	Horizontal stabilizer	204	22	
737	Wing spoilers	—	37	Installed in 1973
756	Ailerons, rudders, elevators, fairings, etc.	3340 (total)	31	
McDonnell-Douglas				
DC-10	Upper rudder	67	26	13 units installed in 1976
DC-10	Vertical stabilizer	834	17	
Lockheed				
L-1011	Aileron	107	23	10 units installed in 1981
L-1011	Vertical stabilizer	622	25	

Apart from these damages, there was no degradation of residual strengths due to either fatigue or environmental exposure. A good correlation was found between the on-ground environmental test program and the performance of the composite components after flight exposure.

Airbus was the first commercial aircraft manufacturer to make extensive use of composites in their A310 aircraft, which was introduced in 1987. The composite components weighed about 10% of the aircraft's weight and included such components as the lower access panels and top panels of the wing leading edge, outer deflector doors, nose wheel doors, main wheel leg fairing doors, engine cowling panels, elevators and fin box, leading and trailing edges of fins, flap track fairings, flap access doors, rear and forward wing–body fairings, pylon fairings, nose radome, cooling air inlet fairings, tail leading edges, upper surface skin panels above the main wheel bay, glide slope antenna cover, and rudder. The composite vertical stabilizer, which is 8.3 m high by 7.8 m wide at the base, is about 400 kg lighter than the aluminum vertical stabilizer previously used [2]. The Airbus A320, introduced in 1988, was the first commercial aircraft to use an all-composite tail, which includes the tail cone, vertical stabilizer, and horizontal stabilizer. Figure 1.2 schematically shows the composite usage in Airbus A380 introduced in 2006. About 25% of its weight is made of composites. Among the major composite components in A380 are the central torsion box (which links the left and right wings under the fuselage), rear-pressure bulkhead (a dome-shaped partition that separates the passenger cabin from the rear part of the plane that is not pressurized), the tail, and the flight control surfaces, such as the flaps, spoilers, and ailerons.

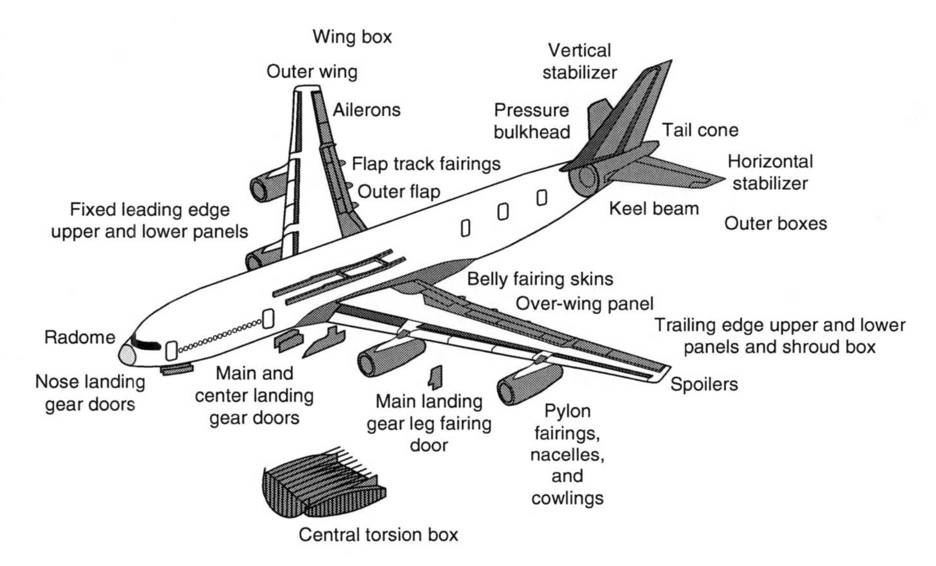

FIGURE 1.2 Use of fiber-reinforced polymer composites in Airbus 380.

Starting with Boeing 777, which was first introduced in 1995, Boeing has started making use of composites in the empennage (which include horizontal stabilizer, vertical stabilizer, elevator, and rudder), most of the control surfaces, engine cowlings, and fuselage floor beams (Figure 1.3). About 10% of Boeing 777's structural weight is made of carbon fiber-reinforced epoxy and about 50% is made of aluminum alloys. About 50% of the structural weight of Boeing's

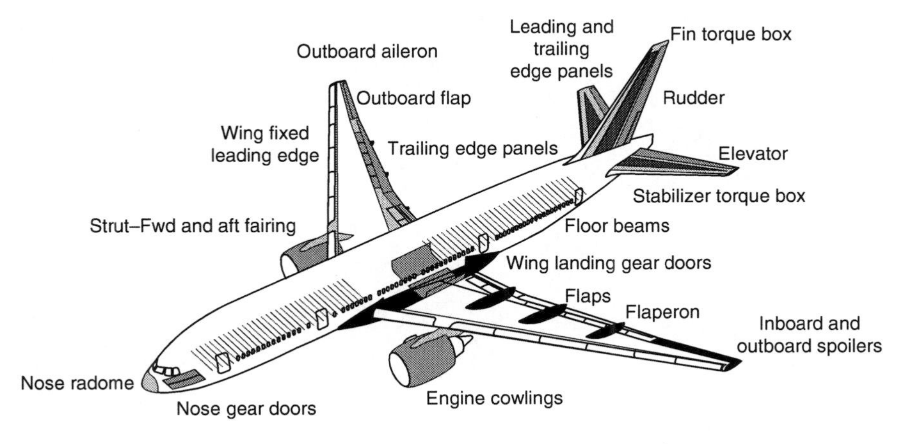

FIGURE 1.3 Use of fiber-reinforced polymer composites in Boeing 777.

next line of airplanes, called the Boeing 787 Dreamliner, will be made of carbon fiber-reinforced polymers. The other major materials in Boeing 787 will be aluminum alloys (20%), titanium alloys (15%), and steel (10%). Two of the major composite components in 787 will be the fuselage and the forward section, both of which will use carbon fiber-reinforced epoxy as the major material of construction.

There are several pioneering examples of using larger quantities of composite materials in smaller aircrafts. One of these examples is the Lear Fan 2100, a business aircraft built in 1983, in which carbon fiber–epoxy and Kevlar 49 fiber–epoxy accounted for ~70% of the aircraft's airframe weight. The composite components in this aircraft included wing skins, main spar, fuselage, empennage, and various control surfaces [3]. Another example is the Rutan Voyager (Figure 1.4), which was an all-composite airplane and made the first-ever nonstop flight around the world in 1986. To travel 25,000 miles without refueling, the Voyager airplane had to be extremely light and contain as much fuel as needed.

Fiber-reinforced polymers are used in many military and commercial helicopters for making baggage doors, fairings, vertical fins, tail rotor spars, and so on. One key helicopter application of composite materials is the rotor blades. Carbon or glass fiber-reinforced epoxy is used in this application. In addition to significant weight reduction over aluminum, they provide a better control over the vibration characteristics of the blades. With aluminum, the critical flopping

FIGURE 1.4 Rutan Voyager all-composite plane.

and twisting frequencies are controlled principally by the classical method of mass distribution [4]. With fiber-reinforced polymers, they can also be controlled by varying the type, concentration, distribution, as well as orientation of fibers along the blade's chord length. Another advantage of using fiber-reinforced polymers in blade applications is the manufacturing flexibility of these materials. The composite blades can be filament-wound or molded into complex airfoil shapes with little or no additional manufacturing costs, but conventional metal blades are limited to shapes that can only be extruded, machined, or rolled.

The principal reason for using fiber-reinforced polymers in aircraft and helicopter applications is weight saving, which can lead to significant fuel saving and increase in payload. There are several other advantages of using them over aluminum and titanium alloys.

1. Reduction in the number of components and fasteners, which results in a reduction of fabrication and assembly costs. For example, the vertical fin assembly of the Lockheed L-1011 has 72% fewer components and 83% fewer fasteners when it is made of carbon fiber-reinforced epoxy than when it is made of aluminum. The total weight saving is 25.2%.
2. Higher fatigue resistance and corrosion resistance, which result in a reduction of maintenance and repair costs. For example, metal fins used in helicopters flying near ocean coasts use an 18 month repair cycle for patching corrosion pits. After a few years in service, the patches can add enough weight to the fins to cause a shift in the center of gravity of the helicopter, and therefore the fin must then be rebuilt or replaced. The carbon fiber-reinforced epoxy fins do not require any repair for corrosion, and therefore, the rebuilding or replacement cost is eliminated.
3. The laminated construction used with fiber-reinforced polymers allows the possibility of aeroelastically tailoring the stiffness of the airframe structure. For example, the airfoil shape of an aircraft wing can be controlled by appropriately adjusting the fiber orientation angle in each lamina and the stacking sequence to resist the varying lift and drag loads along its span. This produces a more favorable airfoil shape and enhances the aerodynamic characteristics critical to the aircraft's maneuverability.

The key limiting factors in using carbon fiber-reinforced epoxy in aircraft structures are their high cost, relatively low impact damage tolerance (from bird strikes, tool drop, etc.), and susceptibility to lightning damage. When they are used in contact with aluminum or titanium, they can induce galvanic corrosion in the metal components. The protection of the metal components from corrosion can be achieved by coating the contacting surfaces with a corrosion-inhibiting paint, but it is an additional cost.

1.3.2 Space Applications

Weight reduction is the primary reason for using fiber-reinforced composites in many space vehicles [5]. Among the various applications in the structure of space shuttles are the mid-fuselage truss structure (boron fiber-reinforced aluminum tubes), payload bay door (sandwich laminate of carbon fiber-reinforced epoxy face sheets and aluminum honeycomb core), remote manipulator arm (ultrahigh-modulus carbon fiber-reinforced epoxy tube), and pressure vessels (Kevlar 49 fiber-reinforced epoxy).

In addition to the large structural components, fiber-reinforced polymers are used for support structures for many smaller components, such as solar arrays, antennas, optical platforms, and so on [6]. A major factor in selecting them for these applications is their dimensional stability over a wide temperature range. Many carbon fiber-reinforced epoxy laminates can be "designed" to produce a CTE close to zero. Many aerospace alloys (e.g., Invar) also have a comparable CTE. However, carbon fiber composites have a much lower density, higher strength, as well as a higher stiffness–weight ratio. Such a unique combination of mechanical properties and CTE has led to a number of applications for carbon fiber-reinforced epoxies in artificial satellites. One such application is found in the support structure for mirrors and lenses in the space telescope [7]. Since the temperature in space may vary between −100°C and 100°C, it is critically important that the support structure be dimensionally stable; otherwise, large changes in the relative positions of mirrors or lenses due to either thermal expansion or distortion may cause problems in focusing the telescope.

Carbon fiber-reinforced epoxy tubes are used in building truss structures for low earth orbit (LEO) satellites and interplanetary satellites. These truss structures support optical benches, solar array panels, antenna reflectors, and other modules. Carbon fiber-reinforced epoxies are preferred over metals or metal matrix composites because of their lower weight as well as very low CTE. However, one of the major concerns with epoxy-based composites in LEO satellites is that they are susceptible to degradation due to atomic oxygen (AO) absorption from the earth's rarefied atmosphere. This problem is overcome by protecting the tubes from AO exposure, for example, by wrapping them with thin aluminum foils.

Other concerns for using fiber-reinforced polymers in the space environment are the outgassing of the polymer matrix when they are exposed to vacuum in space and embrittlement due to particle radiation. Outgassing can cause dimensional changes and embrittlement may lead to microcrack formation. If the outgassed species are deposited on the satellite components, such as sensors or solar cells, their function may be seriously degraded [8].

1.3.3 Automotive Applications

Applications of fiber-reinforced composites in the automotive industry can be classified into three groups: body components, chassis components, and engine components. Exterior body components, such as the hood or door panels, require high stiffness and damage tolerance (dent resistance) as well as a "Class A" surface finish for appearance. The composite material used for these components is E-glass fiber-reinforced sheet molding compound (SMC) composites, in which discontinuous glass fibers (typically 25 mm in length) are randomly dispersed in a polyester or a vinyl ester resin. E-glass fiber is used instead of carbon fiber because of its significantly lower cost. The manufacturing process used for making SMC parts is called compression molding. One of the design requirements for many exterior body panels is the "Class A" surface finish, which is not easily achieved with compression-molded SMC. This problem is usually overcome by in-mold coating of the exterior molded surface with a flexible resin. However, there are many underbody and under-the-hood components in which the external appearance is not critical. Examples of such components in which SMC is used include radiator supports, bumper beams, roof frames, door frames, engine valve covers, timing chain covers, oil pans, and so on. Two of these applications are shown in Figures 1.5 and 1.6.

SMC has seen a large growth in the automotive industry over the last 25 years as it is used for manufacturing both small and large components, such as hoods, pickup boxes, deck lids, doors, fenders, spoilers, and others, in automobiles, light trucks, and heavy trucks. The major advantages of using SMC instead of steel in these components include not only the weight reduction, but also lower tooling cost and parts integration. The tooling cost for compression molding SMC parts can be 40%–60% lower than that for stamping steel parts. An example of parts integration can be found in radiator supports in which SMC is used as a substitution for low carbon steel. The composite

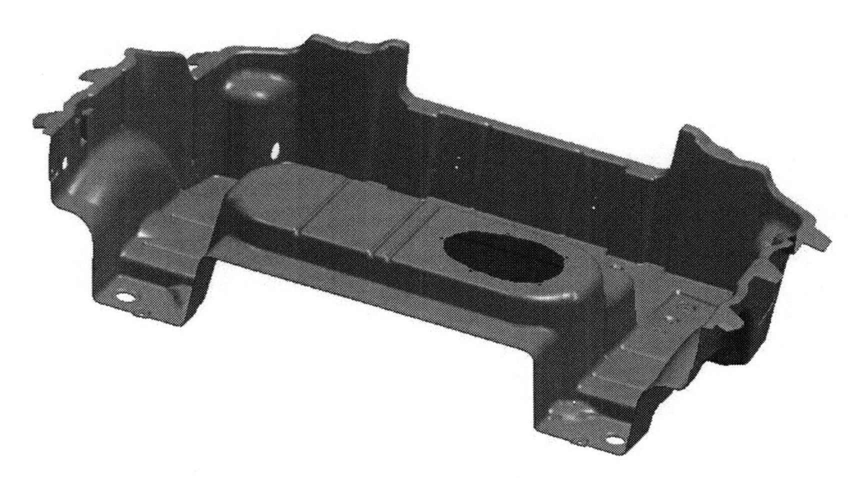

FIGURE 1.5 Compression-molded SMC trunk of Cadillac Solstice. (Courtesy of Molded Fiber Glass and American Composites Alliance. With permission.)

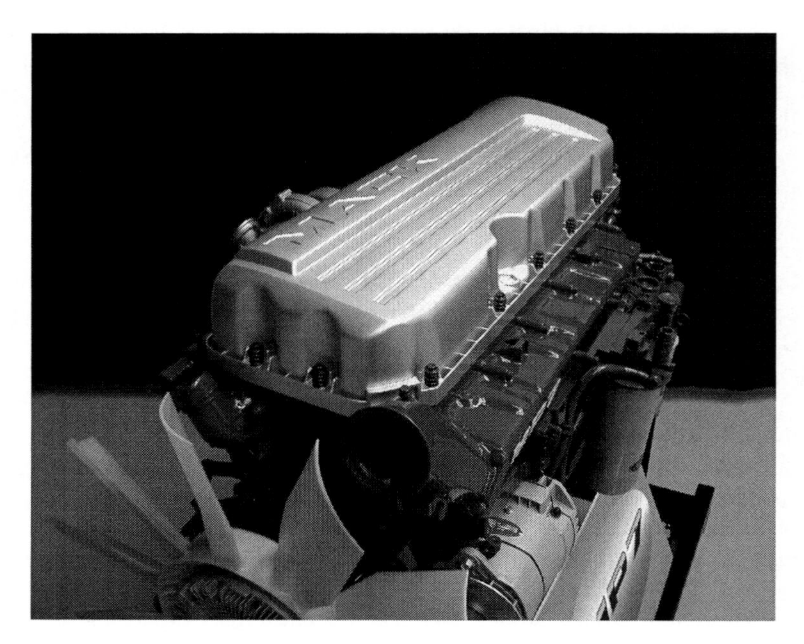

FIGURE 1.6 Compression-molded SMC valve cover for a truck engine. (Courtesy of Ashland Chemicals and American Composites Alliance. With permission.)

radiator support is typically made of two SMC parts bonded together by an adhesive instead of 20 or more steel parts assembled together by large number of screws. The material in the composite radiator support is randomly oriented discontinuous E-glass fiber-reinforced vinyl ester. Another example of parts integration can be found in the station wagon tailgate assembly [9], which has significant load-bearing requirements in the open position. The composite tailgate consists of two pieces, an outer SMC shell and an inner reinforcing SMC piece. They are bonded together using a urethane adhesive. The composite tailgate replaces a seven-piece steel tailgate assembly, at about one-third its weight. The material for both the outer shell and the inner reinforcement is a randomly oriented discontinuous E-glass fiber-reinforced polyester.

Another manufacturing process for making composite body panels in the automotive industry is called the structural reaction injection molding (SRIM). The fibers in these parts are usually randomly oriented discontinuous E-glass fibers and the matrix is a polyurethane or polyurea. Figure 1.7 shows the photograph of a one-piece 2 m long cargo box that is molded using this process. The wall thickness of the SRIM cargo box is 3 mm and its one-piece construction replaces four steel panels that are joined together using spot welds.

Among the chassis components, the first major structural application of fiber-reinforced composites is the Corvette rear leaf spring, introduced first in

FIGURE 1.7 One-piece cargo box for a pickup truck made by the SRIM process.

1981 [10]. Unileaf E-glass fiber-reinforced epoxy springs have been used to replace multileaf steel springs with as much as 80% weight reduction. Other structural chassis components, such as drive shafts and road wheels, have been successfully tested in laboratories and proving grounds. They have also been used in limited quantities in production vehicles. They offer opportunities for substantial weight savings, but so far they have not proven to be cost-effective over their steel counterparts.

The application of fiber-reinforced composites in engine components has not been as successful as the body and chassis components. Fatigue loads at very high temperatures pose the greatest challenge in these applications. Development of high-temperature polymers as well as metal matrix or ceramic matrix composites would greatly enhance the potential for composite usage in this area.

Manufacturing and design of fiber-reinforced composite materials for automotive applications are significantly different from those for aircraft applications. One obvious difference is in the volume of production, which may range from 100 to 200 pieces per hour for automotive components compared with a few hundred pieces per year for aircraft components. Although the labor-intensive hand layup followed by autoclave molding has worked well for fabricating aircraft components, high-speed methods of fabrication, such as compression molding and SRIM, have emerged as the principal manufacturing process for automotive composites. Epoxy resin is the major polymer matrix used in aerospace composites; however, the curing time for epoxy resin is very long, which means the production time for epoxy matrix composites is also very long. For this reason, epoxy is not considered the primary matrix material in automotive composites. The polymer matrix used in automotive applications is either a polyester, a vinyl ester, or polyurethane, all of which require significantly lower curing time than epoxy. The high cost of carbon fibers has prevented their use in the cost-conscious automotive industry. Instead of carbon fibers, E-glass fibers are used in automotive composites because of their significantly lower cost. Even with E-glass fiber-reinforced composites, the cost-effectiveness issue has remained particularly critical, since the basic material of construction in present-day automobiles is low-carbon steel, which is much less expensive than most fiber-reinforced composites on a unit weight basis.

Although glass fiber-reinforced polymers are the primary composite materials used in today's automobiles, it is well recognized that significant vehicle weight reduction needed for improved fuel efficiency can be achieved only with carbon fiber-reinforced polymers, since they have much higher strength–weight and modulus–weight ratios. The problem is that the current carbon fiber price, at $16/kg or higher, is not considered cost-effective for automotive applications. Nevertheless, many attempts have been made in the past to manufacture structural automotive parts using carbon fiber-reinforced polymers; unfortunately most of them did not go beyond the stages of prototyping and structural testing. Recently, several high-priced vehicles have started using carbon fiber-reinforced polymers in a few selected components. One recent example of this is seen in the BMW M6 roof panel (Figure 1.8), which was produced by a process called resin transfer molding (RTM). This panel is twice as thick as a comparable steel panel, but 5.5 kg lighter. One added benefit of reducing the weight of the roof panel is that it slightly lowers the center of gravity of the vehicle, which is important for sports coupe.

Fiber-reinforced composites have become the material of choice in motor sports where lightweight structure is used for gaining competitive advantage of higher speed [11] and cost is not a major material selection decision factor. The first major application of composites in race cars was in the 1950s when glass fiber-reinforced polyester was introduced as replacement for aluminum body panels. Today, the composite material used in race cars is mostly carbon fiber-reinforced epoxy. All major body, chassis, interior, and suspension components in today's Formula 1 race cars use carbon fiber-reinforced epoxy. Figure 1.9 shows an example of carbon fiber-reinforced composite used in the gear box and rear suspension of a Formula 1 race car. One major application of carbon fiber-reinforced epoxy in Formula 1 cars is the survival cell, which protects the driver in the event of a crash. The nose cone located in front of the survival cell is also made of carbon fiber-reinforced epoxy. Its controlled crush behavior is also critical to the survival of the driver.

FIGURE 1.8 Carbon fiber-reinforced epoxy roof panel in BMW M6 vehicle. (Photograph provided by BMW. With permission.)

FIGURE 1.9 Carbon fiber-reinforced epoxy suspension and gear box in a Formula 1 race car. (Courtesy of Bar 1 Formula 1 Racing Team. With permission.)

1.3.4 Sporting Goods Applications

Fiber-reinforced polymers are extensively used in sporting goods ranging from tennis rackets to athletic shoes (Table 1.5) and are selected over such traditional materials as wood, metals, and leather in many of these applications [12]. The advantages of using fiber-reinforced polymers are weight reduction, vibration damping, and design flexibility. Weight reduction achieved by substituting carbon fiber-reinforced epoxies for metals leads to higher speeds and quick maneuvering in competitive sports, such as bicycle races and canoe races. In some applications, such as tennis rackets or snow skis, sandwich constructions of carbon or boron fiber-reinforced epoxies as the skin material and a soft, lighter weight urethane foam as the core material produces a higher weight reduction without sacrificing stiffness. Faster damping of vibrations provided by fiber-reinforced polymers reduces the shock transmitted to the player's arm in tennis or racket ball games and provides a better "feel" for the ball. In archery bows and pole-vault poles, the high stiffness–weight ratio of fiber-reinforced composites is used to store high elastic energy per unit weight, which helps in propelling the arrow over a longer distance or the pole-vaulter to jump a greater height. Some of these applications are described later.

Bicycle frames for racing bikes today are mostly made of carbon fiber-reinforced epoxy tubes, fitted together by titanium fittings and inserts. An example is shown in Figure 1.10. The primary purpose of using carbon fibers is

TABLE 1.5
Applications of Fiber-Reinforced Polymers in Sporting Goods

Tennis rackets
Racket ball rackets
Golf club shafts
Fishing rods
Bicycle frames
Snow and water skis
Ski poles, pole vault poles
Hockey sticks
Baseball bats
Sail boats and kayaks
Oars, paddles
Canoe hulls
Surfboards, snow boards
Arrows
Archery bows
Javelins
Helmets
Exercise equipment
Athletic shoe soles and heels

FIGURE 1.10 Carbon fiber-reinforced epoxy bicycle frame. (Photograph provided by Trek Bicycle Corporation. With permission.)

weight saving (the average racing bicycle weight has decreased from about 9 kg in the 1980s to 1.1 kg in 1990s); however, to reduce material cost, carbon fibers are sometimes combined with glass or Kevlar 49 fibers. Fiber-reinforced polymer wrapped around thin-walled metal tube is not also uncommon. The ancillary components, such as handlebars, forks, seat post, and others, also use carbon fiber-reinforced polymers.

Golf clubs made of carbon fiber-reinforced epoxy are becoming increasingly popular among professional golfers. The primary reason for the composite golf shaft's popularity is its low weight compared with steel golf shafts. The average weight of a composite golf shaft is 65–70 g compared with 115–125 g for steel shafts. Weight reduction in the golf club shaft allows the placement of additional weight in the club head, which results in faster swing and longer drive.

Glass fiber-reinforced epoxy is preferred over wood and aluminum in pole-vault poles because of its high strain energy storage capacity. A good pole must have a reasonably high stiffness (to keep it from flapping excessively during running before jumping) and high elastic limit stress so that the strain energy of the bent pole can be recovered to propel the athlete above the horizontal bar. As the pole is bent to store the energy, it should not show any plastic deformation and should not fracture. The elastic limit of glass fiber-reinforced epoxy is much higher than that of either wood or high-strength aluminum alloys. With glass fiber-reinforced epoxy poles, the stored energy is high enough to clear 6 m or greater height in pole vaulting. Carbon fiber-reinforced epoxy is not used, since it is prone to fracture at high bending strains.

Glass and carbon fiber-reinforced epoxy fishing rods are very common today, even though traditional materials, such as bamboo, are still used. For fly-fishing rods, carbon fiber-reinforced epoxy is preferred, since it produces a smaller tip deflection (because of its higher modulus) and "wobble-free" action during casting. It also dampens the vibrations more rapidly and reduces the transmission of vibration waves along the fly line. Thus, the casting can be longer, quieter, and more accurate, and the angler has a better "feel" for the catch. Furthermore, carbon fiber-reinforced epoxy rods recover their original shape much faster than the other rods. A typical carbon fiber-reinforced epoxy rod of No. 6 or No. 7 line weighs only 37 g. The lightness of these rods is also a desirable feature to the anglers.

1.3.5 Marine Applications

Glass fiber-reinforced polyesters have been used in different types of boats (e.g., sail boats, fishing boats, dinghies, life boats, and yachts) ever since their introduction as a commercial material in the 1940s [13]. Today, nearly 90% of all recreational boats are constructed of either glass fiber-reinforced polyester or glass fiber-reinforced vinyl ester resin. Among the applications are hulls, decks, and various interior components. The manufacturing process used for making a majority of these components is called contact molding. Even though it is a labor-intensive process, the equipment cost is low, and therefore it is affordable to many of the small companies that build these boats. In recent years, Kevlar 49 fiber is replacing glass fibers in some of these applications because of its higher tensile strength–weight and modulus–weight ratios than those of glass fibers. Among the application areas are boat hulls, decks, bulkheads, frames, masts, and spars. The principal advantage is weight reduction, which translates into higher cruising speed, acceleration, maneuverability, and fuel efficiency.

Carbon fiber-reinforced epoxy is used in racing boats in which weight reduction is extremely important for competitive advantage. In these boats, the complete hull, deck, mast, keel, boom, and many other structural components are constructed using carbon fiber-reinforced epoxy laminates and sandwich laminates of carbon fiber-reinforced epoxy skins with either honeycomb core or plastic foam core. Carbon fibers are sometimes hybridized with other lower density and higher strain-to-failure fibers, such as high-modulus polyethylene fibers, to improve impact resistance and reduce the boat's weight.

The use of composites in naval ships started in the 1950s and has grown steadily since then [14]. They are used in hulls, decks, bulkheads, masts, propulsion shafts, rudders, and others of mine hunters, frigates, destroyers, and aircraft carriers. Extensive use of fiber-reinforced polymers can be seen in Royal Swedish Navy's 72 m long, 10.4 m wide Visby-class corvette, which is the largest composite ship in the world today. Recently, the US navy has commissioned a 24 m long combat ship, called Stiletto, in which carbon fiber-reinforced epoxy will be the primary material of construction. The selection of carbon fiber-reinforced epoxy is based on the design requirements of lightweight and high strength needed for high speed, maneuverability, range, and payload capacity of these ships. Their stealth characteristics are also important in minimizing radar reflection.

1.3.6 Infrastructure

Fiber-reinforced polymers have a great potential for replacing reinforced concrete and steel in bridges, buildings, and other civil infrastructures [15]. The principal reason for selecting these composites is their corrosion resistance, which leads to longer life and lower maintenance and repair costs. Reinforced concrete bridges tend to deteriorate after several years of use because of corrosion of steel-reinforcing bars (rebars) used in their construction. The corrosion problem is exacerbated because of deicing salt spread on the bridge road surface in winter months in many parts of the world. The deterioration can become so severe that the concrete surrounding the steel rebars can start to crack (due to the expansion of corroding steel bars) and ultimately fall off, thus weakening the structure's load-carrying capacity. The corrosion problem does not exist with fiber-reinforced polymers. Another advantage of using fiber-reinforced polymers for large bridge structures is their lightweight, which means lower dead weight for the bridge, easier transportation from the production factory (where the composite structure can be prefabricated) to the bridge location, easier hauling and installation, and less injuries to people in case of an earthquake. With lightweight construction, it is also possible to design bridges with longer span between the supports.

One of the early demonstrations of a composite traffic bridge was made in 1995 by Lockheed Martin Research Laboratories in Palo Alto, California. The bridge deck was a 9 m long × 5.4 m wide quarter-scale section and the material selected was E-glass fiber-reinforced polyester. The composite deck was a sandwich laminate of 15 mm thick E-glass fiber-reinforced polyester face sheets and a series of E-glass fiber-reinforced polyester tubes bonded together to form the core. The deck was supported on three U-shaped beams made of E-glass fabric-reinforced polyester. The design was modular and the components were stackable, which simplified both their transportation and assembly.

In recent years, a number of composite bridge decks have been constructed and commissioned for service in the United States and Canada. The Wickwire Run Bridge located in West Virginia, United States is an example of one such construction. It consists of full-depth hexagonal and half-depth trapezoidal profiles made of glass fabric-reinforced polyester matrix. The profiles are supported on steel beams. The road surface is a polymer-modified concrete. Another example of a composite bridge structure is shown in Figure 1.11, which replaced a 73 year old concrete bridge with steel rebars. The replacement was necessary because of the severe deterioration of the concrete deck, which reduced its load rating from 10 to only 4.3 t and was posing safety concerns. In

FIGURE 1.11 Glass fiber-reinforced vinyl ester pultruded sections in the construction of a bridge deck system. (Photograph provided by Strongwell Corporation. With permission.)

the composite bridge, the internal reinforcement for the concrete deck is a two-layer construction and consists primarily of pultruded I-section bars (I-bars) in the width direction (perpendicular to the direction of traffic) and pultruded round rods in the length directions. The material for the pultruded sections is glass fiber-reinforced vinyl ester. The internal reinforcement is assembled by inserting the round rods through the predrilled holes in I-bar webs and keeping them in place by vertical connectors.

Besides new bridge construction or complete replacement of reinforced concrete bridge sections, fiber-reinforced polymer is also used for upgrading, retrofitting, and strengthening damaged, deteriorating, or substandard concrete or steel structures [16]. For upgrading, composite strips and plates are attached in the cracked or damaged areas of the concrete structure using adhesive, wet layup, or resin infusion. Retrofitting of steel girders is accomplished by attaching composite plates to their flanges, which improves the flange stiffness and strength. The strengthening of reinforced concrete columns in earthquake prone areas is accomplished by wrapping them with fiber-reinforced composite jackets in which the fibers are primarily in the hoop direction. They are found to be better than steel jackets, since, unlike steel jackets, they do not increase the longitudinal stiffness of the columns. They are also much easier to install and they do not corrode like steel.

1.4 MATERIAL SELECTION PROCESS

Material selection is one of the most important and critical steps in the structural or mechanical design process. If the material selection is not done properly, the design may show poor performance; may require frequent maintenance, repair, or replacement; and in the extreme, may fail, causing damage, injuries, or fatalities. The material selection process requires the knowledge of the performance requirements of the structure or component under consideration. It also requires the knowledge of

1. Types of loading, for example, axial, bending, torsion, or combination thereof
2. Mode of loading, for example, static, fatigue, impact, shock, and so on
3. Service life
4. Operating or service environment, for example, temperature, humidity conditions, presence of chemicals, and so on
5. Other structures or components with which the particular design under consideration is required to interact
6. Manufacturing processes that can be used to produce the structure or the component
7. Cost, which includes not only the material cost, but also the cost of transforming the selected material to the final product, that is, the manufacturing cost, assembly cost, and so on

The material properties to consider in the material selection process depend on the performance requirements (mechanical, thermal, etc.) and the possible mode or modes of failure (e.g., yielding, brittle fracture, ductile failure, buckling, excessive deflection, fatigue, creep, corrosion, thermal failure due to overheating, etc.). Two basic material properties often used in the preliminary selection of materials for a structural or mechanical design are the modulus and strength. For a given design, the modulus is used for calculating the deformation, and the strength is used for calculating the maximum load-carrying capacity. Which property or properties should be considered in making a preliminary material selection depends on the application and the possible failure modes. For example, yield strength is considered if the design of a structure requires that no permanent deformation occurs because of the application of the load. If, on the other hand, there is a possibility of brittle failure because of the influence of the operating environmental conditions, fracture toughness is the material property to consider.

In many designs the performance requirement may include stiffness, which is defined as load per unit deformation. Stiffness should not be confused with modulus, since stiffness depends not only on the modulus of the material (which is a material property), but also on the design. For example, the stiffness of a straight beam with solid circular cross section depends not only on the modulus of the material, but also on its length, diameter, and how it is supported (i.e., boundary conditions). For a given beam length and support conditions, the stiffness of the beam with solid circular cross section is proportional to Ed^4, where E is the modulus of the beam material and d is its diameter. Therefore, the stiffness of this beam can be increased by either selecting a higher modulus material, or increasing the diameter, or doing both. Increasing the diameter is more effective in increasing the stiffness, but it also increases the weight and cost of the beam. In some designs, it may be possible to increase the beam stiffness by incorporating other design features, such as ribs, or by using a sandwich construction.

In designing structures with minimum mass or minimum cost, material properties must be combined with mass density (ρ), cost per unit mass ($/kg), and so on. For example, if the design objective for a tension linkage or a tie bar is to meet the stiffness performance criterion with minimum mass, the material selection criterion involves not just the tensile modulus of the material (E), but also the modulus–density ratio (E/ρ). The modulus–density ratio is a material index, and the material that produces the highest value of this material index should be selected for minimum mass design of the tension link. The material index depends on the application and the design objective. Table 1.6 lists the material indices for minimum mass design of a few simple structures.

As an example of the use of the material index in preliminary material selection, consider the carbon fiber–epoxy quasi-isotropic laminate in Table 1.1. Thin laminates of this type are considered well-suited for many aerospace applications [1], since they exhibit equal elastic properties (e.g., modulus) in all directions in the plane of load application. The quasi-isotropic laminate in Table 1.1 has an elastic modulus of 45.5 GPa, which is 34% lower than that of the 7178-T6 aluminum alloy and 59% lower than that of the Ti-6 Al-4V titanium alloy. The aluminum and the titanium alloys are the primary metallic alloys used in the construction of civilian and military aircrafts. Even though the quasi-isotropic carbon fiber–epoxy composite laminate has a lower modulus, it is a good candidate for substituting the metallic alloys in stiffness-critical aircraft structures. This is because the carbon fiber–epoxy quasi-isotropic laminate has a superior material index in minimum mass design of stiffness-critical structures. This can be easily verified by comparing the values of the material index $\frac{E^{1/3}}{\rho}$ of all three materials, assuming that the structure can be modeled as a thin plate under bending load.

TABLE 1.6
Material Index for Stiffness and Strength-Critical Designs at Minimum Mass

Structure	Constraints	Design Variable	Material Index: Stiffness-Critical Design	Material Index: Strength-Critical Design
Round tie bar loaded in axial tension	Length fixed	Diameter	$\frac{E}{\rho}$	$\frac{S_f}{\rho}$
Rectangular beam loaded in bending	Length and width fixed	Height	$\frac{E^{1/3}}{\rho}$	$\frac{S_f^{1/2}}{\rho}$
Round bar or shaft loaded in torsion	Length fixed	Diameter	$\frac{G^{1/2}}{\rho}$	$\frac{S_f^{2/3}}{\rho}$
Flat plate loaded in bending	Length and width fixed	Thickness	$\frac{E^{1/3}}{\rho}$	$\frac{S_f^{1/2}}{\rho}$
Round column loaded in compression	Length	Diameter	$\frac{E^{1/2}}{\rho}$	$\frac{S_f}{\rho}$

Source: Adapted from Ashby, M.F., *Material Selection in Mechanical Design*, 3rd Ed., Elsevier, Oxford, UK, 2005.

Note: ρ = mass density, E = Young's modulus, G = shear modulus, and S_f = strength.

Weight reduction is often the principal consideration for selecting fiber-reinforced polymers over metals, and for many applications, they provide a higher material index than metals, and therefore, suitable for minimum mass design. Depending on the application, there are other advantages of using fiber-reinforced composites, such as higher damping, no corrosion, parts integration, control of thermal expansion, and so on, that should be considered as well, and some of these advantages add value to the product that cannot be obtained with metals. One great advantage is the tailoring of properties according to the design requirements, which is demonstrated in the example of load-bearing orthopedic implants [17]. One such application is the bone plate used for bone fracture fixation. In this application, the bone plate is attached to the bone fracture site with screws to hold the fractured pieces in position, reduce the mobility at the fracture interface, and provide the required stress-shielding of the bone for proper healing. Among the biocompatible materials used for orthopedic implants, stainless steel and titanium are the two most common materials used for bone plates. However, the significantly higher modulus of both of these materials than that of bone creates excessive stress-shielding, that is, they share the higher proportion of the compressive stresses during healing than the bone. The advantage of using fiber-reinforced polymers is that they can be designed to match the modulus of bone, and indeed, this is the reason for selecting carbon fiber-reinforced epoxy or polyether ether ketone (PEEK) for such an application [18,19].

REFERENCES

1. C.E. Harris, J.H. Starnes, Jr., and M.J. Shuart, Design and manufacturing of aerospace composite structures, state-of-the-art assessment, *J. Aircraft, 39*:545 (2002).
2. C. Soutis, Carbon fiber reinforced plastics in aircraft applications, *Mater. Sci. Eng., A, 412*:171 (2005).
3. J.V. Noyes, Composites in the construction of the Lear Fan 2100 aircraft, *Composites, 14*:129 (1983).
4. R.L. Pinckney, Helicopter rotor blades, *Appl. Composite Mat., ASTM STP, 524*:108 (1973).
5. N.R. Adsit, Composites for space applications, *ASTM Standardization News*, December (1983).
6. H. Bansemir and O. Haider, Fibre composite structures for space applications—recent and future developments, *Cryogenics, 38*:51 (1998).
7. E.G. Wolff, Dimensional stability of structural composites for spacecraft applications, *Metal Prog., 115*:54 (1979).
8. J. Guthrie, T.B. Tolle, and D.H. Rose, Composites for orbiting platforms, *AMPTIAC Q., 8*:51 (2004).
9. D.A. Riegner, Composites in the automotive industry, *ASTM Standardization News*, December (1983).
10. P. Beardmore, Composite structures for automobiles, *Compos. Struct., 5*:163 (1986).
11. G. Savage, Composite materials in Formula 1 racing, *Metals Mater., 7*:617 (1991).
12. V.P. McConnell, Application of composites in sporting goods, *Comprehensive Composite Materials*, Vol. 6, Elsevier, Amsterdam, pp. 787–809 (2000).
13. W. Chalmers, The potential for the use of composite materials in marine structures, *Mar. Struct., 7*:441 (1994).
14. A.P. Mouritz, E. Gellert, P. Burchill, and K. Challis, Review of advanced composite structures for naval ships and submarines, *Compos. Struct., 53*:21 (2001).
15. L.C. Hollaway, The evolution of and the way forward for advanced polymer composites in the civil infrastructure, *Construct. Build. Mater., 17*:365 (2003).
16. V.M. Karbhari, Fiber reinforced composite bridge systems—transition from the laboratory to the field, *Compos. Struct., 66*:5 (2004).
17. S.L. Evans and P.J. Gregson, Composite technology in load-bearing orthopedic implants, *Biomaterials, 19*:1329 (1998).
18. M.S. Ali, T.A. French, G.W. Hastings, T. Rae, N. Rushton, E.R.S. Ross, and C.H. Wynn-Jones, Carbon fibre composite bone plates, *J. Bone Joint Surg., 72-B*:586 (1990).
19. Z.-M. Huang and K. Fujihara, Stiffness and strength design of composite bone plate, *Compos. Sci. Tech., 65*:73 (2005).

PROBLEMS*

P1.1. The modulus and tensile strength of a SMC composite are reported as 15 GPa and 230 MPa, respectively. Assuming that the density of the SMC is 1.85 g/cm^3, calculate the strength–weight ratio and modulus–weight ratio for the composite.

P1.2. The material of a tension member is changed from AISI 4340 steel to a unidirectional high-modulus (HM) carbon fiber-reinforced epoxy.
1. Calculate the ratio of the cross-sectional areas, axial stiffnesses, and weights of these two members for equal load-carrying capacities
2. Calculate the ratio of the cross-sectional areas, load-carrying capacities, and weights of these two members for equal axial stiffness (*Hint*: Load = strength × cross-sectional area, and axial stiffness = modulus × cross-sectional area)

P1.3. Compare the heights and weights of three rectangular beams made of aluminum alloy 6061-T6, titanium alloy Ti-6Al-4V, and a unidirectional high-strength (HS) carbon fiber-reinforced epoxy designed to posses (a) equal bending stiffness and (b) equal bending moment carrying capacity. Assume that all three beams have the same length and width. (*Hint*: The bending stiffness is proportional to Eh^3 and the bending moment is proportional to Sh^2, where E, S, and h are the modulus, strength, and height, respectively.)

P1.4. Calculate the flexural (bending) stiffness ratio of two cantilever beams of equal weight, one made of AISI 4340 steel and the other made of a unidirectional high-modulus carbon fiber–epoxy composite. Assume that both beams have the same length and width, but different heights. If the beams were simply supported instead of fixed at one end like a cantilever beam, how will the ratio change?

P1.5. In a certain application, a steel beam of round cross section (diameter = 10 mm) is to be replaced by a unidirectional fiber-reinforced epoxy beam of equal length. The composite beam is designed to have a natural frequency of vibration 25% higher than that of the steel beam. Among the fibers to be considered are high-strength carbon fiber, high-modulus carbon fiber, and Kevlar 49 (see Table 1.1). Select one of these fibers on the basis of minimum weight for the beam.

Note that the natural frequency of vibration of a beam is given by the following equation:

---

* Use Table 1.1 for material properties if needed.

$$w_n = C\left(\frac{EI}{mL^4}\right)^{1/2},$$

where

w_n = fundamental natural frequency
C = a constant that depends on the beam support conditions
E = modulus of the beam material
I = moment of inertia of the beam cross section
m = mass per unit length of the beam
L = beam length

P1.6. The material of a thin flat panel is changed from SAE 1010 steel panel (thickness = 1.5 mm) an E-glass fiber-reinforced polyester SMC panel with equal flexural stiffness. Calculate the percentage weight and cost differences between the two panels. Note that the panel stiffness is $\frac{Eh^3}{12(1-\nu^2)}$, where E is the modulus of the panel material, ν is the Poisson's ratio of the panel material, and h is the panel thickness. The following is known for the two materials:

	Steel	SMC
Modulus (GPa)	207	16
Poisson's ratio	0.30	0.30
Density (g/cm^3)	7.87	1.85
Cost ($/kg)	0.80	1.90

Suggest an alternative design approach by which the wall thickness of the flat SMC panel can be reduced without lowering its flexural stiffness.

P1.7. To reduce the material cost, an engineer decides to use a hybrid beam instead of an all-carbon fiber beam. Both beams have the same overall dimensions. The hybrid beam contains carbon fibers in the outer layers and either E-glass or Kevlar 49 fibers in the core. The matrix used is an epoxy. Costs of these materials are as follows:

Carbon fiber–epoxy matrix: $25.00/lb
E-glass fiber–epoxy matrix: $1.20/lb
Kevlar 49 fiber–epoxy matrix: $8.00/lb

The total carbon fiber–epoxy thickness in the hybrid beam is equal to the core thickness. Compare the percentage weight penalty and cost savings for each hybrid beam over an all-carbon fiber beam. Do you expect both all-carbon and hybrid beams to have the same bending stiffness? If the answer is "no," what can be done to make the two stiffnesses equal?

P1.8. The shear modulus (G) of steel and a quasi-isotropic carbon fiber–epoxy is 78 and 17 GPa, respectively. The mean diameter (D) of a thin-walled steel torque tube is 25 mm and its wall thickness (t) is 3 mm. Knowing that the torsional stiffness of a thin-walled tube is proportional to D^3tG, calculate:

1. Mean diameter of a composite tube that has the same torsional stiffness and wall thickness as the steel tube
2. Wall thickness of a composite tube that has the same torsional stiffness and mean diameter as the steel tube
3. Difference in weight (in percentage) between the steel tube and the composite tube in each of the previous cases, assuming equal length for both tubes

P1.9. Using the information in Problem P1.8, design a composite torque tube that is 30% lighter than the steel tube but has the same torsional stiffness. Will the axial stiffnesses of these two tubes be the same?

P1.10. Write the design and material selection considerations for each of the following applications and discuss why fiber-reinforced polymers can be a good candidate material for each application.

1. Utility poles
2. Aircraft floor panels
3. Aircraft landing gear doors
4. Household step ladders
5. Wind turbine blades
6. Suspension arms of an automobile
7. Drive shaft of an automobile
8. Underground gasoline storage tanks
9. Hydrogen storage tanks for fuel cell vehicles
10. Leg prosthetic
11. Flywheel for energy storage

2 Materials

Major constituents in a fiber-reinforced composite material are the reinforcing fibers and a matrix, which acts as a binder for the fibers. Other constituents that may also be found are coupling agents, coatings, and fillers. Coupling agents and coatings are applied on the fibers to improve their wetting with the matrix as well as to promote bonding across the fiber–matrix interface. Both in turn promote a better load transfer between the fibers and the matrix. Fillers are used with some polymeric matrices primarily to reduce cost and improve their dimensional stability.

Manufacturing of a composite structure starts with the incorporation of a large number of fibers into a thin layer of matrix to form a *lamina* (ply). The thickness of a lamina is usually in the range of 0.1–1 mm (0.004–0.04 in.). If continuous (long) fibers are used in making the lamina, they may be arranged either in a unidirectional orientation (i.e., all fibers in one direction, Figure 2.1a), in a bidirectional orientation (i.e., fibers in two directions, usually normal to each other, Figure 2.1b), or in a multidirectional orientation (i.e., fibers in more than two directions, Figure 2.1c). The bi- or multidirectional orientation of fibers is obtained by weaving or other processes used in the textile industry. The bidirectional orientations in the form of fabrics are shown in Appendix A.1. For a lamina containing unidirectional fibers, the composite material has the highest strength and modulus in the longitudinal direction of the fibers. However, in the transverse direction, its strength and modulus are very low. For a lamina containing bidirectional fibers, the strength and modulus can be varied using different amounts of fibers in the longitudinal and transverse directions. For a balanced lamina, these properties are the same in both directions.

A lamina can also be constructed using discontinuous (short) fibers in a matrix. The discontinuous fibers can be arranged either in unidirectional orientation (Figure 2.1c) or in random orientation (Figure 2.1d). Discontinuous fiber-reinforced composites have lower strength and modulus than continuous fiber composites. However, with random orientation of fibers (Figure 2.1e), it is possible to obtain equal mechanical and physical properties in all directions in the plane of the lamina.

The thickness required to support a given load or to maintain a given deflection in a fiber-reinforced composite structure is obtained by stacking several laminas in a specified sequence and then consolidating them to form a *laminate*. Various laminas in a laminate may contain fibers either all in one

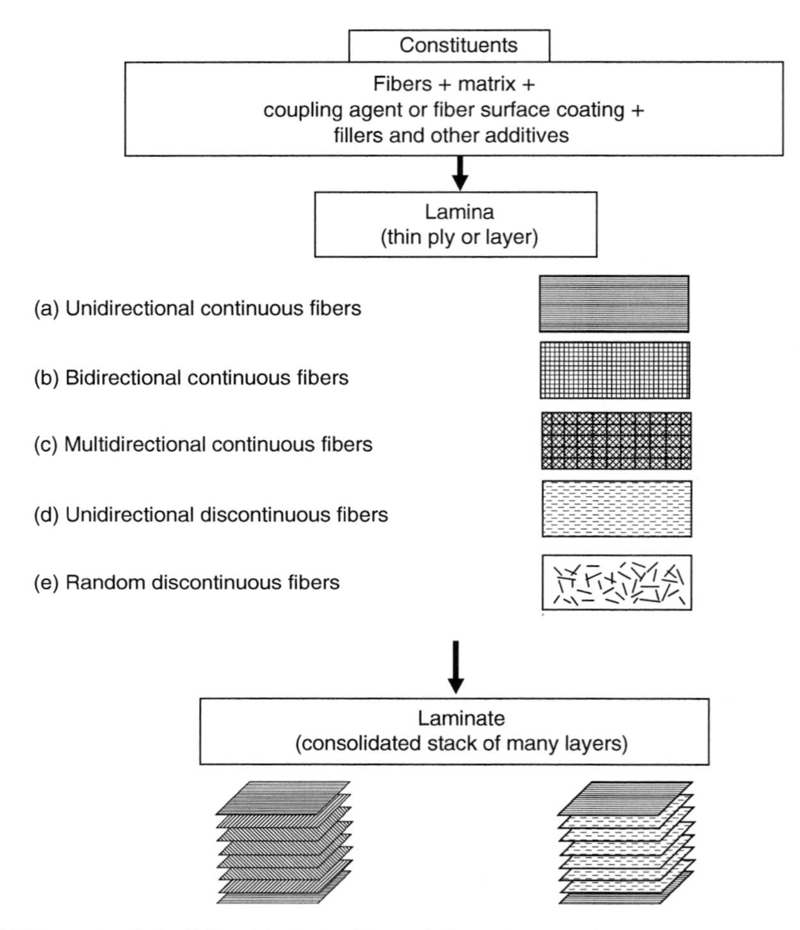

FIGURE 2.1 Basic building blocks in fiber-reinforced composites.

direction or in different directions. It is also possible to combine different kinds of fibers to form either an interply or an intraply hybrid laminate. An *interply* hybrid laminate consists of different kinds of fibers in different laminas, whereas an *intraply* hybrid laminate consists of two or more different kinds of fibers interspersed in the same lamina. Generally, the same matrix is used throughout the laminate so that a coherent interlaminar bond is formed between the laminas.

Fiber-reinforced polymer laminas can also be combined with thin aluminum or other metallic sheets to form metal–composite hybrids, commonly known as *fiber metal laminates* (FML). Two such commercial metal–composite hybrids are ARALL and GLARE. *ARALL* uses alternate layers of aluminum sheets and unidirectional aramid fiber–epoxy laminates

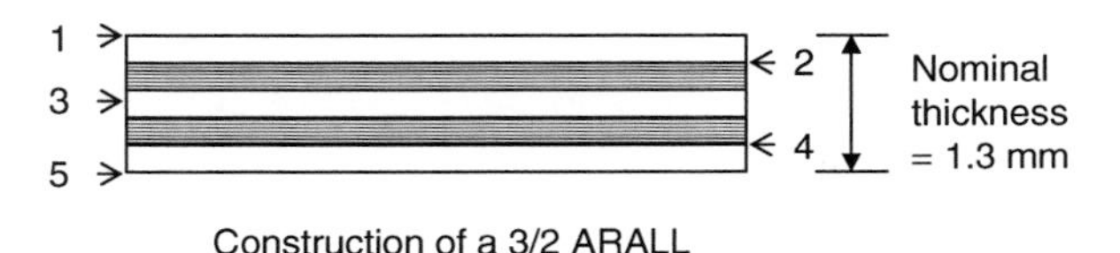

FIGURE 2.2 Construction of an ARALL laminate.

(Figure 2.2). *GLARE* uses alternate layers of aluminum sheets and either unidirectional or bidirectional S-glass fiber–epoxy laminates. Both metal–composite hybrids have been primarily developed for aircraft structures such as wing panels and fuselage sections.

2.1 FIBERS

Fibers are the principal constituents in a fiber-reinforced composite material. They occupy the largest volume fraction in a composite laminate and share the major portion of the load acting on a composite structure. Proper selection of the fiber type, fiber volume fraction, fiber length, and fiber orientation is very important, since it influences the following characteristics of a composite laminate:

1. Density
2. Tensile strength and modulus
3. Compressive strength and modulus
4. Fatigue strength as well as fatigue failure mechanisms
5. Electrical and thermal conductivities
6. Cost

A number of commercially available fibers and their properties are listed in Table 2.1. The first point to note in this table is the extremely small filament diameter for the fibers. Since such small sizes are difficult to handle, the useful form of commercial fibers is a bundle, which is produced by gathering a large number of continuous filaments, either in untwisted or twisted form. The untwisted form is called *strand* or *end* for glass and Kevlar fibers and *tow* for carbon fibers (Figure 2.3a). The twisted form is called *yarn* (Figure 2.3b).

Tensile properties listed in Table 2.1 are the average values reported by the fiber manufacturers. One of the test methods used for determining the tensile properties of filaments is the single filament test. In this test method, designated as ASTM D3379, a single filament is mounted along the centerline of a slotted

TABLE 2.1
Properties of Selected Commercial Reinforcing Fibers

Fiber	Typical Diameter (μm)[a]	Density (g/cm^3)	Tensile Modulus GPa (Msi)	Tensile Strength GPa (ksi)	Strain-to-Failure (%)	Coefficient of Thermal Expansion (10^{-6}/°C)[b]	Poisson's Ratio
Glass							
E-glass	10 (round)	2.54	72.4 (10.5)	3.45 (500)	4.8	5	0.2
S-glass	10 (round)	2.49	86.9 (12.6)	4.30 (625)	5.0	2.9	0.22
PAN carbon							
T-300[c]	7 (round)	1.76	231 (33.5)	3.65 (530)	1.4	−0.6 (longitudinal) 7–12 (radial)	0.2
AS-1[d]	8 (round)	1.80	228 (33)	3.10 (450)	1.32		
AS-4[d]	7 (round)	1.80	248 (36)	4.07 (590)	1.65		
T-40[c]	5.1 (round)	1.81	290 (42)	5.65 (820)	1.8	−0.75 (longitudinal)	
IM-7[d]	5 (round)	1.78	301 (43.6)	5.31 (770)	1.81		
HMS-4[d]	8 (round)	1.80	345 (50)	2.48 (360)	0.7		
GY-70[e]	8.4 (bilobal)	1.96	483 (70)	1.52 (220)	0.38		
Pitch carbon							
P-55[c]	10	2.0	380 (55)	1.90 (275)	0.5	−1.3 (longitudinal)	
P-100[c]	10	2.15	758 (110)	2.41 (350)	0.32	−1.45 (longitudinal)	
Aramid							
Kevlar 49[f]	11.9 (round)	1.45	131 (19)	3.62 (525)	2.8	−2 (longitudinal) 59 (radial)	0.35
Kevlar 149[f]		1.47	179 (26)	3.45 (500)	1.9		
Technora[g]		1.39	70 (10.1)	3.0 (435)	4.6	−6 (longitudinal)	

Extended chain polyethylene								
Spectra 900[h]	38	0.97	117 (17)	2.59 (375)	3.5			
Spectra 1000[h]	27	0.97	172 (25)	3.0 (435)	2.7			
Boron	140 (round)	2.7	393 (57)	3.1 (450)	0.79	5		0.2
SiC								
Monofilament	140 (round)	3.08	400 (58)	3.44 (499)	0.86	1.5		
Nicalon[i] (multifilament)	14.5 (round)	2.55	196 (28.4)	2.75 (399)	1.4			
Al_2O_3								
Nextel 610[j]	10–12 (round)	3.9	380 (55)	3.1 (450)		8		
Nextel 720[j]	10–12	3.4	260 (38)	2.1 (300)		6		
Al_2O_3–SiO_2								
Fiberfrax (discontinuous)	2–12	2.73	103 (15)	1.03–1.72 (150–250)				

[a] 1 μm = 0.0000393 in.
[b] m/m per °C = 0.556 in./in. per °F.
[c] Amoco.
[d] Hercules.
[e] BASF.
[f] DuPont.
[g] Teijin.
[h] Honeywell.
[i] Nippon carbon.
[j] 3-M.

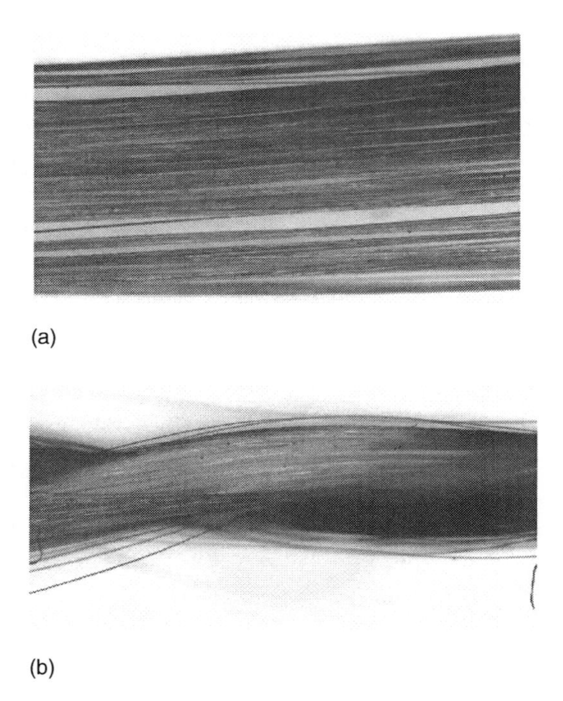

FIGURE 2.3 (a) Untwisted and (b) twisted fiber bundle.

tab by means of a suitable adhesive (Figure 2.4). After clamping the tab in the grips of a tensile testing machine, its midsection is either cut or burned away. The tension test is carried out at a constant loading rate until the filament

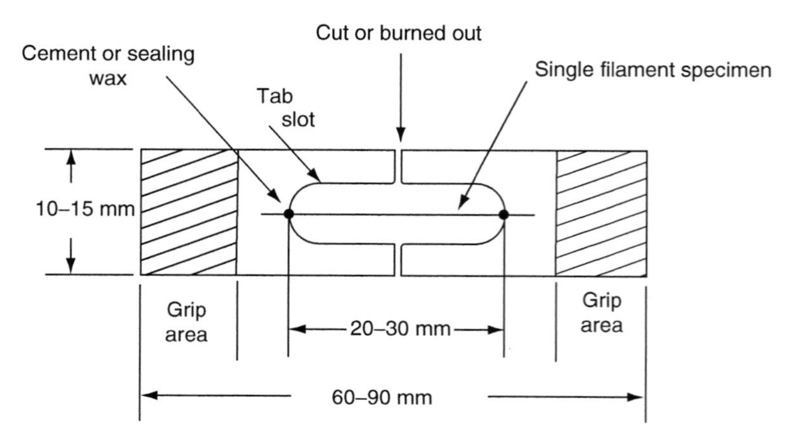

FIGURE 2.4 Mounting tab for tensile testing of single filament.

fractures. From the load-time record of the test, the following tensile properties are determined:

$$\text{Tensile strength } \sigma_{\text{fu}} = \frac{F_{\text{u}}}{A_{\text{f}}} \tag{2.1}$$

and

$$\text{Tensile modulus } E_{\text{f}} = \frac{L_{\text{f}}}{CA_{\text{f}}}, \tag{2.2}$$

where

F_{u} = force at failure
A_{f} = average filament cross-sectional area, measured by a planimeter from the photomicrographs of filament ends
L_{f} = gage length
C = true compliance, determined from the chart speed, loading rate, and the system compliance

Tensile stress–strain diagrams obtained from single filament test of reinforcing fibers in use are almost linear up to the point of failure, as shown in Figure 2.5. They also exhibit very low strain-to-failure and a brittle failure mode. Although the absence of yielding does not reduce the load-carrying capacity of the fibers, it does make them prone to damage during handling as well as during contact with other surfaces. In continuous manufacturing operations, such as filament winding, frequent fiber breakage resulting from such damages may slow the rate of production.

The high tensile strengths of the reinforcing fibers are generally attributed to their filamentary form in which there are statistically fewer surface flaws than in the bulk form. However, as in other brittle materials, their tensile strength data exhibit a large amount of scatter. An example is shown in Figure 2.6.

The experimental strength variation of brittle filaments is modeled using the following Weibull distribution function [1]:

$$f(\sigma_{\text{fu}}) = \alpha\sigma_{\text{o}}^{-\alpha}\sigma_{\text{fu}}^{\alpha-1}\left(\frac{L_{\text{f}}}{L_{\text{o}}}\right)\exp\left[-\left(\frac{L_{\text{f}}}{L_{\text{o}}}\right)\left(\frac{\sigma_{\text{fu}}}{\sigma_{\text{o}}}\right)^{\alpha}\right], \tag{2.3}$$

where

$f(\sigma_{\text{fu}})$ = probability of filament failure at a stress level equal to σ_{fu}
σ_{fu} = filament strength
L_{f} = filament length
L_{o} = reference length
α = shape parameter
σ_{o} = scale parameter (the filament strength at $L_{\text{f}} = L_{\text{o}}$)

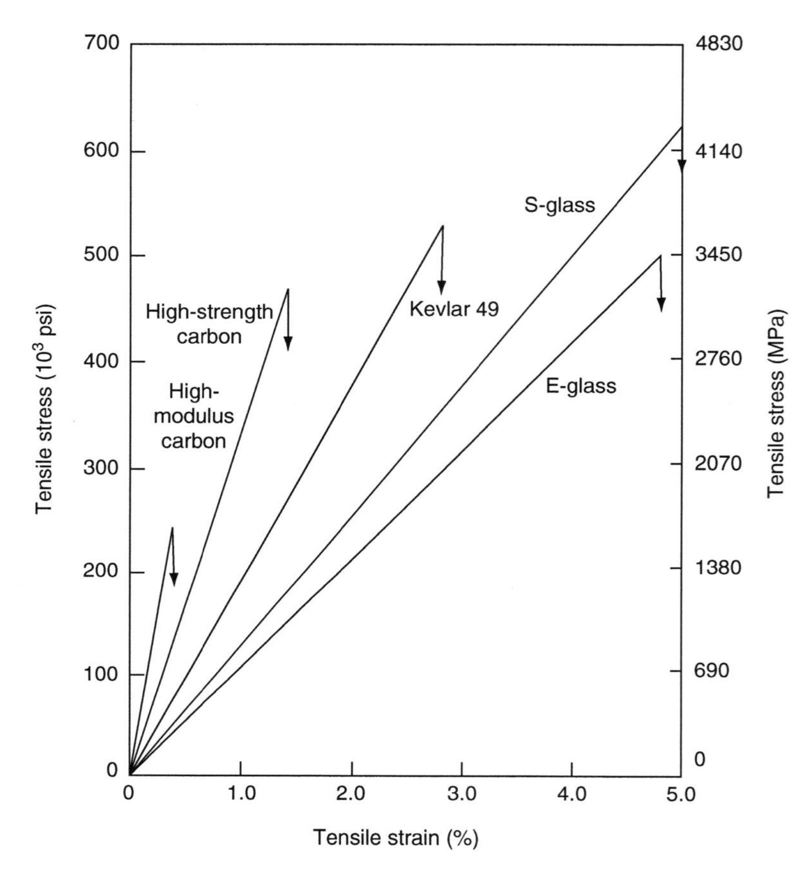

FIGURE 2.5 Tensile stress–strain diagrams for various reinforcing fibers.

The cumulative distribution of strength is given by the following equation:

$$F(\sigma_{fu}) = 1 - \exp\left[-\left(\frac{L_f}{L_o}\right)\left(\frac{\sigma_{fu}}{\sigma_o}\right)^{\alpha}\right], \tag{2.4}$$

where $F(\sigma_{fu})$ represents the probability of filament failure at a stress level lower than or equal to σ_{fu}. The parameters α and σ_o in Equations 2.3 and 2.4 are called the Weibull parameters, and are determined using the experimental data. α can be regarded as an inverse measure of the coefficient of variation. The higher the value of α, the narrower is the distribution of filament strength. The scale parameter σ_o may be regarded as a reference stress level.

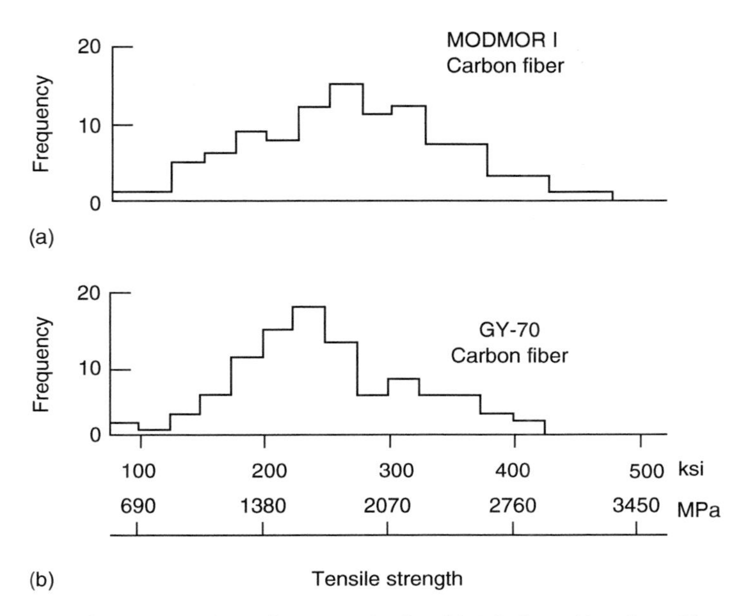

FIGURE 2.6 Histograms of tensile strengths for (a) Modmor I carbon fibers and (b) GY-70 carbon fibers. (After McMahon, P.E., *Analysis of the Test Methods for High Modulus Fibers and Composites, ASTM STP*, 521, 367, 1973.)

The mean filament strength $\overline{\sigma}_{\mathrm{fu}}$ is given by

$$\overline{\sigma}_{\mathrm{fu}} = \sigma_{\mathrm{o}} \left(\frac{L_{\mathrm{f}}}{L_{\mathrm{o}}} \right)^{-1/\alpha} \Gamma\left(1 + \frac{1}{\alpha} \right), \qquad (2.5)$$

where Γ represents a gamma function. Equation 2.5 clearly shows that the mean strength of a brittle filament decreases with increasing length. This is also demonstrated in Figure 2.7.

Tensile properties of fibers can also be determined using fiber bundles. It has been observed that even though the tensile strength distribution of individual filaments follows the Weibull distribution, the tensile strength distribution of fiber bundles containing a large number of parallel filaments follows a normal distribution [1]. The maximum strength, σ_{fm}, that the filaments in the bundle will exhibit and the mean bundle strength, $\overline{\sigma}_{\mathrm{b}}$, can be expressed in terms of the Weibull parameters determined for individual filaments. They are given as follows:

$$\sigma_{\mathrm{fm}} = \sigma_{\mathrm{o}} \left[\left(\frac{L_{\mathrm{f}}}{L_{\mathrm{o}}} \right) \alpha \right]^{-1/\alpha},$$

$$\overline{\sigma}_{\mathrm{b}} = \sigma_{\mathrm{o}} \left[\left(\frac{L_{\mathrm{f}}}{L_{\mathrm{o}}} \right) \alpha \right]^{-1/\alpha} \mathrm{e}^{-1/\alpha}. \qquad (2.6)$$

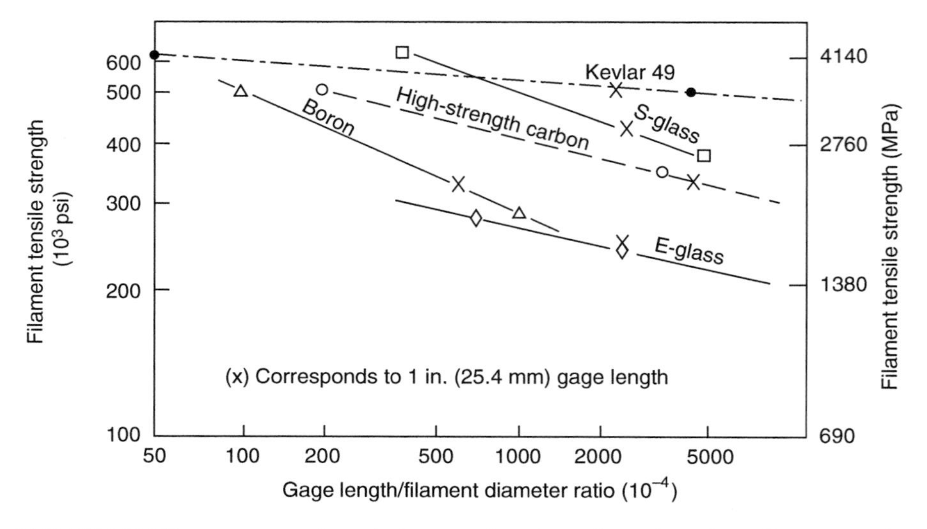

FIGURE 2.7 Filament strength variation as a function of gage length-to-diameter ratio. (After Kevlar 49 Data Manual, E. I. duPont de Nemours & Co., 1975.)

The fiber bundle test method is similar to the single filament test method. The fiber bundle can be tested either in dry or resin-impregnated condition. Generally, the average tensile strength and modulus of fiber bundles are lower than those measured on single filaments. Figure 2.8 shows the stress–strain diagram of a dry glass fiber bundle containing 3000 filaments. Even though a single glass filament shows a linear tensile stress–strain diagram until failure, the glass fiber strand shows not only a nonlinear stress–strain diagram before reaching the maximum stress, but also a progressive failure after reaching the maximum stress. Both nonlinearity and progressive failure occur due to the statistical distribution of the strength of glass filaments. The weaker filaments in the bundle fail at low stresses, and the surviving filaments continue to carry the tensile load; however, the stress in each surviving filament becomes higher. Some of them fail as the load is increased. After the maximum stress is reached, the remaining surviving filaments continue to carry even higher stresses and start to fail, but not all at one time, thus giving the progressive failure mode as seen in Figure 2.8. Similar tensile stress–strain diagrams are observed with carbon and other fibers in fiber bundle tests.

In addition to tensile properties, compressive properties of fibers are also of interest in many applications. Unlike the tensile properties, the compressive properties cannot be determined directly by simple compression tests on filaments or strands. Various indirect methods have been used to determine the compressive strength of fibers [2]. One such method is the loop test in which a filament is bent into the form of a loop until it fails. The compressive strength

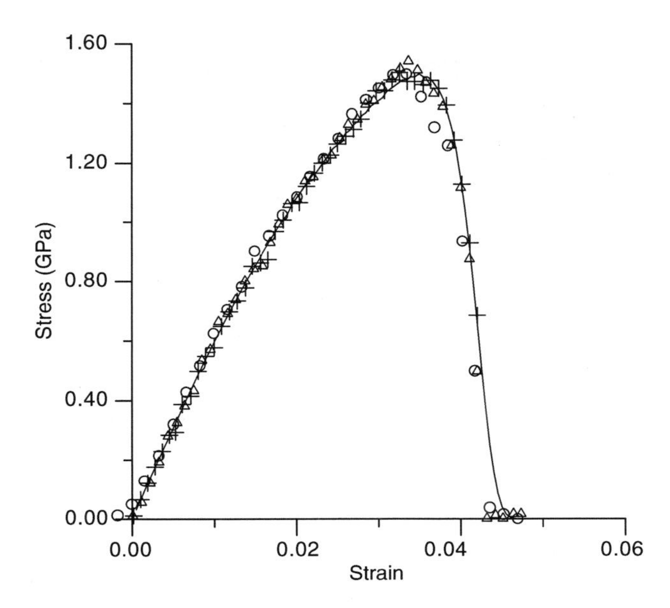

FIGURE 2.8 Tensile stress–strain diagram of an untwisted E-glass fiber bundle containing 3000 filaments.

of the fiber is determined from the compressive strain at the fiber surface. In general, compressive strength of fibers is lower than their tensile strength, as is shown in Table 2.2. The compressive strength of boron fibers is higher than that of carbon and glass fibers. All organic fibers have low compressive

TABLE 2.2
Compressive Strength of a Few Selected Reinforcing Fibers

Fiber	Tensile Strength[a] (GPa)	Compressive Strength[a] (GPa)
E-glass fiber	3.4	4.2
T-300 carbon fiber	3.2	2.7–3.2
AS 4 carbon fiber	3.6	2.7
GY-70 carbon fiber	1.86	1.06
P100 carbon fiber	2.2	0.5
Kevlar 49 fiber	3.5	0.35–0.45
Boron	3.5	5

Source: Adapted from Kozey, V.V., Jiang, H., Mehta, V.R., and Kumar, S., *J. Mater. Res.*, 10, 1044, 1995.

[a] In the longitudinal direction.

strength. This includes Kevlar 49, which has a compressive strength almost 10 times lower than its tensile strength.

2.1.1 Glass Fibers

Glass fibers are the most common of all reinforcing fibers for polymeric matrix composites (PMC). The principal advantages of glass fibers are low cost, high tensile strength, high chemical resistance, and excellent insulating properties. The disadvantages are relatively low tensile modulus and high density (among the commercial fibers), sensitivity to abrasion during handling (which frequently decreases its tensile strength), relatively low fatigue resistance, and high hardness (which causes excessive wear on molding dies and cutting tools).

The two types of glass fibers commonly used in the fiber-reinforced plastics (FRP) industry are E-glass and S-glass. Another type, known as C-glass, is used in chemical applications requiring greater corrosion resistance to acids than is provided by E-glass. E-glass has the lowest cost of all commercially available reinforcing fibers, which is the reason for its widespread use in the FRP industry. S-glass, originally developed for aircraft components and missile casings, has the highest tensile strength among all fibers in use. However, the compositional difference and higher manufacturing cost make it more expensive than E-glass. A lower cost version of S-glass, called S-2-glass, is also available. Although S-2-glass is manufactured with less-stringent nonmilitary specifications, its tensile strength and modulus are similar to those of S-glass.

The chemical compositions of E- and S-glass fibers are shown in Table 2.3. As in common soda-lime glass (window and container glasses), the principal ingredient in all glass fibers is silica (SiO_2). Other oxides, such as B_2O_3 and Al_2O_3, are added to modify the network structure of SiO_2 as well as to improve its workability. Unlike soda-lime glass, the Na_2O and K_2O content in E- and S-glass fibers is quite low, which gives them a better corrosion resistance to water as well as higher surface resistivity. The internal structure of glass fibers is a three-dimensional, long network of silicon, oxygen, and other atoms arranged in a random fashion. Thus, glass fibers are amorphous (noncrystalline) and isotropic (equal properties in all directions).

TABLE 2.3
Typical Compositions of Glass Fibers (in wt%)

Type	SiO_2	Al_2O_3	CaO	MgO	B_2O_3	Na_2O
E-glass	54.5	14.5	17	4.5	8.5	0.5
S-glass	64	26	—	10	—	—

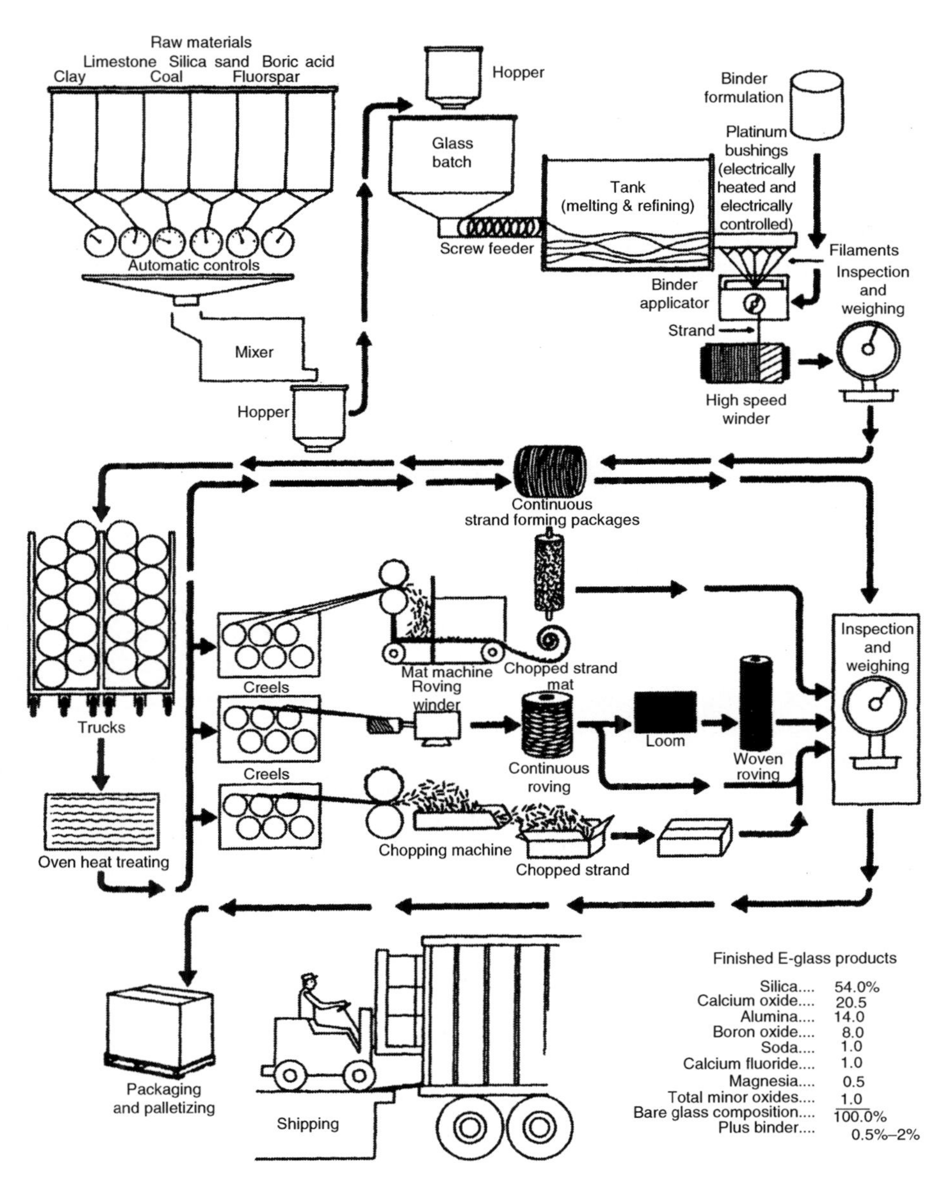

FIGURE 2.9 Flow diagram in glass fiber manufacturing. (Courtesy of PPG Industries. With permission.)

The manufacturing process for glass fibers is depicted in the flow diagram in Figure 2.9. Various ingredients in the glass formulation are first dry-mixed and melted in a refractory furnace at about 1370°C. The molten glass is exuded through a number of orifices contained in a platinum bushing and rapidly drawn into filaments of ~10 μm in diameter. A protective coating (size) is then applied on individual filaments before they are gathered together into a strand and wound on a drum. The coating or size is a mixture of lubricants (which prevent abrasion between the filaments), antistatic agents (which reduce static friction between the filaments), and a binder (which packs the filaments together into a strand). It may also contain small percentages of a coupling agent that promotes adhesion between fibers and the specific polymer matrix for which it is formulated.

The basic commercial form of continuous glass fibers is a *strand*, which is a collection of parallel filaments numbering 204 or more. Other common forms of glass fibers are illustrated in Figure 2.10. A *roving* is a group of untwisted parallel strands (also called *ends*) wound on a cylindrical *forming package*. Rovings are used in continuous molding operations, such as filament winding and pultrusion. They can also be preimpregnated with a thin layer of polymeric resin matrix to form *prepregs*. Prepregs are subsequently cut into required dimensions, stacked, and cured into the final shape in batch molding operations, such as compression molding and hand layup molding.

Chopped strands are produced by cutting continuous strands into short lengths. The ability of the individual filaments to hold together during or after the chopping process depends largely on the type and amount of the size applied during fiber manufacturing operation. Strands of high integrity are called "hard" and those that separate more readily are called "soft." Chopped strands ranging in length from 3.2 to 12.7 mm (0.125–0.5 in.) are used in injection-molding operations. Longer strands, up to 50.8 mm (2 in.) in length, are mixed with a resinous binder and spread in a two-dimensional random fashion to form chopped strand mats (CSMs). These mats are used mostly for hand layup moldings and provide nearly equal properties in all directions in the plane of the structure. *Milled glass fibers* are produced by grinding continuous strands in a hammer mill into lengths ranging from 0.79 to 3.2 mm (0.031–0.125 in.). They are primarily used as a filler in the plastics industry and do not possess any significant reinforcement value.

Glass fibers are also available in woven form, such as *woven roving* or *woven cloth*. Woven roving is a coarse drapable fabric in which continuous rovings are woven in two mutually perpendicular directions. Woven cloth is weaved using twisted continuous strands, called *yarns*. Both woven roving and cloth provide bidirectional properties that depend on the style of weaving as well as relative fiber counts in the length (*warp*) and crosswise (*fill*) directions (See Appendix A.1). A layer of woven roving is sometimes bonded with a layer of CSM to produce a woven roving mat. All of these forms of glass fibers are suitable for hand layup molding and liquid composite molding.

The average tensile strength of freshly drawn glass fibers may exceed 3.45 GPa (500,000 psi). However, surface damage (flaws) produced by abrasion, either by rubbing against each other or by contact with the processing

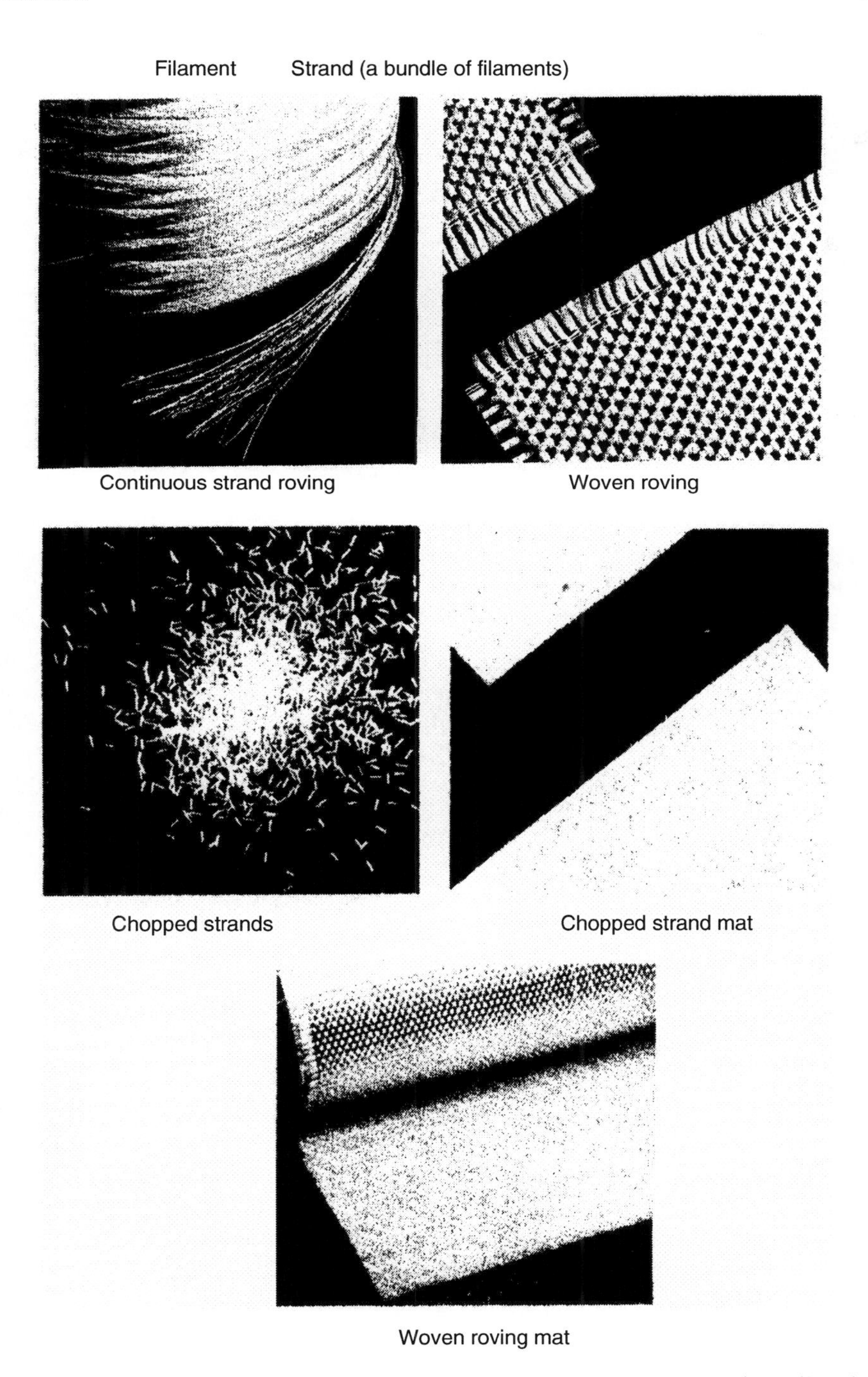

FIGURE 2.10 Common forms of glass fibers. (Courtesy of Owens Corning Fiberglas Corporation.)

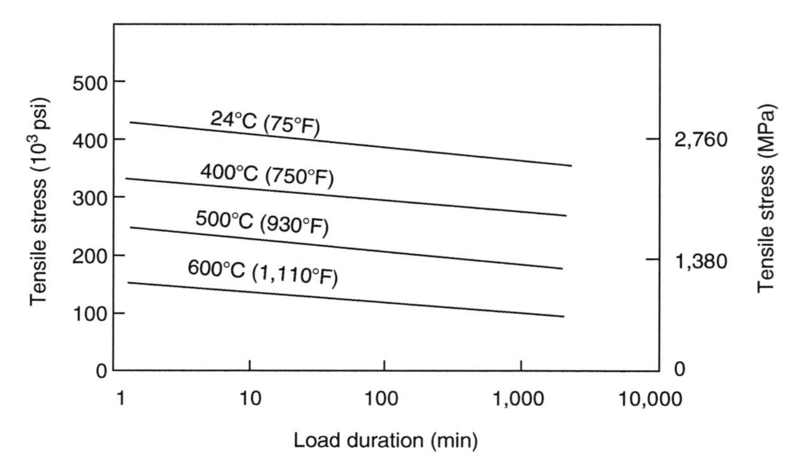

FIGURE 2.11 Reduction of tensile stress in E-glass fibers as a function of time at various temperatures. (After Otto, W.H., Properties of glass fibers at elevated temperatures, Owens Corning Fiberglas Corporation, AD 228551, 1958.)

equipment, tends to reduce it to values that are in the range of 1.72–2.07 GPa (250,000–300,000 psi). Strength degradation is increased as the surface flaws grow under cyclic loads, which is one of the major disadvantages of using glass fibers in fatigue applications. Surface compressive stresses obtained by alkali ion exchange [3] or elimination of surface flaws by chemical etching may reduce the problem; however, commercial glass fibers are not available with any such surface modifications.

The tensile strength of glass fibers is also reduced in the presence of water or under sustained loads (static fatigue). Water bleaches out the alkalis from the surface and deepens the surface flaws already present in fibers. Under sustained loads, the growth of surface flaws is accelerated owing to stress corrosion by atmospheric moisture. As a result, the tensile strength of glass fibers is decreased with increasing time of load duration (Figure 2.11).

2.1.2 Carbon Fibers

Carbon fibers are commercially available with a variety of tensile modulus values ranging from 207 GPa (30×10^6 psi) on the low side to 1035 GPa (150×10^6 psi) on the high side. In general, the low-modulus fibers have lower density, lower cost, higher tensile and compressive strengths, and higher tensile strains-to-failure than the high-modulus fibers. Among the advantages of carbon fibers are their exceptionally high tensile strength–weight ratios as well as tensile modulus–weight ratios, very low coefficient of linear thermal expansion (which provides dimensional stability in such applications as space antennas), high fatigue strengths, and high thermal conductivity (which is even

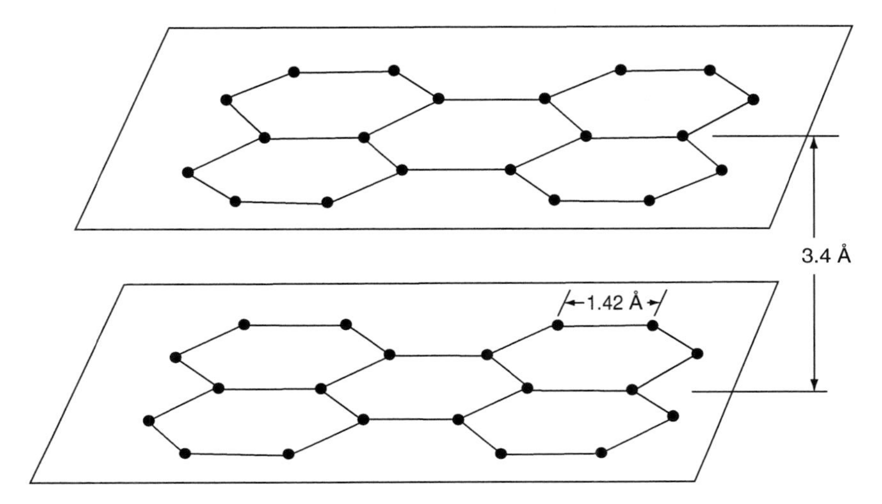

FIGURE 2.12 Arrangement of carbon atoms in a graphite crystal.

higher than that of copper). The disadvantages are their low strain-to-failure, low impact resistance, and high electrical conductivity, which may cause "shorting" in unprotected electrical machinery. Their high cost has so far excluded them from widespread commercial applications. They are used mostly in the aerospace industry, where weight saving is considered more critical than cost.

Structurally, carbon fibers contain a blend of amorphous carbon and graphitic carbon. Their high tensile modulus results from the graphitic form, in which carbon atoms are arranged in a crystallographic structure of parallel planes or layers. The carbon atoms in each plane are arranged at the corners of interconnecting regular hexagons (Figure 2.12). The distance between the planes (3.4 Å) is larger than that between the adjacent atoms in each plane (1.42 Å). Strong covalent bonds exist between the carbon atoms in each plane, but the bond between the planes is due to van der Waals-type forces, which is much weaker. This results in highly anisotropic physical and mechanical properties for the carbon fiber.

The basal planes in graphite crystals are aligned along the fiber axis. However, in the transverse direction, the alignment can be either circumferential, radial, random, or a combination of these arrangements (Figure 2.13). Depending on which of these arrangements exists, the thermoelastic properties, such as modulus (E) and coefficient of thermal expansion (α), in the radial (r) and circumferential (θ) directions of the fiber can be different from those in the axial (a) or longitudinal direction. For example, if the arrangement is circumferential, $E_a = E_\theta > E_r$, and the fiber is said to be circumferentially orthotropic. For the radial arrangement, $E_a = E_r > E_\theta$, and the fiber is radially orthotropic. When there is a random arrangement, $E_a > E_\theta = E_r$, the fiber is transversely isotropic. In commercial fibers, a two-zone structure with circumferential

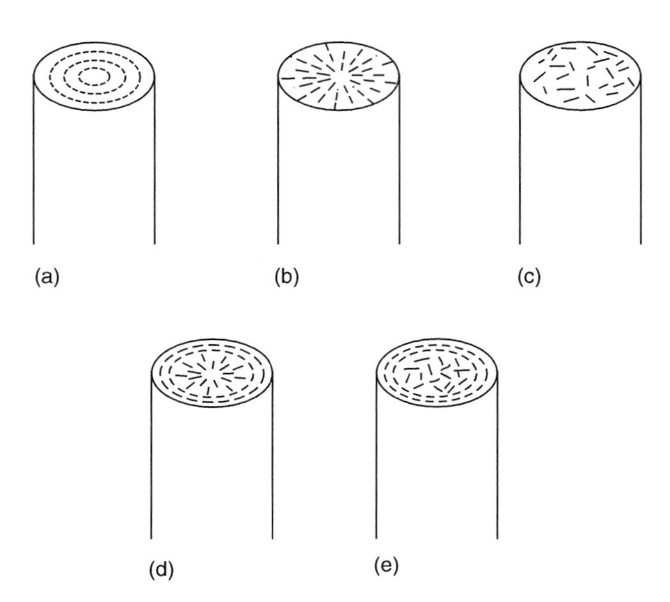

FIGURE 2.13 Arrangement of graphite crystals in a direction transverse to the fiber axis: (a) circumferential, (b) radial, (c) random, (d) radial–circumferential, and (e) random–circumferential.

arrangement in the skin and either radial or random arrangement in the core is commonly observed [4].

Carbon fibers are manufactured from two types of precursors (starting materials), namely, textile precursors and pitch precursors. The manufacturing process from both precursors is outlined in Figure 2.14. The most common textile precursor is polyacrylonitrile (PAN). The molecular structure of PAN, illustrated schematically in Figure 2.15a, contains highly polar CN groups that are randomly arranged on either side of the chain. Filaments are wet spun from a solution of PAN and stretched at an elevated temperature during which the polymer chains are aligned in the filament direction. The stretched filaments are then heated in air at 200°C–300°C for a few hours. At this stage, the CN groups located on the same side of the original chain combine to form a more stable and rigid ladder structure (Figure 2.15b), and some of the CH_2 groups are oxidized. In the next step, PAN filaments are carbonized by heating them at a controlled rate at 1000°C–2000°C in an inert atmosphere. Tension is maintained on the filaments to prevent shrinking as well as to improve molecular orientation. With the elimination of oxygen and nitrogen atoms, the filaments now contain mostly carbon atoms, arranged in aromatic ring patterns in parallel planes. However, the carbon atoms in the neighboring planes are not yet perfectly ordered, and the filaments have a relatively low tensile modulus. As the carbonized filaments are subsequently heat-treated at or above 2000°C,

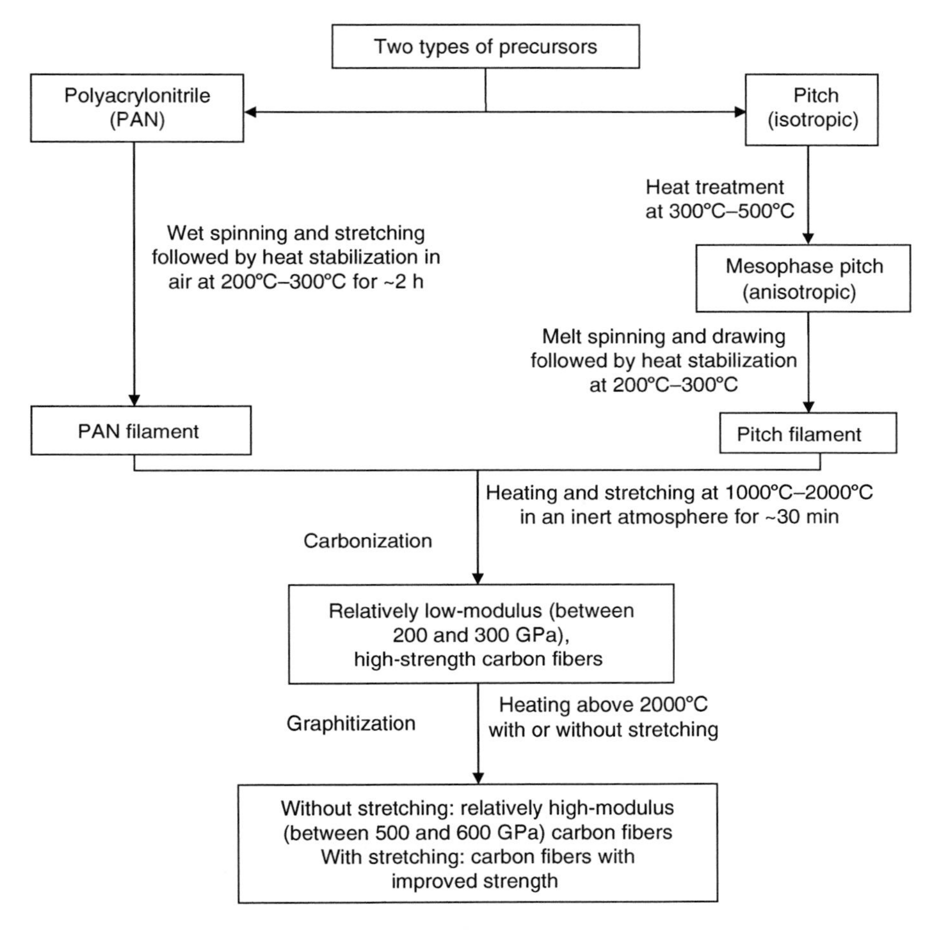

FIGURE 2.14 Flow diagram in carbon fiber manufacturing.

their structure becomes more ordered and turns toward a true graphitic form with increasing heat treatment temperature. The graphitized filaments attain a high tensile modulus, but their tensile strength may be relatively low (Figure 2.16). Their tensile strength can be increased by hot stretching them above 2000°C, during which the graphitic planes are aligned in the filament direction. Other properties of carbon fibers (e.g., electrical conductivity, thermal conductivity, longitudinal coefficient of thermal expansion, and oxidation resistance) can be improved by controlling the amount of crystallinity and eliminating the defects, such as missing carbon atoms or catalyst impurities. Tensile strength and tensile modulus are also affected by the amount of crystallinity and the presence of defects (Table 2.4).

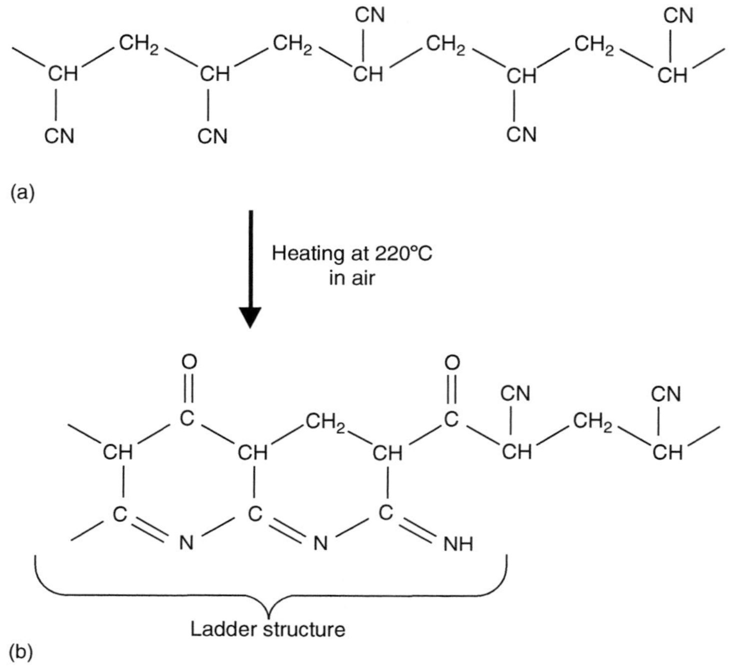

FIGURE 2.15 Ladder structure in an oxidized PAN molecule. (a) Molecular structure of PAN and (b) rigid ladder structure.

Pitch, a by-product of petroleum refining or coal coking, is a lower cost precursor than PAN. The carbon atoms in pitch are arranged in low-molecular-weight aromatic ring patterns. Heating to temperatures above 300°C polymerizes (joins) these molecules into long, two-dimensional sheetlike structures. The highly viscous state of pitch at this stage is referred to as "mesophase." Pitch filaments are produced by melt spinning mesophase pitch through a spinneret (Figure 2.17). While passing through the spinneret die, the mesophase pitch molecules become aligned in the filament direction. The filaments are cooled to freeze the molecular orientation, and subsequently heated between 200°C and 300°C in an oxygen-containing atmosphere to stabilize them and make them infusible (to avoid fusing the filaments together). In the next step, the filaments are carbonized at temperatures around 2000°C. The rest of the process of transforming the structure to graphitic form is similar to that followed for PAN precursors.

PAN carbon fibers are generally categorized into high tensile strength (HT), high modulus (HM), and ultrahigh modulus (UHM) types. The high tensile strength PAN carbon fibers, such as T-300 and AS-4 in Table 2.1, have the

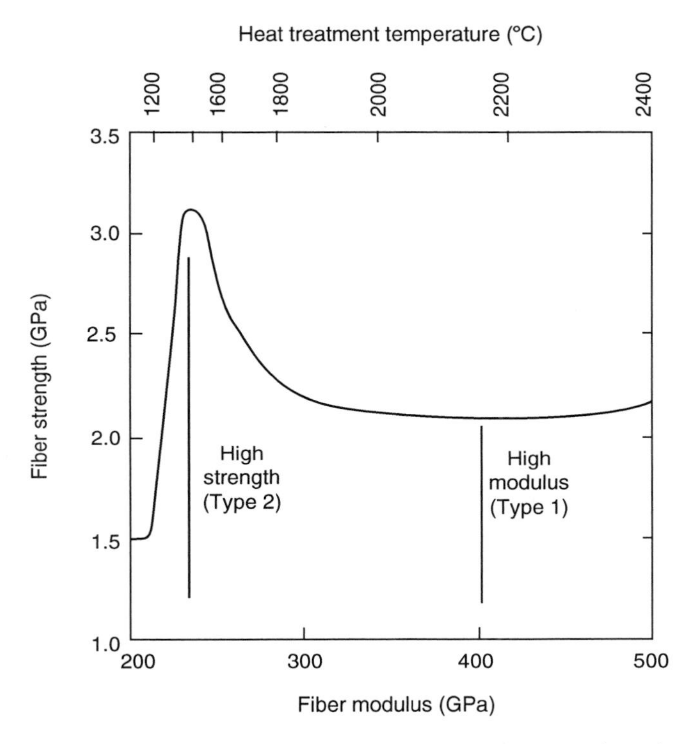

FIGURE 2.16 Influence of heat treatment temperature on strength and modulus of carbon fibers. (After Watt, W., *Proc. R. Soc. Lond.*, A319, 5, 1970.)

lowest modulus, while the ultrahigh-modulus PAN carbon fibers, such as GY-70, have the lowest tensile strength as well as the lowest tensile strain-to-failure. A number of intermediate modulus (IM) high-strength PAN carbon fibers, such as T-40 and IM-7, have also been developed that possess the highest strain-to-failure among carbon fibers. Another point to note in Table 2.1 is that the pitch carbon fibers have very high modulus values, but their tensile strength and strain-to-failure are lower than those of the PAN carbon fibers. The high modulus of pitch fibers is the result of the fact that they are more graphitizable; however, since shear is easier between parallel planes of a graphitized fiber and graphitic fibers are more sensitive to defects and flaws, their tensile strength is not as high as that of PAN fibers.

The axial compressive strength of carbon fibers is lower than their tensile strength. The PAN carbon fibers have higher compressive strength than pitch carbon fibers. It is also observed that the higher the modulus of a carbon fiber, the lower is its compressive strength. Among the factors that contribute to the reduction in compressive strength are higher orientation, higher graphitic order, and larger crystal size [5,6].

TABLE 2.4
Structural Features and Controlling Parameters Affecting the Properties of Carbon Fibers

Structural Feature	Controlling Parameters	Properties	
		Increase	Decrease
Increasing orientation of crystallographic basal planes parallel to the fiber axis	Fiber drawing, fiber structure, restraint against shrinkage during heat treatment	Longitudinal strength and modulus, longitudinal negative CTE, thermal and electrical conductivities	Transverse strength and modulus
Increasing crystallinity (larger and perfect crystals)	Precursor chemistry, heat treatment	Thermal and electrical conductivities, longitudinal negative CTE, oxidation resistance	Longitudinal tensile and compressive strengths, transverse strength and modulus
Decreasing defect content	Precursor purity, fiber handling	Tensile strength, thermal and electrical conductivities, oxidation resistance	

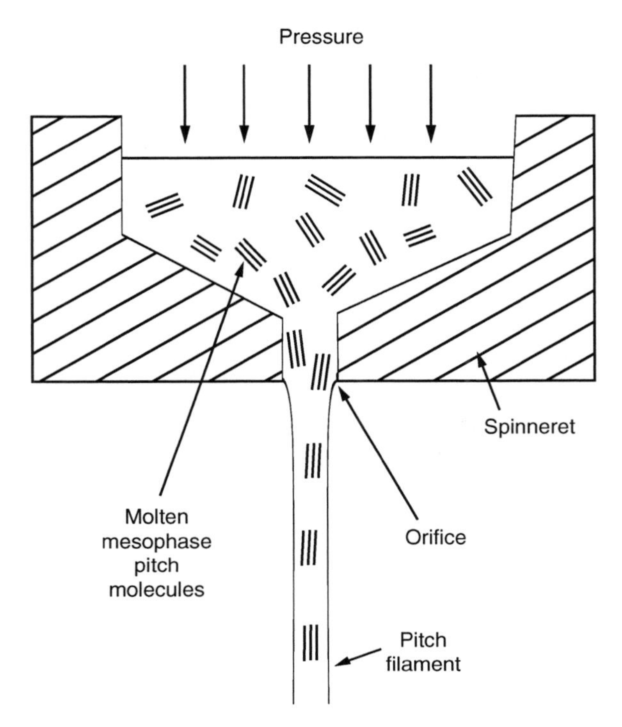

FIGURE 2.17 Alignment of mesophase pitch into a pitch filament. (After *Commercial Opportunities for Advanced Composites, ASTM STP*, 704, 1980.)

The PAN carbon fibers have lower thermal conductivity and electrical conductivity than pitch carbon fibers [6]. For example, thermal conductivity of PAN carbon fibers is in the range of 10–100 W/m °K compared with 20–1000 W/m °K for pitch carbon fibers. Similarly, electrical conductivity of PAN carbon fibers is in the range of 10^4–10^5 S/m compared with 10^5–10^6 S/m for pitch carbon fibers. For both types of carbon fibers, the higher the tensile modulus, the higher are the thermal and electrical conductivities.

Carbon fibers are commercially available in three basic forms, namely, long and continuous tow, chopped (6–50 mm long), and milled (30–3000 μm long). The long and continuous tow, which is simply an untwisted bundle of 1,000–160,000 parallel filaments, is used for high-performance applications. The price of carbon fiber tow decreases with increasing filament count. Although high filament counts are desirable for improving productivity in continuous molding operations, such as filament winding and pultrusion, it becomes increasingly difficult to wet them with the matrix. "Dry" filaments are not conducive to good mechanical properties.

Carbon fiber tows can also be weaved into two-dimensional fabrics of various styles (see Appendix A.1). Hybrid fabrics containing commingled or coweaved carbon and other fibers, such as E-glass, Kevlar, PEEK, PPS, and so on, are also available. Techniques of forming three-dimensional weaves with fibers running in the thickness direction have also been developed.

2.1.3 Aramid Fibers

Aramid fibers are highly crystalline aromatic polyamide fibers that have the lowest density and the highest tensile strength-to-weight ratio among the current reinforcing fibers. Kevlar 49 is the trade name of one of the aramid fibers available in the market. As a reinforcement, aramid fibers are used in many marine and aerospace applications where lightweight, high tensile strength, and resistance to impact damage (e.g., caused by accidentally dropping a hand tool) are important. Like carbon fibers, they also have a negative coefficient of thermal expansion in the longitudinal direction, which is used in designing low thermal expansion composite panels. The major disadvantages of aramid fiber-reinforced composites are their low compressive strengths and difficulty in cutting or machining.

The molecular structure of aramid fibers, such as Kevlar 49 fibers, is illustrated in Figure 2.18. The repeating unit in its molecules contains an amide (–NH) group (which is also found in nylons) and an aromatic ring represented by —⌬— in Figure 2.18. The aromatic ring gives it a higher chain stiffness (modulus) as well as better chemical and thermal stability over other commercial organic fibers, such as nylons.

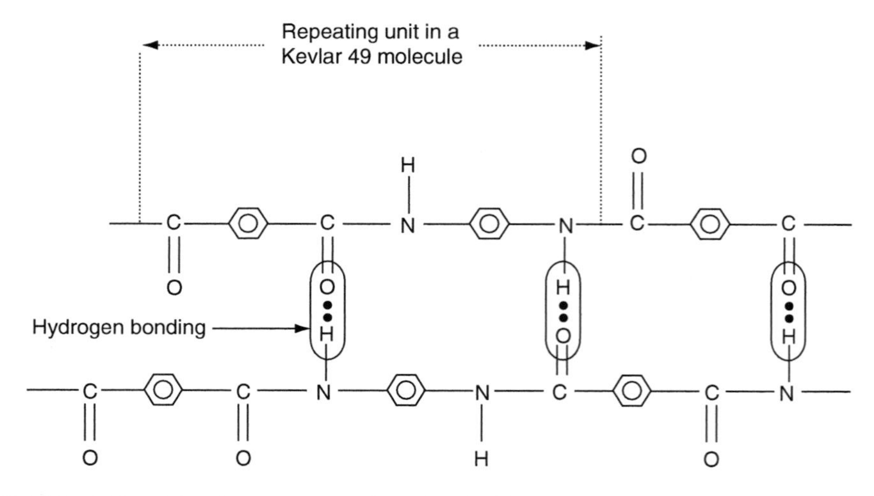

FIGURE 2.18 Molecular structure of Kevlar 49 fiber.

Kevlar 49 filaments are manufactured by extruding an acidic solution of a proprietary precursor (a polycondensation product of terephthaloyol chloride and *p*-phenylene diamine) from a spinneret. During the filament drawing process, Kevlar 49 molecules become highly oriented in the direction of the filament axis. Weak hydrogen bonds between hydrogen and oxygen atoms in adjacent molecules hold them together in the transverse direction. The resulting filament is highly anisotropic, with much better physical and mechanical properties in the longitudinal direction than in the radial direction.

Although the tensile stress–strain behavior of Kevlar 49 is linear, fiber fracture is usually preceded by longitudinal fragmentation, splintering, and even localized drawing. In bending, Kevlar 49 fibers exhibit a high degree of yielding on the compression side. Such a noncatastrophic failure mode is not observed in glass or carbon fibers, and gives Kevlar 49 composites superior damage tolerance against impact or other dynamic loading. One interesting application of this characteristic of Kevlar 49 fibers is found in soft lightweight body armors and helmets used for protecting police officers and military personnel.

Kevlar 49 fibers do not melt or support combustion but will start to carbonize at about 427°C. The maximum long-term use temperature recommended for Kevlar 49 is 160°C. They have very low thermal conductivity, but a very high vibration damping coefficient. Except for a few strong acids and alkalis, their chemical resistance is good. However, they are quite sensitive to ultraviolet lights. Prolonged direct exposure to sunlight causes discoloration and significant loss in tensile strength. The problem is less pronounced in composite laminates in which the fibers are covered with a matrix. Ultraviolet light-absorbing fillers can be added to the matrix to further reduce the problem.

Kevlar 49 fibers are hygroscopic and can absorb up to 6% moisture at 100% relative humidity and 23°C. The equilibrium moisture content (i.e., maximum moisture absorption) is directly proportional to relative humidity and is attained in 16–36 h. Absorbed moisture seems to have very little effect on the tensile properties of Kevlar 49 fibers. However, at high moisture content, they tend to crack internally at the preexisting microvoids and produce longitudinal splitting [7].

A second-generation Kevlar fiber is Kevlar 149, which has the highest tensile modulus of all commercially available aramid fibers. The tensile modulus of Kevlar 149 is 40% higher than that of Kevlar 49; however, its strain-to-failure is lower. Kevlar 149 has the equilibrium moisture content of 1.2% at 65% relative humidity and 22°C, which is nearly 70% lower than that of Kevlar 49 under similar conditions. Kevlar 149 also has a lower creep rate than Kevlar 49.

2.1.4 Extended Chain Polyethylene Fibers

Extended chain polyethylene fibers, commercially available under the trade name Spectra, are produced by gel spinning a high-molecular-weight polyethylene. Gel spinning yields a highly oriented fibrous structure with exceptionally high crystallinity (95%–99%) relative to melt spinning used for conventional polyethylene fibers.

Spectra polyethylene fibers have the highest strength-to-weight ratio of all commercial fibers available to date. Two other outstanding features of Spectra fibers are their low moisture absorption (1% compared with 5%–6% for Kevlar 49) and high abrasion resistance, which make them very useful in marine composites, such as boat hulls and water skis.

The melting point of Spectra fibers is 147°C; however, since they exhibit a high level of creep above 100°C, their application temperature is limited to 80°C–90°C. The safe manufacturing temperature for composites containing Spectra fibers is below 125°C, since they exhibit a significant and rapid reduction in strength as well as increase in thermal shrinkage above this temperature. Another problem with Spectra fibers is their poor adhesion with resin matrices, which can be partially improved by their surface modification with gas plasma treatment.

Spectra fibers provide high impact resistance for composite laminates even at low temperatures and are finding growing applications in ballistic composites, such as armors, helmets, and so on. However, their use in high-performance aerospace composites is limited, unless they are used in conjunction with stiffer carbon fibers to produce hybrid laminates with improved impact damage tolerance than all-carbon fiber laminates.

2.1.5 Natural Fibers

Examples of natural fibers are jute, flax, hemp, remi, sisal, coconut fiber (coir), and banana fiber (abaca). All these fibers are grown as agricultural plants in various parts of the world and are commonly used for making ropes, carpet backing, bags, and so on. The components of natural fibers are cellulose microfibrils dispersed in an amorphous matrix of lignin and hemicellulose [8]. Depending on the type of the natural fiber, the cellulose content is in the range of 60–80 wt% and the lignin content is in the range of 5–20 wt%. In addition, the moisture content in natural fibers can be up to 20 wt%. The properties of some of the natural fibers in use are given in Table 2.5.

TABLE 2.5
Properties of Selected Natural Fibers

Property	Hemp	Flax	Sisal	Jute
Density (g/cm^3)	1.48	1.4	1.33	1.46
Modulus (GPa)	70	60–80	38	10–30
Tensile strength (MPa)	550–900	800–1500	600–700	400–800
Elongation to failure (%)	1.6	1.2–1.6	2–3	1.8

Source: Adapted from Wambua, P., Ivens, J., and Verpoest, I., *Compos. Sci. Tech.*, 63, 1259, 2003.

Recently, natural fiber-reinforced polymers have created interest in the automotive industry for the following reasons. The applications in which natural fiber composites are now used include door inner panel, seat back, roof inner panel, and so on.

1. They are environment-friendly, meaning that they are biodegradable, and unlike glass and carbon fibers, the energy consumption to produce them is very small.
2. The density of natural fibers is in the range of 1.25–1.5 g/cm^3 compared with 2.54 g/cm^3 for E-glass fibers and 1.8–2.1 g/cm^3 for carbon fibers.
3. The modulus–weight ratio of some natural fibers is greater than that of E-glass fibers, which means that they can be very competitive with E-glass fibers in stiffness-critical designs.
4. Natural fiber composites provide higher acoustic damping than glass or carbon fiber composites, and therefore are more suitable for noise attenuation, an increasingly important requirement in interior automotive applications.
5. Natural fibers are much less expensive than glass and carbon fibers.

However, there are several limitations of natural fibers. The tensile strength of natural fibers is relatively low. Among the other limitations are low melting point and moisture absorption. At temperatures higher than 200°C, natural fibers start to degrade, first by the degradation of hemicellulose and then by the degradation of lignin. The degradation leads to odor, discoloration, release of volatiles, and deterioration of mechanical properties.

2.1.6 Boron Fibers

The most prominent feature of boron fibers is their extremely high tensile modulus, which is in the range of 379–414 GPa (55–60 $\times$ 10^6 psi). Coupled with their relatively large diameter, boron fibers offer excellent resistance to buckling, which in turn contributes to high compressive strength for boron fiber-reinforced composites. The principal disadvantage of boron fibers is their high cost, which is even higher than that of many forms of carbon fibers. For this reason, its use is at present restricted to a few aerospace applications.

Boron fibers are manufactured by chemical vapor deposition (CVD) of boron onto a heated substrate (either a tungsten wire or a carbon monofilament). Boron vapor is produced by the reaction of boron chloride with hydrogen:

$$2BCl_3 + 3H_2 = 2B + 6HCl$$

The most common substrate used in the production of boron fibers is tungsten wire, typically 0.0127 mm (0.0005 in.) in diameter. It is continuously pulled through a reaction chamber in which boron is deposited on its surface at 1100°C–1300°C. The speed of pulling and the deposition temperature can be varied to control the resulting fiber diameter. Currently, commercial boron fibers are produced in diameters of 0.1, 0.142, and 0.203 mm (0.004, 0.0056, and 0.008 in.), which are much larger than those of other reinforcing fibers.

During boron deposition, the tungsten substrate is converted into tungsten boride by diffusion and reaction of boron with tungsten. The core diameter increases in diameter from 0.0127 mm (0.0005 in.) to 0.0165 mm (0.00065 in.), placing boron near the core in tension. However, near the outer surface of the boron layer, a state of biaxial compression exists, which makes the boron fiber less sensitive to mechanical damage [9]. The adverse reactivity of boron fibers with metals is reduced by chemical vapor deposition of silicon carbide on boron fibers, which produces borsic fibers.

2.1.7 Ceramic Fibers

Silicon carbide (SiC) and aluminum oxide (Al_2O_3) fibers are examples of ceramic fibers notable for their high-temperature applications in metal and ceramic matrix composites. Their melting points are 2830°C and 2045°C, respectively. Silicon carbide retains its strength well above 650°C, and aluminum oxide has excellent strength retention up to about 1370°C. Both fibers are suitable for reinforcing metal matrices in which carbon and boron fibers exhibit adverse reactivity. Aluminum oxide fibers have lower thermal and electrical conductivities and have higher coefficient of thermal expansion than silicon carbide fibers.

Silicon carbide fibers are available in three different forms [10]:

1. Monofilaments that are produced by chemical vapor deposition of β-SiC on a 10–25 μm diameter carbon monofilament substrate. The carbon monofilament is previously coated with ~1 μm thick pyrolitic graphite to smoothen its surface as well as to enhance its thermal conductivity. β-SiC is produced by the reaction of silanes and hydrogen gases at around 1300°C. The average fiber diameter is 140 μm.
2. Multifilament yarn produced by melt spinning of a polymeric precursor, such as polycarbosilane, at 350°C in nitrogen gas. The resulting polycarbosilane fiber is first heated in air to 190°C for 30 min to cross-link the polycarbosilane molecules by oxygen and then heat-treated to 1000°C–1200°C to form a crystalline structure. The average fiber diameter in the yarn is 14.5 μm and a commercial yarn contains 500 fibers. Yarn fibers have a considerably lower strength than the monofilaments.
3. Whiskers, which are 0.1–1 μm in diameter and around 50 μm in length. They are produced from rice hulls, which contain 10–20 wt% SiO_2. Rice hulls are first heated in an oxygen-free atmosphere to 700°C–900°C to remove the volatiles and then to 1500°C–1600°C for 1 h to produce SiC

> whiskers. The final heat treatment is at 800°C in air, which removes free carbon. The resulting SiC whiskers contain 10 wt% of SiO_2 and up to 10 wt% Si_3N_4. The tensile modulus and tensile strength of these whiskers are reported as 700 GPa and 13 GPa (101.5×10^6 psi and 1.88×10^6 psi), respectively.

Many different aluminum oxide fibers have been developed over the years, but many of them at present are not commercially available. One of the early aluminum oxide fibers, but not currently available in the market, is called the Fiber FP [11]. It is a high-purity (>99%) polycrystalline α-Al_2O_3 fiber, dry spun from a slurry mix of alumina and proprietary spinning additives. The spun filaments are fired in two stages: low firing to control shrinkage, followed by flame firing to produce a suitably dense α-Al_2O_3. The fired filaments may be coated with a thin layer of silica to improve their strength (by healing the surface flaws) as well as their wettability with the matrix. The filament diameter is 20 μm. The tensile modulus and tensile strength of Fiber FP are reported as 379 GPa and 1.9 GPa (55×10^6 psi and 275,500 psi), respectively. Experiments have shown that Fiber FP retains almost 100% of its room temperature tensile strength after 300 h of exposure in air at 1000°C. Borsic fiber, on the other hand, loses 50% of its room temperature tensile strength after only 1 h of exposure in air at 500°C. Another attribute of Fiber FP is its remarkably high compressive strength, which is estimated to be about 6.9 GPa (1,000,000 psi).

Nextel 610 and Nextel 720, produced by 3 M, are two of the few aluminum oxide fibers available in the market now [12]. Both fibers are produced in continuous multifilament form using the sol–gel process. Nextel 610 contains greater than 99% Al_2O_3 and has a single-phase structure of α-Al_2O_3. The average grain size is 0.1 μm and the average filament diameter is 14 μm. Because of its fine-grained structure, it has a high tensile strength at room temperature; but because of grain growth, its tensile strength decreases rapidly as the temperature is increased above 1100°C. Nextel 720, which contains 85% Al_2O_3 and 15% SiO_2, has a lower tensile strength at room temperature, but is able to retain about 85% of its tensile strength even at 1400°C. Nextel 720 also has a much lower creep rate than Nextel 610 and other oxide fibers at temperatures above 1000°C. The structure of Nextel 720 contains α-Al_2O_3 grains embedded in mullite grains. The strength retention of Nextel 720 at high temperatures is attributed to reduced grain boundary sliding and reduced grain growth.

Another ceramic fiber, containing approximately equal parts of Al_2O_3 and silica (SiO_2), is available in short, discontinuous lengths under the trade name Fiberfrax. The fiber diameter is 2–12 μm and the fiber aspect ratio (length to diameter ratio) is greater than 200. It is manufactured either by a melt blowing or by a melt spinning process. Saffil, produced by Saffil Ltd., is also a discontinuous aluminosilicate fiber, containing 95% Al_2O_3 and 5% SiO_2. Its diameter is 1–5 μm. It is produced by blow extrusion of partially hydrolyzed solution of aluminum salts with a small amount of SiO_2. It contains mainly δ-Al_2O_3 grains of 50 nm size, but it also contains some larger size α-Al_2O_3. These two fibers are mostly used for high temperature insulation.

2.2 MATRIX

The roles of the matrix in a fiber-reinforced composite are: (1) to keep the fibers in place, (2) to transfer stresses between the fibers, (3) to provide a barrier against an adverse environment, such as chemicals and moisture, and (4) to protect the surface of the fibers from mechanical degradation (e.g., by abrasion). The matrix plays a minor role in the tensile load-carrying capacity of a composite structure. However, selection of a matrix has a major influence on the compressive, interlaminar shear as well as in-plane shear properties of the composite material. The matrix provides lateral support against the possibility of fiber buckling under compressive loading, thus influencing to a large extent, the compressive strength of the composite material. The interlaminar shear strength is an important design consideration for structures under bending loads, whereas the in-plane shear strength is important under torsional loads. The interaction between fibers and matrix is also important in designing damage-tolerant structures. Finally, the processing and defects in a composite material depend strongly on the processing characteristics of the matrix. For example, for epoxy polymers used as matrix in many aerospace composites, the processing characteristics include the liquid viscosity, the curing temperature, and the curing time.

Table 2.6 lists various matrix materials that have been used either commercially or in research. Among these, thermoset polymers, such as epoxies, polyesters, and vinyl esters, are more commonly used as matrix material in continuous or long fiber-reinforced composites, mainly because of the ease of processing due to their low viscosity. Thermoplastic polymers are more commonly used with short fiber-reinforced composites that are injection-molded; however, the interest in continuous fiber-reinforced thermoplastic matrix is growing. Metallic and ceramic matrices are primarily considered for high-temperature applications. We briefly discuss these three categories of matrix in this section.

2.2.1 Polymer Matrix

A polymer is defined as a long-chain molecule containing one or more repeating units of atoms (Figure 2.19), joined together by strong covalent bonds. A polymeric material (commonly called a plastic) is a collection of a large number of polymer molecules of similar chemical structure (but not of equal length). In the solid state, these molecules are frozen in space, either in a random fashion in amorphous polymers or in a mixture of random fashion

TABLE 2.6
Matrix Materials

Polymeric
Thermoset polymers
- Epoxies: principally used in aerospace and aircraft applications
- Polyesters, vinyl esters: commonly used in automotive, marine, chemical, and electrical applications
- Phenolics: used in bulk molding compounds
- Polyimides, polybenzimidazoles (PBI), polyphenylquinoxaline (PPQ): for high-temperature aerospace applications (temperature range: 250°C–400°C)
- Cyanate ester

Thermoplastic polymers
- Nylons (such as nylon 6, nylon 6,6), thermoplastic polyesters (such as PET, PBT), polycarbonate (PC), polyacetals: used with discontinuous fibers in injection-molded articles
- Polyamide-imide (PAI), polyether ether ketone (PEEK), polysulfone (PSUL), polyphenylene sulfide (PPS), polyetherimide (PEI): suitable for moderately high temperature applications with continuous fibers

Metallic
- Aluminum and its alloys, titanium alloys, magnesium alloys, copper-based alloys, nickel-based superalloys, stainless steel: suitable for high-temperature applications (temperature range: 300°C–500°C)

Ceramic
- Aluminum oxide (Al_2O_3), carbon, silicon carbide (SiC), silicon nitride (Si_3N_4): suitable for high-temperature applications

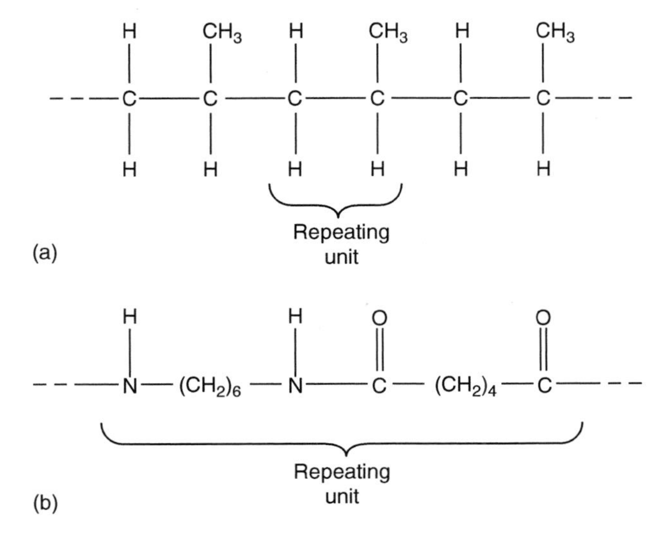

FIGURE 2.19 Examples of repeating units in polymer molecules. (a) A polypropylene molecule. (b) A nylon 6,6 molecule.

and orderly fashion (folded chains) in semicrystalline polymers (Figure 2.20). However, on a submicroscopic scale, various segments in a polymer molecule may be in a state of random excitation. The frequency, intensity, and number of these segmental motions increase with increasing temperature, giving rise to the temperature-dependent properties of a polymeric solid.

2.2.1.1 Thermoplastic and Thermoset Polymers

Polymers are divided into two broad categories: thermoplastics and thermosets. In a thermoplastic polymer, individual molecules are not chemically joined together (Figure 2.21a). They are held in place by weak secondary bonds or intermolecular forces, such as van der Waals bonds and hydrogen bonds. With the application of heat, these secondary bonds in a solid thermoplastic polymer can be temporarily broken and the molecules can now be moved relative to each other or flow to a new configuration if pressure is applied on them. On cooling, the molecules can be frozen in their new configuration and the secondary bonds are restored, resulting in a new solid shape. Thus, a thermoplastic polymer can be heat-softened, melted, and reshaped (or postformed) as many times as desired.

In a thermoset polymer, on the other hand, the molecules are chemically joined together by cross-links, forming a rigid, three-dimensional network structure (Figure 2.21b). Once these cross-links are formed during the polymerization reaction (also called the curing reaction), the thermoset polymer

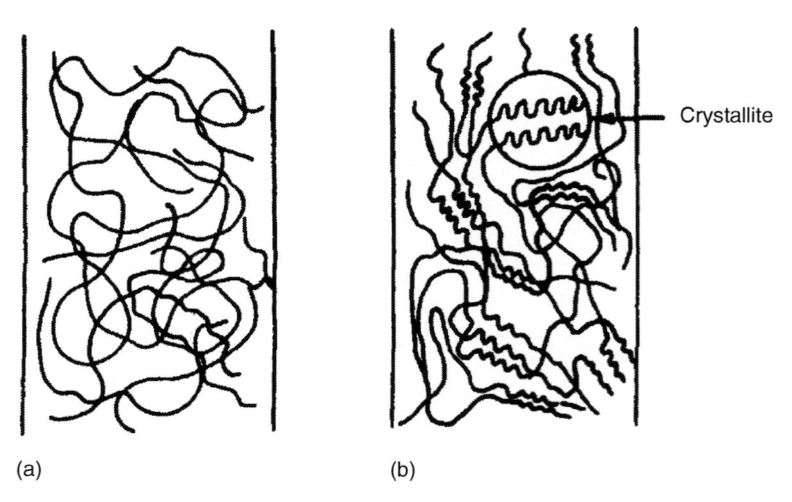

FIGURE 2.20 Arrangement of molecules in (a) amorphous polymers and (b) semicrystalline polymers.

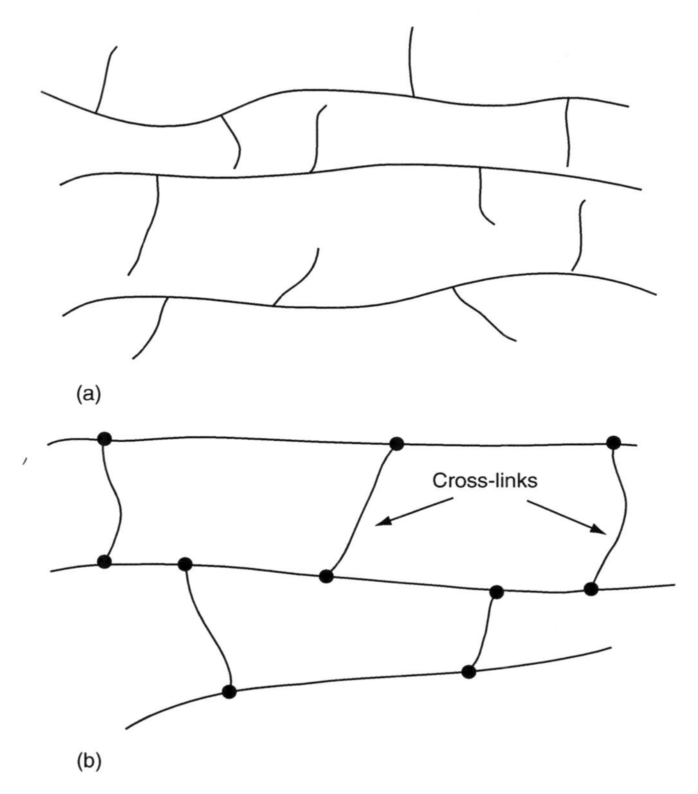

FIGURE 2.21 Schematic representation of (a) thermoplastic polymer and (b) thermoset polymer.

cannot be melted by the application of heat. However, if the number of cross-links is low, it may still be possible to soften them at elevated temperatures.

2.2.1.2 Unique Characteristics of Polymeric Solids

There are two unique characteristics of polymeric solids that are not observed in metals under ordinary conditions, namely, that their mechanical properties depend strongly on both the ambient temperature and the loading rate. Figure 2.22 schematically shows the general trends in the variation of tensile modulus of various types of polymers with temperature. Near the glass transition temperature, denoted by T_g in this diagram, the polymeric solid changes from a hard, sometimes brittle (glass-like) material to a soft, tough (leather-like) material. Over a temperature range around T_g, its modulus is reduced by as much as five orders of magnitude. Near this temperature, the material is also highly viscoelastic. Thus, when an external load is applied, it exhibits an

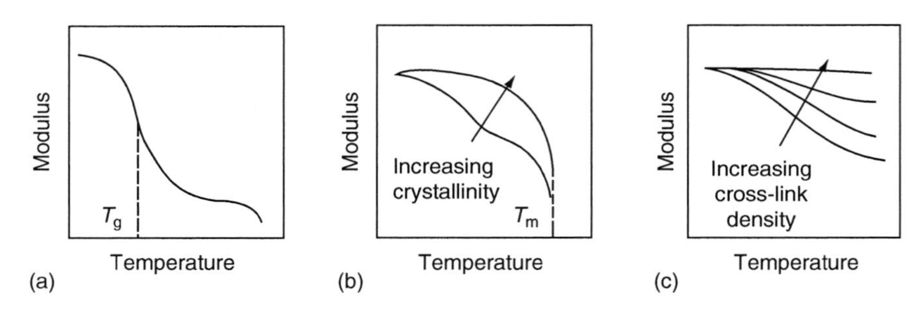

FIGURE 2.22 Variation of tensile modulus with temperature for three different types of polymers: (a) amorphous thermoplastic, (b) semicrystalline thermoplastic, and (c) thermoset.

instantaneous elastic deformation followed by a slow viscous deformation. With increasing temperature, the polymer changes into a rubber-like solid capable of undergoing large, elastic deformations under external loads. As the temperature is increased further, both amorphous and semicrystalline thermoplastics achieve highly viscous liquid states, with the latter showing a sharp transition at the crystalline melting point, denoted by T_m. However, for a thermoset polymer, no melting occurs; instead, it chars and finally burns at very high temperatures. The glass transition temperature of a thermoset polymer can be controlled by varying the amount of cross-linking between the molecules. For very highly cross-linked polymers, the glass transition and the accompanying softening may not be observed.

The mechanical characteristics of a polymeric solid depend on the ambient temperature relative to the glass transition temperature of the polymer. If the ambient temperature is above T_g, the polymeric solid exhibits low surface hardness, low modulus, and high ductility. At temperatures below T_g, the segmental motion in a polymer plays an important role. If the molecular structure of a polymer allows many segmental motions, it behaves in a ductile manner even below T_g. Polycarbonate (PC), polyethylene terephthalate (PET), and various nylons fall into this category. If, on the other hand, the segmental motions are restricted, as in polymethyl methacrylate (PMMA), polystyrene (PS), and many thermoset polymers, it shows essentially a brittle failure.

Figure 2.23 shows the effects of temperature and loading rate on the stress–strain behavior of polymeric solids. At low temperatures, the stress–strain behavior is much like that of a brittle material. The polymer may not exhibit any signs of yielding and the strain-to-failure is low. As the temperature is increased, yielding may occur; but the yield strength decreases with increasing temperature. The strain-to-failure, on the other hand, increases with increasing temperature, transforming the polymer from a brittle material at low temperatures to a ductile material at elevated temperatures.

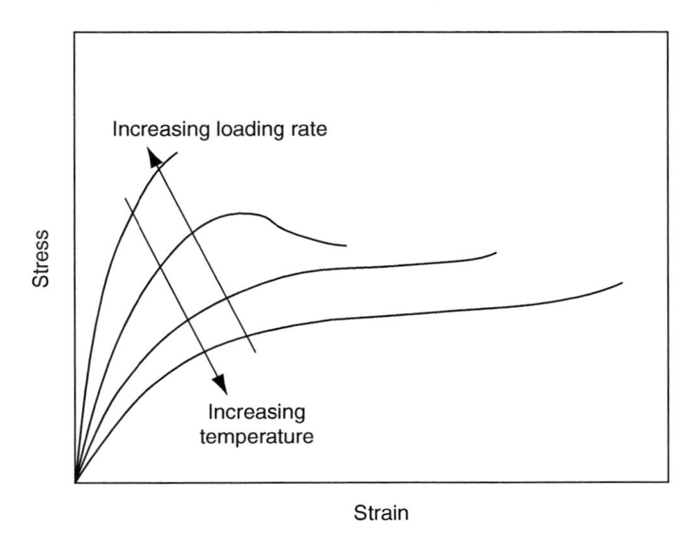

FIGURE 2.23 Effects of loading rate and temperature on the stress–strain behavior of polymeric solids.

The effect of loading rate on the stress–strain behavior is opposite to that due to temperature (Figure 2.23). At low loading rates or long durations of loading, the polymer may behave in a ductile manner and show high toughness. At high loading rates or short durations of loading, the same polymer behaves in a rigid, brittle (glass-like) manner.

2.2.1.3 Creep and Stress Relaxation

The viscoelastic characteristic of a polymeric solid is best demonstrated by creep and stress relaxation tests. In creep tests, a constant stress is maintained on a specimen while its deformation (or strain) is monitored as a function to time. As the polymer creeps, the strain increases with time. In stress relaxation tests, a constant deformation (strain) is maintained while the stress on the specimen is monitored as a function of time. In stress relaxation, stress decreases with time. Both tests are performed at various ambient temperatures of interest. Typical creep and stress relaxation diagrams, shown schematically in Figure 2.24, exhibit an instantaneous elastic response followed by a delayed viscous response. In general, thermoset polymers exhibit lower creep and stress relaxation than thermoplastic polymers.

2.2.1.4 Heat Deflection Temperature

Softening characteristics of various polymers are often compared on the basis of their heat deflection temperatures (HDT). Measurement of HDT is

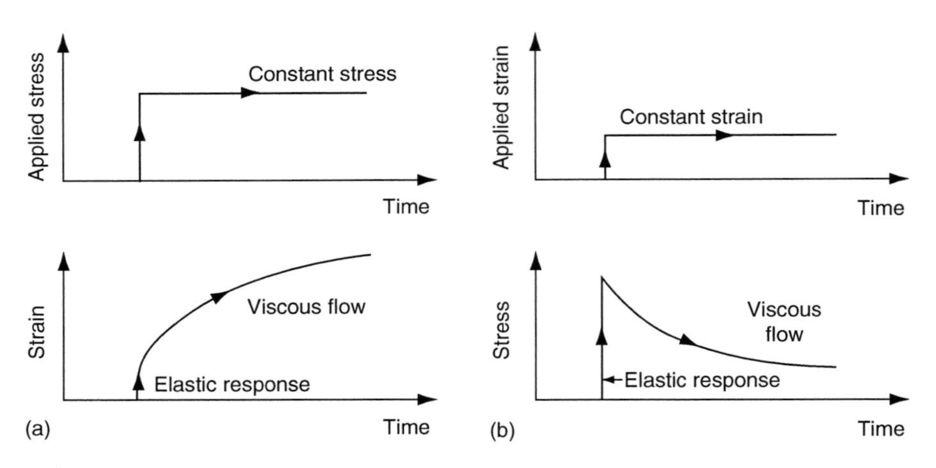

FIGURE 2.24 (a) Creep and (b) stress relaxation in solid polymers.

described in ASTM test method D648. In this test, a polymer bar of rectangular cross section is loaded as a simply supported beam (Figure 2.25) inside a suitable nonreacting liquid medium, such as mineral oil. The load on the bar is adjusted to create a maximum fiber stress of either 1.82 MPa (264 psi) or 0.455 MPa (66 psi). The center deflection of the bar is monitored as the temperature of the liquid medium is increased at a uniform rate of 2 ± 0.2°C/min. The temperature at which the bar deflection increases by 0.25 mm

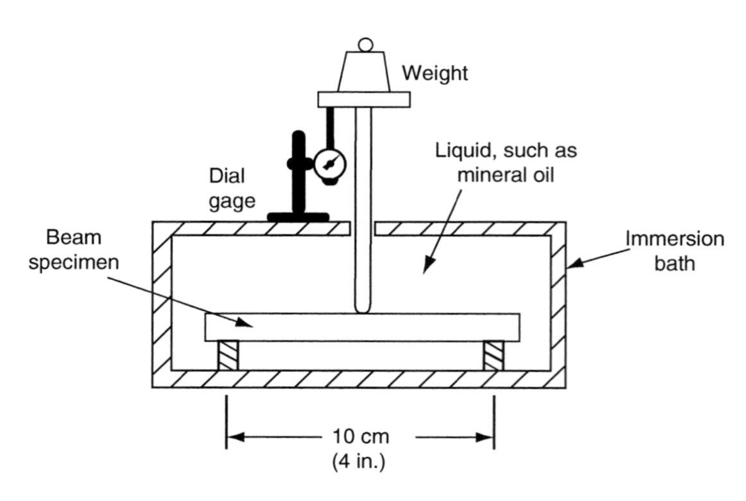

FIGURE 2.25 Test setup for measuring heat deflection temperature (HDT).

(0.01 in.) from its initial room temperature deflection is called the HDT at the specific fiber stress.

Although HDT is widely reported in the plastics product literature, it should not be used in predicting the elevated temperature performance of a polymer. It is used mostly for quality control and material development purposes. It should be pointed out that HDT is not a measure of the glass transition temperature. For glass transition temperature measurements, such methods as differential scanning calorimetry (DSC) or differential thermal analysis (DTA) are used [13].

2.2.1.5 Selection of Matrix: Thermosets vs. Thermoplastics

The primary consideration in the selection of a matrix is its basic mechanical properties. For high-performance composites, the most desirable mechanical properties of a matrix are

1. High tensile modulus, which influences the compressive strength of the composite
2. High tensile strength, which controls the intraply cracking in a composite laminate
3. High fracture toughness, which controls ply delamination and crack growth

For a polymer matrix composite, there may be other considerations, such as good dimensional stability at elevated temperatures and resistance to moisture or solvents. The former usually means that the polymer must have a high glass transition temperature T_g. In practice, the glass transition temperature should be higher than the maximum use temperature. Resistance to moisture and solvent means that the polymer should not dissolve, swell, crack (craze), or otherwise degrade in hot–wet environments or when exposed to solvents. Some common solvents in aircraft applications are jet fuels, deicing fluids, and paint strippers. Similarly, gasoline, motor oil, and antifreeze are common solvents in the automotive environment.

Traditionally, thermoset polymers (also called resins) have been used as a matrix material for fiber-reinforced composites. The starting materials used in the polymerization of a thermoset polymer are usually low-molecular-weight liquid chemicals with very low viscosities. Fibers are either pulled through or immersed in these chemicals before the polymerization reaction begins. Since the viscosity of the polymer at the time of fiber incorporation is very low, it is possible to achieve a good wet-out between the fibers and the matrix without the aid of either high temperature or pressure. Fiber surface wetting is extremely important in achieving fiber–matrix interaction in the composite, an essential requirement for good mechanical performance. Among other

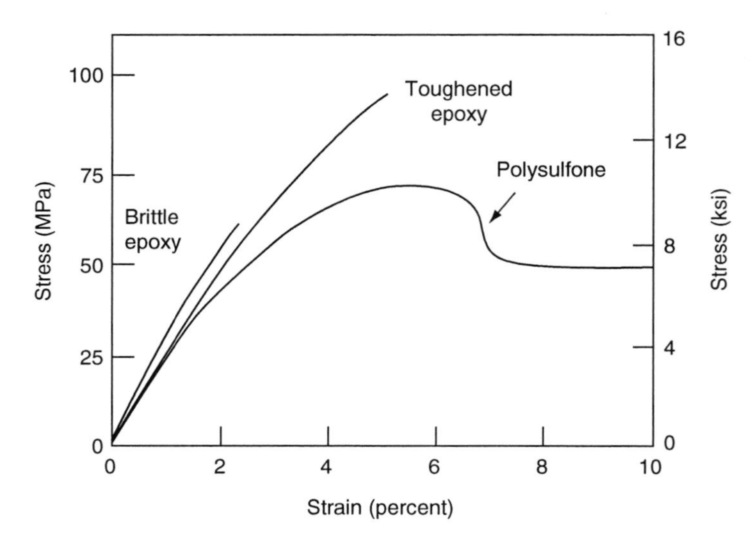

FIGURE 2.26 Tensile stress–strain diagrams of a thermoset polymer (epoxy) and a thermoplastic polymer (polysulfone).

advantages of using thermoset polymers are their thermal stability and chemical resistance. They also exhibit much less creep and stress relaxation than thermoplastic polymers. The disadvantages are

1. Limited storage life (before the final shape is molded) at room temperature
2. Long fabrication time in the mold (where the polymerization reaction, called the curing reaction or simply called *curing*, is carried out to transform the liquid polymer to a solid polymer)
3. Low strain-to-failure (Figure 2.26), which also contributes to their low impact strengths

The most important advantage of thermoplastic polymers over thermoset polymers is their high impact strength and fracture resistance, which in turn impart an excellent damage tolerance characteristic to the composite material. In general, thermoplastic polymers have higher strain-to-failure (Figure 2.25) than thermoset polymers, which may provide a better resistance to matrix microcracking in the composite laminate. Other advantages of thermoplastic polymers are

1. Unlimited storage (shelf) life at room temperature
2. Shorter fabrication time
3. Postformability (e.g., by thermoforming)

TABLE 2.7
Maximum Service Temperature for Selected Polymeric Matrices

Polymer	T_g, °C	Maximum Service Temperature, °C (°F)
Thermoset matrix		
DGEBA epoxy	180	125 (257)
TGDDM epoxy	240–260	190 (374)
Bismaleimides (BMI)	230–290	232 (450)
Acetylene-terminated polyimide (ACTP)	320	280 (536)
PMR-15	340	316 (600)
Thermoplastic matrix		
Polyether ether ketone (PEEK)	143	250 (482)
Polyphenylene sulfide (PPS)	85	240 (464)
Polysulfone	185	160 (320)
Polyetherimide (PEI)	217	267 (512)
Polyamide-imide (PAI)	280	230 (446)
K-III polyimide	250	225 (437)
LARC-TPI polyimide	265	300 (572)

4. Ease of joining and repair by welding, solvent bonding, and so on
5. Ease of handling (no tackiness)
6. Can be reprocessed and recycled

In spite of such distinct advantages, the development of continuous fiber-reinforced thermoplastic matrix composites has been much slower than that of continuous fiber-reinforced thermoset matrix composites. Because of their high melt or solution viscosities, incorporation of continuous fibers into a thermoplastic matrix is difficult. Commercial engineering thermoplastic polymers, such as nylons and polycarbonate, are of very limited interest in structural applications because they exhibit lower creep resistance and lower thermal stability than thermoset polymers. Recently, a number of thermoplastic polymers have been developed that possess high heat resistance (Table 2.7) and they are of interest in aerospace applications.

2.2.2 Metal Matrix

Metal matrix has the advantage over polymeric matrix in applications requiring a long-term resistance to severe environments, such as high temperature [14]. The yield strength and modulus of most metals are higher than those for polymers, and this is an important consideration for applications requiring high transverse strength and modulus as well as compressive strength for the composite. Another advantage of using metals is that they can be plastically deformed and strengthened by a variety of thermal and mechanical treatments. However, metals have a number of disadvantages, namely, they have high densities, high melting points (therefore, high process temperatures), and a tendency toward corrosion at the fiber–matrix interface.

The two most commonly used metal matrices are based on aluminum and titanium. Both of these metals have comparatively low densities and are available in a variety of alloy forms. Although magnesium is even lighter, its great affinity toward oxygen promotes atmospheric corrosion and makes it less suitable for many applications. Beryllium is the lightest of all structural metals and has a tensile modulus higher than that of steel. However, it suffers from extreme brittleness, which is the reason for its exclusion as a potential matrix material. Nickel- and cobalt-based superalloys have also been used as matrix; however, the alloying elements in these materials tend to accentuate the oxidation of fibers at elevated temperatures.

Aluminum and its alloys have attracted the most attention as matrix material in metal matrix composites. Commercially, pure aluminum has been used for its good corrosion resistance. Aluminum alloys, such as 201, 6061, and 1100, have been used for their higher tensile strength–weight ratios. Carbon fiber is used with aluminum alloys; however, at typical fabrication temperatures of 500°C or higher, carbon reacts with aluminum to form aluminum carbide (Al_4C_3), which severely degrades the mechanical properties of the composite. Protective coatings of either titanium boride (TiB_2) or sodium has been used on carbon fibers to reduce the problem of fiber degradation as well as to improve their wetting with the aluminum alloy matrix [15]. Carbon fiber-reinforced aluminum composites are inherently prone to galvanic corrosion, in which carbon fibers act as a cathode owing to a corrosion potential of 1 V higher than that of aluminum. A more common reinforcement for aluminum alloys is SiC.

Titanium alloys that are most useful in metal matrix composites [16] are α, β alloys (e.g., Ti-6Al-9V) and metastable β-alloys (e.g., Ti-10V-2Fe-3Al). These titanium alloys have higher tensile strength–weight ratios as well as better strength retentions at 400°C–500°C over those of aluminum alloys. The thermal expansion coefficient of titanium alloys is closer to that of reinforcing fibers, which reduces the thermal mismatch between them. One of the problems with titanium alloys is their high reactivity with boron and Al_2O_3 fibers at normal fabrication temperatures. Borsic (boron fibers coated with silicon carbide) and silicon carbide (SiC) fibers show less reactivity with titanium. Improved tensile strength retention is obtained by coating boron and SiC fibers with carbon-rich layers.

2.2.3 Ceramic Matrix

Ceramics are known for their high temperature stability, high thermal shock resistance, high modulus, high hardness, high corrosion resistance, and low density. However, they are brittle materials and possess low resistance to crack propagation, which is manifested in their low fracture toughness. The primary reason for reinforcing a ceramic matrix is to increase its fracture toughness.

Structural ceramics used as matrix materials can be categorized as either oxides or nonoxides. Alumina (Al_2O_3) and mullite (Al_2O_3–SiO_2) are the two most commonly used oxide ceramics. They are known for their thermal and chemical stability. The common nonoxide ceramics are silicon carbide (SiC), silicon nitride (Si_3N_4), boron carbide (B_4C), and aluminum nitride (AlN). Of these, SiC has found wider applications, particularly where high modulus is desired. It also has an excellent high temperature resistance. Si_3N_4 is considered for applications requiring high strength and AlN is of interest because of its high thermal conductivity.

The reinforcements used in ceramic matrix composites are SiC, Si_3N_4, AlN, and other ceramic fibers. Of these, SiC has been the most commonly used reinforcement because of its thermal stability and compatibility with a broad range of both oxide and nonoxide ceramic matrices. The forms in which the reinforcement is used in ceramic matrix composites include whiskers (with length to diameter ratio as high as 500), platelets, particulates, and both monofilament and multifilament continuous fibers.

2.3 THERMOSET MATRIX

2.3.1 EPOXY

Starting materials for epoxy matrix are low-molecular-weight organic liquid resins containing a number of epoxide groups, which are three-member rings of one oxygen atom and two carbon atoms:

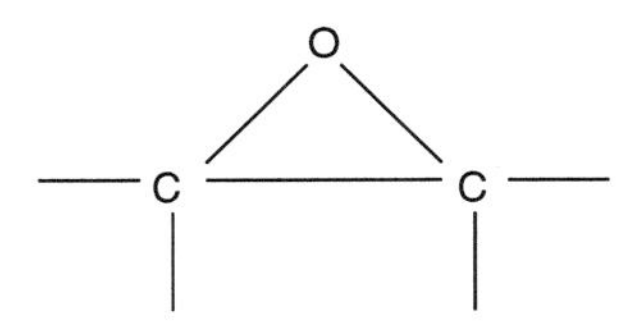

A common starting material is diglycidyl ether of bisphenol A (DGEBA), which contains two epoxide groups, one at each end of the molecule (Figure 2.27a). Other ingredients that may be mixed with the starting liquid are diluents to reduce its viscosity and flexibilizers to improve the impact strength of the cured epoxy matrix.

The polymerization (curing) reaction to transform the liquid resin to the solid state is initiated by adding small amounts of a reactive curing agent just before incorporating fibers into the liquid mix. One such curing agent is diethylene triamine (DETA, Figure 2.27b). Hydrogen atoms in the amine (NH_2) groups of a DETA molecule react with the epoxide groups of DGEBA molecules in the manner illustrated in Figure 2.28a. As the reaction continues, DGEBA molecules form cross-links with each other (Figure 2.28b) and a three-dimensional network structure is slowly formed (Figure 2.28c). The resulting material is a solid epoxy polymer.

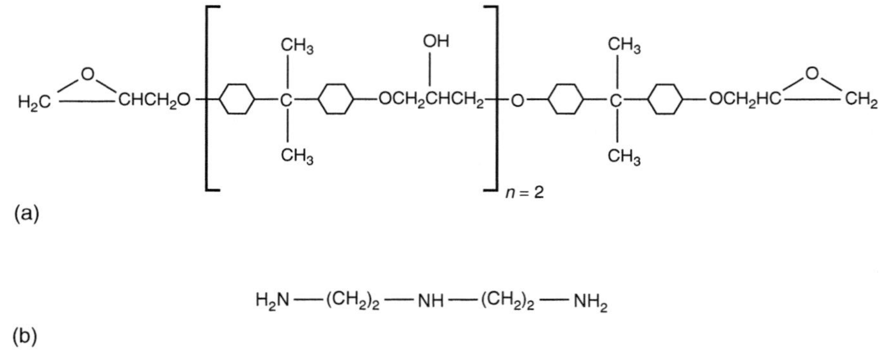

H_2C—(O)—$CHCH_2O$—[—⟨⟩—$C(CH_3)_2$—⟨⟩—$OCH_2CH(OH)CH_2$—]$_{n=2}$—O—⟨⟩—$C(CH_3)_2$—⟨⟩—OCH_2HC—(O)—CH_2

(a)

$H_2N—(CH_2)_2—NH—(CH_2)_2—NH_2$

(b)

FIGURE 2.27 Principal ingredients in the preparation of an epoxy matrix. (a) A molecule of diglycidyl ether of bisphenol A (DGEBA) epoxy resin. (b) A molecule of diethylene triamine (DETA) curing agent.

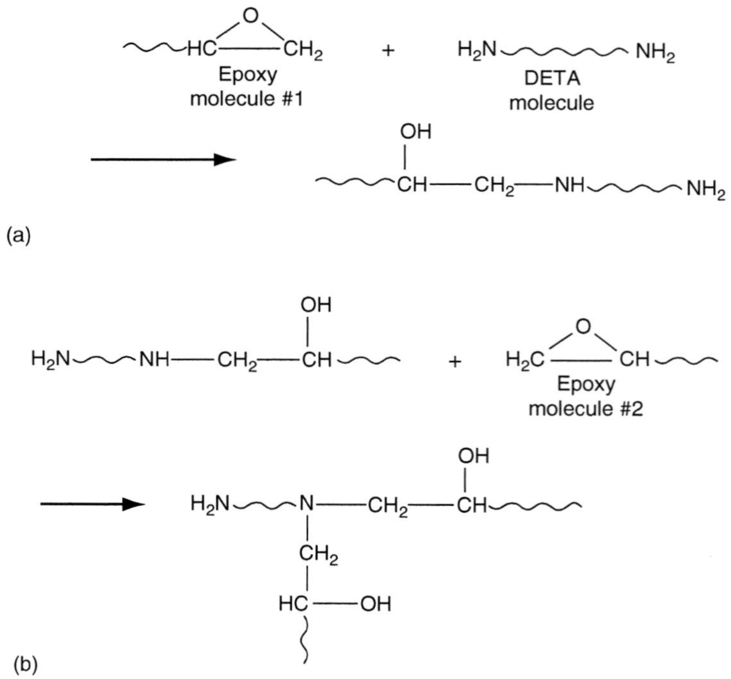

~HC—(O)—CH_2 + H_2N~NH_2

Epoxy molecule #1 DETA molecule

→ ~CH(OH)—CH_2—NH~NH_2

(a)

H_2N~NH—CH_2—CH(OH)~ + H_2C—(O)—CH~

Epoxy molecule #2

→ H_2N~N(CH_2—HC(OH)~)—CH_2—CH(OH)~

(b)

FIGURE 2.28 Schematic representation of a cross-linked epoxy resin. (a) Reaction of epoxide group with DETA molecule; (b) formation of cross-links; and

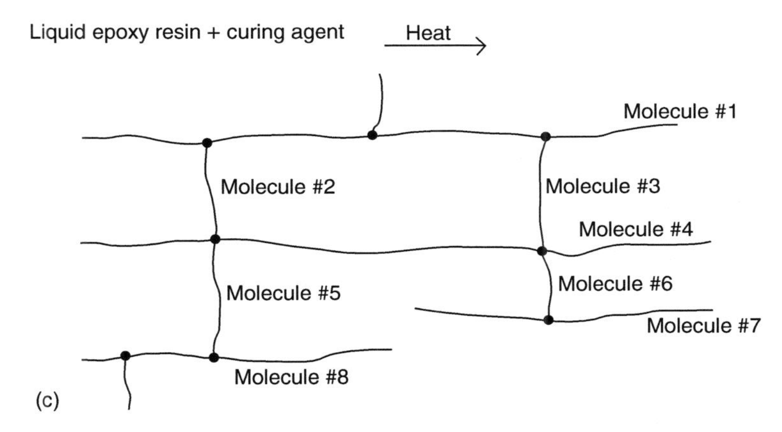

FIGURE 2.28 (continued) (c) three-dimensional network structure of solid epoxy.

If the curing reaction is slowed by external means (e.g., by lowering the reaction temperature) before all the molecules are cross-linked, the resin would exist in B-stage form. At this stage, cross-links have formed at widely spaced points in the reactive mass. Hardness, tackiness, and the solvent reactivity of the B-staged resin depend on the cure advancement or the degree of cure at the end of B-staging. The B-staged resin can be transformed into a hard, insoluble mass by completing the cure at a later time.

Curing time (also called pot life) and temperature to complete the polymerization reaction depend on the type and amount of curing agent. With some curing agents, the reaction initiates and proceeds at room temperature; but with others, elevated temperatures are required. Accelerators are sometimes added to the liquid mix to speed up a slow reaction and shorten the curing time.

The properties of a cured epoxy resin depend principally on the cross-link density (spacing between successive cross-link sites). In general, the tensile modulus, glass transition temperature, and thermal stability as well as chemical resistance are improved with increasing cross-link density, but the strain-to-failure and fracture toughness are reduced. Factors that control the cross-link density are the chemical structure of the starting liquid resin (e.g., number of epoxide groups per molecule and spacing between epoxide groups), functionality of the curing agent (e.g., number of active hydrogen atoms in DETA), and the reaction conditions, such as temperature and time.

The continuous use temperature for DGEBA-based epoxies is 150°C or less. Higher heat resistance can be obtained with epoxies based on novolac and cycloaliphatics, for example, which have a continuous use temperature ranging up to 250°C. In general, the heat resistance of an epoxy is improved if it contains more aromatic (—⬡—) rings in its basic molecular chain.

TABLE 2.8
Typical Properties of Cast Epoxy Resin (at 23°C)

Density (g/cm^3)	1.2–1.3
Tensile strength, MPa (psi)	55–130 (8,000–19,000)
Tensile modulus, GPa (10^6 psi)	2.75–4.10 (0.4–0.595)
Poisson's ratio	0.2–0.33
Coefficient of thermal expansion, 10^{-6} m/m per °C (10^{-6} in./in. per °F)	50–80 (28–44)
Cure shrinkage, %	1–5

Epoxy matrix, as a class, has the following advantages over other thermoset matrices:

1. Wide variety of properties, since a large number of starting materials, curing agents, and modifiers are available
2. Absence of volatile matters during cure
3. Low shrinkage during cure
4. Excellent resistance to chemicals and solvents
5. Excellent adhesion to a wide variety of fillers, fibers, and other substrates

The principal disadvantages are its relatively high cost and long cure time. Typical properties of cast epoxy resins are given in Table 2.8.

One of the epoxy resins used in the aerospace industry is based on tetraglycidal diaminodiphenyl methane (TGDDM). It is cured with diaminodiphenyl sulfone (DDS) with or without an accelerator. The TGDDM–DDS system is used due to its relatively high glass transition temperature (240°C–260°C, compared with 180°C–190°C for DGEBA systems) and good strength retention even after prolonged exposure to elevated temperatures. Prepregs made with this system can be stored for a longer time period due to relatively low curing reactivity of DDS in the "B-staged" resin. Limitations of the TGDDM system are their poor hot–wet performance, low strain-to-failure, and high level of atmospheric moisture absorption (due to its highly polar molecules). High moisture absorption reduces its glass transition temperature as well as its modulus and other mechanical properties.

Although the problems of moisture absorption and hot–wet performance can be reduced by changing the resin chemistry (Table 2.9), brittleness or low strain-to-failure is an inherent problem of any highly cross-linked resin. Improvement in the matrix strain-to-failure and fracture toughness is considered essential for damage-tolerant composite laminates. For epoxy resins, this can be accomplished by adding a small amount of highly reactive carboxyl-terminated butadiene–acrylonitrile (CTBN) liquid elastomer, which forms a

TABLE 2.9
Mechanical Properties of High-Performance Epoxy Resins[a,b]

Property	Epoxy 1	Epoxy 2 (Epon HPT 1072, Shell Chemical)	Epoxy 3 (Tactix 742, Dow Chemical)
T_g, °C	262	261	334
Flexural properties (at room temperature)			
Strength, MPa (ksi)	140.7 (20.4)	111.7 (16.2)	124.1 (18)
Modulus, GPa (Msi)	3.854 (0.559)	3.378 (0.490)	2.965 (0.430)
Flexural properties (hot–wet)[c]			
Strength (% retained)	55	65	—
Modulus (% retained)	64.5	87.3	—
Fracture energy, G_{Ic}, kJ/m² (in. lb/in.²)	0.09 (0.51)	0.68 (3.87)	0.09 (0.51)
Moisture gain, %	5.7	2.6	—

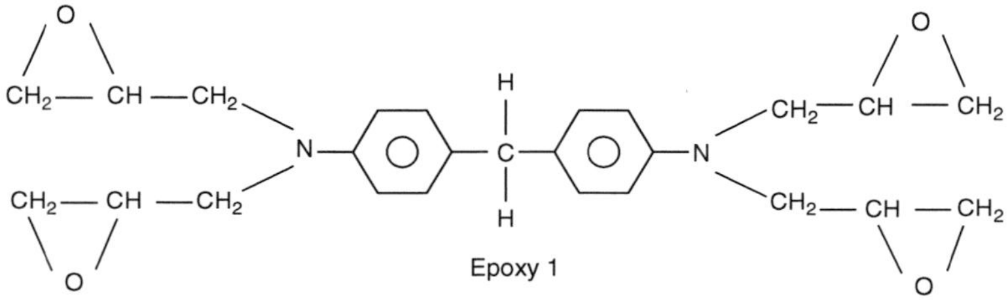

Epoxy 1

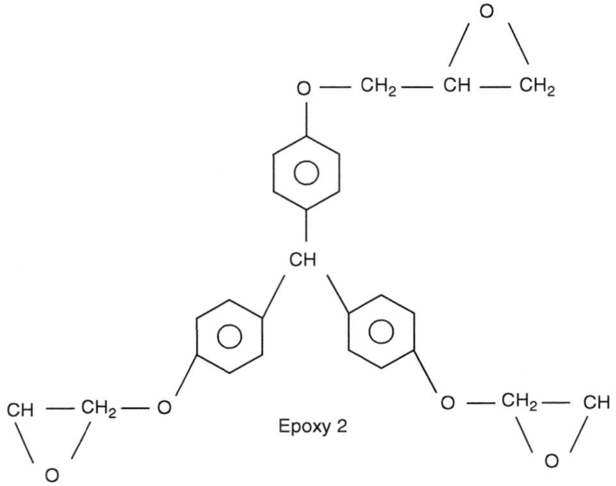

Epoxy 2

(*continued*)

TABLE 2.9 (continued)
Mechanical Properties of High-Performance Epoxy Resins[a,b]

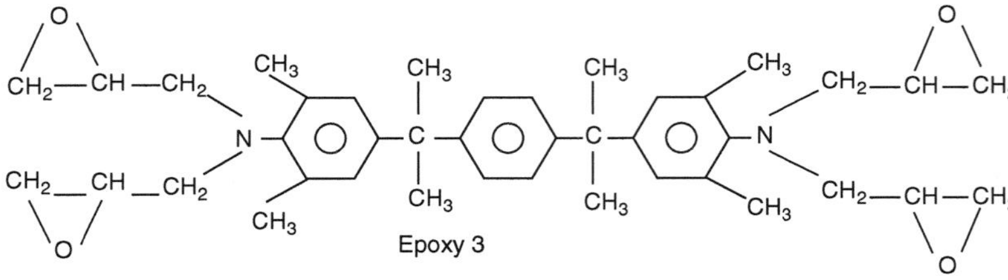

Epoxy 3

[a] All epoxies were cured with DDS.
[b] Molecular structures for Epoxies 1–3 are given in the accompanying figures.
[c] Percent retained with room temperature and dry properties when tested in water at 93°C after 2 week immersion at 93°C.

second phase in the cured matrix and impedes its microcracking. Although the resin is toughened, its glass transition temperature, modulus, and tensile strength as well as solvent resistance are reduced (Table 2.10). This problem is overcome by blending epoxy with a tough thermoplastic resin, such as polyethersulfone, but the toughness improvement depends on properly matching the epoxy and thermoplastic resin functionalities, their molecular weights, and so on [17].

2.3.2 Polyester

The starting material for a thermoset polyester matrix is an unsaturated polyester resin that contains a number of C=C double bonds. It is prepared by the

TABLE 2.10
Effect of CTBN Addition on the Properties of Cast Epoxy Resin

CTBN parts per 100 parts of epoxy[a]	0	5	10	15
Tensile strength, MPa	65.8	62.8	58.4	51.4
Tensile modulus, GPa	2.8	2.5	2.3	2.1
Elongation at break (%)	4.8	4.6	6.2	8.9
Fracture energy, G_{Ic}, kJ/m^2	1.75	26.3	33.3	47.3
HDT, °C (at 1.82 MPa)	80	76	74	71

Source: Adapted from Riew, C.K., Rowe, E.H., and Siebert, A.R., *Toughness and Brittleness of Plastics*, R.D. Deanin and A.M. Crugnola, eds., American Chemical Society, Washington, D.C., 1976.

[a] DGEBA epoxy (Epon 828, Shell) cured with five parts of piperidine at 120°C for 16 h.

reaction of maleic anhydride and ethylene glycol or propylene glycol (Figure 2.29a). Saturated acids, such as isophthalic acid or orthophthalic acid, are also added to modify the chemical structure between the cross-linking sites; however, these acids do not contain any C=C double bonds. The resulting polymeric liquid is dissolved in a reactive (polymerizable) diluent, such as styrene (Figure 2.29b), which reduces its viscosity and makes it easier to handle. The diluent also contains C=C double bonds and acts as a cross-linking agent by bridging the adjacent polyester molecules at their unsaturation points. Trace amounts of an inhibitor, such as hydroquinone or benzoquinone, are added to the liquid mix to prevent premature polymerization during storage.

The curing reaction for polyester resins is initiated by adding small quantities of a catalyst, such as an organic peroxide or an aliphatic azo compound (Figure 2.29c), to the liquid mix. With the application of heat (in the temperature range

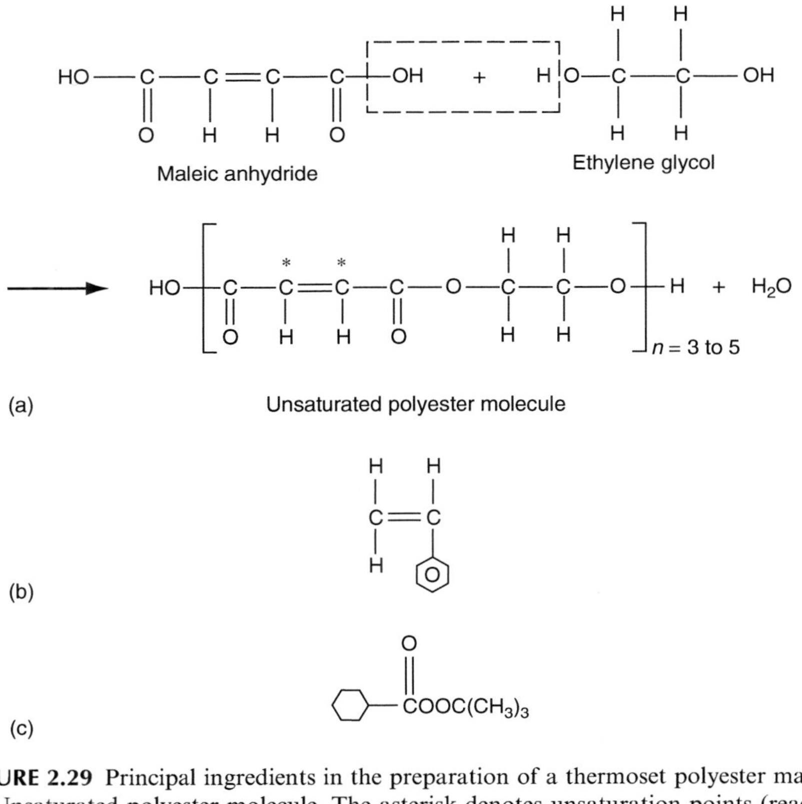

FIGURE 2.29 Principal ingredients in the preparation of a thermoset polyester matrix. (a) Unsaturated polyester molecule. The asterisk denotes unsaturation points (reactive sites) in the unsaturated polyester molecule; (b) styrene molecule; and (c) *t*-butyl perbenzoate molecule (tBPB).

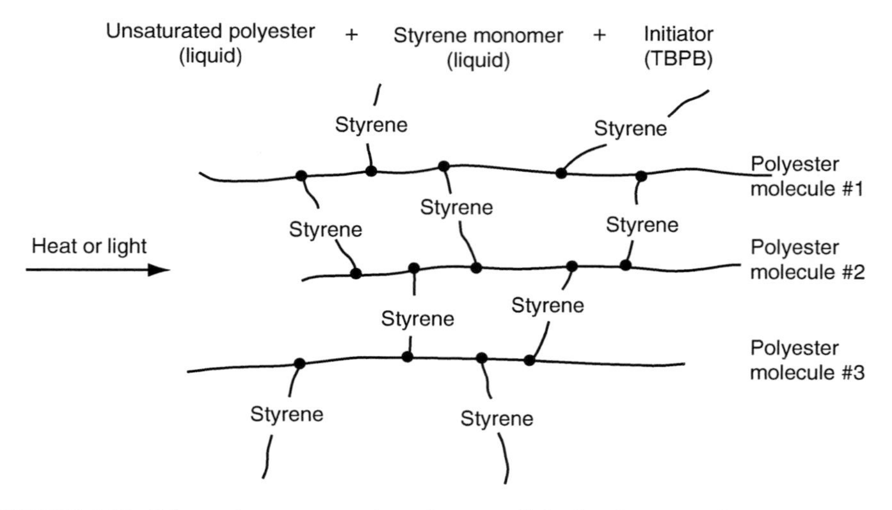

FIGURE 2.30 Schematic representation of a cross-linked polyester resin.

of 107°C–163°C), the catalyst decomposes rapidly into free radicals, which react (mostly) with the styrene molecules and break their C=C bonds. Styrene radicals, in turn, join with the polyester molecules at their unsaturation points and eventually form cross-links between them (Figure 2.30). The resulting material is a solid polyester resin.

The curing time for polyester resins depends on the decomposition rate of the catalyst, which can be increased by increasing the curing temperature. However, for a given resin–catalyst system, there is an optimum temperature at which all of the free radicals generated from the catalyst are used in curing the resin. Above this optimum temperature, free radicals are formed so rapidly that wasteful side reactions occur and deterioration of the curing reaction is observed. At temperatures below the optimum, the curing reaction is very slow. The decomposition rate of a catalyst is increased by adding small quantities of an accelerator, such as cobalt naphthanate (which essentially acts as a catalyst for the primary catalyst).

As in the case of epoxy resins, the properties of polyester resins depend strongly on the cross-link density. The modulus, glass transition temperature, and thermal stability of cured polyester resins are improved by increasing the cross-link density, but the strain-to-failure and impact energy are reduced. The major factor influencing the cross-link density is the number of unsaturation points in an uncured polyester molecule. The simplest way of controlling the frequency of unsaturation points is to vary the weight ratio of various ingredients used for making unsaturated polyesters. For example, the frequency of unsaturation in an isophthalic polyester resin decreases as the weight ratio

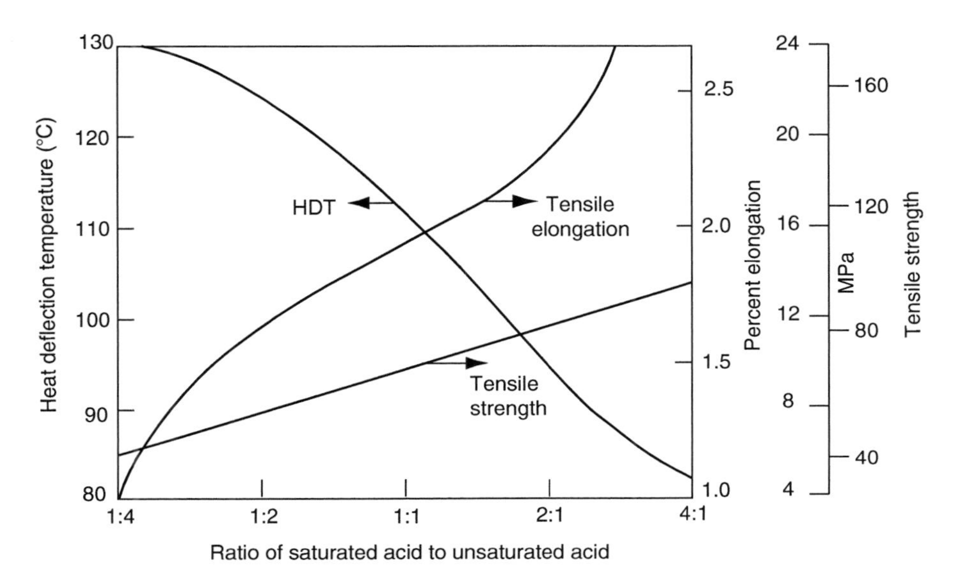

FIGURE 2.31 Effect of unsaturation level on the properties of a thermoset polyester resin. (After How ingredients influence unsaturated polyester properties, Amoco Chemicals Corporation, Bulletin IP-70, 1980.)

of isophthalic acid to maleic anhydride is increased. The effect of such weight ratio variation on various properties of a cured isophthalic polyester resin is shown in Figure 2.31. The type of ingredients also influences the properties and processing characteristics of polyester resins. For example, terephthalic acid generally provides a higher HDT than either isophthalic or orthophthalic acid, but it has the slowest reactivity of the three phthalic acids. Adipic acid, if used instead of any of the phthalic acids, lowers the stiffness of polyester molecules, since it does not contain an aromatic ring in its backbone. Thus, it can be used as a flexibilizer for polyester resins. Another ingredient that can also lower the stiffness is diethylene glycol. Propylene glycol, on the other hand, makes the polyester resin more rigid, since the pendant methyl groups in its structure restrict the rotation of polyester molecules.

The amount and type of diluent are also important factors in controlling the properties and processing characteristics of polyester resins. Styrene is the most widely used diluent because it has low viscosity, high solvency, and low cost. Its drawbacks are flammability and potential (carcinogenic) health hazard due to excessive emissions. Increasing the amount of styrene reduces the modulus of the cured polyester resin, since it increases the space between polyester molecules. Because styrene also contributes unsaturation points, higher styrene content in the resin solution increases the total amount of unsaturation and, consequently, the curing time is increased. An excessive amount of styrene tends to promote self-polymerization (i.e., formation of

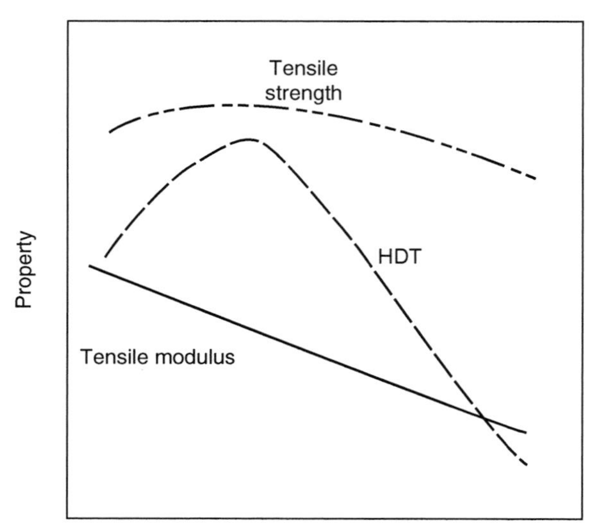

FIGURE 2.32 Effect of increasing styrene content on the properties of a thermoset polyester resin.

polystyrene) and causes polystyrene-like properties to dominate the cured polyester resin (Figure 2.32).

Polyester resins can be formulated in a variety of properties ranging from hard and brittle to soft and flexible. Its advantages are low viscosity, fast cure time, and low cost. Its properties (Table 2.11) are generally lower than those for epoxies. The principal disadvantage of polyesters over epoxies is their high volumetric shrinkage. Although this allows easier release of parts from the mold, the difference in shrinkage between the resin and fibers results in uneven depressions (called *sink marks*) on the molded surface. The sink marks are undesirable for exterior surfaces requiring high gloss and good appearance

TABLE 2.11
Typical Properties of Cast Thermoset Polyester Resins (at 23°C)

Density (g/cm^3)	1.1–1.43
Tensile strength, MPa (psi)	34.5–103.5 (5,000–15,000)
Tensile modulus, GPa (10^6 psi)	2.1–3.45 (0.3–0.5)
Elongation, %	1–5
HDT, °C (°F)	60–205 (140–400)
Cure shrinkage, %	5–12

(e.g., Class A surface quality in automotive body panels, such as hoods). One way of reducing these surface defects is to use low-shrinkage (also called low-profile) polyester resins that contain a thermoplastic component (such as polystyrene or PMMA). As curing proceeds, phase changes in the thermoplastic component allow the formation of microvoids that compensate for the normal shrinkage of the polyester resin.

2.3.3 Vinyl Ester

The starting material for a vinyl ester matrix is an unsaturated vinyl ester resin produced by the reaction of an unsaturated carboxylic acid, such as methacrylic or acrylic acid, and an epoxy (Figure 2.33). The C═C double bonds (unsaturation points) occur only at the ends of a vinyl ester molecule, and therefore, cross-linking can take place only at the ends, as shown schematically in Figure 2.34. Because of fewer cross-links, a cured vinyl ester resin is more flexible and has higher fracture toughness than a cured polyester resin. Another unique characteristic of a vinyl ester molecule is that it contains a number of OH (hydroxyl) groups along its length. These OH groups can form hydrogen bonds with similar groups on a glass fiber surface resulting in excellent wet-out and good adhesion with glass fibers.

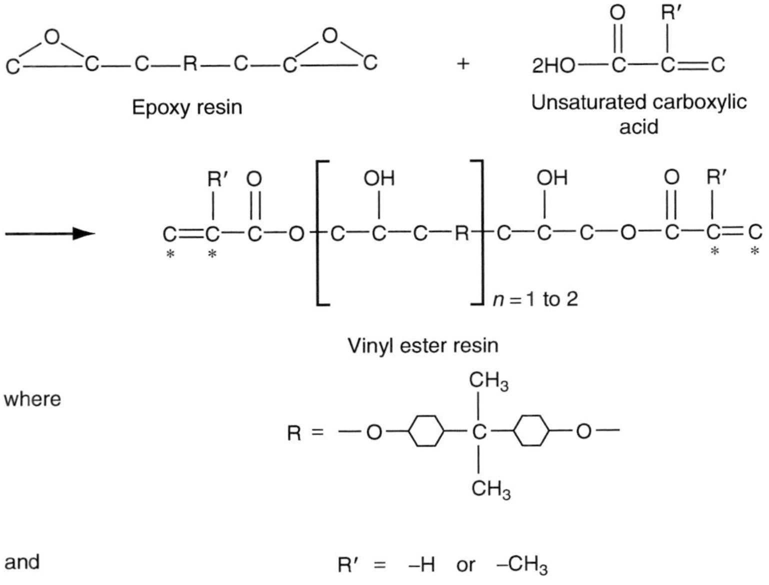

FIGURE 2.33 Chemistry of a vinyl ester resin. The asterisk denotes unsaturation points (reactive sites).

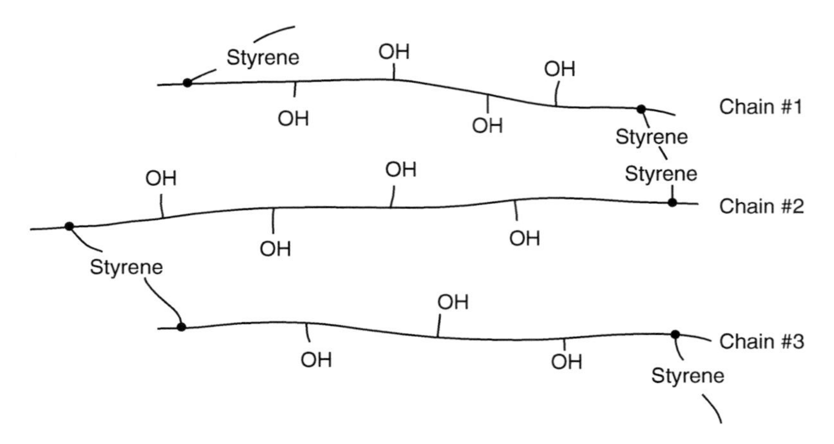

FIGURE 2.34 Schematic representation of a cross-linked vinyl ester resin.

Vinyl ester resins, like unsaturated polyester resins, are dissolved in styrene monomer, which reduces their viscosity. During polymerization, styrene coreacts with the vinyl ester resin to form cross-links between the unsaturation points in adjacent vinyl ester molecules. The curing reaction for vinyl ester resins is similar to that for unsaturated polyesters.

Vinyl ester resins possess good characteristics of epoxy resins, such as excellent chemical resistance and tensile strength, and of unsaturated polyester resins, such as low viscosity and fast curing. However, the volumetric shrinkage of vinyl ester resins is in the range of 5%–10%, which is higher than that of the parent epoxy resins (Table 2.12). They also exhibit only moderate adhesive strengths compared with epoxy resins. The tensile and flexural properties of cured vinyl ester resins do not vary appreciably with the molecular weight and type of epoxy resin or other coreactants. However, the HDT and thermal stability can be improved by using heat-resistant epoxy resins, such as phenolic-novolac types.

TABLE 2.12
Typical Properties of Cast Vinyl Ester Resins (at 23°C)

Density (g/cm^3)	1.12–1.32
Tensile strength, MPa (psi)	73–81 (10,500–11,750)
Tensile modulus, GPa (10^6 psi)	3–3.5 (0.44–0.51)
Elongation, %	3.5–5.5
HDT, °C (°F)	93–135 (200–275)
Cure shrinkage, %	5.4–10.3

2.3.4 Bismaleimides and Other Thermoset Polyimides

Bismaleimide (BMI), PMR-15 (for polymerization of monomer reactants), and acetylene-terminated polyimide (ACTP) are examples of thermoset polyimides (Table 2.13). Among these, BMIs are suitable for applications requiring a service temperature of 127°C–232°C. PMR and ACTP can be used up to 288°C and 316°C, respectively. PMR and ACTP also have exceptional thermo-oxidative stability and show only 20% weight loss over a period of 1000 h at 316°C in flowing air [18].

Thermoset polyimides are obtained by addition polymerization of liquid monomeric or oligomeric imides to form a cross-linked infusible structure. They are available either in solution form or in hot-melt liquid form. Fibers can be coated with the liquid imides or their solutions before the cross-linking reaction. On curing, they not only offer high temperature resistance, but also high chemical and solvent resistance. However, these materials are inherently very brittle due to their densely cross-linked molecular structure. As a result, their composites are prone to excessive microcracking. One useful method of reducing their brittleness without affecting their heat resistance is to combine them with one or more tough thermoplastic polyimides. The combination produces a semi-interpenetrating network (semi-IPN) polymer [19], which retains the easy processability of a thermoset and exhibits the good toughness of a thermoplastic. Although the reaction time is increased, this helps in broadening the processing window, which otherwise is very narrow for some

TABLE 2.13
Properties of Thermoset Polyimide Resins (at 23°C)

	Bismaleimide[a]			
Property	**Without Modifier**	**With[b] Modifier**	**PMR-15[c]**	**ACTP[d]**
Density (g/cm^3)	—	1.28	1.32	1.34
Tensile strength, MPa (ksi)	—	—	38.6 (5.6)	82.7 (12)
Tensile modulus, GPa (Msi)	—	—	3.9 (0.57)	4.1 (0.60)
Strain-to-failure (%)	—	—	1.5	1.5
Flexural strength, MPa (ksi)	60 (8.7)	126.2 (18.3)	176 (25.5)	145 (21)
Flexural modulus, GPa (Msi)	5.5 (0.8)	3.7 (0.54)	4 (0.58)	4.5 (0.66)
Fracture energy, G_{Ic}, J/m^2 (in. lb/in.^{2})	24.5 (0.14)	348 (1.99)	275 (1.57)	—

[a] Compimide 353 (Shell Chemical Co.).

[b] Compimide 353 melt blended with a bis-allylphenyl compound (TM 121), which acts as a toughening modifier (Shell Chemical Co.).

[c] From Ref. [18].

[d] Thermid 600 (National Starch and Chemical Corporation).

of these polyimides and causes problems in manufacturing large or complex composite parts.

BMIs are the most widely used thermoset polyimides in the advanced composite industry. BMI monomers (prepolymers) are prepared by the reaction of maleic anhydride with a diamine. A variety of BMI monomers can be prepared by changing the diamine. One commercially available BMI monomer has the following chemical formula:

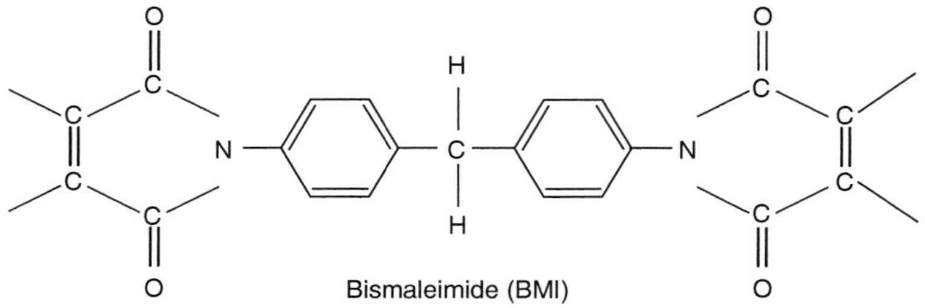

Bismaleimide (BMI)

BMI monomers are mixed with reactive diluents to reduce their viscosity and other comonomers, such as vinyl, acrylic, and epoxy, to improve the toughness of cured BMI. The handling and processing techniques for BMI resins are similar to those for epoxy resins. The curing of BMI occurs through addition-type homopolymerization or copolymerization that can be thermally induced at 170°C–190°C.

2.3.5 Cyanate Ester

Cyanate ester resin has a high glass transition temperature ($T_g = 265°C$), lower moisture absorption than epoxies, good chemical resistance, and good dimensional stability [20]. Its mechanical properties are similar to those of epoxies. The curing reaction of cyanate ester involves the formation of thermally stable triazine rings, which is the reason for its high temperature resistance. The curing shrinkage of cyanate ester is also relatively small. For all these reasons, cyanate ester is considered a good replacement for epoxy in some aerospace applications. Cyanate ester is also considered for printed circuit boards, encapsulants, and other electronic components because of its low dielectric constant and high dielectric breakdown strength, two very important characteristics for many electronic applications.

Cyanate ester is commonly used in blended form with other polymers. For example, it is sometimes blended with epoxy to reduce cost. Blending it with BMI has shown to improve its T_g. Like many other thermoset polymers, cyanate ester has low fracture toughness. Blending it with thermoplastics, such as polyarylsulfone and polyethersulfone, has shown to improve its fracture toughness.

2.4 THERMOPLASTIC MATRIX

Table 2.14 lists the mechanical properties of selected thermoplastic polymers that are considered suitable for high-performance composite applications. The molecules in these polymers contain rigid aromatic rings that give them a relatively high glass transition temperature and an excellent dimensional stability at elevated temperatures. The actual value of T_g depends on the size and flexibility of other chemical groups or linkages in the chain.

2.4.1 Polyether Ether Ketone

Polyether ether ketone (PEEK) is a linear aromatic thermoplastic based on the following repeating unit in its molecules:

TABLE 2.14
Properties of Selected Thermoplastic Matrix Resins (at 23°C)

Property	PEEK[a]	PPS[b]	PSUL[c]	PEI[d]	PAI[e]	K-III[f]	LARC-TPI[g]
Density (g/cm^3)	1.30–1.32	1.36	1.24	1.27	1.40	1.31	1.37
Yield (Y) or tensile (T) strength, MPa (ksi)	100 (14.5) (Y)	82.7 (12) (T)	70.3 (10.2) (Y)	105 (15.2) (Y)	185.5 (26.9) (T)	102 (14.8) (T)	138 (20) (T)
Tensile modulus, GPa (Msi)	3.24 (0.47)	3.3 (0.48)	2.48 (0.36)	3 (0.43)	3.03 (0.44)	3.76 (0.545)	3.45 (0.5)
Elongation-at-break (%)	50	4	75	60	12	14	5
Poisson's ratio	0.4	—	0.37	—	—	0.365	0.36
Flexural strength, MPa (ksi)	170 (24.65)	152 (22)	106.2 (15.4)	150 (21.75)	212 (30.7)	— —	— —
Flexural modulus, GPa (Msi)	4.1 (0.594)	3.45 (0.5)	2.69 (0.39)	3.3 (0.48)	4.55 (0.66)	— —	— —
Fracture energy (G_{Ic}), kJ/m^2	6.6	—	3.4	3.7	3.9	1.9	—
HDT, °C (at 1.82 MPa)	160	135	174	200	274	—	—
CLTE, 10^{-5}/°C	4.7	4.9	5.6	5.6	3.6	—	3.5

[a] Victrex.
[b] Ryton.
[c] Udel.
[d] Ultem.
[e] Torlon.
[f] Avimid.
[g] Durimid.

Ketone Ether Ether

Polyetherether ketone (PEEK)

Continuous carbon fiber-reinforced PEEK composites are known in the industry as aromatic polymer composite or APC.

PEEK is a semicrystalline polymer with a maximum achievable crystallinity of 48% when it is cooled slowly from its melt. Amorphous PEEK is produced if the melt is quenched. At normal cooling rates, the crystallinity is between 30% and 35%. The presence of fibers in PEEK composites tends to increase the crystallinity to a higher level, since the fibers act as nucleation sites for crystal formation [21]. Increasing the crystallinity increases both modulus and yield strength of PEEK, but reduces its strain-to-failure (Figure 2.35).

PEEK has a glass transition temperature of 143°C and a crystalline melting point of 335°C. Melt processing of PEEK requires a temperature range of 370°C–400°C. The maximum continuous use temperature is 250°C.

The outstanding property of PEEK is its high fracture toughness, which is 50–100 times higher than that of epoxies. Another important advantage of PEEK is its low water absorption, which is less than 0.5% at 23°C compared with 4%–5% for conventional aerospace epoxies. As it is semicrystalline, it does not dissolve in common solvents. However, it may absorb some of these

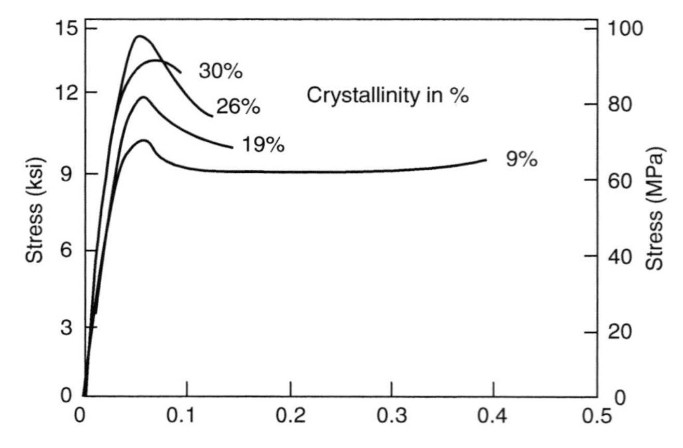

FIGURE 2.35 Tensile stress–strain diagram of PEEK at different crystallinities. (Adapted from Seferis, J.C., *Polym. Compos.*, 71, 58, 1986.)

solvents, most notably methylene chloride. The amount of solvent absorption decreases with increasing crystallinity.

2.4.2 Polyphenylene Sulfide

Polyphenylene sulfide (PPS) is a semicrystalline polymer with the following repeating unit in its molecules:

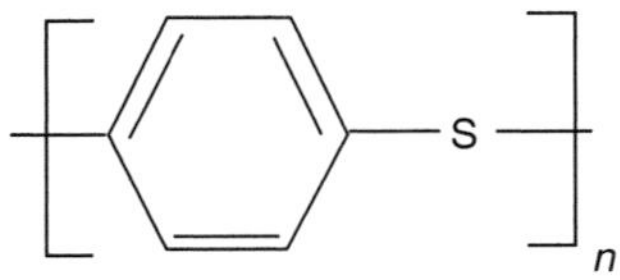

$$\left[-C_6H_4-S- \right]_n$$

PPS is normally 65% crystalline. It has a glass transition temperature of 85°C and a crystalline melting point of 285°C. The relatively low T_g of PPS is due to the flexible sulfide linkage between the aromatic rings. Its relatively high crystallinity is attributed to the chain flexibility and structural regularity of its molecules. Melt processing of PPS requires heating the polymer in the temperature range of 300°C–345°C. The continuous use temperature is 240°C. It has excellent chemical resistance.

2.4.3 Polysulfone

Polysulfone is an amorphous thermoplastic with the repeating unit shown as follows:

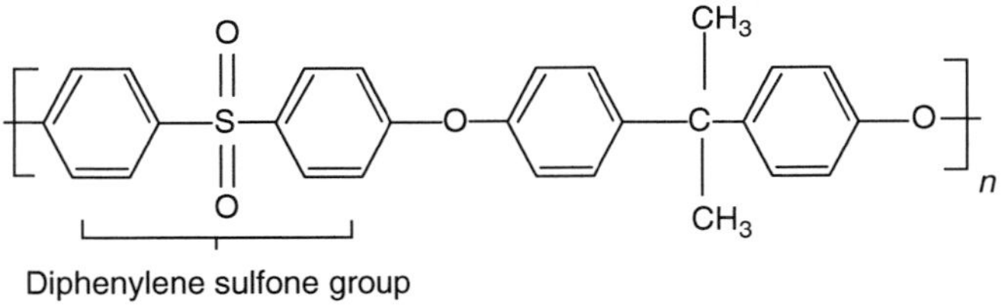

$$\left[-C_6H_4-SO_2-C_6H_4-O-C_6H_4-C(CH_3)_2-C_6H_4-O- \right]_n$$

Diphenylene sulfone group

Polysulfone

Polysulfone has a glass transition temperature of 185°C and a continuous use temperature of 160°C. The melt processing temperature is between 310°C and 410°C. It has a high tensile strain-to-failure (50%–100%) and an excellent hydrolytic stability under hot–wet conditions (e.g., in steam). Although polysulfone has good resistance to mineral acids, alkalis, and salt solutions, it will swell, stress-crack, or dissolve in polar organic solvents such as ketones, chlorinated hydrocarbons, and aromatic hydrocarbons.

2.4.4 Thermoplastic Polyimides

Thermoplastic polyimides are linear polymers derived by condensation polymerization of a polyamic acid and an alcohol. Depending on the types of the polyamic acid and alcohol, various thermoplastic polyimides can be produced. The polymerization reaction takes place in the presence of a solvent and produces water as its by-product. The resulting polymer has a high melt viscosity and must be processed at relatively high temperatures. Unlike thermosetting polyimides, they can be reprocessed by the application of heat and pressure.

Polyetherimide (PEI) and polyamide-imide (PAI) are melt-processable thermoplastic polyimides. Their chemical structures are shown as follows.

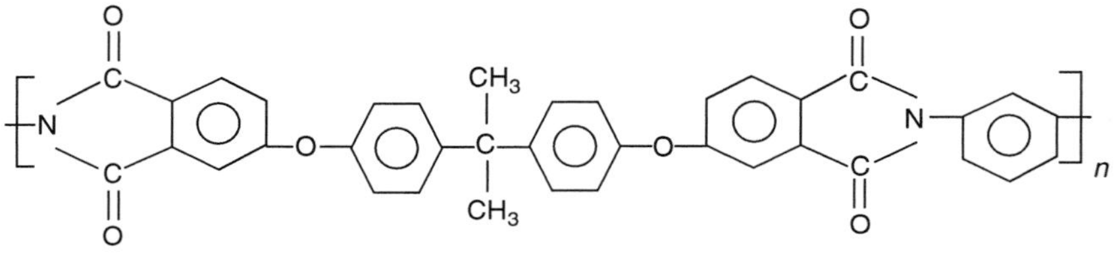

Polyetherimide (PEI)

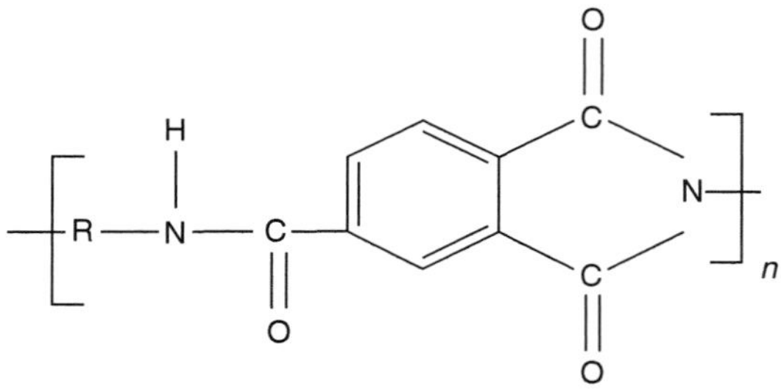

Polyamide-imide (PAI)

Both are amorphous polymers with high glass transition temperatures, 217°C for PEI and 280°C for PAI. The processing temperature is 350°C or above.

Two other thermoplastic polyimides, known as K polymers and Langley Research Center Thermoplastic Imide (LARC-TPI), are generally available as prepolymers dissolved in suitable solvents. In this form, they have low viscosities so that the fibers can be coated with their prepolymers to produce flexible prepregs. Polymerization, which for these polymers means imidization or imide ring formation, requires heating up to 300°C or above.

The glass transition temperatures of K polymers and LARC-TPI are 250°C and 265°C, respectively. Both are amorphous polymers, and offer excellent heat and solvent resistance. Since their molecules are not cross-linked, they are not as brittle as thermoset polymers. They are processed with fibers from low-viscosity solutions much like the thermoset resins; yet after imidization, they can be made to flow and be shape-formed like conventional thermoplastics by heating them over their T_g. This latter characteristic is due to the presence of flexible chemical groups between the stiff, fused-ring imide groups in their backbones. In LARC-TPI, for example, the sources of flexibility are the carbonyl groups and the *meta*-substitution of the phenyl rings in the diamine-derived portion of the chain. The *meta*-substitution, in contrast to *para*-substitution, allows the polymer molecules to bend and flow.

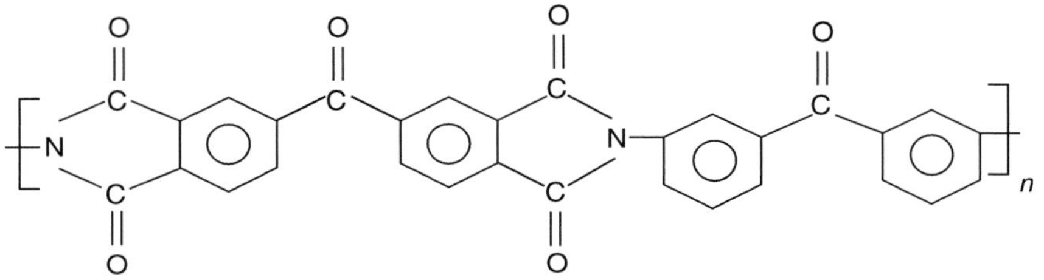

LARC-TPI

2.5 FIBER SURFACE TREATMENTS

The primary function of a fiber surface treatment is to improve the fiber surface wettability with the matrix and to create a strong bond at the fiber–matrix interface. Both are essential for effective stress transfer from the matrix to the fiber and vice versa. Surface treatments for glass, carbon, and Kevlar fibers for their use in polymeric matrices are described in this section.*

2.5.1 Glass Fibers

Chemical coupling agents are used with glass fibers to (1) improve the fiber–matrix interfacial strength through physical and chemical bonds and (2) protect the fiber surface from moisture and reactive fluids.

* Several investigators [27] have suggested that there is a thin but distinct interphase between the fibers and the matrix. The interphase surrounds the fiber and has properties that are different from the bulk of the matrix. It may be created by local variation of the matrix microstructure close to the fiber surface. For example, there may be a variation in cross-link density in the case of a thermosetting matrix, or a variation of crystallinity in the case of a semicrystalline thermoplastic matrix, both of which are influenced by the presence of fibers as well as the fiber surface chemistry. The interphase may also contain microvoids (resulting from poor fiber surface wetting by the matrix or air entrapment) and unreacted solvents or curing agents that tend to migrate toward the fiber surface.

Common coupling agents used with glass fibers are organofunctional silicon compounds, known as silanes. Their chemical structure is represented by R′-Si(OR)$_3$, in which the functional group R′ must be compatible with the matrix resin in order for it to be an effective coupling agent. Some representative commercial silane coupling agents are listed in Table 2.15.

The glass fiber surface is treated with silanes in aqueous solution. When a silane is added to water, it is hydrolyzed to form R′-Si(OH)$_3$:

$$\underset{\text{Silane}}{R' - Si(OR)_3} + \underset{\text{Water}}{3H_2O} \rightarrow R' - Si(OH)_3 + 3HOR.$$

Before treating glass fiber with a coupling agent, its surface must be cleaned from the size applied at the time of forming. The size is burned away by heating the fiber in an air-circulating oven at 340°C for 15–20 h. As the heat-cleaned fibers are immersed into the aqueous solution of a silane, chemical bonds (Si–O–Si) as well as physical bonds (hydrogen bonds) are established between the (OH) groups on the glass fiber surface (which is hydroscopic owing to alkaline content) and $R' - Si(OH)_3$ molecules.

TABLE 2.15
Recommended Silane Coupling Agents for Glass Fiber-Reinforced Thermoset Polymers

With epoxy matrix:

1. γ-Aminopropyltriethoxysilane

   $H_2N - (CH_2)_3 - Si(OC_2H_5)_3$

2. γ-Glycidyloxypropyltrimethoxysilane

   $H_2\overset{O}{C - C}H - CH_2 - O(CH_2)_3 - Si(OCH_3)_3$ (epoxide ring: O bridging H_2C and CH)

3. *N*-β-Aminoethyl-γ-aminopropyltrimethoxysilane

   $H_2N - CH_2 - CH_2 - NH - (CH_2)_3 - Si(OCH_3)_3$

With polyester and vinyl ester matrix:

1. γ-Methacryloxypropyltrimethoxysilane

   $H_2 = \overset{\overset{CH_3}{|}}{C} - \overset{\overset{O}{\|}}{C} - O(CH_2)_3 - Si(OCH_3)_3$

2. Vinyl triethoxysilane

   $H_2C = CH - Si(OC_2H_5)_3$

3. Vinyl tris(β-methoxyethoxy)silane

   $H_2C = CH - Si(OCH_2 - CH_2 - O - CH_3)_3$

Glass fiber surface —OH —OH —OH + $(HO)_3$ Si-R′ → Glass fiber surface (O–H···O, O—Si-R′, O–H···O) Si-R′

When treated glass fibers are incorporated into a resin matrix, the functional group R′ in the silane film reacts with the resin to form chemical coupling between fibers and matrix.

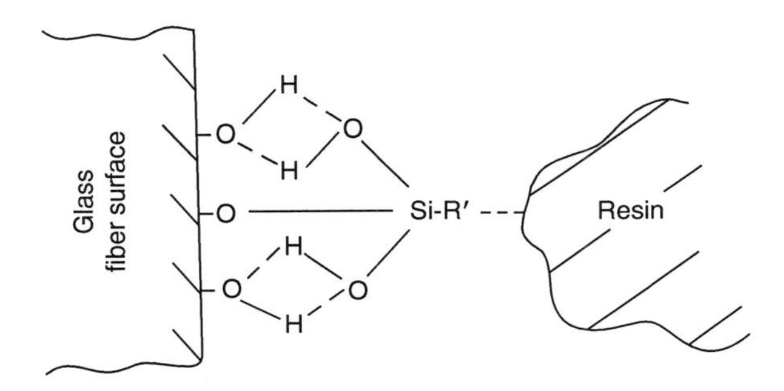

Without a coupling agent, stress transfer between the fibers and the polymer matrix is possible owing to a mechanical interlocking that arises because of higher thermal contraction of the matrix relative to the fibers. Since the coefficient of thermal expansion (and contraction) of the polymer matrix is nearly 10 times higher than that of the fibers, the matrix shrinks considerably more than the fibers as both cool down from the high processing temperature. In addition, polymerization shrinkage in the case of a thermoset polymer and crystallization shrinkage in the case of a semicrystalline polymer contribute to mechanical interlocking. Residual stresses are generated in the fiber as well as the matrix surrounding the fiber as a result of mechanical interlocking. However, at elevated service temperatures or at high applied loads, the difference in expansion of fibers and matrix may relieve this mechanical interlocking and residual stresses. Under extreme circumstances, a microcrack may be formed at the interface, resulting in reduced mechanical properties for the composite. Furthermore, moisture or other reactive fluids that may diffuse through the resin can accumulate at the interface and cause deterioration in fiber properties.

Evidence of fiber–matrix coupling effect can be observed in Figure 2.36, in which fracture surfaces of uncoupled and coupled fiber-reinforced epoxies are compared. In the uncoupled system (Figure 2.36a), the interfacial failure

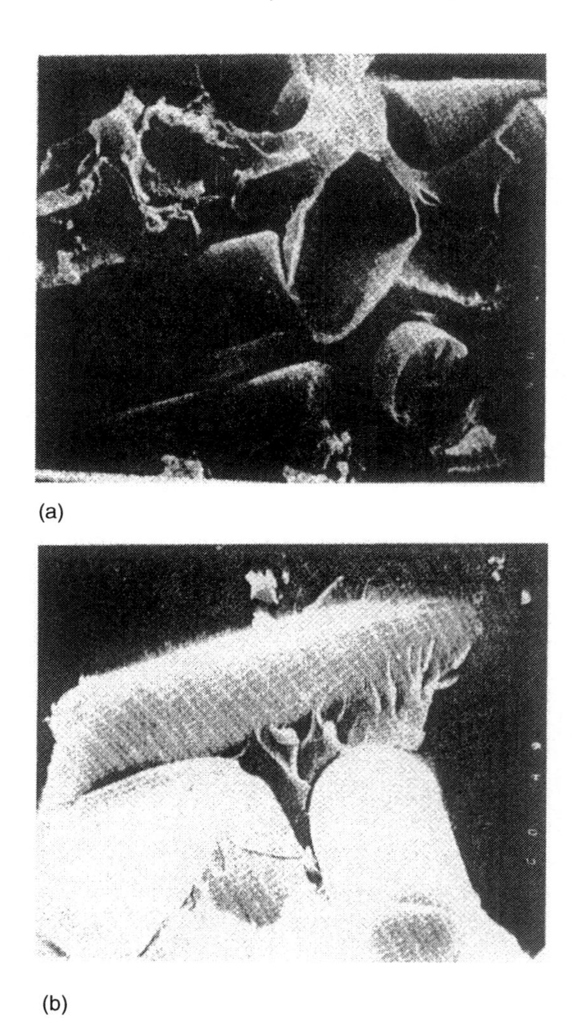

(a)

(b)

FIGURE 2.36 Photomicrographs of fracture surfaces of E-glass–epoxy composites demonstrating (a) poor adhesion with an incompatible silane coupling agent and (b) good adhesion with a compatible silane coupling agent.

is characterized by clean fiber surfaces, thin cracks (debonding) between fibers and matrix, and clear impressions in the matrix from which the fibers have pulled out or separated. In the coupled system (Figure 2.36b), strong interfacial strength is characterized by fiber surfaces coated with thin layers of matrix and the absence of fiber–matrix debonding as well as absence of clear fiber surface impressions in the matrix. In the former, debonding occurs at

TABLE 2.16
Effect of Silane Coupling Agent on the Strength of E-Glass Fiber-Reinforced Polyester Rods

	Strength (MPa)	
Treatment	**Dry**	**Wet[a]**
No silane	916	240
Vinyl silane	740	285
Glycidyl silane	990	380
Methacryl silane	1100	720

[a] After boiling in water at 100°C for 72 h.

the fiber–matrix interface, but in the latter, cohesive failure occurs in the matrix.

The interfacial bond created by silanes or other coupling agents allows a better load stress transfer between fibers and matrix, which in turn improves the tensile strength as well as the interlaminar shear strength of the composite. However, the extent of strength improvement depends on the compatibility of the coupling agent with the matrix resin. Furthermore, it has been observed that although a strong interface produces higher strength, a relatively weaker interface may contribute to higher energy dissipation through debonding at the fiber–matrix interface, which may be beneficial for attaining higher fracture toughness.

The data in Table 2.16 show the improvement in strength achieved by using different silane coupling agents on the glass fiber surface of a glass fiber–polyester composite. The wet strength of the composite, measured after boiling in water for 72 h, is lower than the dry strength; however, by adding the silane coupling agent and creating a stronger interfacial bond between the fibers and the matrix, the wet strength is significantly improved.

2.5.2 Carbon Fibers

Carbon fiber surfaces are chemically inactive and must be treated to form surface functional groups that promote good chemical bonding with the polymer matrix. Surface treatments also increase the surface area by creating micropores or surface pits on already porous carbon fiber surface. Increase in surface area provides a larger number of contact points for fiber–matrix bonding.

Commercial surface treatments for carbon fibers are of two types, oxidative or nonoxidative [4,5]:

1. Oxidative surface treatments produce acidic functional groups, such as carboxylic, phenolic, and hydroxylic, on the carbon fiber surface. They may be carried out either in an oxygen-containing gas (air, oxygen, carbon dioxide, ozone, etc.) or in a liquid (nitric acid, sodium hypochloride, etc.).

   The gas-phase oxidation is conducted at 250°C or above and often in the presence of a catalyst. Oxidation at very high temperatures causes excessive pitting on the carbon fiber surface and reduces the fiber strength.

   Nitric acid is the most common liquid used for the liquid-phase oxidation. The effectiveness of treatment in improving the surface properties depends on the acid concentration, treatment time, and temperature, as well as the fiber type.
2. Several nonoxidative surface treatments have been developed for carbon fibers. In one of these treatments, the carbon fiber surface is coated with an organic polymer that has functional groups capable of reacting with the resin matrix. Examples of polymer coatings are styrene–maleic anhydride copolymers, methyl acrylate–acrylonitrile copolymer, and polyamides. The preferred method of coating the fiber surface is electropolymerization, in which carbon fibers are used as one of the electrodes in an acidic solution of monomers or monomer mixtures [22]. Improved results are obtained if the carbon fiber surface is oxidized before the coating process.

2.5.3 Kevlar Fibers

Similar to carbon fibers, Kevlar 49 fibers also suffer from weak interfacial adhesion with most matrix resins. Two methods have been successful in improving the interfacial adhesion of Kevlar 49 with epoxy resin [7]:

1. Filament surface oxidation or plasma etching, which reduces the fiber tensile strength but tends to improve the off-axis strength of the composite, which depends on better fiber–matrix interfacial strength.
2. Formation of reactive groups, such as amines ($-NH_2$), on the fiber surface. These reactive groups form covalent bonds with the epoxide groups across the interface.

2.6 FILLERS AND OTHER ADDITIVES

Fillers are added to a polymer matrix for one or more of the following reasons:

1. Reduce cost (since most fillers are much less expensive than the matrix resin)
2. Increase modulus

TABLE 2.17
Properties of Calcium Carbonate-Filled Polyester Resin

Property	Unfilled Polyester	Polyester Filled with 30 phr $CaCO_3$
Density, g/cm^3	1.30	1.48
HDT, °C (°F)	79 (174)	83 (181)
Flexural strength, MPa (psi)	121 (17,600)	62 (9,000)
Flexural modulus, GPa (10^6 psi)	4.34 (0.63)	7.1 (1.03)

3. Reduce mold shrinkage
4. Control viscosity
5. Produce smoother surface

The most common filler for polyester and vinyl ester resins is calcium carbonate ($CaCO_3$), which is used to reduce cost as well as mold shrinkage. Examples of other fillers are clay, mica, and glass microspheres (solid as well as hollow). Although fillers increase the modulus of an unreinforced matrix, they tend to reduce its strength and impact resistance. Typical properties obtained with calcium carbonate-filled polyester matrix are shown in Table 2.17.

Impact strength and crack resistance of brittle thermosetting polymers can be improved by mixing them with small amounts of a liquid elastomeric toughener, such as carboxyl-terminated polybutadiene acrylonitrile (CTBN) [23]. In addition to fillers and tougheners, colorants, flame retardants, and ultraviolet (UV) absorbers may also be added to the matrix resin [24].

2.7 INCORPORATION OF FIBERS INTO MATRIX

Processes for incorporating fibers into a polymer matrix can be divided into two categories. In one category, fibers and matrix are processed directly into the finished product or structure. Examples of such processes are filament winding and pultrusion. In the second category, fibers are incorporated into the matrix to prepare ready-to-mold sheets that can be stored and later processed to form laminated structures by autoclave molding or compression molding. In this section, we briefly describe the processes used in preparing these ready-to-mold sheets. Knowledge of these processes will be helpful in understanding the performance of various composite laminates. Methods for manufacturing composite structures by filament winding, pultrusion, autoclave molding, compression molding, and others are described in Chapter 5.

Ready-to-mold fiber-reinforced polymer sheets are available in two basic forms, prepregs and sheet-molding compounds.

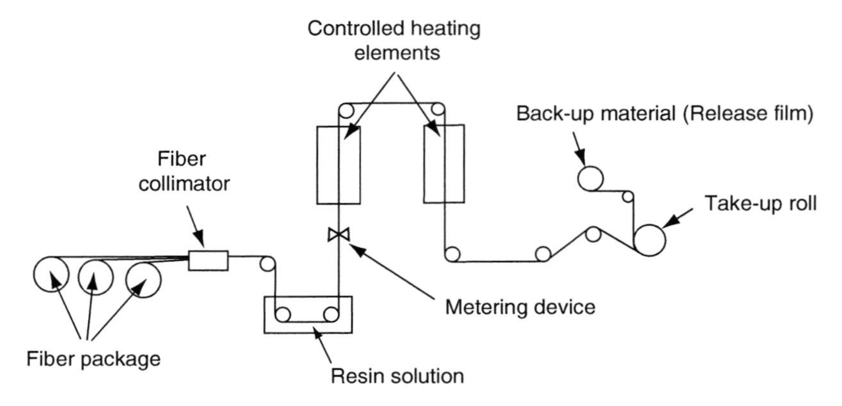

FIGURE 2.37 Schematic of prepreg manufacturing.

2.7.1 Prepregs

These are thin sheets of fibers impregnated with predetermined amounts of uniformly distributed polymer matrix. Fibers may be in the form of continuous rovings, mat, or woven fabric. Epoxy is the primary matrix material in prepreg sheets, although other thermoset and thermoplastic polymers have also been used. The width of prepreg sheets may vary from less than 25 mm (1 in.) to over 457 mm (18 in.). Sheets wider than 457 mm are called broadgoods. The thickness of a ply cured from prepreg sheets is normally in the range of 0.13–0.25 mm (0.005–0.01 in.). Resin content in commercially available prepregs is between 30% and 45% by weight.

Unidirectional fiber-reinforced epoxy prepregs are manufactured by pulling a row of uniformly spaced (collimated) fibers through a resin bath containing catalyzed epoxy resin dissolved in an appropriate solvent (Figure 2.37). The solvent is used to control the viscosity of the liquid resin. Fibers preimpregnated with liquid resin are then passed through a chamber in which heat is applied in a controlled manner to advance the curing reaction to the B-stage. At the end of B-staging, the prepreg sheet is backed up with a release film or waxed paper and wound around a take-up roll. The backup material is separated from the prepreg sheet just before it is placed in the mold to manufacture the composite part. The normal shelf life (storage time before molding) for epoxy prepregs is 6–8 days at 23°C; however, it can be prolonged up to 6 months or more if stored at −18°C.

2.7.2 Sheet-Molding Compounds

Sheet-molding compounds (SMC) are thin sheets of fibers precompounded with a thermoset resin and are used primarily in compression molding process [25]. Common thermoset resins for SMC sheets are polyesters and vinyl esters. The longer cure time for epoxies has limited their use in SMC.

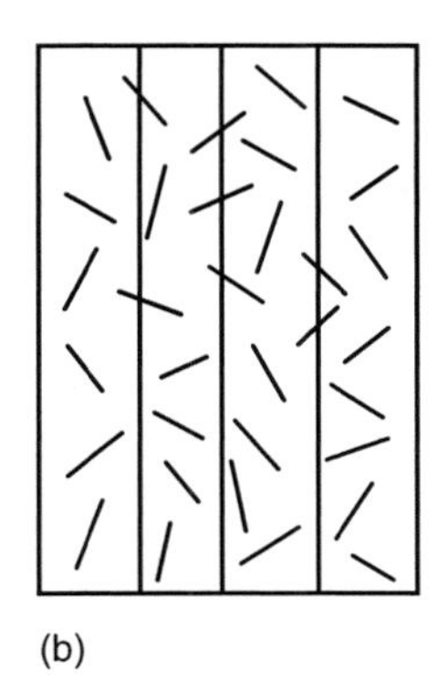

FIGURE 2.38 Various types of sheet-molding compounds (SMC): (a) SMC-R, (b) SMC-CR, and (c) XMC.

The various types of sheet-molding compounds in current use (Figure 2.38) are as follows:

1. SMC-R, containing randomly oriented discontinuous fibers. The nominal fiber content (by weight percent) is usually indicated by two-digit numbers after the letter R. For example, the nominal fiber content in SMC-R30 is 30% by weight.
2. SMC-CR, containing a layer of unidirectional continuous fibers on top of a layer of randomly oriented discontinuous fibers. The nominal fiber contents are usually indicated by two-digit numbers after the letters C and R. For example, the nominal fiber contents in SMC-C40R30 are 40% by weight of unidirectional continuous fibers and 30% by weight of randomly oriented discontinuous fibers.
3. XMC (trademark of PPG Industries), containing continuous fibers arranged in an X pattern, where the angle between the interlaced fibers is between 5° and 7°. Additionally, it may also contain randomly oriented discontinuous fibers interspersed with the continuous fibers.

A typical formulation for sheet-molding compound SMC-R30 is presented in Table 2.18. In this formulation, the unsaturated polyester and styrene are polymerized together to form the polyester matrix. The role of the low shrink additive, which is a thermoplastic polymer powder, is to reduce the polymerization shrinkage. The function of the catalyst (also called the initiator) is to initiate the polymerization reaction, but only at an elevated temperature. The function of the inhibitor is to prevent premature curing (gelation) of the resin that may start by the action of the catalyst while the ingredients are blended together. The mold release agent acts as an internal lubricant, and helps in releasing the molded part from the die. Fillers assist in reducing shrinkage of

TABLE 2.18
Typical Formulation of SMC-R30

Material	Weight (%)	
Resin paste		70%
Unsaturated polyester	10.50	
Low shrink additive	3.45	
Styrene monomer	13.40	
Filler ($CaCO_3$)	40.70	
Thickener (MgO)	0.70	
Catalyst (TBPB)	0.25	
Mold release agent (zinc stearate)	1.00	
Inhibitor (benzoquinone)	<0.005 g	
Glass fiber (25.4 mm, chopped)		30%
Total		100%

the molded part, promote better fiber distribution during molding, and reduce the overall cost of the compound. Typical filler–resin weight ratios are 1.5:1 for SMC-R30, 0.5:1 for SMC-R50, and nearly 0:1 for SMC-R65. The *thickener* is an important component in an SMC formulation since it increases the viscosity of the compound without permanently curing the resin and thereby makes it easier to handle an SMC sheet before molding. However, the thickening reaction should be sufficiently slow to allow proper wet-out and impregnation of fibers with the resin. At the end of the thickening reaction, the compound becomes dry, nontacky, and easy to cut and shape. With the application of heat in the mold, the thickening reaction is reversed and the resin paste becomes sufficiently liquid-like to flow in the mold. Common thickeners used in SMC formulations are oxides and hydroxides of magnesium and calcium, such as MgO, $Mg(OH)_2$, CaO, and $Ca(OH)_2$. Another method of thickening is known as the interpenetrating thickening process (ITP), in which a proprietary polyurethane rubber is used to form a temporary three-dimensional network structure with the polyester or vinyl ester resin.

SMC-R and SMC-CR sheets are manufactured on a sheet-molding compound machine (Figure 2.39). The resin paste is prepared by mechanically blending the various components listed in Table 2.18. It is placed on two moving polyethylene carrier films behind the metering blades. The thickness of the resin paste on each carrier film is determined by the vertical adjustment of the metering blades. Continuous rovings are fed into the chopper arbor, which is commonly set to provide 25.4 mm long discontinuous fibers. Chopped fibers are deposited randomly on the bottom resin paste. For SMC-CR sheets, parallel lines of continuous strand rovings are fed on top of the chopped fiber layer. After covering the fibers with the top resin paste, the carrier films are

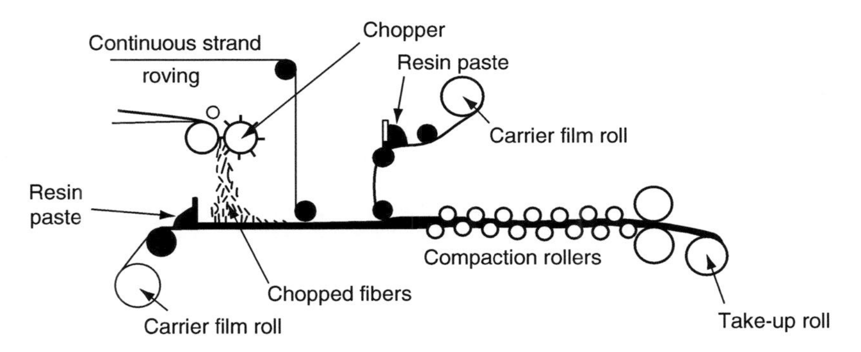

FIGURE 2.39 Schematic of a sheet molding compounding operation.

pulled through a number of compaction rolls to form a sheet that is then wound around a take-up roll. Wetting of fibers with the resin paste takes place at the compaction stage.

XMC sheets are manufactured by the filament winding process (see Chapter 5) in which continuous strand rovings are pulled through a tank of resin paste and wound under tension around a large rotating cylindrical drum. Chopped fibers, usually 25.4 mm long, are deposited on the continuous fiber layer during the time of winding. After the desired thickness is obtained, the built-up material is cut by a knife along a longitudinal slit on the drum to form the XMC sheet.

At the end of manufacturing, SMC sheets are allowed to "mature" (thicken or increase in viscosity) at about 30°C for 1–7 days. The matured sheet can be either compression molded immediately or stored at −18°C for future use.

2.7.3 Incorporation of Fibers into Thermoplastic Resins

Incorporating fibers into high-viscosity thermoplastic resins and achieving a good fiber wet-out are much harder than those in low-viscosity thermoset resins. Nevertheless, several fiber incorporation techniques in thermoplastic resins have been developed, and many of them are now commercially used to produce thermoplastic prepregs. These prepregs can be stored for unlimited time without any special storage facility and, whenever required, stacked and consolidated into laminates by the application of heat and pressure.

1. *Hot-melt impregnation* is used mainly for semicrystalline thermoplastics, such as PEEK and PPS, for which there are no suitable solvents available for solution impregnation. Amorphous polymers are also used for hot-melt impregnation.

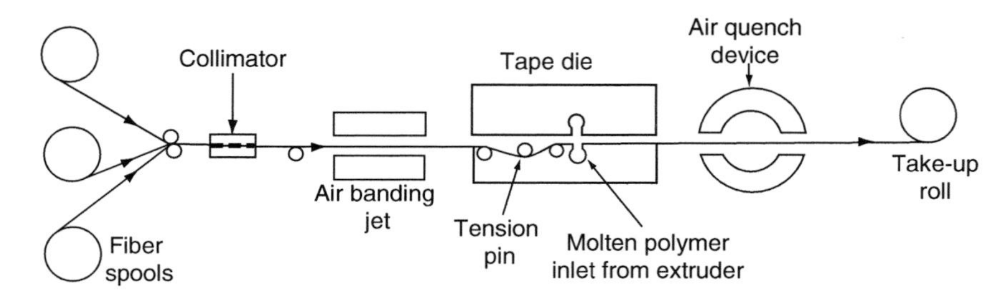

FIGURE 2.40 Hot-melt impregnation of thermoplastic prepregs. (Adapted from Muzzy, J.D., *The Manufacturing Science of Composites*, T.G. Gutowski, ed., ASME, New York, 1988.)

In this process, collimated fiber tows are pulled through a die attached at the end of an extruder, which delivers a fine sheet of hot polymer melt under high pressure to the die. To expose the filaments to the polymer melt, the fiber tows are spread by an air jet before they enter the die (Figure 2.40). The hot prepreg exiting from the die is rapidly cooled by a cold air jet and wound around a take-up roll.

For good and uniform polymer coating on filaments, the resin melt viscosity should be as low as possible. Although the viscosity can be reduced by increasing the melt temperature, there may be polymer degradation at very high temperatures. Hot-melt-impregnated prepregs tend to be stiff, boardy, and tack-free (no stickiness). This may cause problems in draping the mold surface and sticking the prepreg layers to each other as they are stacked before consolidation.

2. *Solution impregnation* is used for polymers that can be dissolved in a suitable solvent, which usually means an amorphous polymer, such as polysulfone and PEI. The choice of solvent depends primarily on the polymer solubility, and therefore, on the chemical structure of the polymer and its molecular weight. The solvent temperature also affects the polymer solubility. In general, a low-boiling-point solvent is preferred, since it is often difficult to remove high-boiling-point solvents from the prepreg.

   Solution impregnation produces drapable and tacky prepregs. However, solvent removal from the prepreg is a critical issue. If the solvent is entrapped, it may create a high void content in the consolidated laminate and seriously affect its properties.

3. *Liquid impregnation* uses low-molecular-weight monomers or prepolymers (precursors) to coat the fibers. This process is commonly used for LARC-TPI and a few other thermoplastic polyimides. In this case, the precursor is dissolved in a solvent to lower its viscosity.

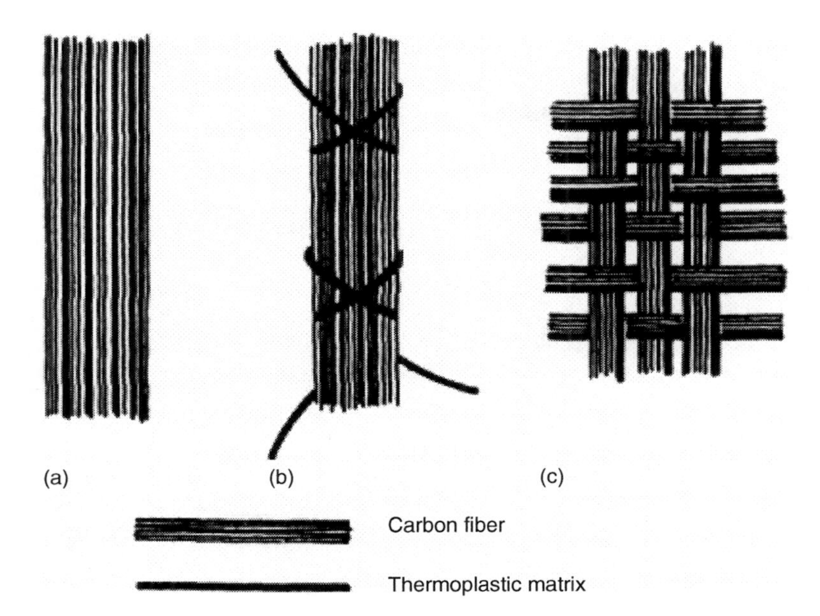

FIGURE 2.41 (a) Commingled, (b) wrapped, and (c) coweaved fiber arrangements.

Liquid-impregnated prepregs are drapable and tacky. However, the removal of residual solvents and reaction by-products from the prepreg during the consolidation stage can be difficult.

4. *Film stacking* is primarily used with woven fabrics or random fiber mats, which are interleaved between unreinforced thermoplastic polymer sheets. The layup is then heated and pressed to force the thermoplastic into the reinforcement layers and thus form a prepregged sheet.
5. *Fiber mixing* is a process of intimately mixing thermoplastic fibers with reinforcement fibers by *commingling*, *wrapping*, or *coweaving* (Figure 2.41). Commingled and wrapped fibers can be woven, knitted, or braided into two- or three-dimensional hybrid fabrics. The thermoplastic fibers in these fabrics can be melted and spread to wet the reinforcement fibers at the consolidation stage during molding.

The principal advantage of using hybrid fabrics is that they are highly flexible and can be draped over a contoured mold, whereas the other thermoplastic prepregs are best suited for relatively flat surfaces. However, fiber mixing is possible only if the thermoplastic polymer is available in filamentary form. Such is the case for PEEK and PPS that are spun into monofilaments with diameters in the range of 16–18 μm. Polypropylene (PP) and polyethylene terephthalate (PET) fibers are also used in making commingled rovings and fabrics.

6. *Dry powder coating* [26] uses charged and fluidized thermoplastic powders to coat the reinforcement fibers. After passing through the fluidized bed, the fibers enter a heated oven, where the polymer coating is melted on the fiber surface.

2.8 FIBER CONTENT, DENSITY, AND VOID CONTENT

Theoretical calculations for strength, modulus, and other properties of a fiber-reinforced composite are based on the fiber volume fraction in the material.

Experimentally, it is easier to determine the fiber weight fraction w_f, from which the fiber volume fraction v_f and composite density ρ_c can be calculated:

$$v_f = \frac{w_f/\rho_f}{(w_f/\rho_f) + (w_m/\rho_m)}, \tag{2.7}$$

$$\rho_c = \frac{1}{(w_f/\rho_f) + (w_m/\rho_m)}, \tag{2.8}$$

where

w_f = fiber weight fraction (same as the fiber mass fraction)
w_m = matrix weight fraction (same as the matrix mass fraction) and is equal to $(1-w_f)$
ρ_f = fiber density
ρ_m = matrix density

In terms of volume fractions, the composite density ρ_c can be written as

$$\rho_c = \rho_f v_f + \rho_m v_m, \tag{2.9}$$

where v_f is the fiber volume fraction and v_m is the matrix volume fraction. Note that v_m is equal to $(1-v_f)$.

The fiber weight fraction can be experimentally determined by either the ignition loss method (ASTM D2854) or the matrix digestion method (ASTM D3171). The ignition loss method is used for PMC-containing fibers that do not lose weight at high temperatures, such as glass fibers. In this method, the cured resin is burned off from a small test sample at 500°C–600°C in a muffle furnace. In the matrix digestion method, the matrix (either polymeric or metallic) is dissolved away in a suitable liquid medium, such as concentrated nitric acid. In both cases, the fiber weight fraction is determined by comparing the weights of the test sample before and after the removal of the matrix. For unidirectional composites containing electrically conductive fibers (such as carbon) in a nonconductive matrix, the fiber volume fraction can be determined directly by comparing the electrical resistivity of the composite with that of fibers (ASTM D3355).

During the incorporation of fibers into the matrix or during the manufacturing of laminates, air or other volatiles may be trapped in the material. The trapped air or volatiles exist in the laminate as microvoids, which may significantly affect some of its mechanical properties. A high void content (over 2% by volume) usually leads to lower fatigue resistance, greater susceptibility to water diffusion, and increased variation (scatter) in mechanical properties. The void content in a composite laminate can be estimated by comparing the theoretical density with its actual density:

$$v_v = \frac{\rho_c - \rho}{\rho_c}, \tag{2.10}$$

where

v_v = volume fraction of voids
ρ_c = theoretical density, calculated from Equation 2.8 or 2.9
ρ = actual density, measured experimentally on composite specimens (which is less than ρ_c due to the presence of voids)

EXAMPLE 2.1

Calculate v_f and ρ_c for a composite laminate containing 30 wt% of E-glass fibers in a polyester resin. Assume $\rho_f = 2.54\ g/cm^3$ and $\rho_m = 1.1\ g/cm^3$.

Solution

Assume a small composite sample of mass 1 g and calculate its volume.

	Fiber	Matrix
Mass (g)	0.3	1 − 0.3 = 0.7
Density (g/cm^3)	2.54	1.1
Volume (cm^3)	$\frac{0.3}{2.54} = 0.118$	$\frac{0.7}{1.1} = 0.636$

Therefore, volume of 1 g of composite is (0.118 + 0.636) or 0.754 cm^3. Now, we calculate

Fiber volume fraction $v_f = \frac{0.118}{0.754} = 0.156$ or 15.6%

Matrix volume fraction $v_m = 1 - v_f = 1 - 0.156 = 0.844$ or 84.4%

$$\text{Composite density } \rho_c = \frac{1 \text{ g}}{0.754 \text{ cm}^3} = 1.326 \text{ g/cm}^3$$

Note: These values can also be obtained using Equations 2.7 and 2.8.

EXAMPLE 2.2

Assume that the fibers in a composite lamina are arranged in a square array as shown in the figure. Determine the maximum fiber volume fraction that can be packed in this arrangement.

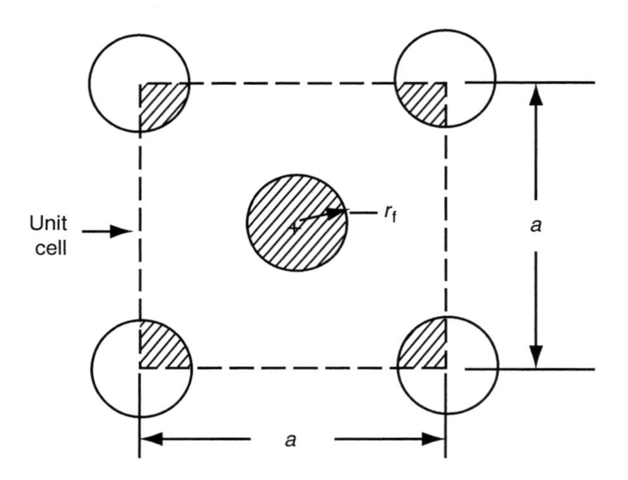

Solution

Number of fibers in the unit cell = 1 + (4) (1/4) = 2
Fiber cross-sectional area in the unit cell = (2) (πr_f^2)
Unit cell area = a^2
Therefore,
Fiber volume fraction (v_f) = $\dfrac{2\pi r_f^2}{a^2}$
from which, we can write

$$a = \frac{\sqrt{2\pi}}{v_f^{1/2}} r_f.$$

Interfiber spacing (R) between the central fiber and each corner fiber is given by

$$R = \frac{a}{\sqrt{2}} - 2r_f = r_f\left[\left(\frac{\pi}{v_f}\right)^{1/2} - 2\right].$$

For maximum volume fraction, $R = 0$, which gives

$$v_{f_{max}} = 0.785 \text{ or } 78.5\%.$$

2.9 FIBER ARCHITECTURE

Fiber architecture is defined as the arrangement of fibers in a composite, which not only influences the properties of the composite, but also its processing. The characteristics of fiber architecture that influence the mechanical properties include fiber continuity, fiber orientation, fiber crimping, and fiber interlocking. During processing, matrix flow through the fiber architecture determines the void content, fiber wetting, fiber distribution, dry area and others in the final composite, which in turn, also affect its properties and performance.

If continuous fibers are used, the fiber architecture can be one-dimensional, two-dimensional, or three-dimensional. The one-dimensional architecture can be produced by the prepregging technique described earlier or by other manufacturing methods, such as pultrusion. The two- and three-dimensional architectures are produced by textile manufacturing processes and are used with liquid composite molding processes, such as resin transfer molding, in which a liquid polymer is injected into the dry fiber preform containing two- or three-dimensional fiber architecture. Each fiber architecture type has its unique characteristics, and if properly used, can provide an opportunity not only to tailor the structural performance of the composite, but also to produce a variety of structural shapes and forms.

In the one-dimensional architecture, fiber strands (or yarns) are oriented all in one direction. The unidirectional orientation of continuous fibers in the composite produces the highest strength and modulus in the fiber direction, but much lower strength and modulus in the transverse to the fiber direction. A multilayered composite laminate can be built using the one-dimensional architecture in which each layer may contain unidirectional continuous fibers, but the angle of orientation from layer to layer can be varied. With proper orientation of fibers in various layers, the difference in strength and modulus values in different directions can be reduced. However, one major problem with many multilayered laminates is that their interlaminar properties can be low and they can be prone to early failure by delamination, in which cracks originated at the interface between the layers due to high interlaminar tensile and shear stresses cause separation of layers.

The two-dimensional architecture with continuous fibers can be either bidirectional or multidirectional. In a bidirectional architecture, fiber yarns (or strands) are either woven or interlaced together in two mutually perpendicular directions (Figure 2.42a). These two directions are called warp and fill directions, and represent 0° and 90° orientations, respectively. The fiber yarns are crimped or undulated as they move up and down to form the interlaced

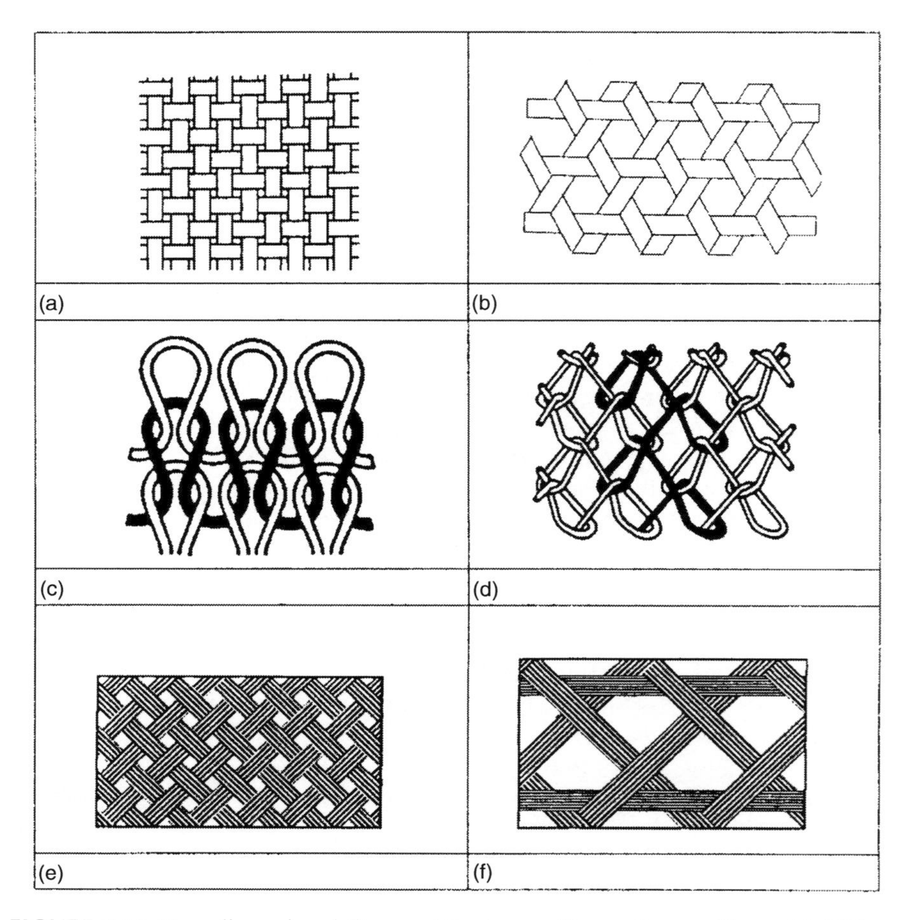

FIGURE 2.42 Two-dimensional fiber architectures with continuous fibers. (a) Bidirectional fabric, (b) Multidirectional fabric, (c) Weft-knitted fabric, (d) Warp-knitted fabric, (e) Biaxial braided fabric, and (f) Triaxial braided fabric.

structure. The nomenclature used for woven fabric architecture is given in Appendix A.1. By changing the number of fiber yarns per unit width in the warp and fill directions, a variety of properties can be obtained in these two directions. If the number of fiber yarns is the same in both warp and fill directions, then the properties are the same in these two directions and the fabric is balanced. However, the properties in other directions are still low. In order to improve the properties in the other directions, fiber yarns can be interlaced in the other directions to produce multidirectional fabrics (Figure 2.42b), such as $0/\pm\theta$ or $0/90/\pm\theta$ fabric. The angle $\pm\theta$ refers to $+\theta$ and $-\theta$ orientation of the bias yarns relative to the warp or 0° direction.

Knitting and braiding are two other textile processes used for making two-dimensional fiber architecture. In a knitted fabric, the fiber yarns are interlooped instead of interlaced (Figure 2.42c and d). If the knitting yarn runs in the cross-machine direction, the fabric is called the weft knit, and if it runs in the machine direction, it is called the warp knit. Knitted fabrics are produced on industrial knitting machines in which a set of closely spaced needles pull the yarns and form the loops. Knitted fabrics are more flexible than woven fabrics and are more suitable for making shapes with tight corners. Biaxial braided fabrics are produced by intertwining two sets of continuous yarns, one in the $+\theta$ direction and the other in the $-\theta$ direction relative to the braiding axis (Figure 2.42e). The angle θ is called the braid angle or the bias angle. Triaxial braids contain a third set of yarns oriented along the braiding axis (Figure 2.42f). Braided construction is most suitable for tubular structures, although it is also used for flat form.

A two-dimensional architecture can also be created using randomly oriented fibers, either with continuous lengths or with discontinuous lengths (Figure 2.43). The former is called the continuous fiber mat (CFM), while the latter is called the chopped strand mat (CSM). In a CFM, the continuous yarns can be either straight or oriented in a random swirl pattern. In a CSM, the fiber yarns are discontinuous (chopped) and randomly oriented. In both mats, the fibers are held in place using a thermoplastic binder. Because of the random orientation of fibers, the composite made from either CFM or CSM displays equal or nearly equal properties in all directions in the plane of the composite and thus, can be considered planar isotropic.

Composites made with one- and two-dimensional fiber architectures are weak in the z-direction (thickness direction) and often fail by delamination. To improve the interlaminar properties, fibers are added in the thickness direction, creating a three-dimensional architecture (Figure 2.44). The simplest form of

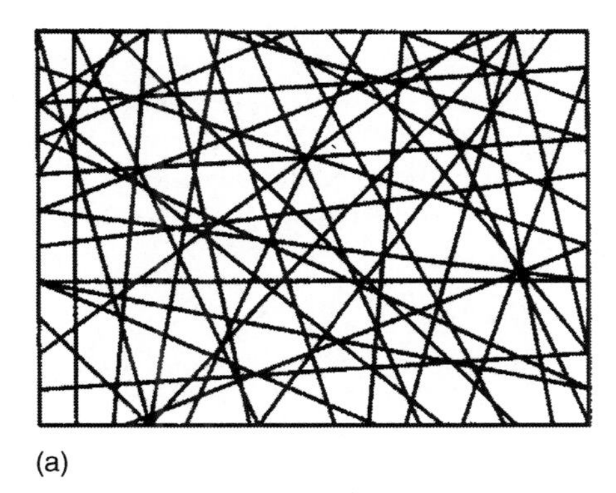

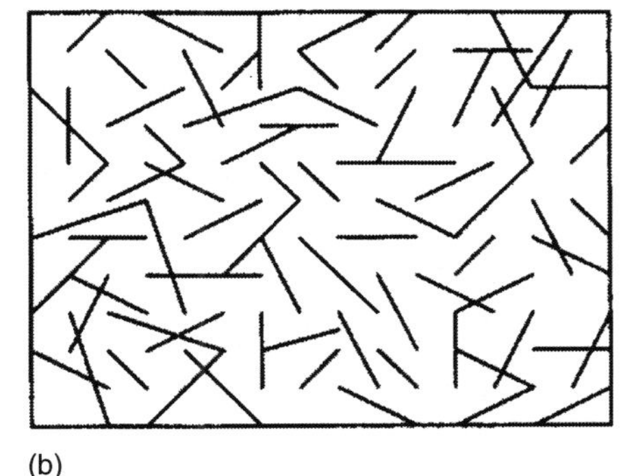

FIGURE 2.43 Two-dimensional fiber architecture with random fibers. (a) Continuous fiber mat (CFM) and (b) Chopped strand mat (CSM).

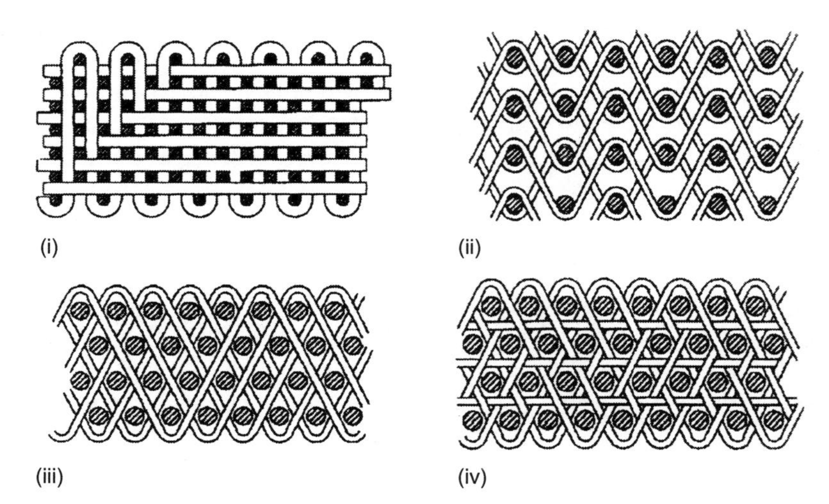

FIGURE 2.44 Examples of three-dimensional fiber architecture. (i) 3D, (ii) 2D or angle interlock, (iii) 3X, and (iv) 3X with warp stuffer yarns. (From Wilson, S., Wenger, W., Simpson, D. and Addis, S., "SPARC" 5 Axis, 3D Woven, Low Crimp Performs, Resin Transfer Molding, *SAMPE Monograph 3, SAMPE*, 1999. With permission.)

three-dimensional architecture can be created by stitching a stack of woven fabrics with stitching threads. The three-dimensional architecture can also be produced by weaving or braiding.

REFERENCES

1. T.-W. Chou, *Microstructural Design of Fiber Composites*, Cambridge University Press, Cambridge, UK (1992).
2. V.V. Kozey, H. Jiang, V.R. Mehta, and S. Kumar, Compressive behavior of materials: Part II. High performance fibers, *J. Mater. Res., 10*:1044 (1995).
3. J.F. Macdowell, K. Chyung, and K.P. Gadkaree, Fatigue resistant glass and glass-ceramic macrofiber reinforced polymer composites, 40th Annual Technical Conference, Society of Plastics Industry, January (1985).
4. J.-B. Donnet and R.C. Bansal, *Carbon Fibers*, Marcel Dekker, New York (1984).
5. S. Chand, Carbon fibres for composites, *J. Mater. Sci., 35*:1303 (2000).
6. M.L. Minus and S. Kumar, The processing, properties, and structure of carbon fibers, *JOM, 57*:52 (2005).
7. R.J. Morgan and R.E. Allred, Aramid fiber reinforcements, *Reference Book for Composites Technology*, Vol. 1 (S.M. Lee, ed.), Technomic Publishing Company, Lancaster, PA (1989).
8. D. Nabi Saheb and J.P. Jog, Natural fiber polymer composites: a review, *Adv. Polym. Tech., 18*:351 (1999).

9. J.G. Morley, Fibre reinforcement of metals and alloys, *Int. Metals Rev., 21*:153 (1976).
10. C.-H. Andersson and R. Warren, Silicon carbide fibers and their potential for use in composite materials, Part 1, *Composites, 15*:16 (1984).
11. A.K. Dhingra, Alumina Fibre FP, *Phil. Trans. R. Soc. Lond., A, 294*:1 (1980).
12. A.R. Bunsell, Oxide fibers for high-temperature reinforcement and insulation, *JOM, 57, 2*:48 (2005).
13. C.D. Armeniades and E. Baer, Transitions and relaxations in polymers, *Introduction to Polymer Science and Technology* (H.S. Kaufman and J.J. Falcetta, eds.), John Wiley & Sons, New York (1977).
14. J.E. Schoutens, Introduction to metal matrix composite materials, MMCIAC Tutorial Series, DOD Metal Matrix Composites Information Analysis Center, Santa Barbara, CA (1982).
15. M.U. Islam and W. Wallace, Carbon fibre reinforced aluminum matrix composites: a critical review, Report No. DM-4, National Research Council, Canada (1984).
16. P.R. Smith and F.H. Froes, Developments in titanium metal matrix composites, *J. Metal*, March (1984).
17. H.G. Recker et al., Highly damage tolerant carbon fiber epoxy composites for primary aircraft structural applications, *SAMPE Q.,* October (1989).
18. D.A. Scola and J.H. Vontell, High temperature polyimides: chemistry and properties, *Polym. Compos., 9*:443 (1988).
19. R.H. Pater, Improving processing and toughness of a high performance composite matrix through an interpenetrating polymer network, Part 6, *SAMPE J., 26* (1990).
20. I. Hamerton and J.H. Nay, Recent technological developments in cyanate ester resins, *High Perform. Polym., 10*:163 (1998).
21. H.X. Nguyen and H. Ishida, Poly(aryl-ether-ether-ketone) and its advanced composites: a review, *Polym. Compos., 8*:57 (1987).
22. J.P. Bell, J. Chang, H.W. Rhee, and R. Joseph, Application of ductile polymeric coatings onto graphite fibers, *Polym. Compos., 8*:46 (1987).
23. E.H. Rowe, Developments in improved crack resistance of thermoset resins and composites, 37th Annual Technical Conference, Society of Plastics Engineers, May (1979).
24. *Modern Plastics Encyclopedia*, McGraw-Hill, New York (1992).
25. P.K. Mallick, *Sheet Molding Compounds, Composite Materials Technology* (P.K. Mallick and S. Newman, eds.), Hanser Publishers, Munich (1990).
26. J.D. Muzzy, Processing of advanced thermoplastic composites, *The Manufacturing Science of Composites* (T.G. Gutowski, ed.), ASME, New York (1988).
27. L.T. Drzal, Composite property dependence on the fiber, matrix, and the interphase, *Tough Composite Materials: Recent Developments*, Noyes Publications, Park Ridge, NJ, p. 207 (1985).

PROBLEMS

P2.1. The linear density of a dry carbon fiber tow is 0.198 g/m. The density of the carbon fiber is 1.76 g/cm^3 and the average filament diameter is 7 μm. Determine the number of filaments in the tow.

P2.2. Glass fiber rovings are commonly designated by the term *yield*, which is the length of the roving per unit weight (e.g., 1275 yd/lb). Estimate the yield for a glass fiber roving that contains 20 ends (strands) per roving. Each strand in the roving is made of 204 filaments, and the average filament diameter is 40×10^{-5} in.

P2.3. The strength of brittle fibers is expressed by the well-known Griffith's formula:

$$\sigma_{\mathrm{fu}} = \left(\frac{2E_{\mathrm{f}}\gamma_{\mathrm{f}}}{\pi c}\right)^{1/2},$$

where
σ_{fu} = fiber tensile strength
E_{f} = fiber tensile modulus
γ_{f} = surface energy
c = critical flaw size

1. The average tensile strength of as-drawn E-glass fibers is 3.45 GPa, and that of commercially available E-glass fibers is 1.724 GPa. Using Griffith's formula, compare the critical flaw sizes in these two types of fibers. Suggest a few reasons for the difference.
2. Assuming that the critical flaw size in the as-drawn E-glass fiber is 10^{-4} cm, estimate the surface energy of E-glass fibers.

P2.4. Scanning electron microscope study of fracture surfaces of carbon fibers broken in tension shows that they fail either at the surface flaws (pits) or at the internal voids. The surface energy of graphite is 4.2 J/m^2. Assuming this as the surface energy of carbon fibers, estimate the range of critical flaw size in these fibers if the observed strength values vary between 1.3 and 4.3 GPa. The fiber modulus in tension is 230 GPa. Use Griffith's formula in Problem P2.3.

P2.5. The filament strength distribution of a carbon fiber is represented by the Weibull distribution function given by Equation 2.4. The following Weibull parameters are known for this particular carbon fiber: $\alpha = 6.58$ and $\sigma_{\mathrm{o}} = 2.56$ GPa at $L_{\mathrm{f}} = 200$ mm. Determine the filament strength at which 99% of the filaments are expected to fail if the filament length is (a) 20 mm, (b) 100 mm, (c) 200 mm, and (d) 500 mm.

P2.6. A carbon fiber bundle containing 2000 parallel filaments is being tested in tension. The filament length is 100 mm. The strength distribution of individual filaments is described in Problem P2.5. Compare the mean filament strength and the mean fiber bundle strength. Schematically show the tensile stress–strain diagram of the carbon fiber bundle and compare it with that of the carbon filament. (*Note*: For $\alpha = 6.58$, $\Gamma\left(1 + \frac{1}{\alpha}\right) \cong 0.93$.)

P2.7. The Weibull parameters for the filament strength distribution of an E-glass fiber are $\alpha = 11.32$ and $\sigma_o = 4.18$ GPa at $L_f = 50$ mm. Assume that the filament stress–strain relationship obeys Hooke's law, that is, $\sigma_f = E_f\varepsilon_f$, where E_f = fiber modulus = 69 GPa, and the tensile load applied on the bundle in a fiber bundle test is distributed uniformly among the filaments, develop the tensile load–strain diagram of the fiber bundle. State any assumption you may make to determine the load–strain diagram.

P2.8. It has been observed that the strength of a matrix-impregnated fiber bundle is significantly higher than that of a dry fiber bundle (i.e., without matrix impregnation). Explain.

P2.9. The strength of glass fibers is known to be affected by the time of exposure at a given stress level. This phenomenon, known as the static fatigue, can be modeled by the following equation:

$$\sigma = A - B \log(t + 1),$$

where A and B are constants and t is the time of exposure (in minutes) under stress. For an as-drawn E-glass fiber, $A = 450{,}000$ psi and $B = 20{,}000$ psi.

Determine the tensile strength of an as-drawn E-glass fiber after 1000 h of continuous tensile loading at 23°C. Using the results of Problem P2.3, determine the rate of increase of the critical flaw size in this fiber at 1, 10, and 1000 h.

P2.10. Kevlar 49 fiber strands are used in many high strength cable applications where its outstanding strength–weight ratio leads to a considerable weight saving over steel cables.

1. Compare the breaking loads and weights of Kevlar 49 and steel cables, each with a 6.4 mm diameter
2. Compare the maximum stresses and elongations in 1000 m long Kevlar 49 and steel cables due to their own weights

P2.11. The smallest radius to which a fiber can be bent or knotted without fracturing is an indication of how easily it can be handled in a manufacturing operation. The handling characteristic of a fiber is particularly important in continuous manufacturing operations, such as filament winding or pultrusion, in which continuous strand rovings are pulled through a number of guides or eyelets with sharp corners. Frequent breakage of fibers at these locations is undesirable since it slows down the production rate.

Using the following relationship between the bending radius r_b and the maximum tensile strain in the fiber,

$$r_b = \frac{d_f}{2\varepsilon_{max}},$$

compare the smallest radii to which various glass, carbon, and Kevlar fibers in Table 2.1 can be bent without fracturing.

P2.12. During a filament winding operation, T-300 carbon fiber tows containing 6000 filaments per tow are pulled over a set of guide rollers 6 mm in diameter. The mean tensile strength of the filaments is 3000 MPa, and the standard deviation is 865 MPa (assuming a standard normal distribution for filament strength). Determine the percentage of filaments that may snap as the tows are pulled over the rollers. The fiber modulus is 345 GPa, and the filament diameter is 7 μm.

P2.13. The energy required to snap (break) a brittle fiber is equal to the strain energy stored (which is equal to the area under the stress–strain diagram) in the fiber at the time of its failure. Compare the strain energies of E-glass, T-300, IM-7, GY-70, Kevlar 49, and Spectra 900 fibers.

P2.14. Assume that the area under the stress–strain diagram of a material is a measure of its toughness. Using the stress–strain diagrams shown in the following figure, compare the toughness values of the three matrix resins considered.

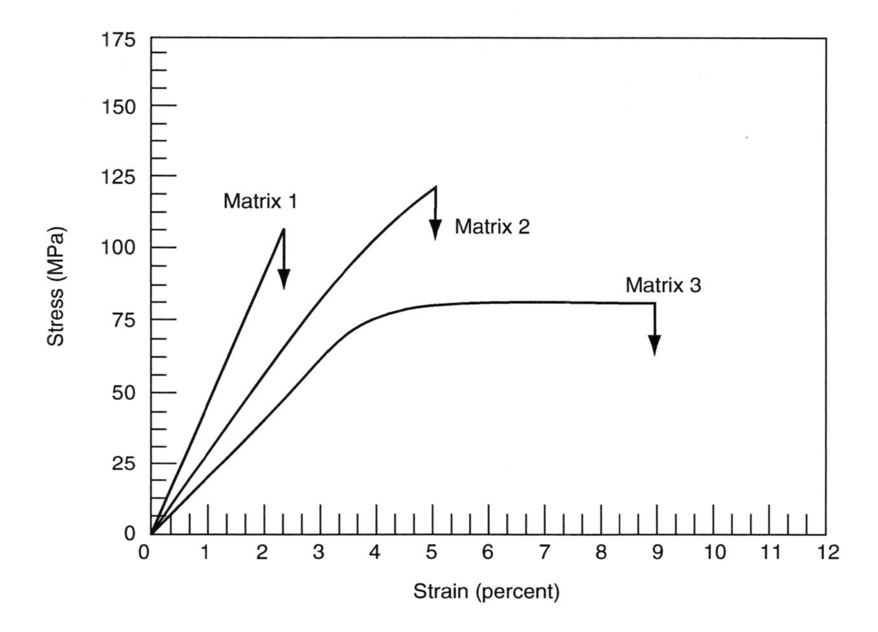

P2.15. Assuming that the unidirectional continuous fibers of round cross section are arranged in a simple square array as shown in the accompanying figure, calculate the theoretical fiber volume fraction in the composite lamina. What is the maximum fiber volume fraction that can be arranged in this fashion?

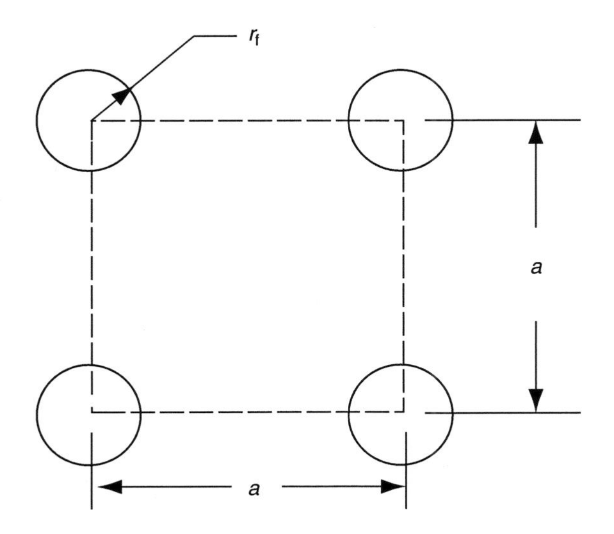

P2.16. Using the simple square arrangement in the earlier figure , show that fibers with square or hexagonal cross sections can be packed to higher fiber volume fractions than fibers with round cross sections. Compare the fiber surface area per unit volume fraction for each cross section. What is the significance of the surface area calculation?

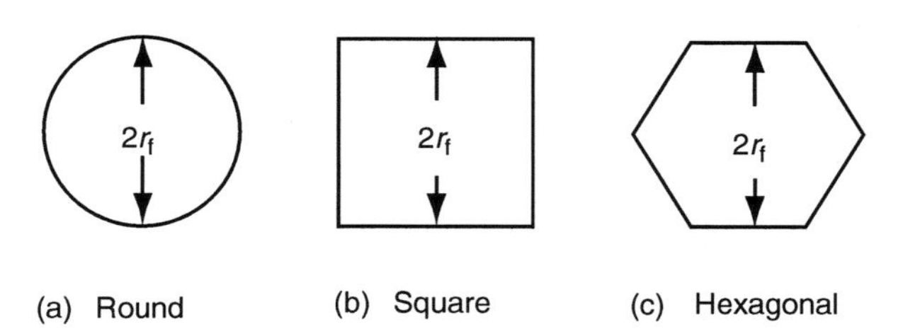

P2.17. Assuming that the fibers in a bundle are arranged in a simple square array, calculate the interfiber spacings in terms of the fiber radius r_f for a fiber volume fraction of 0.6.

P2.18. Assuming that the fibers in a unidirectional continuous fiber composite are arranged in a hexagonal packing as shown in the accompanying figure, show that the fiber volume fraction is given by 3.626 $(r_f/a)^2$. Calculate the maximum volume fraction of fibers that can be packed in this fashion.

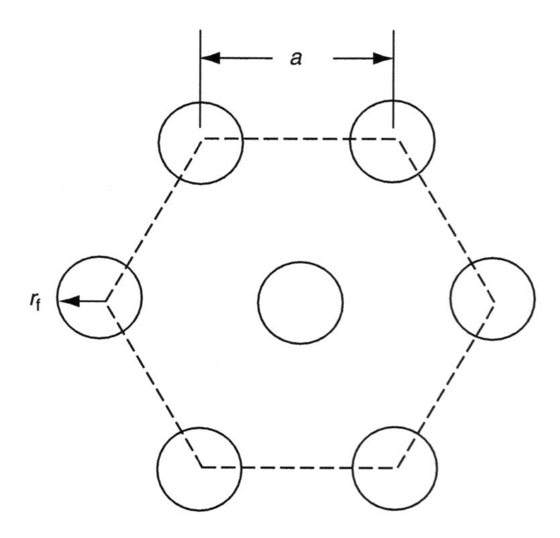

P2.19. Assume that the unit cell of a composite containing commingled filaments of E-glass and T-300 carbon fibers can be represented by a square array shown in the following figure. The diameters of the E-glass filaments and T-300 filaments are 7×10^{-6} m and 10×10^{-6} m, respectively.

1. Determine the unit cell dimension if the fiber volume fraction is 60%
2. Determine the theoretical density of the composite

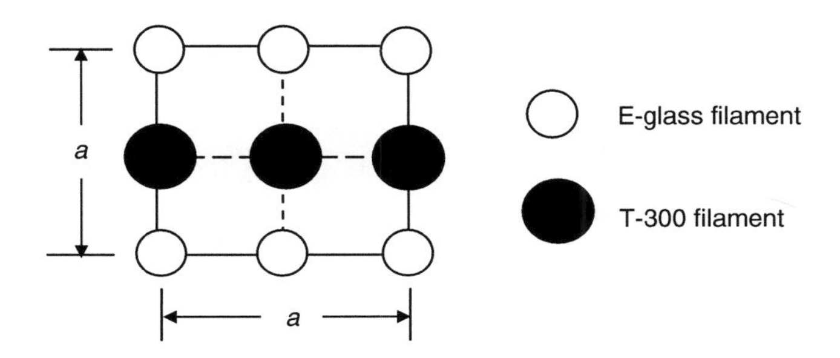

P2.20. The following data were obtained in a resin burn-off test of an E-glass–polyester sample:

Weight of an empty crucible = 10.1528 g
Weight of crucible + sample before burn-off = 10.5219 g
Weight of crucible + sample after burn-off = 10.3221 g

Calculate the fiber weight fraction, the fiber volume fraction, and the density of the composite sample. Assume $\rho_f = 2.54$ g/cm^3 and $\rho_m = 1.25$ g/cm^3. Do you expect the calculated value to be higher or lower than the actual value?

P2.21. An interply hybrid composite contains 30 wt% AS-4 carbon fibers, 30 wt% Kevlar 49 fibers, and 40 wt% epoxy resin. Assume that the density of the epoxy resin is 1.2 g/cm^3. Calculate the density of the composite.

P2.22. The fiber content in an E-glass fiber-reinforced polypropylene is 30% by volume.

1. How many kilograms of E-glass fibers are in the composite for every 100 kg of polypropylene? The density of polypropylene is 0.9 g/cm^3.
2. Assume that half of the E-glass fibers (by weight) in part (1) is replaced with T-300 carbon fibers. How will the density of the new composite compare with the density of the original composite?

P2.23. The density, ρ_m, of a semicrystalline polymer matrix, such as PEEK, can be expressed as

$$\rho_m = \rho_{mc} v_{mc} + \rho_{ma} v_{ma},$$

where ρ_{mc} and ρ_{ma} are the densities of the crystalline phase and the amorphous phase in the matrix, respectively, and, v_{mc} and v_{ma} are the corresponding volume fractions.

The density of an AS-4 carbon fiber-reinforced PEEK is reported as 1.6 g/cm^3. Knowing that the fiber volume fraction is 0.6, determine the volume and weight fractions of the crystalline phase in the matrix. For PEEK, $\rho_{mc} = 1.401$ g/cm^3 and $\rho_{ma} = 1.263$ g/cm^3.

P2.24. A carbon fiber–epoxy plate of thickness t was prepared by curing N prepeg plies of equal thickness. The number of fiber yarns per unit prepreg width is n, and the yarn weight per unit length is W_y. Show that the fiber volume fraction in the plate is

$$v_f = \frac{W_y n N}{t \rho_f g},$$

where
ρ_f is the fiber density
g is the acceleration due to gravity

P2.25. Determine the weight and cost of prepreg required to produce a hollow composite tube (outside diameter = 50 mm, wall thickness = 5 mm, and length = 4 m) if it contains 60 vol% AS-4 carbon fibers (at \$60/kg) in an epoxy matrix (at \$8/kg). The tube is manufactured by wrapping the prepreg around a mandrel, and the cost of prepregging is \$70/kg.

P2.26. The material in a composite beam is AS-1 carbon fiber-reinforced epoxy ($v_f = 0.60$). In order to save cost, carbon fibers are being replaced by equal volume percent of E-glass fibers. To compensate for the lower modulus of E-glass fibers, the thickness of the beam is increased threefold. Assuming that the costs of carbon fibers, E-glass fibers, and epoxy are \$40/kg, \$4/kg, and \$6/kg, respectively, determine the percent material cost saved. The density of epoxy is 1.25 g/cm^3.

P2.27. Calculate the average density of an interply hybrid beam containing m layers of T-300 carbon fiber–epoxy and n layers of E-glass fiber–epoxy. The thickness of each carbon fiber layer is t_c and that of each glass layer is t_g. The fiber volume fractions in carbon and glass layers are v_c and v_g, respectively.

P2.28. An interply hybrid laminate contains 24 layers of GY-70 carbon fiber–epoxy and 15 layers of Kevlar 49 fiber–epoxy. The fiber weight fraction in both carbon and Kevlar 49 layers is 60%. Determine

1. Overall volume fraction of carbon fibers in the laminate
2. Volume fraction of layers containing carbon fibers

3 Mechanics

The mechanics of materials deal with stresses, strains, and deformations in engineering structures subjected to mechanical and thermal loads. A common assumption in the mechanics of conventional materials, such as steel and aluminum, is that they are homogeneous and isotropic continua. For a homogeneous material, properties do not depend on the location, and for an isotropic material, properties do not depend on the orientation. Unless severely cold-worked, grains in metallic materials are randomly oriented so that, on a statistical basis, the assumption of isotropy can be justified. Fiber-reinforced composites, on the other hand, are microscopically inhomogeneous and nonisotropic (orthotropic). As a result, the mechanics of fiber-reinforced composites are far more complex than that of conventional materials.

The mechanics of fiber-reinforced composite materials are studied at two levels:

1. The micromechanics level, in which the interaction of the constituent materials is examined on a microscopic scale. Equations describing the elastic and thermal characteristics of a lamina are, in general, based on micromechanics formulations. An understanding of the interaction between various constituents is also useful in delineating the failure modes in a fiber-reinforced composite material.
2. The macromechanics level, in which the response of a fiber-reinforced composite material to mechanical and thermal loads is examined on a macroscopic scale. The material is assumed to be homogeneous. Equations of orthotropic elasticity are used to calculate stresses, strains, and deflections.

In this chapter, we look into a few basic concepts as well as a number of simple working equations used in the micro- and macromechanics of fiber-reinforced composite materials. Detailed derivations of these equations are given in the references cited in the text.

3.1 FIBER–MATRIX INTERACTIONS IN A UNIDIRECTIONAL LAMINA

We consider the mechanics of materials approach [1] in describing fiber–matrix interactions in a unidirectional lamina owing to tensile and compressive loadings. The basic assumptions in this vastly simplified approach are as follows:

1. Fibers are uniformly distributed throughout the matrix.
2. Perfect bonding exists between the fibers and the matrix.
3. The matrix is free of voids.
4. The applied force is either parallel to or normal to the fiber direction.
5. The lamina is initially in a stress-free state (i.e., no residual stresses are present in the fibers and the matrix).
6. Both fibers and matrix behave as linearly elastic materials.

A review of other approaches to the micromechanical behavior of a composite lamina is given in Ref. [2].

3.1.1 Longitudinal Tensile Loading

In this case, the load on the composite lamina is a tensile force applied parallel to the longitudinal direction of the fibers.

3.1.1.1 Unidirectional Continuous Fibers

Assuming a perfect bonding between fibers and matrix, we can write

$$\varepsilon_f = \varepsilon_m = \varepsilon_c, \tag{3.1}$$

where ε_f, ε_m, and ε_c are the longitudinal strains in fibers, matrix, and composite, respectively (Figure 3.1).

Since both fibers and matrix are elastic, the respective longitudinal stresses can be calculated as

$$\sigma_f = E_f \varepsilon_f = E_f \varepsilon_c, \tag{3.2}$$

$$\sigma_m = E_m \varepsilon_m = E_m \varepsilon_c. \tag{3.3}$$

Comparing Equation 3.2 with Equation 3.3 and noting that $E_f > E_m$, we conclude that the fiber stress σ_f is always greater than the matrix stress σ_m.

The tensile force P_c applied on the composite lamina is shared by the fibers and the matrix so that

$$P_c = P_f + P_m. \tag{3.4}$$

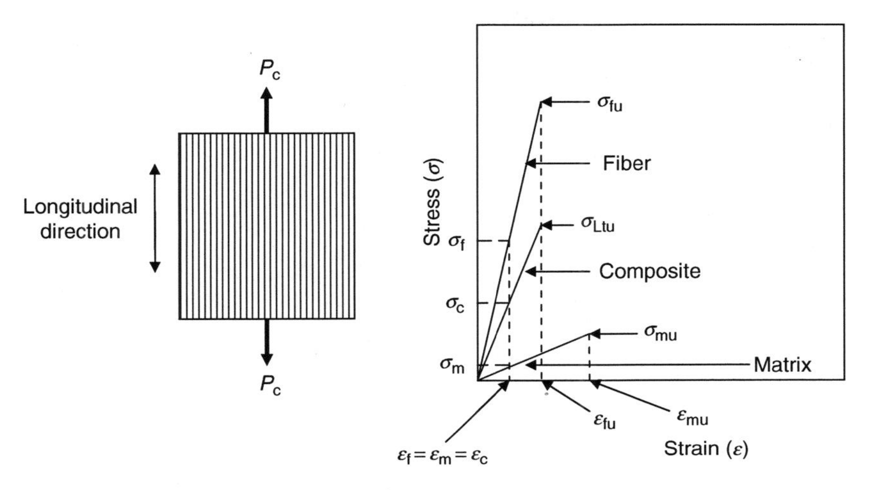

FIGURE 3.1 Longitudinal tensile loading of a unidirectional continuous fiber lamina.

Since force = stress × area, Equation 3.4 can be rewritten as

$$\sigma_c A_c = \sigma_f A_f + \sigma_m A_m$$

or

$$\sigma_c = \sigma_f \frac{A_f}{A_c} + \sigma_m \frac{A_m}{A_c}, \tag{3.5}$$

where

σ_c = average tensile stress in the composite
A_f = net cross-sectional area for the fibers
A_m = net cross-sectional area for the matrix
$A_c = A_f + A_m$

Since $v_f = \frac{A_f}{A_c}$ and $v_m = (1 - v_f) = \frac{A_m}{A_c}$, Equation 3.5 gives

$$\sigma_c = \sigma_f v_f + \sigma_m v_m = \sigma_f v_f + \sigma_m (1 - v_f). \tag{3.6}$$

Dividing both sides of Equation 3.6 by ε_c, and using Equations 3.2 and 3.3, we can write the longitudinal modulus for the composite as

$$E_L = E_f v_f + E_m v_m = E_f v_f + E_m (1 - v_f) = E_m + v_f (E_f - E_m). \tag{3.7}$$

Equation 3.7 is called the *rule of mixtures*. This equation shows that the longitudinal modulus of a unidirectional continuous fiber composite is intermediate between the fiber modulus and the matrix modulus; it increases linearly with increasing fiber volume fraction; and since $E_f > E_m$, it is influenced more by the fiber modulus than the matrix modulus.

The fraction of load carried by fibers in longitudinal tensile loading is

$$\frac{P_f}{P_c} = \frac{\sigma_f v_f}{\sigma_f v_f + \sigma_m(1 - v_f)} = \frac{E_f v_f}{E_f v_f + E_m(1 - v_f)}. \tag{3.8}$$

Equation 3.8 is plotted in Figure 3.2 as a function of $\frac{E_f}{E_m}$ ratio and fiber volume fraction. In polymer matrix composites, the fiber modulus is much greater than the matrix modulus. In most polymer matrix composites, $\frac{E_f}{E_m} > 10$. Thus, even for $v_f = 0.2$, fibers carry >70% of the composite load. Increasing the fiber volume fraction increases the fiber load fraction as well as the composite load. Although cylindrical fibers can be theoretically packed to almost 90% volume fraction, the practical limit is close to ~80%. Over this limit, the matrix will not be able to wet the fibers.

In general, the fiber failure strain is lower than the matrix failure strain, that is, $\varepsilon_{fu} < \varepsilon_{mu}$. Assuming all fibers have the same tensile strength and the tensile rupture of fibers immediately precipitates a tensile rupture of the composite, the

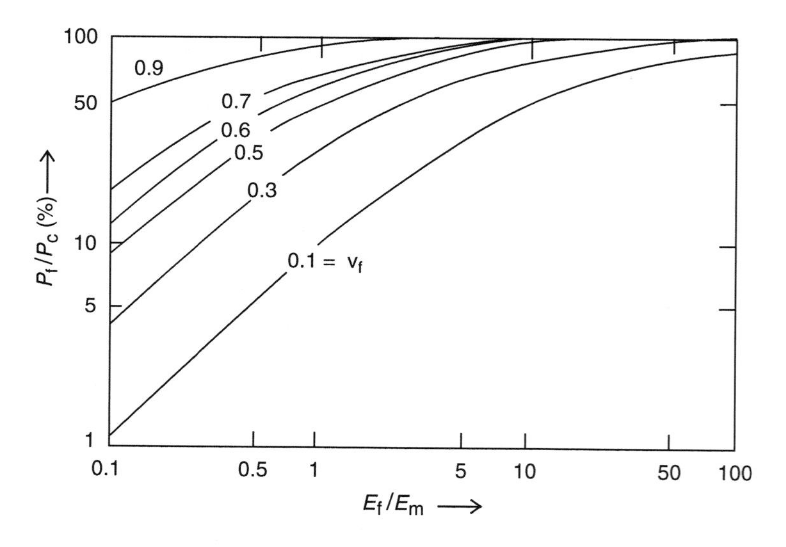

FIGURE 3.2 Fraction of load shared by fibers in longitudinal tensile loading of a unidirectional continuous fiber lamina.

longitudinal tensile strength σ_{Ltu} of a unidirectional continuous fiber composite can be estimated as

$$\sigma_{Ltu} = \sigma_{fu} v_f + \sigma'_m (1 - v_f), \tag{3.9}$$

where

σ_{fu} = fiber tensile strength (assuming a single tensile strength value for all fibers, which is not actually the case)

σ'_m = matrix stress at the fiber failure strain, that is, at $\varepsilon_m = \varepsilon_{fu}$ (Figure 3.1)

For effective reinforcement of the matrix, that is, for $\sigma_{Ltu} > \sigma_{mu}$, the fiber volume fraction in the composite must be greater than a critical value. This *critical fiber volume fraction* is calculated by setting $\sigma_{Ltu} = \sigma_{mu}$. Thus, from Equation 3.9,

$$\text{Critical } v_f = \frac{\sigma_{mu} - \sigma'_m}{\sigma_{fu} - \sigma'_m}. \tag{3.10a}$$

Equation 3.9 assumes that the matrix is unable to carry the load transferred to it after the fibers have failed, and therefore, the matrix fails immediately after the fiber failure. However, at low fiber volume fractions, it is possible that the matrix will be able to carry additional load even after the fibers have failed. For this to occur,

$$\sigma_{mu}(1 - v_f) > \sigma_{fu} v_f + \sigma'_m (1 - v_f),$$

from which the *minimum fiber volume fraction* can be calculated as

$$\text{Minimum } v_f = \frac{\sigma_{mu} - \sigma'_m}{\sigma_{mu} + \sigma_{fu} - \sigma'_m}. \tag{3.10b}$$

If the fiber volume fraction is less than the minimum value given by Equation 3.10b, the matrix will continue to carry the load even after the fibers have failed at $\sigma_f = \sigma_{fu}$. As the load on the composite is increased, the strain in the matrix will also increase, but some of the load will be transferred to the fibers. The fibers will continue to break into smaller and smaller lengths, and with decreasing fiber length, the average stress in the fibers will continue to decrease. Eventually, the matrix will fail when the stress in the matrix reaches σ_{mu}, causing the composite to fail also. The longitudinal tensile strength of the composite in this case will be $\sigma_{mu}(1 - v_f)$.

Figure 3.3 shows the longitudinal strength variation with fiber volume fraction for a unidirectional continuous fiber composite containing an elastic, brittle matrix. Table 3.1 shows critical fiber volume fraction and minimum fiber

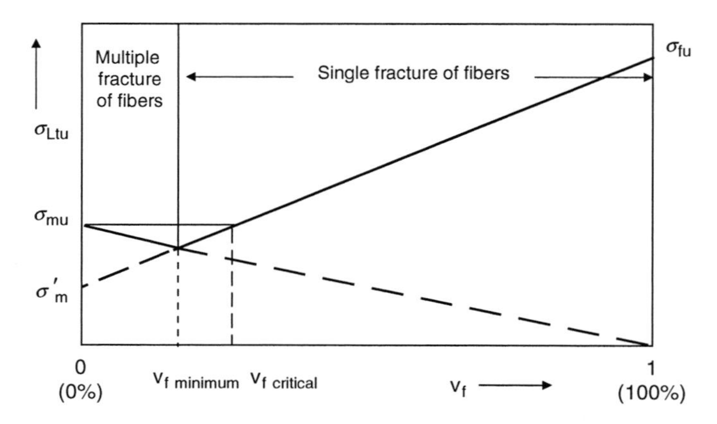

FIGURE 3.3 Longitudinal tensile strength variation with fiber volume fraction in a unidirectional continuous fiber composite in which the matrix failure strain is greater than the fiber failure strain.

volume fraction for unidirectional continuous fiber-reinforced epoxy. For all practical applications, fiber volume fractions are much greater than these values.

There are other stresses in the fibers as well as the matrix besides the longitudinal stresses. For example, transverse stresses, both tangential and radial, may arise due to the difference in Poisson's ratios, ν_f and ν_m, between the fibers and matrix. If $\nu_f < \nu_m$, the matrix tends to contract more in the transverse directions than the fibers as the composite is loaded in tension in the longitudinal direction. This creates a radial pressure at the interface and, as a result, the matrix near the interface experiences a tensile stress in the tangential

TABLE 3.1
Critical and Minimum Fiber Volume Fractions in E-glass, Carbon, and Boron Fiber-Reinforced Epoxy Matrix[a] Composite

Property	E-Glass Fiber	Carbon Fiber	Boron Fiber
E_f	10×10^6 psi	30×10^6 psi	55×10^6 psi
σ_{fu}	250,000 psi	400,000 psi	450,000 psi
$\varepsilon_{fu} = \frac{\sigma_{fu}}{E_f}$	0.025	0.0133	0.0082
$\sigma'_m = E_m\ \varepsilon_{fu}$	2,500 psi	1,330 psi	820 psi
Critical v_f	3.03%	2.17%	2.04%
Minimum v_f	2.9%	2.12%	2%

[a] Matrix properties: $\sigma_{mu} = 10{,}000$ psi, $E_m = 0.1 \times 10^6$ psi, and $\varepsilon_{mu} = 0.1$.

direction and a compressive stress in the radial direction. Tangential and radial stresses in the fibers are both compressive. However, all these stresses are relatively small compared with the longitudinal stresses.

Another source of internal stresses in the lamina is due to the difference in thermal contraction between the fibers and matrix as the lamina is cooled down from the fabrication temperature to room temperature. In general, the matrix has a higher coefficient of thermal expansion (or contraction), and, therefore, tends to contract more than the fibers, creating a "squeezing" effect on the fibers. A three-dimensional state of residual stresses is created in the fibers as well as in the matrix. These stresses can be calculated using the equations given in Appendix A.2.

3.1.1.2 Unidirectional Discontinuous Fibers

Tensile load applied to a discontinuous fiber lamina is transferred to the fibers by a shearing mechanism between fibers and matrix. Since the matrix has a lower modulus, the longitudinal strain in the matrix is higher than that in adjacent fibers. If a perfect bond is assumed between the two constituents, the difference in longitudinal strains creates a shear stress distribution across the fiber–matrix interface. Ignoring the stress transfer at the fiber end cross sections and the interaction between the neighboring fibers, we can calculate the normal stress distribution in a discontinuous fiber by a simple force equilibrium analysis (Figure 3.4).

Consider an infinitesimal length dx at a distance x from one of the fiber ends (Figure 3.4). The force equilibrium equation for this length is

$$\left(\frac{\pi}{4}d_f^2\right)(\sigma_f + d\sigma_f) - \left(\frac{\pi}{4}d_f^2\right)\sigma_f - (\pi d_f\, dx)\tau = 0,$$

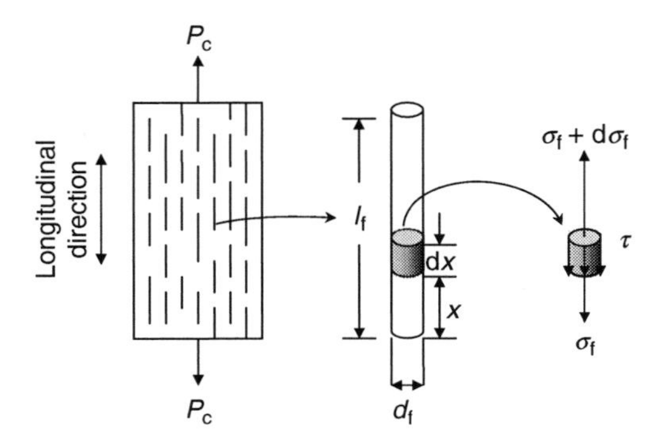

FIGURE 3.4 Longitudinal tensile loading of a unidirectional discontinuous fiber lamina.

which on simplification gives

$$\frac{d\sigma_f}{dx} = \frac{4\tau}{d_f}, \tag{3.11}$$

where
σ_f = longitudinal stress in the fiber at a distance x from one of its ends
τ = shear stress at the fiber–matrix interface
d_f = fiber diameter

Assuming no stress transfer at the fiber ends, that is, $\sigma_f = 0$ at $x = 0$, and integrating Equation 3.11, we determine the longitudinal stress distribution in the fiber as

$$\sigma_f = \frac{4}{d_f}\int_0^x \tau \, dx. \tag{3.12}$$

For simple analysis, let us assume that the interfacial shear stress is constant and is equal to τ_i. With this assumption, integration of Equation 3.12 gives

$$\sigma_f = \frac{4\tau_i}{d_f}x. \tag{3.13}$$

From Equation 3.13, it can be observed that for a composite lamina containing discontinuous fibers, the fiber stress is not uniform. According to Equation 3.13, it is zero at each end of the fiber (i.e., $x = 0$) and it increases linearly with x. The maximum fiber stress occurs at the central portion of the fiber (Figure 3.5). The maximum fiber stress that can be achieved at a given load is

$$(\sigma_f)_{max} = 2\tau_i \frac{l_t}{d_f}, \tag{3.14}$$

where $x = l_t/2$ = load transfer length from each fiber end. Thus, the load transfer length, l_t, is the minimum fiber length in which the maximum fiber stress is achieved.

For a given fiber diameter and fiber–matrix interfacial condition, a critical fiber length l_c is calculated from Equation 3.14 as

$$l_c = \frac{\sigma_{fu}}{2\tau_i} d_f, \tag{3.15}$$

where
σ_{fu} = ultimate tensile strength of the fiber
l_c = minimum fiber length required for the maximum fiber stress to be equal to the ultimate tensile strength of the fiber at its midlength (Figure 3.6b)
τ_i = shear strength of the fiber–matrix interface or the shear strength of the matrix adjacent to the interface, whichever is less

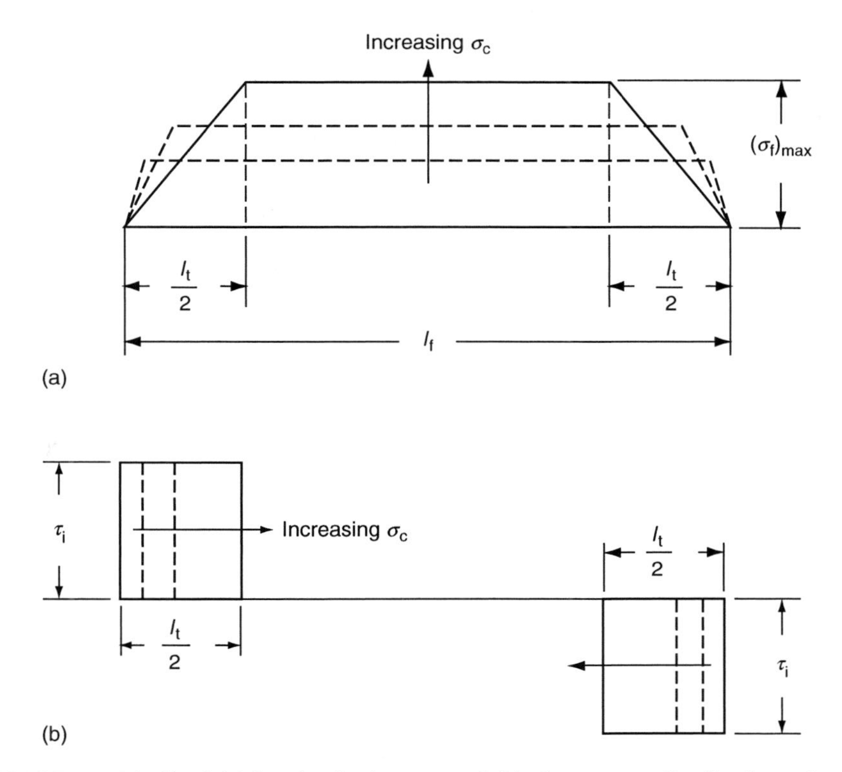

FIGURE 3.5 Idealized (a) longitudinal stress and (b) shear stress distributions along a discontinuous fiber owing to longitudinal tensile loading.

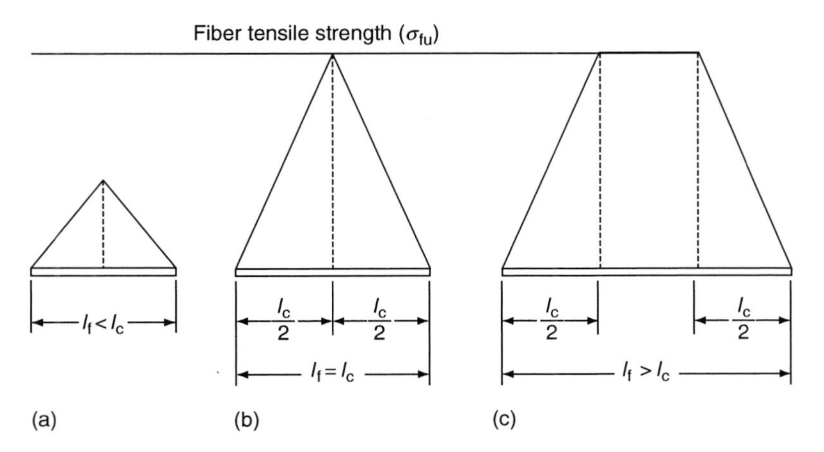

FIGURE 3.6 Significance of critical fiber length on the longitudinal stresses of a discontinuous fiber.

From Equations 3.14 and 3.15, we make the following observations:

1. For $l_f < l_c$, the maximum fiber stress may never reach the ultimate fiber strength (Figure 3.6a). In this case, either the fiber–matrix interfacial bond or the matrix may fail before fibers achieve their potential strength.
2. For $l_f > l_c$, the maximum fiber stress may reach the ultimate fiber strength over much of its length (Figure 3.6c). However, over a distance equal to $l_c/2$ from each end, the fiber remains less effective.
3. For effective fiber reinforcement, that is, for using the fiber to its potential strength, one must select $l_f \gg l_c$.
4. For a given fiber diameter and strength, l_c can be controlled by increasing or decreasing τ_i. For example, a matrix-compatible coupling agent may increase τ_i, which in turn decreases l_c. If l_c can be reduced relative to l_f through proper fiber surface treatments, effective reinforcement can be achieved without changing the fiber length.

Although normal stresses near the two fiber ends, that is, at $x < l_t/2$, are lower than the maximum fiber stress, their contributions to the total load-carrying capacity of the fiber cannot be completely ignored. Including these end stress distributions, an average fiber stress is calculated as

$$\bar{\sigma}_f = \frac{1}{l_f}\int_0^{l_f} \sigma_f \, dx,$$

which gives

$$\bar{\sigma}_f = (\sigma_f)_{max}\left(1 - \frac{l_t}{2l_f}\right). \quad (3.16)$$

Note that the load transfer length for $l_f < l_c$ is $\frac{l_f}{2}$, whereas that for $l_f > l_c$ is $\frac{l_c}{2}$.

For $l_f > l_c$, the longitudinal tensile strength of a unidirectional discontinuous fiber composite is calculated by substituting $(\sigma_f)_{max} = \sigma_{fu}$ and $l_t = l_c$ (Figure 3.6c). Thus,

$$\begin{aligned}\sigma_{Ltu} &= \bar{\sigma}_{fu} v_f + \sigma'_m(1 - v_f) \\ &= \sigma_{fu}\left(1 - \frac{l_c}{2l_f}\right) v_f + \sigma'_m(1 - v_f).\end{aligned} \quad (3.17)$$

In Equation 3.17, it is assumed that all fibers fail at the same strength level of σ_{fu}. Comparison of Equations 3.9 and 3.17 shows that discontinuous fibers always strengthen a matrix to a lesser degree than continuous fibers. However,

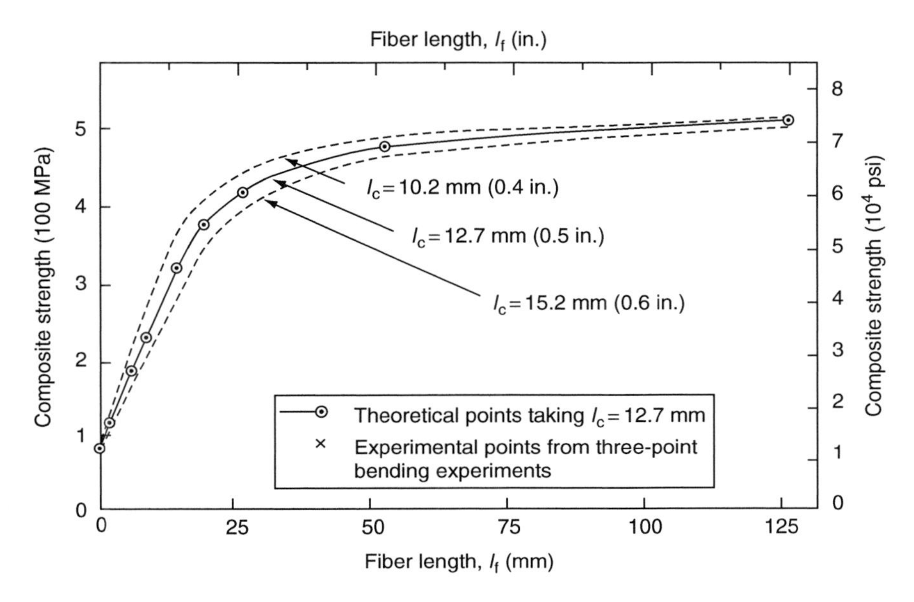

FIGURE 3.7 Variation in the longitudinal strength of a unidirectional discontinuous fiber composite as a function of fiber length. (After Hancock, P. and Cuthbertson, R.C., *J. Mater. Sci.*, 5, 762, 1970.)

for $l_f > 5l_c$, strengthening greater than 90% can be achieved even with discontinuous fibers. An example is shown in Figure 3.7.

For $l_f < l_c$, there will be no fiber failure. Instead, the lamina fails primarily because of matrix tensile failure. Since the average tensile stress in the fiber is $\bar{\sigma}_f = \tau_i \frac{l_f}{d_f}$, the longitudinal tensile strength of the composite is given by

$$\sigma_{Ltu} = \tau_i \frac{l_f}{d_f} v_f + \sigma_{mu}(1 - v_f), \tag{3.18}$$

where σ_{mu} is the tensile strength of the matrix material.

A simple method of determining the fiber–matrix interfacial shear strength is called a single fiber fragmentation test, which is based on the observation that fibers do not break if their length is less than the critical value. In this test, a single fiber is embedded along the centerline of a matrix tensile specimen (Figure 3.8). When the specimen is tested in axial tension, the tensile stress is transferred to the fiber by shear stress at the fiber–matrix interface. The embedded fiber breaks when the maximum tensile stress in the fiber reaches its tensile strength. With increased loading, the fiber breaks into successively shorter lengths until the fragmented lengths become so short that the maximum

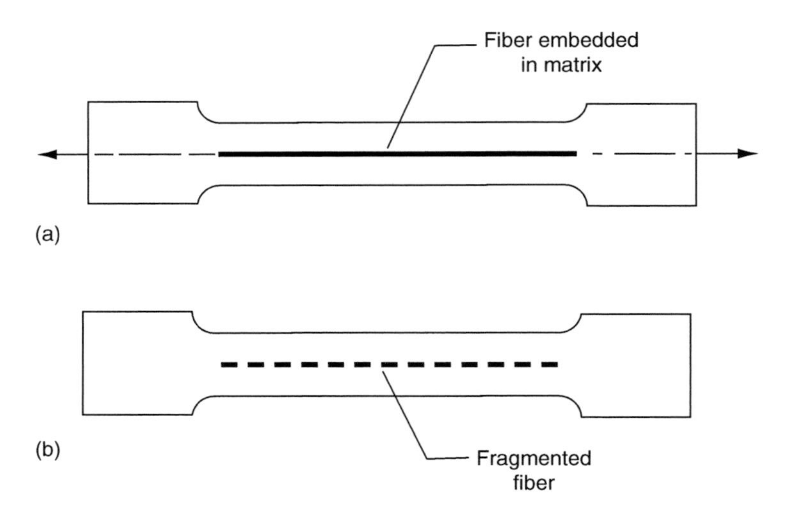

FIGURE 3.8 Single fiber fragmentation test to determine fiber–matrix interfacial shear strength.

tensile stress can no longer reach the fiber tensile strength. The fragmented fiber lengths at this point are theoretically equal to the critical fiber length, l_c. However, actual (measured) fragment lengths vary between $l_c/2$ and l_c. Assuming a uniform distribution for the fragment lengths and a mean value of $\bar{l}$ equal to $0.75l_c$, Equation 3.15 can be used to calculate the interfacial shear strength τ_{im} [3]:

$$\tau_{im} = \frac{3d_f\sigma_{fu}}{8\bar{l}}, \tag{3.19}$$

where $\bar{l}$ is the mean fragment length.

Equation 3.13 was obtained assuming that the interfacial shear stress τ_i is a constant. The analysis that followed Equation 3.13 was used to demonstrate the importance of critical fiber length in discontinuous fiber composites. However, strictly speaking, this analysis is valid only if it can be shown that τ_i is a constant. This will be true in the case of a ductile matrix that yields due to high shear stress in the interfacial zone before the fiber–matrix bond fails and then flows plastically with little or no strain hardening (i.e., the matrix behaves as a perfectly plastic material with a constant yield strength as shown in Figure 3.9). When this occurs, the interfacial shear stress is equal to the shear yield strength of the matrix (which is approximately equal to half of its tensile yield strength) and remains constant at this value. If the fiber–matrix bond fails before matrix yielding, a frictional force may be generated at the interface, which transfers the load from the matrix to the fibers through slippage (sliding). In a polymer matrix composite, the source of this frictional force is the radial pressure on the fiber surface created by the shrinkage of the matrix as it cools down from the

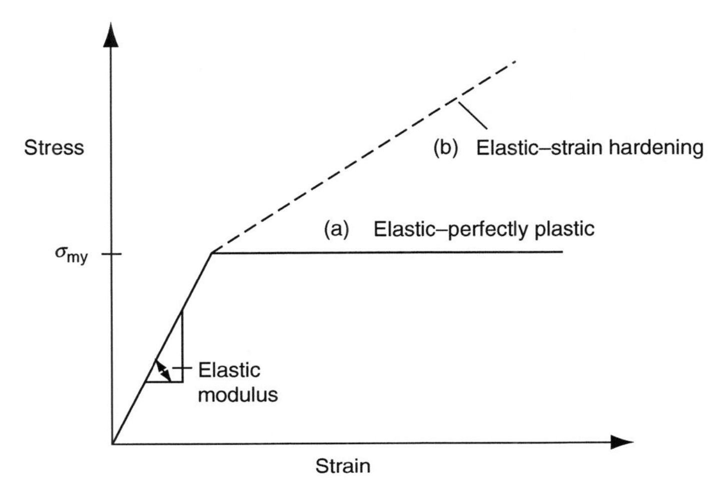

FIGURE 3.9 Stress–strain diagrams of (a) an elastic-perfectly plastic material and (b) an elastic-strain hardening material.

curing temperature. In this case, the interfacial shear stress is equal to the product of the coefficient of sliding friction and the radial pressure.

When the matrix is in the elastic state and the fiber–matrix bond is still unbroken, the interfacial shear stress is not a constant and varies with x. Assuming that the matrix has the same strain as the composite, Cox [4] used a simple shear lag analysis to derive the following expression for the fiber stress distribution along the length of a discontinuous fiber:

$$\sigma_f = E_f\varepsilon_1\left[1 - \frac{\cosh\beta\left(\frac{l_f}{2} - x\right)}{\cosh\frac{\beta l_f}{2}}\right] \quad \text{for } 0 \leq x \leq \frac{l_f}{2}, \tag{3.20}$$

where

σ_f = longitudinal fiber stress at a distance x from its end
E_f = fiber modulus
ε_1 = longitudinal strain in the composite

$$\beta = \sqrt{\frac{2G_m}{E_f r_f^2 \ln(R/r_f)}},$$

where

G_m = matrix shear modulus
r_f = fiber radius
$2R$ = center-to-center distance from a fiber to its nearest neighbor

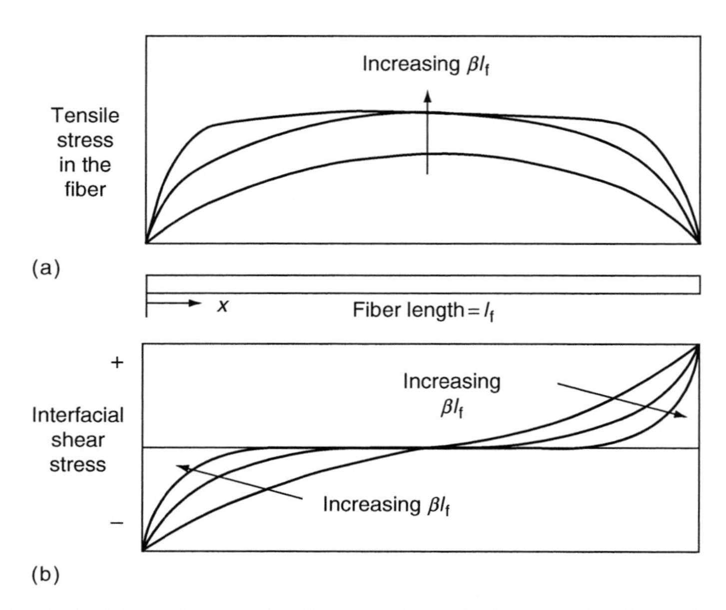

FIGURE 3.10 (a) Normal stress distribution along the length of a discontinuous fiber according to Equation 3.20 and (b) shear stress distribution at the fiber–matrix interface according to Equation 3.21.

Using Equations 3.11 and 3.20, shear stress at the fiber–matrix interface is obtained as:

$$\tau = \frac{1}{2} E_f \varepsilon_1 \beta r_f \frac{\sinh \beta \left(\frac{l_f}{2} - x \right)}{\cosh \frac{\beta l_f}{2}}. \tag{3.21}$$

Equations 3.20 and 3.21 are plotted in Figure 3.10 for various values of βl_f. It shows that the fiber stress builds up over a shorter load transfer length if βl_f is high. This means that not only a high fiber length to diameter ratio (called the *fiber aspect ratio*) but also a high ratio of G_m/E_f is desirable for strengthening a discontinuous fiber composite.

Note that the stress distribution in Figure 3.5 or 3.10 does not take into account the interaction between fibers. Whenever a discontinuity due to fiber end occurs, a stress concentration must arise since the tensile stress normally assumed by the fiber without the discontinuity must be taken up by the surrounding fibers. As a result, the longitudinal stress distribution for each fiber may contain a number of peaks.

EXAMPLE 3.1

A unidirectional fiber composite contains 60 vol% of HMS-4 carbon fibers in an epoxy matrix. Using the fiber properties in Table 2.1 and matrix properties as $E_m = 3.45$ GPa and $\sigma_{my} = 138$ MPa, determine the longitudinal tensile strength of the composite for the following cases:

1. The fibers are all continuous.
2. The fibers are 3.17 mm long and τ_i is (i) 4.11 MPa or (ii) 41.1 MPa.

Solution

Since HMS-4 carbon fibers are linearly elastic, their failure strain is

$$\varepsilon_{fu} = \frac{\sigma_{fu}}{E_f} = \frac{2480 \text{ MPa}}{345 \times 10^3 \text{ MPa}} = 0.0072.$$

Assuming that the matrix behaves in an elastic-perfectly plastic manner, its yield strain can be calculated as

$$\varepsilon_{my} = \frac{\sigma_{my}}{E_m} = \frac{138 \text{ MPa}}{3.45 \times 10^3 \text{ MPa}} = 0.04.$$

Thus, the fibers are expected to break before the matrix yields and the stress in the matrix at the instance of fiber failure is

$$\sigma'_m = E_m \varepsilon_{fu} = (3.45 \times 10^3 \text{ MPa})\,(0.0072) = 24.84 \text{ MPa}.$$

1. Using Equation 3.9, we get

$$\begin{aligned} \sigma_{Ltu} &= (2480)(0.6) + (24.84)(1 - 0.6) \\ &= 1488 + 9.94 = 1497.94 \text{ MPa}. \end{aligned}$$

2. (i) When $\tau_i = 4.11$ MPa, the critical fiber length is

$$l_c = \frac{2480 \text{ MPa}}{(2)(4.11 \text{ MPa})}(8 \times 10^{-3} \text{ mm}) = 2.414 \text{ mm}.$$

Since $l_f > l_c$, we can use Equation 3.17 to calculate

$$\begin{aligned} \sigma_{Ltu} &= (2480)\left[1 - \frac{2.414}{(2)(3.17)}\right](0.6) + (24.84)(1 - 0.6) \\ &= 921.43 + 9.94 = 931.37 \text{ MPa}. \end{aligned}$$

(ii) When $\tau_i = 41.1$ MPa, $l_c = 0.2414$ mm. Thus, $l_f > l_c$.

Equation 3.17 now gives $\sigma_{\mathrm{Ltu}} = 1441.28$ MPa.

This example demonstrates that with the same fiber length, it is possible to achieve a high longitudinal tensile strength for the composite by increasing the interfacial shear stress. Physically, this means that the bonding between the fibers and the matrix must be improved.

3.1.1.3 Microfailure Modes in Longitudinal Tension

In deriving Equations 3.9 and 3.17, it was assumed that all fibers have equal strength and the composite lamina fails immediately after fiber failure. In practice, fiber strength is not a unique value; instead it follows a statistical distribution. Therefore, it is expected that a few fibers will break at low stress levels. Although the remaining fibers will carry higher stresses, they may not fail simultaneously.

When a fiber breaks (Figure 3.11), the normal stress at each of its broken ends becomes zero. However, over a distance of $l_c/2$ from each end, the stress builds back up to the average value by shear stress transfer at the fiber–matrix interface (Figure 3.11c). Additionally, the stress states in a region close to the broken ends contain

1. Stress concentrations at the void created by the broken fiber
2. High shear stress concentrations in the matrix near the fiber ends
3. An increase in the average normal stress in adjacent fibers (Figure 3.11b)

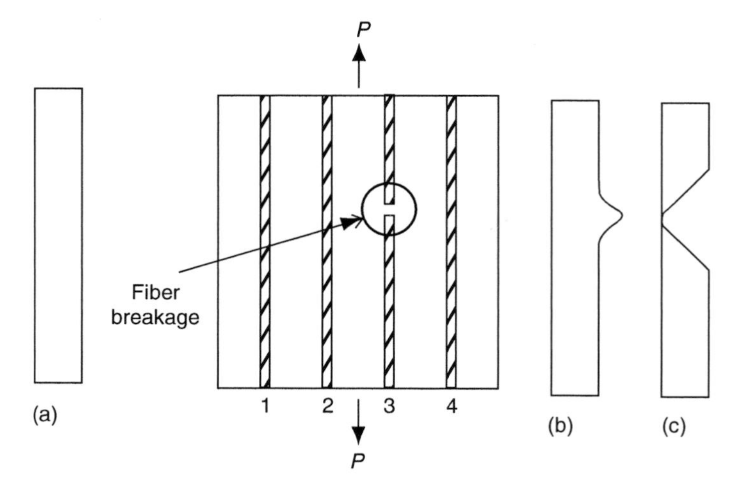

FIGURE 3.11 Longitudinal stress distributions (a) in unidirectional continuous fibers before the failure of fiber 3, (b) in fibers 2 and 4 after the failure of fiber 3, and (c) in fiber 3 after it fails.

Owing to these local stress magnifications, possibilities for several microfailure modes exist:

1. Partial or total debonding of the broken fiber from the surrounding matrix due to high interfacial shear stresses at its ends. As a result, the fiber effectiveness is reduced either completely or over a substantial length (Figure 3.12a).
2. Initiation of a microcrack in the matrix due to high stress concentration at the ends of the void (Figure 3.12b).

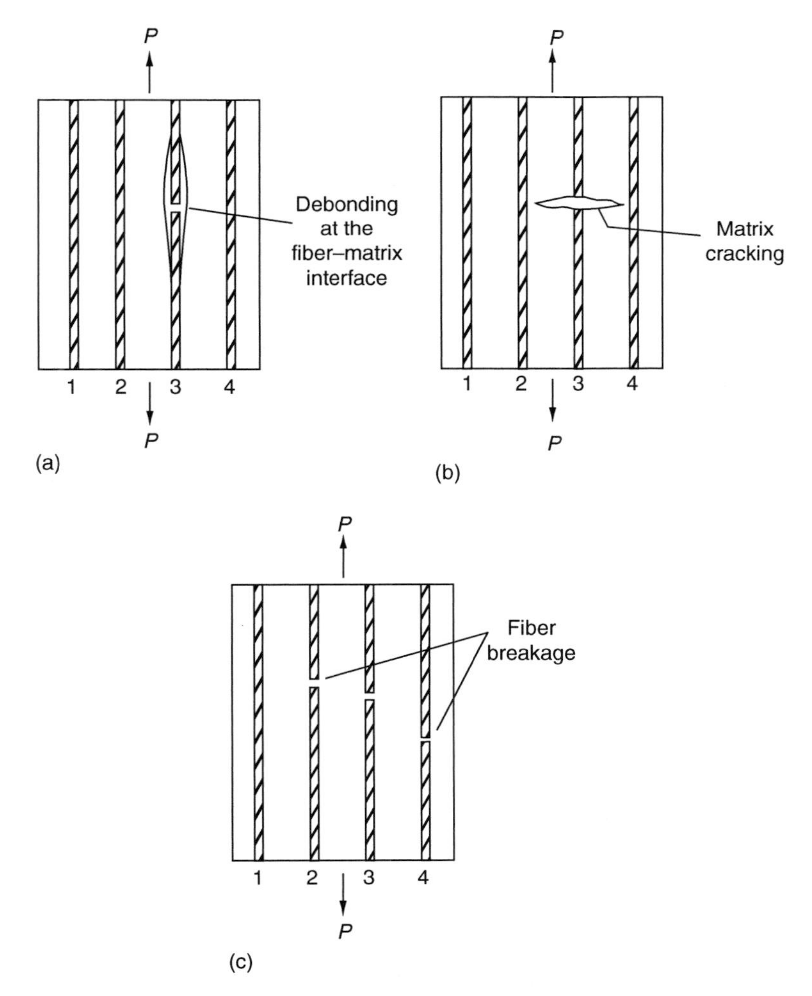

FIGURE 3.12 Possible microfailure modes following the breakage of fiber 3.

3. Plastic deformation (microyielding) in the matrix, particularly if the matrix is ductile.
4. Failure of other fibers in the vicinity of the first fiber break due to high average normal stresses and the local stress concentrations (Figure 3.12c). Each fiber break creates additional stress concentrations in the matrix as well as in other fibers. Eventually, many of these fiber breaks and the surrounding matrix microcracks may join to form a long microcrack in the lamina.

The presence of longitudinal stress (σ_{yy}) concentration at the tip of an advancing crack is well known. Cook and Gordon [5] have shown that the stress components σ_{xx} and τ_{xy} may also reach high values slightly ahead of the crack tip (Figure 3.13a). Depending on the fiber–matrix interfacial strength, these stress components are capable of debonding the fibers from the surrounding matrix even before they fail in tension (Figure 3.13b). Fiber–matrix debonding ahead of the crack tip has the effect of blunting the crack front and reducing the notch sensitivity of the material. High fiber strength and low interfacial strength promote debonding over fiber tensile failure.

With increasing load, fibers continue to break randomly at various locations in the lamina. Because of the statistical distribution of surface flaws, the

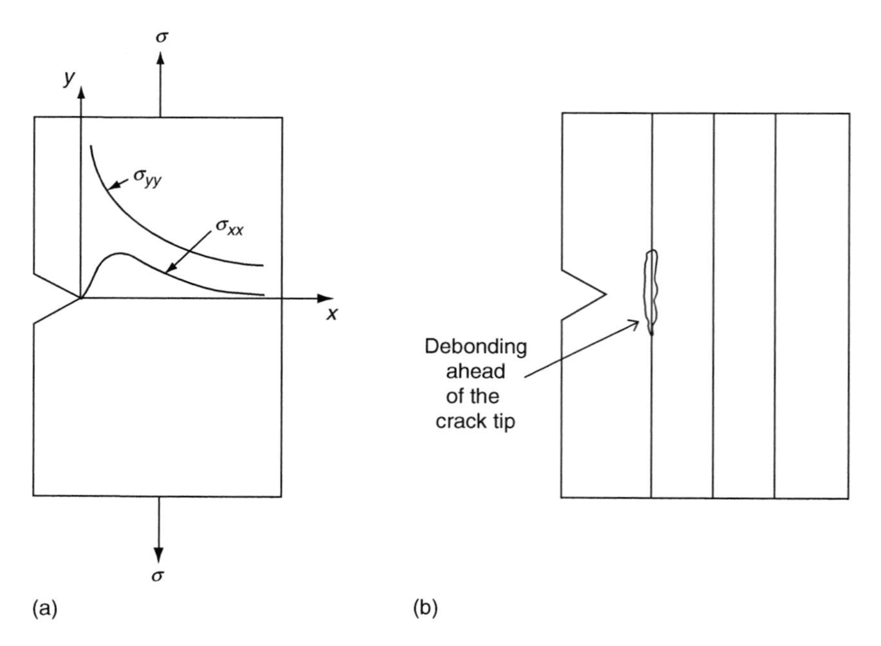

FIGURE 3.13 Schematic representation of (a) normal stress distributions and (b) fiber–matrix debonding ahead of a crack tip.

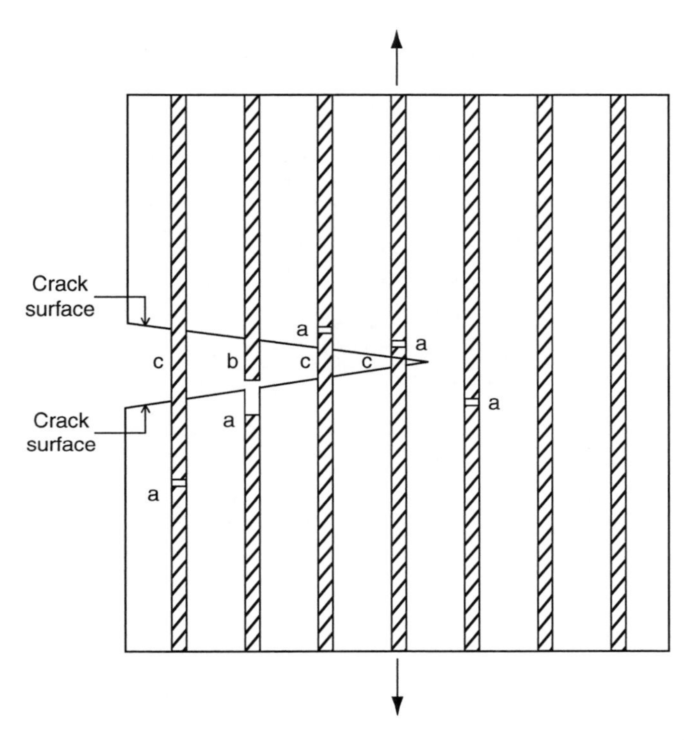

FIGURE 3.14 Schematic representation of fiber pullout and matrix bridging by broken fibers (a) fiber breakage; (b) fiber pullout; and (c) matrix bridging.

fiber failure does not always occur in the crack plane (Figure 3.14). Therefore, the opening of the matrix crack may cause broken fibers to pull out from the surrounding matrix (Figure 3.15), which is resisted by the friction at the fiber–matrix interface. If the interfacial strength is high or the broken fiber lengths are greater than $l_c/2$, the fiber pullout is preceded by either debonding or fiber failure even behind the crack front. Thus, broken fibers act as a bridge between the two faces of the matrix crack. In some instances, multiple parallel cracks are formed in the matrix normal to the fiber direction. If these cracks are bridged by fibers, the volume of matrix between the cracks may deform significantly before rupture.

Fracture toughness of a unidirectional 0° lamina is the sum of the energies consumed by various microfailure processes, namely, fiber fracture, matrix cracking or yielding, debonding, and fiber pullout. Theoretical models to calculate the energy contributions from some of these failure modes are given in Table 3.2. Although the true nature of the fracture process and stress fields are not known, these models can serve to recognize the variables that play

FIGURE 3.15 Fracture surface of a randomly oriented discontinuous fiber composite showing the evidence of fiber pullout.

major roles in the development of high fracture toughness for a fiber-reinforced composite lamina. It should be noted that energy contributions from the fracturing of brittle fibers and a brittle matrix are negligible (<10%) compared with those listed in Table 3.2.

3.1.2 Transverse Tensile Loading

When a transverse tensile load is applied to the lamina, the fibers act as hard inclusions in the matrix instead of the principal load-carrying members. Although the matrix modulus is increased by the presence of fibers, local stresses and strains in the surrounding matrix are higher than the applied stress. Figure 3.16b shows the variation of radial stress (σ_{rr}) and tangential stress ($\sigma_{\theta\theta}$) in a lamina containing a single cylindrical fiber. Near the fiber–matrix interface, the radial stress is tensile and is nearly 50% higher than the applied stress. Because of this radial stress component, cracks normal to the loading direction

TABLE 3.2
Important Energy Absorption Mechanisms During Longitudinal Tensile Loading of a Unidirectional Continuous Fiber Lamina

Stress relaxation energy (energy dissipated owing to reduction in stresses at the ends of a broken fiber [6])	$E_r = \frac{v_f \sigma_{fu}^2 l_c}{6E_f}$
Stored elastic energy in a partially debonded fiber [7]	$E_s = \frac{v_f \sigma_{fu}^2 y}{4E_f}$ (where y = debonded length of the fiber when it breaks)
Fiber pullout energy [8]	$E_{po} = \frac{v_f \sigma_{fu} l_c^2}{12 l_f}$ for $l_f > l_c$ $= \frac{v_f \sigma_{fu} l_f^2}{12 l_c}$ for $l_f < l_c$
Energy absorption by matrix deformation between parallel matrix cracks [9]	$E_{md} = \frac{(1 - v_f)^2}{v_f} \left(\frac{\sigma_{mu} d_f}{4\tau_i} \right) U_m$ (where U_m = energy required in deforming unit volume of the matrix to rupture)

Notes:
1. All energy expressions are on the basis of unit fracture surface area.
2. Debonding of fibers ahead of a crack tip or behind a crack tip is an important energy absorption mechanism. However, no suitable energy expression is available for this mechanism.
3. Energy absorption may also occur because of yielding of fibers or matrix if either of these constituents is ductile in nature.

may develop either at the fiber–matrix interface or in the matrix at $\theta = 90°$ (Figure 3.16c).

In a lamina containing a high volume fraction of fibers, there will be interactions of stress fields from neighboring fibers. Adams and Doner [10] used a finite difference method to calculate the stresses in unidirectional composites under transverse loading. A rectangular packing arrangement of parallel fibers was assumed, and solutions were obtained for various interfiber spacings representing different fiber volume fractions. Radial stresses at the fiber–matrix interface for 55% and 75% fiber volume fractions are shown in Figure 3.17. The maximum principal stress increases with increasing E_f/E_m ratio and fiber volume fraction, as indicated in Figure 3.18. The transverse modulus of the composite has a similar trend. Although an increased transverse modulus is desirable in many applications, an increase in local stress concentrations at high volume fractions and high fiber modulus may reduce the transverse strength of the composite (Table 3.3).

The simplest model used for deriving the equation for the transverse modulus of a unidirectional continuous fiber-reinforced composite is shown in Figure 3.19

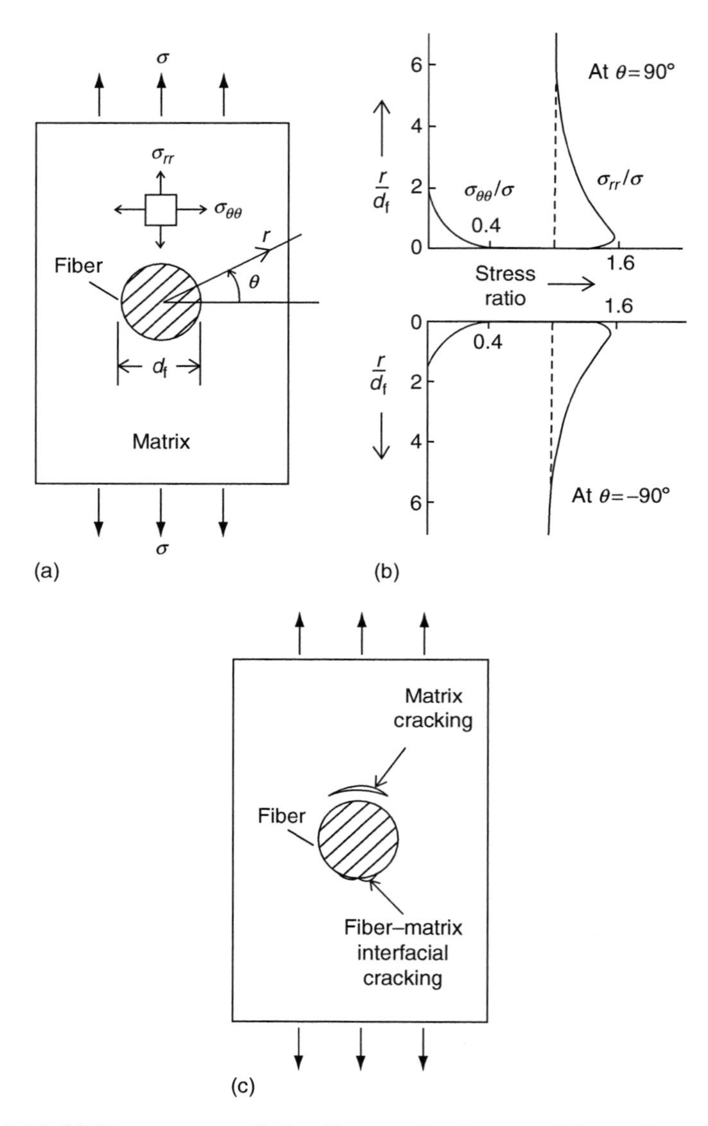

FIGURE 3.16 (a) Transverse tensile loading on a lamina containing a single cylindrical fiber, (b) stress distribution around a single fiber due to transverse tensile loading, and (c) possible microfailure modes.

in which the fibers and the matrix are replaced by their respective "equivalent" volumes and are depicted as two structural elements (slabs) with strong bonding across their interface. The tensile load is acting normal to the fiber direction. The other assumptions made in this simple slab model are as follows.

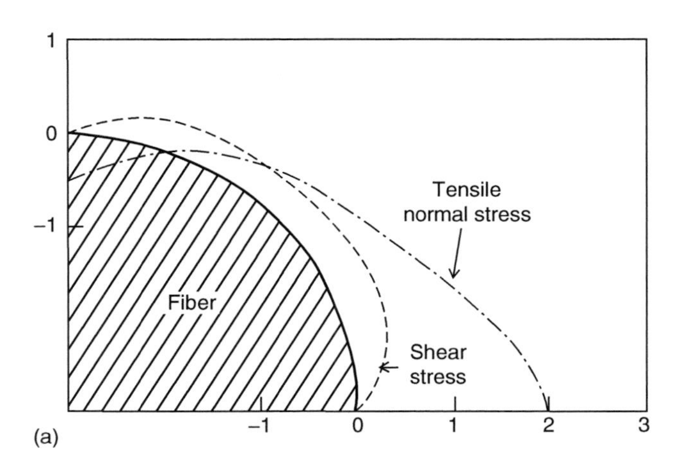

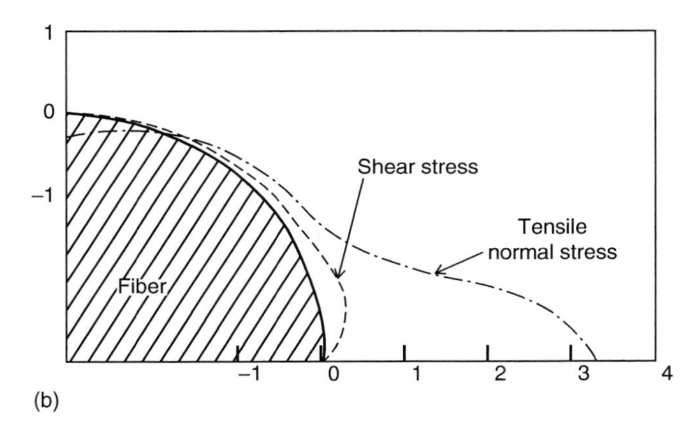

FIGURE 3.17 Variation of shear stress $\tau_{r\theta}$ and normal stress σ_{rr} at the surface of a circular fiber in a square array subjected to an average tensile stress σ transverse to the fiber directions: (a) $v_f = 55\%$ and (b) $v_f = 75\%$. (After Adams, D.F. and Doner, D.R., *J. Compos. Mater.*, 1, 152, 1967.)

1. Total deformation in the transverse direction is the sum of the total fiber deformation and the total matrix deformation, that is, $\Delta W_c = \Delta W_f + \Delta W_m$.
2. Tensile stress in the fibers and the tensile stress in the matrix are both equal to the tensile stress in the composite, that is, $\sigma_f = \sigma_m = \sigma_c$.

Since $\varepsilon_c = \frac{\Delta W_c}{W_c}$, $\varepsilon_f = \frac{\Delta W_f}{W_f}$, and $\varepsilon_m = \frac{\Delta W_m}{W_m}$, the deformation equation $\Delta W_c = \Delta W_f + \Delta W_m$ can be written as

$$\varepsilon_c W_c = \varepsilon_f W_f + \varepsilon_m W_m. \tag{3.22}$$

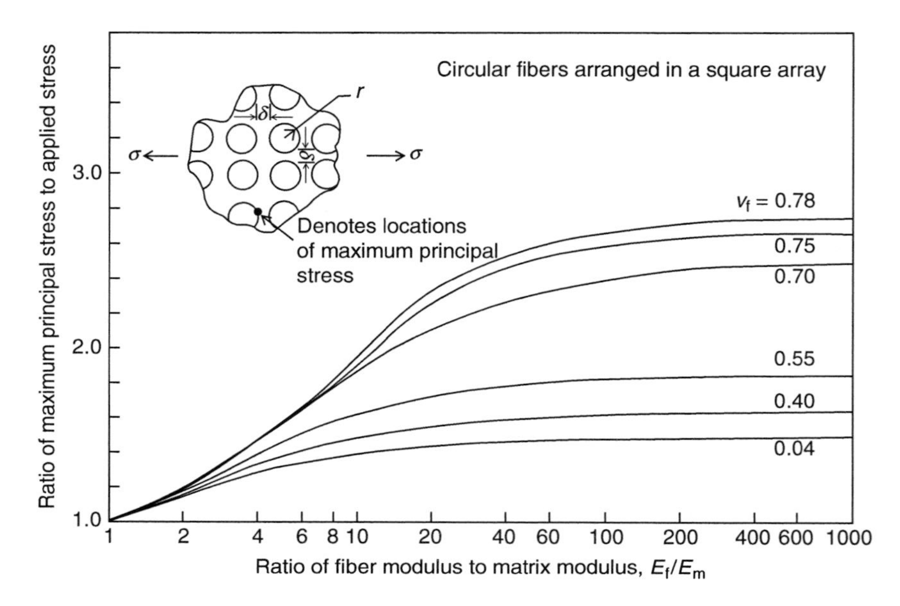

FIGURE 3.18 Ratio of the maximum principal stress in the matrix to the applied transverse stress on the composite for various fiber volume fractions. (After Adams, D.F. and Doner, D.R., *J. Compos. Mater.*, 1, 152, 1967.)

TABLE 3.3
Effect of Transverse Loading in a Unidirectional Composite

Composite Material	$\frac{E_f}{E_m}$	v_f (%)	Transverse Modulus, GPa (Msi)	Transverse Strength, MPa (ksi)
E-glass–epoxy	20	39	8.61 (1.25)	47.2 (6.85)
		67	18.89 (2.74)	30.87 (4.48)
E-glass–epoxy	24	46	8.96 (1.30)	69.1 (10.03)
		57	13.23 (1.92)	77.92 (11.31)
		68	21.91 (3.18)	67.93 (9.86)
		73	25.9 (3.76)	41.27 (5.99)
Boron–epoxy	120	65	23.43 (3.4)	41.96 (6.09)

Source: Adapted from Adams, D.F. and Doner, D.R., *J. Compos. Mater.*, 1, 152, 1967.

Dividing both sides by W_c and noting that $\frac{W_f}{W_c} = v_f$ and $\frac{W_m}{W_c} = v_m$, we can rewrite Equation 3.22 as

$$\varepsilon_c = \varepsilon_f v_f + \varepsilon_m v_m. \tag{3.23}$$

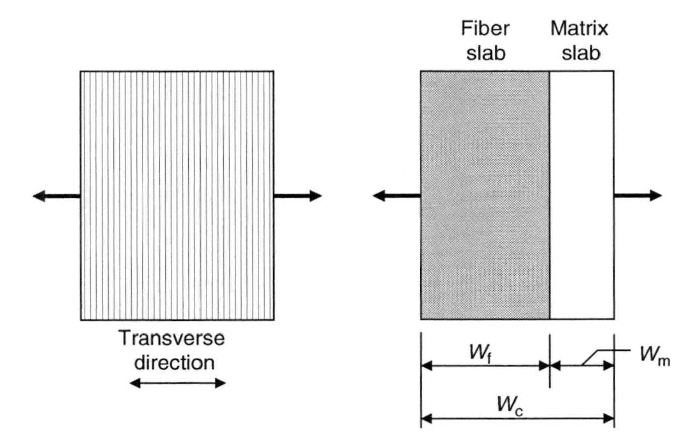

FIGURE 3.19 Transverse loading of a unidirectional continuous fiber lamina and the equivalent slab model.

Since $\varepsilon_c = \frac{\sigma_c}{E_T}$, $\varepsilon_f = \frac{\sigma_f}{E_f}$, and $\varepsilon_m = \frac{\sigma_m}{E_m}$, Equation 3.23 can be written as

$$\frac{\sigma_c}{E_T} = \frac{\sigma_f}{E_f} v_f + \frac{\sigma_m}{E_m} v_m. \tag{3.24}$$

In Equation 3.24, E_T is the transverse modulus of the unidirectional continuous fiber composite.

Finally, since it is assumed that $\sigma_f = \sigma_m = \sigma_c$, Equation 3.24 becomes

$$\frac{1}{E_T} = \frac{v_f}{E_f} + \frac{v_m}{E_m}. \tag{3.25}$$

Rearranging Equation 3.25, the expression for the transverse modulus E_T becomes

$$E_T = \frac{E_f E_m}{E_f v_m + E_m v_f} = \frac{E_f E_m}{E_f - v_f(E_f - E_m)}. \tag{3.26}$$

Equation 3.26 shows that the transverse modulus increases nonlinearly with increasing fiber volume fraction. By comparing Equations 3.7 and 3.26, it can be seen that the transverse modulus is lower than the longitudinal modulus and is influenced more by the matrix modulus than by the fiber modulus.

A simple equation for predicting the transverse tensile strength of a unidirectional continuous fiber lamina [11] is

$$\sigma_{Ttu} = \frac{\sigma_{mu}}{K_\sigma}, \tag{3.27}$$

where

$$K_\sigma = \frac{1 - v_f[1 - (E_m/E_f)]}{1 - (4v_f/\pi)^{1/2}[1 - (E_m/E_f)]}.$$

Equation 3.27 assumes that the transverse tensile strength of the composite is limited by the ultimate tensile strength of the matrix. Note that K_σ represents the maximum stress concentration in the matrix in which fibers are arranged in a square array. The transverse tensile strength values predicted by Equation 3.27 are found to be in reasonable agreement with those predicted by the finite difference method for fiber volume fractions <60% [2]. Equation 3.27 predicts that for a given matrix, the transverse tensile strength decreases with increasing fiber modulus as well as increasing fiber volume fraction.

3.1.3 Longitudinal Compressive Loading

An important function of the matrix in a fiber-reinforced composite material is to provide lateral support and stability for fibers under longitudinal compressive loading. In polymer matrix composites in which the matrix modulus is relatively low compared with the fiber modulus, failure in longitudinal compression is often initiated by localized buckling of fibers. Depending on whether the matrix behaves in an elastic manner or shows plastic deformation, two different localized buckling modes are observed: *elastic microbuckling* and *fiber kinking*.

Rosen [12] considered two possible elastic microbuckling modes of fibers in an elastic matrix as demonstrated in Figure 3.20. The extensional mode of

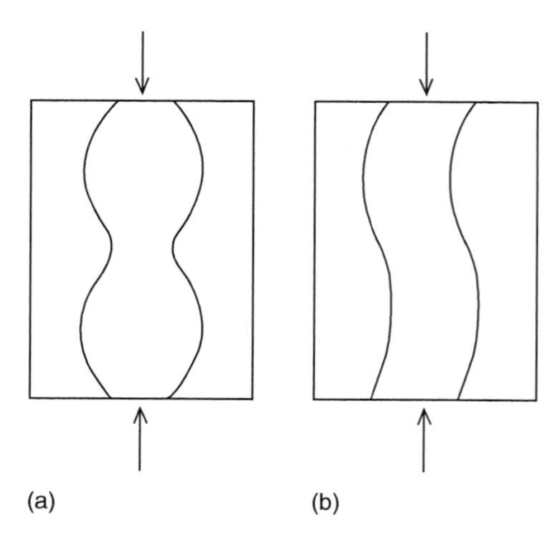

FIGURE 3.20 Fiber microbuckling modes in a unidirectional continuous fiber composite under longitudinal compressive loading: (a) extensional mode and (b) shear mode.

microbuckling occurs at low fiber volume fractions ($v_f < 0.2$) and creates an extensional strain in the matrix because of out-of-phase buckling of fibers. The shear mode of microbuckling occurs at high fiber volume fractions and creates a shear strain in the matrix because of in-phase buckling of fibers. Using buckling theory for columns in an elastic foundation, Rosen [12] predicted the compressive strengths in extensional mode and shear mode as

$$\text{Extensional mode: } \sigma_{\text{Lcu}} = 2v_f \left(\frac{v_f E_m E_f}{3(1 - v_f)} \right)^{1/2}, \tag{3.28a}$$

$$\text{Shear mode: } \sigma_{\text{Lcu}} = \frac{G_m}{(1 - v_f)}, \tag{3.28b}$$

where
G_m is the matrix shear modulus
v_f is the fiber volume fraction

Since most fiber-reinforced composites contain fiber volume fraction >30%, the shear mode is more important than the extensional mode. As Equation 3.28b shows, the shear mode is controlled by the matrix shear modulus as well as fiber volume fraction. The measured longitudinal compressive strengths are generally found to be lower than the theoretical values calculated from Equation 3.28b. Some experimental data suggest that the longitudinal compressive strength follows a rule of mixtures prediction similar to Equation 3.9.

The second important failure mode in longitudinal compressive loading is fiber kinking, which occurs in highly localized areas in which the fibers are initially slightly misaligned from the direction of the compressive loading. Fiber bundles in these areas rotate or tilt by an additional angle from their initial configuration to form kink bands and the surrounding matrix undergoes large shearing deformation (Figure 3.21). Experiments conducted on glass and carbon fiber-reinforced composites show the presence of fiber breakage at the ends of kink bands [13]; however, whether fiber breakage precedes or follows the kink band formation has not been experimentally verified. Assuming an elastic-perfectly plastic shear stress–shear strain relationship for the matrix, Budiansky and Fleck [14] have determined the stress at which kinking is initiated as

$$\sigma_{\text{ck}} = \frac{\tau_{\text{my}}}{\varphi + \gamma_{\text{my}}}, \tag{3.29}$$

where
τ_{my} = shear yield strength of the matrix
γ_{my} = shear yield strain of the matrix
φ = initial angle of fiber misalignment

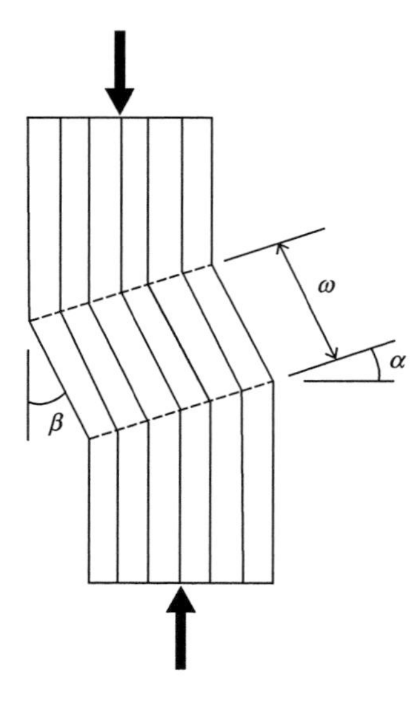

FIGURE 3.21 Kink band geometry. α = Kink band angle, β = Fiber tilt angle, and ω = Kink band width.

Besides fiber microbuckling and fiber kinking, a number of other failure modes have also been observed in longitudinal compressive loading of unidirectional continuous fiber-reinforced composites. They include shear failure of the composite, compressive failure or yielding of the reinforcement, longitudinal splitting in the matrix due to Poisson's ratio effect, matrix yielding, interfacial debonding, and fiber splitting or fibrillation (in Kevlar 49 composites). Factors that appear to improve the longitudinal compressive strength of unidirectional composites are increasing values of the matrix shear modulus, fiber tensile modulus, fiber diameter, matrix ultimate strain, and fiber–matrix interfacial strength. Fiber misalignment or bowing, on the other hand, tends to reduce the longitudinal compressive strength.

3.1.4 Transverse Compressive Loading

In transverse compressive loading, the compressive load is applied normal to the fiber direction, and the most common failure mode observed is the matrix shear failure along planes that are parallel to the fiber direction, but inclined to the loading direction (Figure 3.22). The failure is initiated by fiber–matrix debonding. The transverse compressive modulus and strength are considerably lower than the longitudinal compressive modulus and strength. The transverse compressive modulus is higher than the matrix modulus and is close to the transverse tensile modulus. The transverse compressive strength is found to be nearly independent of fiber volume fraction [15].

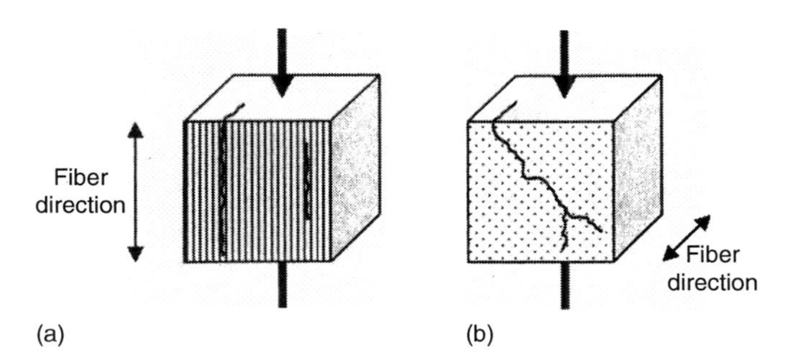

FIGURE 3.22 Shear failure (a) in longitudinal compression (compressive load parallel to the fiber direction) and (b) in transverse compression (compressive load normal to the fiber direction).

3.2 CHARACTERISTICS OF A FIBER-REINFORCED LAMINA

3.2.1 FUNDAMENTALS

3.2.1.1 Coordinate Axes

Consider a thin lamina in which fibers are positioned parallel to each other in a matrix, as shown in Figure 3.23. To describe its elastic properties, we first define two right-handed coordinate systems, namely, the 1-2-z system and the

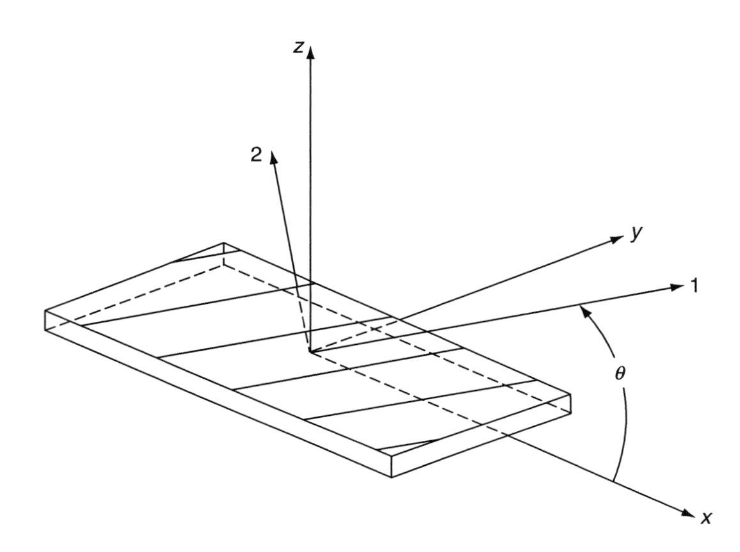

FIGURE 3.23 Definition of principal material axes and loading axes for a lamina.

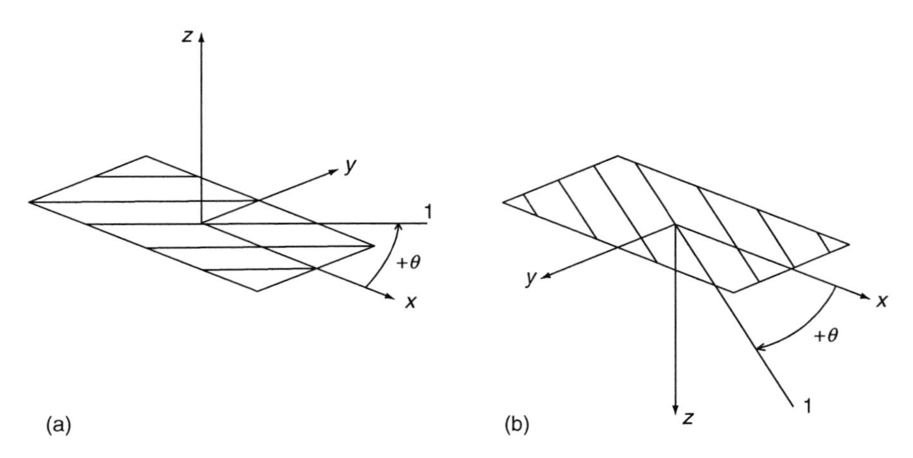

FIGURE 3.24 Right-handed coordinate systems. Note the difference in fiber orientation in (a) and (b).

x-y-z system. Both 1-2 and x-y axes are in the plane of the lamina, and the z axis is normal to this plane. In the 1-2-z system, axis 1 is along the fiber length and represents the longitudinal direction of the lamina, and axis 2 is normal to the fiber length and represents the transverse direction of the lamina. Together they constitute the *principal material directions* in the plane of the lamina. In the xyz system, x and y axes represent the *loading directions*.

The angle between the positive x axis and the 1-axis is called the *fiber orientation angle* and is represented by θ. The sign of this angle depends on the right-handed coordinate system selected. If the z axis is vertically upward to the lamina plane, θ is positive when measured counterclockwise from the positive x axis (Figure 3.24a). On the other hand, if the z axis is vertically downward, θ is positive when measured clockwise from the positive x axis (Figure 3.24b). In a 0° lamina, the principal material axis 1 coincides with the loading axis x, but in a 90° lamina, the principal material axis 1 is at a 90° angle with the loading axis x.

3.2.1.2 Notations

Fiber and matrix properties are denoted by subscripts f and m, respectively. Lamina properties, such as tensile modulus, Poisson's ratio, and shear modulus, are denoted by two subscripts. The first subscript represents the loading direction, and the second subscript represents the direction in which the particular property is measured. For example, ν_{12} represents the ratio of strain in direction 2 to the applied strain in direction 1, and ν_{21} represents the ratio of strain in direction 1 to the applied strain in direction 2.

Stresses and strains are also denoted with double subscripts (Figure 3.25). The first of these subscripts represents the direction of the outward normal to the plane in which the stress component acts. The second subscript represents

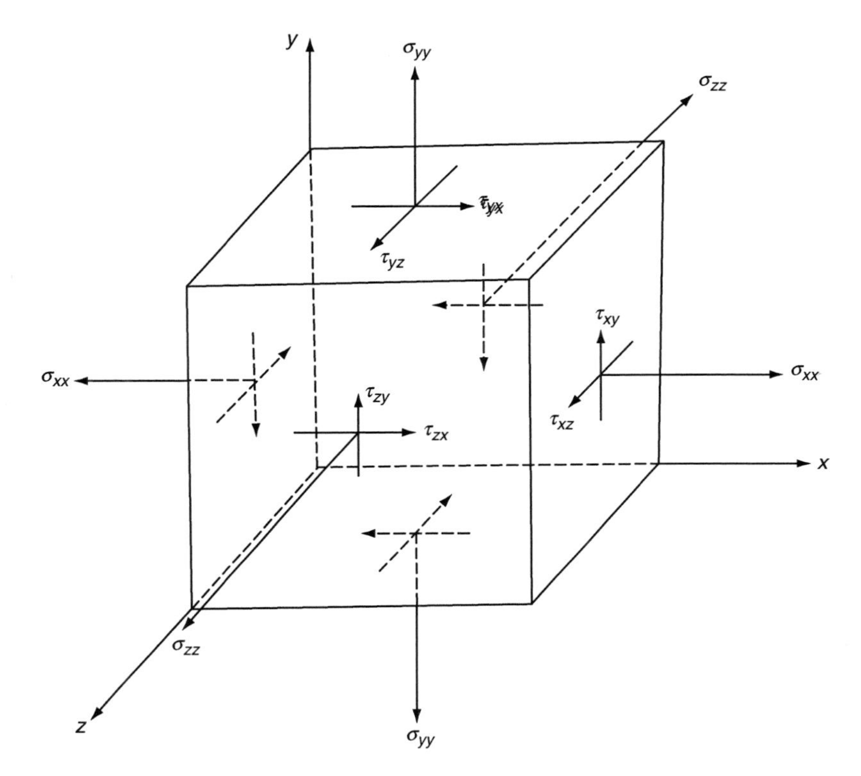

FIGURE 3.25 Normal stress and shear stress components.

the direction of the stress component. Thus, for example, the subscript x in the shear stress component τ_{xy} represents the outward normal to the yz plane and the subscript y represents its direction. The stress components σ_{xx}, σ_{yy}, and τ_{xy} are called *in-plane* (*intralaminar*) *stresses*, whereas σ_{zz}, τ_{xz}, and τ_{yz} are called *interlaminar stresses*.

In order to visualize the direction (sense) of various stress components, we adopt the following sign conventions:

1. If the outward normal to a stress plane is directed in a positive coordinate direction, we call it a positive plane. A negative plane has its outward normal pointing in the negative coordinate direction.
2. A stress component is positive in sign if it acts in a positive direction on a positive plane or in a negative direction on a negative plane. On the other hand, the stress component is negative in sign if it acts in a negative direction on a positive plane or in a positive direction on a negative plane. Thus, all stress components in Figure 3.25 are positive in sign.

3.2.1.3 Stress and Strain Transformations in a Thin Lamina under Plane Stress

In stress analysis of a thin lamina with fiber orientation angle θ, it is often desirable to transform stresses in the xy directions to stresses in the 12 directions. The stress transformation equations are

$$\begin{aligned}\sigma_{11} &= \sigma_{xx}\cos^2\theta + \sigma_{yy}\sin^2\theta + 2\tau_{xy}\cos\theta\sin\theta,\\ \sigma_{22} &= \sigma_{xx}\sin^2\theta + \sigma_{yy}\cos^2\theta - 2\tau_{xy}\cos\theta\sin\theta,\\ \tau_{12} &= (-\sigma_{xx} + \sigma_{yy})\sin\theta\cos\theta + \tau_{xy}(\cos^2\theta - \sin^2\theta).\end{aligned} \tag{3.30}$$

where σ_{xx}, σ_{yy}, and τ_{xy} are applied stresses in the xy directions and σ_{11}, σ_{22}, and τ_{12} are transformed stresses in the 12 directions. Similar equations can also be written for strain transformation by replacing each σ with ε and each τ with $\gamma/2$ in Equation 3.30. Thus, the strain transformation equations are

$$\begin{aligned}\varepsilon_{11} &= \varepsilon_{xx}\cos^2\theta + \varepsilon_{yy}\sin^2\theta + \gamma_{xy}\cos\theta\sin\theta,\\ \varepsilon_{22} &= \varepsilon_{xx}\sin^2\theta + \varepsilon_{yy}\cos^2\theta - \gamma_{xy}\cos\theta\sin\theta,\\ \gamma_{12} &= 2(-\varepsilon_{xx} + \varepsilon_{yy})\sin\theta\cos\theta + \gamma_{xy}(\cos^2\theta - \sin^2\theta).\end{aligned} \tag{3.31}$$

EXAMPLE 3.2

A normal stress σ_{xx} of 10 MPa is applied on a unidirectional angle-ply lamina containing fibers at 30° to the x axis, as shown at the top of the figure. Determine the stresses in the principal material directions.

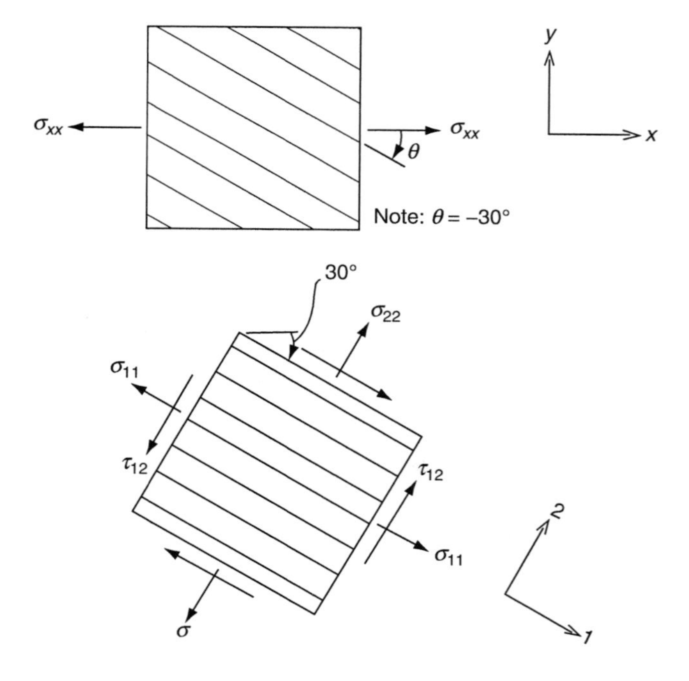

Solution

Since $\sigma_{yy} = \tau_{xy} = 0$, the transformation equations become

$$\begin{aligned}\sigma_{11} &= \sigma_{xx}\cos^2\theta,\\ \sigma_{22} &= \sigma_{xx}\sin^2\theta,\\ \tau_{12} &= -\sigma_{xx}\sin\theta\cos\theta.\end{aligned}$$

In this example, $\sigma_{xx} = +10$ MPa and $\theta = -30°$. Therefore,

$$\begin{aligned}\sigma_{11} &= 7.5 \text{ MPa},\\ \sigma_{22} &= 2.5 \text{ MPa},\\ \tau_{12} &= 4.33 \text{ MPa}.\end{aligned}$$

The stresses in the principal material directions are shown in the figure.

3.2.1.4 Isotropic, Anisotropic, and Orthotropic Materials

In an isotropic material, properties are the same in all directions. Thus, the material contains an infinite number of planes of material property symmetry passing through a point. In an anisotropic material, properties are different in all directions so that the material contains no planes of material property symmetry. Fiber-reinforced composites, in general, contain three orthogonal planes of material property symmetry, namely, the 1–2, 2–3, and 1–3 plane shown in Figure 3.26, and are classified as *orthotropic materials*. The intersections of

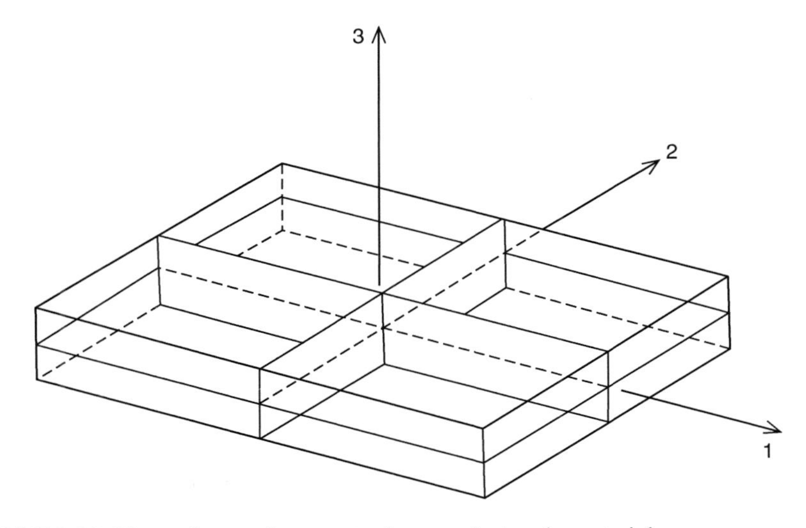

FIGURE 3.26 Three planes of symmetry in an orthotropic material.

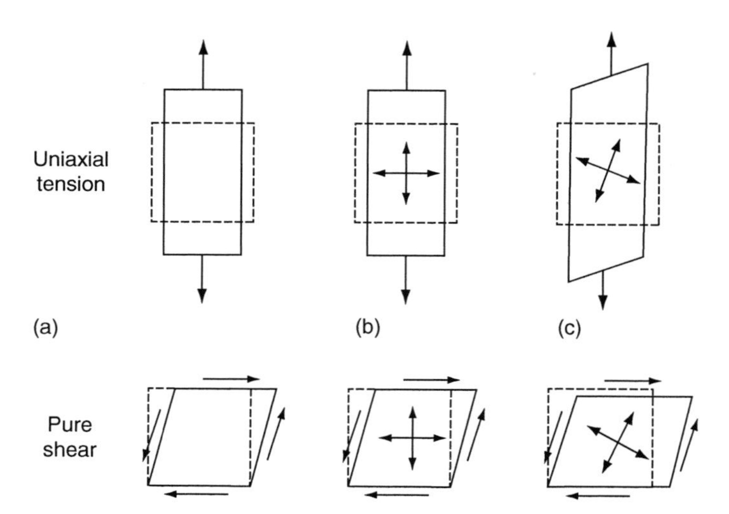

FIGURE 3.27 Differences in the deformations of isotropic, specially orthotropic and anisotropic materials subjected to uniaxial tension ((a) Isotropic, (b) Special orthotropic, and (c) General orthotropic and anisotropic) and pure shear stresses.

these three planes of symmetry, namely, axes 1, 2, and 3, are called the principal material directions.

Differences in the mechanical behavior of isotropic, orthotropic, and anisotropic materials are demonstrated schematically in Figure 3.27. Tensile normal stresses applied in any direction on an isotropic material cause elongation in the direction of the applied stresses and contractions in the two transverse directions. Similar behavior is observed in orthotropic materials only if the normal stresses are applied in one of the principal material directions. However, normal stresses applied in any other direction create both extensional and shear deformations. In an anisotropic material, a combination of extensional and shear deformation is produced by a normal stress acting in any direction. This phenomenon of creating both extensional and shear deformations by the application of either normal or shear stresses is termed *extension-shear coupling* and is not observed in isotropic materials.

The difference in material property symmetry in isotropic, orthotropic, and anisotropic materials is also reflected in the mechanics and design of these types of materials. Two examples are given as follows.

1. The elastic stress–strain characteristics of an isotropic material are described by three elastic constants, namely, Young's modulus E, Poisson's ratio ν, and shear modulus G. Only two of these three elastic

constants are independent since they can be related by the following equation:

$$G = \frac{E}{2(1 + \nu)}. \tag{3.32}$$

The number of independent elastic constants required to characterize anisotropic and orthotropic materials are 21 and 9, respectively [16]. For an orthotropic material, the nine independent elastic constants are E_{11}, E_{22}, E_{33}, G_{12}, G_{13}, G_{23}, ν_{12}, ν_{13}, and ν_{23}.

Unidirectionally oriented fiber composites are a special class of orthotropic materials. Referring to Figure 3.28, which shows a

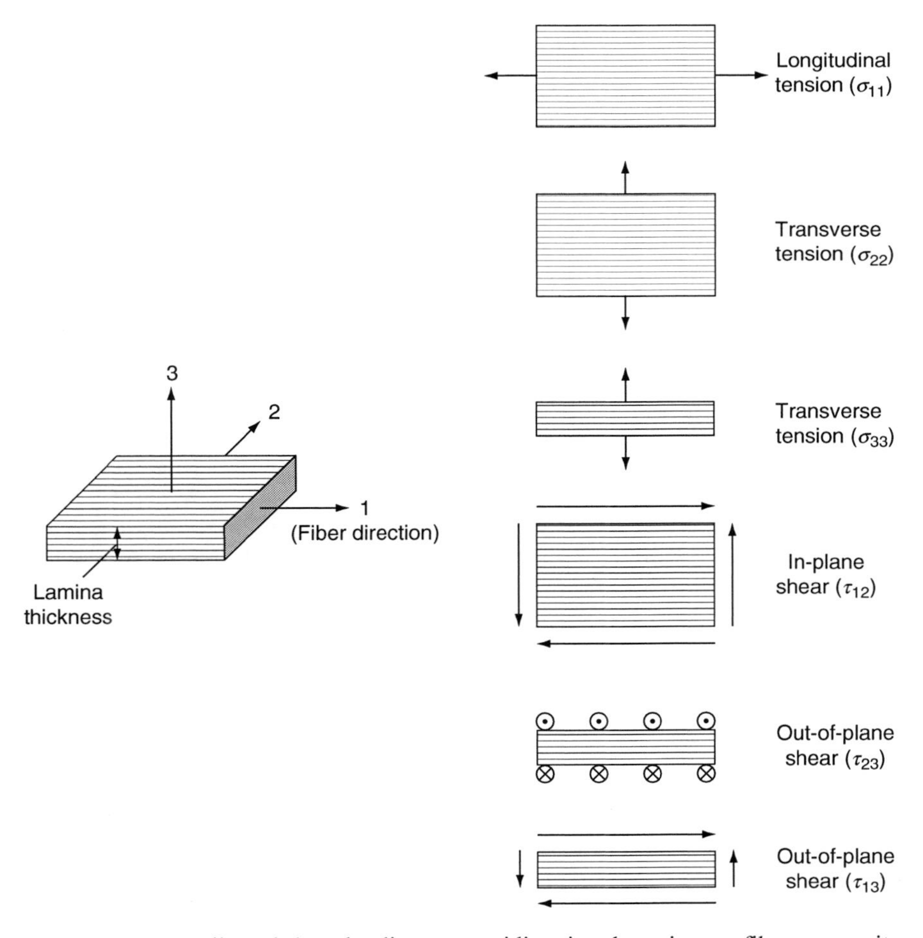

FIGURE 3.28 Tensile and shear loading on a unidirectional continuous fiber composite.

composite in which the fibers are in the 12 plane, it can be visualized that the elastic properties are equal in the 2–3 direction so that $E_{22} = E_{33}$, $\nu_{12} = \nu_{13}$, and $G_{12} = G_{13}$. Furthermore, G_{23} can be expressed in terms of E_{22} and ν_{23} by an expression similar to Equation 3.32.

$$G_{23} = \frac{E_{22}}{2(1 + \nu_{23})}. \quad (3.33)$$

Thus, the number of independent elastic constants for a unidirectionally oriented fiber composite reduces to 5, namely, E_{11}, E_{22}, ν_{12}, G_{12}, and ν_{23}. Such composites are often called *transversely isotropic*.

Note that $\nu_{21} \neq \nu_{12}$ and $\nu_{31} \neq \nu_{13}$, but $\nu_{31} = \nu_{21}$. However, ν_{21} is related to ν_{12} by the following equation, and therefore is not an independent elastic constant.

$$\nu_{21} = \left(\frac{E_{22}}{E_{11}}\right)\nu_{12}. \quad (3.34)$$

Christensen [17] has shown that in the case of unidirectional fiber-reinforced composites with fibers oriented in the 1-direction, ν_{23} can be related to ν_{12} and ν_{21} using the following equation:

$$\nu_{23} = \nu_{32} = \nu_{12}\frac{(1 - \nu_{21})}{(1 - \nu_{12})}. \quad (3.35)$$

Equation 3.35 fits the experimental data within the range of experimental accuracy. Thus, for a unidirectional fiber-reinforced composite, the number of independent elastic constants is reduced from 5 to 4.

2. For an isotropic material, the sign convention for shear stresses and shear strains is of little practical significance, since its mechanical behavior is independent of the direction of shear stress. For an orthotropic or anisotropic material, the direction of shear stress is critically important in determining its strength and modulus [18]. For example, consider a unidirectional fiber-reinforced lamina (Figure 3.29) subjected to states of pure shear of opposite sense. For positive shear (Figure 3.29a), the maximum (tensile) principal stress is parallel to the fiber direction that causes fiber fracture. For negative shear (Figure 3.29b), the maximum (tensile) principal stress is normal to the fiber direction, which causes either a matrix failure or a fiber–matrix interface failure. Obviously, a positive shear condition will favor a higher load-carrying capacity than the negative shear condition. For an isotropic material, shear strength is equal in all directions. Therefore, the direction of shear stress will not influence the failure of the material.

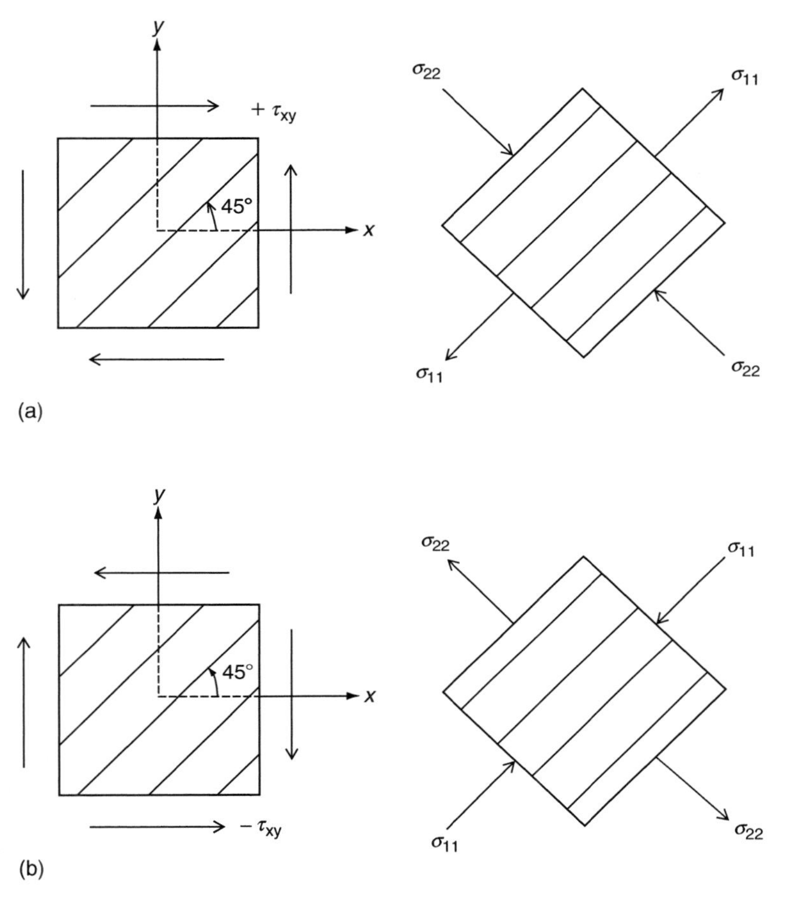

FIGURE 3.29 Normal stress components parallel and perpendicular to the fibers due to (a) positive shear stress and (b) negative shear stress on a 45° lamina.

3.2.2 Elastic Properties of a Lamina

3.2.2.1 Unidirectional Continuous Fiber 0° Lamina

Elastic properties of a unidirectional continuous fiber 0° lamina (Figure 3.30) are calculated from the following equations.

1. Referring to Figure 3.30a in which the tensile stress is applied in the 1-direction,

   Longitudinal modulus:

$$E_{11} = E_f v_f + E_m v_m \tag{3.36}$$

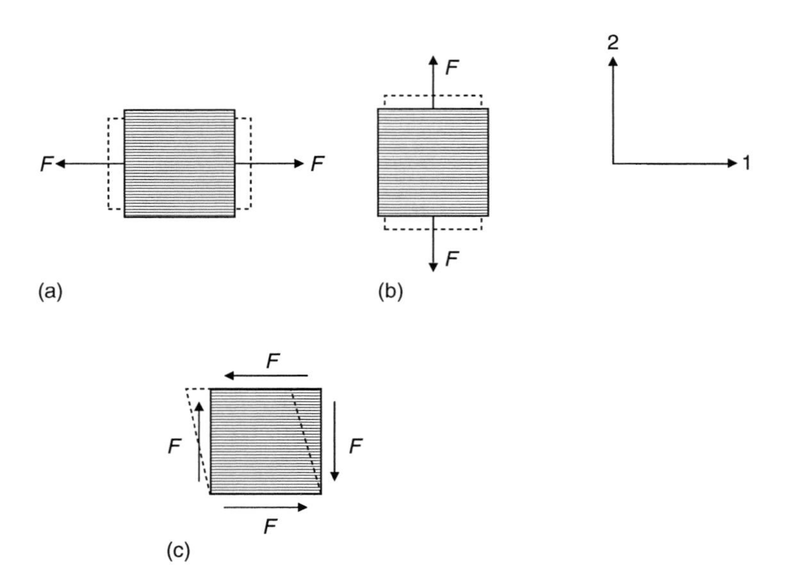

FIGURE 3.30 Applications of (a) longitudinal tensile stress, (b) transverse tensile stress, and (c) in-plane shear stress on a unidirectional continuous fiber 0° lamina.

and

Major Poisson's ratio:

$$\nu_{12} = \nu_f v_f + \nu_m v_m, \tag{3.37}$$

where $\nu_{12} = -\dfrac{\text{Strain in the 2-direction}}{\text{Strain in the 1-direction (i.e., the stress direction)}}$.

2. Referring to Figure 3.30b in which the tensile stress is applied in the 2-direction

Transverse modulus:

$$E_{22} = \frac{E_f E_m}{E_f v_m + E_m v_f} \tag{3.38}$$

and

Minor Poisson's ratio:

$$\nu_{21} = \frac{E_{22}}{E_{11}} \nu_{12}, \tag{3.39}$$

where $\nu_{21} = -\dfrac{\text{Strain in the 1-direction}}{\text{Strain in the 2-direction (i.e., the stress direction)}}$.

3. Referring to Figure 3.30c in which the shear stress is applied in 12 plane

   In-plane shear modulus:

$$G_{12} = G_{21} = \frac{G_f G_m}{G_f v_m + G_m v_f}. \tag{3.40}$$

The following points should be noted from Equations 3.36 through 3.40:

1. The longitudinal modulus (E_{11}) is always greater than the transverse modulus (E_{22}) (Figure 3.31).
2. The fibers contribute more to the development of the longitudinal modulus, and the matrix contributes more to the development of the transverse modulus.
3. The major Poisson's ratio (ν_{12}) is always greater than the minor Poisson's ratio (ν_{21}). Since these Poisson's ratios are related to Equation 3.39, only one of them can be considered independent.
4. As for E_{22}, the matrix contributes more to the development of G_{12} than the fibers.
5. Four independent elastic constants, namely, E_{11}, E_{22}, ν_{12}, and G_{12}, are required to describe the in-plane elastic behavior of a lamina. The ratio E_{11}/E_{22} is often considered a *measure of orthotropy.*

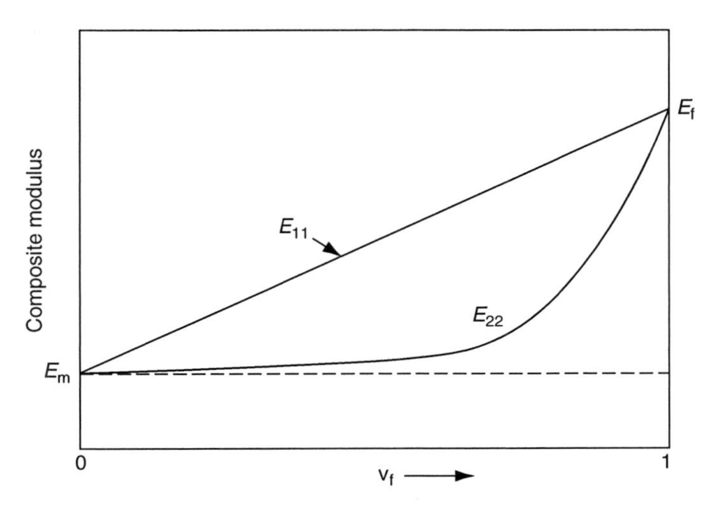

FIGURE 3.31 Variations of longitudinal and transverse modulus of a unidirectional continuous fiber 0° lamina with fiber volume fraction.

Equations 3.36 through 3.40 are derived using the simple mechanics of materials approach along with the following assumptions:

1. Both fibers and matrix are linearly elastic isotropic materials.
2. Fibers are uniformly distributed in the matrix.
3. Fibers are perfectly aligned in the 1-direction.
4. There is perfect bonding between fibers and matrix.
5. The composite lamina is free of voids.

Since, in practice, none of these assumptions is completely valid, these equations provide only approximate values for the elastic properties of a continuous fiber 0° lamina. It has been found that the values of E_{11} and ν_{12} predicted by Equations 3.36 and 3.37 agree well with the experimental data, but values of E_{22} and G_{12} predicted by Equations 3.38 and 3.40 are lower than the experimental data [19]. Both E_{22} and G_{12} are sensitive to void content, fiber anisotropy, and the matrix Poisson's ratio. Since equations based on the theory of elasticity or the variational approach, for example, are difficult to solve, Equations 3.36 through 3.40 or empirically modified versions of these equations (see Appendix A.3) are used frequently for the laminate design.

In Equations 3.36 through 3.40, it is assumed that both fibers and matrix are isotropic materials. While the matrix in most fiber-reinforced composites exhibits isotropic behavior, many reinforcing fibers are not isotropic and their elastic modulus in the longitudinal direction, E_{fL}, is much greater than their elastic modulus in the transverse direction, E_{fT}. Accordingly, Equations 3.36 and 3.38 should be modified in the following manner.

$$E_{11} = E_{fL}v_f + E_m v_m, \tag{3.41}$$

$$E_{22} = \frac{E_{fT}E_m}{E_{fT}v_m + E_m v_f}. \tag{3.42}$$

The Poisson's ratio of the fiber in Equation 3.37 should be represented by ν_{fLT}, and its shear modulus in Equation 3.40 should be represented by G_{fLT}. Since for most of the fibers, E_{fT}, ν_{fLT}, and G_{fLT} are difficult to measure and are not available, Equations 3.36 and 3.40 are commonly used albeit the errors that they can introduce.

EXAMPLE 3.3

To demonstrate the difference between ν_{12} and ν_{21}, consider the following example in which a square composite plate containing unidirectional continuous T-300 carbon fiber-reinforced epoxy is subjected to a uniaxial tensile load of 1000 N. The plate thickness is 1 mm. The length (L_o) and width (W_o) of the plate are 100 mm each.

Consider two loading cases, where

1. Load is applied parallel to the fiber direction
2. Load is applied normal to the fiber direction

Calculate the changes in length and width of the plate in each case. The basic elastic properties of the composite are given in Appendix A.5.

SOLUTION

From Appendix A.5, $E_{11} = 138$ GPa, $E_{22} = 10$ GPa, and $\nu_{12} = 0.21$. Using Equation 3.39, we calculate ν_{21}.

$$\nu_{21} = \frac{E_{22}}{E_{11}}\nu_{12} = \left(\frac{10\text{ GPa}}{138\text{ GPa}}\right)(0.21) = 0.0152.$$

1. Tensile load is applied parallel to the fiber direction, that is, in the 1-direction. Therefore, $\sigma_{11} = \frac{1000\text{ N}}{(100\text{ mm})(1\text{ mm})} = 10$ MPa and $\sigma_{22} = 0$. Now, we calculate the normal strains ε_{11} and ε_{22}.

$$\varepsilon_{11} = \frac{\sigma_{11}}{E_{11}} = \frac{10\text{ MPa}}{138\text{ GPa}} = 0.725 \times 10^{-4},$$

$$\varepsilon_{22} = -\nu_{12}\varepsilon_{11} = -(0.21)\ (0.725 \times 10^{-4}) = -0.152 \times 10^{-4}.$$

Since $\varepsilon_{11} = \frac{\Delta L}{L_o}$ and $\varepsilon_{22} = \frac{\Delta W}{W_o}$,

$$\Delta L = L_o\varepsilon_{11} = (100\text{ mm})\ (0.725 \times 10^{-4}) = 0.00725\text{ mm},$$

$$\Delta W = W_o\varepsilon_{22} = (100\text{ mm})\ (-0.152 \times 10^{-4}) = -0.00152\text{ mm}.$$

2. Tensile load is applied normal to the fiber direction, that is, in the 2-direction. Therefore, $\sigma_{22} = \frac{1000\text{ N}}{(100\text{ mm})(1\text{ mm})} = 10$ MPa and $\sigma_{11} = 0$. The normal strains in this case are

$$\varepsilon_{22} = \frac{\sigma_{22}}{E_{22}} = \frac{10\text{ MPa}}{10\text{ GPa}} = 10 \times 10^{-4},$$

$$\varepsilon_{11} = -\nu_{21}\varepsilon_{22} = -(0.0152)\ (10 \times 10^{-4}) = -0.152 \times 10^{-4}.$$

Since $\varepsilon_{11} = \frac{\Delta L}{L_o}$ and $\varepsilon_{22} = \frac{\Delta W}{W_o}$,

$$\Delta L = L_o\varepsilon_{11} = (100\text{ mm})\ (-0.152 \times 10^{-4}) = -0.00152\text{ mm},$$

$$\Delta W = W_o\varepsilon_{22} = (100\text{ mm})\ (10 \times 10^{-4}) = 0.1\text{ mm}.$$

3.2.2.2 Unidirectional Continuous Fiber Angle-Ply Lamina

The following equations are used to calculate the elastic properties of an angle-ply lamina in which continuous fibers are aligned at an angle θ with the positive x direction (Figure 3.32):

$$\frac{1}{E_{xx}} = \frac{\cos^4\theta}{E_{11}} + \frac{\sin^4\theta}{E_{22}} + \frac{1}{4}\left(\frac{1}{G_{12}} - \frac{2\nu_{12}}{E_{11}}\right)\sin^2 2\theta, \quad (3.43)$$

$$\frac{1}{E_{yy}} = \frac{\sin^4\theta}{E_{11}} + \frac{\cos^4\theta}{E_{22}} + \frac{1}{4}\left(\frac{1}{G_{12}} - \frac{2\nu_{12}}{E_{11}}\right)\sin^2 2\theta, \quad (3.44)$$

$$\frac{1}{G_{xy}} = \frac{1}{E_{11}} + \frac{2\nu_{12}}{E_{11}} + \frac{1}{E_{22}} - \left(\frac{1}{E_{11}} + \frac{2\nu_{12}}{E_{11}} + \frac{1}{E_{22}} - \frac{1}{G_{12}}\right)\cos^2 2\theta, \quad (3.45)$$

$$\nu_{xy} = E_{xx}\left[\frac{\nu_{12}}{E_{11}} - \frac{1}{4}\left(\frac{1}{E_{11}} + \frac{2\nu_{12}}{E_{11}} + \frac{1}{E_{22}} - \frac{1}{G_{12}}\right)\sin^2 2\theta\right], \quad (3.46)$$

$$\nu_{yx} = \frac{E_{yy}}{E_{xx}}\nu_{xy}, \quad (3.47)$$

where E_{11}, E_{22}, ν_{12}, and G_{12} are calculated using Equations 3.36 through 3.40.

Figure 3.33 shows the variation of E_{xx} as a function of fiber orientation angle θ for an angle-ply lamina. Note that at $\theta = 0°$, E_{xx} is equal to E_{11}, and at $\theta = 90°$, E_{xx} is equal to E_{22}. Depending on the shear modulus G_{12}, E_{xx} can be

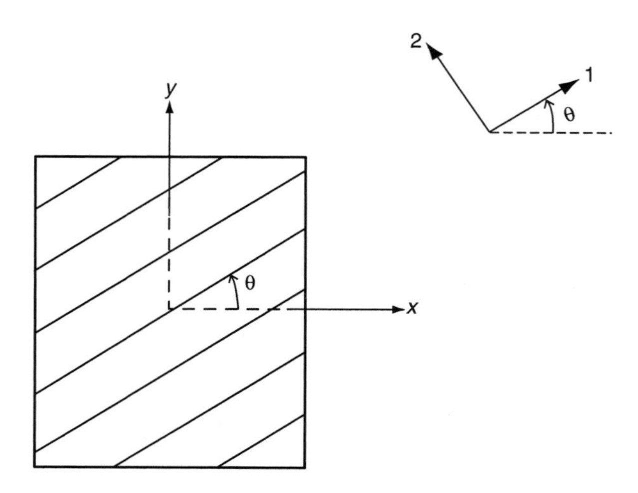

FIGURE 3.32 Unidirectional continuous fiber angle-ply lamina.

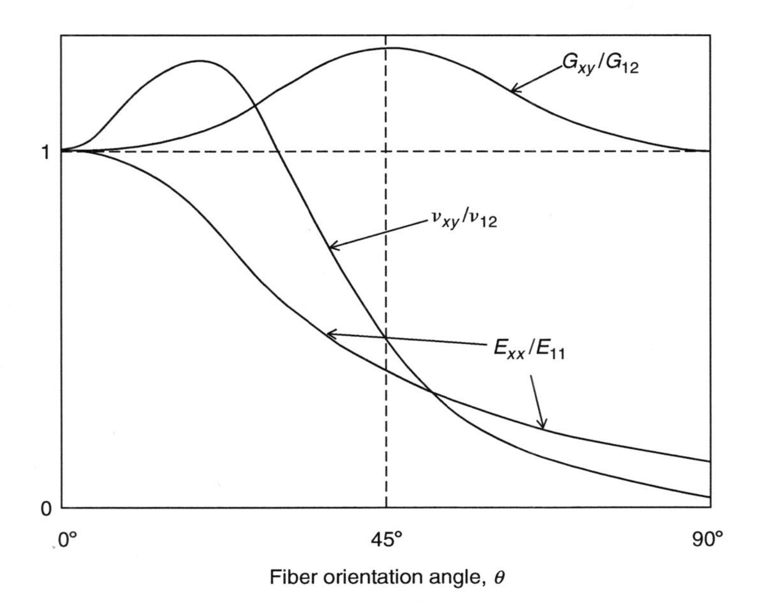

FIGURE 3.33 Variation of elastic constants of continuous E-glass fiber lamina with fiber-orientation angle.

either greater than E_{11} or less than E_{22} at some intermediate values of θ. The range of G_{12} for which E_{xx} is within E_{11} and E_{22} [20] is given by

$$\frac{E_{11}}{2(1+\nu_{12})} > G_{12} > \frac{E_{11}}{2\left(\frac{E_{11}}{E_{22}}+\nu_{12}\right)}. \tag{3.48}$$

For glass fiber–epoxy, high-strength carbon fiber–epoxy, and Kevlar 49 fiber–epoxy composites, G_{12} is within the range given by Equation 3.48, and therefore, for these composite laminas, $E_{22} < E_{xx} < E_{11}$. However, for very high-modulus carbon fiber–epoxy and boron fiber–epoxy composites, G_{12} is less than the lower limit in Equation 3.48, and therefore for a range of angles between 0° and 90°, E_{xx} for these laminas can be lower than E_{22}.

3.2.2.3 Unidirectional Discontinuous Fiber 0° Lamina

Elastic properties of a unidirectional discontinuous fiber 0° lamina are calculated using the following equations (Figure 3.34).

Longitudinal modulus:

$$E_{11} = \frac{1+2(l_\mathrm{f}/d_\mathrm{f})\eta_\mathrm{L}\mathrm{v}_\mathrm{f}}{1-\eta_\mathrm{L}\mathrm{v}_\mathrm{f}}E_\mathrm{m}, \tag{3.49}$$

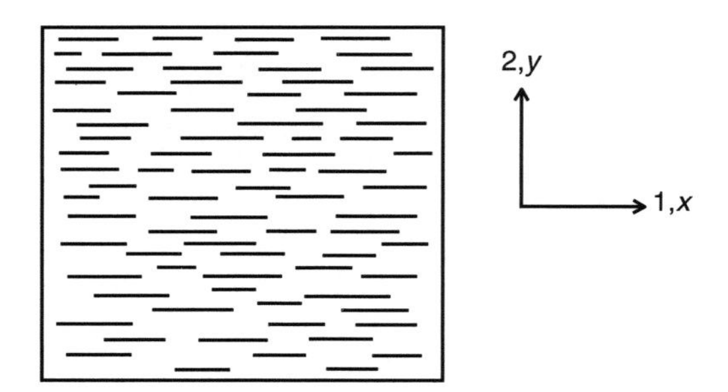

FIGURE 3.34 Unidirectional discontinuous fiber 0° lamina.

Transverse modulus:

$$E_{22} = \frac{1 + 2\eta_{\mathrm{T}} \mathrm{v_f}}{1 - \eta_{\mathrm{T}} \mathrm{v_f}} E_{\mathrm{m}}, \tag{3.50}$$

Shear modulus:

$$G_{12} = G_{21} = \frac{1 + \eta_{\mathrm{G}} \mathrm{v_f}}{1 - \eta_{\mathrm{G}} \mathrm{v_f}} G_{\mathrm{m}}, \tag{3.51}$$

Major Poisson's ratio:

$$\nu_{12} = \nu_{\mathrm{f}} \mathrm{v_f} + \nu_{\mathrm{m}} \mathrm{v_m}, \tag{3.52}$$

Minor Poisson's ratio:

$$\nu_{21} = \frac{E_{22}}{E_{11}} \nu_{12}, \tag{3.53}$$

where

$$\begin{aligned} \eta_{\mathrm{L}} &= \frac{(E_{\mathrm{f}}/E_{\mathrm{m}}) - 1}{(E_{\mathrm{f}}/E_{\mathrm{m}}) + 2(l_{\mathrm{f}}/d_{\mathrm{f}})} \\ \eta_{\mathrm{T}} &= \frac{(E_{\mathrm{f}}/E_{\mathrm{m}}) - 1}{(E_{\mathrm{f}}/E_{\mathrm{m}}) + 2} \\ \eta_{\mathrm{G}} &= \frac{(G_{\mathrm{f}}/G_{\mathrm{m}}) - 1}{(G_{\mathrm{f}}/G_{\mathrm{m}}) + 1} \end{aligned} \tag{3.54}$$

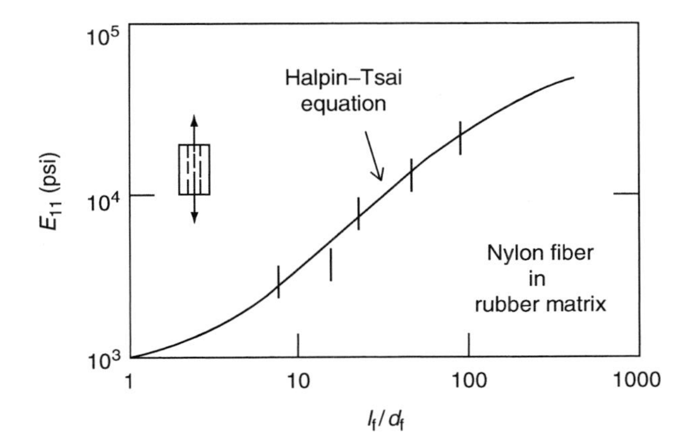

FIGURE 3.35 Variation of longitudinal modulus of a unidirectional discontinuous fiber lamina with fiber length–diameter ratio. (After Halpin, J.C., *J. Compos. Mater.*, 3, 732, 1969.)

Equations 3.49 through 3.53 are derived from the Halpin–Tsai equations (Appendix A.4) with the following assumptions:

1. Fiber cross section is circular.
2. Fibers are arranged in a square array.
3. Fibers are uniformly distributed throughout the matrix.
4. Perfect bonding exists between the fibers and the matrix.
5. Matrix is free of voids.

Fiber aspect ratio, defined as the ratio of average fiber length l_f to fiber diameter d_f, has a significant effect on the longitudinal modulus E_{11} (Figure 3.35). On the other hand, the transverse modulus E_{22} is not affected by the fiber aspect ratio. Furthermore, the longitudinal modulus E_{11} for a discontinuous fiber 0° lamina is always less than that for a continuous fiber 0° lamina.

3.2.2.4 Randomly Oriented Discontinuous Fiber Lamina

A thin lamina containing randomly oriented discontinuous fibers (Figure 3.36) exhibits planar isotropic behavior. The properties are ideally the same in all directions in the plane of the lamina. For such a lamina, the tensile modulus and shear modulus are calculated from

$$E_{\text{random}} = \frac{3}{8}E_{11} + \frac{5}{8}E_{22}, \tag{3.55}$$

FIGURE 3.36 Randomly oriented discontinuous fiber lamina.

$$G_{\text{random}} = \frac{1}{8}E_{11} + \frac{1}{4}E_{22}, \tag{3.56}$$

where E_{11} and E_{22} are the longitudinal and transverse tensile moduli given by Equations 3.49 and 3.50, respectively, for a unidirectional discontinuous fiber 0° lamina of the same fiber aspect ratio and same fiber volume fraction as the randomly oriented discontinuous fiber composite. The Poisson's ratio in the plane of the lamina is

$$\nu_{\text{random}} = \frac{E_{\text{random}}}{2G_{\text{random}}} - 1. \tag{3.57}$$

EXAMPLE 3.4

Consider a sheet molding compound composite, designated SMC-R65, containing E-glass fibers in a thermoset polyester matrix. The following data are known.

For E-glass fiber,

$$E_f = 68.9 \text{ GPa}$$
$$\rho_f = 2.54 \text{ g/cm}^3$$
$$l_f = 25 \text{ mm}$$
$$d_f = 2.5 \text{ mm.}$$

For polyester,

$$E_m = 3.45 \text{ GPa}$$
$$\rho_m = 1.1 \text{ g/cm}^3.$$

Calculate the tensile modulus, shear modulus, and Poisson's ratio for the material.

Solution

Step 1: Calculate the fiber volume fraction v_f.

Fiber weight fraction in SMC-R65 is $w_f = 0.65$. Therefore, from Equation 2.7,

$$v_f = \frac{0.65/2.54}{(0.65/2.54) + (1 - 0.65)/1.1} = 0.446 \text{ or } 44.6\%.$$

Step 2: Calculate E_{11} for a unidirectional lamina containing 44.6 vol% discontinuous fibers of length $l_f = 25$ mm.

$$\frac{E_f}{E_m} = \frac{68.9}{3.45} = 19.97,$$
$$\frac{l_f}{d_f} = \frac{25}{2.5} = 10.$$

Therefore, from Equation 3.54,

$$\eta_L = \frac{19.97 - 1}{19.97 + (2)(10)} = 0.475.$$

Using Equation 3.49, we calculate

$$E_{11} = \frac{1 + (2)(10)(0.475)(0.446)}{1 - (0.475)(0.446)}$$
$$= 22.93 \text{ GPa}.$$

Step 3: Calculate E_{22} for a unidirectional lamina containing 44.6 vol% discontinuous fibers of length $l_f = 25$ mm. From Equation 3.54,

$$\eta_T = \frac{19.97 - 1}{19.97 + 2} = 0.863.$$

Using Equation 3.50, we calculate

$$E_{22} = \frac{1 + (2)(0.863)(0.446)}{1 - (0.863)(0.446)}$$
$$= 9.93 \text{ GPa}.$$

Step 4: Calculate E and G for SMC-R65 using values of E_{11} and E_{22} in Equations 3.55 and 3.56, and then calculate ν using Equation 3.57.

$$E = E_{\text{random}} = \frac{3}{8}E_{11} + \frac{5}{8}E_{22} = 14.81 \text{ GPa},$$
$$G = G_{\text{random}} = \frac{1}{8}E_{11} + \frac{1}{4}E_{22} = 5.35 \text{ GPa},$$
$$\nu = \nu_{\text{random}} = \frac{E}{2G} - 1 = 0.385.$$

3.2.3 Coefficients of Linear Thermal Expansion [21]

For a unidirectional continuous fiber lamina, coefficients of linear thermal expansion in the 0° and 90° directions can be calculated from the following equations:

$$\alpha_{11} = \frac{\alpha_{fl} E_f v_f + \alpha_m E_m v_m}{E_f v_f + E_m v_m} \tag{3.58}$$

and

$$\alpha_{22} = (1 + \nu_f)\frac{(\alpha_{fl} + \alpha_{fr})}{2} v_f + (1 + \nu_m)\alpha_m v_m - \alpha_{11}\nu_{12}, \tag{3.59}$$

where

$\nu_{12} = \nu_f v_f + \nu_m v_m$
α_{fl} = coefficient of linear thermal expansion for the fiber in the longitudinal direction
α_{fr} = coefficient of linear thermal expansion for the fiber in the radial direction
α_m = coefficient of linear thermal expansion for the matrix

Equations 3.58 and 3.59 are plotted in Figure 3.37 as a function of fiber volume fraction for a typical glass fiber-reinforced polymer matrix composite for which $\alpha_m \gg \alpha_f$. It should be noted that the coefficient of linear thermal expansion in such composites is greater in the transverse (90°) direction than in the longitudinal (0°) direction.

If the fibers are at an angle θ with the x direction, the coefficients of thermal expansion in the x and y directions can be calculated using α_{11} and α_{22}:

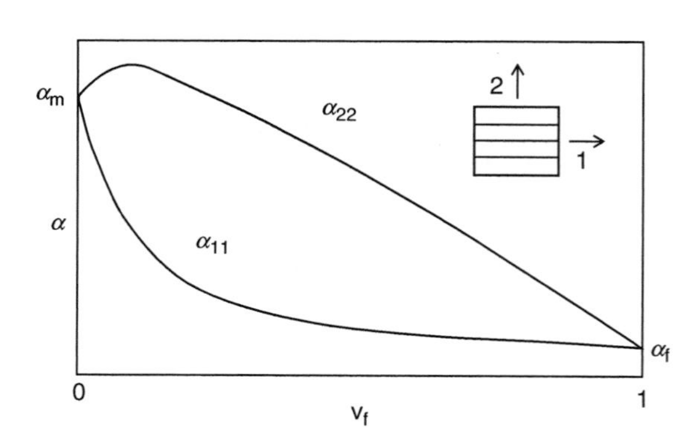

FIGURE 3.37 Variation of longitudinal and transverse coefficients of thermal expansion with fiber volume fraction in a 0° unidirectional continuous E-glass fiber-reinforced epoxy lamina.

$$\alpha_{xx} = \alpha_{11}\cos^2\theta + \alpha_{22}\sin^2\theta,$$
$$\alpha_{yy} = \alpha_{11}\sin^2\theta + \alpha_{22}\cos^2\theta,$$
$$\alpha_{xy} = (2\sin\theta\cos\theta)\,(\alpha_{11} - \alpha_{22}), \tag{3.60}$$

where α_{xx} and α_{yy} are coefficients of linear expansion and α_{xy} is the coefficient of shear expansion. It is important to observe that, unless $\theta = 0°$ or $90°$, a change in temperature produces a shear strain owing to the presence of α_{xy}. The other two coefficients, α_{xx} and α_{yy}, produce extensional strains in the x and y directions, respectively.

3.2.4 Stress–Strain Relationships for a Thin Lamina

3.2.4.1 Isotropic Lamina

For a thin isotropic lamina in plane stress (i.e., $\sigma_{zz} = \tau_{xz} = \tau_{yz} = 0$) (Figure 3.38), the strain–stress relations in the elastic range are

$$\varepsilon_{xx} = \frac{1}{E}(\sigma_{xx} - \nu\sigma_{yy}),$$
$$\varepsilon_{yy} = \frac{1}{E}(-\nu\sigma_{xx} + \sigma_{yy}),$$
$$\gamma_{xy} = \frac{1}{G}\tau_{xy}, \tag{3.61}$$

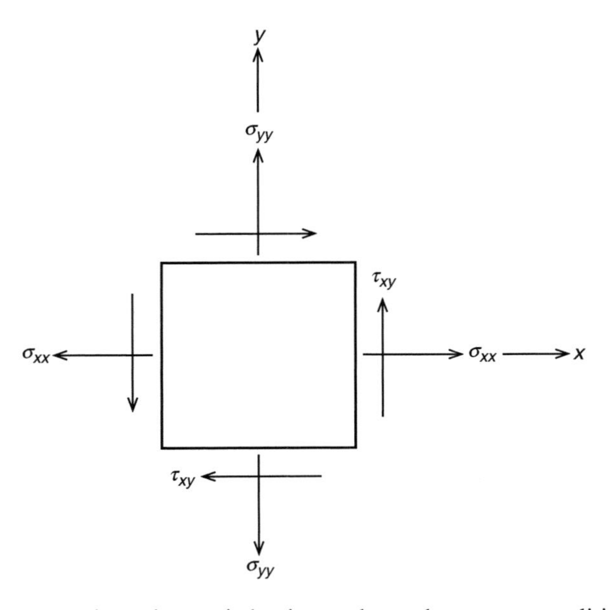

FIGURE 3.38 Stresses in an isotropic lamina under a plane stress condition.

where E, G, and ν represent the Young's modulus, shear modulus, and Poisson's ratio, respectively.

An important point to note in Equation 3.61 is that there is no coupling between the shear stress τ_{xy} and normal stresses σ_{xx} and σ_{yy}. In other words, shear stress τ_{xy} does not influence the normal strains ε_{xx} and ε_{yy} just as the normal stresses σ_{xx} and σ_{yy} do not influence the shear strain γ_{xy}.

3.2.4.2 Orthotropic Lamina

For a thin orthotropic lamina in plane stress ($\sigma_{zz} = \tau_{xz} = \tau_{yz} = 0$) (Figure 3.39), the strain–stress relations in the elastic range are

$$\varepsilon_{xx} = \frac{\sigma_{xx}}{E_{xx}} - \nu_{yx}\frac{\sigma_{yy}}{E_{yy}} - m_x\tau_{xy}, \tag{3.62}$$

$$\varepsilon_{yy} = -\nu_{xy}\frac{\sigma_{xx}}{E_{xx}} + \frac{\sigma_{yy}}{E_{yy}} - m_y\tau_{xy}, \tag{3.63}$$

$$\gamma_{xy} = -m_x\sigma_{xx} - m_y\sigma_{yy} + \frac{\tau_{xy}}{G_{xy}}, \tag{3.64}$$

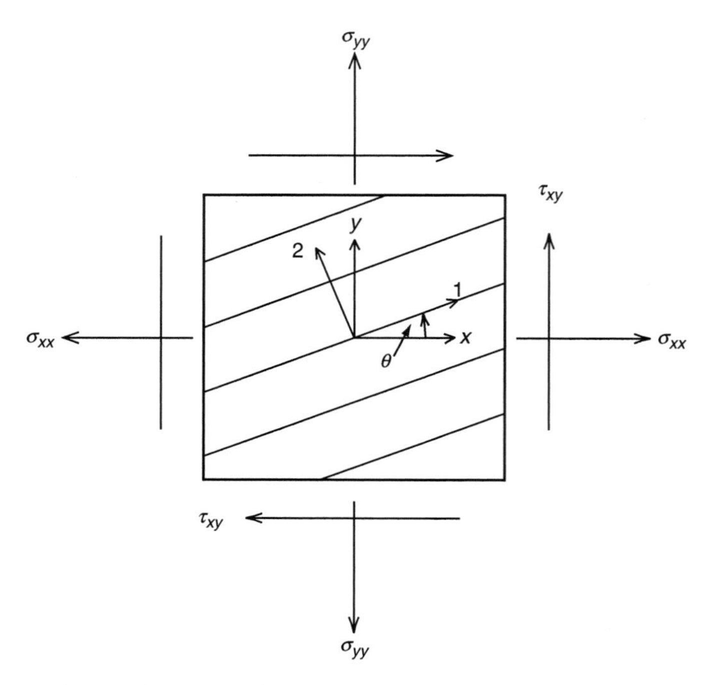

FIGURE 3.39 Stresses in a general orthotropic lamina under a plane stress condition.

where E_{xx}, E_{yy}, G_{xy}, ν_{xy}, and ν_{yx} are elastic constants for the lamina obtained from Equations 3.43 through 3.47 and m_x and m_y are given by the following equations:

$$m_x = (\sin 2\theta)\left[\frac{\nu_{12}}{E_{11}} + \frac{1}{E_{22}} - \frac{1}{2G_{12}} - (\cos^2\theta)\left(\frac{1}{E_{11}} + \frac{2\nu_{12}}{E_{11}} + \frac{1}{E_{22}} - \frac{1}{G_{12}}\right)\right], \quad (3.65)$$

$$m_y = (\sin 2\theta)\left[\frac{\nu_{12}}{E_{11}} + \frac{1}{E_{22}} - \frac{1}{2G_{12}} - (\sin^2\theta)\left(\frac{1}{E_{11}} + \frac{2\nu_{12}}{E_{11}} + \frac{1}{E_{22}} - \frac{1}{G_{12}}\right)\right]. \quad (3.66)$$

The new elastic constants m_x and m_y represent the influence of shear stresses on extensional strains in Equations 3.62 and 3.63 and the influence of normal stresses on shear strain in Equation 3.64. These constants are called *coefficients of mutual influence*.

The following important observations can be made from Equations 3.62 through 3.66:

1. Unlike isotropic lamina, extensional and shear deformations are coupled in a general orthotropic lamina; that is, normal stresses cause both normal strains and shear strains, and shear stress causes both shear strain and normal strains. The effects of such extension-shear coupling phenomena are demonstrated in Figure 3.27c.
2. For $\theta = 0°$ and 90°, both m_x and m_y are zero, and therefore, for these fiber orientations, there is no extension-shear coupling. Such a lamina, in which the principal material axes (1 and 2 axes) coincide with the loading axes (x and y axes), is called *specially orthotropic*. For a specially orthotropic lamina (Figure 3.40), the strain–stress relations are

$$\varepsilon_{xx} = \varepsilon_{11} = \frac{\sigma_{xx}}{E_{11}} - \nu_{21}\frac{\sigma_{yy}}{E_{22}}, \quad (3.67)$$

$$\varepsilon_{yy} = \varepsilon_{22} = -\nu_{12}\frac{\sigma_{xx}}{E_{11}} + \frac{\sigma_{yy}}{E_{22}}, \quad (3.68)$$

$$\gamma_{xy} = \gamma_{yx} = \gamma_{12} = \gamma_{21} = \frac{\tau_{xy}}{G_{12}}. \quad (3.69)$$

3. Both m_x and m_y are functions of the fiber orientation angle θ and exhibit maximum values at an intermediate angle between $\theta = 0°$ and 90° (Figure 3.41).

A critical point to note is that, unlike isotropic materials, the directions of principal stresses and principal strains do not coincide in a general orthotropic lamina. The only exception is found for specially orthotropic lamina in which principal stresses are in the same direction as the material principal axes.

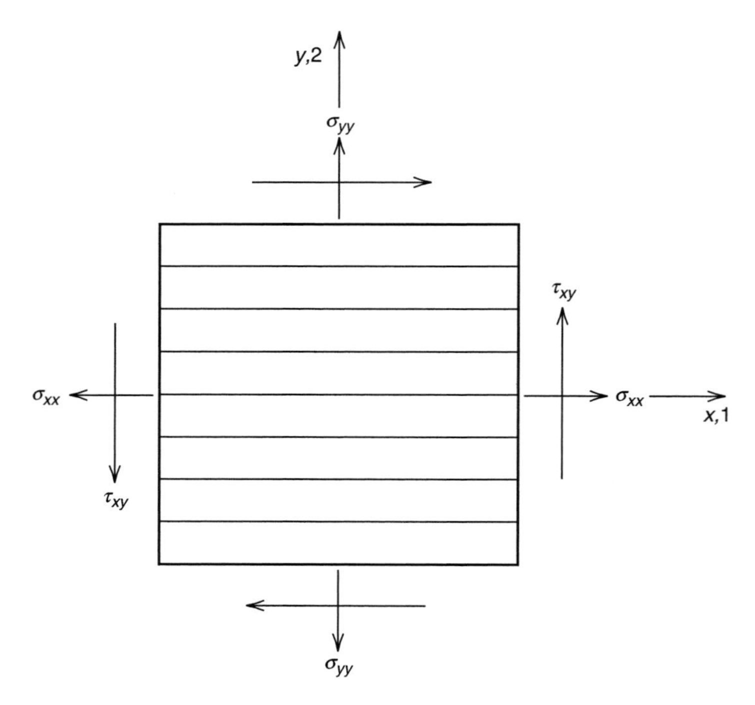

FIGURE 3.40 Stresses in a specially orthotropic lamina under a plane stress condition.

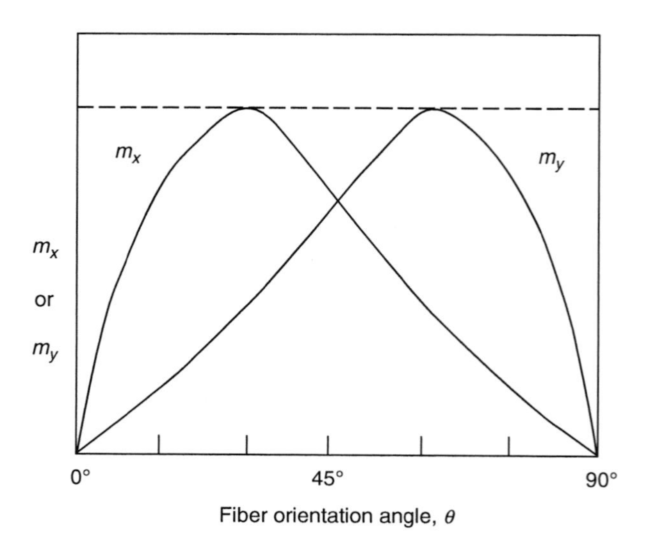

FIGURE 3.41 Variation of coefficients of mutual influence with fiber orientation angle in an E-glass fiber–epoxy lamina.

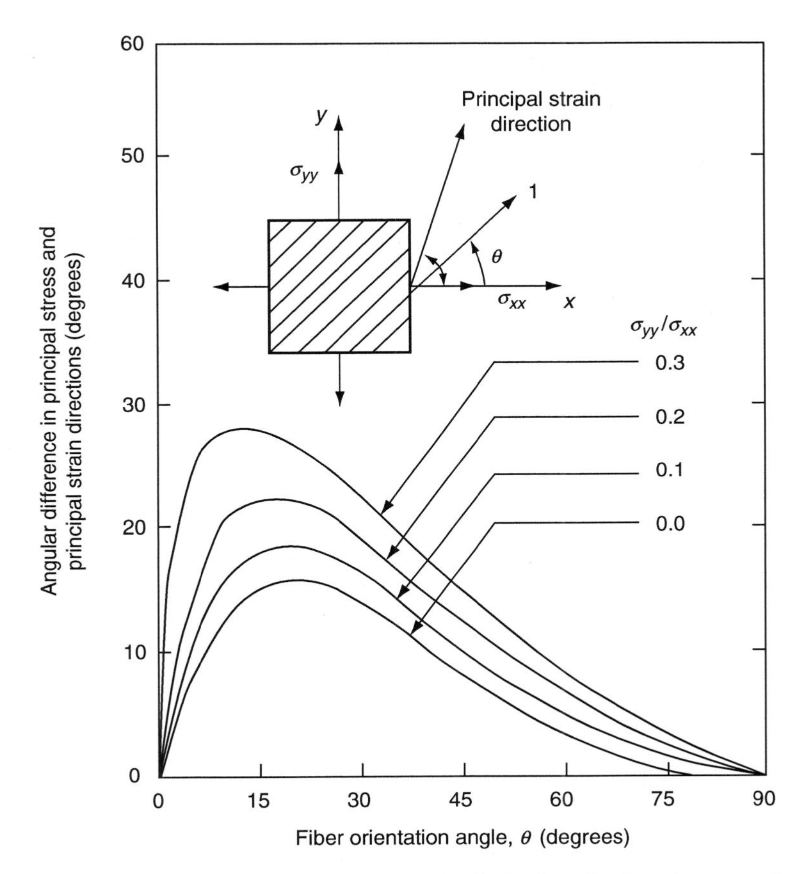

FIGURE 3.42 Difference in principal stress and principal strain directions as a function of fiber orientation angle in an E-glass–epoxy composite ($E_{11}/E_{22} = 2.98$). Note that, for the biaxial normal stress condition shown in this figure, σ_{xx} and σ_{yy} represent the principal stresses σ_1 and σ_2, respectively. (After Greszczuk, L.B., *Orientation Effects in the Mechanical Behavior of Anisotropic Structural Materials, ASTM STP*, 405, 1, 1966.)

Greszczuk [22] has shown that the difference between the principal stress and principal strain directions is a function of the material orthotropy (i.e., the ratio E_{11}/E_{22}) as well as the ratio of the two principal stresses (i.e., the ratio σ_2/σ_1, Figure 3.42).

EXAMPLE 3.5

A thin plate is subjected to a biaxial stress field of $\sigma_{xx} = 1$ GPa and $\sigma_{yy} = 0.5$ GPa. Calculate the strains in the *xy* directions if the plate is made of (a) steel, (b) a 0° unidirectional boron–epoxy composite, and (c) a 45° unidirectional boron–epoxy composite.

Use the elastic properties of the boron–epoxy composite given in Appendix A.5.

Solution

1. Using $E = 207$ GPa and $\nu = 0.33$ for steel in Equation 3.61, we obtain

$$\varepsilon_{xx} = \frac{1}{207}[1 - (0.33)\,(0.5)] = 4.034 \times 10^{-3},$$
$$\varepsilon_{yy} = \frac{1}{207}[-(0.33)\,(1) + 0.5] = 0.821 \times 10^{-3},$$
$$\gamma_{xy} = 0.$$

2. For the 0° unidirectional boron–epoxy (from Appendix A.5):

$$E_{11} = 207 \text{ GPa (same as steel's modulus)}$$
$$E_{22} = 19 \text{ GPa}$$
$$\nu_{12} = 0.21$$
$$G_{12} = 6.4 \text{ GPa.}$$

We first calculate ν_{21}:

$$\nu_{21} = (0.21)\frac{19}{207} = 0.0193.$$

Since 0° unidirectional boron–epoxy is a specially orthotropic lamina, we use Equations 3.67 through 3.69 to obtain

$$\varepsilon_{xx} = \frac{1}{207} - (0.0193)\frac{0.5}{19} = 4.323 \times 10^{-3},$$
$$\varepsilon_{yy} = -(0.21)\frac{1}{207} + \frac{0.5}{19} = 25.302 \times 10^{-3},$$
$$\gamma_{xy} = 0.$$

3. We first need to calculate the elastic constants of the 45° boron–epoxy laminate using Equations 3.43 through 3.47:

$$E_{xx} = E_{yy} = 18.896 \text{ GPa},$$
$$\nu_{xy} = \nu_{yx} = 0.476.$$

Next, we calculate the coefficients of mutual influence using Equations 3.65 and 3.66:

$$m_x = m_y = 0.0239 \text{ GPa}^{-1}.$$

Now, we use Equations 3.62 through 3.64 to calculate:

$$\begin{aligned}
\varepsilon_{xx} &= \frac{1}{18.896} - (0.476)\frac{0.5}{18.896} = 40.326 \times 10^{-3}, \\
\varepsilon_{yy} &= -(0.476)\frac{1}{18.896} + \frac{0.5}{18.896} = 1.270 \times 10^{-3}, \\
\gamma_{xy} &= -(0.0239)\ (1 + 0.5) = -35.85 \times 10^{-3}.
\end{aligned}$$

Note that although the shear stress is zero, there is a shear strain due to extension-shear coupling. This causes a distortion of the plate in addition to the extensions due to ε_{xx} and ε_{yy} as shown in the figure. In addition, note that a negative shear strain means that the initial 90° angle between the adjacent edges of the stress element is increased.

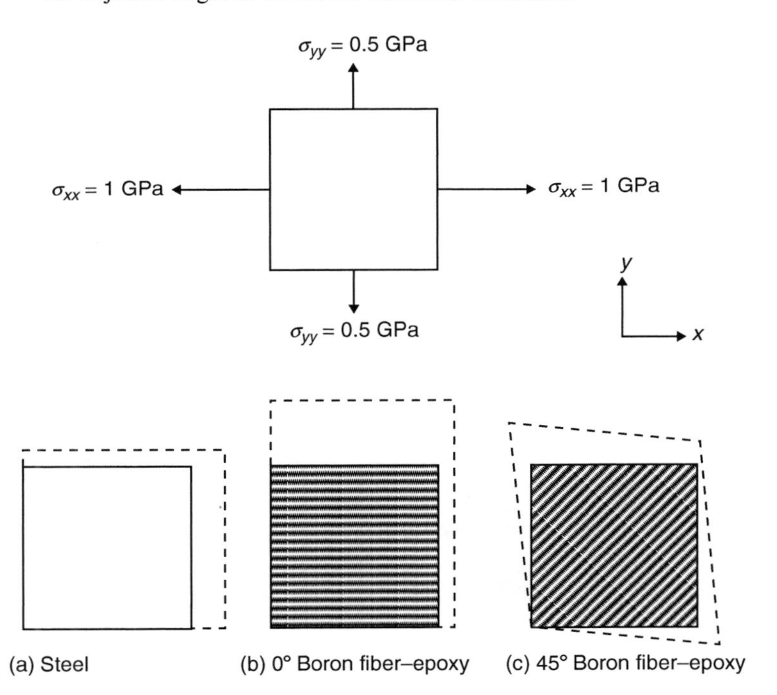

3.2.5 Compliance and Stiffness Matrices

3.2.5.1 Isotropic Lamina

For an isotropic lamina, Equation 3.61 can be written in the matrix form as

$$\begin{bmatrix} \varepsilon_{xx} \\ \varepsilon_{yy} \\ \gamma_{xy} \end{bmatrix} = \begin{bmatrix} \frac{1}{E} & -\frac{\nu}{E} & 0 \\ -\frac{\nu}{E} & \frac{1}{E} & 0 \\ 0 & 0 & \frac{1}{G} \end{bmatrix} \begin{bmatrix} \sigma_{xx} \\ \sigma_{yy} \\ \tau_{xy} \end{bmatrix} = [S] \begin{bmatrix} \sigma_{xx} \\ \sigma_{yy} \\ \tau_{xy} \end{bmatrix}, \tag{3.70}$$

where $[S]$ represents the *compliance matrix* relating strains to known stresses. The inverse of the compliance matrix is called the *stiffness matrix*, which is used in relating stresses to known strains. Thus, the stiffness matrix $[Q]$ for an isotropic lamina is

$$[Q] = [S]^{-1} = \begin{bmatrix} \frac{E}{1-\nu^2} & \frac{\nu E}{1-\nu^2} & 0 \\ \frac{\nu E}{1-\nu^2} & \frac{E}{1-\nu^2} & 0 \\ 0 & 0 & G \end{bmatrix}. \tag{3.71}$$

3.2.5.2 Specially Orthotropic Lamina ($\theta = 0°$ or $90°$)

Arranging Equations 3.67 through 3.69 in matrix form, we can write the strain–stress relation for a specially orthotropic lamina as

$$\begin{bmatrix} \varepsilon_{xx} \\ \varepsilon_{yy} \\ \gamma_{xy} \end{bmatrix} = \begin{bmatrix} S_{11} & S_{12} & 0 \\ S_{21}(=S_{12}) & S_{22} & 0 \\ 0 & 0 & S_{66} \end{bmatrix} \begin{bmatrix} \sigma_{xx} \\ \sigma_{yy} \\ \tau_{xy} \end{bmatrix} = [S] \begin{bmatrix} \sigma_{xx} \\ \sigma_{yy} \\ \tau_{xy} \end{bmatrix}, \tag{3.72}$$

where

$$\begin{aligned} S_{11} &= \frac{1}{E_{11}} \\ S_{12} = S_{21} &= -\frac{\nu_{12}}{E_{11}} = -\frac{\nu_{21}}{E_{22}} \\ S_{22} &= \frac{1}{E_{22}} \\ S_{66} &= \frac{1}{G_{12}} \end{aligned} \tag{3.73}$$

The $[S]$ matrix is the compliance matrix for the specially orthotropic lamina. Inverting Equation 3.72, we can write the stress–strain relations for a specially orthotropic lamina as

$$\begin{bmatrix} \sigma_{xx} \\ \sigma_{yy} \\ \tau_{xy} \end{bmatrix} = \begin{bmatrix} Q_{11} & Q_{12} & 0 \\ Q_{21}(=Q_{12}) & Q_{22} & 0 \\ 0 & 0 & Q_{66} \end{bmatrix} \begin{bmatrix} \varepsilon_{xx} \\ \varepsilon_{yy} \\ \gamma_{xy} \end{bmatrix} = [Q] \begin{bmatrix} \varepsilon_{xx} \\ \varepsilon_{yy} \\ \gamma_{xy} \end{bmatrix}, \tag{3.74}$$

where $[Q]$ represents the stiffness matrix for the specially orthotropic lamina. Various elements in the $[Q]$ matrix are

$$
\begin{aligned}
Q_{11} &= \frac{E_{11}}{1-\nu_{12}\nu_{21}},\\
Q_{22} &= \frac{E_{22}}{1-\nu_{12}\nu_{21}},\\
Q_{12} = Q_{21} &= \frac{\nu_{12}E_{22}}{1-\nu_{12}\nu_{21}} = \frac{\nu_{21}E_{11}}{1-\nu_{12}\nu_{21}},\\
Q_{66} &= G_{12}.
\end{aligned} \tag{3.75}
$$

3.2.5.3 General Orthotropic Lamina ($\theta \neq 0°$ or $90°$)

The strain–stress relations for a general orthotropic lamina, Equations 3.62 through 3.64, can be expressed in matrix notation as

$$
\begin{bmatrix} \varepsilon_{xx} \\ \varepsilon_{yy} \\ \gamma_{xy} \end{bmatrix} = \begin{bmatrix} \bar{S}_{11} & \bar{S}_{12} & \bar{S}_{16} \\ \bar{S}_{12} & \bar{S}_{22} & \bar{S}_{26} \\ \bar{S}_{16} & \bar{S}_{26} & \bar{S}_{66} \end{bmatrix} \begin{bmatrix} \sigma_{xx} \\ \sigma_{yy} \\ \tau_{xy} \end{bmatrix} = [\bar{S}] \begin{bmatrix} \sigma_{xx} \\ \sigma_{yy} \\ \tau_{xy} \end{bmatrix}, \tag{3.76}
$$

where $[\bar{S}]$ represents the compliance matrix for the lamina. Various elements in the $[\bar{S}]$ matrix are expressed in terms of the elements in the $[S]$ matrix for a specially orthotropic lamina. These expressions are

$$
\begin{aligned}
\bar{S}_{11} &= \frac{1}{E_{xx}} = S_{11}\cos^4\theta + (2S_{12}+S_{66})\sin^2\theta\cos^2\theta + S_{22}\sin^4\theta,\\
\bar{S}_{12} &= -\frac{\nu_{xy}}{E_{xx}} = S_{12}(\sin^4\theta+\cos^4\theta) + (S_{11}+S_{22}-S_{66})\sin^2\theta\cos^2\theta,\\
\bar{S}_{22} &= \frac{1}{E_{yy}} = S_{11}\sin^4\theta + (2S_{12}+S_{66})\sin^2\theta\cos^2\theta + S_{22}\cos^4\theta,\\
\bar{S}_{16} &= -m_x = (2S_{11}-2S_{12}-S_{66})\sin\theta\cos^3\theta - (2S_{22}-2S_{12}-S_{66})\sin^3\theta\cos\theta,\\
\bar{S}_{26} &= -m_y = (2S_{11}-2S_{12}-S_{66})\sin^3\theta\cos\theta - (2S_{22}-2S_{12}-S_{66})\sin\theta\cos^3\theta,\\
\bar{S}_{66} &= \frac{1}{G_{xy}} = 2(2S_{11}+2S_{22}-4S_{12}-S_{66})\sin^2\theta\cos^2\theta + S_{66}(\sin^4\theta+\cos^4\theta).
\end{aligned} \tag{3.77}
$$

On substitution for S_{11}, S_{12}, and so on, into Equation 3.77, we obtain the same equations as Equations 3.43 through 3.46 for E_{xx}, E_{yy}, G_{xy}, and ν_{xy}, and Equations 3.65 and 3.66 for m_x and m_y.

Inverting Equation 3.76, the stress–strain relations for a general orthotropic lamina can be written as

$$
\begin{bmatrix} \sigma_{xx} \\ \sigma_{yy} \\ \tau_{xy} \end{bmatrix} = \begin{bmatrix} \bar{Q}_{11} & \bar{Q}_{12} & \bar{Q}_{16} \\ \bar{Q}_{12} & \bar{Q}_{22} & \bar{Q}_{26} \\ \bar{Q}_{16} & \bar{Q}_{26} & \bar{Q}_{66} \end{bmatrix} \begin{bmatrix} \varepsilon_{xx} \\ \varepsilon_{yy} \\ \gamma_{xy} \end{bmatrix} = [\bar{Q}] \begin{bmatrix} \sigma_{xx} \\ \sigma_{yy} \\ \gamma_{xy} \end{bmatrix}, \tag{3.78}
$$

where $[\bar{Q}]$ represents the stiffness matrix for the lamina. Various elements in the $[\bar{Q}]$ matrix are expressed in terms of the elements in the $[Q]$ matrix as

$$\begin{aligned}
\bar{Q}_{11} &= Q_{11}\cos^4\theta + 2(Q_{12} + 2Q_{66})\sin^2\theta\cos^2\theta + Q_{22}\sin^4\theta,\\
\bar{Q}_{12} &= Q_{12}(\sin^4\theta + \cos^4\theta) + (Q_{11} + Q_{22} - 4Q_{66})\sin^2\theta\cos^2\theta,\\
\bar{Q}_{22} &= Q_{11}\sin^4\theta + 2(Q_{12} + 2Q_{66})\sin^2\theta\cos^2\theta + Q_{22}\cos^4\theta,\\
\bar{Q}_{16} &= (Q_{11} - Q_{12} - 2Q_{66})\sin\theta\cos^3\theta + (Q_{12} - Q_{22} + 2Q_{66})\sin^3\theta\cos\theta,\\
\bar{Q}_{26} &= (Q_{11} - Q_{12} - 2Q_{66})\sin^3\theta\cos\theta + (Q_{12} - Q_{22} + 2Q_{66})\sin\theta\cos^3\theta,\\
\bar{Q}_{66} &= (Q_{11} + Q_{22} - 2Q_{12} - 2Q_{66})\sin^2\theta\cos^2\theta + Q_{66}(\sin^4\theta + \cos^4\theta).
\end{aligned} \tag{3.79}$$

In using Equations 3.77 and 3.79, the following points should be noted:

1. Elements $\bar{S}_{16}$ and $\bar{S}_{26}$ in the $[\bar{S}]$ matrix or $\bar{Q}_{16}$ and $\bar{Q}_{26}$ in the $[\bar{Q}]$ matrix represent extension-shear coupling.
2. From Equation 3.77 or 3.79, it appears that there are six elastic constants that govern the stress–strain behavior of a lamina. However, a closer examination of these equations would indicate that $\bar{S}_{16}$ and $\bar{S}_{26}$ (or $\bar{Q}_{16}$ and $\bar{Q}_{26}$) are linear combinations of the four basic elastic constants, namely, $\bar{S}_{11}$, $\bar{S}_{12}$, $\bar{S}_{22}$, and $\bar{S}_{66}$, and therefore are not independent.
3. Elements in both the $[\bar{S}]$ and $[\bar{Q}]$ matrices are expressed in terms of the properties in the principal material directions, namely, E_{11}, E_{22}, G_{12}, and ν_{12}, which can be either experimentally determined or predicted from the constituent properties using Equations 3.36 through 3.40.
4. Elements in the $[\bar{Q}]$ and $[\bar{S}]$ matrices can be expressed in terms of five invariant properties of the lamina, as shown below.

Using trigonometric identities, Tsai and Pagano [23] have shown that the elements in the $[\bar{Q}]$ matrix can be written as

$$\begin{aligned}
\bar{Q}_{11} &= U_1 + U_2\cos 2\theta + U_3\cos 4\theta,\\
\bar{Q}_{12} &= \bar{Q}_{21} = U_4 - U_3\cos 4\theta,\\
\bar{Q}_{22} &= U_1 - U_2\cos 2\theta + U_3\cos 4\theta,\\
\bar{Q}_{16} &= \frac{1}{2}U_2\sin 2\theta + U_3\sin 4\theta,\\
\bar{Q}_{26} &= \frac{1}{2}U_2\sin 2\theta - U_3\sin 4\theta,\\
\bar{Q}_{66} &= U_5 - U_3\cos 4\theta,
\end{aligned} \tag{3.80}$$

where U_1 through U_5 represent angle-invariant stiffness properties of a lamina and are given as

$$
\begin{aligned}
U_1 &= \frac{1}{8}(3Q_{11} + 3Q_{22} + 2Q_{12} + 4Q_{66}),\\
U_2 &= \frac{1}{2}(Q_{11} - Q_{22}),\\
U_3 &= \frac{1}{8}(Q_{11} + Q_{22} - 2Q_{12} - 4Q_{66}),\\
U_4 &= \frac{1}{8}(Q_{11} + Q_{22} + 6Q_{12} - 4Q_{66}),\\
U_5 &= \frac{1}{2}(U_1 - U_4).
\end{aligned} \tag{3.81}
$$

It is easy to observe from Equation 3.80 that for fiber orientation angles θ and $-\theta$,

$$
\begin{aligned}
\bar{Q}_{11}(-\theta) &= \bar{Q}_{11}(\theta),\\
\bar{Q}_{12}(-\theta) &= \bar{Q}_{12}(\theta),\\
\bar{Q}_{22}(-\theta) &= \bar{Q}_{22}(\theta),\\
\bar{Q}_{66}(-\theta) &= \bar{Q}_{66}(\theta),\\
\bar{Q}_{16}(-\theta) &= -\bar{Q}_{16}(\theta),\\
\bar{Q}_{26}(-\theta) &= -\bar{Q}_{26}(\theta).
\end{aligned}
$$

Similar expressions for the elements in the $[\bar{S}]$ matrix are

$$
\begin{aligned}
\bar{S}_{11} &= V_1 + V_2\cos 2\theta + V_3\cos 4\theta,\\
\bar{S}_{12} &= \bar{S}_{21} = V_4 - V_3\cos 4\theta,\\
\bar{S}_{22} &= V_1 - V_2\cos 2\theta + V_3\cos 4\theta,\\
\bar{S}_{16} &= V_2\sin 2\theta + 2V_3\sin 4\theta,\\
\bar{S}_{26} &= V_2\sin 2\theta - 2V_3\sin 4\theta,\\
\bar{S}_{66} &= V_5 - 4V_3\cos 4\theta,
\end{aligned} \tag{3.82}
$$

where

$$
\begin{aligned}
V_1 &= \frac{1}{8}(3S_{11} + 3S_{22} + 2S_{12} + S_{66}),\\
V_2 &= \frac{1}{2}(S_{11} - S_{22}),\\
V_3 &= \frac{1}{8}(S_{11} + S_{22} - 2S_{12} - S_{66}),\\
V_4 &= \frac{1}{8}(S_{11} + S_{22} + 6S_{12} - S_{66}),\\
V_5 &= 2(V_1 - V_4).
\end{aligned} \tag{3.83}
$$

These invariant forms are very useful in computing the elements in $[\bar{Q}]$ and $[\bar{S}]$ matrices for a lamina.

EXAMPLE 3.6

Determine the elements in the stiffness matrix for an angle-ply lamina containing 60 vol% of T-300 carbon fibers in an epoxy matrix. Consider fiber orientation angles of both +45° and −45° for the fiber, $E_f = 220$ GPa and $\nu_f = 0.2$, and for the matrix, $E_m = 3.6$ GPa and $\nu_m = 0.35$.

SOLUTION

Step 1: Calculate E_{11}, E_{22}, ν_{12}, ν_{21}, and G_{12} using Equations 3.36 through 3.40.

$$E_{11} = (220)(0.6) + (3.6)(1 - 0.6) = 133.44 \text{ GPa},$$

$$E_{22} = \frac{(220)(3.6)}{(220)(1 - 0.6) + (3.6)(0.6)} = 8.78 \text{ GPa},$$

$$\nu_{12} = (0.2)(0.6) + (0.35)(1 - 0.6) = 0.26,$$

$$\nu_{21} = \frac{8.78}{133.44}(0.26) = 0.017.$$

To calculate G_{12}, we need to know the values of G_f and G_m. Assuming isotropic relationships, we estimate

$$G_f = \frac{E_f}{2(1 + \nu_f)} = \frac{220}{2(1 + 0.2)} = 91.7 \text{ GPa},$$

$$G_m = \frac{E_m}{2(1 + \nu_m)} = \frac{3.6}{2(1 + 0.35)} = 1.33 \text{ GPa}.$$

Therefore,

$$G_{12} = \frac{(91.7)(1.33)}{(91.7)(1 - 0.6) + (1.33)(0.6)} = 3.254 \text{ GPa}.$$

Note that the T-300 carbon fiber is not isotropic, and therefore, the calculation of G_f based on the isotropic assumption will certainly introduce error. Since the actual value of G_f is not always available, the isotropic assumption is often made to calculate G_f.

Step 2: Calculate Q_{11}, Q_{22}, Q_{12}, Q_{21}, and Q_{66} using Equation 3.75.

$$Q_{11} = \frac{133.44}{1 - (0.26)(0.017)} = 134.03 \text{ GPa},$$

$$Q_{22} = \frac{8.78}{1 - (0.26)(0.017)} = 8.82 \text{ GPa},$$

$$Q_{12} = Q_{21} = \frac{(0.26)(8.78)}{1 - (0.26)(0.017)} = 2.29 \text{ GPa},$$

$$Q_{66} = 3.254 \text{ GPa}.$$

Step 3: Calculate U_1, U_2, U_3, U_4, and U_5 using Equation 3.81.

$$U_1 = \frac{1}{8}[(3)(134.03) + (3)(8.82) + (2)(2.29) + (4)(3.254)] = 55.77 \text{ GPa},$$

$$U_2 = \frac{1}{2}(134.03 - 8.82) = 62.6 \text{ GPa},$$

$$U_3 = \frac{1}{8}[134.03 + 8.82 - (2)(2.29) - (4)(3.254)] = 15.66 \text{ GPa},$$

$$U_4 = \frac{1}{8}[134.03 + 8.82 + (6)(2.29) - (4)(3.259)] = 17.95 \text{ GPa},$$

$$U_5 = \frac{1}{2}(55.77 - 17.95) = 18.91 \text{ GPa}.$$

Step 4: Calculate $\bar{Q}_{11}$, $\bar{Q}_{22}$, $\bar{Q}_{12}$, $\bar{Q}_{16}$, $\bar{Q}_{26}$, and $\bar{Q}_{66}$ using Equation 3.80. For a $\theta = +45°$ lamina,

$$\bar{Q}_{11} = 55.77 + (62.6)\cos 90° + (15.66)\cos 180° = 40.11 \text{ GPa},$$

$$\bar{Q}_{22} = 55.77 - (62.6)\cos 90° + (15.66)\cos 180° = 40.11 \text{ GPa},$$

$$\bar{Q}_{12} = 17.95 - (15.66)\cos 180° = 33.61 \text{ GPa},$$

$$\bar{Q}_{66} = 18.91 - (15.66)\cos 180° = 34.57 \text{ GPa},$$

$$\bar{Q}_{16} = \frac{1}{2}(62.6)\sin 90° + (15.66)\sin 180° = 31.3 \text{ GPa},$$

$$\bar{Q}_{26} = \frac{1}{2}(62.6)\sin 90° - (15.66)\sin 180° = 31.3 \text{ GPa}.$$

Similarly, for a $\theta = -45°$ lamina,

$$\bar{Q}_{11} = 40.11 \text{ GPa},$$

$$\bar{Q}_{22} = 40.11 \text{ GPa},$$

$$\bar{Q}_{12} = 33.61 \text{ GPa},$$

$$\bar{Q}_{66} = 34.57 \text{ GPa},$$

$$\bar{Q}_{16} = -31.3 \text{ GPa},$$

$$\bar{Q}_{26} = -31.3 \text{ GPa}.$$

In the matrix form,

$$[\bar{Q}]_{45°} = \begin{bmatrix} 40.11 & 33.61 & 31.3 \\ 33.61 & 40.11 & 31.3 \\ 31.3 & 31.3 & 34.57 \end{bmatrix} \text{ GPa},$$

$$[\bar{Q}]_{-45°} = \begin{bmatrix} 40.11 & 33.61 & -31.3 \\ 33.61 & 40.11 & -31.3 \\ -31.3 & -31.3 & 34.57 \end{bmatrix} \text{ GPa}.$$

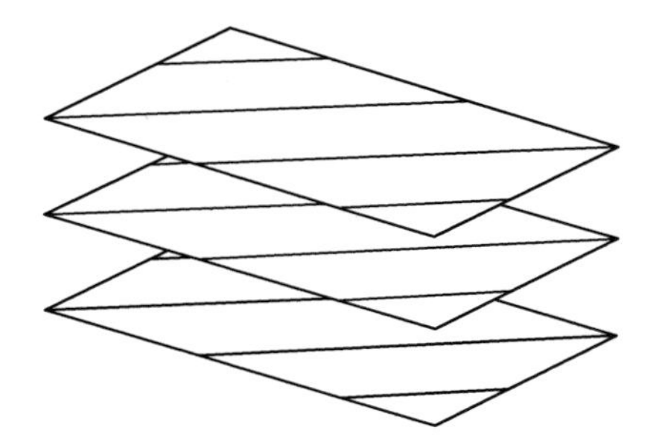

FIGURE 3.43 Unidirectional laminate.

3.3 LAMINATED STRUCTURE

3.3.1 From Lamina to Laminate

A laminate is constructed by stacking a number of laminas in the thickness (z) direction. Examples of a few special types of laminates and the standard lamination code are given as follows:

Unidirectional laminate: In a unidirectional laminate (Figure 3.43), fiber orientation angles are the same in all laminas. In unidirectional 0° laminates, for example, $\theta = 0°$ in all laminas.

Angle-ply laminate: In an angle-ply laminate (Figure 3.44), fiber orientation angles in alternate layers are $/\theta/-\theta/\theta/-\theta/$ when $\theta \neq 0°$ or 90°.

Cross-ply laminate: In a cross-ply laminate (Figure 3.45), fiber orientation angles in alternate layers are $/0°/90°/0°/90°/$.

Symmetric laminate: In a symmetric laminate, the ply orientation is symmetrical about the centerline of the laminate; that is, for each ply above the midplane, there is an identical ply (in material, thickness, and fiber orientation angle) at an equal distance below the midplane. Thus, for a symmetric laminate,

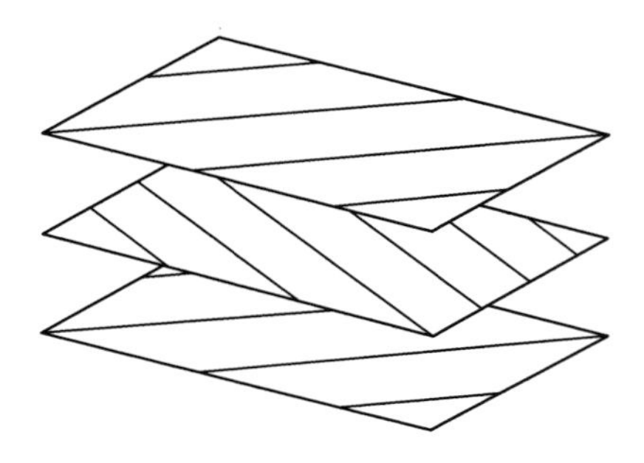

FIGURE 3.44 Angle-ply laminate.

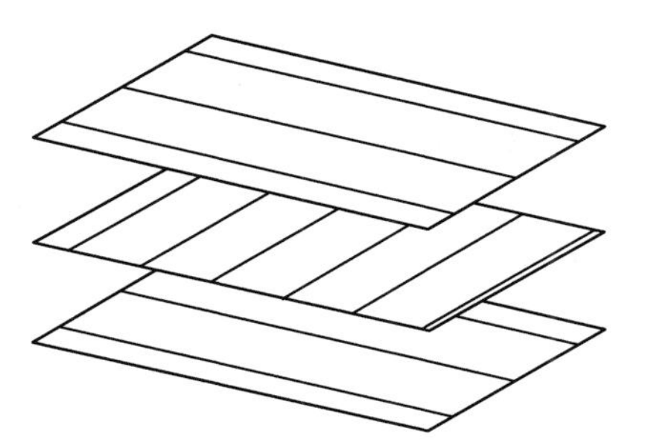

FIGURE 3.45 Cross-ply laminate.

$$\theta(z) = \theta(-z),$$

where z is the distance from the midplane of the laminate. Some examples of symmetric laminates and their codes are listed.

1. $\quad$ 1 2 3 4 5 6
   [0/+45/90/90 +45/0]
   Code: $[0/45/90]_S$

Subscript S in the code indicates symmetry about the midplane.

2. $\quad$ 1 2 3 4 5
   [0/+45/90/+45/0]
   Code: $[0/45/\overline{90}]_S$

The bar over 90 indicates that the plane of symmetry passes midway through the thickness of the 90° lamina.

3. $\quad$ 1 2 3 4 5 6 7
   [0/+45/−45/90/−45/+45/0]
   Code: $[0/\pm45/\overline{90}]_S$

Adjacent +45° and −45° laminas are grouped as ±45°.

4. $\quad$ 1 2 3 4 5 6 7 8 9 10 11 12 13 14
   [0 / 90 / 0 / 0 / 0 / 0 / 45 / 45 / 0 / 0 / 0 / 0 / 90 / 0]
   Code: $[0/90/0_4/45]_S$

Four adjacent 0° plies are grouped together as 0_4.

5. $\quad$ 1 2 3 4 5 6 7 8 9 10
   [0/45/−45/+45/−45/−45/+45/−45/+45/0]
   Code: $[0/(\pm45)_2]_S$

Two adjacent ±45° plies are grouped as $(\pm45)_2$.

6. $[0/45/-45/45/-45/45/-45/0/0/0/0/0/-45/45/-45/45/-45/45/0]$
   Code: $[0/(\pm 45)_3/0_2/\bar{0}]_S$

7. $[\theta/-\theta/\theta/-\theta/-\theta/\theta/-\theta/\theta]$
   Code: $[\theta/-\theta]_{2S}$ or $[\pm\theta]_{2S}$

   Two adjacent $\pm\theta$ plies on each side of the plane of symmetry are denoted by the subscript 2S.

8. Symmetric angle-ply laminate
   $[\theta/-\theta/\theta/-\theta/\theta/-\theta/\theta]$
   Code: $[\pm\theta/\theta/-\bar{\theta}]_S$

   Note that symmetric angle-ply laminates contain an odd number of plies.

9. Symmetric cross-ply laminate
   $[0/90/0/90/0/90/0/90/0]$
   Code: $[(0/90)_2/\bar{0}]_S$

   Note that symmetric cross-ply laminates contain an odd number of plies.

10. Hybrid (interply) laminate.
   $[0_B/0_B/45_C/-45_C/90_G/90_G/-45_C/45_C/0_B/0_B]$
   Code: $[0_{2B}/(\pm 45)_C/90_G]_S$

   where B, C, and G represent boron, carbon, and glass fiber, respectively.

Antisymmetric laminate: In antisymmetric laminates, the ply orientation is antisymmetric about the centerline of the laminate; that is, for each ply of fiber orientation angle θ above the midplane, there is a ply of fiber orientation angle $-\theta$ with identical material and thickness at an equal distance below the midplane. Thus, for an antisymmetric laminate,

$$\theta(z) = -\theta(-z).$$

For example, $\theta/-\theta/\theta/-\theta$ is an antisymmetric laminate. In contrast, $\theta/-\theta/-\theta/\theta$ is symmetric.

Unsymmetric laminate: In unsymmetric laminates, there is no symmetry or antisymmetry. Examples are 0/0/0/90/90/90 and $0/\theta/-\theta/90$.

Quasi-isotropic laminate: These laminates are made of three or more laminas of identical thickness and material with equal angles between each adjacent lamina. Thus, if the total number of laminas is n, the orientation angles of the laminas are at increments of π/n. The resulting laminate exhibits an in-plane isotropic elastic behavior in the xy plane. However, its strength properties may still vary with the direction of loading. Examples of simple quasi-isotropic laminates are [+60/0/−60] and [+45/0/−45/90]. Other combinations of these stacking sequences, such as [0/+60/−60] and [0/+45/−45/90], also exhibit in-plane isotropic elastic behavior. A very common and widely used quasi-isotropic symmetrical stacking sequence is $[0/\pm45/90]_S$.

3.3.2 Lamination Theory

Lamination theory is useful in calculating stresses and strains in each lamina of a thin laminated structure. Beginning with the stiffness matrix of each lamina, the step-by-step procedure in lamination theory includes

1. Calculation of stiffness matrices for the laminate
2. Calculation of midplane strains and curvatures for the laminate due to a given set of applied forces and moments
3. Calculation of in-plane strains ε_{xx}, ε_{yy}, and γ_{xy} for each lamina
4. Calculation of in-plane stresses σ_{xx}, σ_{yy}, and τ_{xy} in each lamina

The derivation of lamination theory is given in Ref. [16]. The principal equations and a number of examples are presented in the following sections.

3.3.2.1 Assumptions

Basic assumptions in the lamination theory are

1. Laminate is thin and wide (width $\gg$ thickness).
2. A perfect interlaminar bond exists between various laminas.
3. Strain distribution in the thickness direction is linear.
4. All laminas are macroscopically homogeneous and behave in a linearly elastic manner.

The geometric midplane of the laminate contains the xy axes, and the z axis defines the thickness direction. The total thickness of the laminate is h, and the thickness of various laminas are represented by t_1, t_2, t_3, and so on. The total number of laminas is N. A sketch for the laminate is shown in Figure 3.46.

3.3.2.2 Laminate Strains

Following assumption 3, laminate strains are linearly related to the distance from the midplane as

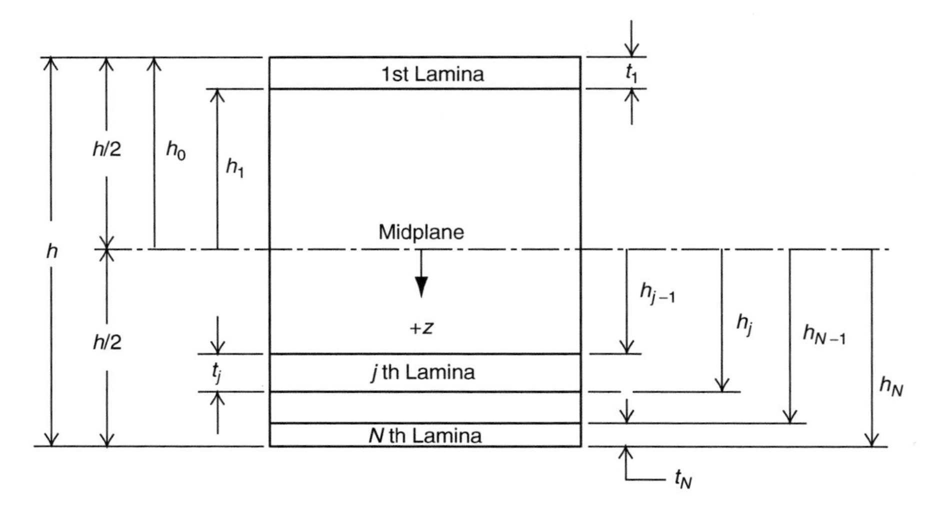

FIGURE 3.46 Laminate geometry.

$$\begin{aligned}
\varepsilon_{xx} &= \varepsilon^{\circ}_{xx} + zk_{xx},\\
\varepsilon_{yy} &= \varepsilon^{\circ}_{yy} + zk_{yy},\\
\gamma_{xy} &= \gamma^{\circ}_{xy} + zk_{xy},
\end{aligned} \tag{3.84}$$

where

ε°_{xx}, ε°_{yy} = midplane normal strains in the laminate
$\gamma^{\circ}{}_{xy}$ = midplane shear strain in the laminate
k_{xx}, k_{yy} = bending curvatures of the laminate
k_{xy} = twisting curvature of the laminate
z = distance from the midplane in the thickness direction

3.3.2.3 Laminate Forces and Moments

Applied force and moment resultant (Figure 3.47) on a laminate are related to the midplane strains and curvatures by the following equations:

$$\begin{aligned}
N_{xx} &= A_{11}\varepsilon^{\circ}_{xx} + A_{12}\varepsilon^{\circ}_{yy} + A_{16}\gamma^{\circ}_{xy} + B_{11}k_{xx} + B_{12}k_{yy} + B_{16}k_{xy},\\
N_{yy} &= A_{12}\varepsilon^{\circ}_{xx} + A_{22}\varepsilon^{\circ}_{yy} + A_{26}\gamma^{\circ}_{xy} + B_{12}k_{xx} + B_{22}k_{yy} + B_{26}k_{xy},\\
N_{xy} &= A_{16}\varepsilon^{\circ}_{xx} + A_{26}\varepsilon^{\circ}_{yy} + A_{66}\gamma^{\circ}_{xy} + B_{16}k_{xx} + B_{26}k_{yy} + B_{66}k_{xy},\\
M_{xx} &= B_{11}\varepsilon^{\circ}_{xx} + B_{12}\varepsilon^{\circ}_{yy} + B_{16}\gamma^{\circ}_{xy} + D_{11}k_{xx} + D_{12}k_{yy} + D_{16}k_{xy},\\
M_{yy} &= B_{12}\varepsilon^{\circ}_{xx} + B_{22}\varepsilon^{\circ}_{yy} + B_{26}\gamma^{\circ}_{xy} + D_{12}k_{xx} + D_{22}k_{yy} + D_{26}k_{xy},\\
M_{xy} &= B_{16}\varepsilon^{\circ}_{xx} + B_{26}\varepsilon^{\circ}_{yy} + B_{66}\gamma^{\circ}_{xy} + D_{16}k_{xx} + D_{26}k_{yy} + D_{66}k_{xy}.
\end{aligned}$$

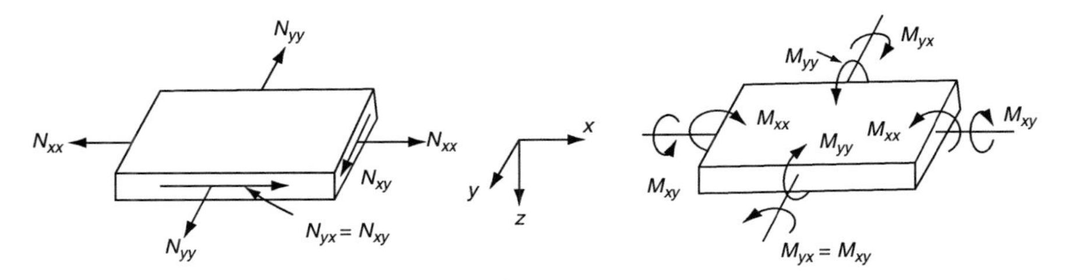

FIGURE 3.47 In-plane, bending, and twisting loads applied on a laminate.

In matrix notation,

$$\begin{bmatrix} N_{xx} \\ N_{yy} \\ N_{xy} \end{bmatrix} = [A]\begin{bmatrix} \varepsilon^{\circ}_{xx} \\ \varepsilon^{\circ}_{yy} \\ \gamma^{\circ}_{xy} \end{bmatrix} + [B]\begin{bmatrix} k_{xx} \\ k_{yy} \\ k_{xy} \end{bmatrix} \tag{3.85}$$

and

$$\begin{bmatrix} M_{xx} \\ M_{yy} \\ M_{xy} \end{bmatrix} = [B]\begin{bmatrix} \varepsilon^{\circ}_{xx} \\ \varepsilon^{\circ}_{yy} \\ \gamma^{\circ}_{xy} \end{bmatrix} + [D]\begin{bmatrix} k_{xx} \\ k_{yy} \\ k_{xy} \end{bmatrix}, \tag{3.86}$$

where

N_{xx} = normal force resultant in the x direction (per unit width)
N_{yy} = normal force resultant in the y direction (per unit width)
N_{xy} = shear force resultant (per unit width)
M_{xx} = bending moment resultant in the yz plane (per unit width)
M_{yy} = bending moment resultant in the xz plane (per unit width)
M_{xy} = twisting moment (torsion) resultant (per unit width)

$[A]$ = extensional stiffness matrix for the laminate (unit: N/m or lb/in.)

$$[A] = \begin{bmatrix} A_{11} & A_{12} & A_{16} \\ A_{12} & A_{22} & A_{26} \\ A_{16} & A_{26} & A_{66} \end{bmatrix}, \tag{3.87}$$

$[B]$ = coupling stiffness matrix for the laminate (unit: N or lb)

$$[B] = \begin{bmatrix} B_{11} & B_{12} & B_{16} \\ B_{12} & B_{22} & B_{26} \\ B_{16} & B_{26} & B_{66} \end{bmatrix}, \tag{3.88}$$

$[D]$ = bending stiffness matrix for the laminate (unit: N m or lb in.)

$$[D] = \begin{bmatrix} D_{11} & D_{12} & D_{16} \\ D_{12} & D_{22} & D_{26} \\ D_{16} & D_{26} & D_{66} \end{bmatrix}. \tag{3.89}$$

Referring to Equation 3.85, it can be observed that

1. A_{16} and A_{26} couple in-plane normal forces to midplane shear strain and in-plane shear force to midplane normal strains.
2. B_{11}, B_{12}, and B_{22} couple in-plane normal forces to bending curvatures and bending moments to midplane normal strains.
3. B_{16} and B_{26} couple in-plane normal forces to twisting curvature and twisting moment to midplane normal strains.
4. B_{66} couples in-plane shear force to twisting curvature and twisting moment to midplane shear strain.
5. D_{16} and D_{26} couple bending moments to twisting curvature and twisting moment to bending curvatures.

The couplings between normal forces and shear strains, bending moments and twisting curvatures, and so on, occur only in laminated structures and not in a monolithic structure. If the laminate is properly constructed, some of these couplings can be eliminated. For example, if the laminate is constructed such that both A_{16} and $A_{26} = 0$, there will be no coupling between in-plane normal forces and midplane shear strains, that is, in-plane normal forces will not cause shear deformation of the laminate. Similarly, if the laminate is constructed such that both D_{16} and $D_{26} = 0$, there will be coupling between bending moments and twisting curvature, that is, bending moments will not cause twisting of the laminate. These special constructions are described in the following section.

3.3.2.4 Elements in Stiffness Matrices

The elements in $[A]$, $[B]$, and $[D]$ matrices are calculated from

$$A_{mn} = \sum_{j=1}^{N} \left(\bar{Q}_{mn}\right)_j \left(h_j - h_{j-1}\right), \tag{3.90}$$

$$B_{mn} = \frac{1}{2} \sum_{j=1}^{N} \left(\bar{Q}_{mn}\right)_j \left(h_j^2 - h_{j-1}^2\right), \tag{3.91}$$

$$D_{\mathrm{mn}} = \frac{1}{3}\sum_{j=1}^{N} (\bar{Q}_{\mathrm{mn}})_j \left(h_j^3 - h_{j-1}^3\right), \tag{3.92}$$

where

N = total number of laminas in the laminate
$(\bar{Q}_{\mathrm{mn}})_j$ = elements in the $[\bar{Q}]$ matrix of the jth lamina
h_{j-1} = distance from the midplane to the top of the jth lamina
h_j = distance from the midplane to the bottom of the jth lamina

For the coordinate system shown in Figure 3.46, h_j is positive below the midplane and negative above the midplane.

The elements of the stiffness matrices [A], [B], and [D] are functions of the elastic properties of each lamina and its location with respect to the midplane of the laminate. The following observations are important regarding these stiffness matrices:

1. If [B] is a nonzero matrix, a normal force, such as N_{xx}, will create extension and shear deformations as well as bending–twisting curvatures. Similarly, a bending moment, such as M_{xx}, will create bending and twisting curvatures as well as extension-shear deformations. Such "extension-bending coupling," represented by the [B] matrix, is unique in laminated structures regardless of whether the layers are isotropic or orthotropic. The coupling occurs because of the stacking of layers.
2. For a *symmetric* laminate, [B] = [0] and there is no extension-bending coupling. To construct a symmetric laminate, every lamina of $+\theta$ orientation above the midplane must be matched with an identical (in thickness and material) lamina of $+\theta$ orientation at the same distance below the midplane (Figure 3.48). Note that a symmetric angle-ply or cross-ply laminate contains an odd number of plies.

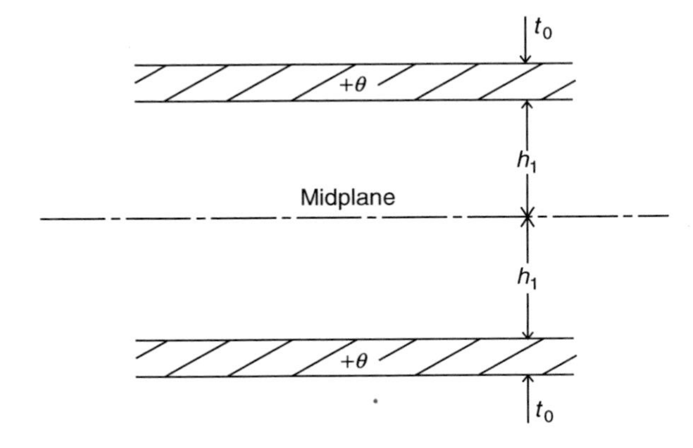

FIGURE 3.48 Symmetric laminate configuration for which [B] = [0], and therefore no extension-bending coupling.

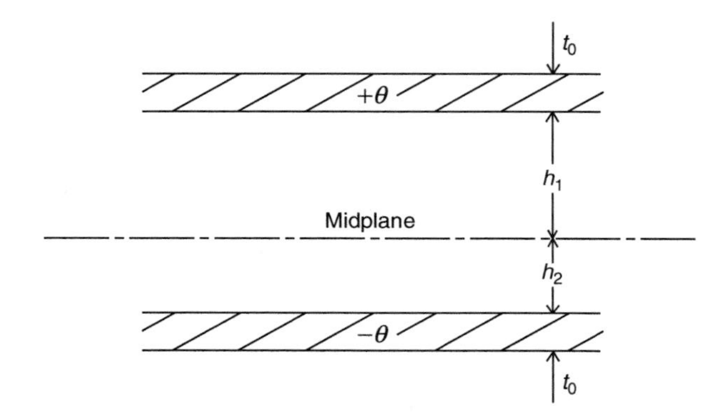

FIGURE 3.49 Balanced laminate configuration for which $A_{16} = A_{26} = 0$, and therefore no extension-shear coupling.

3. If for every lamina of $+\theta$ orientation, there is an identical (equal in thickness and material) lamina of $-\theta$ orientation (Figure 3.49), the normal stress–shear strain coupling (represented by A_{16} and A_{26} in the $[A]$ matrix) for the laminate is zero. The locations of these two laminas are arbitrary. Such a laminate is called *balanced*; for example, [0/+30/−30/+30/−30/0] is a balanced laminate for which $A_{16} = A_{26} = 0$. Note that, with proper positioning of layers, it is possible to prepare a balanced symmetric laminate. For example, [0/+30/−30/−30/+30/0] is a balanced symmetric laminate, for which $A_{16} = A_{26} = 0$ as well as $[B] = [0]$.
4. If for every lamina of $+\theta$ orientation above the midplane, there is an identical lamina (in thickness and material) of $-\theta$ orientation at the same distance below the midplane (Figure 3.50), the bending

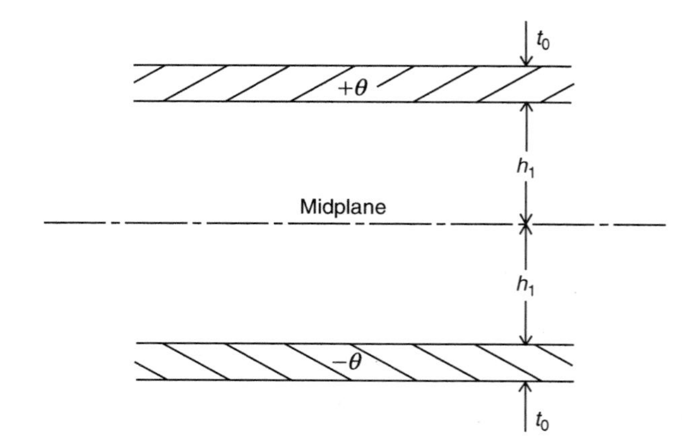

FIGURE 3.50 Laminate configuration for which $D_{16} = D_{26} = 0$, and therefore no bending-twisting coupling.

moment-twisting curvature coupling (represented by D_{16} and D_{26} in the [D] matrix) for the laminate is zero. For example, for a [0/+30/−30/+30/−30/0] laminate, $D_{16} = D_{26} = 0$. Note that the D_{16} and D_{26} terms cannot be zero for a symmetric laminate, unless $\theta = 0°$ and 90°.

EXAMPLE 3.7

Determine [A], [B], and [D] matrices for (a) a [+45/−45] angle-ply laminate, (b) a [+45/−45]$_S$ symmetric laminate, and (c) a [+45/0/−45] unsymmetric laminate. Each lamina is 6 mm thick and contains 60 vol% of T-300 carbon fiber in an epoxy matrix. Use the same material properties as in Example 3.6.

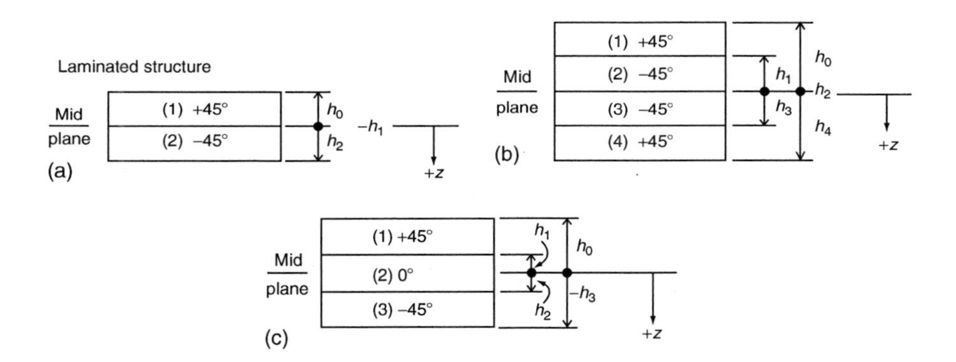

SOLUTION

From Example 3.6, [$\bar{Q}$] matrices for the 0°, +45°, and −45° layers are written as

$$[\bar{Q}]_{0°} = [Q]_{0°} = \begin{bmatrix} 134.03 & 2.29 & 0 \\ 2.29 & 8.82 & 0 \\ 0 & 0 & 3.254 \end{bmatrix} \text{GPa},$$

$$[\bar{Q}]_{+45°} = \begin{bmatrix} 40.11 & 33.61 & 31.3 \\ 33.61 & 40.11 & 31.3 \\ 31.3 & 31.3 & 34.57 \end{bmatrix} \text{GPa},$$

$$[\bar{Q}]_{-45°} = \begin{bmatrix} 40.11 & 33.61 & -31.3 \\ 33.61 & 40.11 & -31.3 \\ -31.3 & -31.3 & 34.57 \end{bmatrix} \text{GPa}.$$

(a) [+45/−45] Angle-ply laminate: From the figure (top left), we note $h_0 = -0.006$ m, $h_1 = 0$, and $h_2 = 0.006$ m. In this laminate, $(\bar{Q}_{mn})_1 = (\bar{Q}_{mn})_{+45°}$ and $(\bar{Q}_{mn})_2 = (\bar{Q}_{mn})_{-45°}$. Therefore,

$$A_{mn} = (\bar{Q}_{mn})_1(h_1 - h_0) + (\bar{Q}_{mn})_2(h_2 - h_1)$$

$$= 6 \times 10^{-3}(\bar{Q}_{mn})_{+45°} + 6 \times 10^{-3}(\bar{Q}_{mn})_{-45°},$$

$$B_{mn} = \frac{1}{2}\left[(\bar{Q}_{mn})_1\left(h_1^2 - h_0^2\right) + (\bar{Q}_{mn})_2\left(h_2^2 - h_1^2\right)\right]$$

$$= -18 \times 10^{-6}(\bar{Q}_{mn})_{+45°} + 18 \times 10^{-6}(\bar{Q}_{mn})_{-45°},$$

$$D_{mn} = \frac{1}{3}\left[(\bar{Q}_{mn})_1\left(h_1^3 - h_0^3\right) + (\bar{Q}_{mn})_2\left(h_2^3 - h_1^3\right)\right]$$

$$= 72 \times 10^{-9}(\bar{Q}_{mn})_{+45°} + 72 \times 10^{-9}(\bar{Q}_{mn})_{-45°}.$$

Substituting for various $(\bar{Q}_{mn})$ values, we calculate

$$[A] = \begin{bmatrix} 481.32 & 403.32 & 0 \\ 403.32 & 481.32 & 0 \\ 0 & 0 & 414.84 \end{bmatrix} \times 10^6 \text{ N/m},$$

$$[B] = \begin{bmatrix} 0 & 0 & -1126.8 \\ 0 & 0 & -1126.8 \\ -1126.8 & -1126.8 & 0 \end{bmatrix} \times 10^3 \text{ N},$$

$$[D] = \begin{bmatrix} 5775.84 & 4839.84 & 0 \\ 4839.84 & 5775.84 & 0 \\ 0 & 0 & 4978.08 \end{bmatrix} \text{N m}.$$

Note that for a [+45/−45] angle-ply laminate, $A_{16} = A_{26} = 0$ (since it is balanced) as well as $D_{16} = D_{26} = 0$.

(b) $[(45/-45)]_S$ Symmetric laminate: From the figure (top right), we note that $h_3 = -h_1 = 0.006$ m, $h_4 = h_0 = 0.012$ m, and $h_2 = 0$. In this laminate,

$$(\bar{Q}_{mn})_4 = (\bar{Q}_{mn})_1 = (\bar{Q}_{mn})_{+45°}$$

and

$$(\bar{Q}_{mn})_3 = (\bar{Q}_{mn})_2 = (\bar{Q}_{mn})_{-45°}.$$

Therefore,

$$\begin{aligned}
A_{mn} &= (\bar{Q}_{mn})_1(h_1 - h_0) + (\bar{Q}_{mn})_2(h_2 - h_1) + (\bar{Q}_{mn})_3(h_3 - h_2) + (\bar{Q}_{mn})_4(h_4 - h_3) \\
&= (\bar{Q}_{mn})_{+45°}(h_1 - h_0 + h_4 - h_3) + (\bar{Q}_{mn})_{-45°}(h_2 - h_1 + h_3 - h_2) \\
&= 12 \times 10^{-3}(\bar{Q}_{mn})_{+45°} + 12 \times 10^{-3}(\bar{Q}_{mn})_{-45°}, \\
B_{mn} &= \frac{1}{2}\left[(\bar{Q}_{mn})_1\left(h_1^2 - h_0^2\right) + (\bar{Q}_{mn})_2\left(h_2^2 - h_1^2\right) + (\bar{Q}_{mn})_3\left(h_3^2 - h_2^2\right) + (\bar{Q}_{mn})_4\left(h_4^2 - h_3^2\right)\right] \\
&= \frac{1}{2}\left[(\bar{Q}_{mn})_{45°}\left(h_1^2 - h_0^2 + h_4^2 - h_3^2\right) + (\bar{Q}_{mn})_{-45°}\left(h_2^2 - h_1^2 + h_3^2 - h_2^2\right)\right] \\
&= 0 \quad \text{since } h_1^2 = h_3^2 \text{ and } h_0^2 = h_4^2, \\
D_{mn} &= \frac{1}{3}\left[(\bar{Q}_{mn})_1\left(h_1^3 - h_0^3\right) + (\bar{Q}_{mn})_2\left(h_2^3 - h_1^3\right) + (\bar{Q}_{mn})_3\left(h_3^3 - h_2^3\right) + (Q_{mn})_4\left(h_4^3 - h_3^3\right)\right] \\
&= \frac{1}{3}\left[(\bar{Q}_{mn})_{+45°}\left(h_1^3 - h_0^3 + h_4^3 - h_3^3\right) + (\bar{Q}_{mn})_{-45°}\left(h_2^3 - h_1^3 + h_3^3 - h_2^3\right)\right] \\
&= 1008 \times 10^{-9}(\bar{Q}_{mn})_{+45°} + 144 \times 10^{-9}(\bar{Q}_{mn})_{-45°}.
\end{aligned}$$

Substituting for various $(\bar{Q}_{mn})$ values, we calculate

$$[A] = \begin{bmatrix} 962.64 & 806.64 & 0 \\ 806.64 & 962.64 & 0 \\ 0 & 0 & 829.68 \end{bmatrix} \times 10^6 \text{ N/m},$$

$$[B] = [0],$$

$$[D] = \begin{bmatrix} 46.21 & 38.72 & 27.04 \\ 38.72 & 46.21 & 27.04 \\ 27.04 & 27.04 & 39.82 \end{bmatrix} \times 10^3 \text{ N m}.$$

Note that $[\pm 45]_S$ is a balanced symmetric laminate in which $A_{16} = A_{26} = 0$ and $[B] = [0]$.

(c) [+45/0/−45] Unsymmetric laminate: From the figure (bottom), we note

$$h_2 = -h_1 = 3 \times 10^{-3} \text{ m},$$
$$h_3 = -h_0 = 9 \times 10^{-3} \text{ m}.$$

In this laminate,

$$(\bar{Q}_{mn})_1 = (\bar{Q}_{mn})_{+45°},$$
$$(\bar{Q}_{mn})_2 = (\bar{Q}_{mn})_{0°},$$
$$(\bar{Q}_{mn})_3 = (\bar{Q}_{mn})_{-45°}.$$

Therefore,

$$\begin{aligned}
A_{mn} &= (\bar{Q}_{mn})_1(h_1 - h_0) + (\bar{Q}_{mn})_2(h_2 - h_1) + (\bar{Q}_{mn})_3(h_3 - h_2) \\
&= 6 \times 10^{-3}(\bar{Q}_{mn})_{+45°} + 6 \times 10^{-3}(\bar{Q}_{mn})_{0°} + 6 \times 10^{-3}(\bar{Q}_{mn})_{-45°}, \\
B_{mn} &= \frac{1}{2}\left[(\bar{Q}_{mn})_1\left(h_1^2 - h_0^2\right) + (\bar{Q}_{mn})_2\left(h_2^2 - h_1^2\right) + (\bar{Q}_{mn})_3\left(h_3^2 - h_2^2\right)\right] \\
&= -36 \times 10^{-6}(\bar{Q}_{mn})_{+45°} + 36 \times 10^{-6}(\bar{Q}_{mn})_{-45°}, \\
D_{mn} &= \frac{1}{3}\left[(\bar{Q}_{mn})_1\left(h_1^3 - h_0^3\right) + (\bar{Q}_{mn})_2\left(h_2^3 - h_1^3\right) + (\bar{Q}_{mn})_3\left(h_3^3 - h_2^3\right)\right] \\
&= 234 \times 10^{-9}(\bar{Q}_{mn})_{+45°} + 18 \times 10^{-9}(\bar{Q}_{mn})_{0°} + 234 \times 10^{-9}(\bar{Q}_{mn})_{-45°}.
\end{aligned}$$

Substituting for $[\bar{Q}_{mn}]$ values, we calculate

$$[A] = \begin{bmatrix} 1285.50 & 417.06 & 0 \\ 417.06 & 534.24 & 0 \\ 0 & 0 & 434.36 \end{bmatrix} \times 10^6 \text{ N/m},$$

$$[B] = \begin{bmatrix} 0 & 0 & -2253.6 \\ 0 & 0 & -2253.6 \\ -2253.6 & -2253.6 & 0 \end{bmatrix} \times 10^3 \text{ N},$$

$$[D] = \begin{bmatrix} 21{,}183.84 & 15{,}770.70 & 0 \\ 15{,}770.70 & 18{,}930.24 & 0 \\ 0 & 0 & 16{,}237.33 \end{bmatrix} \text{N m}.$$

Comparing cases (a) and (c), we note that the addition of a 0° lamina increases the value of A_{11} by a significant amount, but A_{12}, A_{22}, and A_{66} are only marginally improved. Elements in the [D] matrix are improved owing to the presence of the 0° lamina as well as the additional thickness in the [+45/0/−45] laminate.

EXAMPLE 3.8

Compare the stiffness matrices of [0/90/90/0] and [0/90/0/90] laminates. Assume each ply has a thickness of $h/4$.

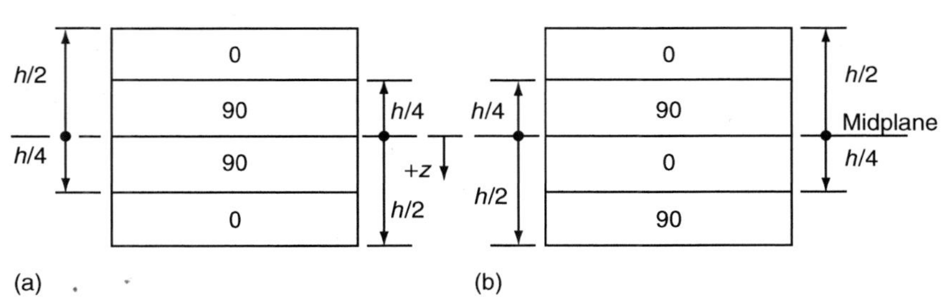

Solution

First, we note that for 0° and 90° plies,

$$(Q_{11})_0 = (Q_{22})_{90},$$
$$(Q_{22})_0 = (Q_{11})_{90},$$
$$(Q_{12})_0 = (Q_{12})_{90},$$
$$(Q_{66})_0 = (Q_{66})_{90},$$
$$(Q_{16})_0 = (Q_{16})_{90} = 0,$$
$$(Q_{26})_0 = (Q_{26})_{90} = 0.$$

For the [0/90/90/0] laminate on the left,

$$A_{ij} = (Q_{ij})_0\left[-\frac{h}{4} - \left(-\frac{h}{2}\right)\right] + (Q_{ij})_{90}\left[0 - \left(-\frac{h}{4}\right)\right] + (Q_{ij})_{90}\left(\frac{h}{4} - 0\right) + (Q_{ij})_0\left(\frac{h}{2} - \frac{h}{4}\right)$$
$$= \frac{h}{2}[(Q_{ij})_0 + (Q_{ij})_{90}],$$

$B_{ij} = 0$ (since this is a symmetric laminate),

$$D_{ij} = \frac{1}{3}\left\{(Q_{ij})_0\left[\left(-\frac{h}{4}\right)^3 - \left(-\frac{h}{2}\right)^3\right] + (Q_{ij})_{90}\left[0 - \left(-\frac{h}{4}\right)^3\right] + (Q_{ij})_{90}\left[\left(\frac{h}{4}\right)^3 - 0\right] + (Q_{ij})_0\left[\left(\frac{h}{2}\right)^3 - \left(\frac{h}{4}\right)^3\right]\right\}$$
$$= \frac{h^3}{96}[7(Q_{ij})_0 + (Q_{ij})_{90}].$$

For the [0/90/0/90] laminate on the right,

$$A_{ij} = (Q_{ij})_0\left[\left(-\frac{h}{4}\right) - \left(-\frac{h}{2}\right)\right] + (Q_{ij})_{90}\left[0 - \left(-\frac{h}{4}\right)\right]$$
$$+ (Q_{ij})_0\left(\frac{h}{4} - 0\right) + (Q_{ij})_{90}\left(\frac{h}{2} - \frac{h}{4}\right)$$
$$= \frac{h}{2}[(Q_{ij})_0 + (Q_{ij})_{90}],$$

$$B_{ij} = \frac{1}{2}\left\{(Q_{ij})_0\left[\left(-\frac{h}{4}\right)^2 - \left(-\frac{h}{2}\right)^2\right] + (Q_{ij})_{90}\left[0 - \left(-\frac{h}{4}\right)^2\right]\right.$$
$$\left. + (Q_{ij})_0\left[\left(\frac{h}{4}\right)^2 - 0\right] + (Q_{ij})_{90}\left[\left(\frac{h}{2}\right)^2 - \left(\frac{h}{4}\right)^2\right]\right\}$$
$$= \frac{h^2}{16}[-(Q_{ij})_0 + (Q_{ij})_{90}],$$

$$D_{ij} = \frac{1}{3}\left\{(Q_{ij})_0\left[\left(-\frac{h}{4}\right)^3 - \left(-\frac{h}{2}\right)^3\right] + (Q_{ij})_{90}\left[0 - \left(-\frac{h}{4}\right)^3\right]\right.$$
$$\left. + (Q_{ij})_0\left[\left(\frac{h}{4}\right)^3 - 0\right] + (Q_{ij})_{90}\left[\left(\frac{h}{2}\right)^3 - \left(\frac{h}{4}\right)^3\right]\right\}$$
$$= \frac{h^3}{24}[(Q_{ij})_0 + (Q_{ij})_{90}].$$

This example demonstrates the influence of stacking sequence on the stiffness matrices and the difference it can make to the elastic response of laminates containing similar plies, but arranged in different orders. In this case, although [A] matrices for the [0/90/90/0] and [0/90/0/90] are identical, their [B] and [D] matrices are different.

3.3.2.5 Midplane Strains and Curvatures

If the normal force and moment resultants acting on a laminate are known, its midplane strains and curvatures can be calculated by inverting Equations 3.85 and 3.86. Thus,

$$\begin{bmatrix} \varepsilon_{xx}^{\circ} \\ \varepsilon_{yy}^{\circ} \\ \gamma_{xy}^{\circ} \end{bmatrix} = [A_1]\begin{bmatrix} N_{xx} \\ N_{yy} \\ N_{xy} \end{bmatrix} + [B_1]\begin{bmatrix} M_{xx} \\ M_{yy} \\ M_{xy} \end{bmatrix} \tag{3.93}$$

and

$$\begin{bmatrix} k_{xx} \\ k_{yy} \\ k_{xy} \end{bmatrix} = [C_1]\begin{bmatrix} N_{xx} \\ N_{yy} \\ N_{xy} \end{bmatrix} + [D_1]\begin{bmatrix} M_{xx} \\ M_{yy} \\ M_{xy} \end{bmatrix}, \tag{3.94}$$

where

$$
\begin{aligned}
[A_1] &= [A^{-1}] + [A^{-1}][B][(D^*)^{-1}][B][A^{-1}] \\
[B_1] &= -[A^{-1}][B][(D^*)^{-1}] \\
[C_1] &= -[(D^*)^{-1}][B][A^{-1}] = [B_1]^T \\
[D^*] &= [D] - [B][A^{-1}][B] \\
[D_1] &= [(D^*)^{-1}]
\end{aligned}
\tag{3.95}
$$

Note that for a symmetric laminate, $[B]=[0]$, and therefore, $[A_1]=[A^{-1}]$, $[B_1]=[C_1]=[0]$, and $[D_1]=[D^{-1}]$. In this case, equations for midplane strains and curvatures become

$$
\begin{bmatrix} \varepsilon_{xx}^{\circ} \\ \varepsilon_{yy}^{\circ} \\ \gamma_{xy}^{\circ} \end{bmatrix} = [A^{-1}] \begin{bmatrix} N_{xx} \\ N_{yy} \\ N_{xy} \end{bmatrix}
\tag{3.96}
$$

and

$$
\begin{bmatrix} k_{xx} \\ k_{yy} \\ k_{xy} \end{bmatrix} = [D^{-1}] \begin{bmatrix} M_{xx} \\ M_{yy} \\ M_{xy} \end{bmatrix}.
\tag{3.97}
$$

Equation 3.96 shows that for a symmetric laminate, in-plane forces cause only in-plane strains and no curvatures. Similarly, Equation 3.97 shows that bending and twisting moments cause only curvatures and no in-plane strains.

EXAMPLE 3.9

Elastic properties of a balanced symmetric laminate: For a balanced symmetric laminate, the extensional stiffness matrix is

$$
[A] = \begin{bmatrix} A_{11} & A_{12} & 0 \\ A_{12} & A_{22} & 0 \\ 0 & 0 & A_{66} \end{bmatrix}
$$

and the coupling stiffness matrix $[B]=[0]$.

The inverse of the $[A]$ matrix is

$$
[A^{-1}] = \frac{1}{A_{11}A_{22} - A_{12}^2} \begin{bmatrix} A_{22} & -A_{12} & 0 \\ -A_{12} & A_{11} & 0 \\ 0 & 0 & \dfrac{(A_{11}A_{22} - A_{12}^2)}{A_{66}} \end{bmatrix}.
$$

Therefore, Equation 3.96 gives

$$\begin{bmatrix} \varepsilon_{xx}^{\circ} \\ \varepsilon_{yy}^{\circ} \\ \gamma_{xy}^{\circ} \end{bmatrix} = \frac{1}{A_{11}A_{22} - A_{12}^2} \begin{bmatrix} A_{22} & -A_{12} & 0 \\ -A_{12} & A_{11} & 0 \\ 0 & 0 & \dfrac{(A_{11}A_{22} - A_{12}^2)}{A_{66}} \end{bmatrix} \begin{bmatrix} N_{xx} \\ N_{yy} \\ N_{xy} \end{bmatrix}. \tag{3.98}$$

Let us assume that the laminate is subjected to a uniaxial tensile stress σ_{xx} in the x direction, and both σ_{yy} and τ_{xy} are zero. If the laminate thickness is h, the tensile force per unit width in the x direction $N_{xx} = h\sigma_{xx}$, $N_{yy} = 0$, and $N_{xy} = 0$. Thus, from Equation 3.98, we obtain

$$\varepsilon_{xx}^{\circ} = \frac{A_{22}}{A_{11}A_{22} - A_{12}^2} h\sigma_{xx},$$

$$\varepsilon_{yy}^{\circ} = -\frac{A_{12}}{A_{11}A_{22} - A_{12}^2} h\sigma_{xx},$$

$$\gamma_{xy}^{\circ} = 0,$$

which give

$$E_{xx} = \frac{\sigma_{xx}}{\varepsilon_{xx}^{\circ}} = \frac{A_{11}A_{22} - A_{12}^2}{hA_{22}}, \tag{3.99}$$

$$\nu_{xy} = -\frac{\varepsilon_{yy}^{\circ}}{\varepsilon_{xx}^{\circ}} = \frac{A_{12}}{A_{22}}. \tag{3.100}$$

In turn, applying N_{yy} and N_{xy} separately, we can determine

$$E_{yy} = \frac{A_{11}A_{22} - A_{12}^2}{hA_{11}}, \tag{3.101}$$

$$\nu_{yx} = \frac{A_{12}}{A_{11}} \left(\text{which is the same as } \nu_{xy}\frac{E_{yy}}{E_{xx}} \right), \tag{3.102}$$

and

$$G_{xy} = \frac{A_{66}}{h}. \tag{3.103}$$

EXAMPLE 3.10

Elastic properties of a symmetric quasi-isotropic laminate: For a symmetric quasi-isotropic laminate,

$$[A] = \begin{bmatrix} A_{11} & A_{12} & 0 \\ A_{12} & A_{22} = A_{11} & 0 \\ 0 & 0 & A_{66} = \dfrac{A_{11} - A_{12}}{2} \end{bmatrix}$$

and $[B] = [0]$

Now using the results of Example 3.9, we obtain

$$E_{xx} = E_{yy} = \frac{A_{11}^2 - A_{12}^2}{hA_{11}},$$

$$\nu_{xy} = \frac{A_{12}}{A_{11}},$$

$$G_{xy} = \frac{A_{11} - A_{12}}{2h},$$

where h is the laminate thickness.

Note that for a quasi-isotropic laminate, $E_{xx} = E_{yy}$ and, from the previous equations, it can be easily shown that

$$G_{xy} = \frac{E_{xx}}{2(1 + \nu_{xy})}.$$

However, $E_{xx} = E_{yy}$ does not necessarily mean quasi-isotropy. For example, E_{xx} and E_{yy} are equal for a $[0/90]_S$ laminate, but it is not a quasi-isotropic laminate. For a quasi-isotropic laminate, elastic modulus at any arbitrary angle in the plane of the laminate is the same as E_{xx} or E_{yy}. That will not be the case with the $[0/90]_S$ laminate.

EXAMPLE 3.11

Elastic properties of symmetric angle-ply laminates: For angled plies with θ and $-\theta$ fiber orientation angles,

$$\bar{Q}_{11}(\theta) = \bar{Q}_{11}(-\theta),$$

$$\bar{Q}_{22}(\theta) = \bar{Q}_{22}(-\theta),$$

$$\bar{Q}_{12}(\theta) = \bar{Q}_{12}(-\theta),$$

$$\bar{Q}_{66}(\theta) = \bar{Q}_{66}(-\theta),$$

$$\bar{Q}_{16}(\theta) = -\bar{Q}_{16}(-\theta),$$

$$\bar{Q}_{26}(\theta) = -\bar{Q}_{26}(-\theta).$$

Referring to the four-layer angle-ply laminate shown in the left side of the figure, we can write the elements in the extensional stiffness matrix $[A]$ of the $[\theta/-\theta]_S$ as

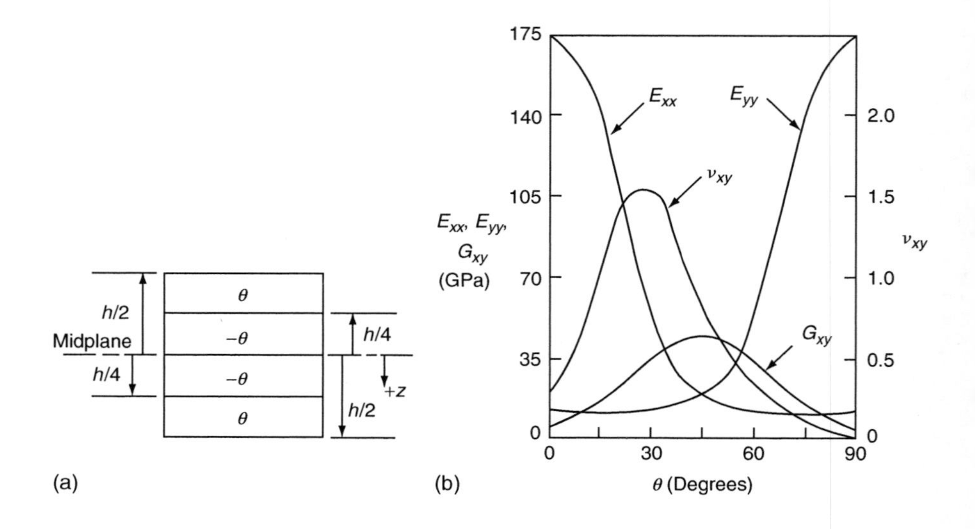

$$
\begin{aligned}
A_{ij} &= (\bar{Q}_{ij})_\theta\left(-\frac{h}{4}+\frac{h}{2}\right) + (\bar{Q}_{ij})_{-\theta}\left(0+\frac{h}{4}\right) \\
&\quad + (\bar{Q}_{ij})_{-\theta}\left(\frac{h}{4}-0\right) + (\bar{Q}_{ij})_\theta\left(\frac{h}{2}-\frac{h}{4}\right) \\
&= \frac{h}{2}\left[(\bar{Q}_{ij})_\theta + (\bar{Q}_{ij})_{-\theta}\right].
\end{aligned}
$$

Thus,

$$
[A] = \begin{bmatrix} h\bar{Q}_{11} & h\bar{Q}_{12} & 0 \\ h\bar{Q}_{12} & h\bar{Q}_{22} & 0 \\ 0 & 0 & h\bar{Q}_{66} \end{bmatrix}.
$$

Now, using Equations 3.99 through 3.103, we can write

$$
\begin{aligned}
E_{xx} &= \frac{\bar{Q}_{11}\bar{Q}_{22} - \bar{Q}_{12}^2}{\bar{Q}_{22}}, \\
E_{yy} &= \frac{\bar{Q}_{11}\bar{Q}_{22} - \bar{Q}_{12}^2}{\bar{Q}_{11}}, \\
\nu_{xy} &= \frac{\bar{Q}_{12}}{\bar{Q}_{22}}, \\
G_{xy} &= \bar{Q}_{66}.
\end{aligned}
$$

Since $\bar{Q}_{11}$, $\bar{Q}_{22}$, $\bar{Q}_{12}$, and $\bar{Q}_{66}$ are functions of the fiber orientation angle θ, the elastic properties of the angle-ply laminate will also be functions of θ. This is illustrated in the right side of the figure. Note that the shear modulus is maximum at $\theta = 45°$, that is, for a $[\pm 45]_S$ laminate. In addition, note the variation in the Poisson's ratio, which has values greater than unity for a range of fiber orientation angles. In an isotropic material, the Poisson's ratio cannot exceed a value of 0.5.

EXAMPLE 3.12

Bending of a balanced symmetric laminate beam specimen: For a balanced symmetric laminate, $[B] = [0]$.

$$[D] = \begin{bmatrix} D_{11} & D_{12} & D_{16} \\ D_{12} & D_{22} & D_{26} \\ D_{16} & D_{26} & D_{66} \end{bmatrix},$$

$$[D^{-1}] = \frac{1}{D_0}\begin{bmatrix} D_{11}^{\circ} & D_{12}^{\circ} & D_{16}^{\circ} \\ D_{12}^{\circ} & D_{22}^{\circ} & D_{26}^{\circ} \\ D_{16}^{\circ} & D_{26}^{\circ} & D_{66}^{\circ} \end{bmatrix},$$

where

$$\begin{aligned}
D_0 &= D_{11}\left(D_{22}D_{66} - D_{26}^2\right) - D_{12}(D_{12}D_{66} - D_{16}D_{26}) + D_{16}(D_{12}D_{26} - D_{22}D_{16}) \\
D_{11}^{\circ} &= \left(D_{22}D_{66} - D_{26}^2\right) \\
D_{12}^{\circ} &= -(D_{12}D_{66} - D_{16}D_{26}) \\
D_{16}^{\circ} &= (D_{12}D_{26} - D_{22}D_{16}) \\
D_{22}^{\circ} &= \left(D_{11}D_{66} - D_{16}^2\right) \\
D_{26}^{\circ} &= -(D_{11}D_{26} - D_{12}D_{16}) \\
D_{66}^{\circ} &= \left(D_{11}D_{12} - D_{12}^2\right)
\end{aligned}$$

If a bending moment is applied in the yz plane so that M_{xx} is present and $M_{yy} = M_{xy} = 0$, the specimen curvatures can be obtained from Equation 3.97:

$$\begin{aligned}
k_{xx} &= \frac{D_{11}^{\circ}}{D_0} M_{xx}, \\
k_{yy} &= \frac{D_{12}^{\circ}}{D_0} M_{xx}, \\
k_{xy} &= \frac{D_{16}^{\circ}}{D_0} M_{xx}.
\end{aligned} \tag{3.104}$$

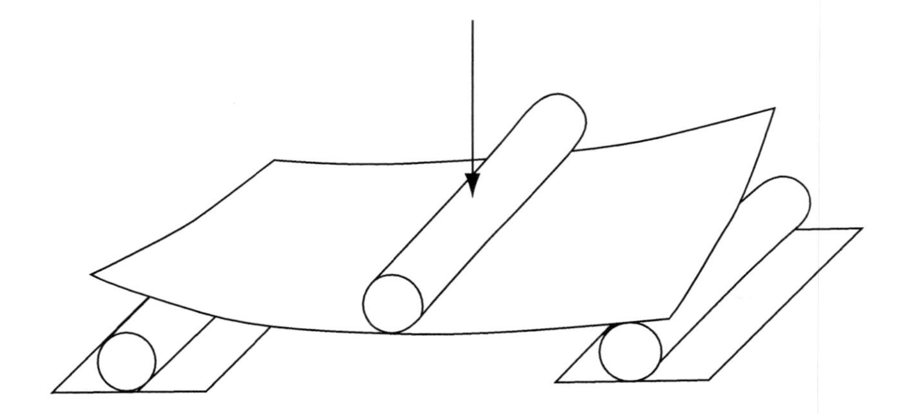

Thus, even though no twisting moment is applied, the specimen would tend to twist unless $D^{\circ}_{16} = (D_{12}D_{26} - D_{22}D_{16}) = 0$. This is possible only if the balanced symmetric laminate contains fibers in the 0° and 90° directions. The twisting phenomenon can be easily demonstrated in a three-point flexural test in which the specimen lifts off the support on opposite corners of its span, as shown in the figure.

3.3.2.6 Lamina Strains and Stresses Due to Applied Loads

Knowing the midplane strains and curvatures for the laminate, strains at the midplane of each lamina can be calculated using the following linear relationships:

$$\begin{bmatrix} \varepsilon_{xx} \\ \varepsilon_{yy} \\ \gamma_{xy} \end{bmatrix}_j = \begin{bmatrix} \varepsilon^{\circ}_{xx} \\ \varepsilon^{\circ}_{yy} \\ \gamma^{\circ}_{xy} \end{bmatrix} + z_j \begin{bmatrix} k_{xx} \\ k_{yy} \\ k_{xy} \end{bmatrix}, \tag{3.105}$$

where z_j is the distance from the laminate midplane to the midplane of the jth lamina.

In turn, stresses in the jth lamina can be calculated using its stiffness matrix. Thus,

$$\begin{bmatrix} \sigma_{xx} \\ \sigma_{yy} \\ \tau_{xy} \end{bmatrix}_j = [\bar{Q}_{\text{mn}}]_j \begin{bmatrix} \varepsilon_{xx} \\ \varepsilon_{yy} \\ \gamma_{xy} \end{bmatrix}_j = [\bar{Q}_{\text{mn}}]_j \begin{bmatrix} \varepsilon^{\circ}_{xx} \\ \varepsilon^{\circ}_{yy} \\ \gamma^{\circ}_{xy} \end{bmatrix} + z_j[\bar{Q}_{\text{mn}}]_j \begin{bmatrix} k_{xx} \\ k_{yy} \\ k_{xy} \end{bmatrix}. \tag{3.106}$$

Figure 3.51 demonstrates schematically the strain and stress distributions in a laminate. Note that the strain distribution is continuous and linearly varies with the distance z from the laminate midplane. The stress distribution is not continuous, although it varies linearly across each lamina thickness. For thin laminas, the strain and stress variation across the thickness of each lamina is small. Therefore, their average values are calculated using the center distance z_j, as shown in Equations 3.105 and 3.106.

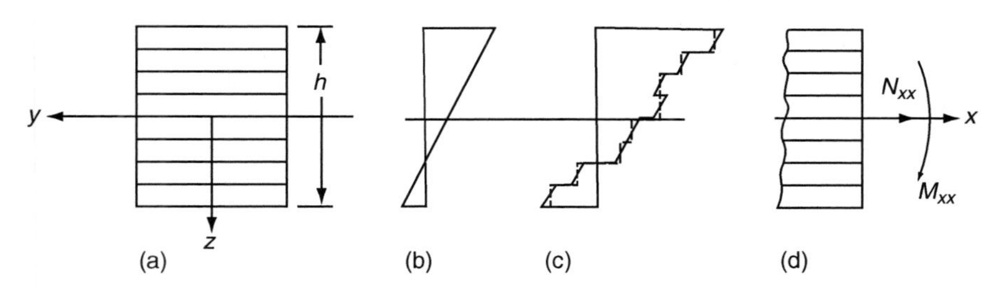

FIGURE 3.51 Strain and stress distributions in a laminate. (a) Laminate; (b) Strain distribution; (c) Stress distribution; and (d) Normal force and bending moment resultants.

EXAMPLE 3.13

Calculate lamina stresses at the midplane of each lamina in the [+45/−45] laminate in Example 3.7 due to $N_{xx} = 100$ kN/m.

SOLUTION

Step 1: From the laminate stiffness matrices [A], [B], and [D], determine [A^{-1}], [D^*], [A_1], [B_1], [C_1], and [D_1].

$$[A^{-1}] = \begin{bmatrix} 0.697 & -0.584 & 0 \\ -0.584 & 0.697 & 0 \\ 0 & 0 & 0.241 \end{bmatrix} \times 10^{-8} \text{ m/N},$$

$$[B][A^{-1}][B] = \begin{bmatrix} 3.06 & 3.06 & 0 \\ 3.06 & 3.06 & 0 \\ 0 & 0 & 2.87 \end{bmatrix} \times 10^{3} \text{ N m},$$

$$[D^*] = [D] - [B][A^{-1}][B] = \begin{bmatrix} 2715.84 & 1779.84 & 0 \\ 1779.84 & 2715.84 & 0 \\ 0 & 0 & 2108.08 \end{bmatrix} \text{N m},$$

$$[D_1] = [(D^*)^{-1}] = \begin{bmatrix} 6.45 & -4.23 & 0 \\ -4.23 & 6.45 & 0 \\ 0 & 0 & 4.74 \end{bmatrix} \times 10^{-4} \frac{1}{\text{N m}},$$

$$[B_1] = -[A^{-1}][B][(D^*)^{-1}] = \begin{bmatrix} 0 & 0 & 603.54 \\ 0 & 0 & 603.54 \\ 602.74 & 602.74 & 0 \end{bmatrix} \times 10^{-9} \frac{1}{\text{N}},$$

$$[C_1] = -[(D^*)^{-1}][B][A^{-1}] = \begin{bmatrix} 0 & 0 & 602.74 \\ 0 & 0 & 602.74 \\ 603.54 & 603.54 & 0 \end{bmatrix} \times 10^{-9} \frac{1}{\text{N}},$$

$$[A_1] = [A^{-1}] + [A^{-1}][B][(D^*)^{-1}][B][A^{-1}] = \begin{bmatrix} 7.7385 & -5.0715 & 0 \\ -5.0715 & 7.7385 & 0 \\ 0 & 0 & 5.683 \end{bmatrix} \times 10^{-9}\ \text{m/N}.$$

Step 2: Using Equations 3.93 and 3.94, calculate the $[\varepsilon^\circ]$ and $[k]$ matrices.

$$[\varepsilon^\circ] = \begin{bmatrix} \varepsilon_{xx}^\circ \\ \varepsilon_{yy}^\circ \\ \gamma_{xy}^\circ \end{bmatrix} = [A_1] \begin{bmatrix} 100 \times 10^3\ \text{N/m} \\ 0 \\ 0 \end{bmatrix}.$$

Therefore,

$$\varepsilon_{xx}^\circ = 77.385 \times 10^{-5}\ \text{m/m},$$
$$\varepsilon_{yy}^\circ = -50.715 \times 10^{-5}\ \text{m/m},$$
$$\gamma_{xy}^\circ = 0.$$

$$[k] = \begin{bmatrix} k_{xx} \\ k_{yy} \\ k_{xy} \end{bmatrix} = [C_1] \begin{bmatrix} 100 \times 10^3\ \text{N/m} \\ 0 \\ 0 \end{bmatrix},$$

Therefore,

$$k_{xx} = 0$$
$$k_{yy} = 0$$
$$k_{xy} = 0.060354\ \text{per m}$$

Step 3: Using Equation 3.105, calculate $\varepsilon^\circ{}_{xx}$, $\varepsilon^\circ{}_{yy}$, and γ_{xy} at the midplane of +45° and −45° laminas.

$$\begin{bmatrix} \varepsilon_{xx} \\ \varepsilon_{yy} \\ \gamma_{xy} \end{bmatrix}_{+45^\circ} = \begin{bmatrix} 77.385 \times 10^{-5} \\ -50.715 \times 10^{-5} \\ 0 \end{bmatrix} + (-3 \times 10^{-3}) \begin{bmatrix} 0 \\ 0 \\ 0.060354 \end{bmatrix}$$
$$= \begin{bmatrix} 77.385 \\ -50.715 \\ -18.106 \end{bmatrix} \times 10^{-5}.$$

Similarly,

$$\begin{bmatrix} \varepsilon_{xx} \\ \varepsilon_{yy} \\ \gamma_{xy} \end{bmatrix}_{-45^\circ} = \begin{bmatrix} 77.385 \\ -50.715 \\ 18.106 \end{bmatrix} \times 10^{-5}.$$

Step 4: Using Equation 3.106, calculate σ_{xx}, σ_{yy}, and τ_{xy} at the midplanes of +45° and −45° laminas.

$$\begin{bmatrix} \sigma_{xx} \\ \sigma_{yy} \\ \tau_{xy} \end{bmatrix}_{+45^\circ} = \begin{bmatrix} 40.11 & 33.61 & 31.3 \\ 33.61 & 40.11 & 31.3 \\ 31.3 & 31.3 & 34.57 \end{bmatrix} \text{GPa} \begin{bmatrix} 77.385 \times 10^{-5} \\ -50.715 \times 10^{-5} \\ -18.106 \times 10^{-5} \end{bmatrix} = \begin{bmatrix} 8.33 \\ 0 \\ 2.09 \end{bmatrix} \text{MPa.}$$

Similarly,

$$\begin{bmatrix} \sigma_{xx} \\ \sigma_{yy} \\ \tau_{xy} \end{bmatrix}_{-45^\circ} = \begin{bmatrix} 8.33 \\ 0 \\ -2.09 \end{bmatrix} \text{MPa.}$$

Using the stress transformation Equation 3.30, we may compute the longitudinal, transverse, and shear stresses in the 1–2 directions, which give the following results:

	45° Layer (MPa)	−45° Layer (MPa)
σ_{11}	6.255	6.255
σ_{22}	2.075	2.075
τ_{12}	−4.165	4.165

3.3.2.7 Thermal Strains and Stresses

If a temperature variation ΔT is involved, lamina strains will be

$$\begin{aligned} \varepsilon_{xx} &= \varepsilon_{xx}^{M} + \varepsilon_{xx}^{T} = \varepsilon_{xx}^{\circ} + zk_{xx}, \\ \varepsilon_{yy} &= \varepsilon_{yy}^{M} + \varepsilon_{yy}^{T} = \varepsilon_{yy}^{\circ} + zk_{yy}, \\ \gamma_{xy} &= \gamma_{xy}^{M} + \gamma_{xy}^{T} = \gamma_{xy}^{\circ} + zk_{xy}, \end{aligned} \tag{3.107}$$

where the superscripts M and T denote the mechanical and thermal strains, respectively.

Thermal strains are due to free expansions (or contractions) caused by temperature variations, but mechanical strains are due to both applied loads and thermal loads. Thermal loads appear due to restrictions imposed by various layers against their free thermal expansion. In many applications involving polymer matrix composites, moisture can also influence the laminate strains owing to volumetric expansion (swelling) or contraction of the matrix caused by moisture absorption or desorption [24]. In such cases, a third term representing hygroscopic strains must be added in the middle column of Equation 3.107.

Modifying Equations 3.85 and 3.86 for thermal effects, we can write

$$\begin{bmatrix} N_{xx} \\ N_{yy} \\ N_{xy} \end{bmatrix} = [A] \begin{bmatrix} \varepsilon_{xx}^{\circ} \\ \varepsilon_{yy}^{\circ} \\ \gamma_{xy}^{\circ} \end{bmatrix} + [B] \begin{bmatrix} k_{xx} \\ k_{yy} \\ k_{xy} \end{bmatrix} - [T^*]\Delta T \tag{3.108}$$

and

$$\begin{bmatrix} M_{xx} \\ M_{yy} \\ M_{xy} \end{bmatrix} = [B] \begin{bmatrix} \varepsilon_{xx}^{\circ} \\ \varepsilon_{yy}^{\circ} \\ \gamma_{xy}^{\circ} \end{bmatrix} + [D] \begin{bmatrix} k_{xx} \\ k_{yy} \\ k_{xy} \end{bmatrix} - [T^{**}]\Delta T, \tag{3.109}$$

where

$$\begin{aligned} [T^*] &= \begin{bmatrix} \sum_{j=1}^{N} \left[(\bar{Q}_{11})_j(\alpha_{xx})_j + (\bar{Q}_{12})_j(\alpha_{yy})_j + (\bar{Q}_{16})_j(\alpha_{xy})_j\right](h_j - h_{j-1}) \\ \sum_{j=1}^{N} \left[(\bar{Q}_{12})_j(\alpha_{xx})_j + (\bar{Q}_{22})_j(\alpha_{yy})_j + (\bar{Q}_{26})_j(\alpha_{xy})_j\right](h_j - h_{j-1}) \\ \sum_{j=1}^{N} \left[(\bar{Q}_{16})_j(\alpha_{xx})_j + (\bar{Q}_{26})_j(\alpha_{yy})_j + (\bar{Q}_{66})_j(\alpha_{xy})_j\right](h_j - h_{j-1}) \end{bmatrix} \\ [T^{**}] &= \frac{1}{2} \begin{bmatrix} \sum_{j=1}^{N} \left[(\bar{Q}_{11})_j(\alpha_{xx})_j + (\bar{Q}_{12})_j(\alpha_{yy})_j + (\bar{Q}_{16})_j(\alpha_{xy})_j\right]\left(h_j^2 - h_{j-1}^2\right) \\ \sum_{j=1}^{N} \left[(\bar{Q}_{12})_j(\alpha_{xx})_j + (\bar{Q}_{22})_j(\alpha_{yy})_j + (\bar{Q}_{26})_j(\alpha_{xy})_j\right]\left(h_j^2 - h_{j-1}^2\right) \\ \sum_{j=1}^{N} \left[(\bar{Q}_{16})_j(\alpha_{xx})_j + (\bar{Q}_{26})_j(\alpha_{yy})_j + (\bar{Q}_{66})_j(\alpha_{xy})_j\right]\left(h_j^2 - h_{j-1}^2\right) \end{bmatrix} \end{aligned} \tag{3.110}$$

Note that even if no external loads are applied, that is, if $[N] = [M] = [0]$, there may be midplane strains and curvatures due to thermal effects, which in turn will create thermal stresses in various laminas. These stresses can be calculated using midplane strains and curvatures due to thermal effects in Equation 3.106.

When a composite laminate is cooled from the curing temperature to room temperature, significant curing (residual) stresses may develop owing to the thermal mismatch of various laminas. In some cases, these curing stresses may be sufficiently high to cause intralaminar cracks [25]. Therefore, it may be prudent to consider them in the analysis of composite laminates.

For example, consider a $[0/90]_S$ laminate being cooled from the curing temperature to room temperature. If the plies were not joined and could contract freely, the 0° ply will contract much less in the x direction than the 90° ply, while the reverse is true in the y direction. Since the plies are joined and must deform together, internal residual stresses are generated to maintain the geometric compatibility between the plies. In $[0/90]_S$ laminate, residual stresses are compressive in the fiber direction, but tensile in the transverse direction in both 0° and 90° plies (see Example 3.16). Thus, when such a laminate is loaded in tension in the x direction, residual tensile stress added to the applied tensile stress can initiate transverse cracks in the 90° plies at relatively low loads.

Equations 3.108 and 3.109 are also useful for calculating the coefficients of thermal expansion and the cured shapes of a laminate. This is demonstrated in the following two examples.

EXAMPLE 3.14

Coefficients of thermal expansion for a balanced symmetric laminate

SOLUTION

For a balanced symmetric laminate, $A_{16}=A_{26}=0$ and $[B]=[0]$. In a thermal experiment, $[N]=[M]=[0]$. Therefore, from Equation 3.108,

$$\begin{bmatrix} 0 \\ 0 \\ 0 \end{bmatrix} = \begin{bmatrix} A_{11} & A_{12} & 0 \\ A_{12} & A_{22} & 0 \\ 0 & 0 & A_{66} \end{bmatrix} \begin{bmatrix} \varepsilon^{\circ}_{xx} \\ \varepsilon^{\circ}_{yy} \\ \gamma^{\circ}_{xy} \end{bmatrix} - \begin{bmatrix} T_1{}^* \\ T_2{}^* \\ T_3{}^* \end{bmatrix} \Delta T,$$

which gives

$$A_{11}\varepsilon^{\circ}_{xx} + A_{12}\varepsilon^{\circ}_{yy} = T_1^*\Delta T,$$
$$A_{12}\varepsilon^{\circ}_{xx} + A_{22}\varepsilon^{\circ}_{yy} = T_2^*\Delta T,$$

and

$$A_{66}\gamma^{\circ}_{xy} = T_3^*\Delta T.$$

From the first two of these equations, we calculate ε°_{xx} and ε°_{yy} as

$$\varepsilon^{\circ}_{xx} = \frac{A_{22}T_1^* - A_{12}T_2^*}{A_{11}A_{22} - A_{12}^2}\Delta T,$$
$$\varepsilon^{\circ}_{yy} = \frac{A_{11}T_2^* - A_{12}T_1^*}{A_{11}A_{22} - A_{12}^2}\Delta T.$$

Following the definitions of thermal expansion coefficients, we write

$$\alpha_{xx} = \frac{\varepsilon^{\circ}_{xx}}{\Delta T} = \frac{A_{22}T_1^* - A_{12}T_2^*}{A_{11}A_{22} - A_{12}^2},$$
$$\alpha_{yy} = \frac{\varepsilon^{\circ}_{yy}}{\Delta T} = \frac{A_{11}T_2^* - A_{12}T_1^*}{A_{11}A_{22} - A_{12}^2},$$
$$\alpha_{xy} = \frac{\gamma^{\circ}_{xy}}{\Delta T} = \frac{T_3^*}{A_{66}}.$$

For a balanced symmetric laminate, elements in the $[T^{**}]$ matrix are zero. Therefore, there will be no curvatures due to temperature variation. However, the same is not true for unsymmetric laminates.

EXAMPLE 3.15

Determine the curvatures of a two-layer unsymmetric [0/90] laminate after it is cooled from the curing temperature to the room temperature. The material is T-300 carbon fiber in an epoxy matrix for which $\alpha_{11} = -0.5 \times 10^{-6}$ m/m per °C and $\alpha_{22} = 12 \times 10^{-6}$ m/m per °C. Other material properties are the same as those in Example 3.6. The thickness of each layer is t, and the temperature drop is ΔT.

Solution

From Example 3.6, the stiffness matrices for the 0° and 90° layers are

$$[\bar{Q}]_{0^\circ} = \begin{bmatrix} 134.03 & 2.29 & 0 \\ 2.29 & 8.82 & 0 \\ 0 & 0 & 3.254 \end{bmatrix} \times 10^9 \text{ N/m}^2,$$

$$[\bar{Q}]_{90^\circ} = \begin{bmatrix} 8.82 & 2.29 & 0 \\ 2.29 & 134.03 & 0 \\ 0 & 0 & 3.254 \end{bmatrix} \times 10^9 \text{ N/m}^2.$$

Step 1: Determine the $[A]$, $[B]$, and $[D]$ matrices for the laminate.

Referring to the figure (top), we note that $h_0 = -t$, $h_1 = 0$, and $h_2 = t$. Following Equations 3.90 through 3.92, the $[A]$, $[B]$, and $[D]$ matrices for the [0/90] laminate can be written.

$$[A] = \begin{bmatrix} 142.85 & 4.58 & 0 \\ 4.58 & 142.85 & 0 \\ 0 & 0 & 6.508 \end{bmatrix} \times 10^9 \; t \text{ N/m},$$

$$[B] = \begin{bmatrix} -62.605 & 0 & 0 \\ 0 & 62.605 & 0 \\ 0 & 0 & 0 \end{bmatrix} \times 10^9 \; t^2 \text{ N},$$

$$[D] = \begin{bmatrix} 47.62 & 1.53 & 0 \\ 1.53 & 47.62 & 0 \\ 0 & 0 & 2.17 \end{bmatrix} \times 10^9 \; t^3 \text{ Nm}.$$

Step 2: Determine the $[T^*]$ and $[T^{**}]$ matrices for the laminate.

The first element in the $[T^*]$ matrix is

$$T_1^* = [(\bar{Q}_{11})_1(\alpha_{xx})_1 + (\bar{Q}_{12})_1(\alpha_{yy})_1](0 + t) + [(\bar{Q}_{11})_2(\alpha_{xx})_2 + (\bar{Q}_{12})_2(\alpha_{yy})_2](t - 0).$$

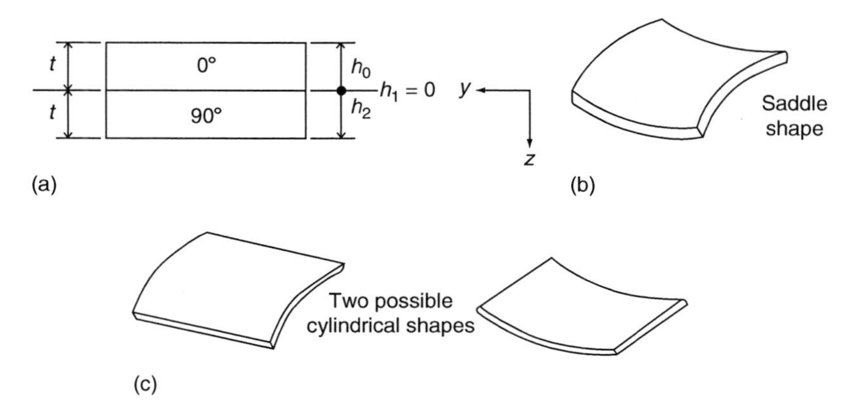

Since $(\alpha_{xx})_1 = (\alpha_{yy})_2 = \alpha_{11} = -0.5 \times 10^{-6}$ m/m per °C and $(\alpha_{xx})_2 = (\alpha_{yy})_1 = 12 \times 10^{-6}$ m/m per °C, we obtain

$$\begin{aligned} T_1^* &= [(134.03)(-0.5) + (2.29)(12)](10^9)(10^{-6})t \\ &\quad + [(8.82)(12) + (2.29)(-0.5)](10^9)(10^{-6})t \\ &= 65.16 \times 10^3 t \text{ N/m°C}. \end{aligned}$$

Using appropriate expressions for other elements in $[T^*]$ and $[T^{**}]$, we obtain

$$[T^*] = \begin{bmatrix} 65.16 \\ 65.16 \\ 0 \end{bmatrix} \times 10^3 \; t \text{ N/m°C},$$

$$[T^{**}] = \begin{bmatrix} 72.12 \\ -72.12 \\ 0 \end{bmatrix} \times 10^3 \; t^2 \text{ N/°C}.$$

Step 3: Determine the laminate curvature matrix.

Substitution of $[T^*]$ and $[T^{**}]$ in Equations 3.108 and 3.109 gives

$$\begin{bmatrix} 0 \\ 0 \\ 0 \end{bmatrix} = [A] \begin{bmatrix} \varepsilon_{xx}^{\circ} \\ \varepsilon_{yy}^{\circ} \\ \gamma_{xy}^{\circ} \end{bmatrix} + [B] \begin{bmatrix} k_{xx} \\ k_{yy} \\ k_{xy} \end{bmatrix} - \begin{bmatrix} 65.16 \\ 65.16 \\ 0 \end{bmatrix} \times 10^3 t \Delta T,$$

$$\begin{bmatrix} 0 \\ 0 \\ 0 \end{bmatrix} = [B] \begin{bmatrix} \varepsilon_{xx}^{\circ} \\ \varepsilon_{yy}^{\circ} \\ \gamma_{xy}^{\circ} \end{bmatrix} + [D] \begin{bmatrix} k_{xx} \\ k_{yy} \\ k_{xy} \end{bmatrix} - \begin{bmatrix} 72.12 \\ -72.12 \\ 0 \end{bmatrix} \times 10^3 t^2 \Delta T,$$

where $[A]$, $[B]$, and $[D]$ are laminate stiffness matrices.

Eliminating the midplane strain matrix from the previous equations, we obtain the following expression relating the laminate curvature matrix to temperature variation ΔT:

$$[k] = [C_1][T^*] + [D_1][T^{**}],$$

where $[C_1]$ and $[D_1]$ are given in Equation 3.95.

In this example,

$$[C_1] = \begin{bmatrix} 0.0218 & 0 & 0 \\ 0 & -0.0218 & 0 \\ 0 & 0 & 0 \end{bmatrix} 10^{-9} t^{-2} \frac{1}{\text{N}},$$

$$[D_1] = \begin{bmatrix} 0.0497 & -0.0016 & 0 \\ -0.0016 & 0.0497 & 0 \\ 0 & 0 & 0.4608 \end{bmatrix} 10^{-9} t^{-3} \frac{1}{\text{N m}}.$$

Therefore, solving for $[k]$, we obtain

$$k_{xx} = -k_{yy} = 5.119 \times 10^{-6} t^{-1} \Delta T \text{ per m},$$
$$k_{xy} = 0.$$

From the expressions for k_{xx} and k_{yy}, we note that both curvatures decrease with increasing layer thickness as well as decreasing temperature variation. Furthermore, since $k_{yy} = -k_{xx}$, the laminate will assume a saddle shape at room temperature, as shown in the figure ((b) on page 207).

Classical lamination theory, such as that used here, predicts the room temperature shapes of all unsymmetric laminates to be a saddle. However, Hyer [26,27] has shown that both cylindrical and saddle shapes are possible, as shown in the figure ((c) on page 207). The cured shape of the laminate depends on the thickness–width ratio as well as the thickness–length ratio. Saddle shapes are obtained for thick laminates, but depending on the relative values of length and width, two different cylindrical shapes (with either k_{xx} or $k_{yy} = 0$) are obtained for thin laminates in which the thickness–length or thickness–width ratios are small. It should be noted that symmetric laminates do not curve (warp) on curing since $[B] = [0]$ as well as $[T^{**}] = [0]$.

EXAMPLE 3.16

Residual stresses generated because of cooling from high curing temperatures: A $[0/90_2]_S$ laminate of AS-4 carbon fiber–epoxy is cured at temperature $T_i = 190°C$ and slowly cooled down to room temperature, $T_f = 23°C$. Determine the residual stresses generated in each layer because of cooling from the curing temperature. Assume each layer in the laminate has a thickness t_0.

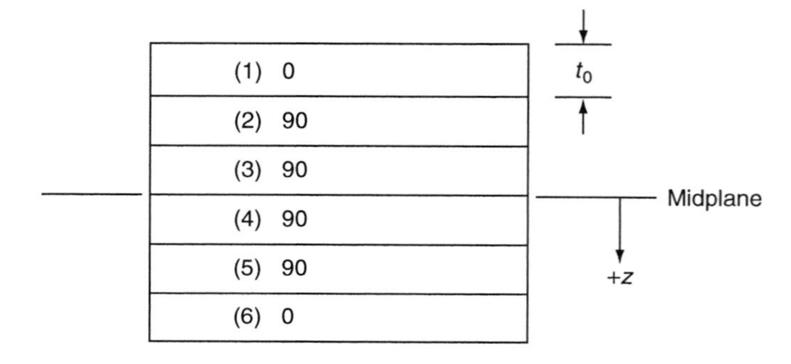

Following material properties are known:

$$E_{11} = 142 \text{ GPa},$$
$$E_{22} = 10.3 \text{ GPa},$$
$$\nu_{12} = 0.27,$$
$$G_{12} = 7.6 \text{ GPa},$$
$$\alpha_{11} = -1.8 \times 10^{-6} \text{ per}\,^{\circ}\text{C},$$
$$\alpha_{22} = 27 \times 10^{-6} \text{ per}\,^{\circ}\text{C}.$$

SOLUTION

Step 1: Using Equation 3.80, determine stiffness matrices for the 0° and 90° layers.

$$[\bar{Q}]_{0^{\circ}} = \begin{bmatrix} 142.77 & 2.796 & 0 \\ 2.796 & 10.356 & 0 \\ 0 & 0 & 7.6 \end{bmatrix} \times 10^{9} \text{ N/m}^{2},$$

$$[\bar{Q}]_{90^{\circ}} = \begin{bmatrix} 10.356 & 2.796 & 0 \\ 2.796 & 142.77 & 0 \\ 0 & 0 & 7.6 \end{bmatrix} \times 10^{9} \text{ N/m}^{2}.$$

Step 2: Determine the $[A]$ matrix for the laminate.

Note that because of symmetry, $[B] = [0]$ and, since $[k] = [0]$, we need not determine the $[D]$ matrix.

For a $[0/90_2]_S$ laminate, $A_{mn} = 2t_0\,[(\bar{Q}_{mn})_0 + 2(\bar{Q}_{mn})_{90}]$. Therefore,

$$[A] = 2t_0 \begin{bmatrix} 163.48 & 8.39 & 0 \\ 8.39 & 295.90 & 0 \\ 0 & 0 & 22.8 \end{bmatrix} \times 10^{9} \text{ N/m}.$$

Step 3: Determine the $[T^*]$ matrix for the laminate.

$$\begin{aligned} T_1^* = 2[&\{(\bar{Q}_{11})_0(\alpha_{xx})_0 + (\bar{Q}_{12})_0(\alpha_{yy})_0 + 0\}(-2t_0 + 3t_0) \\ &+ \{(\bar{Q}_{11})_{90}(\alpha_{xx})_{90} + (\bar{Q}_{12})_{90}(\alpha_{yy})_{90} + 0\}(-t_0 + 2t_0) \\ &+ \{(\bar{Q}_{11})_{90}(\alpha_{xx})_{90} + (\bar{Q}_{12})_{90}(\alpha_{yy})_{90} + 0\}(0 + t_0)]. \end{aligned}$$

Since $(\alpha_{xx})_0 = (\alpha_{yy})_{90} = \alpha_{11} = -1.8 \times 10^{-6}$ per °C and $(\alpha_{yy})_0 = (\alpha_{xx})_{90} = \alpha_{22} = 27 \times 10^{-6}$ per °C, we obtain

$$T_1^* = 735.32 t_0 \times 10^{3} \text{ N/m}^{\circ}\text{C}.$$

Similarly, $T_2^* = -176.82\ t_0 \times 10^3$ N/m °C and $T_3^* = 0$.

Therefore,

$$[T^*] = \begin{bmatrix} 735.32t_0 \\ -176.82t_0 \\ 0 \end{bmatrix} \times 10^3 \text{ N/m°C}.$$

Step 4: Using Equation 3.108, determine the midplane strains.

Since there are no external forces, $[N] = [0]$. Since $[B] = [0]$, we can write Equation 3.108 as

$$[0] = [A] \begin{bmatrix} \varepsilon_{xx}^{\circ} \\ \varepsilon_{yy}^{\circ} \\ \gamma_{xy}^{\circ} \end{bmatrix} + [0] - \begin{bmatrix} T_1^* \\ T_2^* \\ T_3^* \end{bmatrix} \Delta T,$$

where $\Delta T = T_f - T_i$ (which, in this case, has a negative value).

Solving for the strain components gives

$$\begin{bmatrix} \varepsilon_{xx}^{\circ} \\ \varepsilon_{yy}^{\circ} \\ \gamma_{xy}^{\circ} \end{bmatrix} = \begin{bmatrix} 2.267 \\ -0.352 \\ 0 \end{bmatrix} \times 10^{-6} \Delta T \text{ m/m}.$$

Step 5: Determine strains in each layer.

Since $[k] = [0]$, strains in each layer are the same as the midplane strains.

Step 6: Determine the free thermal contraction strains in each layer.

$$\begin{bmatrix} \varepsilon_{xxf} \\ \varepsilon_{yyf} \\ \gamma_{xyf} \end{bmatrix}_{0^\circ} = \begin{bmatrix} -1.8 \\ 27 \\ 0 \end{bmatrix} \times 10^{-6} \Delta T \text{ m/m}$$

and

$$\begin{bmatrix} \varepsilon_{xxf} \\ \varepsilon_{yyf} \\ \gamma_{xyf} \end{bmatrix}_{90^\circ} = \begin{bmatrix} 27 \\ -1.8 \\ 0 \end{bmatrix} \times 10^{-6} \Delta T \text{ m/m}.$$

Step 7: Subtract free thermal contraction strains from strains determined in Step 5 to obtain residual strains in each layer.

$$\begin{bmatrix} \varepsilon_{xxr} \\ \varepsilon_{yyr} \\ \gamma_{xyr} \end{bmatrix}_{0^\circ} = \begin{bmatrix} (2.267 - (-1.8) \times 10^{-6} \Delta T \\ (-0.352 - 27) \times 10^{-6} \Delta T \\ 0 \end{bmatrix} = \begin{bmatrix} 4.067 \\ -27.352 \\ 0 \end{bmatrix} \times 10^{-6} \Delta T \text{ m/m}$$

and

$$\begin{bmatrix} \varepsilon_{xxr} \\ \varepsilon_{yyr} \\ \gamma_{xyr} \end{bmatrix}_{90^\circ} = \begin{bmatrix} (2.267 - 27) \times 10^{-6} \Delta T \\ (-0.352 - (-1.8)) \times 10^{-6} \Delta T \\ 0 \end{bmatrix} = \begin{bmatrix} -24.733 \\ 1.448 \\ 0 \end{bmatrix} \times 10^{-6} \Delta T \text{ m/m}.$$

Step 8: Calculate the residual stresses in each layer

$$\begin{bmatrix} \sigma_{xxr} \\ \sigma_{yyr} \\ \tau_{xyr} \end{bmatrix}_{0^\circ} = [\bar{Q}]_{0^\circ} \begin{bmatrix} \varepsilon_{xxr} \\ \varepsilon_{yyr} \\ \gamma_{xyr} \end{bmatrix}_{0^\circ} = \begin{bmatrix} 504.17 \\ -271.89 \\ 0 \end{bmatrix} \times 10^3 \Delta T \text{ N/m}^2$$

and

$$\begin{bmatrix} \sigma_{xxr} \\ \sigma_{yyr} \\ \tau_{xyr} \end{bmatrix}_{90^\circ} = [\bar{Q}]_{90^\circ} \begin{bmatrix} \varepsilon_{xxr} \\ \varepsilon_{yyr} \\ \gamma_{xyr} \end{bmatrix}_{90^\circ} = \begin{bmatrix} -252.087 \\ 137.550 \\ 0 \end{bmatrix} \times 10^3 \Delta T \text{ N/m}^2.$$

Since, in this case, $\Delta T = 23°\text{C} - 190°\text{C} = -167°\text{C}$, the residual stresses are as follows:

	0° Layer	90° Layer
In the fiber direction	−84.2 MPa	−22.97 MPa
In the transverse direction	45.40 MPa	42.10 MPa

3.4 INTERLAMINAR STRESSES

Load transfer between adjacent layers in a fiber-reinforced laminate takes place by means of interlaminar stresses, such as σ_{zz}, τ_{xz}, and τ_{yz}. To visualize the mechanism of load transfer, let us consider a balanced symmetric $[\pm 45]_S$ laminate under uniaxial tensile load N_{xx} (Figure 3.52). Since $A_{16} = A_{26} = 0$ and $[B] = [0]$ for this laminate, the midplane strains are given by

$$\varepsilon_{xx}^{\circ} = \frac{A_{22}}{A_{11}A_{22} - A_{12}^2} N_{xx},$$

$$\varepsilon_{yy}^{\circ} = -\frac{A_{12}}{A_{11}A_{22} - A_{12}^2} N_{xx},$$

$$\gamma_{xy}^{\circ} = 0.$$

The state of stress in the jth layer is

$$\begin{bmatrix} \sigma_{xx} \\ \sigma_{yy} \\ \tau_{xy} \end{bmatrix}_j = \begin{bmatrix} \bar{Q}_{11} & \bar{Q}_{12} & \bar{Q}_{16} \\ \bar{Q}_{12} & \bar{Q}_{22} & \bar{Q}_{26} \\ \bar{Q}_{16} & \bar{Q}_{26} & \bar{Q}_{66} \end{bmatrix}_j \begin{bmatrix} \varepsilon_{xx}^{\circ} \\ \varepsilon_{yy}^{\circ} \\ 0 \end{bmatrix}.$$

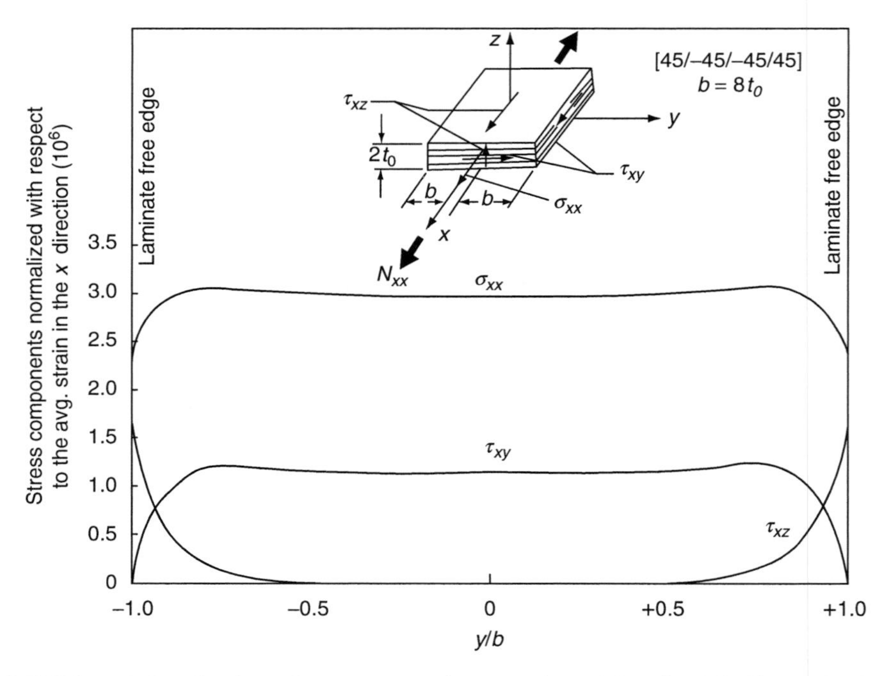

FIGURE 3.52 Interlaminar shear stress τ_{xz} between the +45° and −45° plies at the free edges of a $[\pm 45]_S$ laminate. (After Pipes, R.B. and Pagano, N.J., *J. Compos. Mater.*, 4, 538, 1970.)

Thus, although the shear stress resultant N_{xy} on the laminate is zero, each layer experiences an in-plane shear stress τ_{xy}. Since there is no applied shear stress at the laminate boundary, the in-plane shear stress must diminish from a finite value in the laminate interior to zero at its free edges. The large shear stress gradient at the ends of the laminate width is equilibrated by the development of the interlaminar shear stress τ_{xz} near the free edges, as shown in Figure 3.52. Similar equilibrium arguments can be made to demonstrate the presence of τ_{yz} and σ_{zz} in other laminates.

The principal reason for the existence of interlaminar stresses is the mismatch of Poisson's ratios ν_{xy} and coefficients of mutual influence m_x and m_y between adjacent laminas. If the laminas were not bonded and could deform freely, an axial loading in the x direction would create dissimilar transverse strains ε_{yy} in various laminas because of the difference in their Poisson's ratios. However, in perfect bonding, transverse strains must be identical throughout the laminate. The constraint against free transverse deformations produces normal stress σ_{yy} in each lamina and interlaminar shear stress τ_{yz} at the lamina

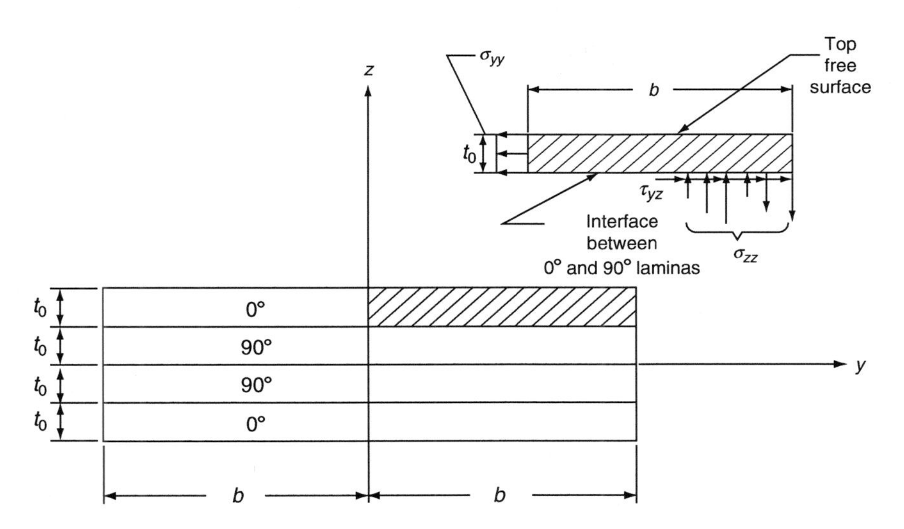

FIGURE 3.53 Source of interlaminar shear stress τ_{yz} and interlaminar normal stress σ_{zz} in a [0/90/90/0] laminate.

interfaces (Figure 3.53). Similarly, the difference in the coefficients of mutual influence m_x would create dissimilar shear strains γ_{xy} in various laminas only if they were not bonded. For a bonded laminate, equal shear strains for all laminas require the development of interlaminar shear stress τ_{zx}. Although the force equilibrium in the y direction is maintained by the action of σ_{yy} and τ_{yz}, the force resultants associated with σ_{yy} and τ_{yz} are not collinear. The moment equilibrium about the x axis is satisfied by the action of the interlaminar normal stress σ_{zz}.

Interlaminar stresses σ_{zz}, τ_{xz}, and τ_{yz} are determined by numerical methods (e.g., finite difference [28] or finite element methods [29,30]), which are beyond the scope of this book. A few approximate methods have also been developed [31,32]. For practical purposes, it may be sufficient to note the following.

1. Interlaminar stresses in laminated composites develop owing to mismatch in the Poisson's ratios and coefficients of mutual influence between various layers. If there is no mismatch of these two engineering properties, there are no interlaminar stresses regardless of the mismatch in elastic and shear moduli.
2. Interlaminar stresses can be significantly high over a region equal to the laminate thickness near the free edges of a laminate. The free edges may be at the boundaries of a laminated plate, around a cutout or hole, or at the ends of a laminated tube. As a result of high interlaminar stresses,

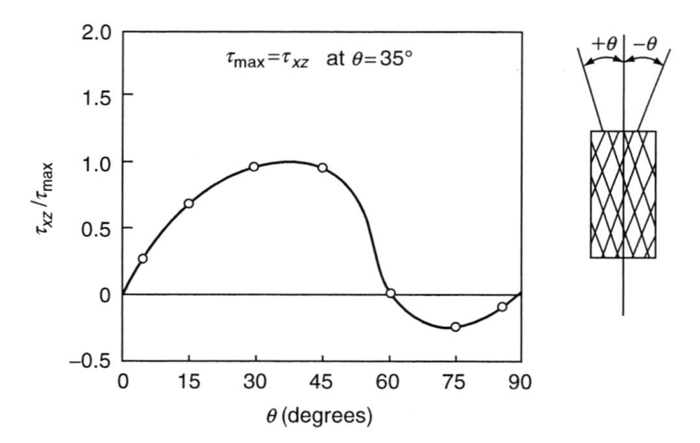

FIGURE 3.54 Variation of interlaminar shear stress τ_{xz} in a $[\pm\theta]_S$ laminate with fiber orientation angle θ. (Adapted from Pipes, R.B. and Pagano, N.J., *J. Compos. Mater.*, 4, 538, 1970.)

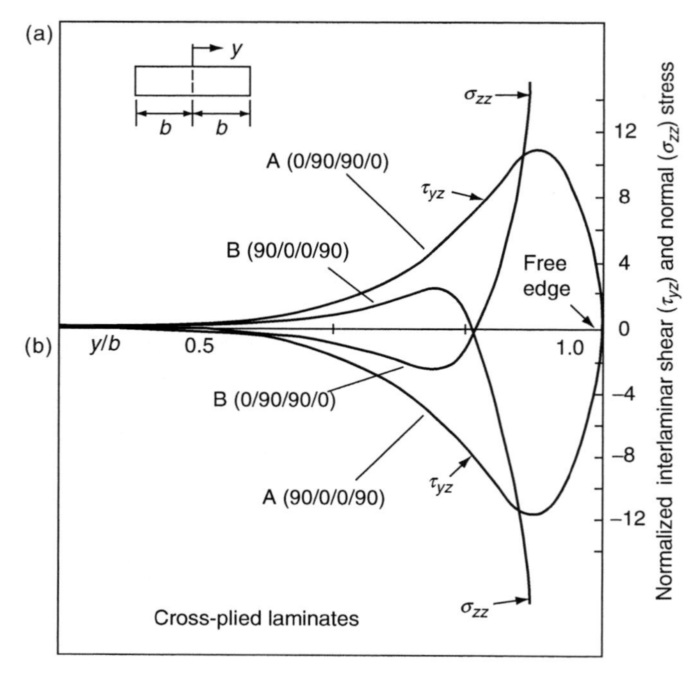

FIGURE 3.55 Distribution of (a) interlaminar shear stress τ_{yz} and (b) interlaminar normal stress σ_{zz} along the width of cross-plied laminates. (Note that the interlaminar stresses are normalized with respect to the average normal strain in the x direction.) (Adapted from Pipes, R.B., *Fibre Sci. Technol.*, 13, 49, 1980.)

delamination (i.e., separation between various laminas) may initiate at the free edges.

3. For an $[\theta/-\theta]$ angle-ply laminate in uniaxial tension, τ_{xz} is the most significant interlaminar stress at the interfaces of the θ and $-\theta$ laminas. Its magnitude and direction depend strongly on the fiber orientation angle θ (Figure 3.54). Furthermore, τ_{xz} has a higher value at the $(\theta/-\theta)$ interfaces in a clustered $[\theta_n/-\theta_n]_S$ laminate than in an alternating $[(\theta/-\theta)_n]_S$ laminate.
4. For a [0/90] type laminate in uniaxial tension, the significant interlaminar stresses are σ_{zz} and τ_{yz}. Their magnitude, locations, and directions depend strongly on the stacking sequence (Figure 3.55). For example, the maximum σ_{zz} at the midplane of a [0/90/90/0] laminate is tensile, but maximum σ_{zz} at the midplane of a [90/0/0/90] laminate is compressive. Thus, delamination is likely in the [0/90/90/0] laminate.
5. For a general laminate, different combinations of τ_{xz}, τ_{yz}, and σ_{zz} may be present between various laminas. For example, consider a [45/−45/0/0/−45/45] laminate in uniaxial tension. In this case, all

TABLE 3.4
Effect of Stacking Sequence on the Critical Interlaminar Stresses in Quasi-Isotropic $[0/90/\pm45]_S$ T-300 Carbon–Epoxy Laminates under Uniaxial Tension[a]

	Max σ_{zz}		Max τ_{xz}	
Laminate	**Value**	**Location**	**Value**	**Location**
$[90/45/0/-45]_S$	−6.8	Midplane	−6.9	0°/−45°
$[0/-45/90/45]_S$	6.2	90° layer	−6.6	90°/45°
$[45/90/0/-45]_S$	6.6	90° layer	5.9	0°/−45°
$[45/90/-45/0]_S$	6.9	90° layer	−6.5	45°/90°
$[45/0/90/-45]_S$	7.6	90° layer	−5.8	90°/−45°
$[45/0/-45/90]_S$	10.4	Midplane	−6.0	0°/−45°
$[90/0/-45/45]_S$	−8.2	−45°/45°	9.0	−45°/45°
$[90/45/-45/0]_S$	−7.4	45°/−45°	−9.2	45°/−45°
$[0/90/45/-45]_S$	−7.6	45° layer	−9.2	45°/−45°
$[0/45/-45/90]_S$	10.0	Midplane	−8.3	45°/−45°
$[45/-45/90/0]_S$	9.0	0° layer	−7.7	45°/−45°
$[45/-45/0/90]_S$	10.9	Midplane	−7.2	45°/−45°

Source: Adapted from Herakovich, C.T., *J. Compos. Mater.*, 15, 336, 1981.

[a] The stress magnitudes are in ksi. To transform to MPa, multiply by 6.89. The (/) indicates interface between adjacent layers.

three interlaminar stress components are present between the 45/−45 layers as well as the 0/−45 layers. However, the interlaminar shear stress τ_{xz} between adjacent 45/−45 laminas is higher than that between adjacent 0/−45 laminas. On the other hand, the interlaminar shear stress τ_{yz} between 0/−45 laminas is higher than that between 45/−45 laminas. However, the maximum σ_{zz} occurs at the laminate midplane.

6. Stacking sequence has a strong influence on the nature, magnitude, and location of interlaminar stresses. This is demonstrated in Table 3.4. Note that laminates with interspersed ±45° layers (separated by 0° or 90° layers) have lower τ_{xz} than those with adjacent ±45° layers, and, therefore, are less likely to delaminate. Among the laminates with

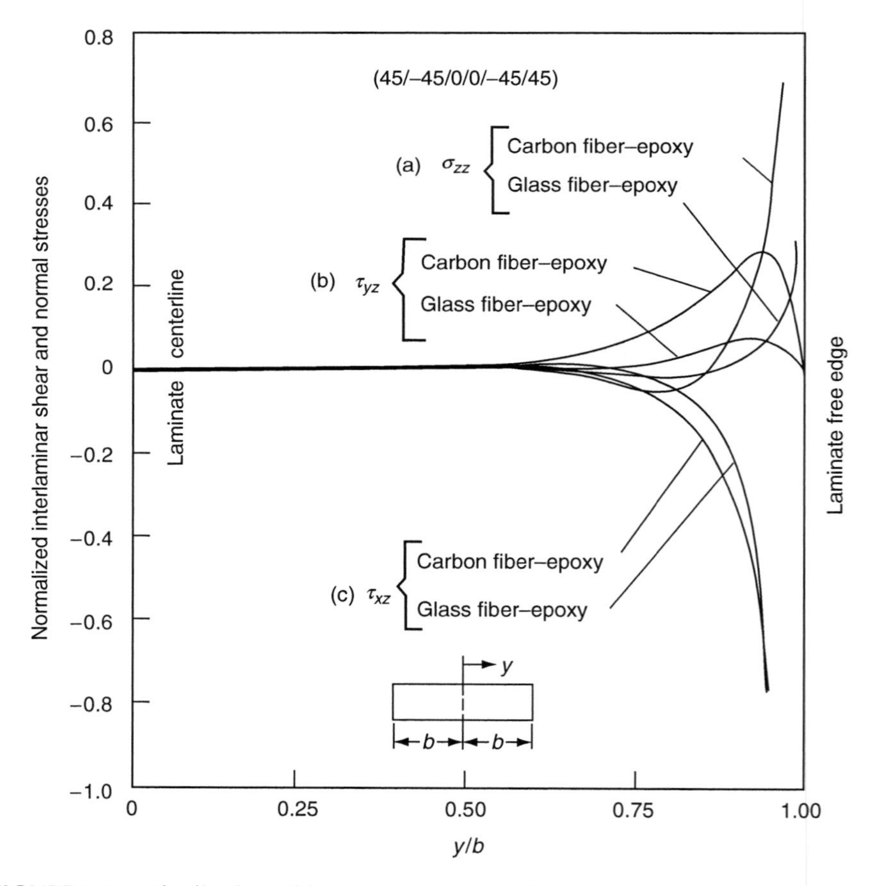

FIGURE 3.56 Distribution of interlaminar stresses in $[\pm 45/0]_S$ laminates with carbon and glass fibers in an epoxy matrix. (Adapted from Pipes, R.B., *Fibre Sci. Technol.*, 13, 49, 1980.)

interspersed ±45° layers, $[90/45/0/-45]_S$ has the most favorable σ_{zz} under a uniaxial tensile load applied on the laminates.

7. Material properties also have a strong influence on the interlaminar shear stresses of a laminate, as shown in Figure 3.56.

REFERENCES

1. G.S. Holister and C. Thomas, *Fibre Reinforced Materials*, Elsevier Publishing Co., London (1966).
2. C.C. Chamis, Micromechanics strength theories, *Composite Materials, Vol. 5, Fracture and Fatigue* (L.J. Broutman, ed.), Academic Press, New York, p. 93 (1974).
3. A.N. Netravali, P. Schwartz, and S.L. Phoenix, Study of interfaces of high performance glass fibers and DGEBA-based epoxy resins using single-fiber-composite test, *Polym. Compos.*, *6*:385 (1989).
4. H.L. Cox, The elasticity and strength of paper and other fibrous materials, *Br. J. Appl. Phys.*, *3*:72 (1952).
5. J. Cook and J.E. Gordon, A mechanism for the control of crack propagation in all brittle systems, *Proc. R. Soc. Lond., A*, *282*:508 (1964).
6. D.C. Phillips, The fracture energy of carbon fibre reinforced glass, *J. Mater. Sci., 7* (1972).
7. J.O. Outwater and M.C. Murphy, The fracture energy of unidirectional laminates, Proc. 24th Annual Technical Conference, Society of the Plastics Industry (1969).
8. A. Kelly, Interface effects and the work of fracture of a fibrous composite, *Proc. R. Soc. Lond., A*, *319*:95 (1970).
9. G.A. Cooper, Micromechanics aspects of fracture and toughness, *Composite Materials, Vol. 5, Fracture and Fatigue* (L.J. Broutman, ed.), Academic Press, New York, p. 415 (1974).
10. D.F. Adams and D.R. Doner, Transverse normal loading of a unidirectional composite, *J. Compos. Mater.*, *1*:152 (1967).
11. L.B. Greszczuk, Theoretical and experimental studies on properties and behavior of filamentary composites, Proc. 21st Annual Technical Conference, Society of the Plastics Industry (1966).
12. B.W. Rosen, Mechanics of composite strengthening, *Fiber Composite Materials*, American Society for Metals, Metals Park, OH, p. 37 (1965).
13. C.R. Schultheisz and A.M. Waas, Compressive failure of composites, Part I: Testing and micromechanical theories, *Prog. Aerospace Sci.*, *32*:1 (1996).
14. B. Budiansky and N.A. Fleck, Compressive failure of fibre composites, *J. Mech. Phys. Solids.*, *41*:183 (1993).
15. T.A. Collings, Transverse compressive behavior of unidirectional carbon fibre reinforced plastics, *Composites*, *5*:108 (1974).
16. R.M. Jones, *Mechanics of Composite Materials*, 2nd Ed., Taylor & Francis, Philadelphia (1999).
17. R.M. Christensen, The numbers of elastic properties and failure parameters for fiber composites, *J. Eng. Mater. Technol.*, *120*:110 (1998).

18. N.J. Pagano and P.C. Chou, The importance of signs of shear stress and shear strain in composites, *J. Compos. Mater.*, *3*:166 (1969).
19. C.C. Chamis and G.P. Sendeckyj, Critique on theories predicting thermoelastic properties of fibrous composites, *J. Compos. Mater.*, *2*:332 (1968).
20. R.M. Jones, Stiffness of orthotropic materials and laminated fiber-reinforced composites, *AIAA J.*, *12*:112 (1974).
21. R.A. Schapery, Thermal expansion coefficients of composite materials based on energy principles, *J. Compos. Mater.*, *2*:280 (1968).
22. L.B. Greszczuk, Effect of material orthotropy on the directions of principal stresses and strains, *Orientation Effects in the Mechanical Behavior of Anisotropic Structural Materials, ASTM STP*, *405*:1 (1966).
23. S.W. Tsai and N.J. Pagano, Invariant properties of composite materials, *Composite Materials Workshop* (S.W. Tsai, J.C. Halpin, and N.J. Pagano, eds.), Technomic Publishing Co., Stamford, CT, p. 233 (1968).
24. R.B. Pipes, J.R. Vinson, and T.W. Chou, On the hygrothermal response of laminated composite systems, *J. Compos. Mater.*, *10*:129 (1976).
25. H.T. Hahn, Residual stresses in polymer matrix composite laminates, *J. Compos. Mater.*, *10*:266 (1976).
26. M.W. Hyer, Some observations on the cured shape of thin unsymmetric laminates, *J. Compos. Mater.*, *15*:175 (1981).
27. M.W. Hyer, Calculations of room-temperature shapes of unsymmetric laminates, *J. Compos. Mater.*, *15*:295 (1981).
28. R.B. Pipes and N.J. Pagano, Interlaminar stresses in composite laminates under uniform axial extension, *J. Compos. Mater.*, *4*:538 (1970).
29. G. Isakson and A. Levy, Finite-element analysis of interlaminar shear in fiberous composites, *J. Compos. Mater.*, *5*:273 (1971).
30. E.F. Rybicki, Approximate three-dimensional solutions for symmetric laminates under inplane loading, *J. Compos. Mater.*, *5*:354 (1971).
31. J.M. Whitney, Free-edge effects in the characterization of composite materials, *Analysis of the Test Methods for High Modulus Fibers and Composites, ASTM STP*, *521*:167 (1973).
32. P. Conti and A.D. Paulis, A simple model to simulate the interlaminar stresses generated near the free edge of a composite laminate, *Delamination and Debonding of Materials, ASTM STP*, *876*:35 (1985).

PROBLEMS

P3.1. Calculate the longitudinal modulus, tensile strength, and failure strain of a unidirectional continuous fiber composite containing 60 vol% of T-800 carbon fibers ($E_f = 294$ GPa and $\sigma_{fu} = 5.6$ GPa) in an epoxy matrix ($E_m = 3.6$ GPa, $\sigma_{mu} = 105$ MPa, and $\varepsilon_{mu} = 3.1\%$). Compare these values with the experimentally determined values of $E_L = 162$ GPa, $\sigma_{Ltu} = 2.94$ GPa, and $\varepsilon_{Ltu} = 1.7\%$. Suggest three possible reasons for the differences. What fraction of load is carried by the fibers in this composite?

P3.2. The material of a tension link is changed from a Ti-6A1-4V (aged) titanium alloy to a unidirectional continuous GY-70 carbon fiber–epoxy. The stress–strain curve of the epoxy resin is shown in the following figure. Calculate the volume fraction of GY-70 fibers required in the composite link to match the modulus of the titanium alloy.

In addition, estimate the tensile strength of the composite link and compare its strength–weight ratio with that of the titanium alloy.

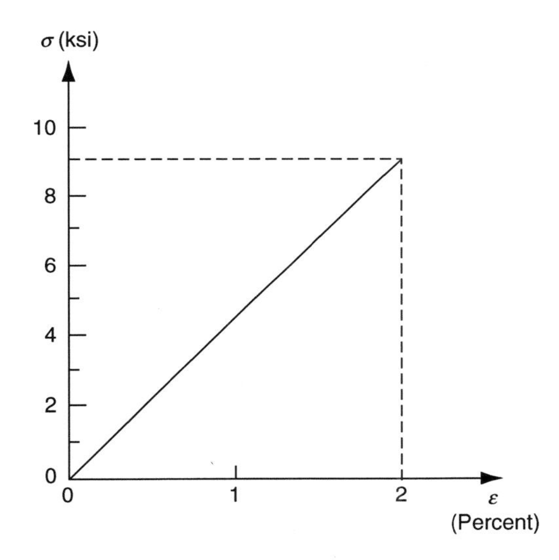

P3.3. To increase the longitudinal tensile modulus of a unidirectional continuous E-glass fiber-reinforced epoxy, some of the E-glass fibers are replaced with T-300 carbon fibers. The total fiber volume fraction is kept unchanged at 60%. Assume that the E-glass and T-300 carbon fibers in the new composite are uniformly distributed.

1. Calculate the volume fraction of T-300 carbon fibers needed in the new composite to double the longitudinal tensile modulus
2. Compare the longitudinal tensile strength of the new composite with that of the original composite
3. Schematically compare the stress–strain diagrams of the fibers, the matrix, and the composite

The tensile modulus and strength of the epoxy are 5 GPa and 50 MPa, respectively. Assume that the tensile stress–strain diagram of the epoxy is linear up to the point of failure.

P3.4. Consider a unidirectional continuous fiber lamina containing brittle, elastic fibers in an elastic-perfectly plastic matrix. The stress–strain diagrams for the fibers and the matrix are shown as follows:

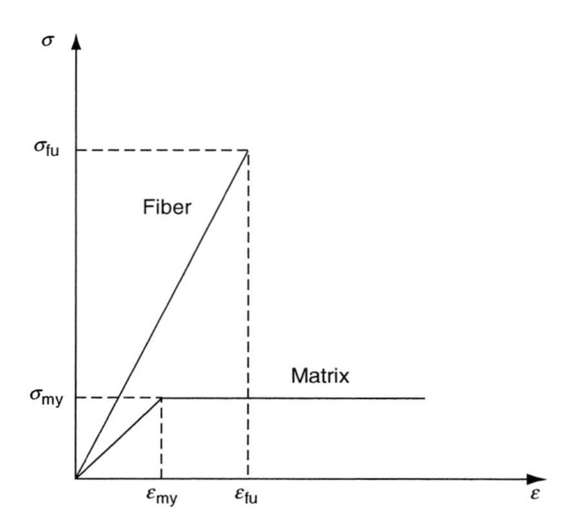

1. Calculate the longitudinal modulus of the composite lamina before and after the matrix yielding
2. Calculate the failure stress for the lamina
3. Draw the stress–strain diagram for the lamina, and explain how it may change if the matrix has the capacity for strain hardening
4. Compare the loads carried by the fibers before and after the matrix yields

P3.5. Compare E_T/E_L vs. v_f of a unidirectional continuous IM-7 carbon fiber-reinforced epoxy and a unidirectional continuous fiber E-glass-reinforced epoxy. Assume $E_m = 2.8$ GPa. What observations will you make from this comparison?

P3.6. A unidirectional continuous fiber lamina is subjected to shear stress as shown in the following figure. Using the "slab" model, show that the shear modulus G_{LT} of the lamina can be represented by the following equation.

$$\frac{1}{G_{LT}} = \frac{v_f}{G_f} + \frac{(1 - v_f)}{G_m}.$$

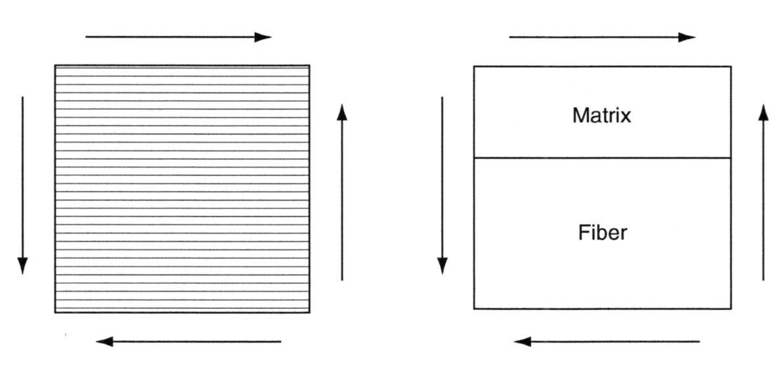

Unidirectional continuous fiber-reinforced composite
and the equivalent slab model in in-plane shear loading

P3.7. Compare the minimum critical fiber aspect ratios for E-glass, T-300 carbon, P-100 carbon, and Kevlar 49 fibers in an epoxy matrix. Assume that the epoxy matrix behaves as an elastic, perfectly plastic material with a tensile yield strength of 10,000 psi.

P3.8. Compare the failure strength of a unidirectional alumina whisker ($l_f/d_f = 200$)-reinforced epoxy with that of a unidirectional continuous alumina fiber-reinforced epoxy. The tensile strength of alumina whiskers is 1,000,000 psi, but that of continuous alumina fibers is 275,000 psi. Assume $v_f = 0.5$ and $\tau_{my} = 4{,}800$ psi.

P3.9. A unidirectional discontinuous E-glass fiber-reinforced vinyl ester composite is required to have a longitudinal tensile strength of 1000 MPa. The fiber volume fraction is 60%. Fiber length and fiber bundle diameter are 12 and 1 mm, respectively. Determine the fiber–matrix interfacial shear strength needed to achieve the required longitudinal tensile strength. The fiber and matrix properties are as follows:
Fiber: Modulus = 72.4 GPa, tensile strength = 2500 MPa
Matrix: Modulus = 2.8 GPa, tensile strength = 110 MPa.

P3.10. Derive an expression for the critical fiber volume fraction in a unidirectional discontinuous fiber composite. On a plot of the composite tensile strength vs. fiber volume fraction, indicate how the critical fiber volume fraction depends on the fiber length.

P3.11. In deriving Equation 3.13, the interfacial shear stress has been assumed constant. Instead, assume that

$$\begin{aligned}\tau_i &= 3000 - 6000\frac{x}{l_t}\,\text{psi for } 0 \le x \le \frac{1}{2}l_t \\ &= 0 \quad \text{for } \frac{1}{2}l_t \le x \le \frac{1}{2}l_f\end{aligned}$$

(a) Show how the fiber stress varies with x, (b) Calculate the critical fiber length, and (c) Calculate the average fiber stress.

P3.12. Using Equation 3.20, derive an expression for the average longitudinal stress in a discontinuous fiber. Assuming a simple square array of AS-1 carbon fibers in an epoxy matrix ($G_m = 1.01$ GPa), plot the average longitudinal fiber stress as a function of l_f/d_f for $v_f = 0.2$, 0.4, and 0.6.

P3.13. The interfacial shear strength of a fiber–matrix joint is often measured by a pullout test. This involves pulling a fiber bundle out of a resin disk cast around a small length of the bundle. A typical load–displacement curve obtained in a pullout test is shown.

1. Calculate the average interfacial shear strength of the joint
2. What must the maximum thickness of the resin disk be so that the fiber bundle pulls out before it breaks within the disk?

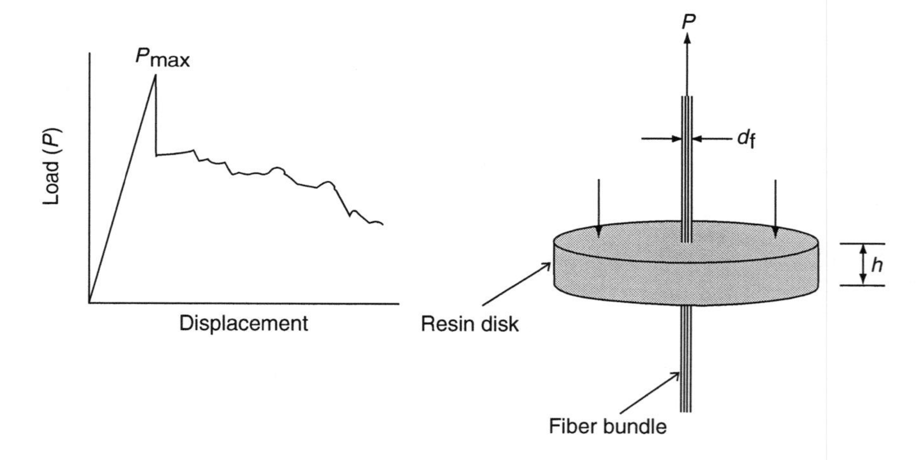

P3.14. Using the equations for the fiber pullout energies (Table 3.2) show that the maximum energy dissipation by fiber pullout occurs at $l_f = l_c$. How do the fiber tensile strength and fiber–matrix interfacial strength affect the pullout energy?

P3.15. Longitudinal tensile tests of single-fiber specimens containing AS-1 carbon fiber in epoxy and HMS-4 carbon fiber in epoxy produce cleavage cracks (normal to the fiber direction) in the matrix adjacent to the fiber rupture. However, the cleavage crack in the AS specimen is longer than in the HMS specimen. Furthermore, the longitudinal tensile strength of the AS specimen increases significantly with increasing matrix ductility, but that of the HMS carbon specimen remains unaffected. Explain both phenomena in terms of the energy released on fiber fracture.

P3.16. A fiber breaks at a location away from the matrix crack plane and pulls out from the matrix with the opening of the matrix crack. Assuming that the embedded fiber length l in the figure is less than half the critical length l_c, show that the work required to pull out the fiber is

$$W_{po} = \frac{\pi}{2} d_f l^2 \tau_i,$$

where τ_i is the interfacial shear stress (assumed constant). What might be expected if the embedded fiber length l is greater than $\frac{1}{2} l_c$?

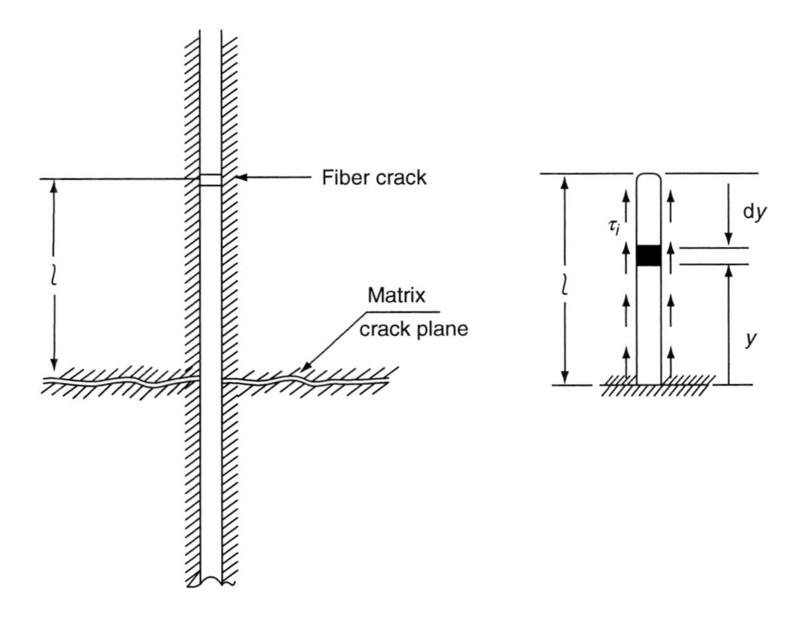

P3.17. Using the rule of mixture approach as was done for longitudinal tensile loading, derive equations for the longitudinal compressive modulus and strength of a unidirectional continuous fiber composite for the following cases:

1. $\varepsilon_{fc} < \varepsilon_{myc}$
2. $\varepsilon_{fc} > \varepsilon_{myc}$

where

ε_{fc} = fiber "fracture" strain in compression
ε_{myc} = matrix yield strain in compression

Compare the rule of mixtures approach with Rosen's microbuckling approach for a carbon fiber-reinforced epoxy composite using the following information: $E_{fc} = 517$ GPa, $\varepsilon_{fc} = 0.25\%$, $E_{mc} = 2.1$ GPa, $\varepsilon_{myc} = 2.85\%$, $\nu_m = 0.39$, and $v_f = 0.4, 0.5, 0.6$.

P3.18. Under longitudinal compressive loads, a unidirectional continuous fiber-reinforced brittle matrix composite often fails by longitudinal matrix cracks running parallel to the fibers. Explain this failure mode in terms of the stress and strain states in the matrix, and, derive an equation for the longitudinal compressive strength of the composite for this failure mode.

P3.19. A 500 mm long × 25 mm wide × 3 mm thick composite plate contains 55% by weight of unidirectional continuous T-300 carbon fibers in an epoxy matrix parallel to its length.

1. Calculate the change in length, width, and thickness of the plate if it is subjected to an axial tensile force of 75 kN in the length direction
2. Calculate the change in length, width, and thickness of the plate if it subjected to an axial tensile force of 75 kN in the width direction

Assume that the density, modulus, and Poisson's ratio of the epoxy matrix are 1.25 g/cm^3, 3.2 GPa, and 0.3, respectively.

P3.20. A round tube (outside diameter = 25 mm, wall thickness = 2.5 mm, and length = 0.5 m) is made by wrapping continuous AS-4 carbon fiber-reinforced epoxy layers, all in the hoop direction. The fiber volume fraction is 60%.

1. Determine the change in length and diameter of the tube if it is subjected to an axial tensile load of 2 kN
2. Determine the maximum axial tensile load that can be applied on the tube?
3. Suppose the tube is used in a torsional application. What will be its torsional stiffness (torque per unit angle of twist)?

The modulus, tensile strength, and Poisson's ratio of the epoxy matrix are 5 GPa, 90 MPa, and 0.34, respectively.

P3.21. The normal stress σ_{xx} of 100 MPa and shear stress τ_{xy} of 25 MPa are applied on a unidirectional angle-ply lamina containing fibers at an angle θ as shown in the figure. Determine the stresses in the principal material directions for $\theta = 0°, 15°, 30°, 45°, 60°, 75°$, and $90°$. Do these stresses remain the same (a) if the direction for the shear stress τ_{xy} is reversed and (b) if the fiber orientation angles are reversed?

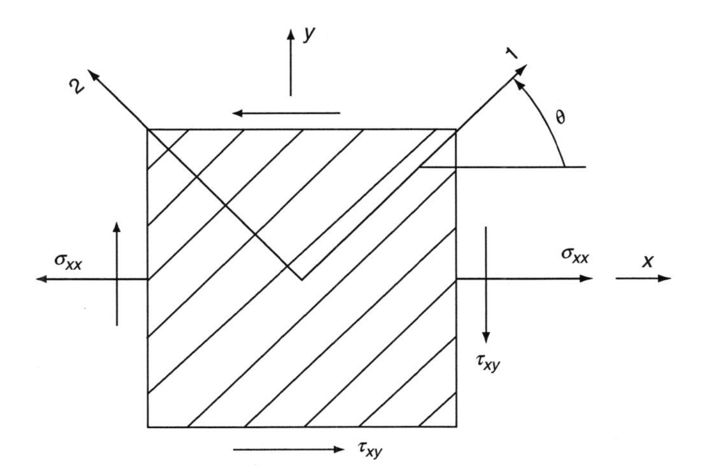

P3.22. A tubular specimen containing fibers at a helix angle α with the tube axis is tested in a combined tension–torsion test. Determine the ratio of σ_{xx} and τ_{xy} as well as the required helix angle α that will create biaxial principal stresses σ_{11} and σ_{22} of ratio m. Note that the shear stress τ_{12} in the principal stress directions is zero.

P3.23. A cylindrical oxygen tank made of an E-glass fiber-reinforced epoxy contains oxygen at a pressure of 10 MPa. The tank has a mean diameter of 300 mm and a wall thickness of 8.9 mm. The fiber orientation angles in various layers of the tank wall are ±55° with its longitudinal axis. Neglecting the interaction between the layers, calculate the stresses in the principal material directions for both fiber orientation angles.

P3.24. The following tensile modulus values were experimentally determined for a unidirectional carbon fiber-reinforced PEEK composite ($v_f = 0.62$):

Fiber orientation angle (degrees)	0	5	10	30	45	60	75	90
Modulus (GPa)	135.2	113.4	72	25.4	11.5	9.65	8.36	9.20

Plot the data as a function of the fiber orientation angle and compare them with the theoretical predictions assuming $E_f = 230$ GPa, $\nu_f = 0.28$, $E_m = 3.45$ GPa, and $\nu_m = 0.4$. Verify the validity of Equation 3.36.

P3.25. Calculate the elastic constants E_{xx}, E_{yy}, ν_{xy}, ν_{yx}, and G_{xy} for a T-300 carbon fiber-reinforced epoxy lamina. The fiber orientation angle is 30°, and the fiber volume fraction is 0.6. For the epoxy matrix, use $E_m = 2.07$ GPa and $\nu_m = 0.45$.

P3.26. A unidirectional discontinuous fiber lamina contains T-300 carbon fiber in an epoxy matrix. The fiber aspect ratio (l_f/d_f) is 50, and the fiber volume fraction is 0.5. Determine the elastic constants E_{11}, E_{22}, ν_{12}, ν_{21}, and G_{12} for the lamina. For the matrix, use $E_m = 2.07$ GPa and $\nu_m = 0.45$. If the fibers are misaligned by 10° with the uniaxial loading direction, how would these elastic constants change?

P3.27. The material used in the transmission gears of an automobile is an injection-molded nylon 6,6 containing 20 wt% of chopped randomly oriented E-glass fibers. The tensile modulus of this material is 1.25×10^6 psi.

In a more demanding application for the transmission gears, the modulus of the material must be 50% higher. An engineer wants to accomplish this by replacing the E-glass fibers with carbon fibers. If the fiber weight fraction remains the same, calculate the length of carbon fibers that must be used to obtain the desired modulus.

Use the following information in your calculations. (a) For the carbon fiber, $\rho_f = 1.8$ g/cm^3, $E_f = 30 \times 10^6$ psi, and $d_f = 0.0006$ in. and (b) for nylon 6,6, $\rho_m = 1.14$ g/cm^3 and $E_m = 0.4 \times 10^6$ psi.

P3.28. A unidirectional discontinuous E-glass fiber-reinforced polyphenylene sulfide (PPS) composite needs to be developed so that its longitudinal tensile modulus is at least 25 GPa and its longitudinal tensile strength is at least 950 MPa. Through the use of proper coupling agent on the glass fiber surface, it would be possible to control the interfacial shear strength between 10 and 30 MPa. The fiber bundle diameter is 0.30 mm and the fiber weight fraction is 60%. Determine the fiber length required for this composite.

The matrix properties are: $\rho_m = 1.36$ g/cm^3, $E_m = 3.5$ GPa, and $S_{mu} = 165$ MPa.

P3.29. A unidirectional continuous fiber lamina contains carbon fibers in an epoxy matrix. The fiber volume fraction is 0.55. The coefficient of longitudinal thermal expansion for the lamina is measured as -0.61×10^{-6} per °C, and that for the matrix at the same temperature is 54×10^{-6} per °C. Estimate the coefficient of thermal expansion for the fiber. The longitudinal modulus of the lamina is 163.3 GPa and the matrix modulus is 3.5 GPa.

P3.30. Coefficients of axial and transverse thermal expansion of 0° unidirectional Spectra 900 fiber-reinforced epoxy composite ($v_f = 60\%$) are

-9×10^{-6} and 100×10^{-6} per °C, respectively. For the same composite, the major Poisson's ratio is 0.32. The matrix properties are $E_m = 2.8$ GPa, $\nu_m = 0.38$, and $\alpha_m = 60 \times 10^{-6}$ per °C. Using these values, estimate (a) the Poisson's ratio of the fiber, (b) coefficients of thermal expansion of the fiber in longitudinal and radial directions, and (c) the fiber volume fraction at which the composite has a zero CTE.

P3.31. A 1 m long thin-walled composite tube has a mean diameter of 25 mm and its wall thickness is 2 mm. It contains 60 vol% E-glass fibers in a vinyl ester matrix. Determine the change in length and diameter of the tube if the temperature is increased by 50°C. The matrix properties are $E_m = 3.5$ GPa, $\nu_m = 0.35$, $\alpha_m = 70 \times 10^{-6}$ per °C.

P3.32. An E-glass fiber–epoxy laminate has the following construction:

$$[0/30/-30/45/-45/90/-45/45/-30/30/0].$$

The following are known: $v_f = 0.60$, $E_f = 10 \times 10^6$ psi, $E_m = 0.34 \times 10^6$ psi, $\nu_f = 0.2$, $\nu_m = 0.35$, $\alpha_f = 5 \times 10^{-6}$ per °C, and $\alpha_m = 60 \times 10^{-6}$ per °C. Determine the coefficients of thermal expansion in the x and y directions for each lamina.

P3.33. Consider a unidirectional continuous fiber lamina. Applying σ_{11}, σ_{22}, and τ_{12} separately, show that the engineering elastic constants E_{11}, E_{22}, ν_{12}, ν_{21}, and G_{12} can be expressed in terms of the elements in the lamina stiffness matrix as

$$E_{11} = Q_{11} - \frac{Q_{12}^2}{Q_{22}},$$

$$E_{22} = Q_{22} - \frac{Q_{12}^2}{Q_{11}},$$

$$\nu_{12} = \frac{Q_{12}}{Q_{22}},$$

$$\nu_{21} = \frac{Q_{12}}{Q_{11}},$$

$$G_{12} = Q_{66}.$$

P3.34. A T-300 carbon fiber–epoxy lamina ($v_f = 0.60$) with a fiber orientation angle of 45° is subjected to a biaxial stress state of $\sigma_{xx} = 100$ MPa and $\sigma_{yy} = -50$ MPa. Determine (a) the strains in the x–y directions, (b) the strains in the 1–2 directions, and (c) the stresses in the 1–2 directions. Use the material property data of Example 3.6.

P3.35. Plot and compare the coefficients of mutual influence as functions of fiber orientation angle θ in T-300 carbon fiber–epoxy laminas containing fibers at the $+\theta$ and $-\theta$ orientations. For what fiber orientation angle θ do the coefficients of mutual influence have the maximum values? Use the material property data of Example 3.6.

P3.36. The elastic constants of a 0° unidirectional carbon fiber-reinforced PEEK lamina are $E_{11} = 132.2$ GPa, $E_{22} = 9.2$ GPa, $G_{12} = 4.90$ GPa, and $\nu_{12} = 0.35$. Write the compliance and stiffness matrices for the same material if the fiber orientation angle is (a) 30°, (b) −30°, (c) 60°, and (d) 90°.

P3.37. A T-300 carbon fiber–epoxy lamina ($v_f = 0.6$) is subjected to a uniaxial normal stress σ_{xx}. Compare the strains in the x–y directions as well as in the 1–2 directions for $\theta = 0°$, +45°, −45°, and 90°. Use the material property data of Example 3.6 and Problem P3.35.

P3.38. Compare the stiffness matrices of three-layered [0/60/−60], [−60/0/60], and [−60/60/−60] laminates. Which of these laminates can be considered quasi-isotropic, and why? Assume that each layer has the same thickness t_0.

P3.39. Compare the stiffness matrices of two-layered, three-layered, and four-layered angle-ply laminates containing alternating θ and $-\theta$ laminas. Assume that each layer has the same thickness t_0.

P3.40. Show that the extensional stiffness matrices for quasi-isotropic $[0/\pm60]_S$, $[\pm60/0]_S$, and $[60/0/-60]_S$ laminates are identical, while their bending stiffness matrices are different.

P3.41. The modulus of a $[0_m/90_n]$ laminate can be calculated using the following "averaging" equation.

$$E_{xx} = \frac{m}{m+n} E_{11} + \frac{n}{m+n} E_{22}.$$

Suppose a $[0/90/0]_{3S}$ laminate is constructed using continuous T-300 fibers in an epoxy matrix. Verify that the modulus of the laminate calculated by the averaging equation is the same as calculated by the lamination theory. Use the material properties given in Example 3.6.

P3.42. The $[A]$ matrix for a boron fiber–epoxy $[\pm45]_S$ laminate of thickness h is

$$[A] = \begin{bmatrix} 0.99h & 0.68h & 0 \\ 0.68h & 0.99h & 0 \\ 0 & 0 & 0.72h \end{bmatrix} \times 10^7 \text{ lb/in.}$$

1. Calculate the engineering elastic constants for the laminate
2. Calculate the strains in the +45° and −45° laminas owing to average laminate stresses $N_{xx}/h = N_{yy}/h = p$ and $N_{xy}/h = 0$. Assume that each lamina has a thickness of $h/4$

P3.43. The elastic properties of unidirectional carbon fiber–epoxy lamina are $E_{11} = 181.3$ GPa, $E_{22} = 10.27$ GPa, $G_{12} = 7.17$ GPa, and $\nu_{12} = 0.28$. Compare the engineering elastic constants of the $[\pm 45/0]_S$ and $[\pm 45/0/90]_S$ laminates manufactured from this carbon fiber–epoxy material.

P3.44. Show that the shear modulus of a thin $[\pm 45]_{nS}$ plate is given by

$$G_{xy} = \frac{1}{4}\left[\frac{E_{11} + E_{22} - 2\nu_{12}E_{22}}{1 - \nu_{12}\nu_{21}}\right].$$

P3.45. Show that the elements in the bending stiffness matrix of $[0/-60/60]_S$ and $[0/90/45/-45]_S$ laminates are given by

$$D_{ij} = \frac{h^3}{12}\left[\frac{(\bar{Q}_{ij})_{60°} + 7(\bar{Q}_{ij})_{-60°} + 19(\bar{Q}_{ij})_{0°}}{27}\right]$$

and

$$D_{ij} = \frac{h^3}{12}\left[\frac{(\bar{Q}_{ij})_{-45°} + 7(\bar{Q}_{ij})_{45°} + 19(\bar{Q}_{ij})_{90°} + 37(\bar{Q}_{ij})_{0°}}{64}\right],$$

respectively. Here, h represents the laminate thickness.

P3.46. A torsional moment M_{xy} applied to a symmetric laminated plate creates a bending curvature as well as a twisting curvature. Find an expression for the additional bending moment M_{xx} that must be applied to the plate to create a pure twisting curvature k_{xy}.

P3.47. An ARALL-4 laminate contains three layers of 2024-T8 aluminum alloy sheet (each 0.3 mm thick) and two layers of 0° unidirectional Kevlar 49-epoxy in an alternate sequence, $[\text{Al}/0_K/\text{Al}/0_K/\text{Al}]$. Elastic properties of the aluminum alloy are $E = 73$ GPa and $\nu = 0.32$, whereas those for the Kevlar layers are $E_{11} = 56.2$ GPa, $E_{22} = 4.55$ GPa, $\nu_{12} = 0.456$, and $G_{12} = 1.85$ GPa. The nominal thickness of the laminate is 1.3 mm. Calculate the elastic properties of the ARALL-4 laminate.

P3.48. Using the basic ply level properties of Example 3.6, determine the stresses in each layer of a $[\pm 45]_{32S}$ laminate subjected to $N_{xx} = 0.1$ N/mm. The ply thickness is 0.013 mm.

P3.49. Using the material properties in Example 3.16, determine the residual thermal stresses in each lamina of (i) a $[90_2/0]_S$ and (ii) a $[0/90/0]_S$ laminate. Both laminates are slowly cooled down from a curing temperature of 190°C to 23°C.

P3.50. The following thermomechanical properties are known for a carbon fiber–epoxy composite: $E_{11} = 145$ GPa, $E_{22} = 9$ GPa, $G_{12} = 4.5$ GPa, $\nu_{12} = 0.246$, $\alpha_{11} = -0.25 \times 10^{-6}$ per °C, and $\alpha_{22} = 34.1 \times 10^{-6}$ per °C. Determine the coefficients of thermal expansion of a $[45/-45]_S$ laminate of this material.

P3.51. Using the ply level thermomechanical properties given in Problem 3.49, determine the coefficients of thermal expansion of a $[0/45/-45/90]_S$ laminate of this material.

P3.52. An approximate expression for the maximum interlaminar shear stress τ_{xz} in a $[\theta/-\theta]_{nS}$ class of laminates* is

$$\text{Max}\,\tau_{xz} = \frac{1}{2n}\left[\frac{A_{22}\bar{Q}_{16} - A_{12}\bar{Q}_{26}}{A_{11}A_{22} - A_{12}^2}\right]\frac{N_{xx}}{h},$$

where N_{xx}/h is the average tensile stress on the laminate in the x direction.

Using this expression, compare the maximum interlaminar shear stress τ_{xz} in $[15/-15]_{8S}$ and $[45/-45]_{8S}$ T-300 carbon–epoxy laminates. Use Example 3.6 for the basic material property data.

P3.53. Following is an approximate expression* for the maximum interlaminar normal stress, σ_{zz}, at an interface position z from the midplane of a symmetric $[0/90]_S$ type laminate:

* J.M. Whitney, I.M. Daniel, and R.B. Pipes, *Experimental Mechanics of Fiber Reinforced Composite Materials*, Society for Experimental Mechanics, Brookfield Center, CT (1984).

$$\text{Max}\,\sigma_{zz}(z) = \frac{90\bar{\sigma}_{xx}}{7Ah}\sum_{j}\left[A_{22}\left(\bar{Q}_{12}\right)_j - A_{12}\left(\bar{Q}_{22}\right)_j\right]t_j(\eta_j - z),$$

where

σ_{xx} = applied normal stress in the x direction
A = $A_{11}A_{22} - A_{12}^2$
h = laminate thickness
η_j = distance from the midplane of the laminate to the midplane of the jth lamina
t_j = thickness of the jth lamina

and the summation extends over all the laminas above the interface position z. This equation is valid for thin laminas in which the variation of in-plane stresses is assumed to be negligible over the thickness of each lamina.

Using this approximate expression, compare the maximum interlaminar normal stresses at the midplanes of [0/90/90/0] and [90/0/0/90] laminates.

4 Performance

The performance of an engineering material is judged by its properties and behavior under tensile, compressive, shear, and other static or dynamic loading conditions in both normal and adverse test environments. This information is essential for selecting the proper material in a given application as well as designing a structure with the selected material. In this chapter, we describe the performance of fiber-reinforced polymer composites with an emphasis on the general trends observed in their properties and behavior. A wealth of property data for continuous fiber thermoset matrix composites exists in the published literature. Continuous fiber-reinforced thermoplastic matrix composites are not as widely used as continuous fiber-reinforced thermoset matrix composites and lack a wide database.

Material properties are usually determined by conducting mechanical and physical tests under controlled laboratory conditions. The orthotropic nature of fiber-reinforced composites has led to the development of standard test methods that are often different from those used for traditional isotropic materials. These unique test methods and their limitations are discussed in relation to many of the properties considered in this chapter. The effects of environmental conditions, such as elevated temperature or humidity, on the physical and mechanical properties of composite laminates are presented near the end of the chapter. Finally, long-term behavior, such as creep and stress rupture, and damage tolerance are also discussed.

4.1 STATIC MECHANICAL PROPERTIES

Static mechanical properties, such as tensile, compressive, flexural, and shear properties, of a material are the basic design data in many, if not most, applications. Typical mechanical property values for a number of 0° laminates and sheet-molding compound (SMC) laminates are given in Appendix A.5 and Appendix A.6, respectively.

4.1.1 Tensile Properties

4.1.1.1 Test Method and Analysis

Tensile properties, such as tensile strength, tensile modulus, and Poisson's ratio of flat composite laminates, are determined by static tension tests in accordance

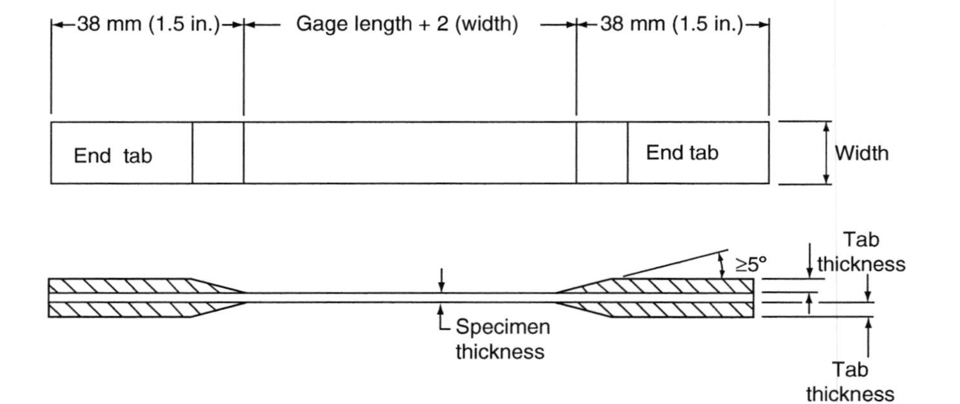

FIGURE 4.1 Tensile test specimen configuration.

with ASTM D3039. The tensile specimen is straight-sided and has a constant cross section with beveled tabs adhesively bonded at its ends (Figure 4.1). A compliant and strain-compatible material is used for the end tabs to reduce stress concentrations in the gripped area and thereby promote tensile failure in the gage section. Balanced [0/90] cross-ply tabs of nonwoven E-glass–epoxy have shown satisfactory results. Any high-elongation (tough) adhesive system can be used for mounting the end tabs to the test specimen.

The tensile specimen is held in a testing machine by wedge action grips and pulled at a recommended cross-head speed of 2 mm/min (0.08 in./min). Longitudinal and transverse strains are measured employing electrical resistance strain gages that are bonded in the gage section of the specimen. Longitudinal tensile modulus E_{11} and the major Poisson's ratio ν_{12} are determined from the tension test data of 0° unidirectional laminates. The transverse modulus E_{22} and the minor Poisson's ratio ν_{21} are determined from the tension test data of 90° unidirectional laminates.

For an off-axis unidirectional specimen ($0° < \theta < 90°$), a tensile load creates both extension and shear deformations (since A_{16} and $A_{26} \neq 0$). Since the specimen ends are constrained by the grips, shear forces and bending couples are induced that create a nonuniform S-shaped deformation in the specimen (Figure 4.2). For this reason, the experimentally determined modulus of an off-axis specimen is corrected to obtain its true modulus [1]:

$$E_{\text{true}} = (1 - \eta)\, E_{\text{experimental}},$$

where

$$\eta = \frac{3\bar{S}_{16}^2}{\bar{S}_{11}^2[3(\bar{S}_{66}/\bar{S}_{11}) + 2(L/w)^2]}, \tag{4.1}$$

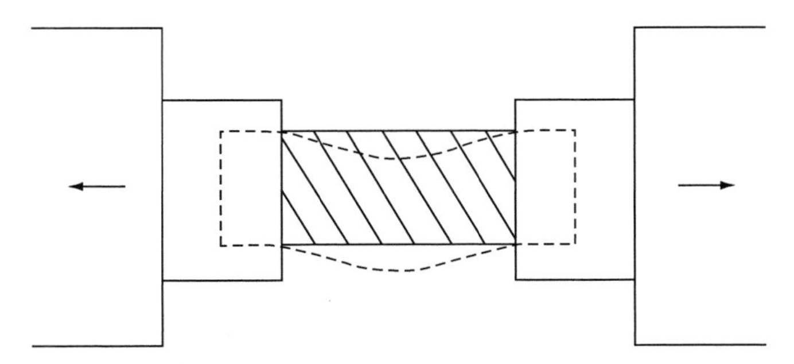

FIGURE 4.2 Nonuniform deformation in a gripped off-axis tension specimen.

where

L is the specimen length between grips
w is the specimen width
$\bar{S}_{11}$,$\bar{S}_{16}$, and $\bar{S}_{66}$ are elements in the compliance matrix (see Chapter 3)

The value of η approaches zero for large values of L/w. Based on the investigation performed by Rizzo [2], L/w ratios >10 are recommended for the tensile testing of off-axis specimens.

The inhomogeneity of a composite laminate and the statistical nature of its constituent properties often lead to large variation in its tensile strength. Assuming a normal distribution, the average strength, standard deviation, and coefficient of variation are usually reported as

$$\text{Average strength} = \sigma_{\text{ave}} = \sum \frac{\sigma_i}{n},$$

$$\text{Standard deviation} = d = \sqrt{\frac{\sum (\sigma_i - \sigma_{\text{ave}})^2}{(n-1)}},$$

$$\text{Coefficient of variation} = \frac{100d}{\sigma_{\text{ave}}}, \tag{4.2}$$

where

n is the number of specimens tested
σ_i is the tensile strength of the ith specimen

Instead of a normal distribution, a more realistic representation of the tensile strength variation of a composite laminate is the Weibull distribution. Using two-parameter Weibull statistics, the cumulative density function for the composite laminate strength is

$$F(\sigma) = \text{Probability of surviving stress } \sigma = \exp\left[-\left(\frac{\sigma}{\sigma_0}\right)^{\alpha}\right], \tag{4.3}$$

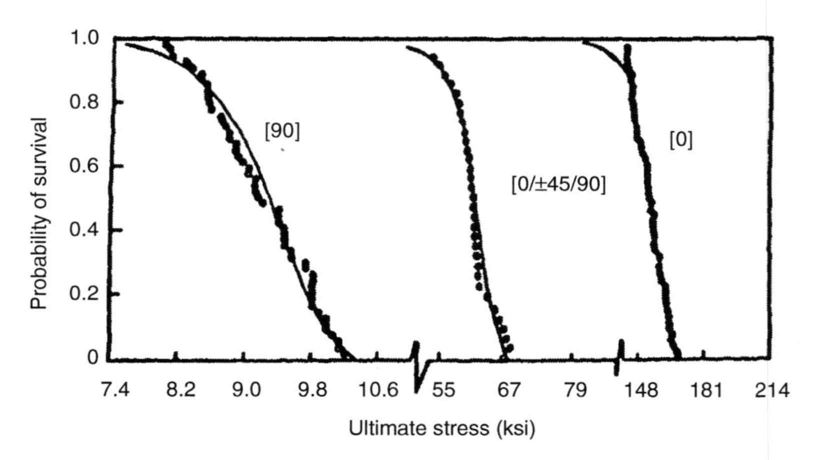

FIGURE 4.3 Tensile strength distribution in various carbon fiber–epoxy laminates. (Adapted from Kaminski, B.E., *Analysis of the Test Methods for High Modulus Fibers and Composites, ASTM STP*, 521, 181, 1973.)

where
 α is a dimensionless shape parameter
 σ_0 is the location parameter (MPa or psi)

The mean tensile strength and variance of the laminates are

$$\bar{\sigma} = \sigma_0 \Gamma\left(\frac{1+\alpha}{\alpha}\right),$$

$$s^2 = \sigma_0^2\left[\Gamma\left(\frac{2+\alpha}{\alpha}\right) - \Gamma^2\left(\frac{1+\alpha}{\alpha}\right)\right], \tag{4.4}$$

where Γ represents a gamma function.

Figure 4.3 shows typical strength distributions for various composite laminates. Typical values of α and σ_0 are shown in Table 4.1. Note that the decreasing value of the shape parameter α is an indication of greater scatter in the tensile strength data.

EXAMPLE 4.1

Static tension test results of 22 specimens of a 0° carbon–epoxy laminate shows the following variations in its longitudinal tensile strength (in MPa): 57.54, 49.34, 68.67, 50.89, 53.20, 46.15, 71.49, 72.84, 58.10, 47.14, 67.64, 67.10, 72.95, 50.78, 63.59, 54.87, 55.96, 65.13, 47.93, 60.67, 57.42, and 67.51. Plot the Weibull distribution curve, and determine the Weibull parameters α and σ_0 for this distribution.

TABLE 4.1
Typical Weibull Parameters for Composite Laminates

Material	Laminate	Shape Parameter, α	Location Parameter, σ_0 MPa (ksi)	
Boron–epoxy[a]	[0]	24.3	1324.2	(192.0)
	[90]	15.2	66.1	(9.6)
	$[0_2/\pm45]_S$	18.7	734.5	(106.6)
	$[0/\pm45/90]_S$	19.8	419.6	(60.9)
	$[90_2/45]_S$	19.8	111.9	(16.1)
T-300 Carbon–epoxy[b]	$[0_8]$	17.7	1784.5	(259)
	$[0_{16}]$	18.5	1660.5	(241)
E-glass–polyester SMC[c]	SMC-R25	7.6	74.2	(10.8)
	SMC-R50	8.7	150.7	(21.9)

[a] From B.E. Kaminski, *Analysis of the Test Methods for High Modulus Fibers and Composites, ASTM STP*, 521, 181, 1973.
[b] From R.E. Bullock, *J. Composite Mater.*, 8, 200, 1974.
[c] From C.D. Shirrell, *Polym. Compos.*, 4, 172, 1983.

Solution

Step 1: Starting with the smallest number, arrange the observed strength values in ascending order and assign the following probability of failure value for each strength.

$$P = \frac{i}{n+1},$$

where
i = 1, 2, 3, ..., n
n = total number of specimens tested

i	σ	P
1	46.15	1/23 = 0.0435
2	47.14	2/23 = 0.0869
3	47.94	3/23 = 0.1304
...	...	...
...	...	...
21	72.84	21/23 = 0.9130
22	72.95	22/23 = 0.9565

Step 2: Plot P vs. tensile strength σ to obtain the Weibull distribution plot (see the following figure).

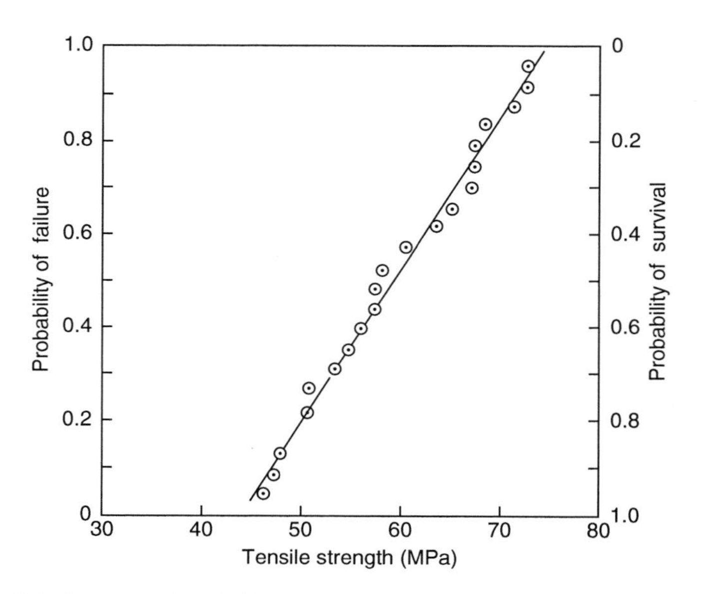

Step 3: Calculate $Y_P = \ln\{\ln[1/(1-P)]\}$ for each strength value, and plot Y_P vs. $\ln \sigma$. Use a linear least-squares method to fit a straight line to the data. The slope of this line is equal to α, and its intersection with the $\ln \sigma$ axis is equal to $\ln \sigma_0$. In our example, $\alpha = 7.62$ and $\ln \sigma_0 = 4.13$, which gives $\sigma_0 = 62.1$ MPa.

4.1.1.2 Unidirectional Laminates

For unidirectional polymer matrix laminates containing fibers parallel to the tensile loading direction (i.e., $\theta = 0°$), the tensile stress–strain curve is linear up to the point of failure (Figure 4.4). These specimens fail by tensile rupture of fibers, which is followed or accompanied by longitudinal splitting (debonding along the fiber–matrix interface) parallel to the fibers. This gives a typical broom-type appearance in the failed area of 0° specimens (Figure 4.5a). For off-axis specimens with $0° < \theta < 90°$, the tensile stress–strain curves may exhibit nonlinearity. For 90° specimens in which the fibers are 90° to the tensile loading direction, tensile rupture of the matrix or the fiber–matrix interface causes the ultimate failure. For intermediate angles, failure may occur by a combination of fiber–matrix interfacial shear failure, matrix shear failure, and matrix tensile rupture. For many of these off-axis specimens (including 90°), matrix craze marks parallel to the fiber direction may appear throughout the gage length at low loads. Representative failure profiles for these specimens are shown in Figure 4.5b and c.

Both tensile strength and modulus for unidirectional specimens depend strongly on the fiber orientation angle θ (Figure 4.6). The maximum tensile strength and modulus are at $\theta = 0°$. With increasing fiber orientation angle, both tensile strength and modulus are reduced. The maximum reduction is observed near $\theta = 0°$ orientations.

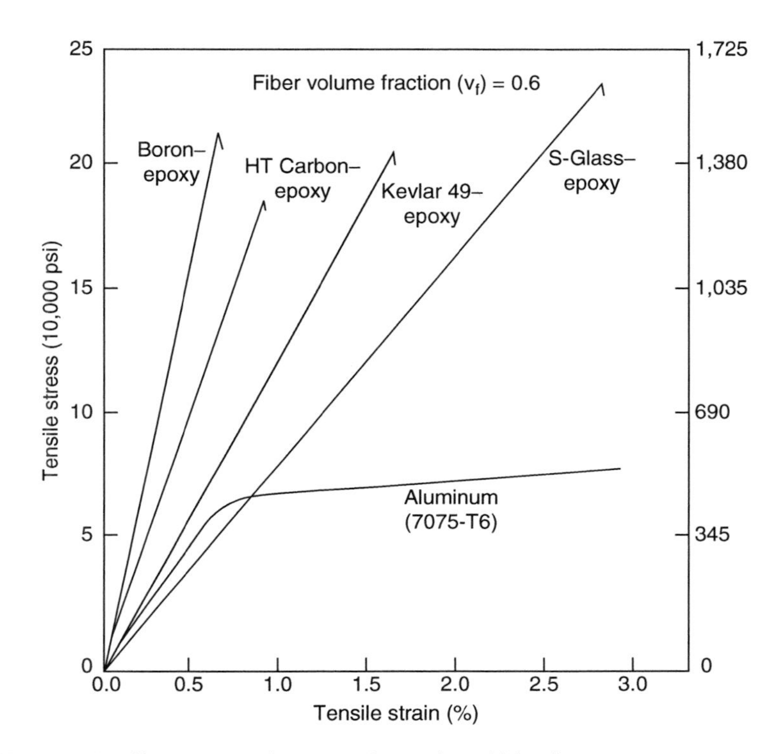

FIGURE 4.4 Tensile stress–strain curves for various 0° laminates.

4.1.1.3 Cross-Ply Laminates

The tensile stress–strain curve for a cross-ply $[0/90]_S$ laminate tested at $\theta = 0°$ direction is slightly nonlinear; however, it is commonly approximated as a bilinear curve (Figure 4.7). The point at which the two linear sections intersect is called the knee and represents the failure of 90° plies. Ultimate failure of the

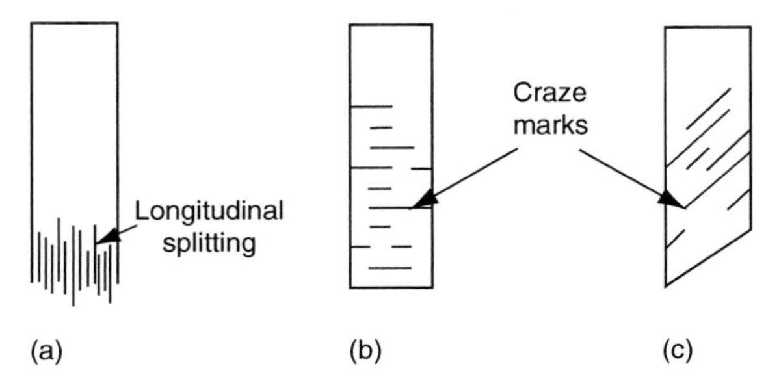

FIGURE 4.5 Schematic failure modes in unidirectional laminates: (a) $\theta = 0°$, (b) $\theta = 90°$, and (c) $0 < \theta < 90°$.

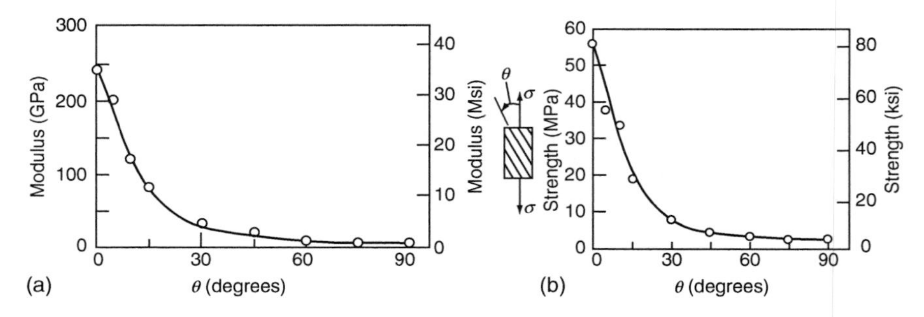

FIGURE 4.6 Variations of tensile modulus and tensile strength of a unidirectional carbon fiber–epoxy laminate with fiber orientation angle. (After Chamis, C.C. and Sinclair, J.H., Mechanical behavior and fracture characteristics of off-axis fiber composites, II—Theory and comparisons, NASA Technical Paper 1082, 1978.)

laminate occurs at the fracture strain of 0° plies. The change in slope of the stress–strain curve at the knee can be reasonably predicted by assuming that all 90° plies have failed at the knee and can no longer contribute to the laminate modulus.

Denoting the moduli of the 0° and 90° plies as E_{11} and E_{22}, respectively, the initial (primary) modulus of the cross-ply laminate can be approximated as

$$E = \frac{A_0}{A} E_{11} + \frac{A_{90}}{A} E_{22}, \tag{4.5}$$

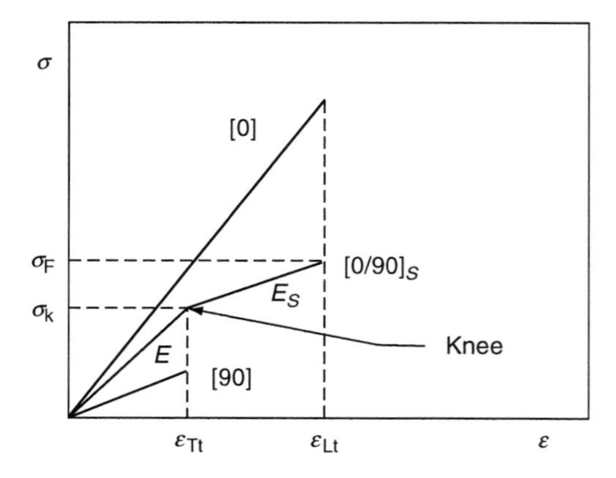

FIGURE 4.7 Schematic tensile stress–strain diagram for a $[0/90]_S$ cross-plied laminate tested at $\theta = 0°$ direction.

where

A_0 = net cross-sectional area of the 0° plies
A_{90} = net cross-sectional area of the 90° plies
$A = A_0 + A_{90}$

At the knee, the laminate strain is equal to the ultimate tensile strain ε_{TU} of the 90° plies. Therefore, the corresponding stress level in the laminate is

$$\sigma_k = E\varepsilon_{TU}, \tag{4.6}$$

where σ_k is the laminate stress at the knee.

If 90° plies are assumed to be completely ineffective after they fail, the secondary modulus (slope after the knee) E_s of the laminate can be approximated as

$$E_s = \frac{A_0}{A} E_{11}. \tag{4.7}$$

Failure of the laminate occurs at the ultimate tensile strain ε_{LU} of the 0° plies. Therefore, the laminate failure stress σ_F is

$$\sigma_F = \sigma_k + E_s(\varepsilon_{LU} - \varepsilon_{TU}). \tag{4.8}$$

Unloading of the cross-ply laminate from a stress level σ_L above the knee follows a path AB (Figure 4.8) and leaves a small residual strain in the laminate. Reloading takes place along the same path until the stress level σ_L is recovered. If the load is increased further, the slope before unloading is also

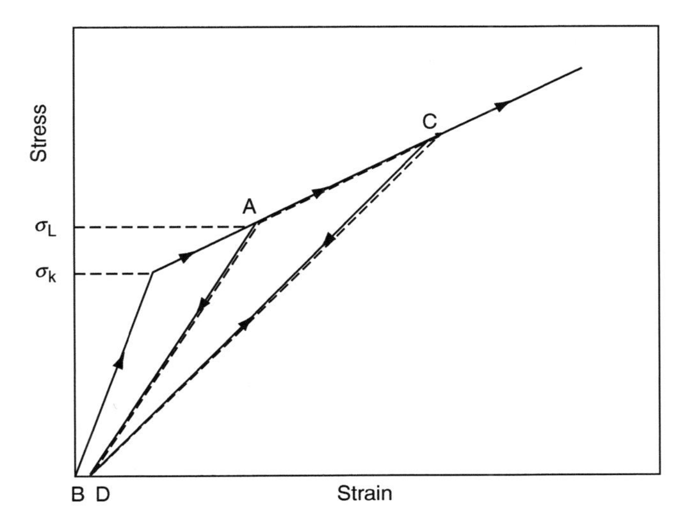

FIGURE 4.8 Unloading and reloading of a $[0/90]_S$ laminate.

recovered. Unloading from a higher stress level follows a path CD, which has a smaller slope than AB. The difference in slope between the two unloading paths AB and CD is evidence that the 90° plies fail in a progressive manner. Neglecting the small residual strains after unloading, Hahn and Tsai [6] predicted the elastic modulus E_D of the damaged laminate as

$$E_D = \frac{E}{1 + [(AE/A_0E_{11}) - 1](1 - \sigma_K/\sigma_L)}. \tag{4.9}$$

4.1.1.4 Multidirectional Laminates

Tensile stress–strain curves for laminates containing different fiber orientations in different laminas are in general nonlinear. A few examples are shown in Figure 4.9. For the purposes of analysis, these curves are approximated by a number of linear portions that have different slopes. When these linear portions are extended, a number of knees, similar to that observed in a cross-ply laminate, can be identified. The first knee in these diagrams is called the first ply failure (FPF) point. Many laminates retain a significant load-carrying capacity beyond the FPF point, but for some laminates with high notch sensitivity, failure occurs just after FPF (Table 4.2). Furthermore, cracks appearing at the FPF may increase the possibility of environmental damage (such as moisture pickup) as well as fatigue failure. For all these reasons, the FPF point has special importance in many laminate designs.

Angle-ply laminates containing [$\pm\theta$] layups exhibit two kinds of stress–strain nonlinearity (Figure 4.10). At values of θ closer to 0°, a stiffening effect

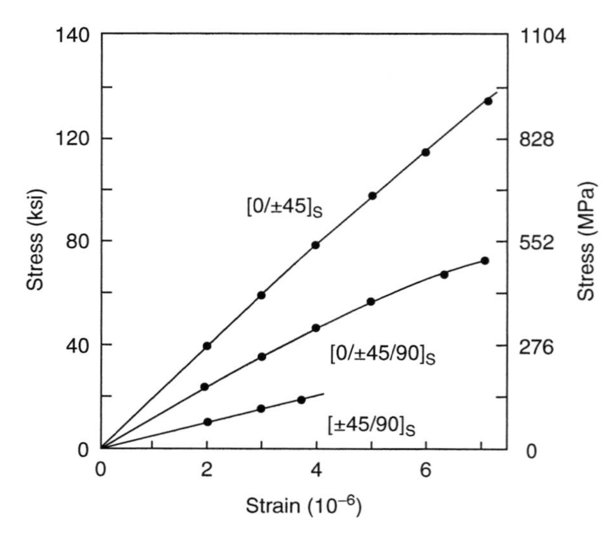

FIGURE 4.9 Typical tensile stress–strain diagrams for multidirectional laminates.

TABLE 4.2
Tensile Strengths and First-Ply Failure (FPF) Stresses in High-Strength Carbon–Epoxy Symmetric Laminates[a]

Laminate	UTS, MPa (ksi)		Estimated FPF Stress, MPa (ksi)		Tensile Modulus, GPa (Msi)	Initial Tensile Strain (%)
	Resin 1	Resin 2	Resin 1	Resin 2		
$[0]_S$	1378 (200)	1378 (200)	—	—	151.6 (22)	0.3
$[90]_S$	41.3 (6)	82.7 (12)	—	—	8.96 (1.3)	0.5–0.9
$[\pm 45]_S$	137.8 (20)	89.6 (13)	89.6 (13)	89.6 (13)	17.2 (2.5)	1.5–4.5
$[0/90]_S$	447.8 (65)	757.9 (110)	413.4 (60)	689 (100)	82.7 (12)	0.5–0.9
$[0_2/\pm 45]_S$	599.4 (87)	689 (110)	592.5 (86)	689 (100)	82.7 (12)	0.8–0.9
$[0/\pm 60]_S$	461.6 (67)	551.2 (80)	323.8 (47)	378.9 (55)	62 (9)	0.8–0.9
$[0/90/\pm 45]_S$	385.8 (56)	413.4 (60)	275.6 (40)	413.4 (60)	55.1 (8)	0.8–0.9

Source: Adapted from Freeman, W.T. and Kuebeler, G.C., *Composite Materials: Testing and Design* (*Third Conference*), *ASTM STP*, 546, 435, 1974.

[a] Resin 2 is more flexible than resin 1 and has a higher strain-to-failure.

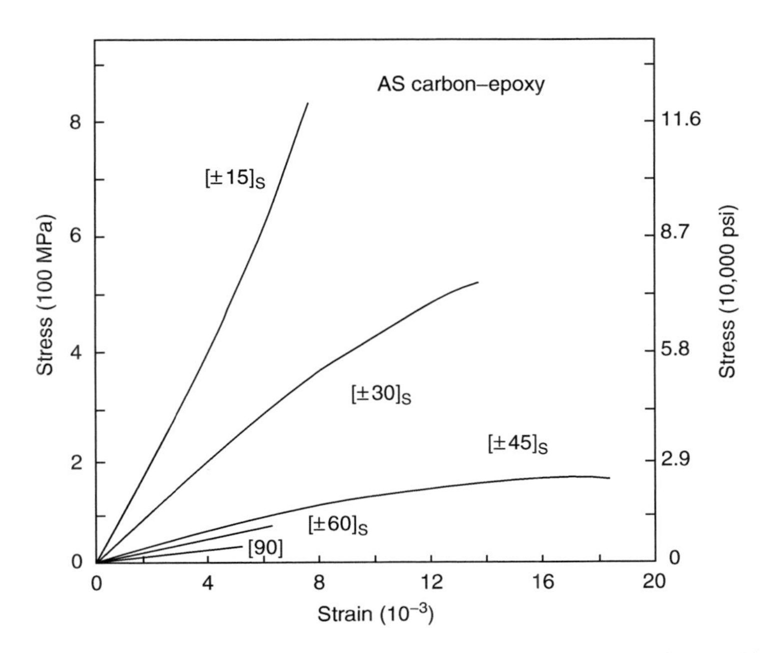

FIGURE 4.10 Typical tensile stress–strain diagrams for angle-ply laminates. (Adapted from Lagace, P.A., *AIAA J.*, 23, 1583, 1985.)

is observed so that the modulus increases with increasing load. At larger values of θ, a softening effect is observed so that the modulus decreases with the increasing load [7]. The stiffening effect is attributed to the longitudinal tensile stresses in various plies, whereas the softening effect is attributed to the shear stresses. Stiffening laminates do not exhibit residual strain on unloading. Softening laminates, on the other hand, exhibit a residual strain on unloading and a hysteresis loop on reloading. However, the slope of the stress–strain curve during reloading does not change from the slope of the original stress–strain curve.

The tensile failure mode and the tensile strength of a multidirectional laminate containing laminas of different fiber orientations depend strongly on the lamina stacking sequence. An example of the stacking sequence effect is observed in the development of cracks in $[0/\pm45/90]_S$ and $[0/90/\pm45]_S$ laminates (Figure 4.11). In both laminates, intralaminar transverse cracks (parallel to fibers) appear in the 90° plies. However, they are arrested at the

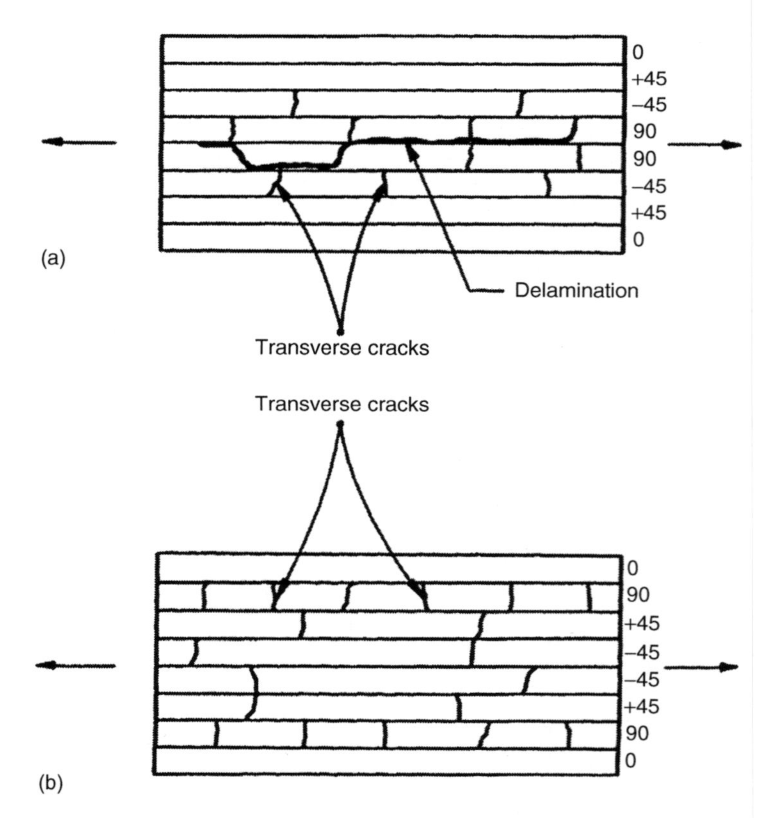

FIGURE 4.11 Damage development in (a) $[0/\pm45/90]_S$ and (b) $[0/90/\pm45]_S$ laminates subjected to static tension loads in the 0° direction.

lamina interfaces and do not immediately propagate into the adjacent plies. The number of transverse cracks in the 90° plies increases until uniformly spaced cracks are formed throughout the specimen length [8]; however, these transverse cracks are more closely spaced in $[0/90/\pm45]_S$ laminates than $[0/\pm45/90]_S$ laminates. Increasing the tensile load also creates a few intralaminar cracks parallel to the fiber directions in both −45° and +45° plies. Apart from these intralaminar crack patterns, subsequent failure modes in these two apparently similar laminates are distinctly different. In $[0/\pm45/90]_S$ laminates, longitudinal interlaminar cracks grow between the 90° plies, which join together to form continuous edge delaminations with occasional jogging into the 90/−45 interfaces. With increasing load, the edge delamination extends toward the center of the specimen; however, the specimen fails by the rupture of 0° fibers before the entire width is delaminated. In contrast to the $[0/\pm45/90]_S$ laminate, there is no edge delamination in the $[0/90/\pm45]_S$ laminate; instead, transverse cracks appear in both +45° and −45° plies before the laminate failure. The difference in edge delamination behavior between the $[0/\pm45/90]_S$ and $[0/90/\pm45]_S$ laminates can be explained in terms of the interlaminar normal stress σ_{zz}, which is tensile in the former and compressive in the latter.

Table 4.3 presents the tensile test data and failure modes observed in several multidirectional carbon fiber–epoxy laminates. If the laminate contains 90° plies, failure begins with transverse microcracks appearing in these plies. With increasing stress level, the number of these transverse microcracks increases until a saturation number, called the characteristic damage state (CDS), is reached. Other types of damages that may follow transverse microcracking are delamination, longitudinal cracking, and fiber failure.

4.1.1.5 Woven Fabric Laminates

The principal advantage of using woven fabric laminates is that they provide properties that are more balanced in the 0° and 90° directions than unidirectional laminates. Although multilayered laminates can also be designed to produce balanced properties, the fabrication (layup) time for woven fabric laminates is less than that of a multilayered laminate. However, the tensile strength and modulus of a woven fabric laminate are, in general, lower than those of multilayered laminates. The principal reason for their lower tensile properties is the presence of fiber undulation in woven fabrics as the fiber yarns in the fill direction cross over and under the fiber yarns in the warp direction to create an interlocked structure. Under tensile loading, these crimped fibers tend to straighten out, which creates high stresses in the matrix. As a result, microcracks are formed in the matrix at relatively low loads. This is also evidenced by the appearance of one or more knees in the stress–strain diagrams of woven fabric laminates (Figure 4.12). Another factor to consider is that the fibers in woven fabrics are subjected to additional mechanical handling during the weaving process, which tends to reduce their tensile strength.

TABLE 4.3
Tensile Test Data and Failure Modes of Several Symmetric Carbon Fiber-Reinforced Epoxy Laminates

Laminate Type	Secant Modulus at Low Strain, GPa	Failure Stress, MPa	Failure Strain	Transverse Ply Strain Cracking	Failure Modes (in Sequence)
$[0_4/90]_S$	122	1620	0.0116	0.0065	Small transverse ply cracks in 90° plies, transverse cracks growing in number as well as in length up to 0° plies, delamination at 0/90 interfaces, 0° ply failure
$[0_4/90_2]_S$	109	1340	0.011	0.004	
$[0_4/90_4]_S$	93	1230	0.0114	0.0035	
$[0_4/90_8]_S$	72	930	0.0115	0.003	
$[\pm 45]_S$	17.3	126	0.017	—	Edge crack formation, edge cracks growing across the width parallel to fiber direction, delamination at the +45/−45 interfaces, single or multiple ply failure
$[+45_2/-45_2]_S$	19	135	0.0117	—	
$[+45_3/-45_3]_S$	14	89	0.01	—	
$[+45/-45_2/45]_S$	18.2	152	0.016	—	
$[(+45/-45)_2]_S$	17	125	0.014	—	
$[\pm 45/90_2/0_2]_S$	64.2	690	0.014	0.0028	Transverse microcracks in 90° plies, longitudinal or angled cracks in 90° plies in the first three laminates, a few edge cracks in 45° plies, delamination (45/90, 0/90, ±45, and 45/0 interfaces in ascending order of threshold strain), longitudinal ply failure
$[\pm 45/0_2/90_2]_S$	61.2	630	0.014	0.0022	
$[0_2/\pm 45/90_2]_S$	56.4	640	0.012	0.0016	
$[0_2/90_2/\pm 45]_S$	59.1	670	0.012	0.0035	

Source: Adapted from Harrison, R.P. and Bader, M.G., *Fibre Sci. Technol.*, 18, 163, 1983.

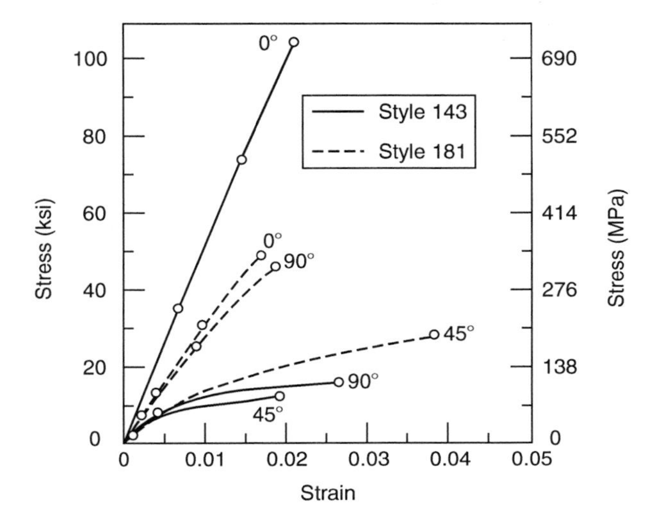

FIGURE 4.12 Stress–strain diagrams of woven glass fabric-epoxy laminates with fabric style 143 (crowfoot weave with 49 × 30 ends) and fabric style 181 (8-harness satin weave with 57 × 54 ends).

Tensile properties of woven fabric laminates can be controlled by varying the yarn characteristics and the fabric style (see Appendix A.1). The yarn characteristics include the number of fiber ends, amount of twist in the yarn, and relative number of yarns in the warp and fill directions. The effect of fiber ends can be seen in Table 4.4 when the differences in the 0° and 90° tensile properties of the parallel laminates with 181 fabric style and 143 fabric style are compared. The difference in the tensile properties of each of these laminates in the 0° and 90° directions reflects the difference in the number of fiber ends in the warp and fill

TABLE 4.4
Tensile Properties of Glass Fabric Laminates

	Tensile Strength, MPa Direction of Testing			Tensile Modulus, GPa Direction of Testing		
Fabric Style[a]	**0° (Warp)**	**90° (Fill)**	**45°**	**0° (Warp)**	**90° (Fill)**	**45°**
181 Parallel lamination	310.4	287.7	182.8	21.4	20.34	15.5
143 Parallel lamination	293.1	34.5	31.0	16.5	6.9	5.5
143 Cross lamination	327.6	327.6	110.3	23.4	23.4	12.2

Source: Adapted from Broutman, L.J., in *Modern Composite Materials*, L.J. Broutman and R.H. Krock, eds., Addison-Wesley, Reading, MA, 1967.

[a] Style 181: 8-harness satin weave, 57 (warp) × 54 (fill) ends, Style 143: Crowfoot weave, 49 (warp) × 30 (fill) ends.

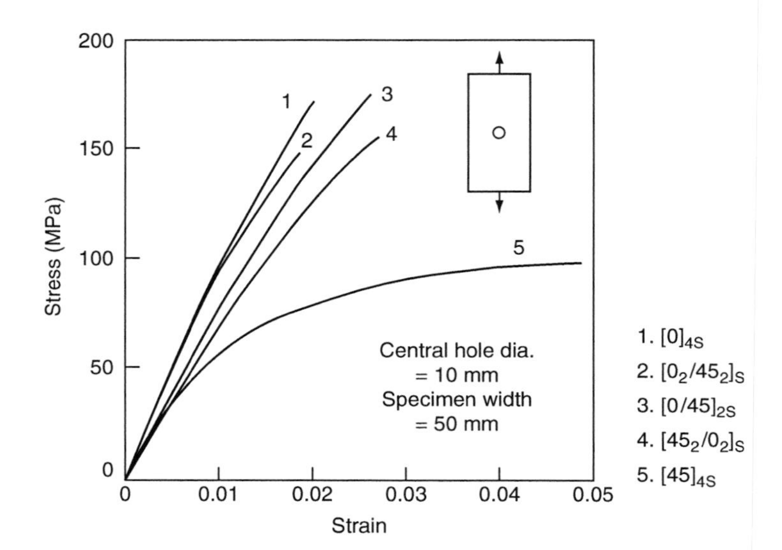

FIGURE 4.13 Effect of stacking sequence on the tensile properties of woven fabric laminates with a central hole. (Adapted from Naik, N.K., Shembekar, P.S., and Verma, M.K., *J. Compos. Mater.*, 24, 838, 1990.)

directions, which is smaller for the 181 fabric style than for the 143 fabric style. Tensile properties of fabric-reinforced laminates can also be controlled by changing the lamination pattern (see, e.g., parallel lamination vs. cross lamination of the laminates with 143 fabric style in Table 4.4) and stacking sequence (Figure 4.13).

4.1.1.6 Sheet-Molding Compounds

Figure 4.14 shows the typical tensile stress–strain diagram for a random fiber SMC (SMC-R) composite containing randomly oriented chopped fibers in a $CaCO_3$-filled polyester matrix. The knee in this diagram corresponds to the development of craze marks in the specimen [9]. At higher loads, the density of craze marks increases until failure occurs by tensile cracking in the matrix and fiber pullout. Both tensile strength and tensile modulus increase with fiber volume fraction. The stress at the knee is nearly independent of fiber volume fractions >20%. Except for very flexible matrices (with high elongations at failure), the strain at the knee is nearly equal to the matrix failure strain. In general, SMC-R composites exhibit isotropic properties in the plane of the laminate; however, they are capable of exhibiting large scatter in strength values from specimen to specimen within a batch or between batches. The variation in strength can be attributed to the manufacturing process for SMC-R composites. They are compression-molded instead of the carefully controlled hand layup technique used for many continuous fiber laminates. A discussion of process-induced defects in compression-molded composites is presented in Chapter 5.

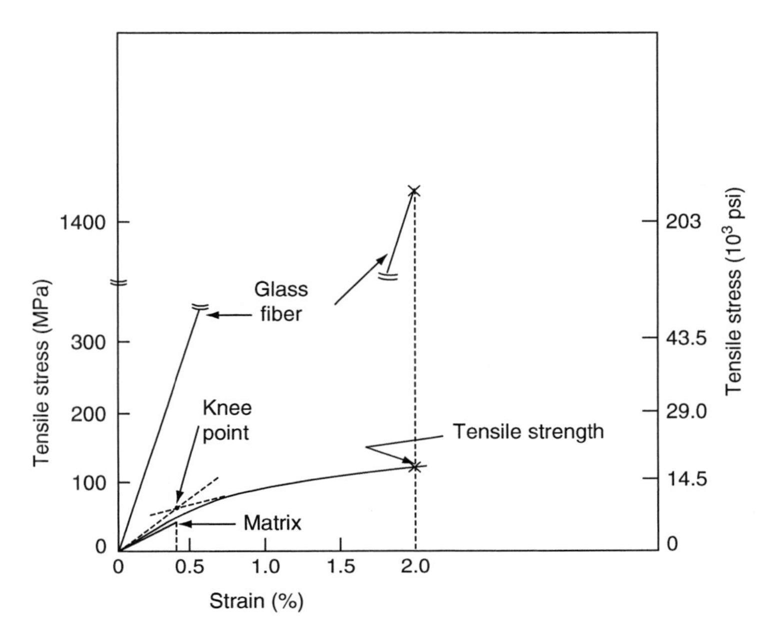

FIGURE 4.14 Tensile stress–strain diagram of an SMC-R laminate. (After Watanabe, T. and Yasuda, M., *Composites*, 13, 54, 1982.)

Tensile stress–strain diagrams for SMC composites containing both continuous and randomly oriented fibers (SMC-CR and XMC) are shown in Figure 4.15. As in the case of SMC-R composites, these stress–strain diagrams are also bilinear. Unlike SMC-R composites, the tensile strength and modulus of SMC-CR and XMC composites depend strongly on the fiber orientation angle of continuous fibers relative to the tensile loading axis. Although the longitudinal tensile strength and modulus of SMC-CR and XMC are considerably higher than those of SMC-R containing equivalent fiber volume fractions, they decrease rapidly to low values as the fiber orientation angle is increased (Figure 4.16). The macroscopic failure mode varies from fiber failure and longitudinal splitting at $\theta = 0°$ to matrix tensile cracking at $\theta = 90°$. For other orientation angles, a combination of fiber–matrix interfacial shear failure and matrix tensile cracking is observed.

4.1.1.7 Interply Hybrid Laminates

Interply hybrid laminates are made of separate layers of low-elongation (LE) fibers, such as high-modulus carbon fibers, and high-elongation (HE) fibers, such as E-glass or Kevlar 49, both in a common matrix. When tested in tension, the interply hybrid laminate exhibits a much higher ultimate strain at failure than the LE fiber composites (Figure 4.17). The strain at which the LE fibers

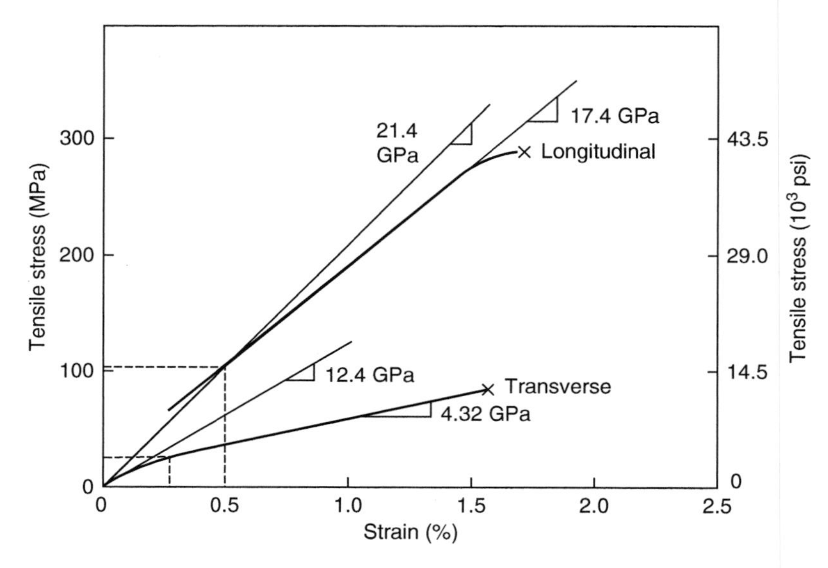

FIGURE 4.15 Tensile stress–strain diagrams for an SMC-C20R30 laminate in the longitudinal (0°) and transverse (90°) directions. (After Riegner, D.A. and Sanders, B.A., A characterization study of automotive continuous and random glass fiber composites, *Proceedings National Technical Conference*, Society of Plastics Engineers, November 1979.)

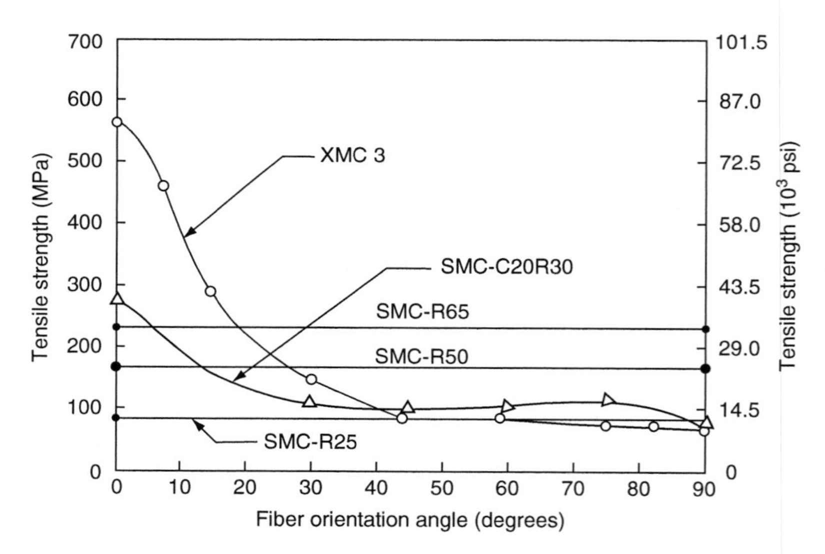

FIGURE 4.16 Variation of tensile strength of various SMC laminates with fiber orientation angle. (After Riegner, D.A. and Sanders, B.A., A characterization study of automotive continuous and random glass fiber composites, *Proceedings National Technical Conference*, Society of Plastics Engineers, November 1979.)

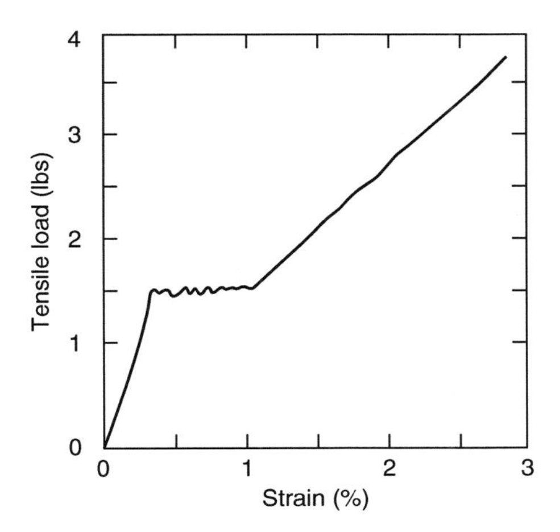

FIGURE 4.17 Tensile stress–strain diagram for a GY-70 carbon/S glass–epoxy interply hybrid laminate. (After Aveston, J. and Kelly, A., *Phil. Trans. R. Soc. Lond., A*, 294, 519, 1980.)

in the hybrid begin to fail is either greater than or equal to the ultimate tensile strain of the LE fibers. Furthermore, instead of failing catastrophically, the LE fibers now fail in a controlled manner, giving rise to a step or smooth inflection in the tensile stress–strain diagram. During this period, multiple cracks are observed in the LE fiber layers [10].

The ultimate strength of interply hybrid laminates is lower than the tensile strengths of either the LE or the HE fiber composites (Figure 4.18). Note that

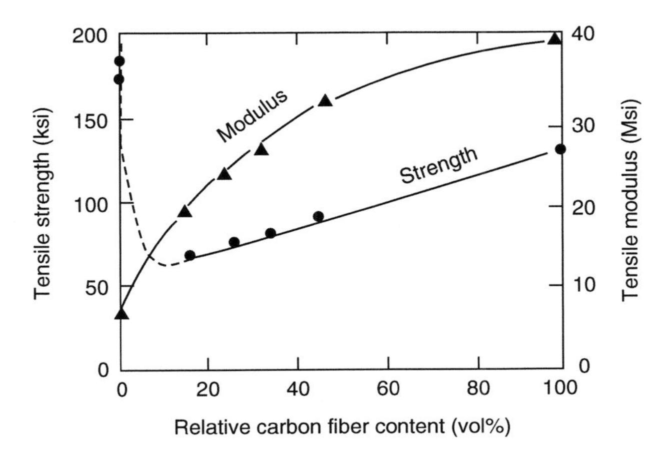

FIGURE 4.18 Variations of tensile strength and modulus of a carbon/glass–epoxy interply hybrid laminate with carbon fiber content. (After Kalnin, L.E., *Composite Materials: Testing and Design (Second Conference), ASTM STP*, 497, 551, 1972.)

the ultimate strain of interply hybrid laminates is also lower than that of the HE fiber composite. The tensile modulus of the interply hybrid laminate falls between the tensile modulus values of the LE and HE fiber composites. Thus, in comparison to the LE fiber composite, the advantage of an interply hybrid laminate subjected to tensile loading is the enhanced strain-to-failure. However, this enhancement of strain, referred to as the hybrid effect, is achieved at the sacrifice of tensile strength and tensile modulus.

The strength variation of hybrid laminates as a function of LE fiber content was explained by Manders and Bader [11]. Their explanation is demonstrated in Figure 4.19, where points A and D represent the tensile strengths of an all-HE fiber composite and an all-LE fiber composite, respectively. If each type of fiber is assumed to have its unique failure strain, the first failure event in the interply hybrid composite will occur when the average tensile strain in it exceeds the failure strain of the LE fibers. The line BD represents the stress in the interply hybrid composite at which failure of the LE fibers occurs. The line AE represents the stress in the interply hybrid composite assuming that the LE fibers carry no load. Thus, if the LE fiber content is less than v_c, the ultimate tensile strength of the interply hybrid laminate is controlled by the HE fibers. Even though the LE fibers have failed at stress levels given by the line BC, the HE fibers continue to sustain increasing load up to the level given by the line AC. For LE fiber contents greater than v_c, the HE fibers fail almost immediately after the failure of the LE fibers. Thus, the line ACD represents the tensile strength of the interply hybrid laminate. For comparison, the rule of mixture prediction, given by the line AD, is also shown in Figure 4.19.

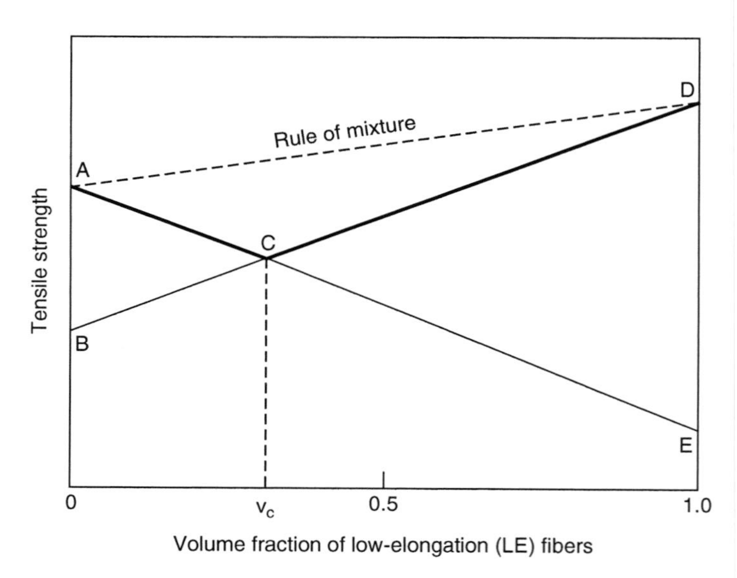

FIGURE 4.19 Model for tensile strength variation in interply hybrid laminates.

4.1.2 Compressive Properties

Compressive properties of thin composite laminates are difficult to measure owing to sidewise buckling of specimens. A number of test methods and specimen designs have been developed to overcome the buckling problem [12]. Three of these test methods are described as follows.

Celanese test: This was the first ASTM standard test developed for testing fiber-reinforced composites in compression; however, because of its several deficiencies, it is no longer a standard test. It employs a straight-sided specimen with tabs bonded at its ends and 10° tapered collet-type grips that fit into sleeves with a matching inner taper (Figure 4.20). An outer cylindrical shell is used for ease of assembly and alignment. As the compressive load is applied at

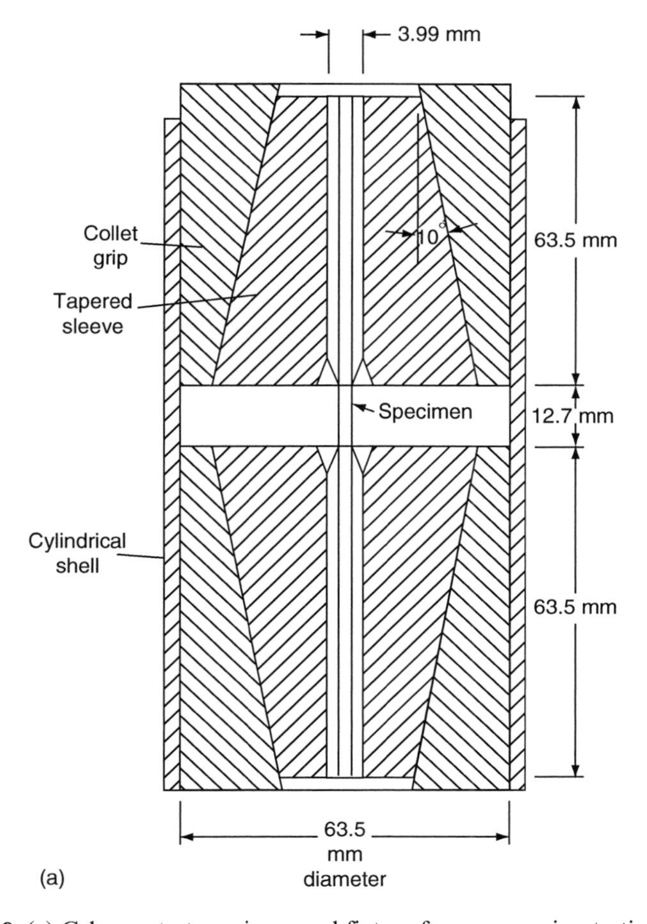

FIGURE 4.20 (a) Celanese test specimen and fixture for compression testing.

(continued)

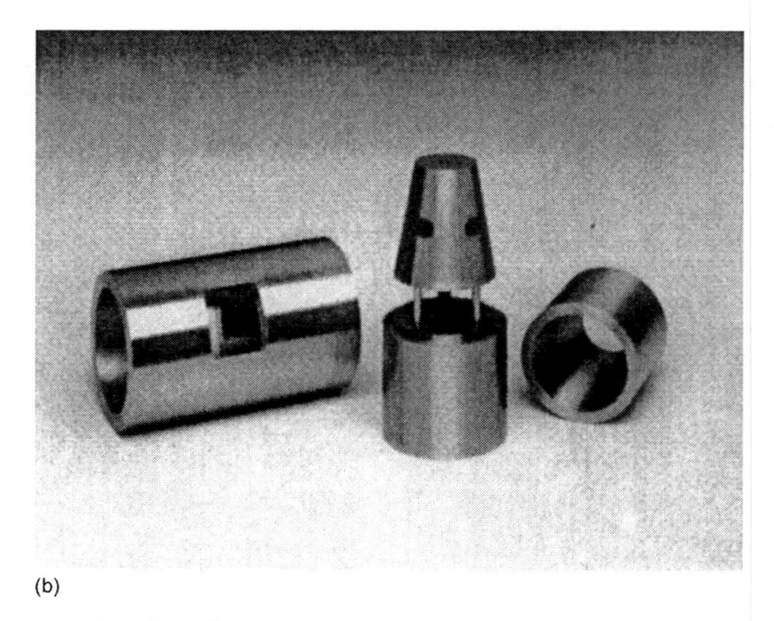

(b)

FIGURE 4.20 (continued) (b) Celanese compression test fixture. (Courtesy of MTS Systems Corporation. With permission.)

the ends of the tapered sleeves, the grip on the specimen tightens and the gage section of the specimen is compressed by the frictional forces transmitted through the end tabs. Strain gages are mounted in the gage section to measure longitudinal and transverse strain data from which compressive modulus and Poisson's ratio are determined.

IITRI test: The IITRI test was first developed at the Illinois Institute of Technology Research Institute and was later adopted as a standard compression test for fiber-reinforced composites (ASTM D3410). It is similar to the Celanese test, except it uses flat wedge grips instead of conical wedge grips (Figure 4.21). Flat wedge surfaces provide a better contact between the wedge and the collet than conical wedge surfaces and improve the axial alignment. Flat wedge grips can also accommodate variation in specimen thickness. The IITRI test fixture contains two parallel guide pins in its bottom half that slide into two roller bushings that are located in its top half. The guide pins help maintain good lateral alignment between the two halves during testing. The standard specimen length is 140 mm, out of which the middle 12.7 mm is unsupported and serves as the gage length. Either untabbed or tabbed specimens can be used; however, tabbing is preferred, since it prevents surface damage and end crushing of the specimen if the clamping force becomes too high.

Sandwich edgewise compression test: In this test, two straight-sided specimens are bonded to an aluminum honeycomb core that provides the necessary

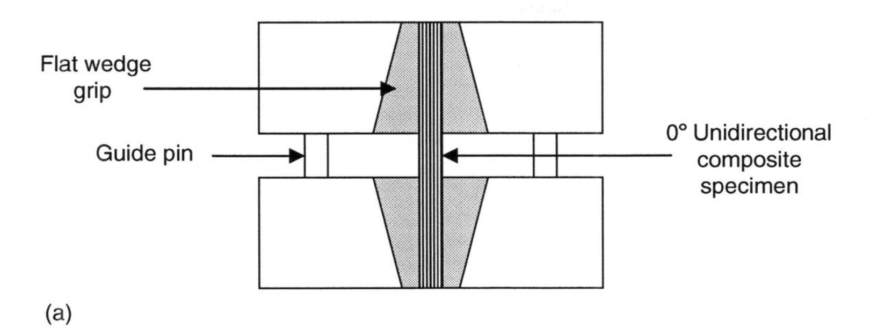

FIGURE 4.21 IITRI compression test fixture. (Courtesy of MTS Systems Corporation. With permission.)

support for lateral stability (Figure 4.22). Compressive load is applied through the end caps, which are used for supporting the specimen as well as preventing end crushing. The average compressive stress in the composite laminate is calculated assuming that the core does not carry any load. Table 4.5 shows representative compressive properties for carbon fiber–epoxy and boron fiber–epoxy laminates obtained in a sandwich edgewise compression test. The data in this table show that the compressive properties depend strongly on the fiber type as well as the laminate configuration.

Compressive test data on fiber-reinforced composites are limited. From the available data on 0° laminates, the following general observations can be made.

1. Unlike ductile metals, the compressive modulus of a 0° laminate is not equal to its tensile modulus.
2. Unlike tensile stress–strain curves, compressive stress–strain curves of 0° laminates may not be linear.

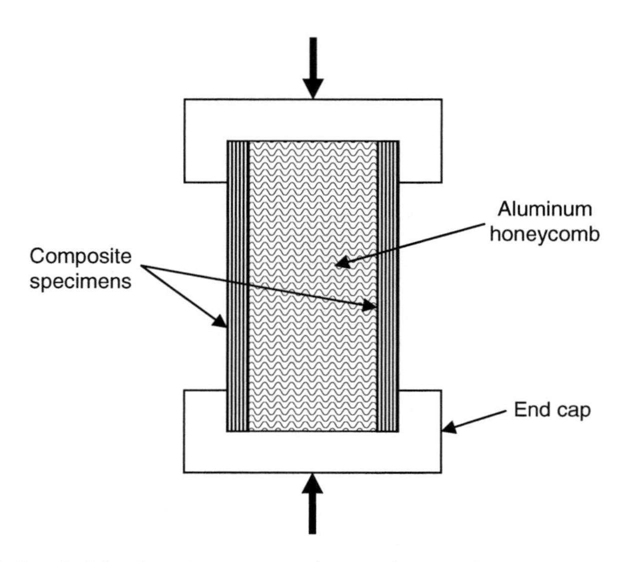

FIGURE 4.22 Sandwich edgewise compression testing specimen.

3. The longitudinal compressive strength of a 0° laminate depends on the fiber type, fiber volume fraction, matrix yield strength, fiber length–diameter ratio, fiber straightness, fiber alignment as well as fiber–matrix interfacial shear strength. The effects of some of these variables on the compressive properties of unidirectional fiber-reinforced polyester composites have been studied by Piggott and Harris and are described in Ref. [4].

TABLE 4.5
Compressive Properties of Carbon and Boron Fiber-Reinforced Epoxy Composites

	Carbon–Epoxy		Boron–Epoxy	
Laminate	Strength, MPa (ksi)	Modulus, GPa (Msi)	Strength, MPa (ksi)	Modulus, GPa (Msi)
[0]	1219.5 (177)	110.9 (16.1)	2101.4 (305)	215.6 (31.3)
[±15]	799.2 (116)	95.8 (13.9)	943.9 (137)	162.9 (23.65)
[±45]	259.7 (37.7)	15.6 (2.27)	235.6 (34.2)	17.4 (2.53)
[90]	194.3 (28.2)	13.1 (1.91)	211.5 (30.7)	20.5 (2.98)
[0/90]	778.6 (113)	60.6 (8.79)	1412.4 (205)	118.3 (17.17)
[0/±45/90]	642.8 (93.3)	46.4 (6.74)	1054.2 (153)	79.0 (11.47)

Source: Adapted from Weller, T., Experimental studies of graphite/epoxy and boron/epoxy angle ply laminates in compression, NASA Report No. NASA-CR-145233, September 1977.

4. Among the commercially used fibers, the compressive strength and modulus of Kevlar 49-reinforced composites are much lower than their tensile strength and modulus. Carbon and glass fiber-reinforced composites exhibit slightly lower compressive strength and modulus than their respective tensile values, and boron fiber-reinforced composites exhibit virtually no difference between the tensile and compressive properties.

4.1.3 Flexural Properties

Flexural properties, such as flexural strength and modulus, are determined by ASTM test method D790. In this test, a composite beam specimen of rectangular cross section is loaded in either a three-point bending mode (Figure 4.23a) or a four-point bending mode (Figure 4.23b). In either mode, a large span–thickness (L/h) ratio is recommended. We will consider only the three-point flexural test for our discussion.

The maximum fiber stress at failure on the tension side of a flexural specimen is considered the flexural strength of the material. Thus, using a homogeneous beam theory, the flexural strength in a three-point flexural test is given by

$$\sigma_{\mathrm{UF}} = \frac{3P_{\max} L}{2bh^2}, \tag{4.10}$$

where

$P_{\max}$ = maximum load at failure
b = specimen width
h = specimen thickness
L = specimen length between the two support points

Flexural modulus is calculated from the initial slope of the load–deflection curve:

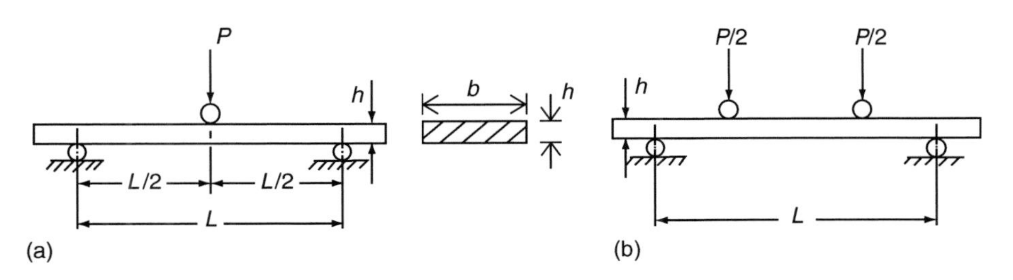

FIGURE 4.23 Flexural test arrangements in (a) three-point bending and (b) four-point bending modes.

$$E_F = \frac{mL^3}{4bh^3}, \tag{4.11}$$

where m is the initial slope of the load–deflection curve.

Three-point flexural tests have received wide acceptance in the composite material industry because the specimen preparation and fixtures are very simple. However, the following limitations of three-point flexural tests should be recognized.

1. The maximum fiber stress may not always occur at the outermost layer in a composite laminate. An example is shown in Figure 4.24. Thus, Equation 4.10 gives only an apparent strength value. For more accurate values, lamination theory should be employed.
2. In the three-point bending mode, both normal stress σ_{xx} and shear stress τ_{xz} are present throughout the beam span. If contributions from both stresses are taken into account, the total deflection at the midspan of the beam is

$$\Delta = \underbrace{\frac{PL^3}{4Ebh^3}}_{\text{normal}} + \underbrace{\frac{3PL}{10Gbh}}_{\text{shear}} = \frac{PL^3}{4Ebh^3}\left[1 + \frac{12}{10}\left(\frac{E}{G}\right)\left(\frac{h}{L}\right)^2\right]. \tag{4.12}$$

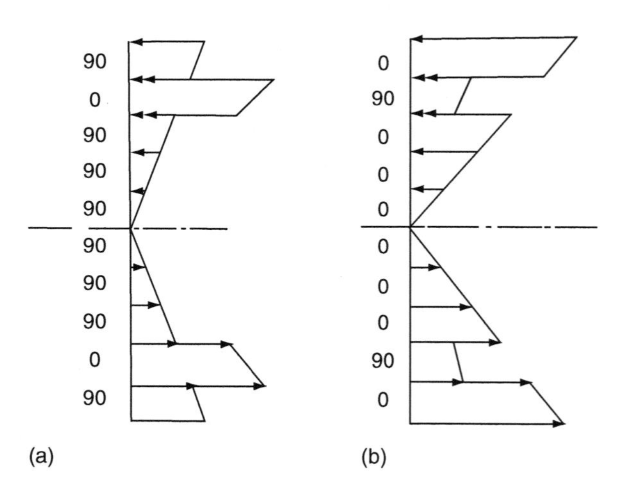

FIGURE 4.24 Normal stress (σ_{xx}) distributions in various layers of (a) [90/0/(90)$_6$/0/90] and (b) [0/90/(0)$_6$/90/0] laminates under flexural loading.

This equation shows that the shear deflection can be quite significant in a composite laminate, since the E/G ratio for fiber-reinforced composites is often quite large. The shear deflection can be reduced employing a high span–thickness (L/h) ratio for the beam. Based on data of Zweben et al. [13], L/h ratios of 60:1 are recommended for the determination of flexural modulus.

3. Owing to large deflection at high L/h ratios, significant end forces are developed at the supports. This in turn affects the flexural stresses in a beam. Unless a lower L/h ratio, say 16:1, is used, Equation 4.10 must be corrected for these end forces in the following way:

$$\sigma_{\max} = \frac{3P_{\max}L}{2bh^2}\left[1 + 6\left(\frac{\Delta}{L}\right)^2 - 4\left(\frac{h}{L}\right)\left(\frac{\Delta}{L}\right)\right], \tag{4.13}$$

where Δ is given by Equation 4.12.

4. Although the flexural strength value is based on the maximum tensile stress in the outer fiber, it does not reflect the true tensile strength of the material. The discrepancy arises owing to the difference in stress distributions in flexural and tensile loadings. Flexural loads create a nonuniform stress distribution along the length, but a tensile load creates a uniform stress distribution. Using a two-parameter Weibull distribution for both tensile strength and flexural strength variations, the ratio of the median flexural strength to the median tensile strength can be written as

$$\frac{\sigma_{\mathrm{UF}}}{\sigma_{\mathrm{UT}}} = \left[2(1+\alpha)^2 \frac{V_{\mathrm{T}}}{V_{\mathrm{F}}}\right]^{1/\alpha}, \tag{4.14}$$

where

α = shape parameter in the Weibull distribution function (assumed to be equal in both tests)

V_{T} = volume of material stressed in a tension test

V_{F} = volume of material stressed in a three-point flexural test

Assuming $V_{\mathrm{T}} = V_{\mathrm{F}}$ and using typical values of $\alpha = 15$ and 25 for 0° E-glass–epoxy and 0° carbon–epoxy laminates, respectively [12], Equation 4.14 shows that

$\sigma_{\mathrm{UF}} = 1.52\sigma_{\mathrm{UT}}$ for 0° E-glass–epoxy laminates

$\sigma_{\mathrm{UF}} = 1.33\sigma_{\mathrm{UT}}$ for 0° carbon–epoxy laminates

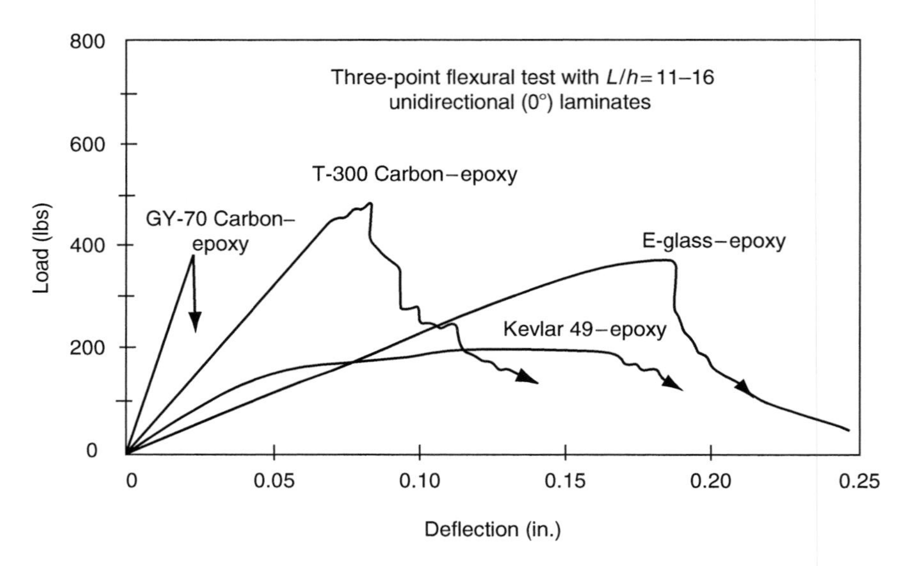

FIGURE 4.25 Load–deflection diagrams for various 0° unidirectional laminates in three-point flexural tests.

Thus, the three-point flexural strength of a composite laminate can be significantly higher than its tensile strength. The experimental data presented by Bullock [14] as well as Whitney and Knight [15] verify this observation.

Figure 4.25 shows the flexural load-deflection diagrams for four unidirectional 0° laminates. The materials of construction are an ultrahigh-modulus carbon (GY-70), a high-strength carbon (T-300), Kevlar 49, and E-glass fiber-reinforced epoxies. The difference in slope in their load–deflection diagrams reflects the difference in their respective fiber modulus. The GY-70 laminate exhibits a brittle behavior, but other laminates exhibit a progressive failure mode consisting of fiber failure, debonding (splitting), and delamination. The Kevlar 49 laminate has a highly nonlinear load–deflection curve due to compressive yielding. Fiber microbuckling damages are observed on the compression side of both E-glass and T-300 laminates. Since high contact stresses just under the loading point create such damage, it is recommended that a large loading nose radius be used.

The flexural modulus is a critical function of the lamina stacking sequence (Table 4.6), and therefore, it does not always correlate with the tensile modulus, which is less dependent on the stacking sequence. In angle-ply laminates, a bending moment creates both bending and twisting curvatures. Twisting curvature causes the opposite corners of a flexural specimen to lift off its supports. This also influences the measured flexural modulus. The twisting curvature is reduced with an increasing length–width (L/b) ratio and a decreasing degree of orthotropy (i.e., decreasing E_{11}/E_{22}).

TABLE 4.6
Tensile and Flexural Properties of Quasi-Isotropic Laminates

	Tension Test		Flexural Test[a]	
Laminate Configuration[b]	Strength, MPa (ksi)	Modulus, GPa (Msi)	Strength, MPa (ksi)	Modulus, GPa (Msi)
$[0/\pm45/90]_S$	506.4 (73.5)	48.23 (7)	1219.5 (177)	68.9 (10)
$[90/\pm45/0]_S$	405.8 (58.9)	45.47 (6.6)	141.2 (20.5)	18.6 (2.7)
$[45/0/-45/90]_S$	460.9 (66.9)	46.85 (6.8)	263.9 (38.3)	47.54 (6.9)

Source: Adapted from Whitney, J.M., Browning, C.E., and Mair, A., *Composite Materials: Testing and Design (Third Conference), ASTM STP*, 546, 30, 1974.

[a] Four-point flexural test with $L/h = 32$ and $L/b = 4.8$.
[b] Material: AS carbon fiber–epoxy composite, $v_f = 0.6$, eight plies.

4.1.4 In-Plane Shear Properties

A variety of test methods [16,17] have been used for measuring in-plane shear properties, such as the shear modulus G_{12} and the ultimate shear strength τ_{12U} of unidirectional fiber-reinforced composites. Three common in-plane shear test methods for measuring these two properties are described as follows.

±45 *Shear test*: The ±45 shear test (ASTM D3518) involves uniaxial tensile testing of a $[+45/-45]_{nS}$ symmetric laminate (Figure 4.26). The specimen dimensions, preparation, and test procedure are the same as those described in the tension test method ASTM D3039. A diagram of the shear stress τ_{12} vs. the shear strain γ_{12} is plotted using the following equations:

$$\tau_{12} = \frac{1}{2}\sigma_{xx},$$
$$\gamma_{12} = \varepsilon_{xx} - \varepsilon_{yy}, \tag{4.15}$$

where σ_{xx}, ε_{xx}, and ε_{yy} represent tensile stress, longitudinal strain, and transverse strain, respectively, in the $[\pm45]_{nS}$ tensile specimen. A typical tensile stress–tensile strain response of a $[\pm45]_S$ boron–epoxy laminate and the corresponding shear stress–shear strain diagram are shown in Figure 4.27.

10° *Off-axis test*: The 10° off-axis test [18] involves uniaxial tensile testing of a unidirectional laminate with fibers oriented at 10° from the tensile loading direction (Figure 4.28). The shear stress τ_{12} is calculated from the tensile stress σ_{xx} using the following expression:

$$\tau_{12} = \frac{1}{2}\sigma_{xx}\sin 2\theta\big|_{\theta=10^\circ} = 0.171\sigma_{xx}. \tag{4.16}$$

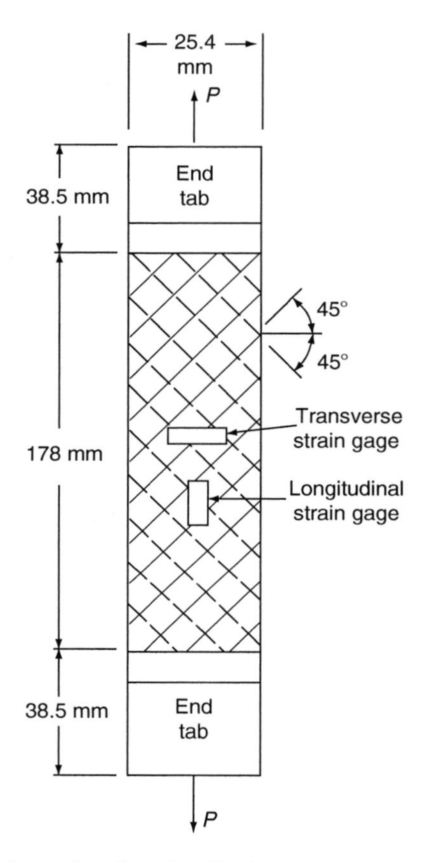

FIGURE 4.26 Test configuration for a [±45]$_S$ shear test.

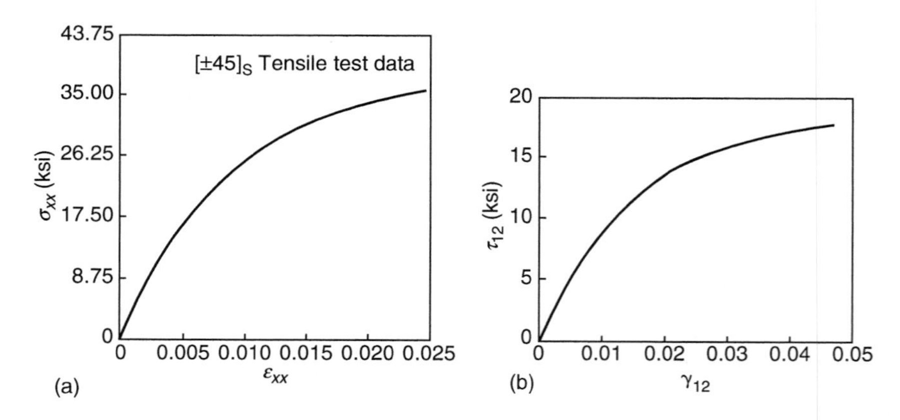

FIGURE 4.27 (a) Tensile stress–strain diagram for a [±45]$_S$ boron–epoxy specimen and (b) the corresponding shear stress–shear strain diagram. (Adapted from the data in Rosen, B.M., *J. Compos. Mater.*, 6, 552, 1972.)

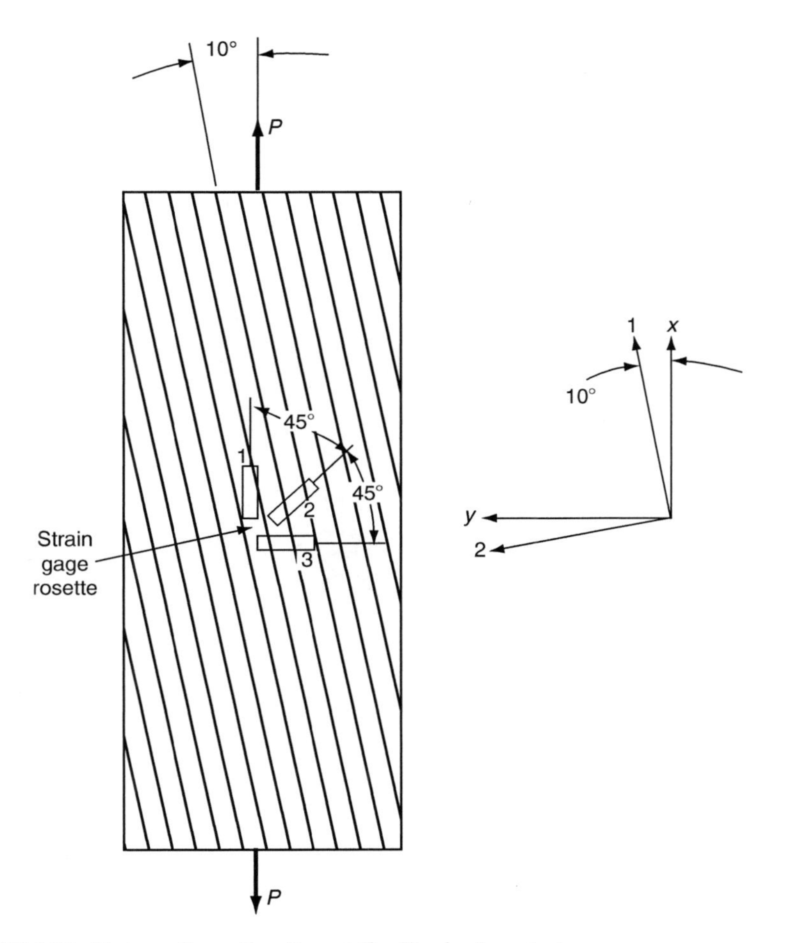

FIGURE 4.28 Test configuration for a 10° off-axis shear test.

Calculation of the shear strain γ_{12} requires measurements of three normal strains using either a rectangular strain gage rosette or a 60° Δ-strain gage rosette. If a rectangular strain gage rosette is used (Figure 4.28), the expression for shear strain γ_{12} is

$$\gamma_{12} = 0.5977\varepsilon_{g1} - 1.8794\varepsilon_{g2} + 1.2817\varepsilon_{g3}, \tag{4.17}$$

where ε_{g1}, ε_{g2}, and ε_{g3} are normal strains in gage 1, 2, and 3, respectively.

Iosipescu shear test: The Iosipescu shear test (ASTM D5379) was originally developed by Nicolai Iosipescu for shear testing of isotropic materials and was later adopted by Walrath and Adams [19] for determining the shear strength and modulus of fiber-reinforced composites. It uses a double V-notched test specimen, which is tested in a four-point bending fixture (Figure 4.29). A uniform transverse shear force is created in the gage section of the specimen, while the bending moment at the notch plane is zero. Various analyses have shown that except at the close vicinity of the notch roots, a state of pure

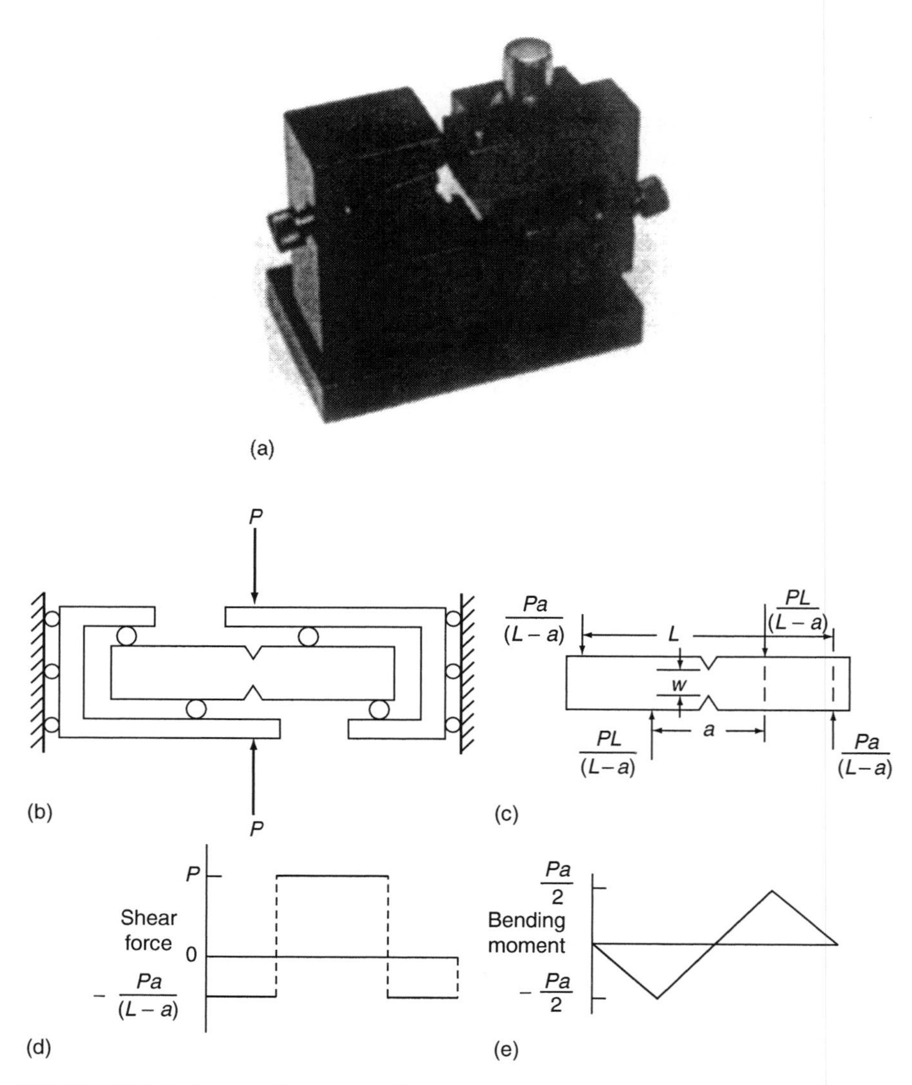

FIGURE 4.29 Iosipescu shear test: (a) test fixture (Courtesy of MTS System Corporation), (b) schematic representation, (c) free body, (d) shear force, and (e) bending moment distribution.

shear exists at the notch plane. The presence of notch creates a shear stress concentration at the notch root, which reduces with increasing notch angle and notch root radius, but increases with increasing orthotropy, that is, increasing (E_{11}/E_{22}). Typical Iosipescu specimens use a 90° notch angle, notch depth equal to 20% of the specimen width, and notch root radius of 1.3 mm.

The shear stress in an Iosipescu shear test is calculated as

$$\tau_{12} = \frac{P}{wh}, \tag{4.18}$$

where

P = applied load
w = distance between the notches
h = specimen thickness

A ±45° strain rosette, centered in the gage section of the specimen, is used to measure the shear strain at the midsection between the notches. The shear strain is given as

$$\gamma_{12} = \varepsilon_{+45^\circ} - \varepsilon_{-45^\circ}. \tag{4.19}$$

Based on a round robin test conducted by the ASTM [20], it is recommended that 0° specimens be used for measuring shear strength τ_{12U} and shear modulus G_{12} of a continuous fiber-reinforced composite material. The 90° specimens show evidence of failure due to transverse tensile stresses that exist outside the notch plane. A schematic of the acceptable and unacceptable failure modes in 0° and 90° specimens is shown in Figure 4.30.

In-plane shear properties τ_{xyU} and G_{xy} of a general laminate are commonly determined by either a two-rail or a three-rail edgewise shear test method (ASTM D4255). In a two-rail shear test, two pairs of steel rails are fastened along the long edges of a 76.2 mm wide × 152.4 mm long rectangular specimen, usually by three bolts on each side (Figure 4.31a). At the other two edges, the specimen remains free.

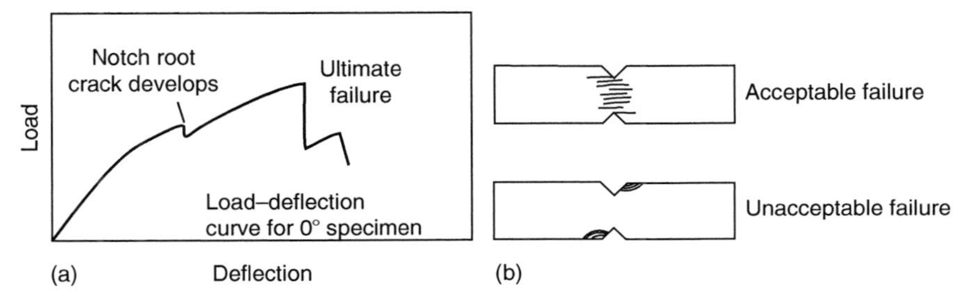

FIGURE 4.30 (a) Load-deflection diagram and (b) acceptable and unacceptable failure modes in an Iosipescu shear test.

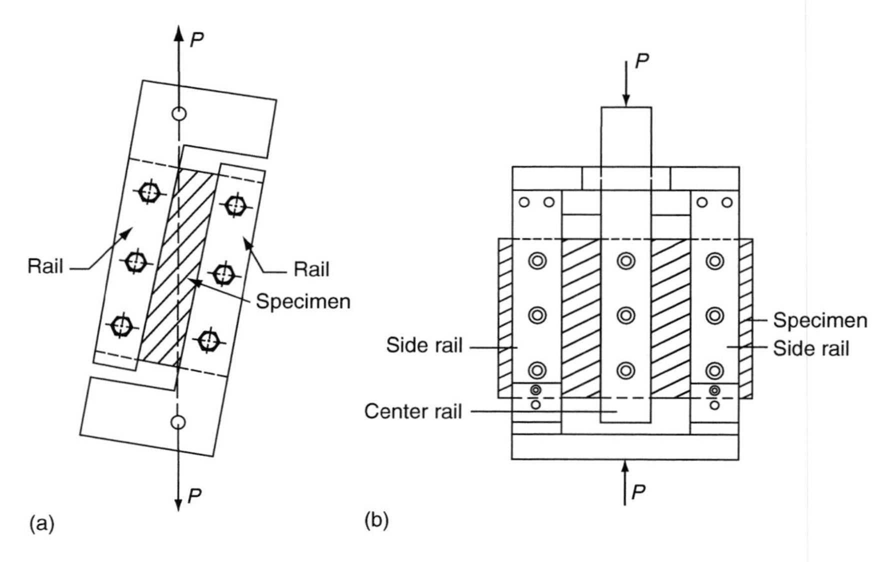

FIGURE 4.31 Test configuration for (a) two-rail and (b) three-rail shear tests.

A tensile load applied to the rails tends to displace one rail relative to the other, thus inducing an in-plane shear load on the test laminate. In a three-rail shear test, three pairs of steel rails are fastened to a 136 mm wide × 152 mm long rectangular specimen, two along its long edges, and one along its centerline (Figure 4.31b). A compressive load applied at the center rail creates an in-plane shear load in the specimen. The shear stress τ_{xy} is calculated from the applied load P as

$$\text{Two-rail:} \quad \tau_{xy} = \frac{P}{Lh},$$
$$\text{Three-rail:} \quad \tau_{xy} = \frac{P}{2Lh}, \tag{4.20}$$

where
L is the specimen length
h is the specimen thickness

The shear strain γ_{xy} is determined using a strain gage mounted in the center of the test section at 45° to the specimen's longitudinal axis. The shear strain γ_{xy} is calculated from the normal strain in the 45° direction:

$$\gamma_{xy} = 2\varepsilon_{45^\circ}. \tag{4.21}$$

In order to assure a uniform shear stress field at a short distance away from the free edges of a rail shear specimen [21], the length–width ratio must be greater than 10:1. A low effective Poisson's ratio for the laminate is also desirable, since

TABLE 4.7
In-Plane Shear Properties of [0] and $[\pm 45]_S$ Laminates[a]

Material ($v_f = 60\%$)	Shear Strength, MPa (ksi)		Shear Modulus, GPa (Msi)	
	[0]	$[\pm 45]_S$	[0]	$[\pm 45]_S$
Boron–epoxy	62 (9)	530.5 (77)	4.82 (0.7)	54.4 (7.9)
Carbon–epoxy	62 (9)	454.7 (66)	4.48 (0.65)	37.9 (5.5)
Kevlar 49–epoxy	55.1 (8)	192.9 (28)	2.07 (0.3)	20.7 (3)
S-Glass–epoxy	55.1 (8)	241.1 (35)	5.51 (0.8)	15.1 (2.2)

[a] For comparison, the shear modulus of steel = 75.8 GPa (11 Msi) and that of aluminum alloys = 26.9 GPa (3.9 Msi).

the shear stress distribution in laminates of a high Poisson's ratio is irregular across the width. For shear properties of unidirectional laminates, either 0° or 90° orientation (fibers parallel or perpendicular to the rails) can be used. However, normal stress concentration near the free edges is transverse to the fibers in a 0° orientation and parallel to the fibers in the 90° orientation. Since normal stresses may cause premature failure in the 0° laminate, it is recommended that a 90° laminate be used for determining τ_{12U} and G_{12} [22].

Although the results from various in-plane shear tests do not always correlate, several general conclusions can be made:

1. The shear stress–strain response for fiber-reinforced composite materials is nonlinear.
2. Even though 0° laminates have superior tensile strength and modulus, their shear properties are poor.

The shear strength and modulus depend on the fiber orientation angle and laminate configuration. The highest shear modulus is obtained with $[\pm 45]_S$ symmetric laminates (Table 4.7). The addition of 0° layers reduces both the shear modulus and the shear strength of $[\pm 45]_S$ laminates.

4.1.5 Interlaminar Shear Strength

Interlaminar shear strength (ILSS) refers to the shear strength parallel to the plane of lamination. It is measured in a short-beam shear test in accordance with ASTM D2344. A flexural specimen of small span–depth (L/h) ratio is tested in three-point bending to produce a horizontal shear failure between the laminas. To explain the short-beam shear test, let us consider the following homogeneous beam equations:

$$\text{Maximum normal stress } \sigma_{xx} = \frac{3PL}{2bh^2} = \frac{3P}{2bh}\left(\frac{L}{h}\right), \tag{4.22a}$$

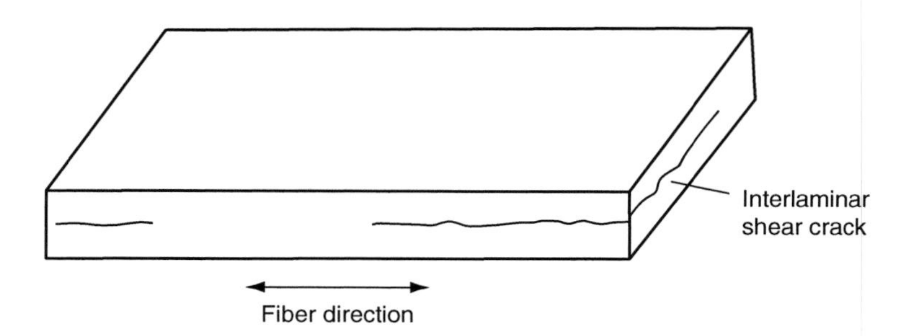

FIGURE 4.32 Interlaminar shear failure in a 0° laminate in a short-beam shear test.

$$\text{Maximum shear stress } \tau_{xz} = \frac{3P}{4bh}. \qquad (4.22b)$$

From Equation 4.22, it can be seen that the maximum normal stress in the beam decreases with decreasing L/h ratio and the maximum shear stress (at the neutral axis) is not affected by the L/h ratio. Thus, for sufficiently small L/h ratios, the maximum shear stress in the beam will reach the ILSS of the material even though the maximum normal stress is still quite low. Thus, the beam will fail in the interlaminar shear mode by cracking along a horizontal plane between the laminas (Figure 4.32). The recommended L/h ratios for short-beam shear tests are between 4 and 5. However, testing a few specimens at various L/h ratios is usually needed before the proper L/h ratio for interlaminar shear failure is found. For very small L/h ratios a compressive failure may occur on the top surface of the specimen, whereas for large L/h ratios a tensile failure may occur at the bottom surface of the specimen [23]. Knowing the maximum load at failure, the ILSS is determined using Equation 4.22b.

Because of its simplicity, the short-beam shear test is widely accepted for material screening and quality control purposes [24]. However, it does not provide design data for the following reasons:

1. Equation 4.22b is based on homogeneous beam theory for long slender beams, which predicts a continuous parabolic shear stress distribution in the thickness direction (Figure 4.33). Such symmetrical shear stress distribution may not occur in a short-beam shear test [25]. Additionally, it may also contain discontinuities at lamina interfaces. Therefore, Equation 4.22b is only an approximate formula for ILSS.
2. In the homogeneous beam theory, maximum shear stress occurs at the neutral plane where normal stresses are zero. In short-beam shear tests of many laminates, maximum shear stress may occur in an area where other stresses may exist. As a result, a combination of failure modes, such as fiber rupture, microbuckling, and interlaminar shear cracking, are observed. Interlaminar shear failure may also not take place at the laminate midplane.

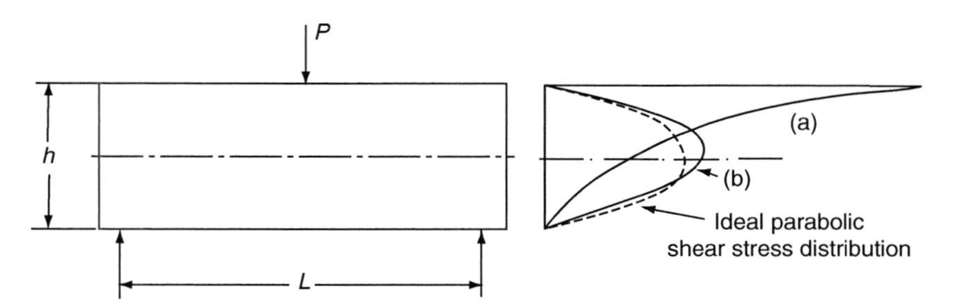

FIGURE 4.33 Shear stress distributions in a short-beam shear specimen: (a) near the support points and (b) near the midspan.

For these reasons, it is often difficult to interpret the short-beam shear test data and compare the test results for various materials.

The ILSS, τ_{xzU} is not the same as the in-plane shear strength, τ_{xyU}. Furthermore, the short-beam shear test should not be used to determine the shear modulus of a material.

Despite the limitations of the short-beam shear test, interlaminar shear failure is recognized as one of the critical failure modes in fiber-reinforced composite laminates. ILSS depends primarily on the matrix properties and fiber–matrix interfacial shear strengths rather than the fiber properties. The ILSS is improved by increasing the matrix tensile strength as well as the matrix volume fraction. Because of better adhesion with glass fibers, epoxies in general produce higher ILSS than vinyl ester and polyester resins in glass fiber-reinforced composites. The ILSS decreases, often linearly, with increasing void content. Fabrication defects, such as internal microcracks and dry strands, also reduce the ILSS.

4.2 FATIGUE PROPERTIES

The fatigue properties of a material represent its response to cyclic loading, which is a common occurrence in many applications. It is well recognized that the strength of a material is significantly reduced under cyclic loads. Metallic materials, which are ductile in nature under normal operating conditions, are known to fail in a brittle manner when they are subjected to repeated cyclic stresses (or strains). The cycle to failure depends on a number of variables, such as stress level, stress state, mode of cycling, process history, material composition, and environmental conditions.

Fatigue behavior of a material is usually characterized by an *S–N* diagram, which shows the relationship between the stress amplitude or maximum stress and number of cycles to failure on a semilogarithmic scale. This diagram is obtained by testing a number of specimens at various stress levels under sinusoidal loading conditions. For a majority of materials, the number of cycles to failure increases continually as the stress level is reduced. For low-carbon steel and a few other alloys, a fatigue limit or endurance limit is observed between 10^5 and 10^6 cycles. For low-carbon steels, the fatigue limit is $\cong$50% of its ultimate tensile strength. Below the fatigue limit, no fatigue failure occurs so that the material has essentially an infinite life. For many fiber-reinforced composites, a fatigue limit may not be observed; however, the slope of the *S–N* plot is markedly reduced at low stress levels. In these situations, it is common practice to specify the fatigue strength of the material at very high cycles, say, 10^6 or 10^7 cycles.

4.2.1 Fatigue Test Methods

The majority of fatigue tests on fiber-reinforced composite materials have been performed with uniaxial tension–tension cycling (Figure 4.34). Tension–compression and compression–compression cycling are not commonly used since failure by compressive buckling may occur in thin laminates. Completely reversed tension–compression cycling is achieved by flexural fatigue tests. In addition, a limited number of interlaminar shear fatigue and in-plane shear fatigue tests have also been performed.

The tension–tension fatigue cycling test procedure is described in ASTM D3479. It uses a straight-sided specimen with the same dimensions and end tabs as in static tension tests. At high cyclic frequencies, polymer matrix composites may generate appreciable heat due to internal damping, which is turn increases the specimen temperature. Since a frequency-induced temperature rise can affect the fatigue performance adversely, low cyclic frequencies (<10 Hz) are

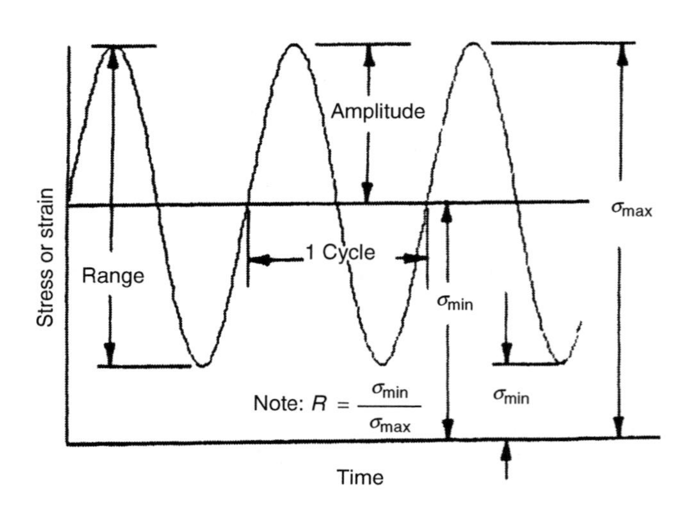

FIGURE 4.34 Stress vs. time diagram in a fatigue test.

preferred. Both stress-controlled and strain-controlled tests are performed. In a stress-controlled test, the specimen is cycled between specified maximum and minimum stresses so that a constant stress amplitude is maintained. In a strain-controlled test, the specimen is cycled between specified maximum and minimum strains so that a constant strain amplitude is maintained.

A unique feature of a fiber-reinforced composite material is that it exhibits a gradual softening or loss in stiffness due to the appearance of microscopic damages long before any visible damage occurs. As a result, the strain in the specimen increases in load-controlled tests, but the stress in the specimen decreases in strain-controlled tests (Figure 4.35). Microscopic damages also cause a loss in residual strength of the material. Instead of specimen separation,

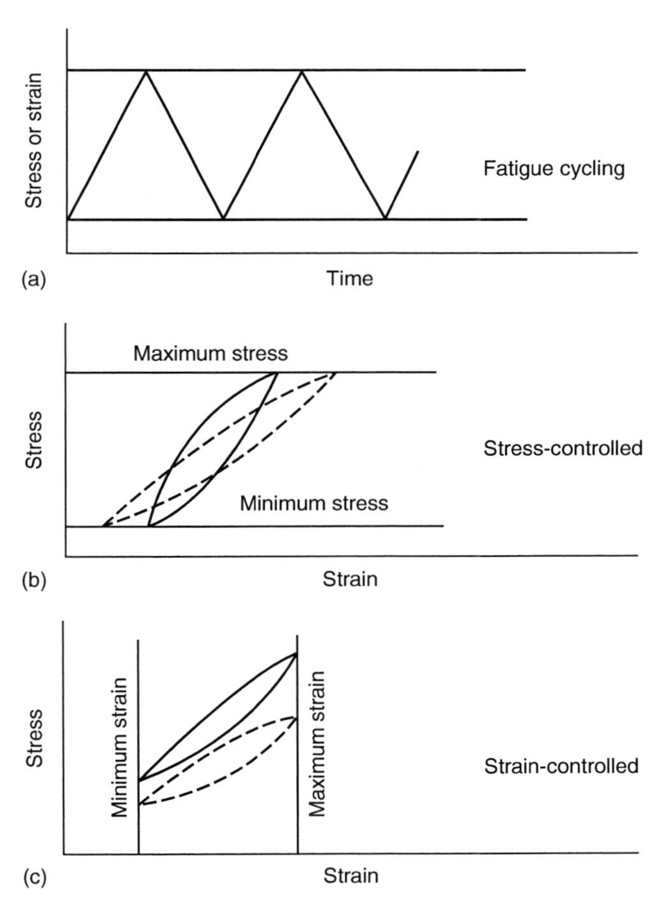

FIGURE 4.35 (a) Fatigue cycling in stress-controlled or strain-controlled fatigue tests. Differences in (b) stress-controlled test and (c) strain-controlled fatigue test of polymer matrix composites.

many fatigue tests are performed until the specimen stiffness or residual strength decreases to a predetermined level. Thus, cycles to failure may not always represent the specimen life at complete fracture.

Many investigators have attempted to describe the S–log N plot for various fiber-reinforced composites by a straight line:

$$S = \sigma_U(m \log N + b), \tag{4.23}$$

where
S = maximum fatigue stress
N = number of cycles to failure
σ_U = average static strength
m, b = constants

Values of m and b for a few epoxy matrix composites are given in Table 4.8.

A power-law representation for the S–N plot is also used:

$$\frac{S}{\sigma_U} N^d = c, \tag{4.24}$$

where c and d are constants. Similar expressions can be written for ε–N plots obtained in strain-controlled fatigue tests.

The number of cycles to failure, also called the fatigue life, usually exhibits a significant degree of scatter. Following a two-parameter Weibull distribution, the probability of fatigue life exceeding L can be written as

$$F(L) = \exp\left[-\left(\frac{L}{L_0}\right)^{\alpha_f}\right], \tag{4.25}$$

TABLE 4.8
Constants in S–N Representation of Composite Laminates

Material	Layup	R	Constants in Equation 4.23		References
			m	b	
E-glass–ductile epoxy	0°	0.1	−0.1573	1.3743	[26]
T-300 Carbon–ductile epoxy	0°	0.1	−0.0542	1.0420	[26]
E-glass–brittle epoxy	0°	0.1	−0.1110	1.0935	[26]
T-300 Carbon–brittle epoxy	0°	0.1	−0.0873	1.2103	[26]
E-glass–epoxy	$[0/\pm45/90]_S$	0.1	−0.1201	1.1156	[27]
E-glass–epoxy	$[0/90]_S$	0.05	−0.0815	0.934	[28]

Note: R represents the ratio of the minimum stress and the maximum stress in fatigue cycling.

where
α_f is the shape parameter in fatigue
L_0 is the location parameter for the fatigue life distribution (cycles)

Comparing the static strength data and fatigue life data of unidirectional 0° E-glass–epoxy, Hahn and Kim [29] proposed the following correlation between the static strength and fatigue data:

$$\frac{L}{L_0} = \left(\frac{S}{\sigma_U}\right)^{\alpha/\alpha_f}. \tag{4.26}$$

Equation 4.26 implies that the higher the static strength of a specimen, the longer would be its fatigue life.

4.2.2 Fatigue Performance

4.2.2.1 Tension–Tension Fatigue

Tension–tension fatigue tests on unidirectional 0° ultrahigh-modulus carbon fiber-reinforced thermoset polymers produce *S–N* curves that are almost horizontal and fall within the static scatter band (Figure 4.36). The fatigue effect is slightly greater for relatively lower modulus carbon fibers. Unidirectional 0° boron and Kevlar 49 fiber composites also exhibit exceptionally good fatigue strength in tension–tension loading (Figure 4.37). Other laminates, such as $[0/\pm45/90]_S$ carbon, $[0/90]_S$ carbon, $[0/\pm30]_S$ carbon, and $[0/\pm45]_S$ boron

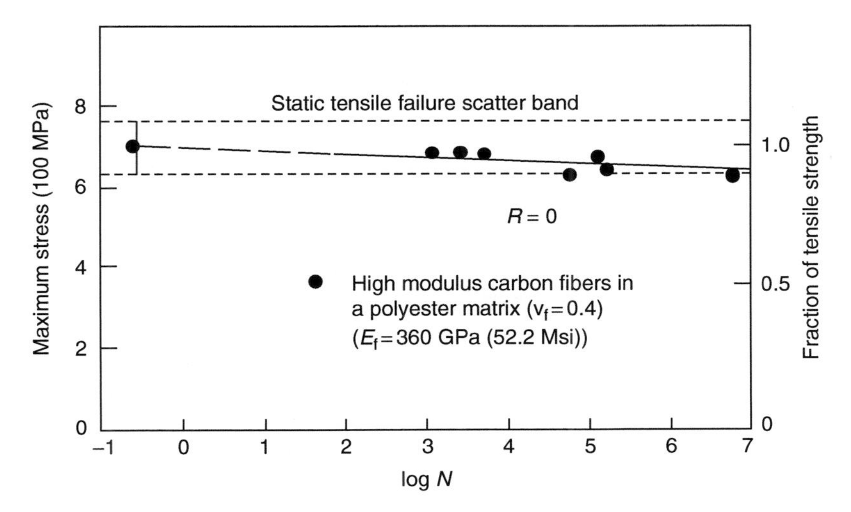

FIGURE 4.36 Tension–tension *S–N* diagram for a 0° ultrahigh-modulus carbon fiber–polyester composite. (After Owen, M.J. and Morris, S., *Carbon Fibres: Their Composites and Applications*, Plastics Institute, London, 1971.)

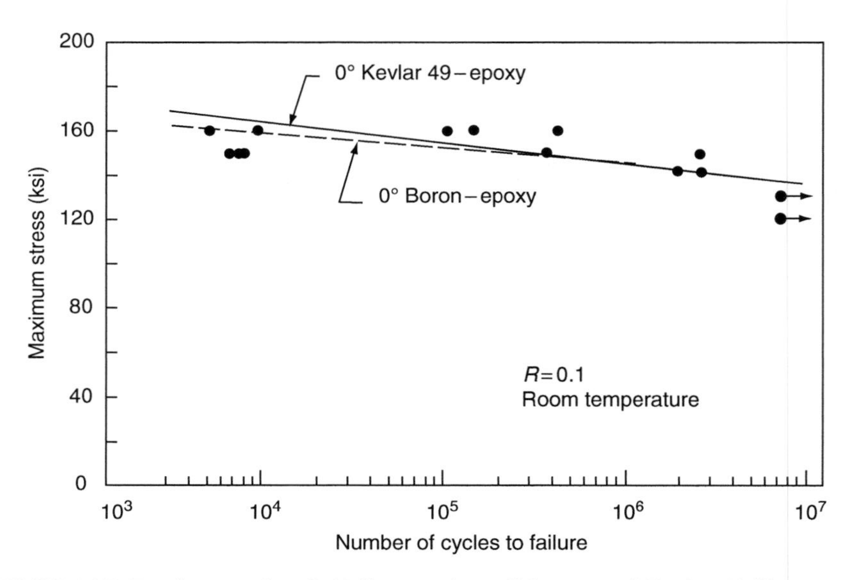

FIGURE 4.37 Tension–tension *S–N* diagram for a 0° boron and Kevlar 49 fiber–epoxy composites. (After Miner, L.H., Wolfe, R.A., and Zweben, C.H., *Composite Reliability, ASTM STP*, 580, 1975.)

(Figure 4.38), show very similar *S–N* diagrams, although the actual fatigue effect depends on the proportion of fibers aligned with the loading axis, stacking sequence, and mode of cycling. The effect of cycling mode is demonstrated in Figure 4.39, in which a tension–compression cycling ($R = -1.6$) produces a steeper *S–N* plot than the tension–tension cycling ($R = 0.1$) and the compression–compression cycling ($R = 10$) gives the lowest *S–N* plot.

The fatigue performances of both E- and S-glass fiber-reinforced composites are inferior to those of carbon, boron, and Kevlar 49 fiber-reinforced

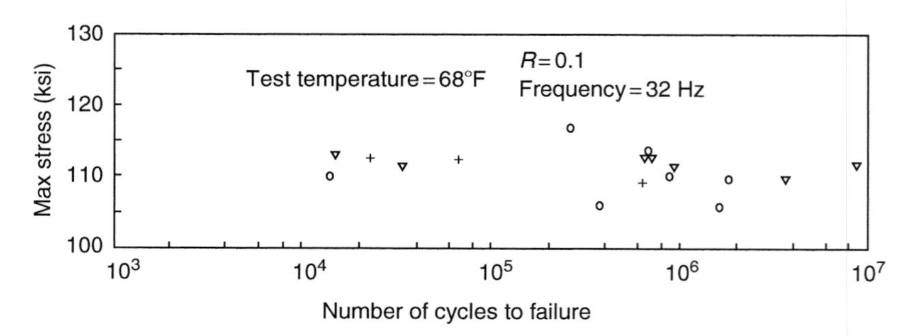

FIGURE 4.38 Tension–tension *S–N* diagram for a $[0/\pm45]_S$ boron fiber–epoxy laminate. (After Donat, R.C., *J. Compos. Mater.*, 4, 124, 1970.)

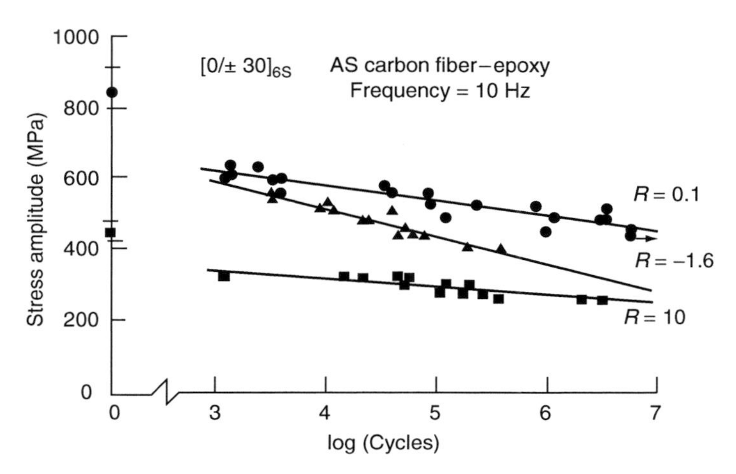

FIGURE 4.39 S–N diagrams for $[0/\pm 30]_{6S}$ AS carbon fiber–epoxy laminates at various fatigue stress ratios (*Note*: $R = 0.1$ in tension–tension cycling, $R = -1.6$ in tension–compression cycling, and $R = 10$ in compression–compression cycling). (After Ramani, S. V. and Williams, D.P., *Failure Mode in Composites III*, AIME, 1976.)

composites. Both types of fibers produce steep S–N plots for unidirectional 0° composites (Figures 4.40 and 4.41). An improvement in their fatigue performance can be achieved by hybridizing them with other high-modulus fibers, such as T-300 carbon (Figure 4.42).

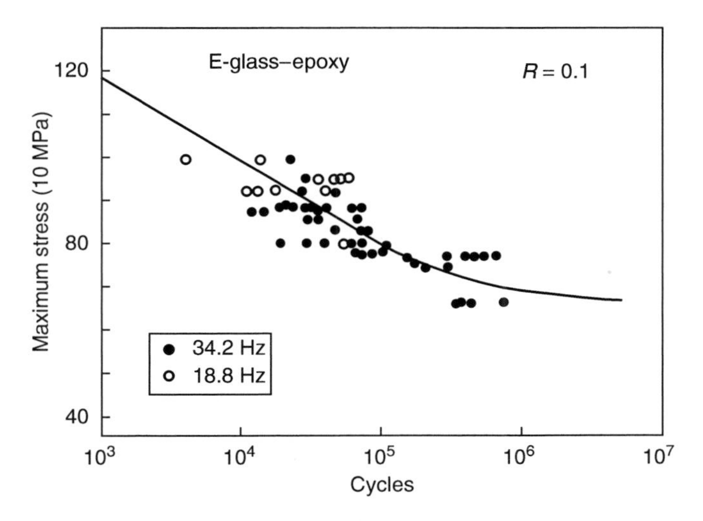

FIGURE 4.40 Tension–tension S–N diagram for a 0° E-glass–epoxy laminate. (After Hashin, Z. and Rotem, A., *J. Compos. Mater.*, 7, 448, 1973.)

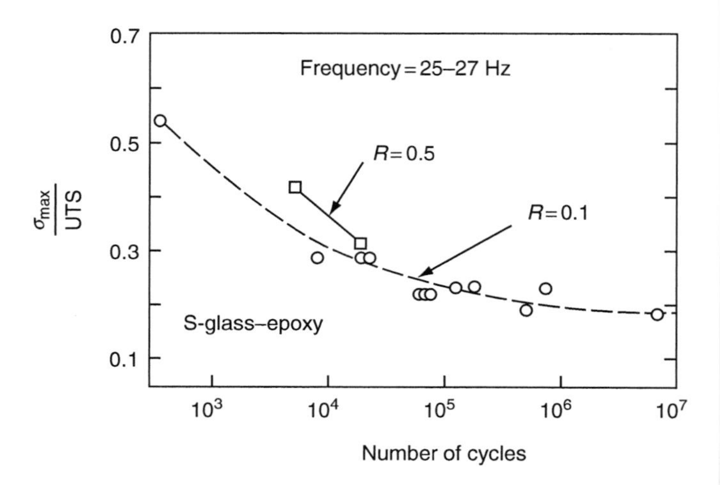

FIGURE 4.41 Tension–tension *S–N* diagram for a 0° S-glass–epoxy laminate at various fatigue stress ratios. (After Tobler, R.L. and Read, D.L., *J. Compos. Mater.*, 10, 32, 1976.)

The tension–tension fatigue properties of SMC composites have been reported by several investigators [30–32]. SMC materials containing E-glass fiber-reinforced polyester or vinyl ester matrix also do not exhibit fatigue limit. Their fatigue performance depends on the proportion of continuous and chopped fibers in the laminate (Figure 4.43).

4.2.2.2 Flexural Fatigue

The flexural fatigue performance of fiber-reinforced composite materials is in general less satisfactory than the tension–tension fatigue performance. This can

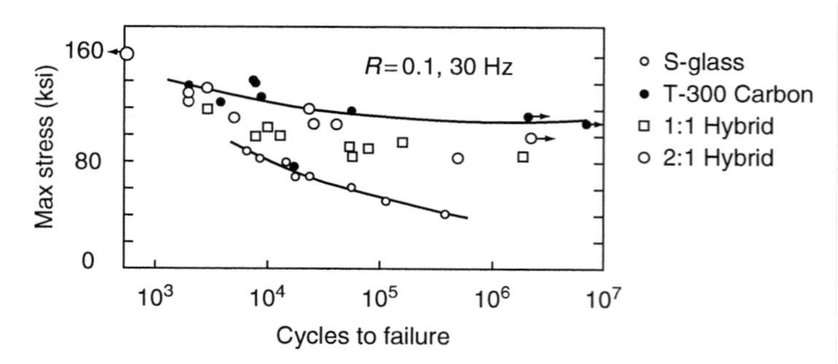

FIGURE 4.42 Tension–tension *S–N* diagram for a 0° S-glass, T-300 carbon, and S-glass/T-300 carbon interply hybrid laminates. (After Hofer, K.E., Jr., Stander, M., and Bennett, L.C., *Polym. Eng. Sci.*, 18, 120, 1978.)

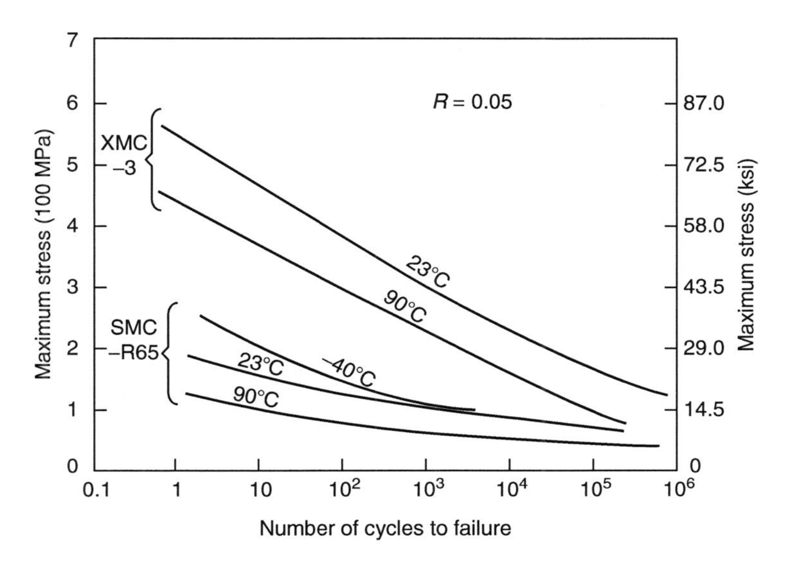

FIGURE 4.43 Tension–tension S–N diagrams for SMC laminates.

be observed in Figure 4.44, where the slope of the flexural S–N curve is greater than that of the tension–tension S–N curve for high-modulus carbon fibers. The lower fatigue strength in flexure is attributed to the weakness of composites on the compression side.

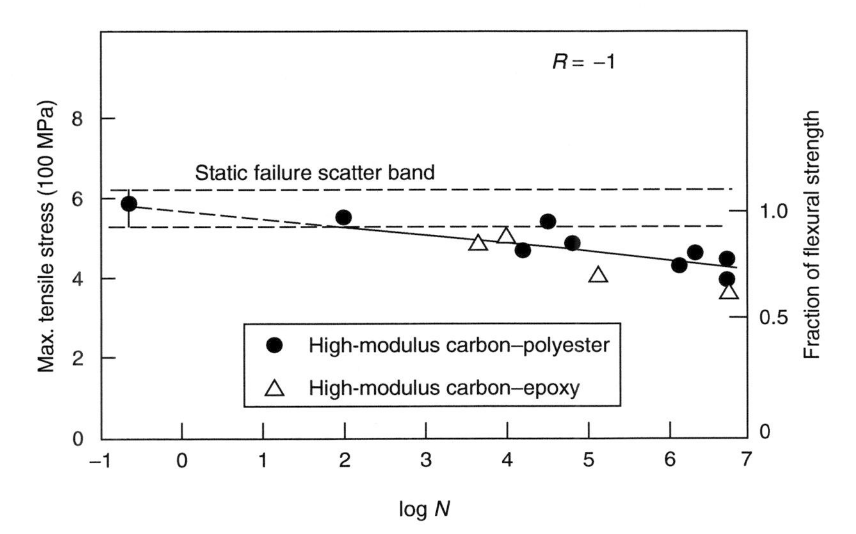

FIGURE 4.44 Flexural S–N diagram for 0° carbon fiber–epoxy and polyester laminates. (After Hahn, H.T. and Kim, R.Y., *J. Compos. Mater.*, 10, 156, 1976.)

4.2.2.3 Interlaminar Shear Fatigue

Fatigue characteristics of fiber-reinforced composite materials in the interlaminar shear (τ_{xz}) mode have been studied by Pipes [33] and several other investigators [34,35]. The interlaminar shear fatigue experiments were conducted using short-beam shear specimens. For a unidirectional 0° carbon fiber-reinforced epoxy, the interlaminar shear fatigue strength at 10^6 cycles was reduced to <55% of its static ILSS even though its tension–tension fatigue strength was nearly 80% of its static tensile strength (Figure 4.45). The interlaminar shear fatigue performance of a unidirectional 0° boron–epoxy system was similar to that of unidirectional 0° carbon–epoxy system. However, a reverse trend was observed for a unidirectional 0° S-glass-reinforced epoxy. For this material, the interlaminar shear fatigue strength at 10^6 cycles was ~60% of its static ILSS, but the tension–tension fatigue strength at 10^6 cycles was <40% of its static tensile strength. Unlike the static interlaminar strengths, fiber volume fraction [34] and fiber surface treatment [35] did not exhibit any significant influence on the high cycle interlaminar fatigue strength.

Wilson [36] has studied the interlaminar shear fatigue behavior of an SMC-R50 laminate. His results show that the interlaminar shear fatigue strength of this material at 10^6 cycles and 26°C is equal to 64% of its static ILSS. The interlaminar shear fatigue strength at 10^6 cycles and 90°C is between 45% and 50% of the corresponding ILSS.

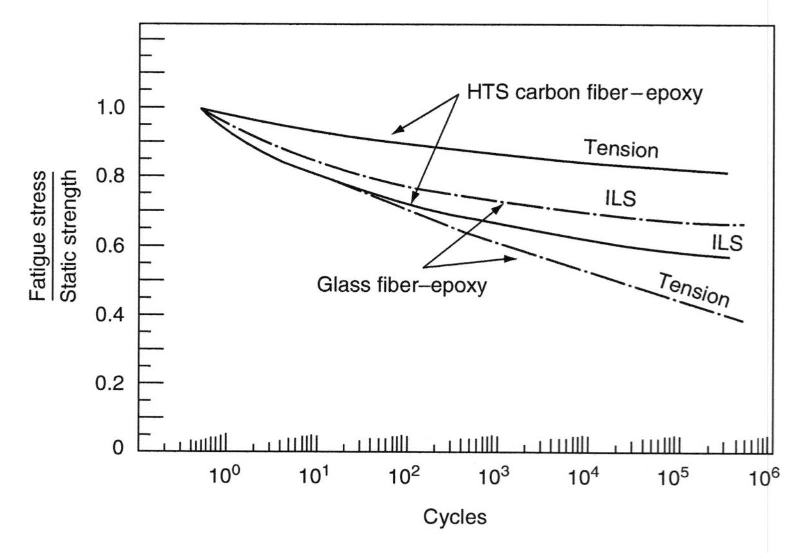

FIGURE 4.45 Interlaminar shear *S–N* diagrams for 0° carbon and glass fiber–epoxy laminates. (After Pipes, R.B., *Composite Materials: Testing and Design (Third Conference)*, *ASTM STP*, 546, 419, 1974.)

4.2.2.4 Torsional Fatigue

The torsional fatigue behavior of carbon fiber-reinforced epoxy thin tubes is shown in Figure 4.46 for 0° and ±45° orientations. On a log–log scale, the *S–N* plot in alternating ($R = -1$) torsional fatigue exhibits linear behavior. The torsional fatigue strength of ±45° specimens is ~3.7–3.8 times higher than that of the 0° specimens at an equivalent number of cycles. The data for [0/±45] tubes fall between the 0° and ±45° lines. The 0° specimens failed by a few longitudinal cracks (cracks parallel to fibers), and the ±45° specimens failed by cracking along the ±45° lines and extensive delamination. Although the 0° specimens exhibited a lower torsional fatigue strength than ±45° specimens, they retained a much higher postfatigue static torsional strength.

Torsional fatigue data for a number of unidirectional 0° fiber-reinforced composites are compared in Figure 4.47. The data in this figure were obtained by shear strain cycling of solid rod specimens [37]. Fatigue testing under pure shear conditions clearly has a severe effect on unidirectional composites, all failing at ~10^3 cycles at approximately half the static shear strain to failure. Short-beam interlaminar shear fatigue experiments do not exhibit such rapid deterioration.

4.2.2.5 Compressive Fatigue

Compression–compression fatigue *S–N* diagram of various E-glass fiber-reinforced polyester and epoxy composites is shown in Figure 4.48. Similar trends are also observed for T-300 carbon fiber-reinforced epoxy systems [38].

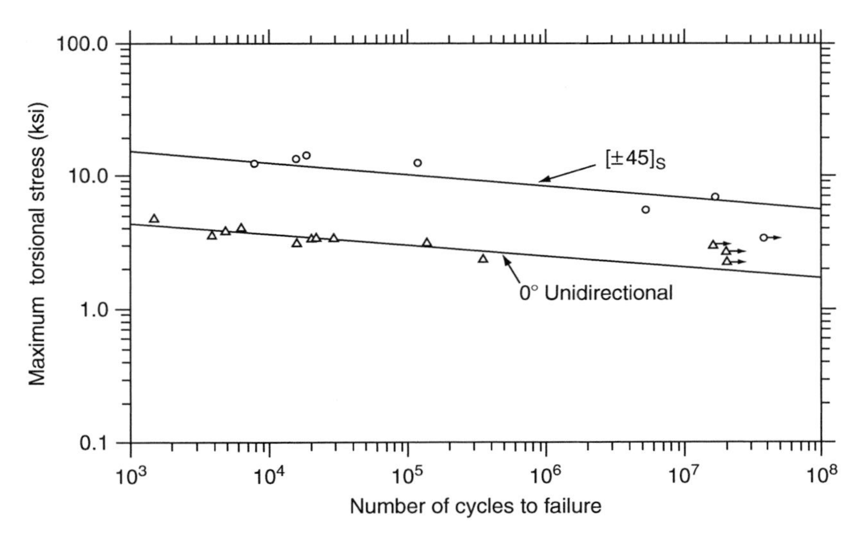

FIGURE 4.46 Torsional *S–N* diagrams for a 0° and $[\pm 45]_S$ high tensile strength carbon fiber–epoxy composites. (After Fujczak, B.R., Torsional fatigue behavior of graphite–epoxy cylinders, U.S. Army Armament Command, Report No. WVT-TR-74006, March 1974.)

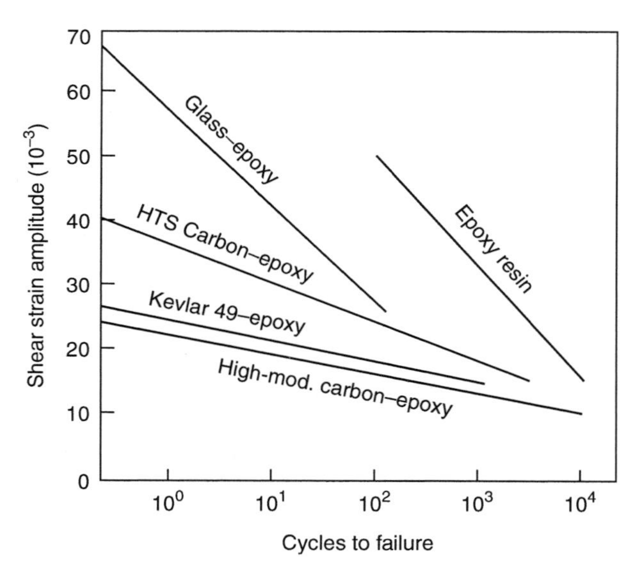

FIGURE 4.47 Torsional shear strain-cycle diagrams for various 0° fiber-reinforced composites. (After Phillips, D.C. and Scott, J.M., *Composites*, 8, 233, 1977.)

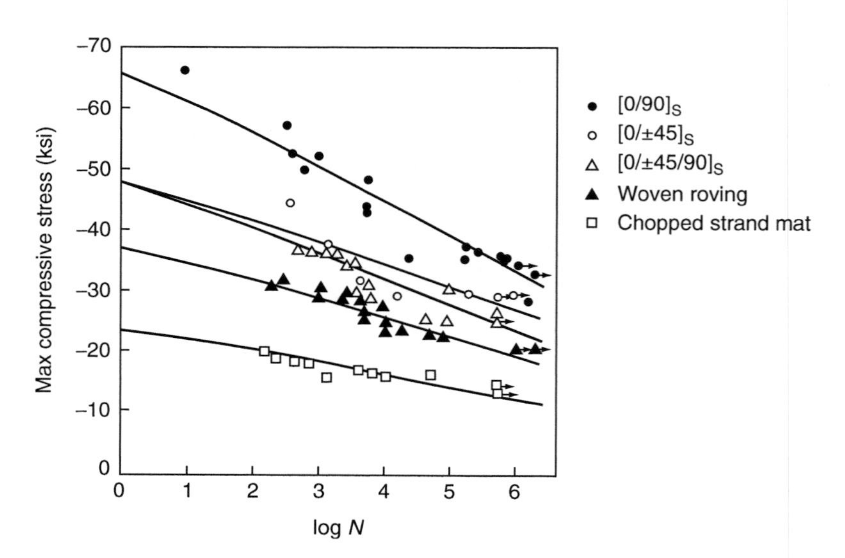

FIGURE 4.48 Compression–compression *S–N* diagrams for various composite laminates. (After Conners, J.D., Mandell, J.F., and McGarry, F.J., Compressive fatigue in glass and graphite reinforced composites, *Proceedings 34th Annual Technical Conference*, Society of the Plastics Industry, 1979.)

4.2.3 Variables in Fatigue Performance

4.2.3.1 Effect of Material Variables

Fatigue tests on unidirectional composites containing off-axis fibers (i.e., $\theta \neq 0°$) show a steady deterioration in fatigue strength with increasing fiber orientation angle [39]. Analogous to static tests, the fatigue failure mode in off-axis composites changes from progressive fiber failure at $\theta < 5°$ to matrix failure or fiber–matrix interface failure at $\theta > 5°$. However, a laminate containing alternate layers of ±5° fibers has higher fatigue strength than a 0° laminate (Figure 4.49). The fatigue performance of 0° laminates is also improved by the addition of a small percentage of 90° plies, which reduce the tendency of splitting (cracks running parallel to fibers in the 0° laminas) due to low transverse strengths of 0° laminas [40]. However, as the percentage of 90° plies increases, the fatigue strength is reduced.

Figure 4.50 shows the zero-tension fatigue data of two carbon fiber-reinforced PEEK laminates, namely, $[-45/0/45/90]_{2S}$ and $[\pm 45]_{4S}$. Higher fatigue strength of the $[-45/0/45/90]_{2S}$ is due to the presence of 0° fibers.

Experiments by Boller [40] and Davis et al. [41] have also shown that the fatigue performance of laminates containing woven fabrics or randomly oriented fibers is lower than that of unidirectional or nonwoven cross-ply laminates (Figure 4.51). Fatigue performance of laminates containing combinations

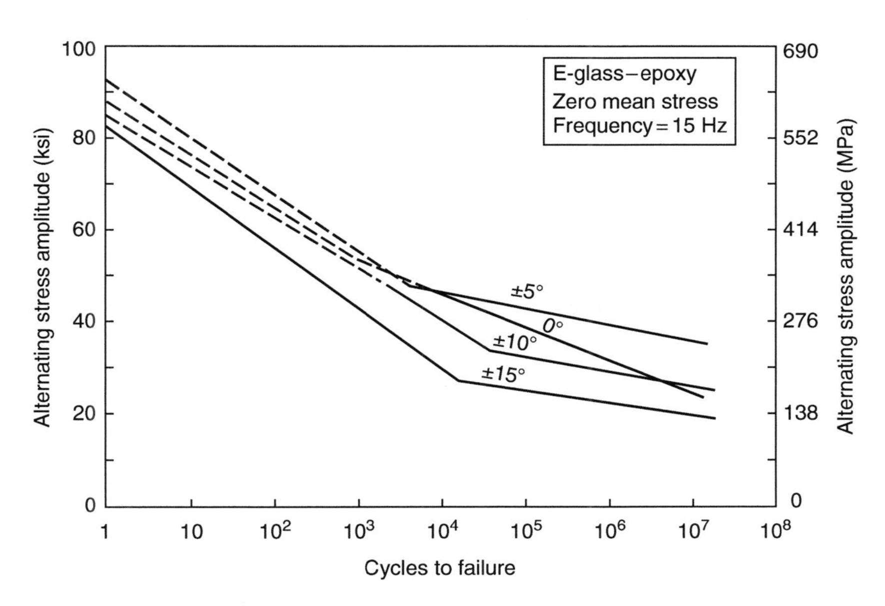

FIGURE 4.49 Effect of fiber orientation angles on the fatigue performance of E-glass–epoxy composites. (After Boller, K.H., *Mod. Plast.*, 41, 145, 1964.)

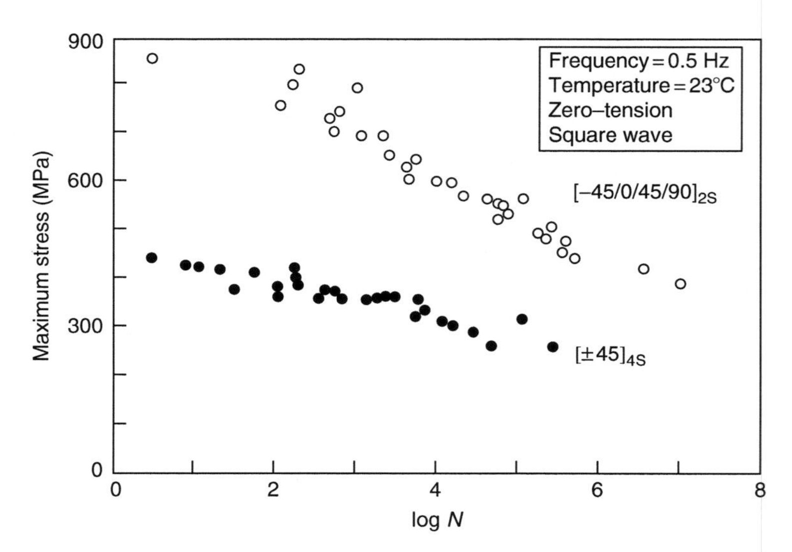

FIGURE 4.50 Zero-tension fatigue data of carbon fiber-reinforced PEEK laminates. (After Carlile, D.R., Leach, D.C., Moore, D.R., and Zahlan, N., *Advances in Thermoplastic Matrix Composite Materials, ASTM STP*, 1044, 199, 1989.)

of fiber orientations, such as $[0/90]_S$ and $[0/\pm45/90]_S$, are particularly sensitive to laminate configuration, since the signs of interlaminar stresses may be reversed by simple variations in stacking sequence (Figure 4.52).

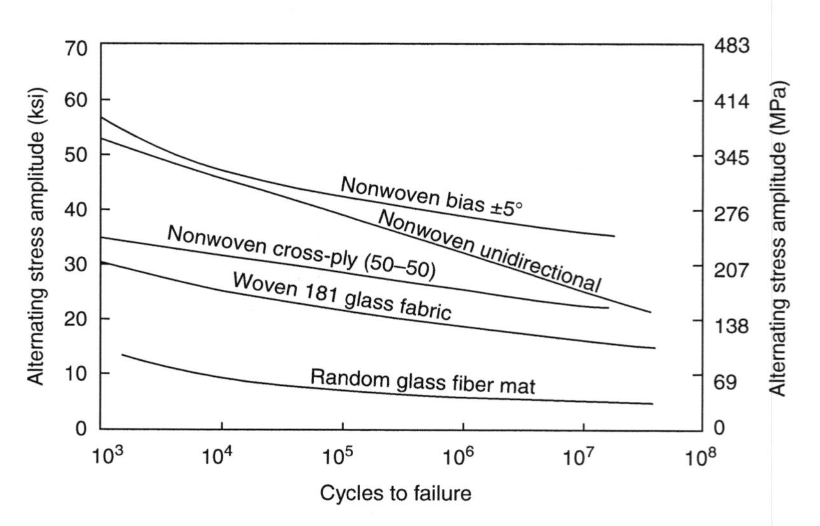

FIGURE 4.51 Fatigue performance of various woven fabric, nonwoven fabric, and mat-reinforced composite laminates. (After Davis, J.W., McCarthy, J.A., and Schrub, J.N., *Materials in Design Engineering*, 1964.)

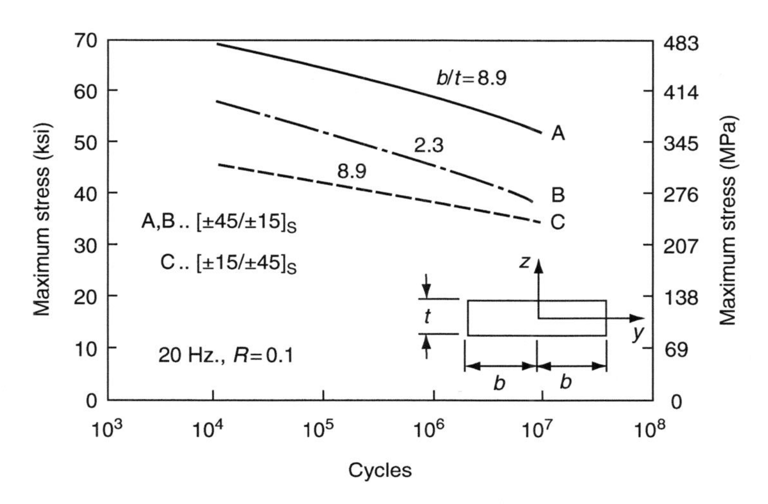

FIGURE 4.52 Effect of laminate stacking sequence on the tension–tension fatigue performance of carbon fiber–epoxy laminates.

A systematic study of the effects of resin type and coupling agents on the fatigue performance of fiber-reinforced polymer composites is lacking. Early work by Boller [40] on balanced E-glass fabric-reinforced laminates has shown the superiority of epoxies over polyesters and other thermoset resins. Mallick [32] has shown that vinyl ester resin provides a better fatigue damage resistance than polyester resin in an SMS-R65 laminate. However, within the same resin category, the effects of compositional differences (e.g., low reactivity vs. high reactivity in polyester resins or hard vs. flexible in epoxy resins) on the long-life fatigue performance are relatively small. In zero-tension fatigue ($R = 0$) experiments with chopped E-glass strand mat–polyester laminates, Owen and Rose [42] have shown that the principal effect of flexibilizing the resin is to delay the onset of fatigue damage. The long-term fatigue lives are not affected by the resin flexibility.

Investigations by Tanimoto and Amijima [43] as well as Dharan [44] have shown that, analogous to static tensile strength, the fatigue strength also increases with increasing fiber volume fraction. An example of the effect of fiber volume fraction is shown in Figure 4.53.

4.2.3.2 Effect of Mean Stress

The effect of tensile mean stress on the fatigue properties of fiber-reinforced composite materials was first studied by Boller [45]. For 0° and ±15° E-glass–epoxy laminates, the stress amplitude at a constant life tends to decrease with increasing tensile mean stress (Figure 4.54). This behavior is

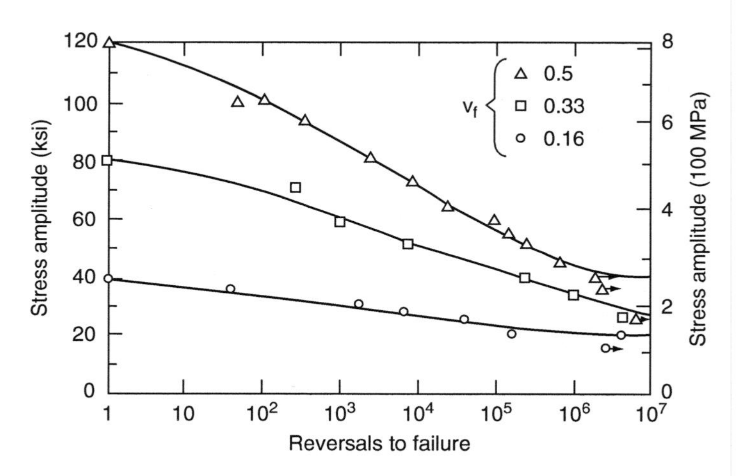

FIGURE 4.53 Effect of fiber volume fraction on the fatigue performance of 0° E-glass–epoxy laminates. (After Dharan, C.K.H., *J. Mater. Sci.*, 10, 1665, 1975.)

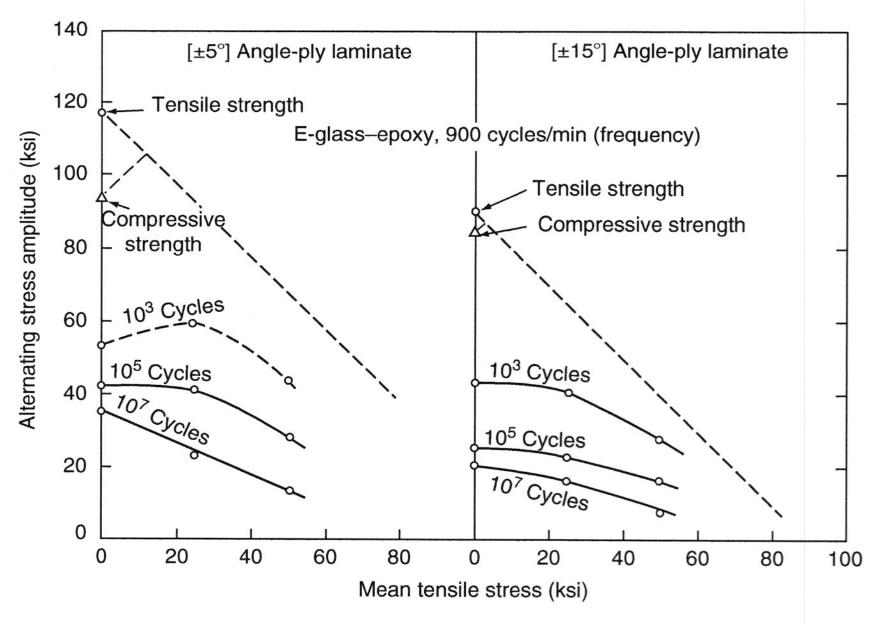

FIGURE 4.54 Effect of tensile mean stress on the fatigue strengths of two E-glass–epoxy laminates. (After Boller, K.H., Effect of tensile mean stresses on fatigue properties of plastic laminates reinforced with unwoven glass fibers, U.S. Air Force Materials Laboratory, Report No. ML-TDR-64–86, June 1964.)

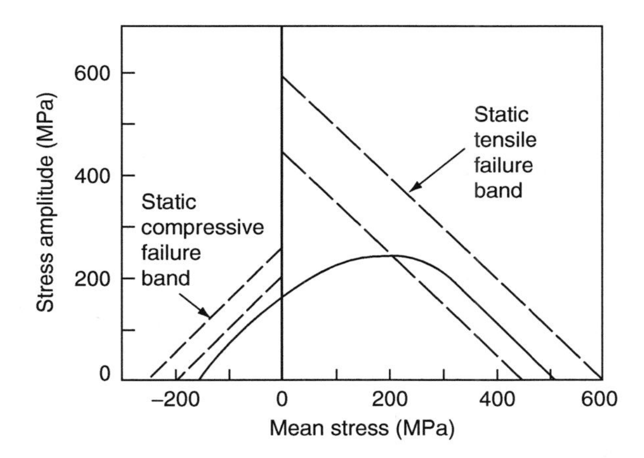

FIGURE 4.55 Effect of mean stress on the fatigue strength of a cross-ply carbon fiber–epoxy laminate. (After Owen, M.J. and Morris, S., *Carbon Fibres: Their Composites and Applications*, Plastic Institute, London, 1971.)

similar to that of ductile metals. However, the Goodman equation,* which is commonly used for ductile metals, may not be applicable for fiber-reinforced composite materials.

Figure 4.55 shows the relationship between mean stress and stress amplitude at 10^6 cycles for a cross-ply high-modulus carbon–epoxy composite. At high tensile mean stresses, the fatigue curve lies within the static tensile strength scatter band. However, low tensile mean stresses as well as compressive mean stresses have a significant adverse effect on the fatigue strength of this material. Similar behavior was also observed for a $[0/\pm30]_{6S}$ carbon–epoxy composite [46].

Smith and Owen [47] have reported the effect of mean stresses on the stress amplitude for chopped E-glass strand mat–polyester composites. Their data (Figure 4.56) show that a small compressive mean stress may have a beneficial effect on the fatigue performance of random fiber composites.

* The Goodman equation used for taking into account the effect of tensile mean stresses for high cycle fatigue design of metals is given by

$$\frac{\sigma_a}{S_f} + \frac{\sigma_m}{S_{ut}} = 1,$$

where

σ_a = alternating stress
σ_m = mean stress
S_f = fatigue strength with $\sigma_m = 0$
S_{ut} = ultimate tensile strength

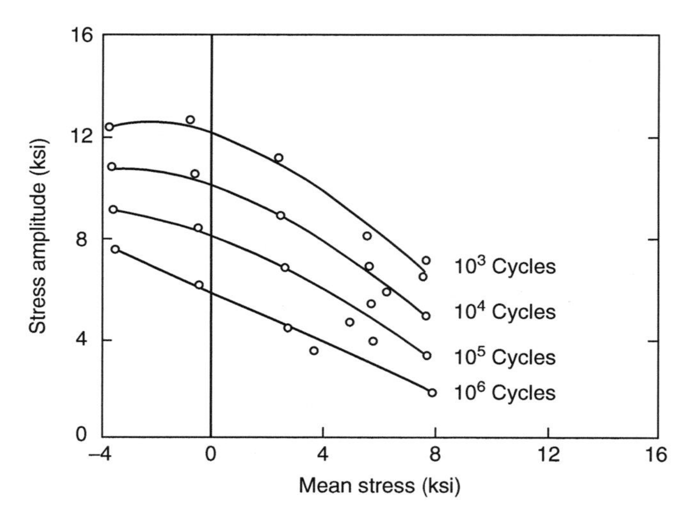

FIGURE 4.56 Effect of mean stress on the fatigue stress amplitude of an E-glass mat-reinforced polyester laminate. (After Smith, T.R. and Owen, M.J., *Mod. Plast.*, 46, 124, 1969.)

4.2.3.3 Effect of Frequency

The viscoelastic nature of polymers causes a phase difference between cyclic stresses and strains in polymer matrix composites, which is exemplified by hysteresis loops even at low stress levels. This results in energy accumulation in the form of heat within the material. Owing to the low thermal conductivity of the material, the heat is not easily dissipated, which in turn creates a temperature difference between the center and surfaces of a polymer matrix laminate. At a constant stress level, the temperature difference due to viscoelastic heating increases with increasing frequency of cycling (Figure 4.57). Depending on the frequency, it may attain a steady-state condition after a few cycles and then rise rapidly to high values shortly before the specimen failure.

In spite of the heating effect at high cyclic frequencies, Dally and Broutman [48] found only a modest decrease in fatigue life with increasing frequency up to 40 Hz for cross-ply as well as quasi-isotropic E-glass–epoxy composites. On the other hand, Mandell and Meier [49] found a decrease in the fatigue life of a cross-ply E-glass–epoxy laminate as the cyclic frequency was reduced from 1 to 0.1 Hz. For a $[\pm 45]_{2S}$ T-300 carbon–epoxy laminate containing a center hole, Sun and Chan [50] have reported a moderate increase in fatigue life to a peak value between 1 and 30 Hz (Figure 4.58). The frequency at which the peak life was observed shifted toward a higher value as the load level was decreased. Similar results were observed by Saff [51] for $[\pm 45]_{2S}$ AS carbon–epoxy laminates with center holes and Reifsnider et al. [52] for $[0/\pm 45/0]_S$ boron–aluminum laminates with center holes.

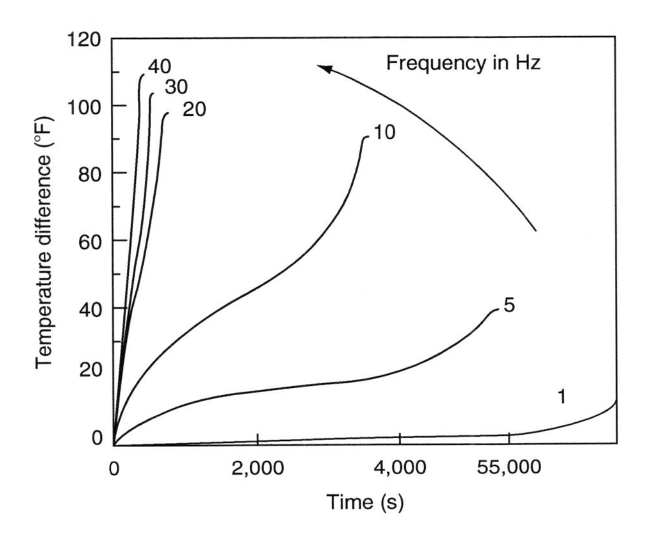

FIGURE 4.57 Difference in temperatures at the center and outside surfaces of cross-ply E-glass–epoxy laminates during fatigue cycling. (After Dally, J.W. and Broutman, L.J., *J. Compos. Mater.*, 1, 424, 1967.)

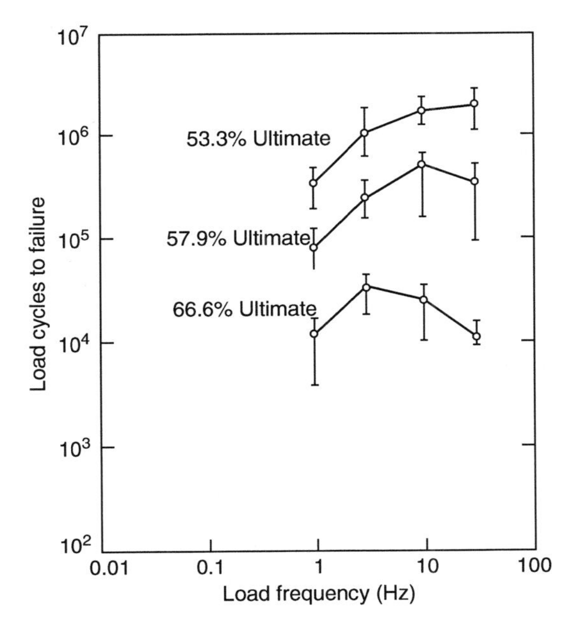

FIGURE 4.58 Effect of cyclic frequency on the fatigue life of a $[\pm 45]_{2S}$ carbon fiber–epoxy laminate containing a center hole. (After Sun, C.T. and Chan, W.S., *Composite Materials, Testing and Design* (*Fifth Conference*), *ASTM STP*, 674, 418, 1979.)

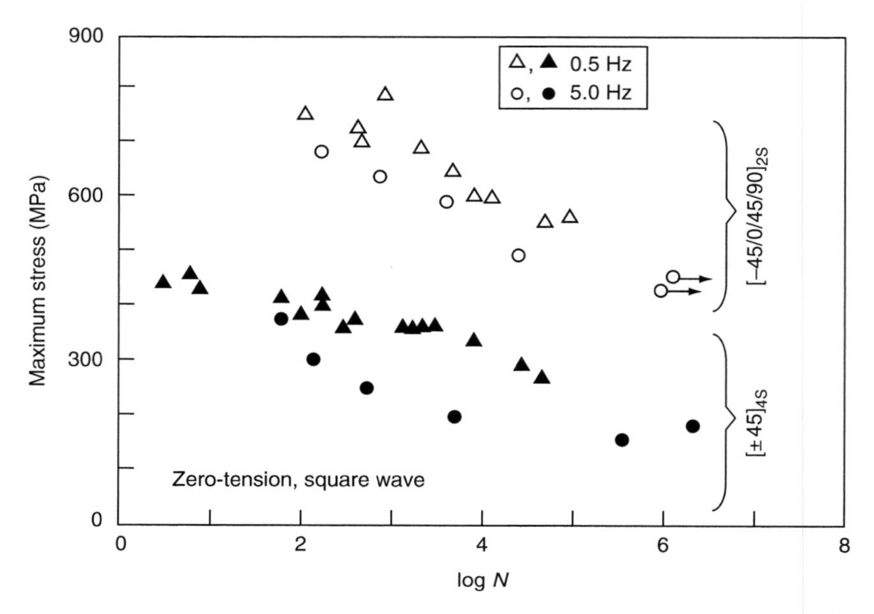

FIGURE 4.59 Influence of test frequency on the fatigue performance of $[-45/0/45/90]_{2S}$ and $[\pm 45]_{4S}$ carbon fiber-reinforced PEEK laminates. (After Carlile, D.R., Leach, D.C., Moore, D.R., and Zahlan, N., *Advances in Thermoplastic Matrix Composite Materials, ASTM STP*, 1044, 199, 1989.)

Frequency-dependent temperature rise was also detected in zero-tension fatigue testing of carbon fiber-reinforced PEEK laminates. Temperature rise was found to be dependent on the laminate configuration. For example, the temperature rise in $[-45/0/45/90]_{2S}$ laminates was only 20°C above the ambient temperature, but it was up to 150°C in $[\pm 45]_{4S}$ laminates, when both were fatigue-tested at 5 Hz. Higher temperature rise in the latter was attributed to their matrix-dominated layup. The difference in the fatigue response at 0.5 and 5 Hz for these laminate configurations is shown in Figure 4.59.

4.2.3.4 Effect of Notches

The fatigue strength of a fiber-reinforced polymer decreases with increasing notch depth (Figure 4.60) as well as increasing notch tip sharpness (Figure 4.61). Stacking sequence also plays an important role in the notch effect in fiber-reinforced polymers. Underwood and Kendall [53] have shown that multidirectional laminates containing 0° layers in the outer surfaces have a much longer fatigue life than either 90° or off-axis layers in the outer surfaces.

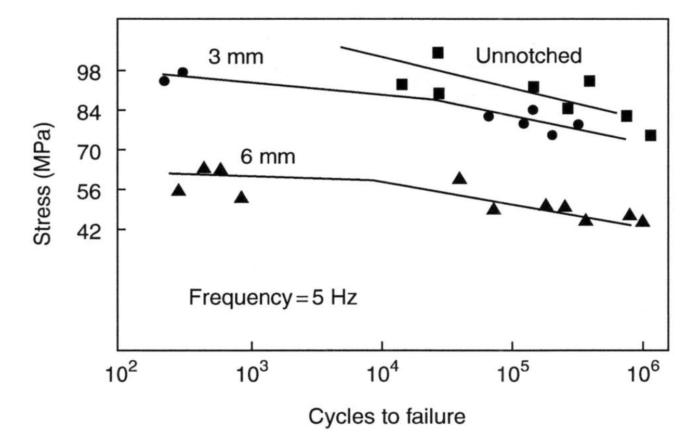

FIGURE 4.60 Effect of notch depth on the fatigue performance of a cross-ply E-glass–epoxy laminate. (After Carswell, W.S., *Composites*, 8, 251, 1977.)

Many fiber-reinforced polymers exhibit less notch sensitivity in fatigue than conventional metals. Fatigue damage created at the notch tip of these laminates tends to blunt the notch severity and increases their residual strengths. A comparison of fatigue strength reductions due to notching in a 2024-T4 aluminum alloy with those in a HT carbon–epoxy laminate (Table 4.9) demonstrates this important advantage of composite materials.

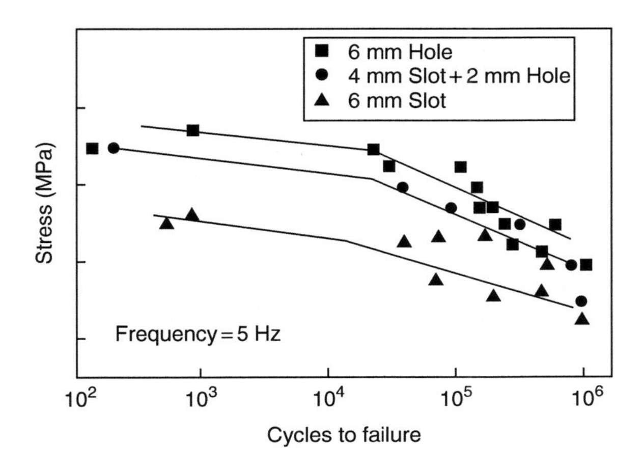

FIGURE 4.61 Effect of notch sharpness on the fatigue performance of a cross-ply E-glass–epoxy laminate. (After Carswell, W.S., *Composites*, 8, 251, 1977.)

TABLE 4.9
Comparison of Unnotched and Notched Fatigue Strengths[a]

Material	Lamination Configuration	UTS, Mpa (ksi)	Ratio of Fatigue Strength to UTS: Unnotched At 10^4 Cycles	Unnotched At 10^7 Cycles	Notched ($K_t = 3$)[b] At 10^4 Cycles	Notched ($K_t = 3$)[b] At 10^7 Cycles
High tensile strength carbon fiber–epoxy ($\rho = 1.57$ g/cm^3)	0°	1138 (165)	0.76	0.70	0.42	0.36
	$[0/90]_S$	759 (110)	0.71	0.59	0.57	0.55
	$[0/90/\pm45]_S$	400 (58)	0.83	0.78	0.55	0.52
2024-T4 aluminum ($\rho = 2.77$ g/cm^3)	—	531 (77)	0.83	0.55	0.65	0.23

Source: Adapted from Freeman, W.T. and Kuebeler, G.C., *Composite Materials: Testing and Design* (*Third Conference*), *ASTM STP*, 546, 435, 1974.

[a] Fatigue stress ratio $R = 0.1$ for all experiments.
[b] K_t is the theoretical stress concentration factor.

4.2.4 Fatigue Damage Mechanisms in Tension–Tension Fatigue Tests

4.2.4.1 Continuous Fiber 0° Laminates

Depending on the maximum stress level, fiber type, and matrix fatigue properties, fatigue damage in continuous fiber 0° laminates is dominated either by fiber breakage or by matrix microcracking [54–56]. At very high fatigue stress levels, the fiber stress may exceed the lower limit of the fiber strength scatter band. Thus, on the first application on the maximum stress, the weakest fibers break. The locations of fiber breakage are randomly distributed in the volume of the composite (Figure 4.62a). High stress concentration at the broken fiber ends initiates more fiber breakage in the nearby areas. Rapidly increasing zones of fiber failure weaken the composite severely, leading eventually to catastrophic failure in a few hundred cycles.

At lower fatigue loads, the fiber stress may be less than the lower limit of the fiber strength scatter band, but the matrix strain may exceed the cyclic strain limit of the matrix. Thus failure initiation takes place by matrix microcracking (Figure 4.62b) instead of fiber failure. High stress concentrations at the ends of matrix microcracks may cause debonding at the fiber–matrix interface and occasional fiber failure. Since the propagation of matrix microcracks is frequently interrupted by debonding, the fatigue failure in this region is progressive and, depending on the stress level, may span over 10^6 cycles.

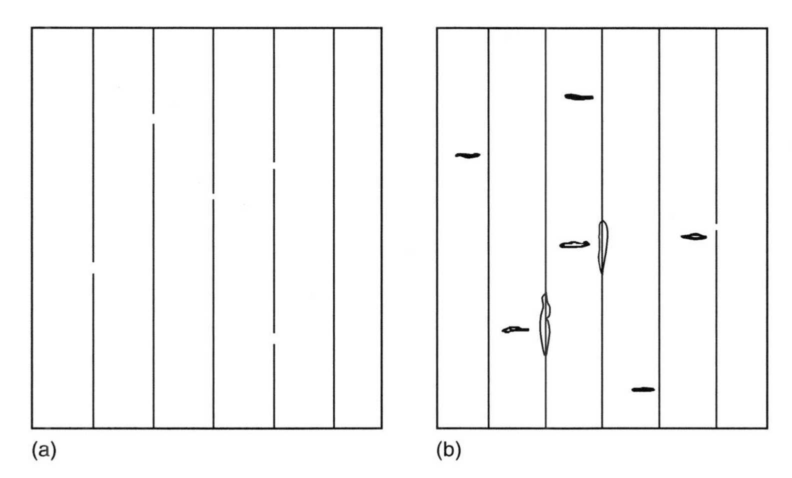

FIGURE 4.62 Damage development during tension–tension fatigue cycling of a 0° laminate. (a) Fiber breakage at high stress levels and (b) Matrix microcracks followed by debonding at low stress levels.

An important factor in determining the fatigue failure mechanism and the nature of the fatigue life diagram in 0° laminates is the fiber stiffness [56], which also controls the composite stiffness. For 0° composites, the composite fracture strain ε_{cu} in the longitudinal direction is equal to the fiber fracture strain ε_{fu}. Their scatter bands are also similar. Now, consider the tensile stress–strain diagrams (Figure 4.63) of a high-modulus fiber composite and a low-modulus fiber composite. For high-modulus fibers, such as GY-70 fibers, ε_{cu} is less than

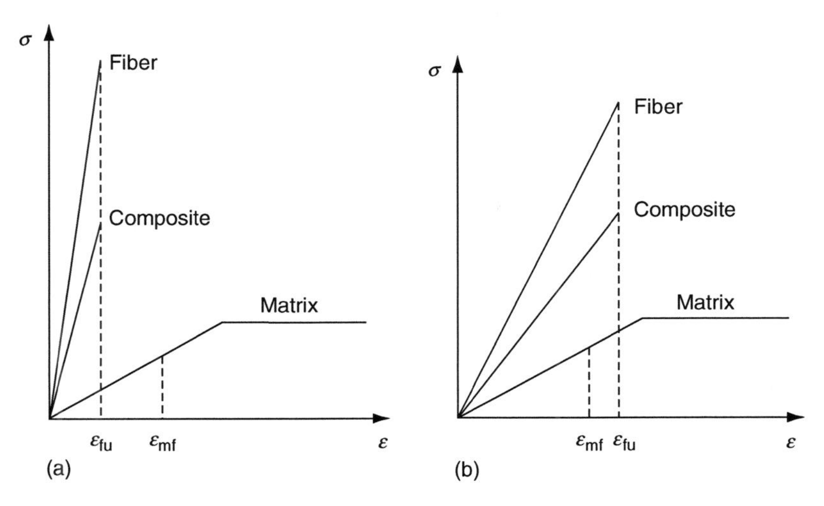

FIGURE 4.63 Schematic longitudinal tensile stress–strain diagrams for (a) high-modulus and (b) low-modulus 0° fiber-reinforced composite laminates. Note that ε_{mf} is the fatigue strain limit of the matrix.

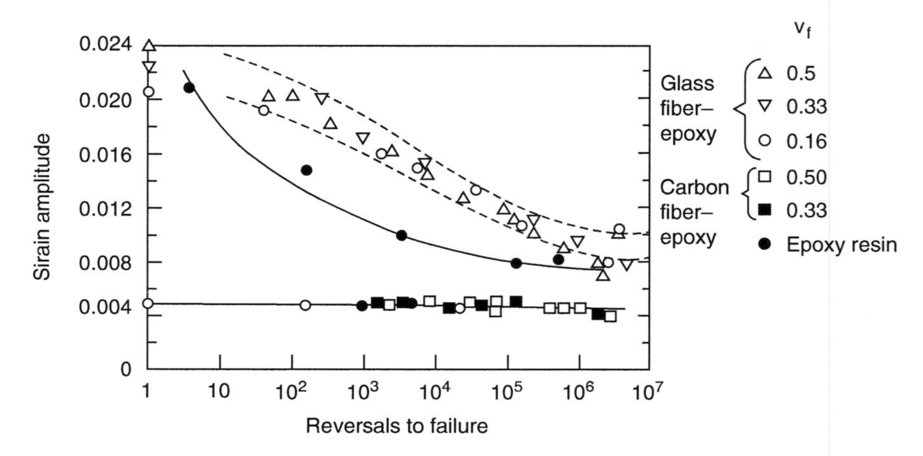

FIGURE 4.64 Cyclic strain amplitude vs. reversals-to-failure. (After Dharan, C.K.H., *J. Mater. Sci.*, 10, 1665, 1975.)

the fatigue strain limit of the matrix ε_{mf}. In this case, catastrophic fatigue failure, initiated by fiber breakage, is expected if the maximum fatigue strain due to applied load, ε_{max}, falls within the fiber fracture scatter band. The fatigue life diagram for such composites is nearly horizontal (as seen for the carbon fiber–epoxy composite in Figure 4.64) and the fatigue strength values are restricted within the fiber strength scatter band, as seen in Figure 4.36. No fatigue failure is expected below this scatter band.

For low-modulus fibers, such as T-300 carbon or E-glass, ε_{mf} falls below the lower bound of ε_{cu}. Thus, if ε_{max} is such that the matrix is strained above ε_{mf}, fatigue failure will be initiated by matrix microcracking and will continue in a progressive manner. The fatigue life diagram in this region will show a sloping band. At very low strain levels, where ε_{max} is less than ε_{mf}, there will be no fatigue failure. On the other hand, if ε_{max} exceeds the lower bound of ε_{cu} there will still be a catastrophic failure dominated by fiber breakage. The entire fatigue strain-life diagram for such composites shows three distinct regions, as seen for the glass fiber–epoxy composite in Figure 4.64:

1. Region I (high strain levels): catastrophic fatigue failure, low cycles to failure, nearly horizontal
2. Region II (intermediate strain levels): progressive fatigue failure, intermediate to high cycles to failure, steep slope
3. Region III (low strain levels): infinite life, horizontal

4.2.4.2 Cross-Ply and Other Multidirectional Continuous Fiber Laminates

Fatigue failure in cross-ply $[0/90]_S$ laminates begins with the formation of transverse microcracks at the fiber–matrix interface in the 90° layers (Figure 4.65).

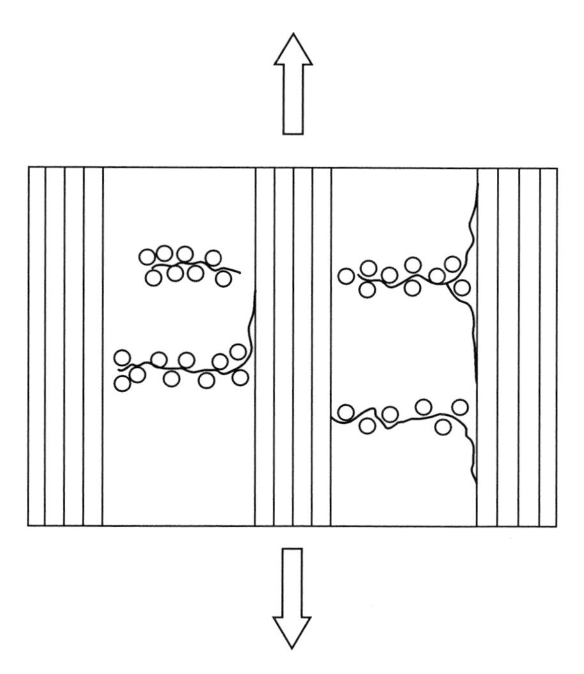

FIGURE 4.65 Damage development during tension–tension fatigue tests of a $[0/90]_S$ laminate.

As the cycling continues, these microcracks propagate across the 90° layers until they reach the adjacent 0° layers. Some of the microcracks are then deflected parallel to the 0° layers, causing delaminations between the 0° and the 90° layers. Depending on the stress level, a number of these transverse microcracks may appear at random locations in the first cycle; however as noted by Broutman and Sahu [57], the transverse crack density (number of microcracks per unit area) becomes nearly constant in a few cycles after their first appearance. It has been found by Agarwal and Dally [28] that delaminations do not propagate for nearly 95% of the fatigue life at a given stress level. It is only during the last 5% of the fatigue life that delaminations propagate rapidly across the 0°/90° interfaces before fiber failure in the 0° layers.

The sequence of damage development events in other multidirectional laminates containing off-axis fibers is similar to that found in cross-ply laminates. Reifsnider et al. [58] have divided this sequence into three regions (Figure 4.66). Region I usually involves matrix microcracking through the thickness of off-axis or 90° layers. These microcracks are parallel to fiber direction in these layers and develop quite early in the life cycle, but they quickly stabilize to a nearly uniform pattern with fixed spacing. The crack pattern developed in region I is called the characteristic damage state (CDS) and is similar to that

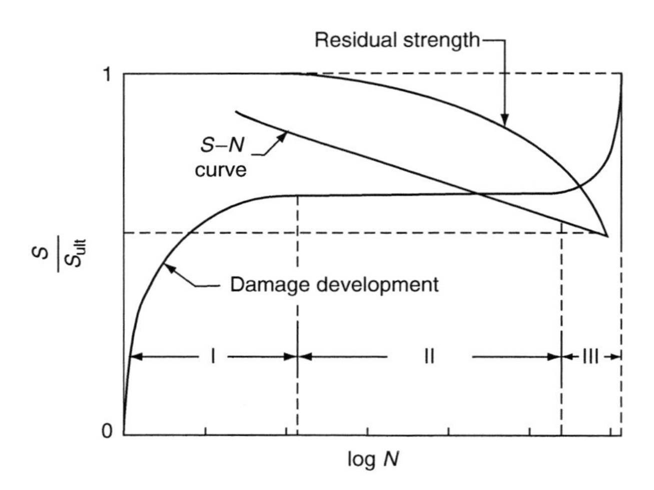

FIGURE 4.66 Three stages of fatigue damage development in multidirectional laminates. (After Reifsnider, K., Schultz, K., and Duke, J.C., *Long-Term Behavior of Composites, ASTM STP*, 813, 136, 1983.)

observed in quasi-static loadings. The CDS is a laminate property in the sense that it depends on the properties of individual layers, their thicknesses, and the stacking sequence. The CDS is found to be independent of load history, environment, residual stresses, or stresses due to moisture absorption.

Region II involves coupling and growth of matrix microcracks that ultimately lead to debonding at fiber–matrix interfaces and delaminations along layer interfaces. Both occur owing to high normal and shear stresses created at the tips of matrix microcracks. Edge delamination may also occur in some laminates (e.g., in $[0/\pm45/90]_S$ laminates) because of high interlaminar stresses between various layers. As a result of delamination, local stresses in the 0° layers increase, since the delaminated off-axis plies cease to share the 0° load. Additional stresses in 0° layers, in turn, cause fiber failure and accelerate the fatigue failure process.

The principal failure mechanism in region III is the fiber fracture in 0° layers followed by debonding (longitudinal splitting) at fiber–matrix interfaces in these layers. These fiber fractures usually develop in local areas adjacent to the matrix microcracks in off-axis plies. It should be noted that fiber fracture occurs in both regions II and III; however, the rate of fiber fracture is much higher in region III, which leads quickly to laminate failure.

4.2.4.3 SMC-R Laminates

The inhomogeneous fiber distribution and random fiber orientation in SMC-R laminates give rise to a multitude of microscopic cracking modes, such as

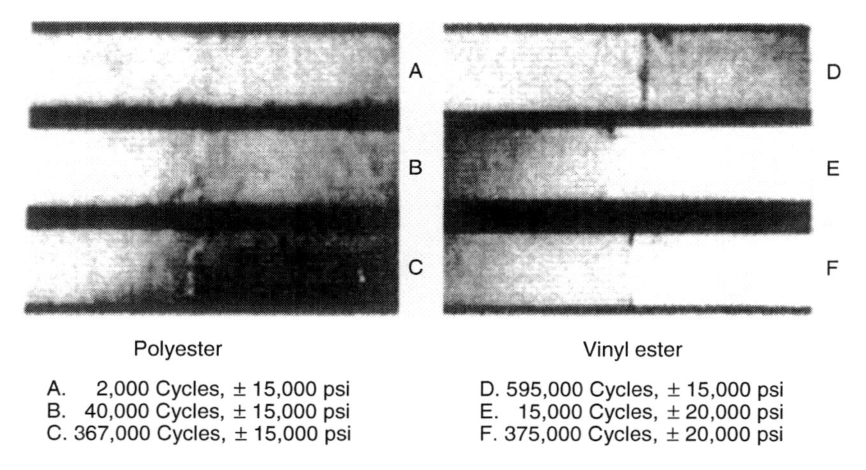

FIGURE 4.67 Damage development during flexural fatigue testing of SMC-R65 laminates containing polyester and vinyl ester matrix. Dark lines at the center of each specimen indicate the formation of microcracks nearly normal to the loading direction. Note that more microcracks were formed in polyester laminates than in vinyl ester laminates.

matrix cracking, fiber–matrix interfacial debonding, and fiber failure [59]. In matrix-rich areas containing sparsely dispersed fibers, matrix cracks are formed normal to the loading direction. Matrix cracks also develop in fiber-rich areas; however, these cracks are shorter in length owing to close interfiber spacing. Furthermore, if the fibers are at an angle with the loading direction, fiber–matrix debonding is observed. Fiber fracture is rarely observed in SMC-R laminates.

Crack density as well as average crack length in SMC-R laminates depends strongly on the stress level. At high fatigue stress levels (>60% of the ultimate tensile strength), the crack density is high but the average crack length is small. At lower stress levels, cracks have a lower density and longer lengths. Another important parameter in controlling crack density and length is the resin type [32]. This is demonstrated in Figure 4.67.

4.2.5 Fatigue Damage and Its Consequences

Unlike metals, fiber-reinforced composites seldom fail along a self-similar and well-defined crack path. Instead, fatigue damage in fiber-reinforced composites occurs at multiple locations in the form of fiber breakage, delamination, debonding, and matrix cracking or yielding. Depending on the stress level, fiber length, fiber orientation, constituent properties, and lamination configuration, some of these failure modes may appear either individually or in combination quite early in the life cycle of a composite. However, they

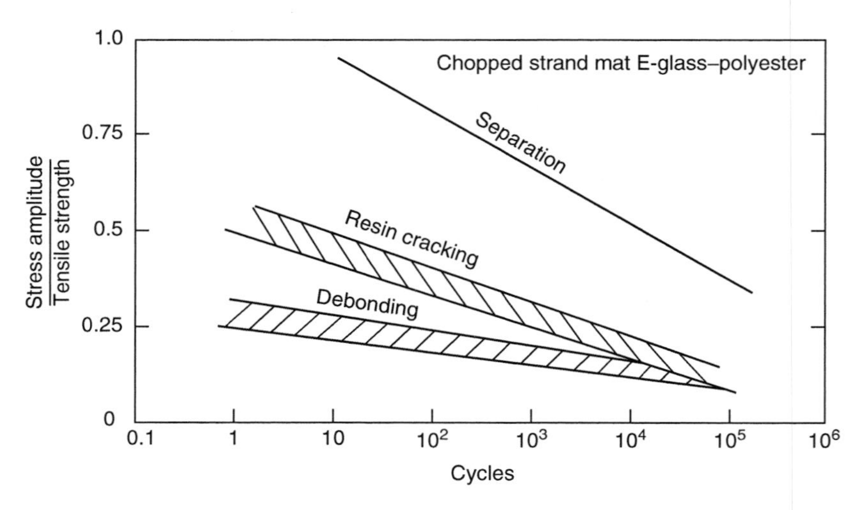

FIGURE 4.68 Various stages of damage growth during tension–tension fatigue cycling of E-glass mat-reinforced polyester laminates. (After Smith, T.R. and Owen, M.J., *Mod. Plast.*, 46, 128, 1969.)

may not immediately precipitate a failure in the material. The damage grows in size or intensity in a progressive manner until the final rupture takes place. Figure 4.68 shows the *S–N* curve of a randomly oriented chopped E-glass fiber strand mat–polyester composite in which the onset of debonding and matrix cracking was observed two to three decades earlier than its final rupture [60].

As the fatigue damage accumulates with cycling, the dynamic modulus of the material is continuously decreased. This cyclic "softening" phenomenon causes an increase in strain level if a stress-controlled test is used and a decrease in stress level if a strain-controlled test is used. Because of less damage development in a continuously reducing stress field, a strain-controlled test is expected to produce a higher fatigue life than a stress-controlled test.

Many investigators have used the reduction in dynamic modulus as a method of monitoring the fatigue damage development in a composite. Since dynamic modulus can be measured frequently during fatigue cycling without discontinuing the test or affecting the material, it is a potential nondestructive test parameter and can be used to provide adequate warning before a structure becomes totally ineffective [61]. Figure 4.69 shows the dynamic modulus loss of $[\pm 45]_S$ boron fiber–epoxy laminates in strain-controlled tension–tension fatigue tests. Reduction in the dynamic modulus of an SMC-R in deflection-controlled flexural fatigue tests is shown in Figure 4.70. These examples demonstrate that the reduction in dynamic modulus depends on the stress or strain level, resin type, and lamination configuration. Other factors that may influence the dynamic modulus reduction are fiber type, fiber orientation angle, temperature,

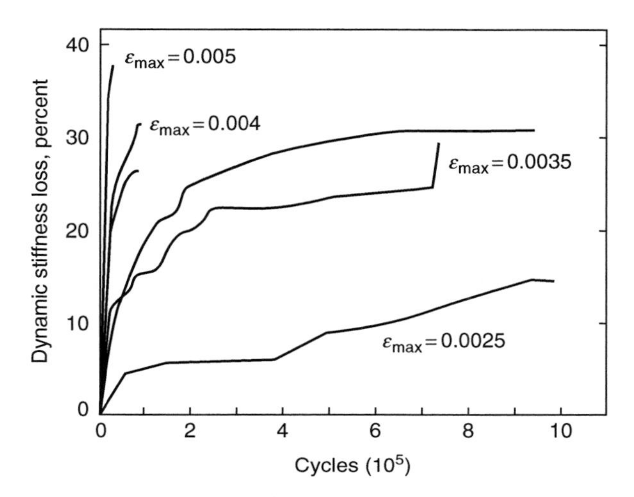

FIGURE 4.69 Reduction in dynamic modulus during strain-controlled fatigue testing of $[\pm 45]_S$ boron–epoxy laminates. (After O'Brien, T.K. and Reifsnider, K.L., *J. Compos. Mater.*, 15, 55, 1981.)

humidity, frequency, and test control mode. The significance of the fiber orientation angle is observed in unidirectional laminates in which the dynamic modulus reduction for $\theta = 0°$ orientation is considerably smaller than that for $\theta = 90°$ orientation.

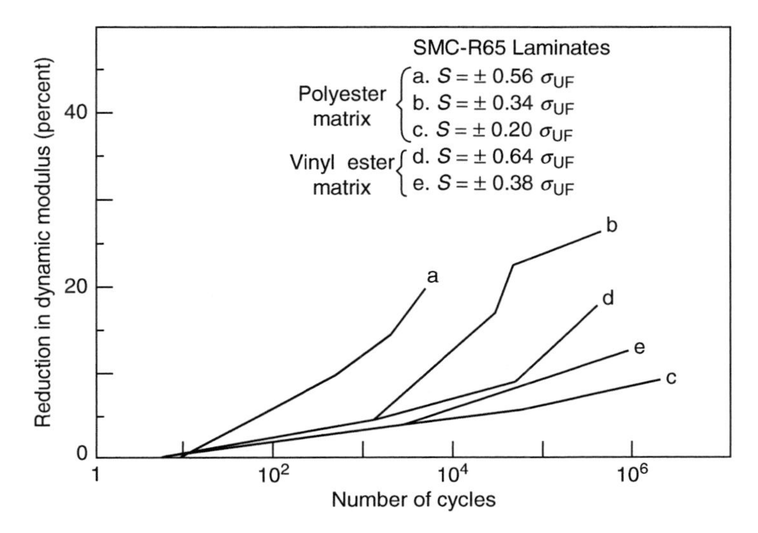

FIGURE 4.70 Reduction in dynamic modulus during flexural fatigue cycling of SMC-R65 laminates. (After Mallick, P.K., *Polym. Compos.*, 2, 18, 1981.)

Several phenomenological models have been proposed to describe the modulus reduction due to fatigue loading. One of them is due to Whitworth [62], who suggested the following equation for calculating the residual modulus E_N after N fatigue cycles in a stress-controlled fatigue test with a maximum stress level S:

$$E_N^a = E_0^a - H(E_0 - E_{\text{ff}})^a \left(\frac{N}{N_{\text{f}}}\right), \tag{4.27}$$

where

E_0 = initial modulus
E_{ff} = modulus at the time of fatigue failure
a, H = damage parameters determined by fitting Equation 4.27 to experimental data (Figure 4.71)
N_{f} = number of cycles to failure when the maximum stress level is $S(N_{\text{f}} > N)$

If the fatigue test is conducted between fixed maximum and minimum stress levels, then the strain level increases with fatigue cycling due to increasing damage accumulation in the material. Assuming that the stress–strain response of the material is linear and the fatigue failure occurs when the maximum strain level reaches the static ultimate strain,

$$E_{\text{ff}} = E_0 \frac{S}{\sigma_{\text{U}}} = E_0 \bar{S}, \tag{4.28}$$

where

σ_{U} = ultimate static strength
$\bar{S} = \frac{S}{\sigma_{\text{U}}}$

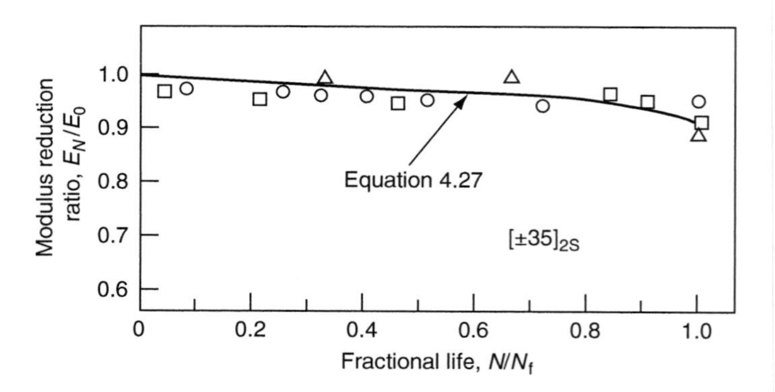

FIGURE 4.71 Dynamic modulus reduction ratio as a function of fractional life for $[\pm 35]_{2S}$ carbon fiber–epoxy laminates (after Whitworth, H.W., *ASME J. Eng. Mater. Technol.*, 112, 358, 1990.)

Thus, Equation 4.27 can be rewritten as

$$E_N^a = E_0^a - HE_0^a(1 - \bar{S})^a \left(\frac{N}{N_f}\right). \tag{4.29}$$

Whitworth defined the fractional damage D_i in a composite laminate after N_i cycles of fatigue loading with a maximum stress level S_i as

$$D_i = \frac{E_0^a - E_{N_i}^a}{E_0^a - E_{ffi}^a}. \tag{4.30}$$

Combining Equations 4.28 through 4.30, the fractional damage can be written as

$$D_i = r_i \frac{N_i}{N_{fi}}, \tag{4.31}$$

where

$$r_i = \frac{H(1 - \bar{S}_i)^a}{1 - \bar{S}_i^a}.$$

In a variable amplitude stress loading, the total damage can be expressed as

$$D = \sum_{i=1}^{m} \frac{r_i}{r_m} \frac{N_i}{N_{fi}}, \tag{4.32}$$

where

$\frac{r_i}{r_m} = \frac{(1 - \bar{S}_i)(1 - \bar{S}_m^a)}{(1 - \bar{S}_i^a)(1 - \bar{S}_m)}$

N_i = number of cycles endured at a maximum stress level S_i

N_{fi} = fatigue life at S_i

m = number of the stress level

In a variable amplitude stress loading, failure occurs when the sum of the fractional damages equals 1, that is, $D = 1$. In addition, note that for $a = 1$,

Equation 4.32 transforms into the linear damage rule (Miner's rule), which is frequently used for describing damage in metals:

$$\text{Miner's rule: } D = \sum_{i=1}^{m} \frac{N_i}{N_{fi}}. \tag{4.33}$$

EXAMPLE 4.2

A quasi-isotropic T-300 carbon fiber–epoxy laminate is subjected to 100,000 cycles at 382 MPa after which the stress level is increased to 437 MPa. Estimate the number of cycles the laminate will survive at the second stress level. Median fatigue lives at 382 and 437 MPa are 252,852 and 18,922 cycles, respectively. From the modulus vs. fatigue cycle data, the constant a and H in Equation 4.27 have been determined as 1.47 and 1.66, respectively (see Ref. [62]). The ultimate static strength of this material is 545 MPa.

Solution

Using Equation 4.32, we can write, for failure to occur at the end of second stress cycling,

$$D = 1 = \frac{r_1}{r_2}\frac{N_1}{N_{f1}} + \frac{r_2}{r_2}\frac{N_2}{N_{f2}} \text{ which gives,}$$

$$N_2 = N_{f2}\left(1 - \frac{r_1}{r_2}\frac{N_1}{N_{f1}}\right).$$

In this example,

$$\begin{aligned}
\sigma_U &= 545 \text{ MPa} \\
S_1 &= 382 \text{ MPa} \\
N_{f1} &= 252{,}852 \text{ cycles} \\
N_1 &= 100{,}000 \text{ cycles} \\
S_2 &= 437 \text{ MPa} \\
N_{f2} &= 18{,}922 \text{ cycles}
\end{aligned}$$

Therefore,

$$r_1 = (1.66)\frac{\left(1 - \frac{382}{545}\right)^{1.47}}{1 - \left(\frac{382}{545}\right)^{1.47}} = 0.6917.$$

Similarly, $r_2 = 0.5551$

Thus,

$$N_2 = (18{,}922)\left[1 - \left(\frac{0.6917}{0.5551}\right)\left(\frac{100{,}000}{252{,}852}\right)\right] \cong 9{,}600 \text{ cycles.}$$

This compares favorably with the experimental median value of 10,800 cycles [62].

4.2.6 Postfatigue Residual Strength

The postfatigue performance of a fiber-reinforced composite is studied by measuring its static modulus and strength after cycling it for various fractions of its total life. Both static modulus and strength are reduced with increasing number of cycles. Broutman and Sahu [57] reported that much of the static strength of a $[0/90]_S$ E-glass fiber–epoxy composite is reduced rapidly in the first 25% of its fatigue life, which is then followed by a much slower rate of strength reduction until the final rupture occurs (Figure 4.72). Tanimoto and Amijima [43] also observed similar behavior for a woven E-glass cloth–polyester laminate; however, in this case the reduction in strength takes place only after a slight increase at <2% of the total life.

An initial increase in static strength was also observed by Reifsnider et al. [52] for a $[0/\pm 45/0]_S$ boron fiber–epoxy laminate containing a center hole. This unique postfatigue behavior of a composite material was explained by means of a wear-in and wear-out mechanism in damage development. The wear-in process takes place in the early stages of fatigue cycling. During this process,

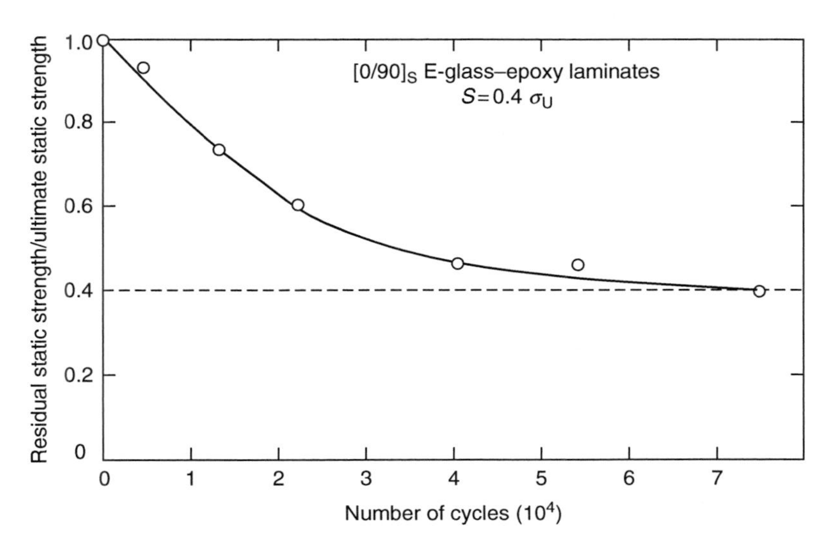

FIGURE 4.72 Reduction in residual static strengths of $[0/90]_S$ laminates after fatigue testing to various cycles. (After Broutman, L.J. and Sahu, S., Progressive damage of a glass reinforced plastic during fatigue, *Proceedings 20th Annual Technical Conference*, Society of the Plastics Industry, 1969.)

the damage developed locally around the center hole reduces the stress concentration in the vicinity of the hole, thus resulting in increased strength. This beneficial stage of fatigue cycling is followed by the wear-out process, which comprises large-scale and widespread damage development leading to strength reduction. Thus, the residual strength of a composite after a period of fatigue cycling is modeled as

$$\sigma_{\text{residual}} = \sigma_{\text{U}} + \sigma_{\text{wear-in}} - \sigma_{\text{wear-out}},$$

where

σ_{U} = ultimate static strength
$\sigma_{\text{wear-in}}$ = change in static strength due to wear-in
$\sigma_{\text{wear-out}}$ = change in static strength due to wear-out

The effect of wear-in is more pronounced at high fatigue load levels. Since fatigue life is longer at low load levels, there is a greater possibility of developing large-scale damage throughout the material. Thus, at low load levels, the effect of wear-out is more pronounced.

A number of phenomenological equations have been proposed to predict the residual static strength of a fatigued composite laminate [63,64]. The simplest among them is based on the assumption of linear strength reduction [63] and is given as

$$\sigma_{r_i} = \sigma_{U_{i-1}} - (\sigma_{U_0} - S_i)\frac{N_i}{N_{\text{f}i}}, \tag{4.34}$$

where

σ_{r_i} = residual strength after N_i cycles at the ith stress level S_i
$N_{\text{f}i}$ = fatigue life at S_i
σ_{U_0} = ultimate static strength of the original laminate
$\sigma_{U_{i-1}}$ = ultimate static strength before being cycled at S_i

In Equation 4.34, the ratio $N_i/N_{\text{f}i}$ represents the fractional fatigue life spent at S_i.

Assuming a nonlinear strength degradation model, Yang [64] proposed the following equation for the residual strength for a composite laminate after N fatigue cycles:

$$\sigma_{\text{res}}^c = \sigma_{\text{U}}^c - \sigma_0^c K S^d N, \tag{4.35}$$

where

c = damage development parameter
σ_0 = location parameter in a two-parameter Weibull function for the static strength distribution of the laminate
S = stress range in the fatigue test = $\sigma_{\max} - \sigma_{\min}$
K, d = parameters used to describe the S–N diagram as $KS^d N_{\text{f}} = 1$

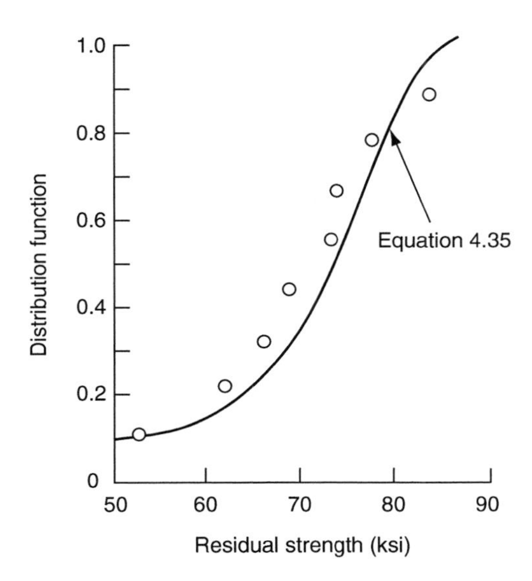

FIGURE 4.73 Residual strength distribution in postfatigue tension tests. (After Yang, J.N., *J. Compos. Mater.*, 12, 19, 1978.)

Procedures for determining c, d, and K are given in Ref. [64]. It is worth noting that, just as the static strengths, the residual strengths at the end of a prescribed number of cycles also follow the Weibull distribution (Figure 4.73).

4.3 IMPACT PROPERTIES

The impact properties of a material represent its capacity to absorb and dissipate energies under impact or shock loading. In practice, the impact condition may range from the accidental dropping of hand tools to high-speed collisions, and the response of a structure may range from localized damage to total disintegration. If a material is strain rate sensitive, its static mechanical properties cannot be used in designing against impact failure. Furthermore, the fracture modes in impact conditions can be quite different from those observed in static tests.

A variety of standard impact test methods are available for metals (ASTM E23) and unreinforced polymers (ASTM D256). Some of these tests have also been adopted for fiber-reinforced composite materials. However, as in the case of metals and unreinforced polymers, the impact tests do not yield basic material property data that can be used for design purposes. They are useful in comparing the failure modes and energy absorption capabilities of two different materials under identical impact conditions. They can also be used in the areas of quality control and materials development.

4.3.1 Charpy, Izod, and Drop-Weight Impact Test

Charpy and Izod impact tests are performed on commercially available machines in which a pendulum hammer is released from a standard height to contact a beam specimen (either notched or unnotched) with a specified kinetic energy. A horizontal simply supported beam specimen is used in the Charpy test (Figure 4.74a), whereas a vertical cantilever beam specimen is used in the Izod test (Figure 4.74b). The energy absorbed in breaking the specimen, usually indicated by the position of a pointer on a calibrated dial attached to the testing machine, is equal to the difference between the energy of the pendulum hammer at the instant of impact and the energy remaining in the pendulum hammer after breaking the specimen.

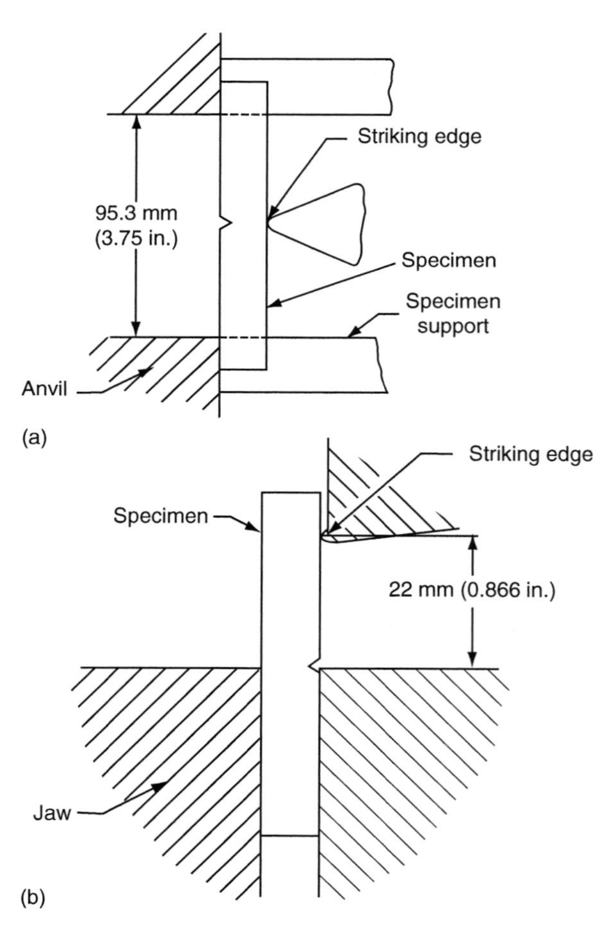

FIGURE 4.74 Schematic arrangements for (a) Charpy and (b) Izod impact tests.

TABLE 4.10
Standard V-Notched Charpy and Izod Impact Energies of Various Materials

Material	Impact Energy, kJ/m^2 (ft lb/in.^{2})	
	Charpy	Izod
S-glass–epoxy, 0°, v_f = 55%	734 (348)	—
Boron–epoxy, 0°, v_f = 55%	109–190 (51.5–90)	—
Kevlar 49–epoxy, 0°, v_f = 60%	317 (150)	158 (75)
AS carbon–epoxy, 0°, v_f = 60%	101 (48)	33 (15.5)
HMS carbon–epoxy, 0°, v_f = 60%	23 (11)	7.5 (3.6)
T-300 carbon–epoxy, 0°, v_f = 60%	132 (62.6)	67.3 (31.9)
4340 Steel (R_c = 43–46)	214 (102)	—
6061-T6 aluminum alloy	153 (72.5)	—
7075-T6 aluminum alloy	67 (31.7)	—

Table 4.10 compares the longitudinal Charpy and Izod impact energies of a number of unidirectional 0° laminates and conventional metals. In general, carbon and boron fiber-reinforced epoxies have lower impact energies than many metals. However, the impact energies of glass and Kevlar 49 fiber-reinforced epoxies are equal to or better than those of steel and aluminum alloys. Another point to note in this table is that the Izod impact energies are lower than the Charpy impact energies.

The drop-weight impact test uses the free fall of a known weight to supply the energy to break a beam or a plate specimen (Figure 4.75). The specimen can be either simply supported or fixed. The kinetic energy of the falling weight is adjusted by varying its drop height. The impact load on the specimen is measured by instrumenting either the striking head or the specimen supports. Energy absorbed by the specimen is calculated as

$$U_t = \frac{W}{2g}\left(u_1^2 - u_2^2\right), \tag{4.36}$$

where

W = weight of the striking head
u_1 = velocity of the striking head just before impact $(=\sqrt{2gH})$
u_2 = measured velocity of the striking head just after impact
g = acceleration due to gravity
H = drop height

A comparison of drop-weight impact energies of carbon, Kevlar 49, and E-glass fiber-reinforced epoxy laminates is given in Table 4.11.

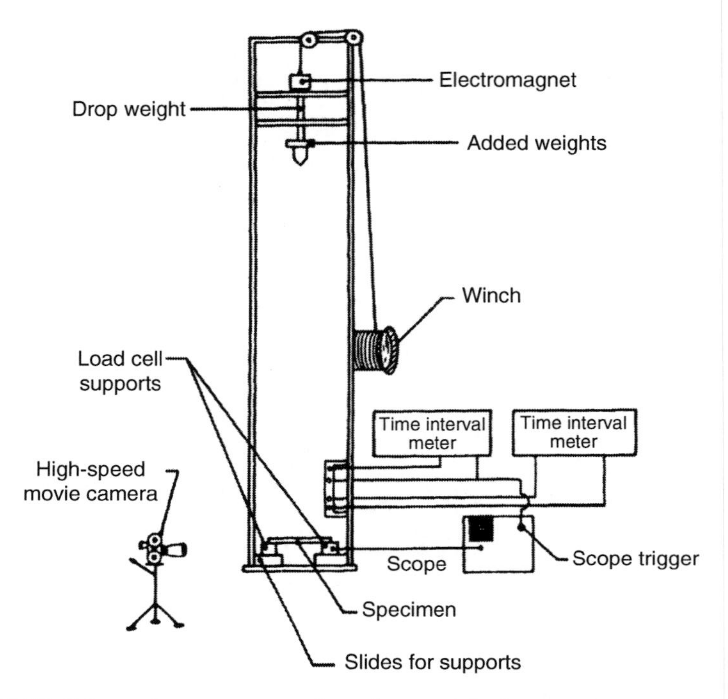

FIGURE 4.75 Schematic arrangement for a drop-weight impact test.

TABLE 4.11
Drop-Weight Impact Force and Energy Values[a]

Material	Force per Unit Thickness, kN/mm (lb/in.)		Energy per Unit Thickness, J/mm (ft lb/in.)	
AS-4 carbon–epoxy				
10-ply cross-ply	1.07	(6,110)	0.155	(2.92)
10-ply fabric	1.21	(6,910)	0.209	(3.94)
Kevlar 49–epoxy				
10-ply cross-ply	1.16	(6,630)	0.284	(5.36)
10-ply fabric	0.91	(5,200)	0.233	(4.39)
E-glass–epoxy				
6-ply cross-ply	2.83	(16,170)	0.739	(13.95)
10-ply cross-ply	2.90	(16,570)	0.789	(14.89)
10-ply fabric	0.99	(5,660)	0.206	(3.89)

Source: Adapted from Winkel, J.D. and Adams, D.F., *Composites*, 16, 268, 1985.

[a] Simply supported 127 mm^2 plate specimens were used in these experiments.

The impact energy measured in all these tests depends on the ratio of beam length to effective depth. Below a critical value of this ratio, there is a considerable increase in impact energy caused by extensive delamination [65]. The effect of notch geometry has relatively little influence on the impact energy because delamination at the notch root at low stresses tends to reduce its severity.

4.3.2 Fracture Initiation and Propagation Energies

The impact energy measured in either of the impact tests does not indicate the fracture behavior of a material unless the relative values of fracture initiation and propagation energies are known. Thus, for example, a high-strength brittle material may have a high fracture initiation energy but a low fracture propagation energy, and the reverse may be true for a low-strength ductile material. Even though the sum of these two energies may be the same, their fracture behavior is completely different. Unless the broken specimens are available for fracture mode inspection, the toughness of a material cannot be judged by the total impact energy alone.

Fracture initiation and propagation energies are determined from the measurements of the dynamic load and striking head velocity during the time of contact. Through proper instrumentation, the load and velocity signals are integrated to produce the variation of cumulative energy as a function of time. Both load–time and energy–time responses are recorded and are then used for energy absorption analysis.

The load–time response during the impact test of a unidirectional composite (Figure 4.76) can be conveniently divided into three regions [66]:

Preinitial fracture region: The preinitial fracture behavior represents the strain energy in the beam specimen before the initial fracture occurs. In unidirectional 0° specimens, strain energy is stored principally by the fibers. The contribution from the matrix is negligible. The fiber strain energy U_f is estimated as

$$U_f = \frac{\sigma_f^2}{6E_f} v_f, \tag{4.37}$$

where

σ_f = longitudinal stress at the outermost fibers in the beam specimen
E_f = fiber modulus
v_f = fiber volume fraction

This equation indicates that the energy absorption in this region can be increased by using a low-modulus fiber and a high fiber volume fraction.

Initial fracture region: Fracture initiation at or near the peak load occurs either by the tensile failure of the outermost fibers or by interlaminar shear

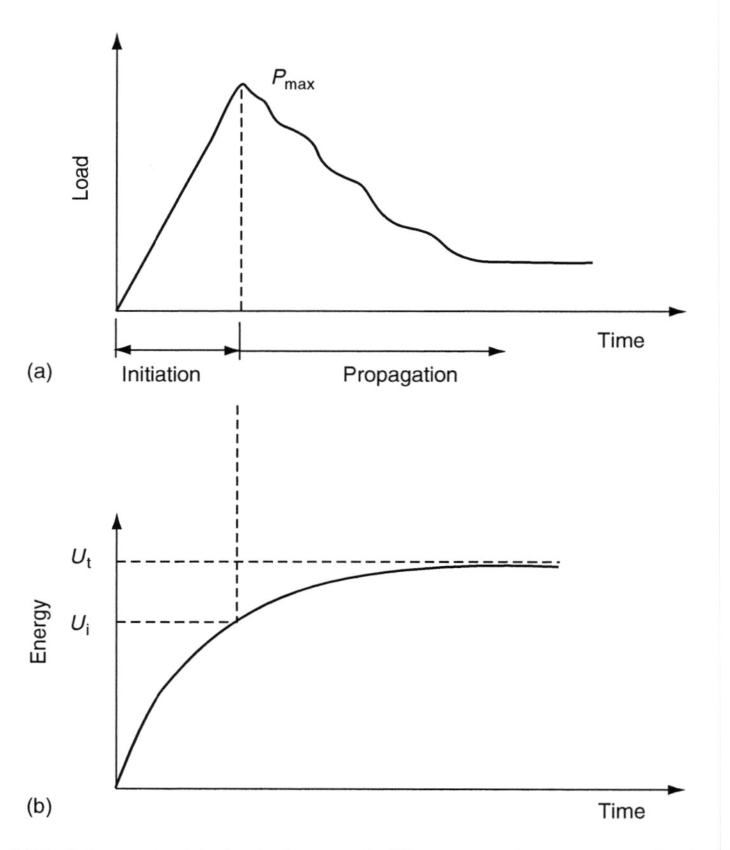

FIGURE 4.76 Schematic (a) load–time and (b) energy–time curves obtained in an instrumented impact test.

failure. In many cases, fiber microbuckling is observed at the location of impact (i.e., on the compression side of the specimen) before reaching the peak load. Compressive yielding is observed in Kevlar 49 composites.

Interlaminar shear failure precedes fiber tensile failure if either the specimen length-to-depth ratio is low or the ILSS is lower than the tensile strength of the material. If the ILSS is high and fibers have either a low tensile strength or a low tensile strain-to-failure, shear failure would not be the first event in the fracture process. Instead, fiber tensile failure would occur on the nonimpacting side as the peak load is reached.

Postinitial fracture region: The postinitial fracture region represents the fracture propagation stage. In unidirectional composites containing low strain-to-failure fibers (e.g., GY-70 carbon fiber), a brittle failure mode is observed. On the other hand, many other fiber-reinforced composites, including unidirectional E-glass, S-glass, T-300 or AS carbon, and Kevlar 49, fail in a sequential manner starting with fiber failure, which is followed by debonding and fiber pullout within each layer and delamination between various layers. Additionally, Kevlar 49 composites exhibit considerable yielding and may not even fracture in an impact test.

Progressive delamination is the most desirable fracture mode in high-energy impact situations. High shear stress ahead of the crack tip causes delamination between adjacent layers, which in turn arrests the advancing crack and reduces its severity as it reaches the delaminated interface. Thus, the specimen continues to carry the load until the fibers in the next layer fail in tension. Depending on the material and lamination configuration, this process is repeated several times until the crack runs through the entire thickness. Energy absorbed in delamination depends on the interlaminar shear fracture energy and the length of delamination, as well as the number of delaminations. Owing to progressive delamination, the material exhibits a "ductile" behavior and absorbs a significant amount of impact energy.

Referring to Figure 4.76b, the energy corresponding to the peak load is called the fracture initiation energy U_i. The remaining energy is called the fracture propagation energy U_p, where

$$U_p = U_t - U_i. \tag{4.38}$$

These energy values are often normalized by dividing them either by the specimen width or by specimen cross-sectional area (effective cross-sectional area in notched specimens).

The fracture initiation and propagation energies of a number of unidirectional fiber-reinforced epoxies are compared in Table 4.12. With the exception of GY-70, other composites in this table fail in a progressive manner. An E-glass–epoxy composite has a much higher fracture initiation energy than other composites owing to higher strain energy. The fracture propagation energy for E-glass and Kevlar 49 fiber composites are higher than that for a T-300 carbon fiber composite. Thus, both E-glass and Kevlar 49 fiber composites have higher impact toughness than carbon fiber composites.

4.3.3 Material Parameters

The primary factor influencing the impact energy of a unidirectional 0° composite is the fiber type. E-glass fiber composites have higher impact energy due to the relatively high strain-to-failure of E-glass fibers. Carbon and boron fiber composites have low strain-to-failure that leads to low impact energies for these composites. Increasing the fiber volume fraction also leads to higher impact energy, as illustrated in Figure 4.77.

TABLE 4.12
Static and Impact Properties of Unidirectional 0° Fiber-Reinforced Epoxy Composites

		Static Flexure Test		Unnotched Charpy Impact Test				
Fiber Type	**Fiber Strain Energy Index**	L/h	σ_{max}, MPa (ksi)	L/h	σ_{max}, MPa (ksi)	U_i, kJ/m^2 (ft lb/in.^{2})	U_p, kJ/m^2 (ft lb/in.^{2})	U_t, kJ/m^2 (ft lb/in.^{2})
E-Glass	82	15.8	1641 (238)	16.1	1938 (281)	466.2 (222)	155.4 (74)	621.6 (296)
Kevlar 49	29	11	703 (102)	10.5	676 (98)	76 (36.2)	162.5 (77.4)	238.5 (113.6)
T-300 Carbon	10.7	14.6	1572 (228)	14.6	1579 (229)	85.7 (40.8)	101.2 (48.2)	186.9 (89)
GY-70 Carbon	2.8	12.8	662 (96)	14.6	483 (70)	12.3 (5.85)	0	12.3 (5.85)

Source: Adapted from Mallick, P.K. and Broutman, L.J., *J. Test. Eval.*, 5, 190, 1977.

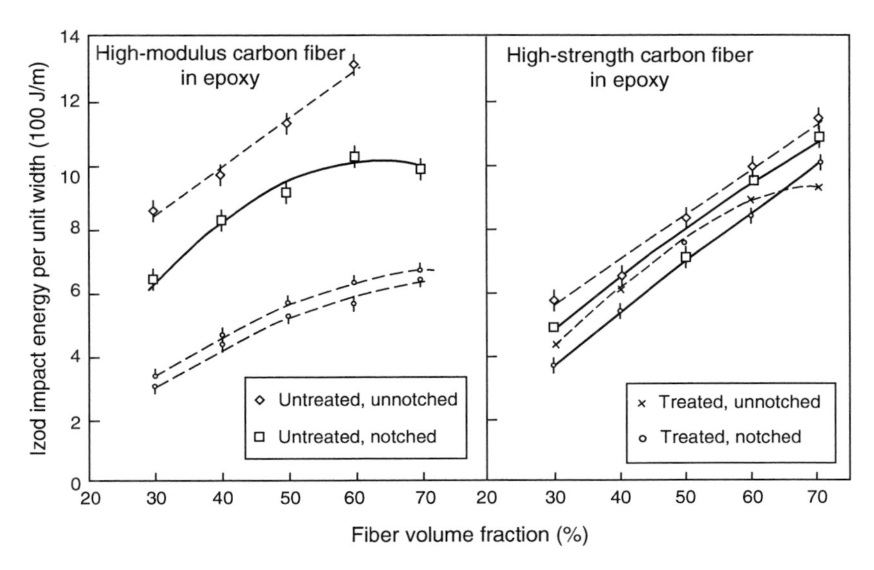

FIGURE 4.77 Variation of unnotched Izod impact energy with fiber volume fraction in 0° carbon–epoxy laminates. (After Hancox, N.L., *Composites*, 3, 41, 1971.)

The next important factor influencing the impact energy is the fiber–matrix interfacial shear strength. Several investigators [69–71] have reported that impact energy is reduced when fibers are surface-treated for improved adhesion with the matrix. At high levels of adhesion, the failure mode is brittle and relatively little energy is absorbed. At very low levels of adhesion, multiple delamination may occur without significant fiber failure. Although the energy absorption is high, failure may take place catastrophically. At intermediate levels of adhesion, progressive delamination occurs, which in turn produces a high impact energy absorption.

Yeung and Broutman [71] have shown that a correlation exists between the impact energy and ILSS of a composite laminate (Figure 4.78). Different coupling agents were used on E-glass woven fabrics to achieve various ILSSs in short-beam shear tests. It was observed that the fracture initiation energy increases modestly with increasing ILSS. However, the fracture propagation energy as well as the total impact energy decrease with increasing ILSS, exhibit a minimum, and appear to level off to intermediate values. The principal failure mode at low ILSSs was delamination. At very high ILSSs, fiber failure was predominant.

The strain energy contribution from the matrix in the development of impact energy is negligible. However, the matrix can influence the impact damage mechanism since delamination, debonding, and fiber pullout energies depend on the fiber–matrix interfacial shear strength. Since epoxies have better

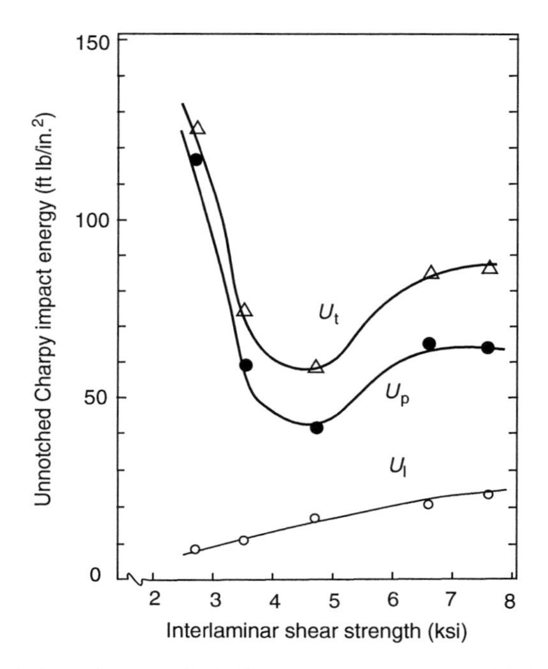

FIGURE 4.78 Variation of unnotched Charpy impact energy with interlaminar shear strength in E-glass fabric–polyester laminates. (After Yeung, P. and Broutman, L.J., *Polym. Eng. Sci.*, 18, 62, 1978.)

adhesion with E-glass fibers than polyesters, E-glass–epoxy composites exhibit higher impact energies than E-glass–polyester composites when the failure mode is a combination of fiber failure and delamination.

In unidirectional composites, the greatest impact energy is exhibited when the fibers are oriented in the direction of the maximum stress, that is, at 0° fiber orientation. Any variation from this orientation reduces the load-carrying capacity as well as the impact energy of the composite laminate. Figure 4.79 shows an example of the effect of fiber orientation on the drop-weight impact energy of $[0/90/0_4/0]_S$ and $[(0/90)_3/0]_S$ laminates [72]. In both cases, a minimum impact energy was observed at an intermediate angle between $\theta = 0°$ and 90°. Furthermore, fracture in off-axis specimens took place principally by interfiber cleavage parallel to the fiber direction in each layer.

The most efficient way of improving the impact energy of a low strain-to-failure fiber composite is to hybridize it with high strain-to-failure fiber laminas. For example, consider the GY-70 carbon fiber composite in Table 4.12 that exhibits a brittle failure mode and a low impact energy. Mallick and Broutman [67] have shown that a hybrid sandwich composite containing GY-70 fiber laminas in the outer skins and E-glass fiber laminas in the core has

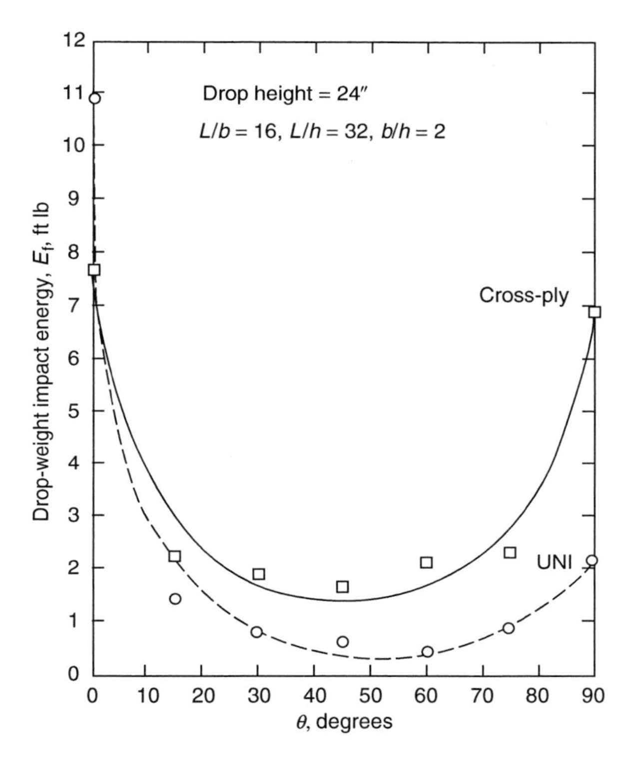

FIGURE 4.79 Variation of drop-weight impact energy with fiber orientation angle. (After Mallick, P.K. and Broutman, L.J., *Eng. Fracture Mech.*, 8, 631, 1976.)

a 35 times higher impact energy than the GY-70 fiber composite. This is achieved without much sacrifice in either the flexural strength or the flexural modulus. The improvement in impact energy is due to delamination at the GY-70/E-glass interface, which occurs after the GY-70 skin on the tension side has failed. Even after the GY-70 skin sheds owing to complete delamination, the E-glass laminas continue to withstand higher stresses, preventing brittle failure of the whole structure. By varying the lamination configuration as well as the fiber combinations, a variety of impact properties can be obtained (Table 4.13). Furthermore, the impact energy of a hybrid composite can be varied by controlling the ratio of various fiber volume fractions (Figure 4.80).

4.3.4 Low-Energy Impact Tests

Low-energy impact tests are performed to study localized damage without destroying the specimen. Two types of low-energy impact tests are performed, namely, the ballistic impact test and the low-velocity drop-weight impact test. In ballistic impact tests, the specimen surface is impinged with very low

TABLE 4.13
Unnotched Charpy Impact Energies of Various Interply Hybrid Laminates

	Impact Energy, kJ/m² (ft lb/in.²)		
Laminate	U_i	U_p	U_t
GY-70/E-glass: $(0_{5G}/0_{5E})_S$[a]	6.7 (3.2)	419.6 (199.8)	426.3 (203)
T-300/E-glass: $(0_{5T}/0_{5E})_S$[a]	60.3 (28.7)	374.4 (178.3)	434.7 (207)
GY-70/Kevlar 49: $(0_{5G}/0_{5K})_S$[a]	7.3 (3.5)	86.9 (41.4)	94.2 (44.9)
GY-70/E-glass: $[(0_G/0_E)_5/0_G]_S$[a]	13.9 (6.6)	204 (97.2)	217.9 (103.8)
T-300/E-glass: $[(0_T/0_E)_5/0_T]_S$[a]	59.4 (28.3)	80.2 (38.2)	139.6 (66.5)
Modmor II carbon/E-glass:[b] $[(0_E/(45/90/0/-45)_M)_7/0_E/(45/90)_M]_S$	29.4 (14)	264.8 (126)	294.2 (140)
For comparison: All Modmor II carbon:[b] $[(0/45/90/0/-45)_7/0/45/90]_S$	27.3 (13)	56.8 (27)	84.1 (40)

[a] From P.K. Mallick and L.J. Broutman, *J. Test. Eval.*, 5, 190, 1977.
[b] From J.L. Perry and D.F. Adams, *Composites*, 7, 166, 1975.

mass spherical balls at high speeds. In low-velocity drop-weight impact tests, a relatively heavier weight or ball is dropped from small heights onto the specimen surface. After the impact test, the specimen is visually as well as nondestructively inspected for internal and surface damages and then tested

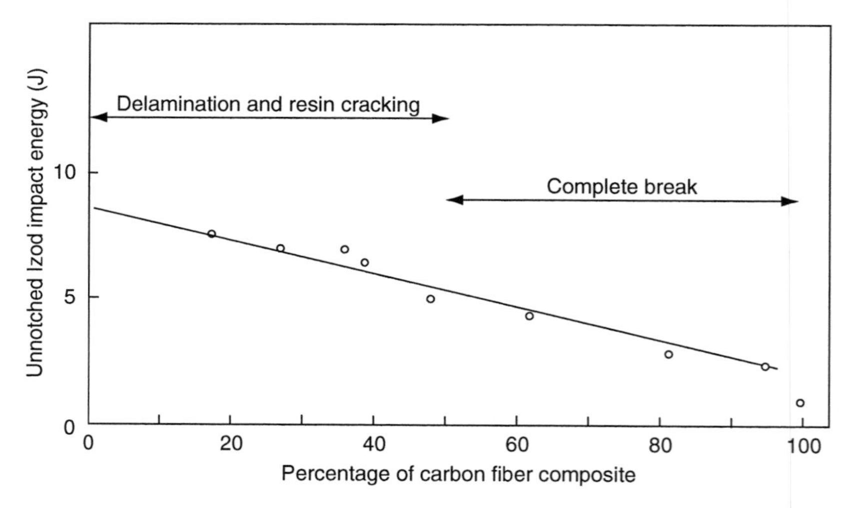

FIGURE 4.80 Effect of carbon fiber content on the unnotched Izod impact energy of 0° carbon/E-glass–epoxy interply hybrid laminates. (After Hancox, N.L. and Wells, H., The Izod impact properties of carbon fibre/glass fibre sandwich structures, U.K. Atomic Energy Research Establishment Report AERE-R7016, 1972.)

in static tension or compression modes to determine its postimpact residual properties.

Sidey and Bradshaw [73] performed ballistic impact experiments on both unidirectional 0° and $[(0/90)]_{2S}$ carbon fiber–epoxy composites. Steel balls, 3 mm in diameter, were impacted on 3 mm thick rectangular specimens. The impact velocity ranged from 70 to 300 m/s. Failure mode in unidirectional composites was longitudinal splitting (through-the-thickness cracks running parallel to the fibers) and subsurface delamination. In cross-ply laminates, the 90° layers prevented the longitudinal cracks from running through the thickness and restricted them to the surface layers only. Delamination was more pronounced with untreated fibers.

Rhodes et al. [74] performed similar ballistic impact experiments on carbon fiber–epoxy composites containing various arrangements of 0°, 90°, and ±45° laminas. Aluminum balls, 12.7 mm in diameter, were impacted on 5–8 mm thick rectangular specimens at impact velocities ranging from 35 to 125 m/s. Their experiments showed that, over a threshold velocity, appreciable internal damage appeared in the impacted area even though the surfaces remained undamaged. The principal internal damage was delamination, which was pronounced at interfaces between 0° and 90° or 0° and 45° laminas. The damaged specimens exhibited lower values of critical buckling loads and strains than the unimpacted specimens.

Ramkumar [75] studied the effects of low-velocity drop-weight impact tests on the static and fatigue strengths of two multidirectional AS carbon fiber–epoxy composites. His experiments indicate that impact-induced delaminations, with or without visible surface damages, can severely reduce the static compressive strengths. Static tensile strengths were affected only if delaminations were accompanied with surface cracks. Fatigue strengths at 10^6 cycles were reduced considerably more in compression–compression and tension–compression fatigue tests than in tension–tension fatigue tests. The growth of impact-induced delaminations toward the free edges was the predominant failure mechanism in these fatigue tests.

Morton and Godwin [76] compared the low-velocity impact damage in carbon fiber-reinforced $[0_2/\pm45]_{2S}$ and $[\pm45/0_3/\pm45/0]_S$ laminates containing either a toughened epoxy or PEEK as the matrix. They observed that the incident impact energy level to produce barely visible impact damage was approximately equal for both toughened epoxy and PEEK composites; however, energy to produce perforation was significantly higher in PEEK composites. Nondestructive inspection of impacted laminates showed that the PEEK laminates had less damage at or near perforation energy. Both epoxy and PEEK laminates showed matrix cracking and ply delamination, but the latter also exhibited local permanent deformation. Morton and Godwin [76] also observed that the stacking sequence with 45° fibers in the outside layers provided a higher residual strength after low-energy impact than that with 0° layers in the outside layers.

4.3.5 Residual Strength After Impact

If a composite laminate does not completely fail by impact loading, it may still be able to carry loads even though it has sustained internal as well as surface damages. The load-carrying capacity of an impact-damaged laminate is measured by testing it for residual strength in a uniaxial tension test.

The postimpact residual strength as well as the damage growth with increasing impact velocity is shown schematically in Figure 4.81. For small impact velocities, no strength degradation is observed (region I). As the damage appears, the residual tensile strength is reduced with increasing impact velocity (region II) until a minimum value is reached just before complete perforation (region III). Higher impact velocities produce complete perforation, and the hole diameter becomes practically independent of impact velocity (region IV). The residual strength in this region remains constant and is equal to the notched tensile strength of the laminate containing a hole of the same diameter as the impacting ball. Husman et al. [77] proposed the following relationship between the residual tensile strength in region II and the input kinetic energy:

$$\sigma_R = \sigma_U \sqrt{\frac{U_s - kU_{KE}}{U_s}}, \tag{4.39}$$

where

σ_R = residual tensile strength after impact

σ_U = tensile strength of an undamaged laminate

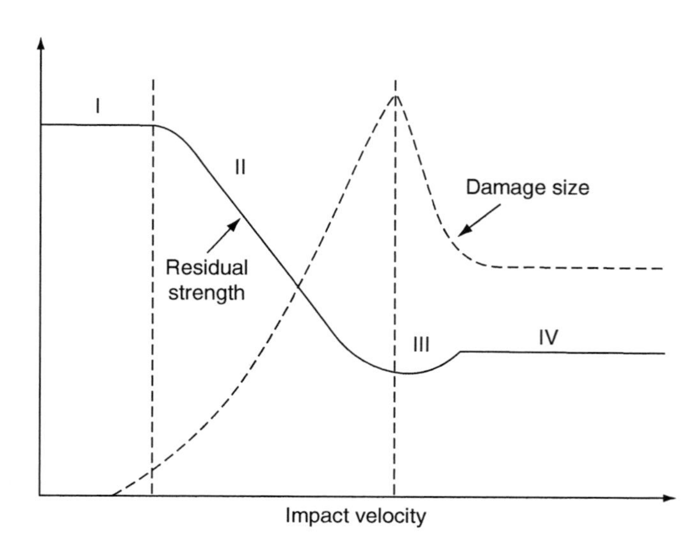

FIGURE 4.81 Schematic representation of the residual static strength in impact-damaged laminates. (After Awerbuch, J. and Hahn, H.T., *J. Compos. Mater.*, 10, 231, 1976.)

U_s = area under the stress–strain curve for an undamaged laminate
U_{KE} = input kinetic energy per unit laminate thickness
k = a constant that depends on the laminate stacking sequence and boundary conditions (e.g., one end fixed vs. both ends fixed)

Two experiments are required to determine the value of k, namely, a static tension test on an undamaged specimen and a static tension test on an impact-damaged specimen. Knowing the preimpact kinetic energy, the value of k can be calculated using Equation 4.39. Although the value of k is not significantly affected by the laminate thickness, it becomes independent of laminate width only for wide specimens. Residual strength measurements on $[0/90]_{3S}$ laminates of various fiber–matrix combinations have shown reasonable agreement with Equation 4.39.

4.3.6 Compression-After-Impact Test

The compression-after-impact test is used for assessing the nonvisible or barely visible impact damage in composite laminates. An edge-supported quasi-isotropic laminated plate, 153 mm × 102 mm × 3–5 mm thick, is impacted at the center with an energy level of 6.7 J/mm (1500 in. lb/in.). After nondestructively examining the extent of impact damage (e.g., by ultrasonic C-scan), the plate is compression-tested in a fixture with antibuckling guides (Figure 4.82).

The compressive strength of an impact-damaged laminate is lower than the undamaged compressive strength. Failure modes observed in compression-after-impact tests are shear crippling of fibers and ply delamination. In brittle epoxy laminates, delamination is the predominant failure mode, while in toughened epoxy matrix composites, significant shear crippling occurs before failure by ply delamination.

Postimpact compressive strength (PICS) of a laminate can be improved by reducing the impact-induced delamination. One way of achieving this is by increasing the interlaminar fracture toughness of the laminate. Figure 4.83 shows that the PICS of carbon fiber-reinforced laminates increases considerably

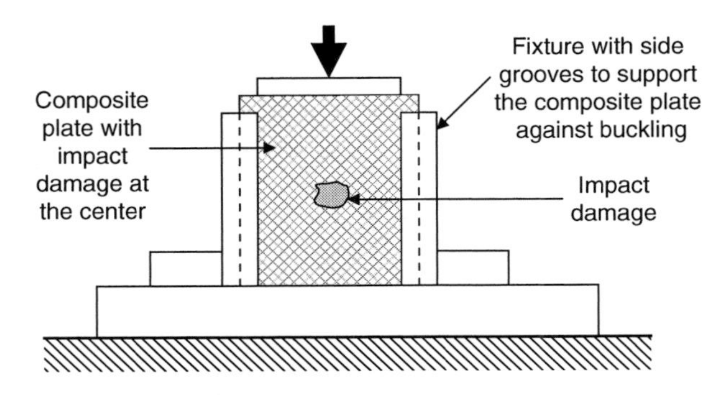

FIGURE 4.82 Test fixture for compression test after impact.

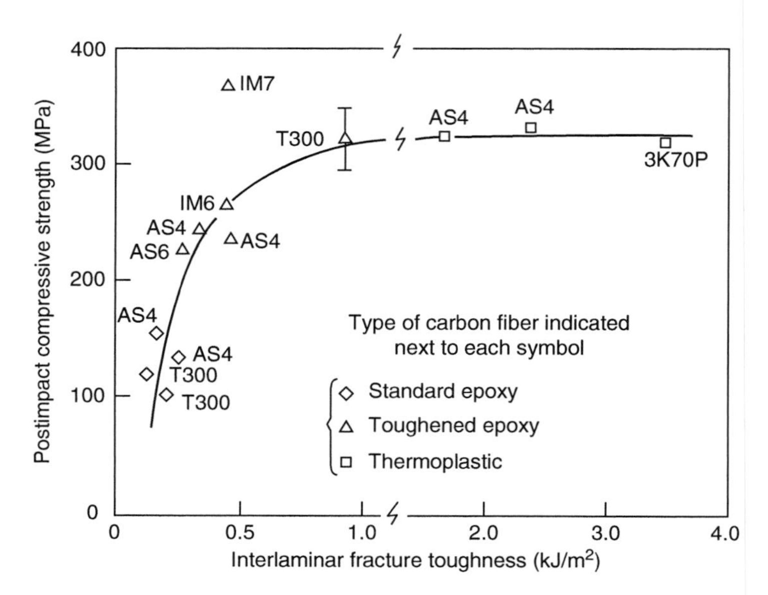

FIGURE 4.83 Postimpact compressive strength of carbon fiber-reinforced laminates as a function of their interlaminar fracture toughness (impact energy $= 6.7$ J/mm). (Adapted from Leach, D., Tough and damage tolerant composites, *Symposium on Damage Development and Failure Mechanisms in Composite Materials*, Leuven, Belgium, 1987.)

when their interlaminar fracture toughness is increased from 200 to 500 J/m^2; however, above 1000 J/m^2, PICS is nearly independent of the interlaminar fracture toughness. In this case, higher interlaminar fracture toughness was obtained by increasing the fracture toughness of the matrix either by toughening it or by changing the matrix from the standard epoxy to a thermoplastic. Other methods of increasing the interlaminar fracture toughness and reducing ply delamination are discussed in Section 4.7.3.

4.4 OTHER PROPERTIES

4.4.1 Pin-Bearing Strength

Pin-bearing strength is an important design parameter for bolted joints and has been studied by a number of investigators. It is obtained by tension testing a pin-loaded hole in a flat specimen (Figure 4.84). The failure mode in pin-bearing tests depends on a number of geometric variables [78]. Generally, at low w/d ratios, the failure is by net tension with cracks originating at the hole boundary, and at low e/d ratios, the failure is by shear-out. The load-carrying capacity of the laminate is low if either of these failure modes occurs instead of bearing failure.

For bearing failure, relatively high values of w/d and e/d ratios are required. The minimum values of w/d and e/d ratios needed to develop full bearing

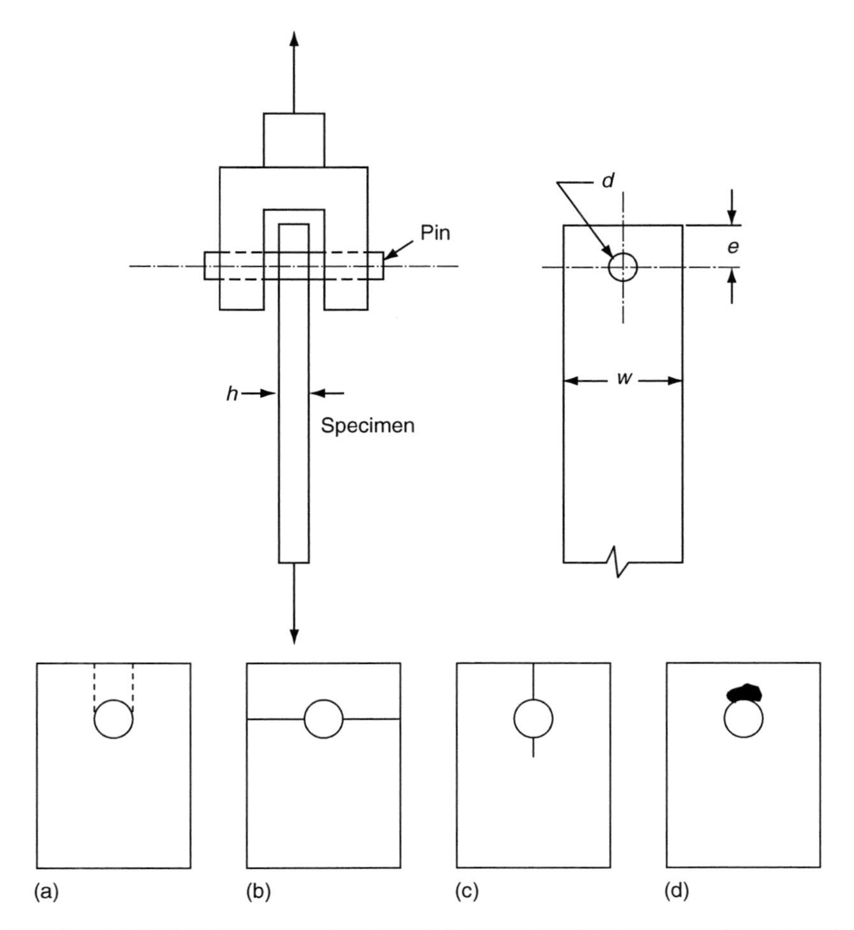

FIGURE 4.84 Pin-bearing test and various failure modes: (a) shear-out, (b) net tension, (c) cleavage, and (d) bearing failure (accompanied by hole elongation).

strength depend on the material and fiber orientation as well as on the stacking sequence. Another geometric variable controlling the bearing strength is the d/h ratio of the specimen. In general, bearing strength is decreased at higher d/h ratios, and a tendency toward shear failure is observed at low d/h ratios. A d/h ratio between 1 and 1.2 is recommended for developing the full bearing strength. A few representative pin-bearing strengths are given in Table 4.14.

For 0° laminates, failure in pin-bearing tests occurs by longitudinal splitting, since such laminates have poor resistance to in-plane transverse stresses at the loaded hole. The bearing stress at failure for 0° laminates is also quite low. Inclusion of 90° layers [79], ±45° layers, or ±60° layers [80] at or near the surfaces improves the bearing strength significantly. However, $[\pm 45]_S$, $[\pm 60]_S$, or $[90/\pm 45]_S$ laminates have lower bearing strengths than $[0/\pm 45]_S$ and $[0/\pm 60]_S$

TABLE 4.14
Representative Pin-Bearing Strength of Various Laminates

Laminates	Tightening Torque, N m (in. lb)	e/d	Pin-Bearing Strength, MPa (ksi)	References
E-glass–vinyl ester SMC-R50	0	3	325 (47.1)	[79]
E-glass–vinyl ester SMC-C40R30	0	3	400 (58)	[79]
E-glass–epoxy				
$[0/90]_S$	1.85 (16.4)	6	600 (87)	[78]
$[0/\pm45]_S$	1.85 (16.4)	4.5	725 (105.1)	[78]
$[\pm45]_S$	1.85 (16.4)	5	720 (104.4)	[78]
HTS carbon–epoxy				
$[0/90]_S$	3.40 (30.2)	6	800 (116)	[80]
$[0/\pm45]_S$	3.40 (30.2)	3	900 (130)	[80]
$[\pm45]_S$	3.40 (30.2)	5	820 (118.9)	[80]

laminates. A number of other observations on the pin-bearing strength of composite laminates are listed as follows.

1. Stacking sequence has a significant influence on the pin-bearing strength of composite laminates. Quinn and Matthews [81] have shown that a $[90/\pm45/0]_S$ layup is nearly 30% stronger in pin-bearing tests than a $[0/90/\pm45]_S$ layup.
2. The number of $\pm\theta$ layers present in a $[0/\pm\theta]_S$ laminate has a great effect on its pin-bearing strength. Collings [80] has shown that a $[0/\pm45]_S$ laminate attains its maximum pin-bearing strength when the ratio of 0° and 45° layers is 60:40.
3. Fiber type is an important material parameter for developing high pin-bearing strength in $[0/\pm\theta]_S$ laminates. Kretsis and Matthews [78] have shown that for the same specimen geometry, the bearing strength of a $[0/\pm45]_S$ carbon fiber-reinforced epoxy laminate is nearly 20% higher than a $[0/\pm45]_S$ E-glass fiber-reinforced epoxy.
4. The pin-bearing strength of a composite laminate can be increased significantly by adhesively bonding a metal insert (preferably an aluminum insert) at the hole boundary [82].
5. Lateral clamping pressure distributed around the hole by washers can significantly increase the pin-bearing strength of a laminate [83]. The increase is attributed to the lateral restraint provided by the washers as well as frictional resistance against slip. The lateral restraint contains the shear cracks developed at the hole boundary within the washer perimeter and allows the delamination to spread over a wider area before final failure occurs. The increase in pin-bearing strength levels off at high clamping pressure. If the clamping pressure is too high, causing the washers to dig into the laminate, the pin-bearing strength may decrease.

4.4.2 Damping Properties

The damping property of a material represents its capacity to reduce the transmission of vibration caused by mechanical disturbances to a structure. The measure of damping of a material is its damping factor η. A high value of η is desirable for reducing the resonance amplitude of vibration in a structure. Table 4.15 compares the typical damping factors for a number of materials. Fiber-reinforced composites, in general, have a higher damping factor than metals. However, its value depends on a number of factors, including fiber and resin types, fiber orientation angle, and stacking sequence.

4.4.3 Coefficient of Thermal Expansion

The coefficient of thermal expansion (CTE) represents the change in unit length of a material due to unit temperature rise or drop. Its value is used for calculating dimensional changes as well as thermal stresses caused by temperature variation.

The CTE of unreinforced polymers is higher than that of metals. The addition of fibers to a polymer matrix generally lowers its CTE. Depending on the fiber type, orientation, and fiber volume fraction, the CTE of fiber-reinforced polymers can vary over a wide range of values. In unidirectional 0° laminates, the longitudinal CTE, α_{11}, reflects the fiber characteristics. Thus, both carbon and Kevlar 49 fibers produce a negative CTE, and glass and boron fibers produce a positive CTE in the longitudinal direction. As in the case of elastic properties, the CTEs for unidirectional 0° laminates are different in longitudinal and transverse directions (Table 4.16). Compared with carbon fiber-reinforced epoxies, Kevlar 49 fiber-reinforced epoxies exhibit a greater anisotropy in their CTE due to greater anisotropy in the CTE of Kevlar 49

TABLE 4.15
Representative Damping Factors of Various Polymeric Laminates

Material	Fiber Orientation	Modulus (10^6 psi)	Damping Factor η
Mild steel	—	28	0.0017
6061 Al alloy	—	10	0.0009
E-glass–epoxy	0°	5.1	0.0070
Boron–epoxy	0°	26.8	0.0067
Carbon–epoxy	0°	27.4	0.0157
	22.5°	4.7	0.0164
	90°	1.0	0.0319
	$[0/22.5/45/90]_S$	10.0	0.0201

Source: Adapted from Friend, C.A., Poesch, J.G., and Leslie, J.C., Graphite fiber composites fill engineering needs, *Proceedings 27th Annual Technical Conference*, Society of the Plastics Industry, 1972.

TABLE 4.16
Coefficients of Thermal Expansion of Various Laminates[a]

Material	Coefficient of Thermal Expansion, 10^{-6} m/m per °C (10^{-6} in./in. per °F)		
	Unidirectional (0°)		
	Longitudinal	Transverse	Quasi-Isotropic
S-glass–epoxy	6.3 (3.5)	19.8 (11)	10.8 (6)
Kevlar 49–epoxy	−3.6 (−2)	54 (30)	−0.9 to 0.9 (−0.5 to 0.5)
Carbon–epoxy			
High-modulus carbon	−0.9 (−0.5)	27 (15)	0 to 0.9 (0 to 0.5)
Ultrahigh-modulus carbon	−1.44 (−0.8)	30.6 (17)	−0.9 to 0.9 (−0.5 to 0.5)
Boron–epoxy	4.5 (2.5)	14.4 (8)	3.6 to 5.4 (2 to 3)
Aluminum		21.6 to 25.2 (12 to 14)	
Steel		10.8 to 18 (6 to 10)	
Epoxy		54 to 90 (30 to 50)	

Source: Adapted from Freeman, W.T. and Kuebeler, G.C., *Composite Materials: Testing and Design* (*Third Conference*), *ASTM STP*, 546, 435, 1974.

[a] The fiber content in all composite laminates is 60% by volume.

TABLE 4.17
Coefficients of Thermal Expansion of Various E-Glass–Epoxy Laminates

Laminate	Fiber Volume Fraction (%)	Direction of Measurement	Coefficient of Thermal Expansion, 10^{-6} m/m per °C (10^{-6} in./in. per °F)
Unidirectional	63	0°	7.13 (3.96)
		15°	9.45 (5.25)
		30°	13.23 (7.35)
		45°	30.65 (12.08)
		60°	30.65 (17.03)
		75°	31.57 (17.54)
		90°	32.63 (18.13)
$[\pm 30/90]_{7S}$	60	In-plane	15.66 (8.7)
$[(0/90/)_9/(\pm 45)_2]_S$	71	In-plane	12.6 (7.0)

Source: Adapted from Raghava, R., *Polym. Compos.*, 5, 173, 1984.

fibers [84]. The anisotropic nature of the CTE of a unidirectional laminate is further demonstrated in Table 4.17.

In quasi-isotropic laminates as well as randomly oriented discontinuous fiber laminates, the CTEs are equal in all directions in the plane of the laminate. Furthermore, with proper fiber type and lamination configuration, CTE in the plane of the laminate can be made close to zero. An example is shown in Figure 4.85, in which the proportions of fibers in 0°, 90°, and ±45° layers

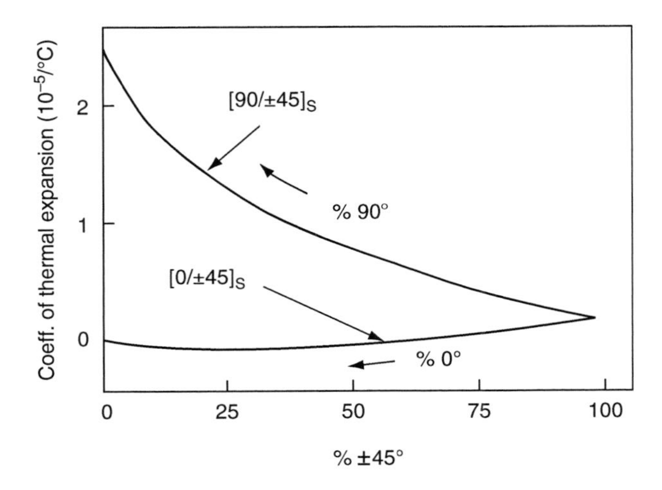

FIGURE 4.85 Coefficients of thermal expansion of $[0/\pm 45]_S$ and $[90/\pm 45]_S$ carbon fiber-epoxy laminates. (After Parker, S.F.H., Chandra, M., Yates, B., Dootson, M., and Walters, B.J., *Composites*, 12, 281, 1981.)

were controlled to obtain a variety of CTEs in the $[0/\pm45]_S$ and $[90/\pm45]_S$ laminates [85].

4.4.4 Thermal Conductivity

The thermal conductivity of a material represents its capacity to conduct heat. Polymers in general have low thermal conductivities, which make them useful as insulating materials. However, in some circumstances, they may also act as a heat sink with little ability to dissipate heat efficiently. As a result, there may be a temperature rise within the material.

The thermal conductivity of a fiber-reinforced polymer depends on the fiber type, orientation, fiber volume fraction, and lamination configuration. A few representative values are shown in Table 4.18. With the exception of carbon fibers, fiber-reinforced polymers in general have low thermal conductivities. Carbon fiber-reinforced polymers possess relatively high thermal conductivities due to the highly conductive nature of carbon fibers. For unidirectional 0° composites, the longitudinal thermal conductivity is controlled by the fibers and the transverse thermal conductivity is controlled by the matrix. This is reflected in widely different values of thermal conductivities in these two directions.

The electrical conductivities of fiber-reinforced polymers are similar in nature to their thermal counterparts. For example, E-glass fiber-reinforced polymers are poor electrical conductors and tend to accumulate static electricity. For protection against static charge buildup and the resulting electromagnetic interference (EMI) or radio frequency interference (RFI), small quantities of conductive fibers, such as carbon fibers, aluminum flakes, or aluminum-coated glass fibers, are added to glass fiber composites.

TABLE 4.18
Thermal Conductivities of Various Composite Laminates

	Thermal Conductivity, W/m per °C (Btu/h ft per °F)		
	Unidirectional (0°)		
Material	Longitudinal	Transverse	Quasi-Isotropic
S-glass–epoxy	3.46 (2)	0.35 (0.2)	0.346 (0.2)
Kevlar 49–epoxy	1.73 (1)	0.173 (0.1)	0.173 (0.1)
Carbon–epoxy			
High modulus	48.44–60.55 (28–35)	0.865 (0.5)	10.38–20.76 (6–12)
Ultrahigh modulus	121.1–29.75 (70–75)	1.04 (0.6)	24.22–31.14 (14–18)
Boron–epoxy	1.73 (1)	1.04 (0.6)	1.384 (0.8)
Aluminum		138.4–216.25 (80–125)	
Steel		15.57–46.71 (9–27)	
Epoxy		0.346 (0.2)	

Source: Adapted from Freeman, W.T. and Kuebeler, G.C., *Composite Materials: Testing and Design* (*Third Conference*), *ASTM STP*, 546, 435, 1974.

4.5 ENVIRONMENTAL EFFECTS

The influence of environmental factors, such as elevated temperatures, high humidity, corrosive fluids, and ultraviolet (UV) rays, on the performance of polymer matrix composites is of concern in many applications. These environmental conditions may cause degradation in the mechanical and physical properties of a fiber-reinforced polymer because of one or more of the following reasons:

1. Physical and chemical degradation of the polymer matrix, for example, reduction in modulus due to increasing temperature, volumetric expansion due to moisture absorption, and scission or alteration of polymer molecules due to chemical attack or ultraviolet rays. However, it is important to note that different groups of polymers or even different molecular configurations within the same group of polymers would respond differently to the same environment.
2. Loss of adhesion or debonding at the fiber–matrix interface, which may be followed by diffusion of water or other fluids into this area. In turn, this may cause a reduction in fiber strength due to stress corrosion. Many experimental studies have shown that compatible coupling agents are capable of either slowing down or preventing the debonding process even under severe environmental conditions, such as exposure to boiling water.
3. Reduction in fiber strength and modulus. For a short-term or intermittent temperature rise up to 150°C–300°C, reduction in the properties of most commercial fibers is insignificant. However, depending on the fiber type, other environmental conditions may cause deterioration in fiber properties. For example, moisture is known to accelerate the static fatigue in glass fibers. Kevlar 49 fibers are capable of absorbing moisture from the environment, which reduces its tensile strength and modulus. The tensile strength of Kevlar 49 fibers is also reduced with direct exposure to ultraviolet rays.

In this section, we consider the effect of elevated temperature and high humidity on the performance of composite laminates containing polymer matrix.

4.5.1 ELEVATED TEMPERATURE

When a polymer specimen is tension-tested at elevated temperatures, its modulus and strength decrease with increasing temperature because of thermal softening. In a polymeric matrix composite, the matrix-dominated properties are more affected by increasing temperature than the fiber-dominated properties. For example, the longitudinal strength and modulus of a unidirectional 0° laminate remain virtually unaltered with increasing temperature, but its transverse and off-axis properties are significantly reduced as the temperature approaches the T_g of the polymer. For a randomly oriented discontinuous fiber composite, strength and modulus are reduced in all directions. Reductions in modulus as a function of increasing test temperature are shown for unidirectional continuous and randomly oriented discontinuous fiber laminates in Figures 4.86 and 4.87, respectively.

Thermal aging due to long-term exposure to elevated temperatures without load can cause deterioration in the properties of a polymer matrix composite. Kerr and Haskins [86] reported the effects of 100–50,000 h of thermal aging on the tensile strength of AS carbon fiber–epoxy and HTS carbon fiber–polyimide unidirectional and cross-ply laminates. For the AS carbon–epoxy systems, thermal aging at 121°C produced no degradation for the first 10,000 h. Matrix degradation began between 10,000 and 25,000 h and was severe after 50,000 h. After 5000 h, the matrix was severely embrittled. Longitudinal tensile strength was considerably reduced for aging times of 5000 h or longer. The HTS carbon–polyimide systems were aged at higher temperatures but showed less degradation than the AS carbon–epoxy systems.

Devine [87] reported the effects of thermal aging on the flexural strength retention in SMC-R laminates containing four different thermoset polyester resins and a vinyl ester resin. At 130°C, all SMC-R laminates retained >80% of their respective room temperature flexural strengths even after thermal aging for 12 months. At 180°C, all SMC-R laminates showed deterioration;

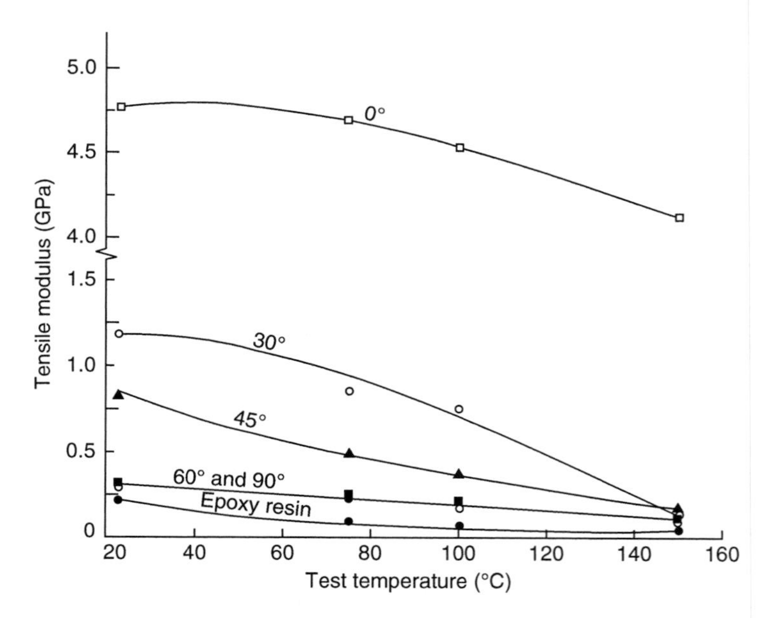

FIGURE 4.86 Effect of increasing test temperature on the static tensile modulus of unidirectional E-glass–epoxy laminates. (After Marom, G. and Broutman, L.J., *J. Adhes.*, 12, 153, 1981.)

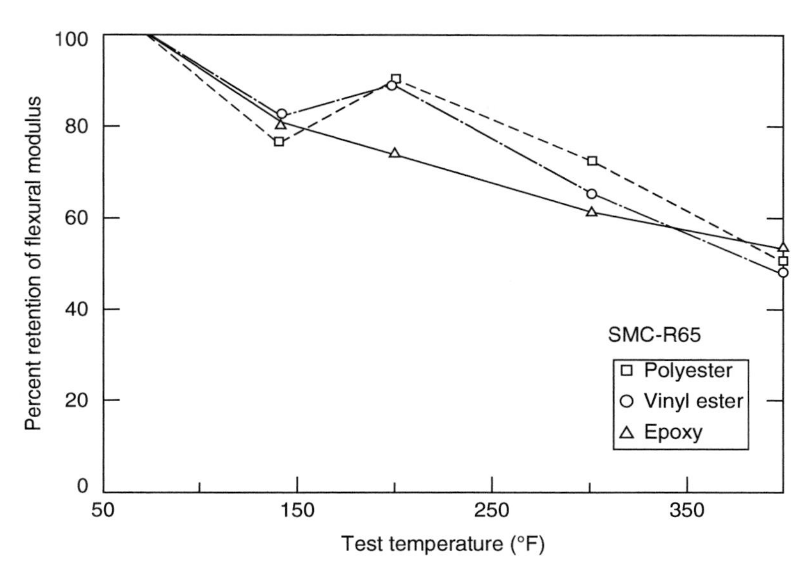

FIGURE 4.87 Effect of increasing test temperature on the static flexural modulus of E-glass-SMC-R65 laminates. (After Mallick, P.K., *Polym. Compos.*, 2, 18, 1981.)

however, vinyl ester laminates had higher strength retention than all polyester laminates.

The concern for the reduction in mechanical properties of thermoplastic matrix composites at elevated temperatures is more than the thermoset matrix composites, since the properties of thermoplastic polymers reduce significantly at or slightly above their glass transition temperatures. As in thermoset matrix composites, the effect of increasing temperature is more pronounced for matrix-dominated properties than for fiber-dominated properties (Figure 4.88).

4.5.2 Moisture

When exposed to humid air or water environments, many polymer matrix composites absorb moisture by instantaneous surface absorption followed by diffusion through the matrix. Analysis of moisture absorption data for epoxy and polyester matrix composites shows that the moisture concentration increases initially with time and approaches an equilibrium (saturation) level after several days of exposure to humid environment (Figure 4.89). The rate at which the composite laminate attains the equilibrium moisture concentration is determined by its thickness as well as the ambient temperature. On drying, the moisture concentration is continually reduced until the composite laminate returns to the original as-dry state. In general, the rate of desorption is higher than the rate of absorption, although for the purposes of analysis they are assumed to be equal.

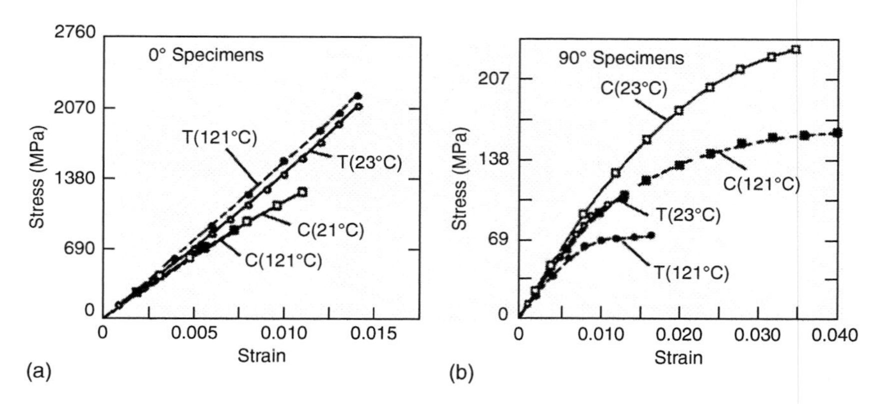

FIGURE 4.88 Tensile (T) and compressive (C) stress–strain diagrams of 0° and 90° carbon fiber-reinforced PEEK laminates at 23°C and 121°C. (After Fisher, J.M., Palazotto, A.N., and Sandhu, R.S., *J. Compos. Technol. Res.*, 13, 152, 1991.)

4.5.2.1 Moisture Concentration

The moisture concentration M, averaged over the thickness, of a composite laminate at any time during its exposure to humid environment at a given temperature can be calculated from the following equation [88]:

$$M = M_i + G(M_m - M_i), \tag{4.40}$$

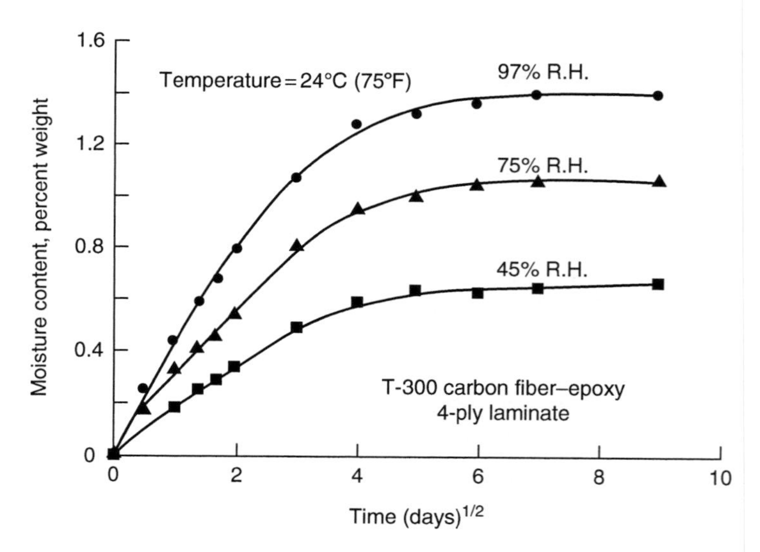

FIGURE 4.89 Moisture absorption in a carbon–epoxy laminate at 24°C (75°F). (After Shen, C.H. and Springer, G.S., *J. Compos. Mater.*, 10, 2, 1976.)

where

- M_i = initial moisture concentration, which is equal to zero if the material is completely dried
- M_m = equilibrium (maximum) moisture concentration in the saturated condition
- G = time-dependent dimensionless parameter related to the diffusion coefficient of the material

For a material immersed in water, the equilibrium moisture concentration M_m is a constant. If the material is exposed to humid air, the equilibrium moisture concentration M_m increases with increasing relative humidity of the surrounding air (Table 4.19); however, it is found that M_m is relatively insensitive to the ambient temperature. For the humid air environment, M_m is expressed as

$$M_m = A(\mathrm{RH})^B, \tag{4.41}$$

where RH is the relative humidity (percent) of the surrounding air, and A and B are constants that depend primarily on the type of polymer; the exponent B has a value between 1 and 2.

Assuming a Fickian diffusion through the laminate thickness, the time-dependent parameter G can be approximated as

$$G \approx 1 - \frac{8}{\pi^2}\exp\left(-\frac{\pi^2 D_z t}{c^2}\right), \tag{4.42}$$

TABLE 4.19
Equilibrium Moisture Content in Various Composite Laminates

Material	Laminate	RH (%)	Temperature (°C)	M_m (%)
T-300 carbon–epoxy[a] ($v_f = 68\%$)	Unidirectional (0°) and quasi-isotropic	50	23	0.35
		75	23	0.7875
		100	23	1.4
		Fully submerged in water	23	1.8
E-glass–polyester[b] ($w_f = 50\%$)	SMC-R50	50	23	0.10
		100	23	1.35
E-glass–vinyl ester[b] ($w_f = 50\%$)	SMC-R50	50	23	0.13
		100	23	0.63

[a] Adapted from C.H. Shen and G.S. Springer, *J. Composite Matls.*, 10, 2, 1976.
[b] Adapted from G.S. Springer, B.A. Sanders and R.W. Tung, *J. Composite Matls.*, 14, 213, 1980.

where

D_z = diffusion coefficient (mm^2/s) of the material in the direction normal to the surface (moisture diffusion is in the thickness direction)

c = laminate thickness h if both sides of the laminate are exposed to humid environment; for exposure on one side, $c = 2h$

t = time (s)

Equation 4.42 is valid at sufficiently large values of t. For shorter times, the average moisture concentration increases linearly with $t^{1/2}$, and the parameter G can be approximated as

$$G = 4\left(\frac{D_z t}{\pi c^2}\right)^{1/2}. \tag{4.43}$$

The diffusion coefficient D_z is related to the matrix diffusion coefficient D_m by the following equation:

$$D_z = D_{11}\cos^2\phi + D_{22}\sin^2\phi, \tag{4.44}$$

where

$D_{11} = D_m(1 - v_f)$

$D_{12} = D_m\left(1 - 2\sqrt{\frac{v_f}{\pi}}\right)$ } Assuming fiber diffusivity (D_f) $\ll$ matrix diffusivity (D_m)

ϕ = fiber angle with the z direction ($\phi = 90°$ for fibers parallel to the laminate surface)

v_f = fiber volume fraction

Equations 4.40 through 4.44 can be used to estimate the moisture concentration in a polymer matrix composite. However, the following internal and external parameters may cause deviations from the calculated moisture concentrations.

Void content: The presence of voids has a dramatic effect on increasing the equilibrium moisture concentration as well as the diffusion coefficient.

Fiber type: Equation 4.44 assumes that the fiber diffusivity is negligible compared with the matrix diffusivity. This assumption is valid for glass, carbon, and boron fibers. However, Kevlar 49 fibers are capable of absorbing and diffusing significant amounts of moisture from the environment. As a result, Kevlar 49 fiber-reinforced composites absorb much more moisture than other composites.

Resin type: Moisture absorption in a resin depends on its chemical structure and the curing agent as well as the degree of cure. Analysis of the water absorption data of various epoxy resin compositions shows that the weight gain due to water absorption may differ by a factor of 10 or more between different resin chemical structures and by a factor of 3 or more for the same resin that has different curing formulations [90]. For many resin systems,

TABLE 4.20
Diffusion Coefficients for Absorption and Desorption in an Epoxy Resin at 100% Relative Humidity

	Diffusion Coefficient (10^{-8} mm^2/s)	
Temperature (°C)	**Absorption**	**Desorption**
0.2	3	3
25	21	17
37	41	40
50	102	88
60	179	152
70	316	282
80	411	489
90	630	661

Source: After Wright, W.W., *Composites*, 12, 201, 1981.

the water absorption process may continue for a long time and equilibrium may not be attained for months or even years.

Temperature: Moisture diffusion in a polymer is an energy-activated process, and the diffusion coefficient depends strongly on the temperature (Table 4.20). In general, the temperature dependence can be predicted from an Arrhenius-type equation:

$$D_z = D_{z0} \exp\left(-\frac{E}{RT}\right), \tag{4.45}$$

where
E = activation energy (cal/g mol)
R = universal gas constant = 1.987 cal/(g mol K)
T = absolute temperature (K)
D_{z0} = a constant (mm^2/s)

Stress level: Gillat and Broutman [91] have shown that increasing the applied stress level on a T-300 carbon–epoxy cross-ply laminate produces higher diffusion coefficients but does not influence the equilibrium moisture content. Similar experiments by Marom and Broutman [92] show that the moisture absorption is a function of fiber orientation angle relative to the loading direction. The maximum effect is observed at $\theta = 90°$.

Microcracks: The moisture concentration in a laminate may exceed the equilibrium moisture concentration if microcracks develop in the material. Moisture absorption is accelerated owing to capillary action at the microcracks as well as exposed fiber–matrix interfaces at the laminate edges. On the other hand, there may be an "apparent" reduction in moisture concentration if there is a loss of material from leaching or cracking.

Thermal spikes: Diffusion characteristics of composite laminates may alter significantly if they are rapidly heated to high temperatures followed by rapid cooling to the ambient condition, a process known as thermal spiking. McKague et al. [93] have shown that the moisture absorption in specimens exposed to 75% relative humidity at 24°C and occasional (twice weekly) thermal spikes (rapid heating to 149°C followed by rapid cooling to 24°C) is twice that of specimens not exposed to spikes. Additionally, thermally spiked specimens exhibit a permanent change in their moisture absorption characteristics. The increased diffusion rate and higher moisture absorption are attributed to microcracks formed owing to stress gradients caused by thermal cycling and resin swelling. The service temperature in a spike environment should be limited to the glass transition temperature T_g of the resin, since spike temperatures above T_g cause much higher moisture absorption than those below T_g.

Reverse thermal effect: Adamson [94] has observed that cast-epoxy resins or epoxy-based laminates containing an equilibrium moisture concentration exhibit a rapid rate of moisture absorption when the ambient temperature is reduced. For example, an AS carbon fiber-reinforced epoxy laminate attained an equilibrium moisture concentration of 2.3 wt% after 140 days of exposure at 74°C. When the exposure temperature was reduced to 25°C, the equilibrium moisture concentration increased to 2.6% within 40 days. This inverse temperature dependence of moisture absorption is called the reverse thermal effect.

4.5.2.2 Physical Effects of Moisture Absorption

Moisture absorption produces volumetric changes (swelling) in the resin, which in turn cause dimensional changes in the material. Assuming that the swollen volume of the resin is equal to the volume of absorbed water, the resulting volume change can be computed from the following relationship:

$$\frac{\Delta V(t)}{V_0} = \frac{\rho_m}{\rho_w} M, \tag{4.46}$$

where

ρ_m = matrix density
ρ_w = water density (≈ 1 kg/mm^3)
M = moisture content at time t

The corresponding dilatational (volumetric) strain in the resin is

$$\varepsilon_m = \frac{1}{3}\frac{\Delta V}{V_0} = \frac{1}{3}\frac{\rho_m}{\rho_w} M = \beta_m M, \tag{4.47}$$

where

$\beta_m = \frac{1}{3}\frac{\rho_m}{\rho_w}$

β_m is called the swelling coefficient

In practice, swelling is negligible until a threshold moisture concentration M_0 is exceeded. Therefore, the dilatational strain in the resin is

$$\begin{aligned} \varepsilon_m &= 0 \quad \text{for } M < M_0, \\ &= \beta_m(M - M_0) \quad \text{for } M > M_0 \end{aligned} \tag{4.48}$$

The threshold moisture concentration M_0 represents the amount of water absorbed in the free volume as well as microvoids present in the resin. For a variety of cast-epoxy resins, the measured swelling coefficient ranges from 0.26 to 0.33 and the threshold moisture concentration is in the range of 0.3%–0.7% [95].

The dilatational strain in a unidirectional 0° composite laminate due to moisture absorption can be calculated as

$$\text{Longitudinal: } \varepsilon_{mL} = 0, \tag{4.49a}$$

$$\text{Transverse: } \varepsilon_{mT} = \beta_T(M - M_v), \tag{4.49b}$$

where

$\beta_T = (1 + \nu_m)\beta_m(\rho_m/\rho_c)$

ρ_c = composite density

ν_m = matrix Poisson's ratio

$M_v = v_v(\rho_w/\rho_c)$

v_v = void volume fraction

Another physical effect of moisture absorption is the reduction in glass transition temperature of the resin (Figure 4.90). Although the room-temperature performance of a resin may not change with a reduction in T_g, its elevated-temperature properties are severely affected. For example, the modulus of an epoxy resin at 150°C decreases from 2,070 MPa (300,000 psi) to 20.7 MPa (3,000 psi) as its T_g is reduced from 215°C to 127°C. Similar effects may be expected for the matrix-dominated properties of a polymer matrix composite.

Finally, the dilatational expansion of the matrix around the fiber reduces the residual compressive stresses at the fiber–matrix interface caused by curing shrinkage. As a result, the mechanical interlocking between the fiber and the matrix may be relieved.

4.5.2.3 Changes in Performance Due to Moisture and Temperature

From the available data on the effects of temperature and moisture content on the tensile strength and modulus of carbon and boron fiber-reinforced epoxy laminates [96,97], the following conclusions can be made.

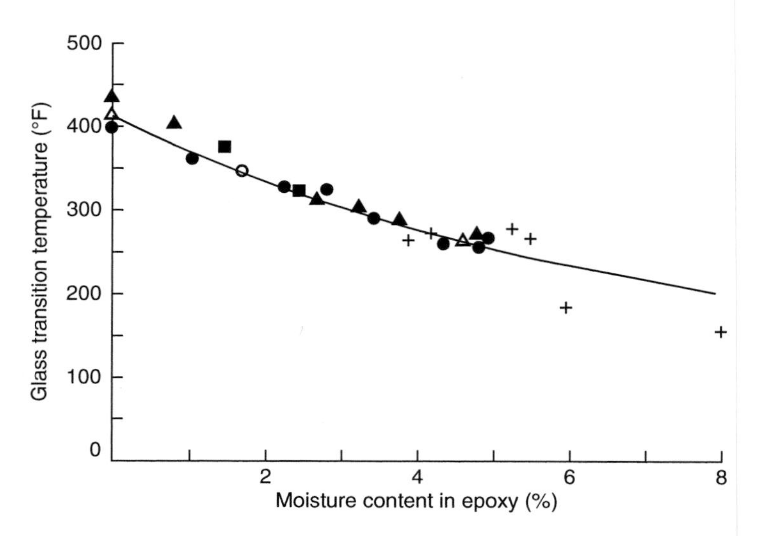

FIGURE 4.90 Variation of glass transition temperature of various epoxy matrices and their composites with moisture content. (After Shirrell, C.D., Halpin, J.C., and Browning, C.E., Moisture—an assessment of its impact on the design of resin based advanced composites, NASA Technical Report, NASA-44-TM-X-3377, April 1976.)

For 0° and $[0/\pm 45/90]_S$ quasi-isotropic laminates, changes in temperature up to 107°C (225°F) have negligible effects on tensile strength and modulus values regardless of the moisture concentration in the material. Although the effect on modulus is negligible up to 177°C (350°F), there may be up to a 20% decrease in tensile strength as the temperature increases from 107°C (225°F) to 177°C (350°F).

For 0° and $[0/\pm 45/90]_S$ laminates, the tensile strength and modulus are not affected by moisture absorption below 1% moisture concentration. Although the modulus is not affected by even higher moisture concentration, the tensile strength may decrease by as much as 20% for moisture concentrations above 1%.

For 90° laminates, increasing temperature and moisture concentration reduce both the tensile strength and the modulus by significant amounts. Depending on the temperature and moisture concentration, the reduction may range as high as 60%–90% of the room temperature properties under dry conditions.

The ILSS of composite laminates is also reduced by increasing moisture absorption. For example, short-beam shear tests of a unidirectional carbon fiber–epoxy show nearly a 10% reduction in ILSS at a moisture concentration of 1.2 wt% that was attained after 33 days of exposure to humid air of 95% relative humidity at 50°C. Immersion in boiling water reduced the ILSS by 35%

for the same exposure time [98]. Experiments by Gillat and Broutman [91] on cross-ply carbon fiber–epoxy show nearly a 25% reduction in ILSS as the moisture concentration increased by 1.5 wt%.

Jones et al. [99] reported the effect of moisture absorption on the tension–tension fatigue and flexural fatigue properties of $[0/90]_S$ cross-ply epoxy matrix composites reinforced with E-glass, HTS carbon, and Kevlar 49 fibers. Conditioning treatments included exposure to humid air (65% relative humidity) and immersion in boiling water. The fatigue resistance of carbon fiber–epoxy was found to be unaffected by the conditioning treatment. Exposure to humid air also did not affect the fatigue response of E-glass fiber–epoxy composites; however, immersion in boiling water reduced the fatigue strength by significant amounts, principally due to the damage incurred on the glass fibers by boiling water. On the other hand, the fatigue response of Kevlar 49–epoxy composites was improved owing to moisture absorption, although at high cycles there appears to be a rapid deterioration as indicated by the sharp downward curvature of the *S–N* curve (Figure 4.91).

Curtis and Moore [100] reported the effect of moisture absorption on the zero tension and zero compression fatigue performance of two matrix-dominated laminates, namely, $[(90/\pm30)_3]_S$ and $[0_2/-45_2/90_2/+45_2]_S$ layups of carbon fibers in an epoxy matrix. Conditioning was performed in humid air of 95% humidity at 70°C. Despite the matrix-dominated behavior of these laminates, moisture absorption had very little effect on their fatigue lives.

Chamis et al. [101] proposed the following empirical equation for estimating the hygrothermal effect on the matrix properties, which can subsequently be used in modifying the matrix-dominated properties of a unidirectional lamina:

$$\frac{P_{wT}}{P_0} = \left(\frac{T_{gw} - T}{T_{gd} - T_0}\right)^{1/2}, \tag{4.50}$$

where

P_{wT} = matrix property at the use temperature T and moisture content M
P_0 = matrix property at a reference temperature T_0
T_{gd} = glass transition temperature in the dry condition
T_{gw} = glass transition temperature in the wet condition with a moisture content M

The glass transition temperature in the wet condition, T_{gw} is calculated using the following equation:

$$T_{gw} = (0.005M^2 - 0.1M + 1)T_{gd} \quad \text{for } M \leq 10\%. \tag{4.51}$$

Equations 4.50 and 4.51 have been used to estimate the hygrothermal effect on epoxy matrix composites, but need experimental validation for other polymer matrix systems.

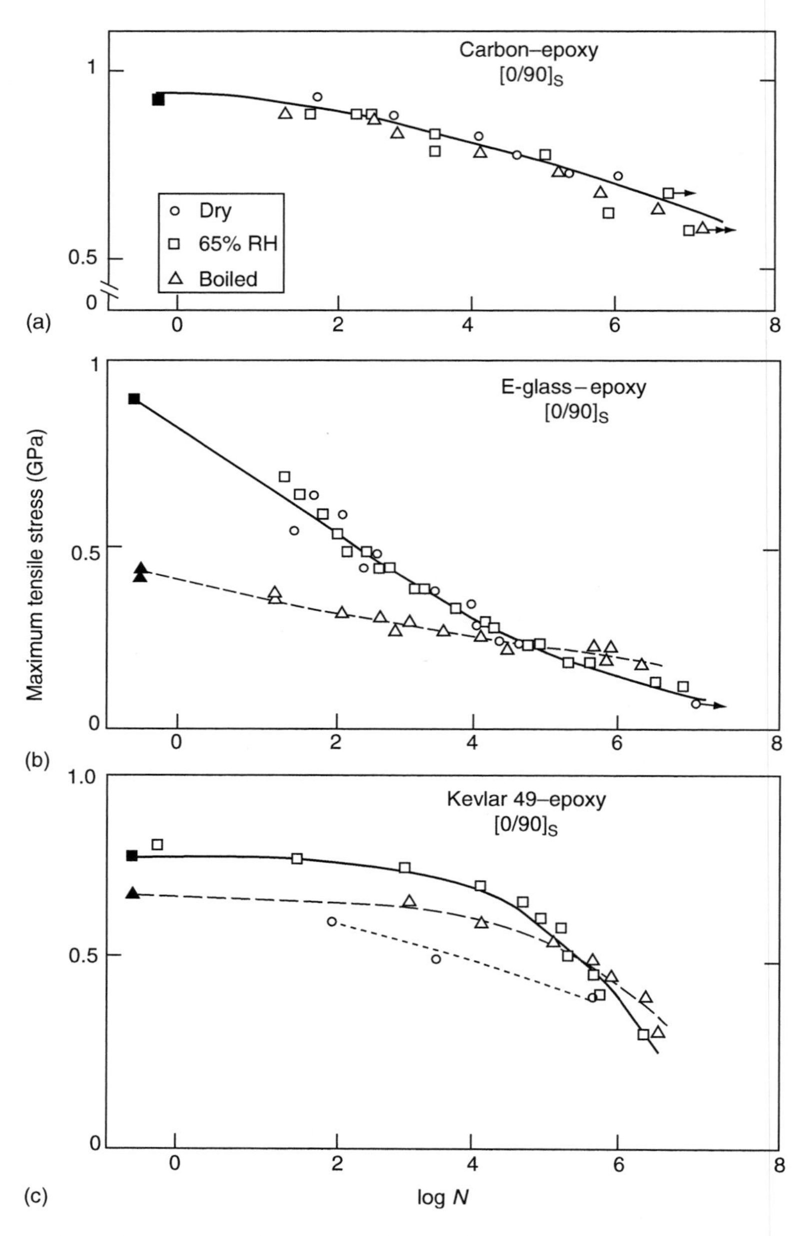

FIGURE 4.91 Effect of moisture absorption on the fatigue behavior of epoxy composites with (a) carbon, (b) E-glass, and (c) Kevlar 49. (After Jones, C.J., Dickson, R.F., Adam, T., Reiter, H., and Harris, B., *Composites*, 14, 288, 1983.)

EXAMPLE 4.3

The tensile strength and modulus of an epoxy matrix at 23°C and dry conditions are 100 MPa and 3.45 GPa, respectively. Estimate its tensile strength and modulus at 100°C and 0.5% moisture content. The glass transition temperature of this epoxy matrix in the dry condition is 215°C.

SOLUTION

Using Equation 4.51, estimate the glass transition temperature at 0.5% moisture content:

$$\begin{aligned} T_{gw} &= [(0.005)(0.5)^2 - (0.1)(0.5) + 1](215) \\ &= 204.5°\text{C}. \end{aligned}$$

Now, using Equation 4.50, estimate P_{wT}:

$$P_{wT} = \left(\frac{204.5 - 100}{215 - 23}\right)^{1/2} P_0 = 0.738 P_0.$$

Thus, the tensile strength and modulus of the epoxy matrix at 100°C and 0.5% moisture content are estimated as:

$$\begin{aligned} \sigma_{mu} &= (0.738)(100\ \text{MPa}) = 73.8\ \text{MPa}, \\ E_m &= (0.738)(3.45\ \text{GPa}) = 2.546\ \text{GPa}. \end{aligned}$$

These values can now be used to estimate the transverse modulus and strength of a unidirectional 0° composite using Equations 3.26 and 3.27, respectively.

4.6 LONG-TERM PROPERTIES

4.6.1 CREEP

Creep is defined as the increase in strain with time at a constant stress level. In polymers, creep occurs because of a combination of elastic deformation and viscous flow, commonly known as viscoelastic deformation. The resulting creep strain increases nonlinearly with increasing time (Figure 4.92). When the stress

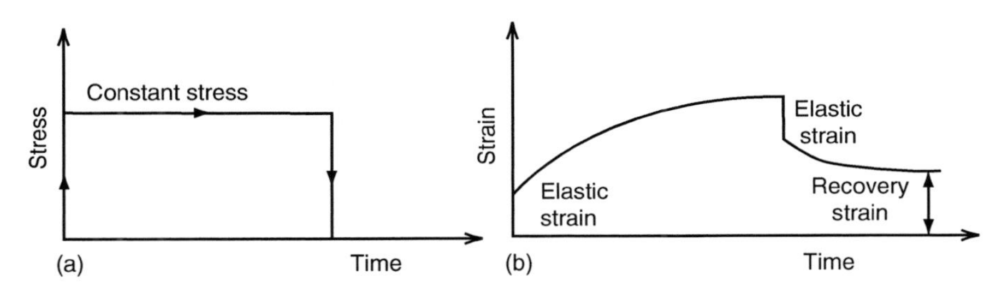

FIGURE 4.92 Schematic representation of creep strain and recovery strain in a polymer.

is released after a period of time, the elastic deformation is immediately recovered. The deformation caused by the viscous flow recovers slowly to an asymptotic value called recovery strain.

Creep strain in polymers and polymer matrix composites depends on the stress level and temperature. Many polymers can exhibit large creep strains at room temperature and at low stress levels. At elevated temperatures or high stress levels, the creep phenomenon becomes even more critical. In general, highly cross-linked thermoset polymers exhibit lower creep strains than thermoplastic polymers. With the exception of Kevlar 49 fibers, commercial reinforcing fibers, such as glass, carbon, and boron, do not creep [102].

4.6.1.1 Creep Data

Under uniaxial stress, the creep behavior of a polymer or a polymer matrix composite is commonly represented by creep compliance, defined as

$$\text{Creep compliance} = D(t) = \frac{\varepsilon(t)}{\sigma}, \tag{4.52}$$

where
σ is the constant stress level in a creep experiment
$\varepsilon(t)$ is the strain measured as a function of time

Figure 4.93 shows typical creep curves for an SMC-R25 laminate at various stress levels. Creep compliances are determined from the slopes of these curves. In general, creep compliance increases with time, stress level, and temperature. For unidirectional fiber-reinforced polymers, it is also a function of fiber orientation angle θ. For $\theta = 0°$ creep compliance is nearly constant, which indicates that creep in the longitudinal direction of a unidirectional 0° laminate is negligible. However, at other fiber orientation angles creep strain can be quite significant.

Fiber orientation angle also influences the temperature dependence of creep compliance. If the fibers are in the loading direction, creep in the composite is governed by the creep in fibers. Thus, with the exception of Kevlar 49 fibers, little temperature dependence is expected in the fiber direction. For other fiber orientations, creep in the matrix becomes the controlling factor. As a result, creep compliance for off-axis laminates increases with increasing temperature (Table 4.21). Creep in SMC-R laminates [103] containing randomly oriented discontinuous fibers is also largely controlled by the matrix creep.

Creep in multidirectional laminates depends on the laminate construction. For example, room temperature creep strains of [±45] and [90/±45] laminates are nearly an order of magnitude different (Figure 4.94), even though the static mechanical properties of these two laminates are similar. The addition of 90° layers to a ±45° construction tends to restrain the rotational tendency of ±45° fibers toward the loading direction and reduces the creep strain significantly.

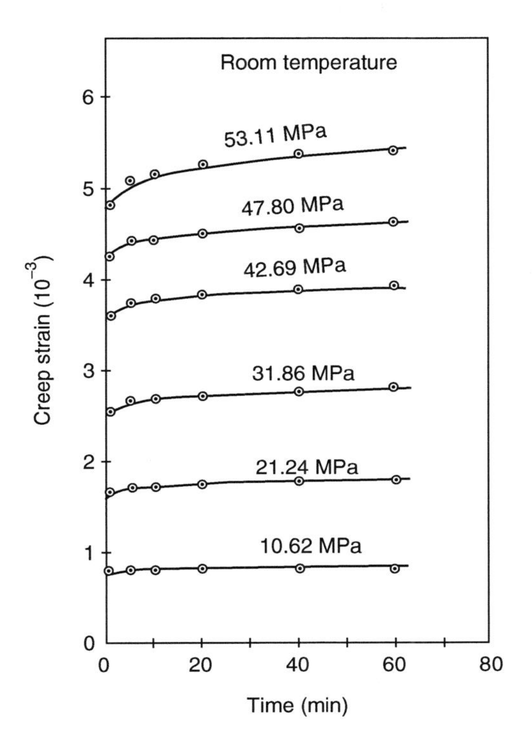

FIGURE 4.93 Tensile creep curves for SMC-R25 polyester laminates. (After Cartner, J.S., Griffith, W.I., and Brinson, H.F., in *Composite Materials in the Automotive Industry*, S.V. Kulkarni, C.H. Zweben, and R.B. Pipes, eds., American Society of Mechanical Engineers, New York, 1978.)

4.6.1.2 Long-Term Creep Behavior

Creep data for a material are generated in the laboratory by conducting either a tensile creep test or a flexural creep test over a period of a few hours to a few hundred hours. Long-term creep behavior of a polymer composite can be predicted from such short-term creep data by the time–temperature superposition method.

The modulus of a polymer at time t and a reference temperature T_0 can be related to its modulus at time t_1 and temperature T_1 by the following equation:

$$E(t,T_0) = \frac{\rho_1 T_1}{\rho_0 T_0} E(t_1,T_1), \quad (4.53)$$

where ρ_1 and ρ_0 are the densities of the polymer at absolute temperatures T_1 and T_0, respectively.

$$t = \left(a_T|_{\text{at } T=T_1} \right) t_1 \quad (4.54)$$

TABLE 4.21
Creep Compliance[a] of Unidirectional E-Glass–Epoxy Laminates

Temperature	Property	Fiber Orientation Angle 30°	45°	60°	90°
28°C	Tensile strength (MPa)	278	186.4	88.8	55.2
	1 h compliance (10^{-6} per MPa)	0.0883	0.2065	0.3346	0.5123
	10 h compliance (10^{-6} per MPa)	0.1300	0.3356	0.6295	1.4390
75°C	Tensile strength (MPa)	230	162.8	74.8	43.2
	1 h compliance (10^{-6} per MPa)	0.1217	0.2511	0.4342	0.6689
	10 h compliance (10^{-6} per MPa)	0.1461	0.3841	0.6539	1.5377
100°C	Tensile strength (MPa)	206	134.4	67.7	40.2
	1 h compliance (10^{-6} per MPa)	0.1460	0.3586	0.6931	0.7728
	10 h compliance (10^{-6} per MPa)	0.1751	0.5739	1.2084	1.9031

Source: Sturgeon, J.B., in *Creep of Engineering Materials*, C.D. Pomeroy, ed., Mechanical Engineering Publishing Ltd., London, 1978.

[a] All compliance values are at stress levels equal to 40% of the tensile strength at the corresponding temperature.

where a_T is the *horizontal shift factor*. For most solids, the variation of density with temperature is negligible so that $\rho_1 = \rho_0$. The horizontal shift factor a_T represents the distance along the time scale over which the modulus value at (t_1, T_1) is shifted to create an equivalent response at the reference temperature T_0. Note that a_T is a function of temperature and is determined from short-term creep test data.

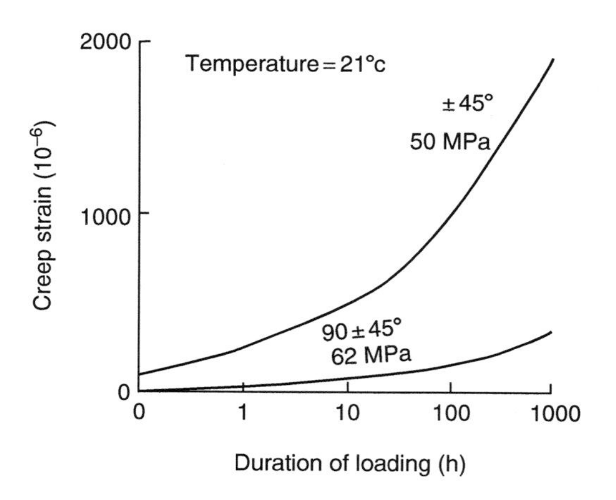

FIGURE 4.94 Comparison of creep curves for $[\pm 45]_S$ and $[90/\pm 45]_S$ laminates. (After Sturgeon, J.B., in *Creep of Engineering Materials*, C.D. Pomeroy, ed., Mechanical Engineering Publishing Ltd., London, 1978.)

The procedure for using the time–temperature superposition method is given as follows.

1. Perform short-term (15 min–1 h) creep tests at various temperatures.
2. Plot creep modulus (or compliance) vs. log (time) for these experiments (Figure 4.95).
3. Select a reference temperature from among the test temperatures used in Step 1.
4. Displace the modulus curves at temperatures other than T_0 horizontally and vertically to match these curves with the modulus curve at T_0.

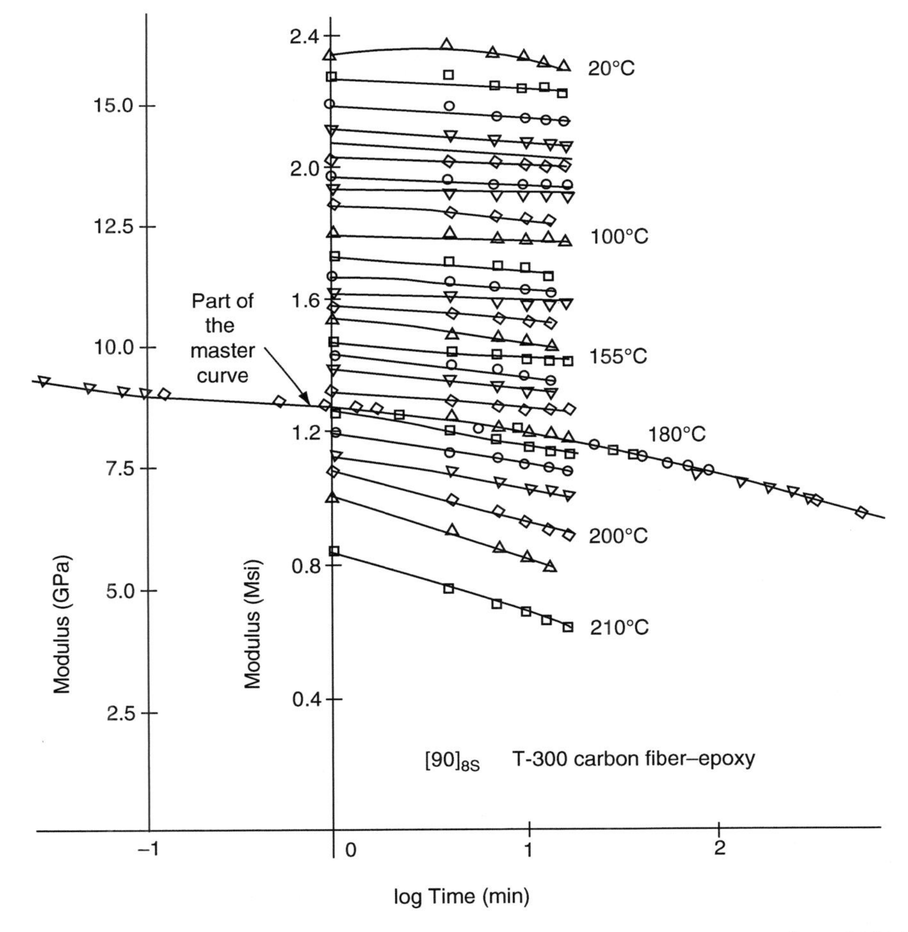

FIGURE 4.95 Creep compliance curves and a portion of the master curve for a T-300 carbon–epoxy laminate. (After Yeow, Y.T., Morris, D.H., and Brinson, H.F., *Composite Materials: Testing and Design (Fifth Conference), ASTM STP*, 674, 263, 1979.)

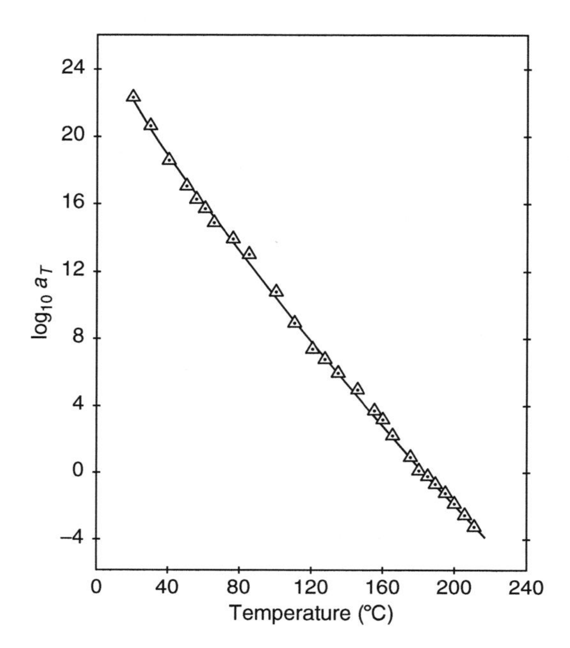

FIGURE 4.96 Shift factor a_T vs. temperature for the time–temperature superposition used in Figure 4.95. (After Yeow, Y.T., Morris, D.H., and Brinson, H.F., *Composite Materials: Testing and Design (Fifth Conference), ASTM STP*, 674, 263, 1979.)

The modulus curves below the reference temperature are shifted to the left; those above the reference temperatures are shifted to the right. The modulus curve thus obtained is called the *master curve* at the selected reference temperature and may extend over several decades of time (Figure 4.95).

5. Plot the horizontal displacements (which are equal to log a_T) as a function of the corresponding temperature. This shift factor curve (Figure 4.96) can now be used to determine the value of a_T at any other temperature within the range of test temperatures used.
6. The master curve at any temperature within the range of test temperatures used can be determined by multiplying the reference master curve with the appropriate a_T value obtained from the shift factor curve.

The time–temperature superposition method described earlier has been used by Yeow et al. [104] to generate master curves for T-300 carbon fiber–epoxy composites.

4.6.1.3 Schapery Creep and Recovery Equations

For the case of uniaxial loading at a constant temperature, Schapery and his coworker [105] have developed the following constitutive equation relating creep strain to the applied stress σ_0:

$$\varepsilon(t) = \left[g_0 D(0) + C \frac{g_1 g_2 t^n}{a_\sigma^n} \right] \sigma_0, \quad (4.55)$$

where

$D(0)$ = initial compliance

g_0, g_1, g_2, a_σ = stress-dependent constants

C, n = constants (independent of stress level)

During recovery after $t > t_1$, the recovery strain ε_r is

$$\varepsilon_r(t) = \frac{\Delta\varepsilon_1}{g_1} [(1 + a_\sigma \lambda)^n - (a_\sigma \lambda)^n], \quad (4.56)$$

where

$$\Delta\varepsilon_1 = \frac{g_1 g_2 C t_1^n}{a_\sigma^n} \sigma_0$$

$$\lambda = \frac{t - t_1}{t_1}$$

where t_1 is the time at which the stress is released. Detailed derivations of Equations 4.55 and 4.56 as well as the assumptions involved are given in Ref. [105].

At low stress levels, the constants g_0, g_1, g_2, and a_σ are equal to unity. Creep data obtained at low stress levels are used to determine the remaining three constants $D(0)$, C, and n. At high stress levels, the constants $g_0 \neq g_1 \neq g_2 \neq a_\sigma \neq 1$; however, the constants $D(0)$, C, and n do not change.

Determination of g_0, g_1, g_2, and a_σ requires creep and creep recovery tests at high stress levels. Lou and Schapery [105] have presented a graphical technique to reduce the high-stress creep data to determine these constants. Computer-based routines have been developed by Brinson and his coworkers and are described in Ref. [104]. Tuttle and Brinson [106] have also presented a scheme for predicting the long-term creep behavior of a general laminate. In this scheme, the lamination theory (see Chapter 3) and the Schapery theory are combined to predict the long-term creep compliance of the general laminate from the long-term creep compliances of individual laminas.

4.6.2 Stress Rupture

Stress rupture is defined as the failure of a material under sustained constant load. It is usually performed by applying a constant tensile stress to a specimen until it fractures completely. The time at which the fracture occurs is termed the lifetime or stress rupture time. The objective of this test is to determine a range of applied stresses and lifetimes within which the material can be considered "safe" in long-term static load applications.

The data obtained from stress rupture tests are plotted with the applied stress on a linear scale along the ordinate and the lifetime on a logarithmic scale along the abscissa. The functional relationship between the applied stress level and lifetime [107] is often represented as

$$\frac{\sigma}{\sigma_U} = A - B \log t \tag{4.57}$$

where σ_U is the static tensile strength, and A and B are constants. Since the stress-rupture data show a wide range of scatter, the constants are determined by the linear regression method.

Glass, Kevlar 49, and boron fibers and their composites exhibit failure by stress rupture. Carbon fibers, on the other hand, are relatively less prone to stress rupture failure. Chiao and his coworkers [108,109] have gathered the most extensive stress rupture data on epoxy-impregnated S-glass and Kevlar 49 fiber strands. For both materials, the lifetime at a stress level varied over a wide range. However, the rate of degradation under sustained tensile load was lower in Kevlar 49 strands than in S-glass strands. The data were analyzed using a two-parameter Weibull distribution in which the Weibull parameters α and σ_0 were the functions of both stress level σ and test temperature T. An example of the maximum likelihood estimates of lifetimes for Kevlar 49 strands is shown in Figure 4.97. To illustrate the use of this figure, consider the first percentile

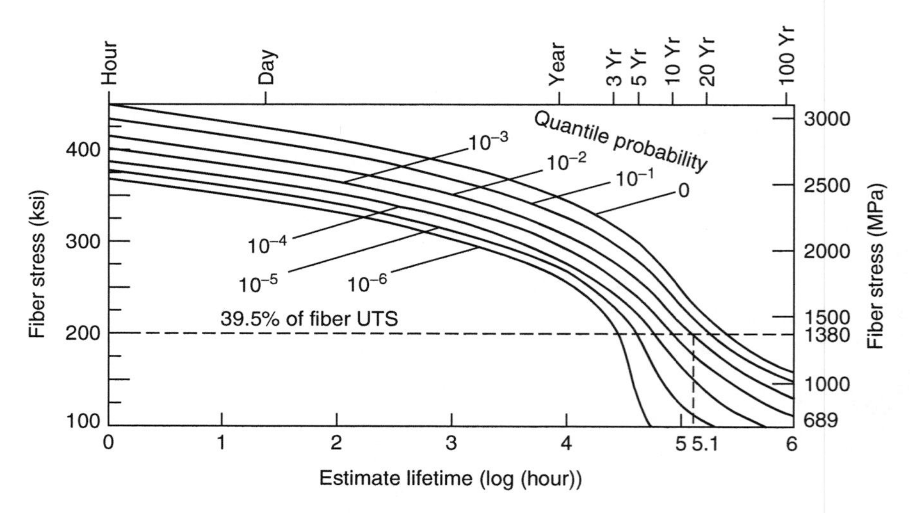

FIGURE 4.97 Maximum likelihood estimates of lifetimes for Kevlar 49–epoxy strands (under ambient conditions with ultraviolet light) for quantile probabilities. (After Glaser, R.E., Moore, R.L., and Chiao, T.T., *Compos. Technol. Rev.*, 6, 26, 1984.)

lifetime estimate of Kevlar 49 strands at 1380 MPa (200 ksi). Corresponding to the first percentile (10^{-2}) curve, $\log t \cong 5.1$, which gives a maximum likelihood estimate for the lifetime as $10^{5.1} = 126{,}000$ h = 14.4 years.

4.7 FRACTURE BEHAVIOR AND DAMAGE TOLERANCE

The fracture behavior of materials is concerned with the initiation and growth of critical cracks that may cause premature failure in a structure. In fiber-reinforced composite materials, such cracks may originate at manufacturing defects, such as microvoids, matrix microcracks, and ply overlaps, or at localized damages caused by in-service loadings, such as subsurface delaminations due to low-energy impacts and hole-edge delaminations due to static or fatigue loads. The resistance to the growth of cracks that originate at the localized damage sites is frequently referred to as the damage tolerance of the material.

4.7.1 Crack Growth Resistance

Many investigators [110–112] have used the linear elastic fracture mechanics (LEFM) approach for studying the crack growth resistance of fiber-reinforced composite materials. The LEFM approach, originally developed for metallic materials, is valid for crack growth with little or no plastic deformation at the crack tip. It uses the concept of stress intensity factor K_I, which is defined as

$$K_I = \sigma_o \sqrt{\pi a} Y, \tag{4.58}$$

where

K_I = Mode I stress intensity factor (Mode I refers to the opening mode of crack propagation due to an applied tensile stress normal to the crack plane)
σ_o = applied stress
a = crack length
Y = geometric function that depends on the crack length, crack location, and mode of loading

Equation 4.58 shows that the stress intensity factor increases with both applied stress and crack length. An existing crack in a material may propagate rapidly in an unstable manner (i.e., with little or no plastic deformation), when the K_I value reaches a critical level. The critical stress intensity factor, K_{Ic}, also called the fracture toughness, indicates the resistance of the material to unstable crack growth.

The critical stress intensity factor of metals is determined by standard test methods, such as ASTM E399. No such standard test method is currently available for fiber-reinforced composite materials. Most investigators have

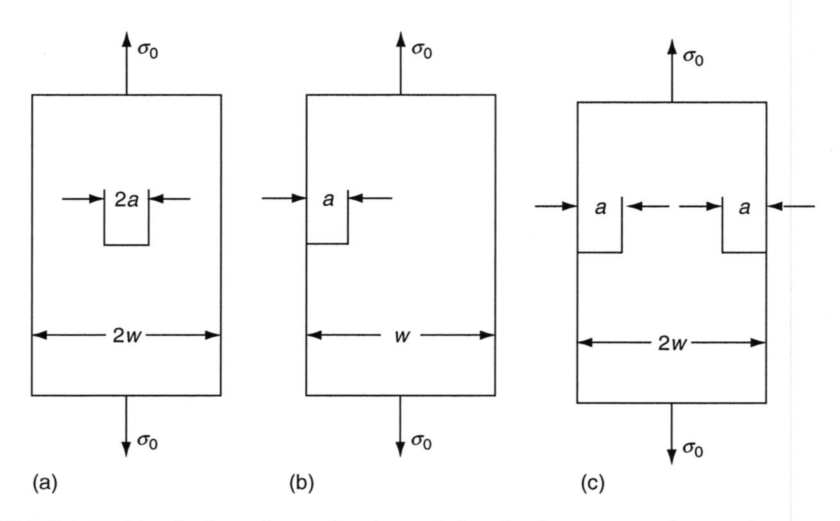

FIGURE 4.98 Notched specimens for determining the fracture toughness of a material: (a) center notched, (b) single-edge notched, and (c) double-edge notched.

used static tensile testing of prenotched straight-sided specimens to experimentally determine the stress intensity factor of fiber-reinforced composite laminates. Three types of specimens, namely, center-notched (CN), single-edge notched (SEN), and double-edge notched (DEN) specimens, are commonly used (Figure 4.98). Load vs. crack opening displacement records (Figure 4.99) obtained in these tests are initially linear. However, they become increasingly nonlinear or even discontinuous as irreversible subcritical damages appear in

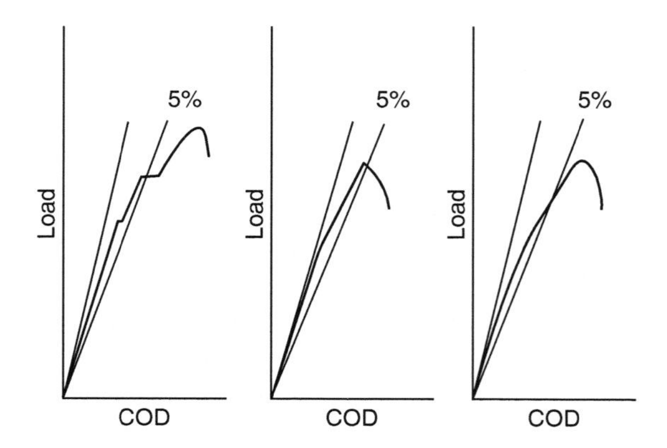

FIGURE 4.99 Typical load vs. crack opening displacement (COD) records obtained in tension testing of center-notched specimens. (After Harris, C.E. and Morris, D.H., *J. Compos. Technol. Res.*, 7, 77, 1985.)

the vicinity of the notch tip. Since the load–displacement curve deviates from linearity, it becomes difficult to determine the load at which crack growth begins in an unstable manner. The critical stress intensity factor calculated on the basis of the maximum load tends to depend on the notch size, laminate thickness, and laminate stacking sequence. Instead of using the maximum load, Harris and Morris [110] calculated the stress intensity factor on the basis of the load where a line drawn through the origin with 95% of the initial slope (i.e., 5% offset from the initial slope) intercepts the load–displacement curve. Physically, this stress intensity factor, denoted as K_5, has been associated with the onset of significant notch tip damage. Harris and Morris found the K_5 value to be relatively insensitive to the geometric variables, such as notch length, laminate thickness, and laminate stacking sequence (Table 4.22), and called it the fracture toughness of the composite material.

TABLE 4.22
Fracture Toughness of T-300 Carbon Fiber–Epoxy Laminates

Laminate Type	Fracture Toughness, K_5 Values	
	(MPa m$^{1/2}$)	(ksi in.$^{1/2}$)
$[0/90/\pm45]_S$	37.0	33.7
$[0/\pm45/90]_S$	33.5	30.5
$[\pm45/0/90]_S$	31.4	28.6
$[0/\pm45/90]_{8S}$	29.8	27.1
$[0/\pm45/90]_{15S}$	29.9	27.2
$[\pm45/0]_S$	32.3	29.4
$[0/\pm45]_S$	32.0	29.1
$[0/\pm45]_{10S}$	32.2	29.3
$[0/\pm45]_{20S}$	31.6	28.8
$[45/0/-45]_S$	27.5	25.0
$[0/90]_S$	27.8	25.3
$[0/90]_{16S}$	28.7	26.1
$[0/90]_{30S}$	26.6	24.2
$[\pm30/90]_S$	35.7	32.5
$[90/\pm30]_S$	36.4	33.1
$[30/90/-30]_S$	37.6	34.2
$[90/\pm30]_{16S}$	28.2	25.7
$[60/0/-60]_S$	26.4	24.0
$[0/\pm60]_S$	33.8	30.8
$[\pm60/0]_S$	35.9	32.7
$[0/\pm60]_S$	27.4	24.9

Source: Adapted from Harris, C.E. and Morris, D.H., *J. Compos. Technol. Res.*, 7, 77, 1985.

Harris and Morris [110,113] have also observed that the fracture process in continuous fiber laminates depends on the laminate type and laminate thickness. For example, in thin $[0/\pm45]_{nS}$ laminates, massive delaminations at the crack tip create uncoupling between the $+45°$ and $-45°$ plies, which is followed by an immediate failure of the laminate. The nonlinearity in the load–displacement diagram of this laminate ensues at or near the maximum load. As the laminate thickness increases, the thickness constraint provided by the outer layers prevents ply delaminations at the interior layers and the notched laminate strength is increased. In contrast, thin $[0/\pm45/90]_{nS}$ laminates develop minor delaminations as well as matrix microcracks at the crack tip at lower than the maximum load; however, the damage developed at the crack tip tends to relieve the stress concentration in its vicinity. As a result, the load–displacement diagram for $[0/\pm45/90]_{nS}$ laminates is more nonlinear and their notched laminate strength is also higher than that of $[0/\pm45]_{nS}$ laminates. With increasing thickness, the size of the crack tip damage in $[0/\pm45/90]_{nS}$ decreases and there is less stress relief in the crack tip region, which in turn lowers the value of their notched tensile strength.

In laminates containing randomly oriented fibers, crack tip damage contains matrix microcracks, fiber–matrix interfacial debonding, fiber breakage, and so on. This damage may start accumulating at load levels as low as 50%–60% of the maximum load observed in a fracture toughness test. Thus the load–displacement diagrams are also nonlinear for these laminates. From the load–displacement records for various initial crack lengths, Gaggar and Broutman [112] developed a crack growth resistance curve similar to that given in Figure 4.100. This curve can be used to predict the stress intensity factor at the point of unstable crack growth.

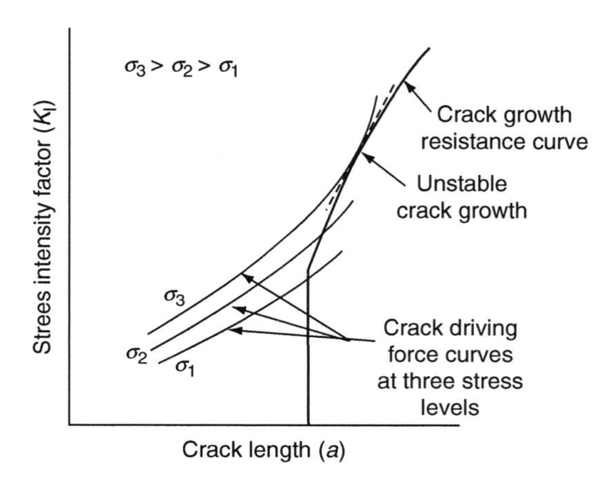

FIGURE 4.100 Crack growth resistance curve of a random fiber laminate. (After Gaggar, S.K. and Broutman, L.J., *J. Compos. Mater.*, 9, 216, 1975.)

4.7.2 Delamination Growth Resistance

Interply delamination is considered the most critical failure mode limiting the long-term fatigue life of certain composite laminates [114]. In general, delamination develops at the free edges of a laminate where adverse interlaminar stresses may exist owing to its particular stacking sequence. Once developed, the delaminated areas may grow steadily with increasing number of cycles, which in turn reduces the effective modulus of the laminate. The presence of delamination also causes a redistribution of stresses within the laminate, which may influence the initiation of fiber breakage in the primary load-bearing plies and reduce the fatigue life of the laminate. In recognition of this problem, a number of test methods [115] have been developed to measure the interlaminar fracture toughness of composite laminates.

The interlaminar fracture toughness, measured in terms of the critical strain energy release rate, is defined as the amount of strain energy released in propagating delamination by a unit length. This LEFM parameter is frequently used for comparing the resistance of various resin systems against the growth of delamination failure.

Delamination may occur in Mode I (opening mode or tensile mode of crack propagation), Mode II (sliding mode or in-plane shear mode of crack propagation), Mode III (tearing mode or antiplane shear mode of crack propagation), or a combination of these modes. The test methods used for determining the critical strain energy release rate in Mode I and Mode II delaminations are briefly described here. In Mode I, crack propagation occurs as the crack surfaces pull apart due to a normal stress perpendicular to the crack plane. In Mode II, crack propagation occurs as the crack surfaces slide over each other due to a shear stress parallel to the crack plane. The compliance method is used in the calculation of the strain energy release rate for each mode, which is related to the specimen compliance by the following equation:

$$G_{\mathrm{I}} \text{ or } G_{\mathrm{II}} = \frac{P^2}{2w}\frac{\mathrm{d}C}{\mathrm{d}a}, \tag{4.59}$$

where

P = applied load
w = specimen width
a = crack length (measured by a traveling microscope during the test)
C = specimen compliance (slope of the load–displacement curve for each crack length) (see Figure 4.102)
$\frac{\mathrm{d}C}{\mathrm{d}a}$ = slope of the compliance vs. crack length curve

The interlaminar fracture toughness, G_{Ic} in Mode I and G_{IIc} in Mode II, are calculated using the critical load P_{c} and $\frac{\mathrm{d}C}{\mathrm{d}a}$ corresponding to the crack length at the onset of delamination propagation.

4.7.2.1 Mode I Delamination

Two commonly used Mode I interlaminar fracture energy tests are the double-cantilever beam (DCB) test and the edge delamination tension (EDT) test.

DCB test: The DCB test is used for determining the strain energy release rate G_I for delamination growth under Mode I loading. It commonly uses a straight-sided 0° unidirectional specimen (Figure 4.101) in which an initial crack (delamination) is created by inserting a thin Teflon film (typically 0.013 mm thick) at the midplane before molding the laminate. Hinged metal tabs are bonded at the delaminated end of the specimen. Load is applied through the metal tabs until the initial crack grows slowly by a predetermined length. The specimen is unloaded and then reloaded until the new crack in the specimen grows slowly by another predetermined length. This process is repeated several times with the same specimen to obtain a series of load–displacement records, as shown in Figure 4.102. For each crack length, a specimen compliance value is calculated using the slope of the loading portion of the corresponding load–displacement record. A typical compliance vs. crack length curve is shown in Figure 4.103.

EDT test: The EDT test uses a straight-sided 11-ply $[(\pm 30)_2/90/\overline{90}]_S$ laminate that exhibits free-edge delaminations at both −30°/90° interfaces under a tensile load. It involves determining the laminate stiffness E_{LAM} and

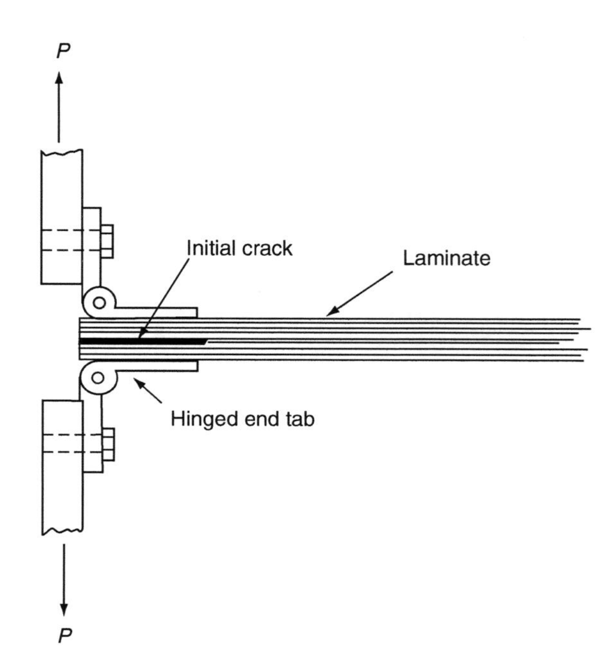

FIGURE 4.101 A double-cantilever beam (DCB) specimen.

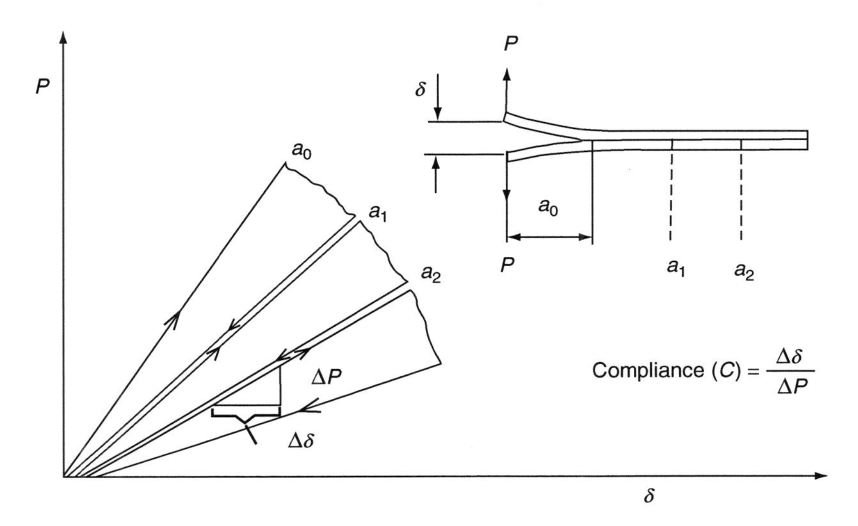

FIGURE 4.102 Typical load–displacement records obtained in a DCB test.

the normal strain ε_c at the onset of delamination from the stress–strain diagram (Figure 4.104) obtained in a static tension test of the $[(\pm 30)_2/90/\overline{90}]_S$ laminate. The critical strain energy release rate at the onset of delamination is then calculated using the following equation [116]:

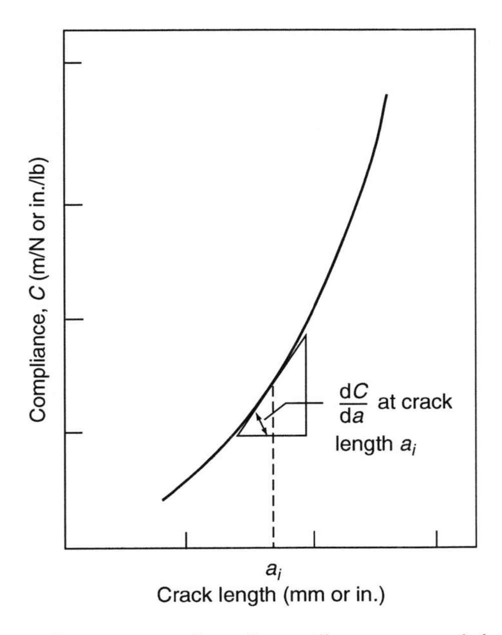

FIGURE 4.103 Schematic representation of compliance vs. crack length in a DCB test.

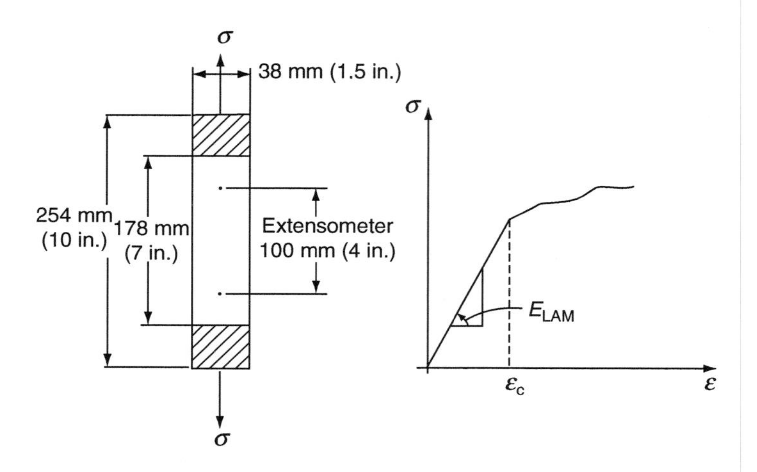

FIGURE 4.104 An edge delamination tension (EDT) test.

$$G_{\text{Ic}} = \frac{\varepsilon_{\text{c}}^2 t}{2}(E_{\text{LAM}} - E^*), \tag{4.60}$$

where

h = specimen thickness
ε_{c} = strain at the onset of delamination
E_{LAM} = initial laminate stiffness
E^* = $[8E_{(30)}+ + 3E_{(90)}]/11$

The E^* term in Equation 4.60 represents a simple rule of mixture calculation for the laminate stiffness after complete delamination has taken place at both $-30°/90°$ interfaces.

Both DCB and EDT tests are useful in qualitatively ranking various matrix materials for their role in delamination growth resistance of a laminate. Both tests produce comparable results (Figure 4.105). In general, the delamination growth resistance increases with increasing fracture toughness of the matrix; however, improving the fracture toughness of a matrix may not translate into an equivalent increase in the delamination growth of a laminate.

4.7.2.2 Mode II Delamination

End notched flexure tests are used to determine the interlaminar fracture toughness in Mode II delamination of [0] unidirectional laminates. There are two types of end notched flexure tests: (a) 3-point or 3-ENF test and (b) 4-point or 4-ENF test. The specimen and loading configurations for these two tests are shown in Figure 4.106. In both specimens, a starter crack is created at one end of the specimen by placing a thin Teflon film (typically 0.013 mm thick) at the

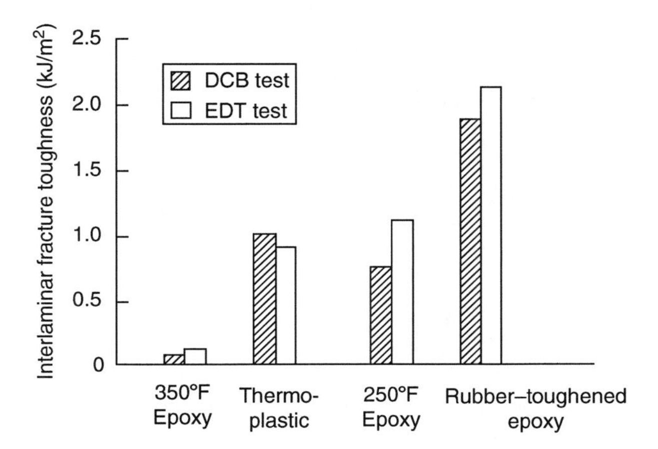

FIGURE 4.105 Comparison of the interlaminar fracture toughness of various resin systems. (After O'Brien, T.K., *Tough Composite Materials: Recent Developments*, Noyes Publications, Park Ridge, NJ, 1985.)

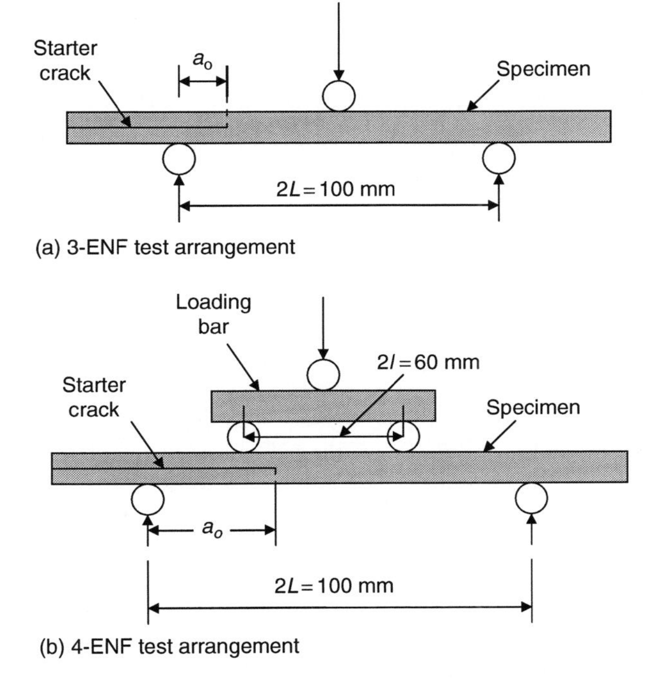

FIGURE 4.106 Schematic of the (a) 3-ENF and (b) 4-ENF specimens and tests.

midplane of the laminate before molding the laminate. The test is conducted at a constant displacement rate of the loading point and the crack growth is monitored. The load–displacement diagram is also recorded during the test.

In the 3-ENF test, it is difficult to obtain stable crack growth, and therefore, multiple specimens with different initial crack lengths are required to plot the compliance vs. crack length curve. On the other hand, the 4-ENF test produces a stable crack growth, and therefore, one specimen is sufficient to determine the Mode II strain energy release rate. In the 4-ENF test, the specimen is unloaded after every 2–3 mm stable crack growth and then reloaded until the new crack grows slowly by another 2–3 mm. As with the DCB specimen, the specimen compliance, C, is determined from the slope of the load–displacement curve corresponding to each new crack length. From the compliance vs. crack length curve dc/da is calculated, which is then used in Equation 4.59 for calculating G_{IIc}. Schuecker and Davidson [117] have shown that if crack length and compliance are measured accurately, the 3-ENF and 4-ENF tests yield similar G_{IIc}.

4.7.3 Methods of Improving Damage Tolerance

Damage tolerance of laminated composites is improved if the initiation and growth of delamination can be either prevented or delayed. Efforts to control delamination have focused on both improving the interlaminar fracture toughness and reducing the interlaminar stresses by means of laminate tailoring. Material and structural parameters that influence the damage tolerance are matrix toughness, fiber–matrix interfacial strength, fiber orientation, stacking sequence, laminate thickness, and support conditions. Some of these parameters have been studied by many investigators and are discussed in the following section.

4.7.3.1 Matrix Toughness

The fracture toughness of epoxy resins commonly used in the aerospace industry is 100 J/m^2 or less. Laminates using these resins have an interlaminar (Mode I delamination) fracture toughness in the range of 100–200 J/m^2. Increasing the fracture toughness of epoxy resins has been shown to increase the interlaminar fracture toughness of the composite. However, the relative increase in the interlaminar fracture toughness of the laminate is not as high as that of the resin itself.

The fracture toughness of an epoxy resin can be increased by adding elastomers (e.g., CTBN), reducing cross-link density, increasing the resin chain flexibility between cross-links, or a combination of all three (Table 4.23). Addition of rigid thermoplastic resins also improves its fracture toughness. Another alternative is to use a thermoplastic matrix, such as PEEK, PPS, PAI, and so on, which has a fracture toughness value in the range of 1000 J/m^2, 10-fold higher than that of conventional epoxy resins.

TABLE 4.23
Mode I Interlaminar Fracture Toughness as Influenced by the Matrix Composition

General Feature of the Base Resin	Matrix Properties: Tensile Strain-to-Failure (%)	Matrix Properties: G_{Ic} (J/m^2)	Composite Properties[a]: v_f (%)	Composite Properties[a]: Transverse Tensile Strain-to-Failure (%)	Composite Properties[a]: G_{Ic} (J/m^2)
Rigid TGMDA/DDS epoxy	1.34	70	76	0.632	190
Moderately cross-linked with rigid backbone between cross-links	1.96	167	54	0.699	335
Same as above plus CTBN rubber particles	3.10	730	60	0.580	1015
			69		520
			71		615
Epoxy with low cross-link density and soft backbone between cross-links	3.24	460	58	0.538	455
Same as above plus CTBN rubber particles	18.00	5000	57	0.538	1730

Source: Adapted from Jordan, W.M., Bradley, W.L., and Moulton, R.J., *J. Compos. Mater.*, 23, 923, 1989.

[a] 0° Unidirectional carbon fiber-reinforced epoxy composites.

4.7.3.2 Interleaving

A second approach of enhancing the interlaminar fracture toughness is to add a thin layer of tough, ductile polymer or adhesive between each consecutive plies in the laminate [118] (Figure 4.107). Although the resin-rich interleaves increase the interlaminar fracture toughness, fiber-dominated properties, such as tensile

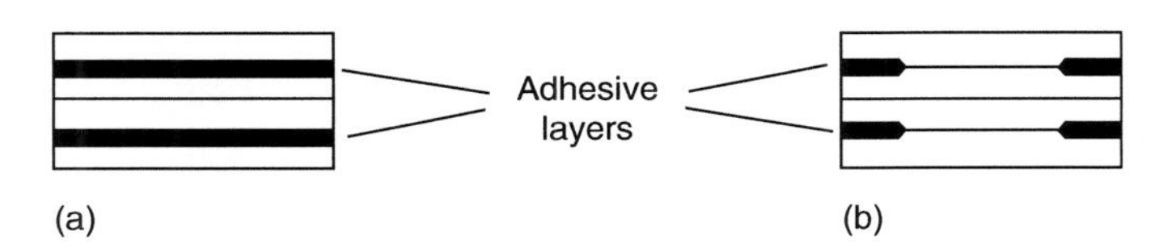

FIGURE 4.107 Interleaving with adhesive layers (a) along the whole laminate width and (b) along the free edges only.

strength and modulus, may decrease due to a reduction in overall fiber volume fraction.

4.7.3.3 Stacking Sequence

High interlaminar normal and shear stresses at the free edges of a laminate are created due to mismatch of Poisson's ratios and coefficients of mutual influence between adjacent layers. Changing the ply stacking sequence may change the interlaminar normal stress from tensile to compressive so that opening mode delamination can be suppressed. However, the growth of delamination may require the presence of an interrupted load path. For example, Lee and Chan [119] used discrete 90° plies at the midplanes of $[30/-30_2/30]_S$ and $[\pm35/0]_S$ laminates to reduce delamination in these laminates. Delamination was arrested at the boundaries of these discrete plies.

4.7.3.4 Interply Hybridization

This can also be used to reduce the mismatch of Poisson's ratios and coefficients of mutual influence between consecutive layers, and thus reduce the possibility of interply edge delamination. For example, replacing the 90° carbon fiber–epoxy plies in $[\pm45/0_2/90_2]_S$ AS-4 carbon fiber–epoxy laminates with 90° E-glass fiber–epoxy plies increases the stress level for the onset of edge delamination (due to interlaminar normal stress, σ_{zz}) from 324 to 655 MPa. The ultimate tensile strength is not affected, since it is controlled mainly by the 0° plies, which are carbon fiber–epoxy for both laminates [120].

4.7.3.5 Through-the-Thickness Reinforcement

Resistance to interlaminar delamination can be improved by means of through-the-thickness reinforcement that can be in the form of stitches, metallic wires and pins, or three-dimensional fabric structures.

Mignery et al. [121] used fine Kevlar thread to stitch the layers in $[\pm30/0]_S$, $[\pm30/90]_S$, and $[\pm45/0_2/90]_S$ AS-4 carbon fiber–epoxy laminates. Stitches parallel to the 0° direction were added after layup with an industrial sewing machine at a distance of 1.3–2.5 mm from the free edges in 32 mm wide test specimens. Although stitching did not prevent the occurrence of free-edge delamination in uniaxial tensile tests, it substantially reduced the rate of delamination growth into the interior of the latter two laminates. No visible edge delamination occurred in either unstitched or stitched $[\pm30/0]_S$ laminates before complete fracture.

Figure 4.108 shows the construction of a three-dimensional composite containing alternate 0/90 layers in the laminate plane (xy plane) and vertical through-the-thickness fibers interlocked with the in-plane layers. Gillespie and his coworkers [122] reported a 10-fold increase in the Mode I interlaminar

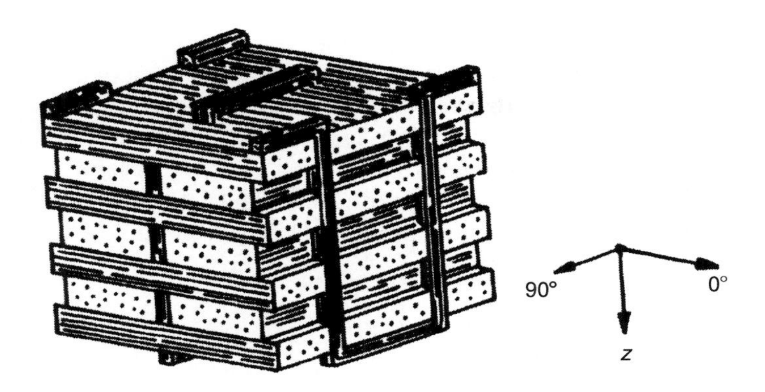

FIGURE 4.108 Construction of a three-dimensional laminate.

fracture toughness of such three-dimensional composites over the two-dimensional 0/90 laminates; however, the in-plane stiffness properties are decreased.

4.7.3.6 Ply Termination

High interlaminar stresses created by mismatching plies in a narrow region near the free edge are reduced if they are terminated away from the free edge. Chan and Ochoa [123] tension tested $[\pm 35/0/09]_S$ laminates in which the 90° layers were terminated at ~3.2 mm away from the free edges of the tension specimen and found no edge delamination up to the point of laminate failure. The ultimate tensile strength of the laminate with 90° ply terminations was 36% higher than the baseline laminate without ply termination. In the baseline laminate, free-edge delamination between the central 90° layers was observed at 49% of the ultimate load.

4.7.3.7 Edge Modification

Sun and Chu [124] introduced a series of narrow and shallow (1.6–3.2 mm deep) notches (Figure 4.109) along the laminate edges and observed a significant

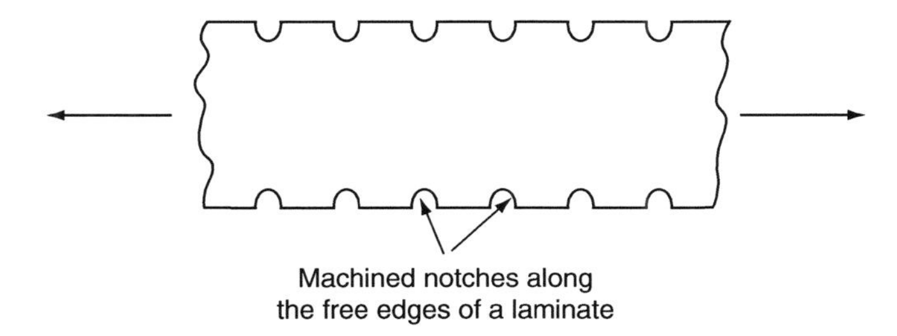

FIGURE 4.109 Edge notched laminate.

increase (25% or higher) in tensile failure load for laminates that are prone to interlaminar shear failure. Delamination was either eliminated or delayed in laminates that are prone to opening mode delamination, but there was no improvement in tensile failure load. The presence of notches disrupts the load path near the free edges and reduces the interlaminar stresses. However, they also introduce high in-plane stress concentration. Thus, suppression of delamination by edge notching may require proper selection of notch size and spacing.

REFERENCES

1. N.J. Pagano and J.C. Halpin, Influence of end constraint in the testing of anisotropic bodies, *J. Compos. Mater., 2*:18 (1968).
2. R.R. Rizzo, More on the influence of end constraints on off-axis tensile tests, *J. Compos. Mater., 3*:202 (1969).
3. B.E. Kaminski, Effects of specimen geometry on the strength of composite materials, *Analysis of the Test Methods for High Modulus Fibers and Composites, ASTM STP, 521*:181 (1973).
4. M.R. Piggott and B. Harris, Compression strength of carbon, glass and Kevlar 49 fibre reinforced polyester resins, *J. Mater. Sci., 15*:2523 (1980).
5. C.D. Shirrell, Variability in static strengths of sheet molding compounds (SMC), *Polym. Compos., 4*:172 (1983).
6. H.T. Hahn and S.W. Tsai, On the behavior of composite laminates after initial failure, *J. Compos. Mater., 8*:288 (1974).
7. P.A. Lagace, Nonlinear stress–strain behavior of graphite/epoxy laminates, *AIAA J., 23*:1583 (1985).
8. J.E. Masters and K.L. Reifsnider, An investigation of cumulative damage development in quasi-isotropic graphite/epoxy laminates, *Damage in Composite Materials, ASTM STP, 775*:40 (1982).
9. T. Watanabe and M. Yasuda, Fracture behavior of sheet molding compounds. Part 1: Under tensile load, *Composites, 13*:54 (1982).
10. J. Aveston and A. Kelly, Tensile first cracking strain and strength of hybrid composites and laminates, *Phil. Trans. R. Soc. Lond., A, 294*:519 (1980).
11. P.W. Manders and M.G. Bader, The strength of hybrid glass/carbon fiber composites. Part 1: Failure strain enhancement and failure mode, *J. Mater. Sci., 16*:2233 (1981).
12. J.M. Whitney, I.M. Daniel, and R.B. Pipes, *Experimental Mechanics of Fiber Reinforced Composite Materials*, Society for Experimental Mechanics, Brookfield Center, CT (1984).
13. C. Zweben, W.S. smith, and M.M. Wardle, Test methods for fiber tensile strength, composite flexural modulus, and properties of fabric-reinforced laminates, *Composite Materials: Testing and Design* (*Fifth Conference*), *ASTM STP, 674*:228 (1979).
14. R.E. Bullock, Strength ratios of composite materials in flexure and in tension, *J. Compos. Mater., 8*:200 (1974).
15. J.M. Whitney and M. Knight, The relationship between tensile strength and flexure strength in fiber-reinforced composites, *Exp. Mech., 20*:211 (1980).

16. Y.T. Yeow and H.F. Brinson, A comparison of simple shear characterization methods for composite laminates, *Composites, 9*:49 (1978).
17. C.C. Chiao, R.L. Moore, and T.T. Chiao, Measurement of shear properties of fibre composites. Part 1: Evaluation of test methods, *Composites, 8*:161 (1977).
18. C.C. Chamis and J.J. Sinclair, Ten-deg off-axis test for shear properties in fiber composites, *Exp. Mech., 17*:339 (1977).
19. D.E. Walrath and D.F. Adams, The Iosipescu shear test as applied to composite materials, *Exp. Mech., 23*:105 (1983).
20. D.W. Wilson, Evaluation of V-notched beam shear test through an interlaboratory study, *J. Compos. Technol. Res., 12*:131 (1990).
21. J.M. Whitney, D.L. Stansberger, and H.B. Howell, Analysis of the rail shear test—applications and limitations, *J. Compos. Mater., 5*:24 (1971).
22. R. Garcia, T.A. Weisshaar, and R.R. McWithey, An experimental and analytical investigation of the rail shear-test method as applied to composite materials, *Exp. Mech., 20*:273 (1980).
23. B.K. Daniels, N.K. Harakas, and R.C. Jackson, Short beam shear tests of graphite fiber composites, *Fibre Sci. Technol., 3*:187 (1971).
24. W.W. Stinchcomb, E.G. Henneke, and H.L. Price, Use of the short-beam shear test for quality control of graphite-polyimide laminates, *Reproducibility and Accuracy of Mechanical Tests, ASTM STP, 626*:96 (1977).
25. C.A. Berg, T. Tirosh, and M. Israeli, Analysis of the short beam bending of fiber reinforced composites, *Composite Materials: Testing and Design (Second Conference), ASTM STP, 497*:206 (1972).
26. L.L. Lorenzo and H.T. Hahn, Fatigue failure mechanisms in unidirectional composites, *Composite Materials—Fatigue and Fracture, ASTM STP, 907*: 210 (1986).
27. H.T. Hahn and R.Y. Kim, Fatigue behavior of composite laminates, *J. Compos. Mater., 10*:156 (1976).
28. B.D. Agarwal and J.W. Dally, Prediction of low-cycle fatigue behavior of GFRP: an experimental approach, *J. Mater. Sci., 10*:193 (1975).
29. H.T. Hahn and R.Y. Kim, Proof testing of composite materials, *J. Compos. Mater., 9*:297 (1975).
30. D.A. Riegner and B.A. Sanders, A characterization study of automotive continuous and random glass fiber composites, *Proceedings National Technical Conference*, Society of Plastics Engineers, November (1979).
31. R.A. Heimbuch and B.A. Sanders, Mechanical properties of automotive chopped fiber reinforced plastics, *Composite Materials in the Automobile Industry*, ASME, New York (1978).
32. P.K. Mallick, Fatigue characteristics of high glass content sheet molding compound (SMC) materials, *Polym. Compos., 2*:18 (1981).
33. R.B. Pipes, Interlaminar shear fatigue characteristics of fiber reinforced composite materials, *Composite Materials: Testing and Design (Third Conference), ASTM STP, 546*:419 (1974).
34. C.K.H. Dharan, Interlaminar shear fatigue of pultruded graphite fibre–polyester composites, *J. Mater. Sci., 13*:1243 (1978).
35. M.J. Owen and S. Morris, Some interlaminar-shear fatigue properties of carbon-fibre-reinforced plastics, *Plast. Polym., 4*:209 (1972).

36. D.W. Wilson, Characterization of the interlaminar shear fatigue properties of SMC-R50, Technical Report No. 80–05, Center for Composite Materials, University of Delaware, Newark, DE (1980).
37. D.C. Phillips and J.M. Scott, The shear fatigue of unidirectional fibre composites, *Composites, 8*:233 (1977).
38. J.D. Conners, J.F. Mandell, and F.J. McGarry, Compressive fatigue in glass and graphite reinforced composites, *Proceedings 34th Annual Technical Conference*, Society of the Plastics Industry (1979).
39. Z. Hashin and A. Rotem, A fatigue failure criterion for fiber reinforced materials, *J. Compos. Mater., 7*:448 (1973).
40. K.H. Boller, Fatigue characteristics of RP laminates subjected to axial loading, *Mod. Plast., 41*:145 (1964).
41. J.W. Davis, J.A. McCarthy, and J.N. Schrub, The fatigue resistance of reinforced plastics, *Materials in Design Engineering*, p. 87 (1964).
42. M.J. Owen and R.G. Rose, Polyester flexibility versus fatigue behavior of RP, *Mod. Plast., 47*:130 (1970).
43. T. Tanimoto and S. Amijima, Progressive nature of fatigue damage of glass fiber reinforced plastics, *J. Compos. Mater., 9* (1975).
44. C.K.H. Dharan, Fatigue failure in graphite fibre and glass fibre-polymer composites, *J. Mater. Sci., 10*:1665 (1975).
45. K.H. Boller, Effect of tensile mean stresses on fatigue properties of plastic laminates reinforced with unwoven glass fibers, U.S. Air Force Materials Laboratory, Report No. ML-TDR-64–86, June (1964).
46. S.V. Ramani and D.P. Williams, Axial fatigue of $[0/\pm30]_{6S}$ graphite/epoxy, *Failure Mode in Composites III*, AIME, p. 115 (1976).
47. T.R. Smith and M.J. Owen, Fatigue properties of RP, *Mod. Plast., 46*:124 (1969).
48. J.W. Dally and L.J. Broutman, Frequency effects on the fatigue of glass reinforced plastics, *J. Compos. Mater., 1*:424 (1967).
49. J.F. Mandell and U. Meier, Effects of stress ratio, frequency, and loading time on the tensile fatigue of glass-reinforced epoxy, *Long-Term Behavior of Composites, ASTM STP, 813*:55 (1983).
50. C.T. Sun and W.S. Chan, Frequency effect on the fatigue life of a laminated composite, *Composite Materials, Testing and Design (Fifth Conference), ASTM STP, 674*:418 (1979).
51. C.R. Saff, Effect of load frequency and lay-up on fatigue life of composites, *Long-Term Behavior of Composites, ASTM STP, 813*:78 (1983).
52. K.L. Reifsnider, W.W. Stinchcomb, and T.K. O'Brien, Frequency effects on a stiffness-based fatigue failure criterion in flawed composite specimens, *Fatigue of Filamentary Composite Materials, ASTM STP, 636*: 171 (1977).
53. J.H. Underwood and D.P. Kendall, Fatigue damage in notched glass–epoxy sheet, *Proceedings International Conference on Composite Materials*, Vol. 2, AIME, p. 1122 (1975).
54. C.K.H. Dharan, Fatigue failure mechanisms in a unidirectionally reinforced composite material, *Fatigue of Composite Materials, ASTM STP, 569*:171 (1975).
55. H.C. Kim and L.J. Ebert, Axial fatigue failure sequence and mechanisms in unidirectional fiberglass composites, *J. Compos. Mater., 12*:139 (1978).
56. R. Talreja, Fatigue of composite materials: damage mechanisms and fatigue-life diagrams, *Proc. R. Soc. Lond., A, 378*:461 (1981).

57. L.J. Broutman and S. Sahu, Progressive damage of a glass reinforced plastic during fatigue, *Proceedings 20th Annual Technical Conference*, Society of the Plastics Industry (1969).
58. K. Reifsnider, K. Schultz, and J.C. Duke, Long-term fatigue behavior of composite materials, *Long-Term Behavior of Composites, ASTM STP, 813*:136 (1983).
59. S.S. Wang and E.S.M. Chim, Fatigue damage and degradation in random short-fiber SMC composite, *J. Compos. Mater., 17*:114 (1983).
60. T.R. Smith and M.J. Owen, Progressive nature of fatigue damage in RP, *Mod. Plast., 46*:128 (1969).
61. P.K. Mallick, A fatigue failure warning method for fiber reinforced composite structures, *Failure Prevention and Reliability—1983* (G. Kurajian, ed.), ASME, New York (1983).
62. H.W. Whitworth, Cumulative damage in composites, *ASME J. Eng. Mater. Technol., 112*:358 (1990).
63. L.J. Broutman and S. Sahu, A new theory to predict cumulative fatigue damage in fiberglass reinforced plastics, *Composite Materials: Testing and Design (Second Conference), ASTM STP, 497*:170 (1972).
64. J.N. Yang, Fatigue and residual strength degradation for graphite/epoxy composites under tension–compression cyclic loadings, *J. Compos. Mater., 12*:19 (1978).
65. M.G. Bader and R.M. Ellis, The effect of notches and specimen geometry on the pendulum impact strength of uniaxial CFRP, *Composites, 6*:253 (1974).
66. R.H. Toland, Failure modes in impact-loaded composite materials, *Symposium on Failure Modes in Composites*, AIME Spring Meeting, May (1972).
67. P.K. Mallick and L.J. Broutman, Static and impact properties of laminated hybrid composites, *J. Test. Eval., 5*:190 (1977).
68. J.L. Perry and D.F. Adams, Charpy impact experiments on graphite/epoxy hybrid composites, *Composites, 7*:166 (1975).
69. N.L. Hancox, Izod impact testing of carbon-fibre-reinforced plastics, *Composites, 3*:41 (1971).
70. M.G. Bader, J.E. Bailey, and I. Bell, The effect of fibre–matrix interface strength on the impact and fracture properties of carbon-fibre-reinforced epoxy resin composites, *J. Phys. D: Appl. Phys., 6*:572 (1973).
71. P. Yeung and L.J. Broutman, The effect of glass-resin interface strength on the impact strength of fiber reinforced plastics, *Polym. Eng. Sci., 18*:62 (1978).
72. P.K. Mallick and L.J. Broutman, Impact properties of laminated angle ply composites, *Eng. Fracture Mech., 8*:631 (1976).
73. G.R. Sidey and F.J. Bradshaw, Some investigations on carbon-fibre-reinforced plastics under impact loading, and measurements of fracture energies, *Carbon Fibres: Their Composites and Applications*, Plastics Institute, London, p. 208 (1971).
74. M.D. Rhodes, J.G. Williams, and J.H. Starnes, Jr., Low-velocity impact damage in graphite-fiber reinforced epoxy laminates, *Proceedings 34th Annual Technical Conference*, Society of the Plastics Industry (1979).
75. R.L. Ramkumar, Effect of low-velocity impact damage on the fatigue behavior of graphite/epoxy laminates, *Long-Term Behavior of Composites, ASTM STP, 813*:116 (1983).
76. J. Morton and E.W. Godwin, Impact response of tough carbon fibre composites, *Compos. Struct., 13*:1 (1989).

77. G.E. Husman, J.M. Whitney, and J.C. Halpin, Residual strength characterization of laminated composites subjected to impact loading, *Foreign Object Impact Damage to Composites, ASTM STP, 568*:92 (1975).
78. G. Kretsis and F.L. Matthews, The strength of bolted joints in glass fibre/epoxy laminates, *Composites, 16*:92 (1985).
79. P.K. Mallick and R.E. Little, Pin bearing strength of fiber reinforced composite laminates, *Proceedings Advanced Composites*, American Society for Metals (1985).
80. T.A. Collings, The strength of bolted joints in multi-directional cfrp laminates, *Composites, 8*:43 (1977).
81. W.J. Quinn and F.L. Matthews, The effect of stacking sequence on the pin-bearing strength in glass fibre reinforced plastic, *J. Compos. Mater., 11*:139 (1977).
82. S. Nilsson, Increasing strength of graphite/epoxy bolted joints by introducing an adhesively bonded metallic insert, *J. Compos. Mater., 23*:641 (1989).
83. J.H. Stockdale and F.L. Matthews, The effect of clamping pressure on bolt bearing loads in glass fibre-reinforced plastics, *Composites, 7*:34 (1976).
84. J.R. Strife and K.M. Prewo, The thermal expansion behavior of unidirectional and bidirectional Kevlar/epoxy composites, *J. Compos. Mater., 13*:264 (1979).
85. S.F.H. Parker, M. Chandra, B. Yates, M. Dootson, and B.J. Walters, The influence of distribution between fibre orientations upon the thermal expansion characteristics of carbon fibre-reinforced plastics, *Composites, 12*:281 (1981).
86. J.R. Kerr and J.F. Haskins, Effects of 50,000 h of thermal aging on graphite/epoxy and graphite/polyimide composites, *AIAA J., 22*:96 (1982).
87. F.E. Devine, Polyester moulding materials in automotive underbonnet environments, *Composites, 14*:353 (1983).
88. C.H. Shen and G.S. Springer, Moisture absorption and desorption of composite materials, *J. Compos. Mater., 10*:2 (1976).
89. G.S. Springer, B.A. Sanders, and R.W. Tung, Environmental effects on glass fiber reinforced polyester and vinyl ester composites, *J. Compos. Mater., 14*:213 (1980).
90. W.W. Wright, The effect of diffusion of water into epoxy resins and their carbon-fibre reinforced composites, *Composites, 12*:201 (1981).
91. O. Gillat and L.J. Broutman, Effect of an external stress on moisture diffusion and degradation in a graphite-reinforced epoxy laminate, *Advanced Composite Materials—Environmental Effects, ASTM STP, 658*:61 (1978).
92. G. Marom and L.J. Broutman, Moisture in epoxy resin composites, *J. Adhes., 12*:153 (1981).
93. E.L. McKague, Jr., J.E. Halkias, and J.D. Reynolds, Moisture in composites: the effect of supersonic service on diffusion, *J. Compos. Mater., 9*:2 (1975).
94. M.J. Adamson, A conceptual model of the thermal-spike mechanism in graphite/epoxy laminates, *Long-Term Behavior of Composites, ASTM STP, 813*:179 (1983).
95. H.T. Hahn, Hygrothermal damage in graphite/epoxy laminates, *J. Eng. Mater. Technol., 109*:1 (1987).
96. C.H. Shen and G.S. Springer, Effects of moisture and temperature on the tensile strength of composite materials, *J. Compos. Mater., 11*:2 (1977).
97. C.H. Shen and G.S. Springer, Environmental effects on the elastic moduli of composite materials, *J. Compos. Mater., 11*:250 (1977).
98. O.K. Joshi, The effect of moisture on the shear properties of carbon fibre composites, *Composites, 14*:196 (1983).
99. C.J. Jones, R.F. Dickson, T. Adam, H. Reiter, and B. Harris, Environmental fatigue of reinforced plastics, *Composites, 14*:288 (1983).

100. P.T. Curtis and B.B. Moore, The effects of environmental exposure on the fatigue behavior of CFRP laminates, *Composites, 14*:294 (1983).
101. C.C. Chamis, R.R. Lark, and J.H. Sinclair, Integrated theory for predicting the hygrothermomechanical response of advanced composite structural components, *Advanced Composite Material—Environmental Effects, ASTM STP, 658*:160 (1978).
102. J.B. Sturgeon, Creep of fibre reinforced thermosetting resins, *Creep of Engineering Materials* (C.D. Pomeroy, ed.), Mechanical Engineering Publishing Ltd., London (1978).
103. J.S. Cartner, W.I. Griffith, and H.F. Brinson, The viscoelastic behavior of composite materials for automotive applications, *Composite Materials in the Automotive Industry* (S.V. Kulkarni, C.H. Zweben, and R.B. Pipes, eds.), American Society of Mechanical Engineers, New York (1978).
104. Y.T. Yeow, D.H. Morris, and H.F. Brinson, Time–temperature behavior of a unidirectional graphite/epoxy composite, *Composite Materials: Testing and Design (Fifth Conference), ASTM STP, 674*:263 (1979).
105. Y.C. Lou and R.A. Schapery, Viscoelastic characterization of a non-linear fiber-reinforced plastic, *J. Compos. Mater., 5*:208 (1971).
106. M.E. Tuttle and H.F. Brinson, Prediction of the long-term creep compliance of general composite laminates, *Exp. Mech., 26*:89 (1986).
107. M.G. Phillips, Prediction of long-term stress-rupture life for glass fibre-reinforced polyester composites in air and in aqueous environments, *Composites, 14*:270 (1983).
108. R.E. Glaser, R.L. Moore, and T.T. Chiao, Life estimation of an S-glass/epoxy composite under sustained tensile loading, *Compos. Technol. Rev., 5*:21 (1983).
109. R.E. Glaser, R.L. Moore, and T.T. Chiao, Life estimation of aramid/epoxy composites under sustained tension, *Compos. Technol. Rev., 6*:26 (1984).
110. C.E. Harris and D.H. Morris, A damage tolerant design parameter for graphite/epoxy laminated composites, *J. Compos. Technol. Res., 7*:77 (1985).
111. J.F. Mandell, S.S. Wang, and F.J. McGarry, The extension of crack tip damage zones in fiber reinforced plastic laminates, *J. Compos. Mater., 9*:266 (1975).
112. S.K. Gaggar and L.J. Broutman, Crack growth resistance of random fiber composites, *J. Compos. Mater., 9*:216 (1975).
113. C.E. Harris and D.H. Morris, Role of delamination and damage development on the strength of thick notched laminates, *Delamination and Debonding of Materials, ASTM STP, 876*:424 (1985).
114. D.J. Wilkins, J.R. Eisenmann, R.A. Camin, W.S. Margolis, and R.A. Benson, Characterizing delamination growth in graphite–epoxy, *Damage in Composite Materials, ASTM STP, 775*:168 (1980).
115. D.F. Adams, L.A. Carlsson, and R.B. Pipes, *Experimental Characterization of Advanced Composite Materials*, 3rd Ed., CRC Press, Boca Raton, FL (2003).
116. T.K. O'Brien, Characterization of delamination onset and growth in a composite laminate, *Damage in Composite Materials, ASTM STP, 775*:140 (1982).
117. C. Schuecker and B.D. Davidson, Evaluation of the accuracy of the four-point end-notched flexure test for mode II delamination toughness determination, *Compos. Sci. Tech., 60*:2137 (2000).
118. O. Ishai, H. Rosenthal, N. Sela, and E. Drukker, Effect of selective adhesive interleaving on interlaminar fracture toughness of graphite/epoxy composite laminates, *Composites, 19*:49 (1988).
119. E.W.Y. Lee and W.S. Chan, Delamination arrestment by discretizing the critical ply in a laminate, AIAA Paper No. 89-1403, *Amer. Inst. Aero. Astro.*, 1989.

120. C.T. Sun, Intelligent tailoring of composite laminates, *Carbon*, *27*:679 (1989).
121. L.A. Mignery, T.M. Tan, and C.T. Sun, The use of stitching to suppress delamination in laminated composites, *Delamination and Debonding, ASTM STP, 876*:371 (1985).
122. J.H. Byun, J.W. Gillespie, and T.W. Chou, Mode I delamination of a three-dimensional fabric composite, *J. Compos. Mater., 24*:497 (1990).
123. W.S. Chan and O.O. Ochoa, Suppression of edge delamination by terminating a critical ply near edges in composite laminates, *Key Engineering Materials Series*, Vol. 37, Transtech Publications, p. 285 (1989).
124. C.T. Sun and G.D. Chu, Reducing free edge effect on laminate strength by edge modification, *J. Compos. Mater., 25*:142 (1991).

PROBLEMS

P4.1. The load–strain data obtained in a tension test of a unidirectional carbon fiber–epoxy composite are given in the following table. Specimen dimensions are length = 254 mm, width = 12.7 mm, and thickness = 1.4 mm.

1. Determine the tensile modulus and Poisson's ratio for each fiber orientation
2. Using Equation 3.45, determine the shear modulus, G_{12} of this material and then verify the validity of Equation 3.46 for the 45° orientation

	Load (N)			Transverse Strain, %		
Axial Strain, %	**0°**	**45°**	**90°**	**0°**	**45°**	**90°**
0.05	2130	130	67	−0.012	−0.00113	−0.0004
0.10	4270	255	134	−0.027	−0.0021	−0.001
0.15	6400	360	204	−0.041	−0.0029	−0.0014
0.20	8620	485	333	−0.054	−0.0038	−0.0019
0.25	—	565	396	—	−0.0048	−0.0025

P4.2. The following tensile modulus (E_{xx}) values were calculated from the tensile stress–strain diagrams of 30° carbon fiber–epoxy off-axis specimens:

L/w	E_{xx}(10^6 psi)
2	2.96
4	2.71
8	2.55

Using the following material properties, calculate the corrected tensile modulus values at each L/w and compare them with the theoretical modulus: $E_{11} = 20 \times 10^6$ psi, $E_{22} = 1 \times 10^6$ psi, $\nu_{12} = 0.25$, and $G_{12} = 0.6 \times 10^6$ psi.

P4.3. The following longitudinal tensile strength data (in MPa) were obtained for a $[0/\pm45/90]_S$ E-glass fiber–epoxy laminate: 520.25, 470.27, 457.60, 541.18, 566.35, 489.82, 524.55, 557.87, 490.00, 498.99, 496.95, 510.84, and 558.76.

(a) Determine the average tensile strength, the standard deviation, and coefficient of variation
(b) Determine the Weibull parameters for the given strength distribution
(c) Using the Weibull parameters in (b), determine the mean strength for the material

P4.4. The tensile stress–strain diagram for a $[(0/90)_2/0/\overline{90}]_S$ E-glass fiber–epoxy laminate is shown in the figure. Determine the initial modulus, the stress at the knee point, the secondary modulus, and the failure strength for the laminate. Compare these experimental values with the theoretical estimates.

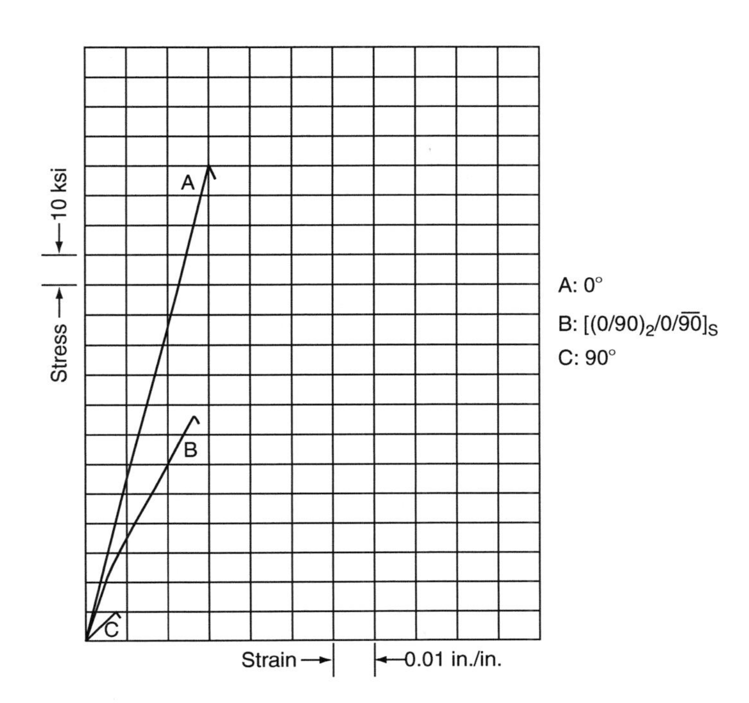

P4.5. Tensile stress–strain diagram of a $[0/90_4]_S$ AS-4 carbon fiber–epoxy laminate is shown in the figure. The longitudinal modulus and transverse modulus of a 0° unidirectional laminate of the same material are 142 and 10.3 GPa, respectively.

1. Determine the initial axial modulus of the $[0/90_4]_S$ laminate and compare it with the theoretical value. How would this value change if the 90° layers are at the outside or the laminate construction is changed to $[0_2/90_3]_S$?
2. The knee in the stress–strain diagram is at a strain of 0.005 mm/mm. However, the ultimate longitudinal and transverse strains of the 0° unidirectional laminate are at 0.0146 and 0.006 mm/mm, respectively. Explain what might have caused a lower strain at the knee
3. Describe the reason for the nonlinear portion of the stress–strain diagram

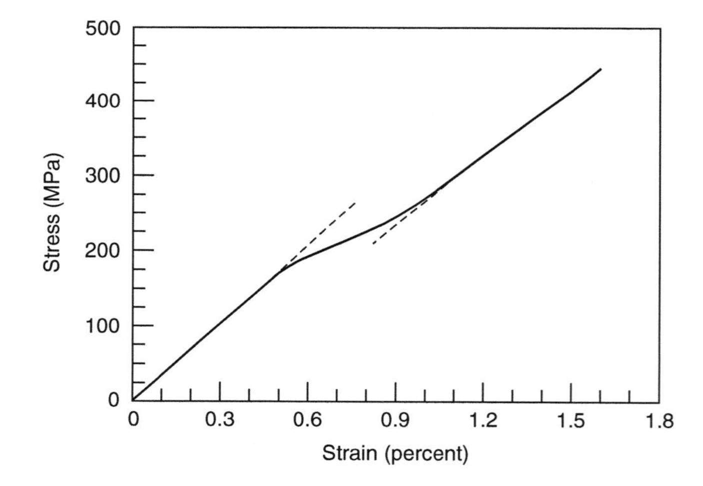

P4.6. The longitudinal compressive strength of a unidirectional 0° carbon fiber–epoxy laminate is estimated as 190 ksi. Using the Euler buckling formula for homogeneous materials, design the minimum thickness of a Celanese compression specimen that will be required to avoid lateral buckling.

P4.7. The following compressive strength data were obtained for a 0° unidirectional surface-treated high-strength carbon fiber–epoxy composite tested in the longitudinal (fiber) direction.* The composite specimens failed by shear cracks at 45° to the axis of loading.

* N.L. Hancox, The compression strength of unidirectional carbon fibre reinforced plastic, *J. Mater. Sci., 10*:234 (1975).

v_f (%)	σ_{LCU} (MPa)
10	325
20	525
30	712
40	915
50	1100
60	1250
70	1750

Plot the data and verify that the compressive strength is linear with v_f. Assuming that the rule of mixture applies, determine the longitudinal compressive strength of the carbon fiber. How may the compressive strength and failure mode of the composite be affected by matrix strength and fiber surface treatment?

P4.8. The following tensile stress–strain values were obtained in uniaxial tensile testing of a $[\pm 45]_S$ laminate:

σ_{xx} (MPa)	ε_{xx} (mm/mm)	ε_{yy} (mm/mm)
27.5	0.001	−0.00083
54.4	0.002	−0.00170
82.7	0.003	−0.00250
96.5	0.004	−0.00340
115.7	0.005	−0.00700
132.5	0.006	−0.00811
161.0	0.007	−0.00905
214.0	0.014	−0.01242

Plot the data and determine the E_{xx} and ν_{xy} of the laminate. In addition, reduce the data to plot τ_{12} vs. γ_{12} for the material and determine G_{12}.

P4.9. Using the stress and strain transformation Equations 3.30 and 3.31, verify Equations 4.16 and 4.17 for the 10° off-axis test shown in Figure 4.28.

P4.10. The following table* gives the stress–strain data for $[0/90/\pm 45]_S$ and $[0/90]_{2S}$ AS-4 carbon fiber–epoxy laminates obtained in a three-rail shear test. Plot the data and determine the shear modulus for each laminate. Compare these values with those predicted by the lamination

* S. Tan and R. Kim, Fracture of composite laminates containing cracks due to shear loading, *Exp. Mech.*, *28*:364 (1988).

theory based on the following material properties: $E_{11} = 143.92$ GPa, $E_{22} = 11.86$ GPa, $G_{12} = 6.68$ GPa, and $\nu_{12} = 0.326$.

	Shear Stress (MPa)	
Shear Strain (%)	**$[0/90/\pm45]_S$**	**$[0/90]_{2S}$**
0.2	40	13.6
0.4	91	24.5
0.6	128.6	37
0.8	155	47.3
1.0	178	54.5
1.2	208	62
1.4	232	65.4
1.6	260	70

P4.11. The following figure shows the schematic of an asymmetric four-point bend (AFPB) test developed for measuring the shear properties of composite laminates. As in the Iosipescu shear test, it uses a V-notched beam specimen. Derive an expression for the shear stress at the center of the specimen and compare it with that in the Iosipescu shear test.

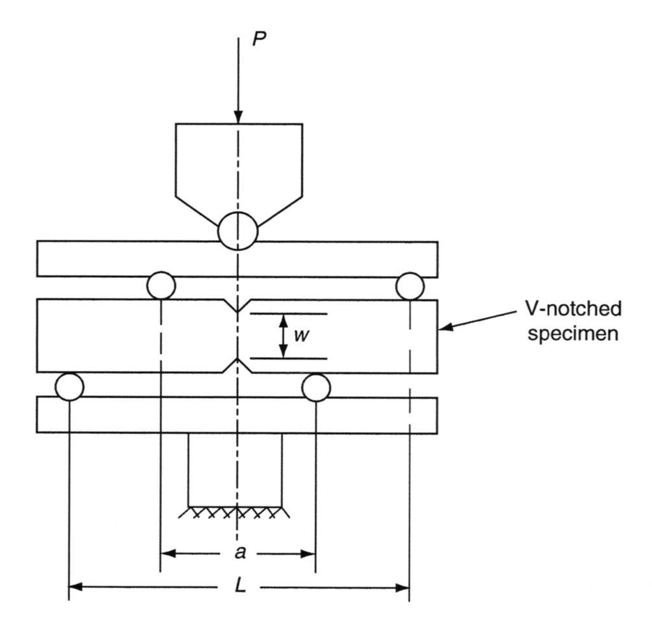

P4.12. Torsion of a thin-walled tube is considered the best method of creating a pure shear stress state in a material. Describe some of the practical problems that may arise in the torsion test of a laminated composite tube.

P4.13. Using Equation 4.12, develop a correction factor for the modulus measurements in a three-point flexure test with a small span–thickness ratio and plot the ratio of corrected modulus and measured modulus as a function of the span–thickness ratio.

P4.14. Using the homogeneous beam theory, develop equations for flexural strength and modulus calculations from the load–deflection diagram in a four-point static flexure test.

P4.15. Using the lamination theory, develop the flexural load–deflection diagram (up to the point of first failure) of a sandwich hybrid beam containing 0° T-300 carbon fiber-reinforced epoxy in the two outer layers and 0° E-glass fiber-reinforced epoxy in the core. Describe the failure mode expected in flexural loading of such a beam.

P4.16. Unidirectional 0° Kevlar 49 composites exhibit a linear stress–strain curve in a longitudinal tension test; however, their longitudinal compressive stress–strain curve is similar to that of an elastic, perfectly plastic metal (see the figure). Furthermore, the compressive proportional limit for a Kevlar 49 composite is lower than its tensile strength. Explain how these two behaviors may affect the stress distribution across the thickness of a unidirectional Kevlar 49 composite beam as the transversely applied load on the beam is increased. Estimate the transverse load at which the flexural load–deflection diagram of a Kevlar 49 beam becomes nonlinear. (*Hint*: Assume that the strain distribution through the thickness of the beam remains linear at all load levels.)

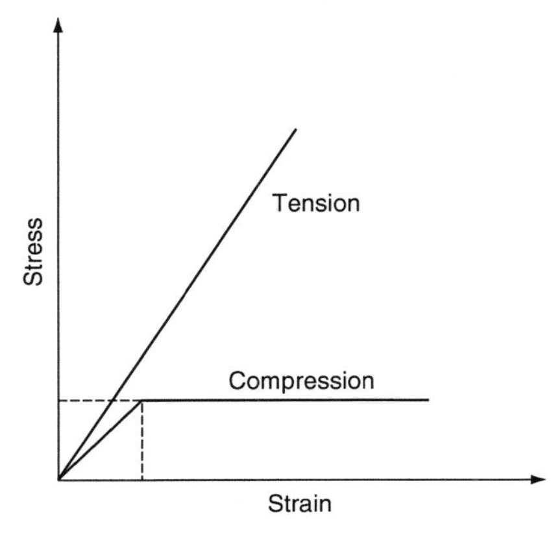

P4.17. The flexural strength of a unidirectional 0° E-glass–vinyl ester beam is 95 ksi, and the estimated ILSS for the material is 6 ksi. Determine the maximum span–thickness ratio for a short-beam shear specimen in which interlaminar shear failure is expected.

P4.18. Some investigators have proposed using a four-point flexure test for ILSS measurements. Using the homogeneous beam theory, develop equations for ILSS in a four-point flexure test and discuss the merits of such a test over a three-point short-beam shear test.

P4.19. Using Table 4.8, compare the fatigue strengths of T-300 carbon–epoxy and E-glass–epoxy composites at 10^4 and 10^7 cycles.

P4.20. The Weibull parameters for the tension–tension fatigue life distribution of a $[0/90/\pm45]_S$ T-300 carbon fiber–epoxy composite are as follows: at $\sigma_{max} = 340$ MPa, $L_0 = 1{,}000{,}000$, and $\alpha_f = 1.6$. At $\sigma_{max} = 410$ MPa, $L_0 = 40{,}000$, and $\alpha_f = 1.7$.

1. Determine the mean fatigue lives at 340 and 410 MPa
2. What is the expected probability of surviving 50,000 cycles at 410 MPa?

P4.21. Fatigue lives (numbers of cycles to failure) of twelve 0° Kevlar 49 fiber-reinforced composite specimens in tension–tension cycling at 90% UTS, $R = 0.1$, 0.5 Hz frequency, and 23°C are reported as 1,585; 25; 44,160; 28,240; 74,140; 47; 339,689; 50,807; 320,415; 865; 5,805; and 4,930. Plot the Weibull distribution curve for these fatigue lives and determine the Weibull parameters for this distribution.

P4.22. For the major portion of the fatigue life of a composite laminate, the instantaneous modulus can be modeled as

$$E = E_0(1 - c \log N)$$

where

E_0 = initial modulus at $N = 1$
N = number of fatigue cycles
c = an experimentally determined damage-controlling factor

Using this model, determine the number of cycles that a simply supported centrally loaded beam can endure before its deflection becomes 50% higher than its initial value.

P4.23. For the quasi-isotropic laminate described in Example 4.2, suppose the loading sequence is 10,000 cycles at 437 MPa followed by cycling to failure at 382 MPa. Estimate the total life expected in this high–low stress sequence using (a) the Whitworth model and (b) the Miner's rule.

P4.24. The mean fatigue lives of $[0/90]_S$ E-glass fiber–epoxy laminates at maximum stress levels of 56,000 and 35,000 psi are 1,500 and 172,000, respectively. The static tensile strength of this material is 65,000 psi.

1. Determine the residual static strength of this material after 50% of the fatigue life at each stress level
2. After cycling for 50% of the fatigue life at 35,000 psi, the maximum stress level in a fatigue specimen is increased to 56,000 psi. Estimate the number of cycles the specimen would survive at 56,000 psi
3. If the first stress level is 56,000 psi, which is then followed by 35,000 psi, estimate the number of cycles at the second stress level

P4.25. Adam et al.* have found that, depending on the fiber type, the post-fatigue residual strength of [0/90] cross-plied composites may show either a gradual decrease with increasing number of cycles (usually after a rapid reduction in the first few cycles) or virtually no change until they fail catastrophically (sudden death). They proposed the following empirical equation to predict the residual strength:

$$\sigma_{res} = \sigma_{max}(\sigma_U - \sigma_{max})(1 - r^x)^{1/y},$$

where

$$r = \frac{\log N - \log 0.5}{\log N_f - \log 0.5}$$

x, y = parameters obtained by fitting the residual strength equation to the experimental data (both vary with material and environmental conditions)

Graphically compare the residual strength after fatigue for the following $[(0/90)_2/0/90]_S$ epoxy laminates at $\sigma_{max} = 0.9$, 0.7, and 0.5 σ_U:

Fiber	σ_U (MPa)	x	y
HTS carbon	944	1.8	23.1
E-glass	578	1.5	4.8
Kevlar 49	674	2.1	8.8

P4.26. Poursartip et al.† have proposed the following empirical equation to represent the fatigue damage (delamination) growth rate at $R = 0.1$ in $[\pm 45/90/45/0]_S$ carbon fiber–epoxy laminate:

$$\frac{dD}{dN} = k_1 \left(\frac{\Delta\sigma}{\sigma_U}\right)^{k_2},$$

* T. Adam, R.F. Dickson, C.J. Jones, H. Reiter, and B. Harris, A power law fatigue damage model for fibre-reinforced plastic laminates, *Proc. Inst. Mech. Eng., 200*:155 (1986).

† A. Poursartip, M.F. Ashby, and P.W.R. Beaumont, The fatigue damage mechanics of fibrous composites, *Polymer NDE* (K.H.G. Ashbee, ed.), Technomic Pub. Co. (1986).

where

D = damage (ratio of delaminated area to total surface area)
$\Delta\sigma$ = stress range
σ_U = ultimate tensile strength
k_1, k_2 = constants determined by fitting a least-square regression line to damage growth rate vs. stress range data

Assuming that the damage can be represented by

$$D = k_3(1 - E/E_0),$$

where

E is the modulus after N cycles
E_0 is the initial modulus
k_3 is a constant

Find expressions for the terminal damage and the number of cycles to failure in terms of $\Delta\sigma$.

P4.27. The following figure shows the load–time curve recorded during the instrumented Charpy impact testing of an unnotched 0° T-300 carbon–epoxy specimen. The pendulum velocity just before impacting the specimen was 16.8 ft./s. The specimen dimensions were as follows: length between specimen supports = 1.6 in.; thickness = 0.125 in.; and width = 0.5 in. Calculate the dynamic flexural strength, dynamic flexural modulus, initiation energy, propagation energy, and total impact energy for the specimen. State your assumptions and critically evaluate your answers.

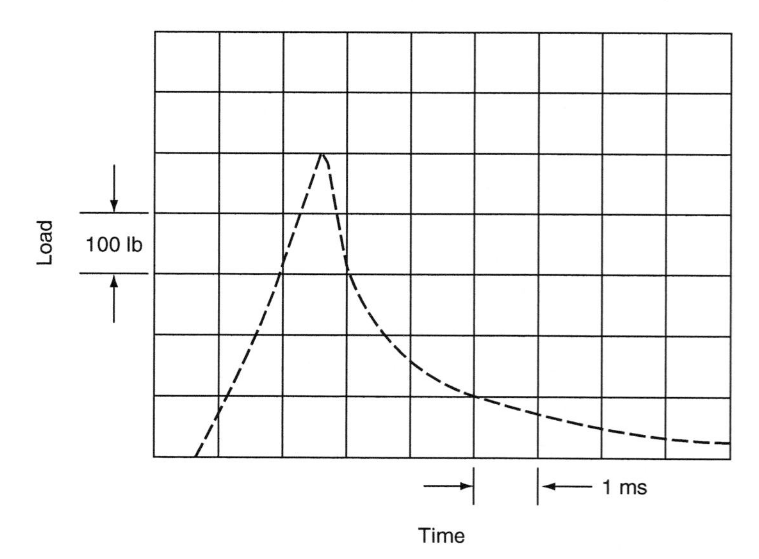

P4.28. In a drop-weight impact test, a 1 in. diameter spherical ball weighing 0.15 lb is dropped freely on 0.120 in. thick $[0/\pm45/90]_{3S}$ carbon fiber–epoxy beam specimens and the rebound heights are recorded. The beam specimens are simply supported, 0.5 in. wide and 6 in. long between the supports. The drop heights, rebound heights, and specimen deflections in three experiments are as follows:

Drop Height (ft)	Rebound Height (ft)	Measured Maximum Deflection (in.)
1	0.72	0.056
4	2.20	0.138
6	3.02	0.150

(a) Calculate the energy lost by the ball in each case.
(b) Assuming that all of the energy lost by the ball is transformed into strain energy in the beam, calculate the maximum deflection of the specimen in each case. The flexural modulus of the laminate is 20×10^6 psi.
(c) Explain why the measured maximum deflections are less than those calculated in (b).

P4.29. The following table* gives the tensile strength and Charpy impact energy data of a unidirectional hybrid composite containing intermingled carbon and Spectra 1000 polyethylene fibers in an epoxy matrix. The polyethylene fibers were either untreated or surface-treated to improve their adhesion with the epoxy matrix. Plot the data as a function of the polyethylene fiber volume fraction and compare them with the rule of mixture predictions. Explain the differences in impact energy in terms of the failure modes that might be expected in untreated and treated fiber composites. Will you expect differences in other mechanical properties of these composites?

* A.A.J.M. Peijs, P. Catsman, L.E. Govaert, and P.J. Lemstra, Hybrid composites based on polyethylene and carbon fibres, Part 2: Influence of composition and adhesion level of polyethylene fibers on mechanical properties, *Composites, 21*:513 (1990).

Polyethylene Fiber Volume Fraction (%)	Tensile Strength (MPa)		Charpy Impact Energy (kJ/m^2)	
	Untreated	Treated	Untreated	Treated
0	1776	1776	90.8	98.8
20	1623	1666	206	100
40	1367	1384	191	105.1
60	1222	1282	171.2	118.2
80	928	986	140.8	124.8
100	1160	1273	116.7	145.3

P4.30. Show that the time required for a material to attain at least 99.9% of its maximum possible moisture content is given by

$$t = \frac{0.679c^2}{D_z},$$

where
t is time in seconds
c is the laminate thickness

P4.31. The maximum possible moisture content in a $[0/90/\pm45]_S$ T-300 carbon fiber–epoxy composite ($v_f = 0.6$) is given by

$$M_m = 0.000145(\text{RH})^{1.8}.$$

A 6.25 mm thick panel of this material is exposed on both sides to air with 90% relative humidity at 25°C. The initial moisture content in the panel is 0.01%.

(a) Estimate the time required for the moisture content to increase to the 0.1% level
(b) If the panel is painted on one side with a moisture-impervious material, what would be the moisture content in the panel at the end of the time period calculated in (a)?

P4.32. Following Equation 4.43, design an experiment for determining the diffusion coefficient, D_z of a composite laminate. Be specific about any calculations that may be required in this experiment.

P4.33. Estimate the transverse tensile strength, transverse modulus, and shear modulus of a 0° unidirectional T-300 carbon fiber-reinforced epoxy composite ($v_f = 0.55$) at 50°C and 0.1% moisture content. The matrix tensile strength, modulus, and Poisson's ratio at 23°C and dry conditions are 56 MPa, 2.2 GPa, and 0.43, respectively; T_{gd} for the matrix is 177°C.

P4.34. Shivakumar and Crews* have proposed the following equation for the bolt clamp-up force F_t between two resin-based laminates:

$$F_t = \frac{F_0}{1 + 0.1126(t/a_{th})^{0.20}},$$

where
F_0 = initial (elastic) clamp-up force
t = elapsed time (weeks)
a_{th} = a shift factor corresponding to a specific steady-state temperature or moisture

1. At 23°C, the a_{th} values for an epoxy resin are 1, 0.1, 0.01, and 0.001 for 0%, 0.5%, 1%, and 1.5% moisture levels, respectively. Using these a_{th} values, estimate the time for 20%, 50%, 80%, and 100% relaxation in bolt clamp-up force.
2. At 66°C and 0.5% moisture level, the bolt clamp-up force relaxes to 80% of its initial value in 3 weeks. Calculate the a_{th} value for this environmental condition.

P4.35. A quasi-isotropic $[0/\pm45/90]_{8S}$ panel of T-300 carbon fiber–epoxy develops a 12 mm long sharp crack at its center. The panel width and thickness are 200 and 8 mm, respectively. The unnotched tensile strength of the laminate is 565 MPa. Determine the safe load that can be applied normal to the crack plane before an unstable fracture occurs. Describe the possible fracture modes for the cracked panel.

P4.36. The load–displacement record shown in the figure was obtained in a double-cantilever beam (DCB) test of a unidirectional 0° carbon fiber–epoxy laminate. The initial crack length was 25 mm and the specimen was unloaded–reloaded after crack extension of every 5 mm. Plot (a) compliance vs. crack length and (b) strain energy release rate G_I vs. crack length for this specimen. What is the Mode I interlaminar fracture toughness of this material? The specimen width was 25 mm.

---

* K.N. Shivakumar and J.H. Crews, Jr., Bolt clamp-up relaxation in a graphite/epoxy laminate, *Long-Term Behavior of Composites, ASTM STP, 813*:5 (1983).

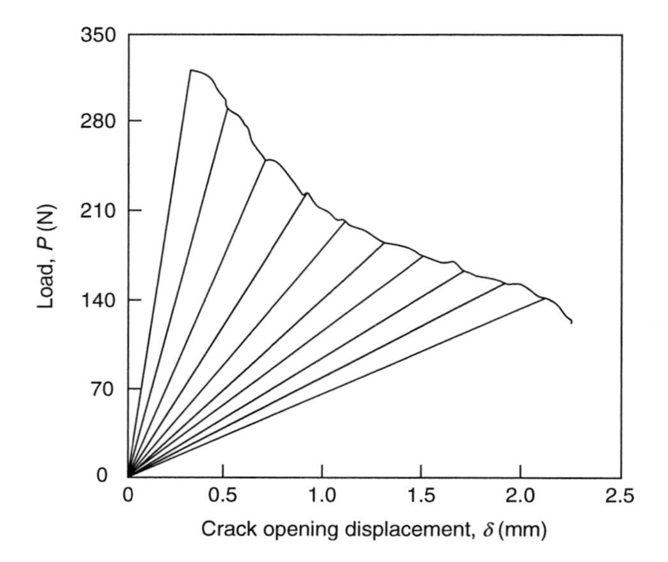

P4.37. The compliance data obtained in a 4-ENF test of a 0° unidirectional carbon fiber–epoxy laminate are given in the following table. The specimen width was 25.4 mm and the initial crack length was 50 mm. The load at which the initial crack was observed to grow was 700 N. Determine the Mode II interlaminar fracture toughness of the material.

Crack Length (mm)	Compliance (mm/N)
60	5.85×10^{-3}
65	6.20×10^{-3}
70	6.60×10^{-3}
75	6.96×10^{-3}

5 Manufacturing

A key ingredient in the successful production application of a material or a component is a cost-effective and reliable manufacturing method. Cost-effectiveness depends largely on the rate of production, and reliability requires a uniform quality from part to part.

The early manufacturing method for fiber-reinforced composite structural parts used a hand layup technique. Although hand layup is a reliable process, it is by nature very slow and labor-intensive. In recent years, particularly due to the interest generated in the automotive industry, there is more emphasis on the development of manufacturing methods that can support mass production rates. Compression molding, pultrusion, and filament winding represent three such manufacturing processes. Although they have existed for many years, investigations on their basic characteristics and process optimization started mostly in the mid-1970s. Resin transfer molding (RTM) is another manufacturing process that has received significant attention in both aerospace and automotive industries for its ability to produce composite parts with complex shapes at relatively high production rates. With the introduction of automation, fast-curing resins, new fiber forms, high-resolution quality control tools, and so on, the manufacturing technology for fiber-reinforced polymer composites has advanced at a remarkably rapid pace.

This chapter describes the basic characteristics of major manufacturing methods used in the fiber-reinforced polymer industry. Emphasis is given to process parameters and their relation to product quality. Quality inspection methods and cost issues are also discussed in this chapter.

5.1 FUNDAMENTALS

Transformation of uncured or partially cured fiber-reinforced thermoset polymers into composite parts or structures involves curing the material at elevated temperatures and pressures for a predetermined length of time. High cure temperatures are required to initiate and sustain the chemical reaction that transforms the uncured or partially cured material into a fully cured solid. High pressures are used to provide the force needed for the flow of the highly viscous resin or fiber–resin mixture in the mold, as well as for the consolidation of individual unbonded plies into a bonded laminate. The magnitude of these two important process parameters, as well as their duration, significantly affects the quality and performance of the molded product. The length of time required to properly cure a part is called the cure cycle. Since the cure cycle determines the production rate for a part, it is desirable to achieve the proper cure in the shortest amount of time. It should be noted that the cure cycle depends on a number of factors, including resin chemistry, catalyst reactivity, cure temperature, and the presence of inhibitors or accelerators.

5.1.1 Degree of Cure

A number of investigators [1–3] have experimentally measured the heat evolved in a curing reaction and related it to the degree of cure achieved at any time during the curing process. Experiments are performed in a differential scanning calorimeter (DSC) in which a small sample, weighing a few milligrams, is heated either isothermally (i.e., at constant temperature) or dynamically (i.e., with uniformly increasing temperature). The instrumentation in DSC monitors the rate of heat generation as a function of time and records it. Figure 5.1 schematically illustrates the rate of heat generation curves for isothermal and dynamic heating.

The total heat generation to complete a curing reaction (i.e., 100% degree of cure) is equal to the area under the rate of heat generation–time curve obtained in a dynamic heating experiment. It is expressed as

$$H_{\mathrm{R}} = \int_0^{t_{\mathrm{f}}} \left(\frac{\mathrm{d}Q}{\mathrm{d}t}\right)_{\mathrm{d}} \mathrm{d}t, \tag{5.1}$$

where

H_{R} = heat of reaction
$(\mathrm{d}Q/\mathrm{d}t)_{\mathrm{d}}$ = rate of heat generation in a dynamic experiment
t_{f} = time required to complete the reaction

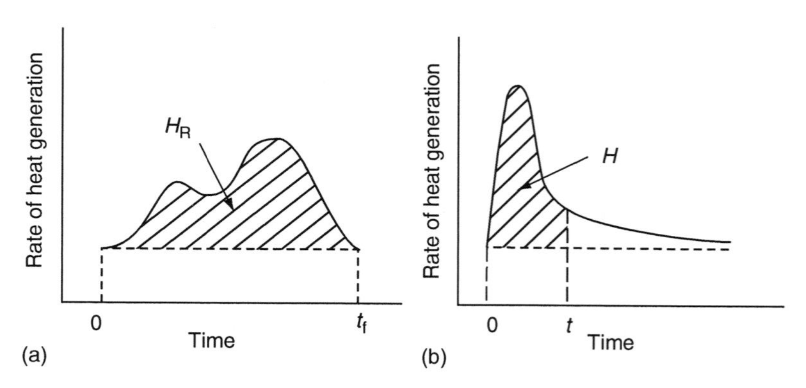

FIGURE 5.1 Schematic representation of the rate of heat generation in (a) dynamic and (b) isothermal heating of a thermoset polymer in a differential scanning calorimeter (DSC).

The amount of heat released in time t at a constant curing temperature T is determined from isothermal experiments. The area under the rate of heat generation–time curve obtained in an isothermal experiment is expressed as

$$H = \int_0^t \left(\frac{dQ}{dt}\right)_i dt, \tag{5.2}$$

where H is the amount of heat released in time t and $(dQ/dt)_i$ is the rate of heat generation in an isothermal experiment conducted at a constant temperature T.

The degree of cure α_c at any time t is defined as

$$\alpha_c = \frac{H}{H_R}. \tag{5.3}$$

Figure 5.2 shows a number of curves relating the degree of cure α_c to cure time for a vinyl ester resin at various cure temperatures. From this figure, it can be seen that α_c increases with both time and temperature; however, the rate of cure, $d\alpha_c/dt$, is decreased as the degree of cure attains asymptotically a maximum value. If the cure temperature is too low, the degree of cure may not reach a 100% level for any reasonable length of time. The rate of cure $d\alpha_c/dt$, obtained from the slope of α_c vs. t curve and plotted in Figure 5.3, exhibits

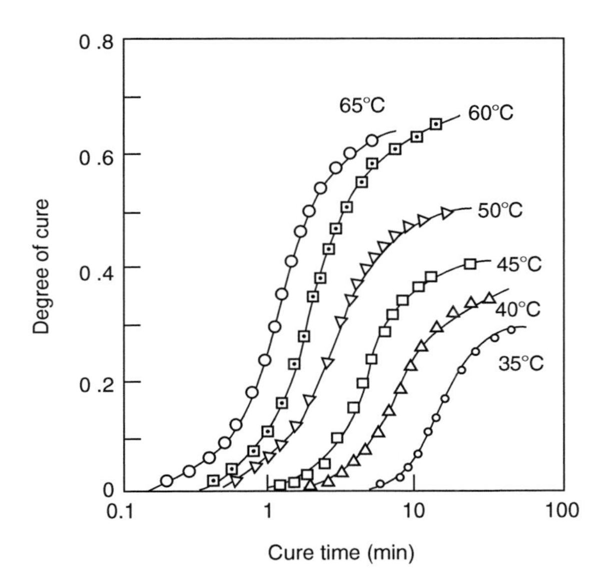

FIGURE 5.2 Degree of cure for a vinyl ester resin at various cure temperatures. (After Han, C.D. and Lem, K.W., *J. Appl. Polym. Sci.*, 29, 1878, 1984.)

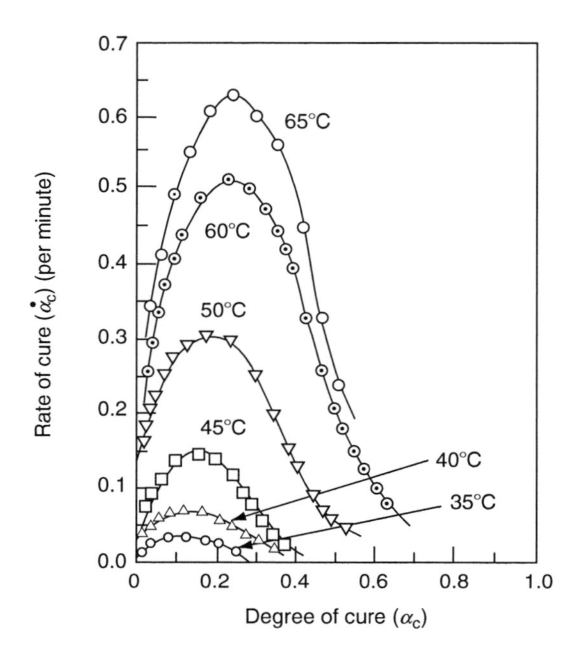

FIGURE 5.3 Rate of cure for a vinyl ester resin at various cure temperatures. (After Han, C.D. and Lem, K.W., *J. Appl. Polym. Sci.*, 29, 1878, 1984.)

a maximum value at 10%–40% of the total cure achieved. Higher cure temperatures increase the rate of cure and produce the maximum degree of cure in shorter periods of time. On the other hand, the addition of a low-profile agent, such as a thermoplastic polymer, to a polyester or a vinyl ester resin decreases the cure rate.

Kamal and Sourour [4] have proposed the following expression for the isothermal cure rate of a thermoset resin:

$$\frac{d\alpha_c}{dt} = (k_1 + k_2\alpha_c^m)(1 - \alpha_c)^n, \tag{5.4}$$

where k_1 and k_2 are reaction rate constants and m and n are constants describing the order of reaction. The parameters m and n do not vary significantly with the cure temperature, but k_1 and k_2 depend strongly on the cure temperature. With the assumption of a second-order reaction (i.e., $m + n = 2$), Equation 5.4 has been used to describe the isothermal cure kinetics of epoxy, unsaturated polyester, and vinyl ester resins. The values of k_1, k_2, m, and n are determined by nonlinear least-squares curve fit to the $d\alpha_c/dt$ vs. α_c data. Typical values of these constants for a number of resins are listed in Table 5.1.

TABLE 5.1
Kinetic Parameters for Various Resin Systems

Resin	Temperature, °C (°F)	Kinetic Parameters in Equation 5.4			
		k_1 (per min)	k_2 (per min)	m	n
Polyester	45 (113)	0.0131	0.351	0.23	1.77
	60 (140)	0.0924	1.57	0.40	1.60
Low-profile polyester	45 (113)	0.0084	0.144	0.27	1.73
(with 20% polyvinyl acetate)	60 (140)	0.0264	0.282	0.27	1.73
Vinyl ester	45 (113)	0.0073	0.219	0.33	1.76
	60 (140)	0.0624	1.59	0.49	1.51

Source: Adapted from Lem, K.W. and Han, C.D., *Polym. Eng. Sci.*, 24, 175, 1984.

5.1.2 Viscosity

Viscosity of a fluid is a measure of its resistance to flow under shear stresses. Low-molecular-weight fluids, such as water and motor oil, have low viscosities and flow readily. High-molecular-weight fluids, such as polymer melts, have high viscosities and flow only under high stresses.

The two most important factors determining the viscosity of a fluid are the temperature and shear rate. For all fluids, the viscosity decreases with increasing temperature. Shear rate does not have any influence on the viscosity of low-molecular-weight fluids, whereas it tends to either increase (shear thickening) or decrease (shear thinning) the viscosity of a high-molecular-weight fluids (Figure 5.4). Polymer melts, in general, are shear-thinning fluids since their viscosity decreases with increasing intensity of shearing.

The starting material for a thermoset resin is a low-viscosity fluid. However, its viscosity increases with curing and approaches a very large value as it transforms into a solid mass. Variation of viscosity during isothermal curing of an epoxy resin is shown in Figure 5.5. Similar viscosity–time curves are also observed for polyester [3] and vinyl ester [5] resins. In all cases, the viscosity increases with increasing cure time and temperature. The rate of viscosity increase is low at the early stage of curing. After a threshold degree of cure is achieved, the resin viscosity increases at a very rapid rate. The time at which this occurs is called the *gel time*. The gel time is an important molding parameter, since the flow of resin in the mold becomes increasingly difficult at the end of this time period.

A number of important observations can be made from the viscosity data reported in the literature:

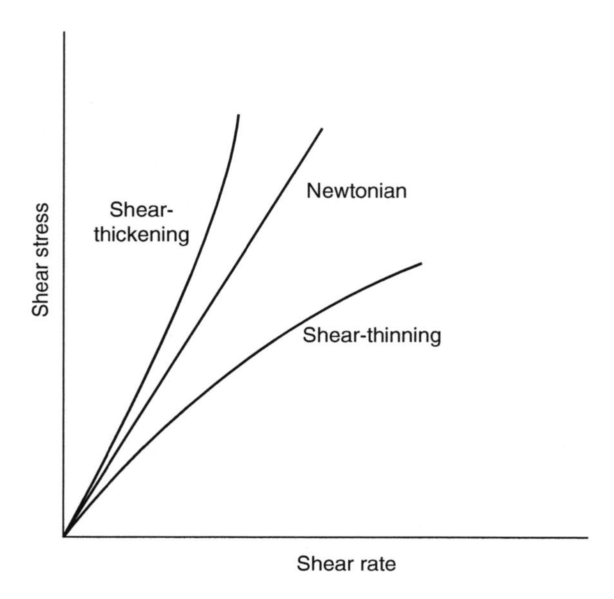

FIGURE 5.4 Schematic shear stress vs. shear rate curves for various types of liquids. Note that the viscosity is defined as the slope of the shear stress–shear rate curve.

1. A B-staged or a thickened resin has a much higher viscosity than the neat resin at all stages of curing.
2. The addition of fillers, such as $CaCO_3$, to the neat resin increases its viscosity as well as the rate of viscosity increase during curing. On the

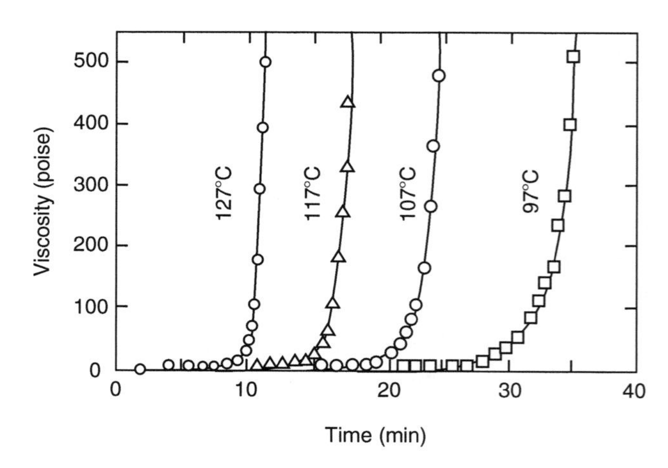

FIGURE 5.5 Variation of viscosity during isothermal curing of an epoxy resin. (After Kamal, M.R., *Polym. Eng. Sci.*, 14, 231, 1974.)

other hand, the addition of thermoplastic additives (such as those added in low-profile polyester and vinyl ester resins) tends to reduce the rate of viscosity increase during curing.

3. The increase in viscosity with cure time is less if the shear rate is increased. This phenomenon, known as shear thinning, is more pronounced in B-staged or thickened resins than in neat resins. Fillers and thermoplastic additives also tend to increase the shear-thinning phenomenon.
4. The viscosity η of a thermoset resin during the curing process is a function of cure temperature T, shear rate $\dot{\gamma}$, and the degree of cure α_c

$$\eta = \eta(T, \dot{\gamma}, \alpha_c). \tag{5.5}$$

The viscosity function for thermosets is significantly different from that for thermoplastics. Since no in situ chemical reaction occurs during the processing of a thermoplastic polymer, its viscosity depends on temperature and shear rate.

5. At a constant shear rate and for the same degree of cure, the η vs. $1/T$ plot is linear (Figure 5.6). This suggests that the viscous flow of a

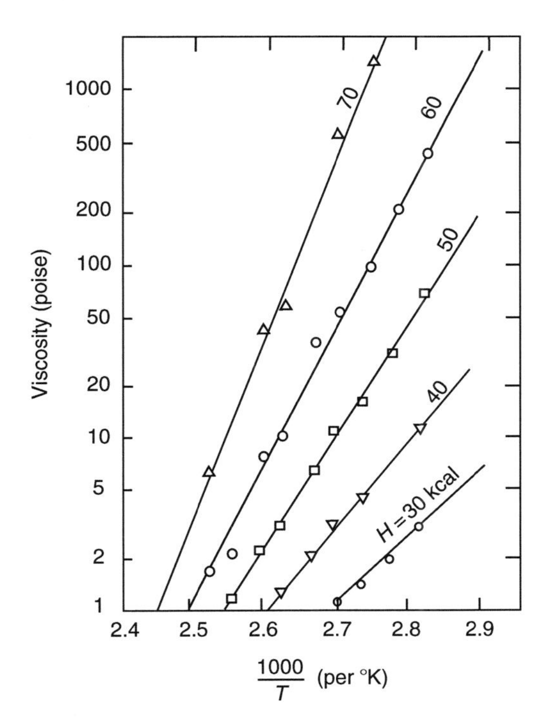

FIGURE 5.6 Viscosity–temperature relationships for an epoxy resin at different levels of cure. (After Kamal, M.R., *Polym. Eng. Sci.*, 14, 231, 1974.)

thermoset polymer is an energy-activated process. Thus, its viscosity as a function of temperature can be written as

$$\eta = \eta_o \exp\left(\frac{E}{RT}\right), \tag{5.6}$$

where

η = viscosity* (Pa s or poise)
E = flow activation energy (cal/g mol)
R = universal gas constant
T = cure temperature (°K)
η_o = constant

The activation energy for viscous flow increases with the degree of cure and approaches a very high value near the gel point.

5.1.3 Resin Flow

Proper flow of resin through a dry fiber network (in liquid composite molding [LCM]) or a prepreg layup (in bag molding) is critical in producing void-free parts and good fiber wet-out. In thermoset resins, curing may take place simultaneously with resin flow, and if the resin viscosity rises too rapidly due to curing, its flow may be inhibited, causing voids and poor interlaminar adhesion.

Resin flow through fiber network has been modeled using Darcy's equation, which was derived for flow of Newtonian fluids through a porous medium. This equation relates the volumetric resin-flow rate q per unit area to the pressure gradient that causes the flow to occur. For one-dimensional flow in the x direction,

$$q = -\frac{P_0}{\eta}\left(\frac{\mathrm{d}p}{\mathrm{d}x}\right), \tag{5.7}$$

where

q = volumetric flow rate per unit area (m/s) in the x direction
P_0 = permeability (m^2)
η = viscosity ($N\,s/m^2$)
$\frac{dp}{dx}$ = pressure gradient (N/m^3), which is negative in the direction of flow (positive x direction)

* Unit of viscosity: 1 Pa s = 1 $N\,s/m^2$ = 10 poise (P) = 1000 centipoise (cP).

The *permeability* is determined by the following equation known as the Kozeny-Carman equation:

$$P_0 = \frac{d_f^2}{16K} \frac{(1 - v_f)^3}{v_f^2}, \tag{5.8}$$

where
d_f = fiber diameter
v_f = fiber volume fraction
K = Kozeny constant

Equations 5.7 and 5.8, although simplistic, have been used by many investigators in modeling resin flow from prepregs in bag-molding process and mold filling in RTM. Equation 5.8 assumes that the porous medium is isotropic, and the pore size and distribution are uniform. However, fiber networks are non-isotropic and therefore, the Kozeny constant, K, is not the same in all directions. For example, for a fiber network with unidirectional fiber orientation, the Kozeny constant in the transverse direction (K_{22}) is an order of magnitude higher than the Kozeny constant in the longitudinal direction (K_{11}). This means that the resin flow in the transverse direction is much lower than that in the longitudinal direction. Furthermore, the fiber packing in a fiber network is not uniform, which also affects the Kozeny constant, and therefore the resin flow.

Equation 5.8 works well for predicting resin flow in the fiber direction. However, Equation 5.8 is not valid for resin flow in the transverse direction, since according to this equation resin flow between the fibers does not stop even when the fiber volume fraction reaches the maximum value at which the fibers touch each other and there are no gaps between them. Gebart [6] derived the following permeability equations in the fiber direction and normal to the fiber direction for unidirectional continuous fiber network with regularly arranged, parallel fibers.

$$\text{In the fiber direction: } P_{11} = \frac{2d_f^2}{C_1} \frac{(1 - v_f^3)}{v_f^2}, \tag{5.9a}$$

$$\text{Normal to the fiber direction: } P_{22} = C_2 \left(\sqrt{\frac{v_{f,\max}}{v_f}} - 1 \right)^{5/2} \frac{d_f^2}{4}, \tag{5.9b}$$

where
C_1 = hydraulic radius between the fibers
C_2 = a constant
$v_{f,max}$ = maximum fiber volume fraction (i.e., at maximum fiber packing)

The parameters C_1, C_2, and $v_{f,max}$ depend on the fiber arrangement in the network. For a square arrangement of fibers, $C_1 = 57$, $C_2 = 0.4$, and $v_{f,max} = 0.785$. For a hexagonal arrangement of fibers (see Problem P2.18), $C_1 = 53$, $C_2 = 0.231$, and $v_{f,max} = 0.906$. Note that Equation 5.9a for resin flow parallel to the fiber direction has the same form as the Kozeny-Carman equation 5.8. According to Equation 5.9b, which is applicable for resin flow transverse to the flow direction, $P_{22} = 0$ at $v_f = v_{f,max}$, and therefore, the transverse resin flow stops at the maximum fiber volume fraction.

The permeability equations assume that the fiber distribution is uniform, the gaps between the fibers are the same throughout the network, the fibers are perfectly aligned, and all fibers in the network have the same diameter. These assumptions are not valid in practice, and therefore, the permeability predictions using Equation 5.8 or 5.9 can only be considered approximate.

5.1.4 Consolidation

Consolidation of layers in a fiber network or a prepreg layup requires good resin flow and compaction; otherwise, the resulting composite laminate may contain a variety of defects, including voids, interply cracks, resin-rich areas, or resin-poor areas. Good resin flow by itself is not sufficient to produce good consolidation [7].

Both resin flow and compaction require the application of pressure during processing in a direction normal to the dry fiber network or prepreg layup. The pressure is applied to squeeze out the trapped air or volatiles, as the liquid resin flows through the fiber network or prepreg layup, suppresses voids, and attains uniform fiber volume fraction. Gutowski et al. [8] developed a model for consolidation in which it is assumed that the applied pressure is shared by the fiber network and the resin so that

$$p = \sigma + \bar{p}_r, \tag{5.10}$$

where

p = applied pressure
σ = average effective stress on the fiber network
$\bar{p}_r$ = average pressure on the resin

The average effective pressure on the fiber network increases with increasing fiber volume fraction and is given by

$$\sigma = A \frac{1 - \sqrt{\frac{v_f}{v_o}}}{\left(\sqrt{\frac{v_a}{v_f}} - 1\right)^4}, \tag{5.11}$$

where
  A is a constant
  v_o is the initial fiber volume fraction in the fiber network (before compaction)
  v_f is the fiber volume fraction at any instant during compaction
  v_a is the maximum possible fiber volume fraction

The constant A in Equation 5.11 depends on the fiber stiffness and the fiber waviness, and is a measure of the deformability of the fiber network. Since the fiber volume fraction, v_f, increases with increasing compaction, Equation 5.11 predicts that σ also increases with increasing compaction, that is, the fiber network begins to take up an increasing amount of the applied pressure. On the other hand, the average pressure on the resin decreases with increasing compaction, which can lead to void formation.

5.1.5 Gel-Time Test

The curing characteristics of a resin–catalyst combination are frequently determined by the gel-time test. In this test, a measured amount (10 g) of a thoroughly mixed resin–catalyst combination is poured into a standard test tube. The temperature rise in the material is monitored as a function of time by means of a thermocouple while the test tube is suspended in a 82°C (180°F) water bath.

A typical temperature–time curve (also known as *exotherm curve*) obtained in a gel-time test is illustrated in Figure 5.7. On this curve, point A indicates the time required for the resin–catalyst mixture to attain the bath temperature. The beginning of temperature rise indicates the initiation of the curing reaction. As the curing reaction begins, the liquid mix begins to transform into a gel-like mass. Heat generated by the exothermic curing reaction increases the mix temperature, which in turn causes the catalyst to decompose at a faster rate and the reaction to proceed at a progressively increasing speed. Since the rate of heat generation is higher than the rate of heat loss to the surrounding medium, the temperature rises rapidly to high values. As the curing reaction nears completion, the rate of heat generation is reduced and a decrease in temperature follows. The exothermic peak temperature observed in a gel-time test is a function of the resin chemistry (level of unsaturation) and the resin–catalyst ratio. The slope of the exotherm curve is a measure of cure rate, which depends primarily on the catalyst reactivity.

Shortly after the curing reaction begins at point A, the resin viscosity increases very rapidly owing to the increasing number of cross-links formed by the curing reaction. The time at which a rapid increase in viscosity ensues is called the gel time and is indicated by point B in Figure 5.7. According to one standard, the time at which the exotherm temperature increases by 5.5°C (10°F) above the bath temperature is considered the gel time. It is sometimes measured by probing the surface of the reacting mass with a clean wooden applicator stick every 15 s until the reacting material no longer adheres to the end of a clean stick.

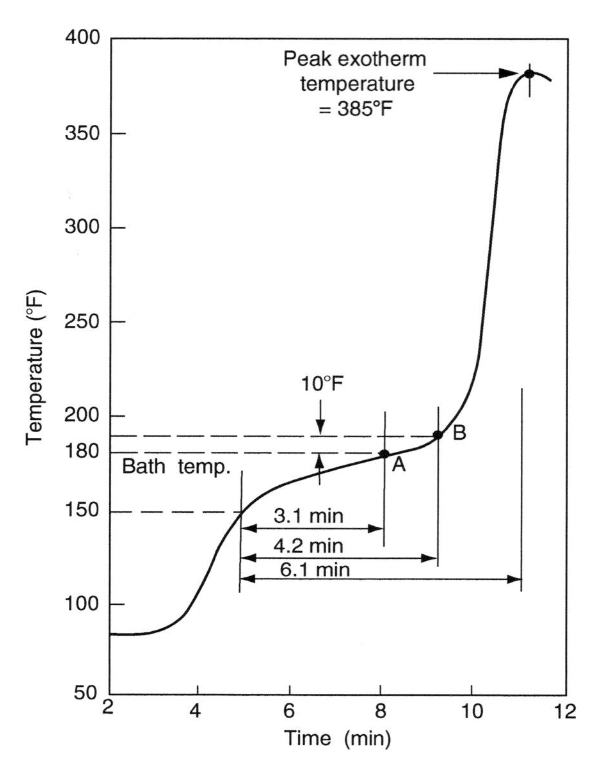

FIGURE 5.7 Typical temperature–time curve obtained in a gel-time test.

5.1.6 Shrinkage

Shrinkage is the reduction in volume or linear dimensions caused by curing as well as thermal contraction. Curing shrinkage occurs because of the rearrangement of polymer molecules into a more compact mass as the curing reaction proceeds. The thermal shrinkage occurs during the cooling period that follows the curing reaction and may take place both inside and outside the mold.

The volumetric shrinkage for cast-epoxy resins is of the order of 1%–5% and that for polyester and vinyl ester resins may range from 5% to 12%. The addition of fibers or fillers reduces the volumetric shrinkage of a resin. However, in the case of unidirectional fibers, the reduction in shrinkage in the longitudinal direction is higher than in the transverse direction.

High shrinkage in polyester or vinyl ester resins can be reduced significantly by the addition of low shrink additives (also called low-profile agents), which are thermoplastic polymers, such as polyethylene, polymethyl acrylate, polyvinyl acetate, and polycaprolactone (see Chapter 2). These thermoplastic additives are usually mixed in styrene monomer during blending with the liquid resin. On curing, the thermoplastic polymer becomes incompatible with the cross-linked resin and forms a dispersed second phase in the cured resin. High resin shrinkage is desirable for easy release of the part from the mold surface; however, at the same time, high resin shrinkage can contribute to many molding defects, such as warpage and sink marks. These defects are described in Section 5.3.

5.1.7 Voids

Among the various defects produced during the molding of a composite laminate, the presence of voids is considered the most critical defect in influencing its mechanical properties. The most common cause for void formation is the inability of the resin to displace air from the fiber surface during the time fibers are coated with the liquid resin. The rate at which the fibers are pulled through the liquid resin, the resin viscosity, the relative values of fiber and resin surface energies, and the mechanical manipulation of fibers in the liquid resin affect air entrapment at the fiber–resin interface. Voids may also be caused by air bubbles and volatiles entrapped in the liquid resin. Solvents used for resin viscosity control, moisture, and chemical contaminants in the resin, as well as styrene monomer, may remain dissolved in the resin mix and volatilize during elevated temperature curing. In addition, air is also entrapped between various layers during the lamination process.

Much of the air or volatiles entrapped at the premolding stages can be removed by (1) degassing the liquid resin, (2) applying vacuum during the molding process, and (3) allowing the resin mix to flow freely in the mold, which helps in carrying the air and volatiles out through the vents in the mold. The various process parameters controlling the resin flow are described in later sections.

The presence of large volume fractions of voids in a composite laminate can significantly reduce its tensile, compressive, and flexural strengths. Large reductions in interlaminar shear strength are observed even if the void content is only 2%–3% by volume (Figure 5.8). The presence of voids generally increases the rate and amount of moisture absorption in a humid environment, which in turn increases the physical dimensions of the part and reduces its matrix-dominated properties.

5.2 BAG-MOLDING PROCESS

The bag-molding process is used predominantly in the aerospace industry where high production rate is not an important consideration. The starting material for bag-molding processes is a prepreg that contains fibers in a partially cured (B-staged) epoxy resin. Typically, a prepreg contains 42 wt% of resin. If this prepreg is allowed to cure without any resin loss, the cured laminate would contain 50 vol% of fibers. Since nearly 10 wt% of resin flows out during the molding process, the actual fiber content in the cured laminate is

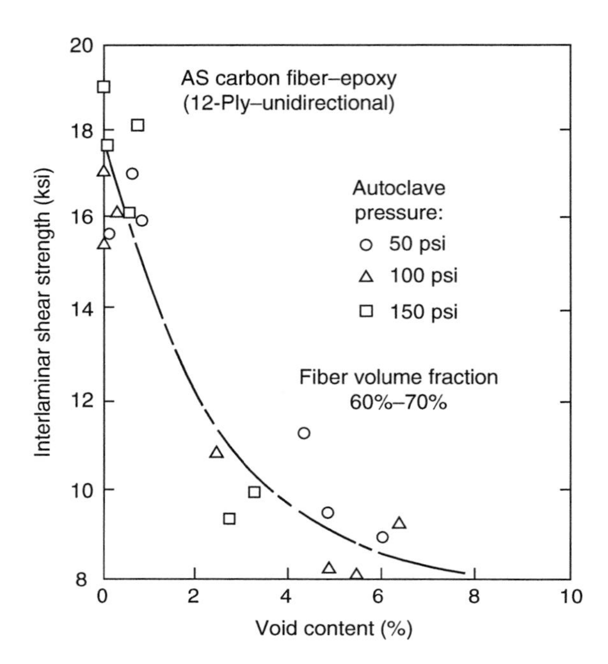

FIGURE 5.8 Effect of void volume fraction on the interlaminar shear strength of a composite laminate. (After Yokota, M.J., *SAMPE J.*, 11, 1978.)

60 vol% which is considered an industry standard for aerospace applications. The excess resin flowing out from the prepreg removes the entrapped air and residual solvents, which in turn reduces the void content in the laminate. However, the recent trend is to employ a near-net resin content, typically 34 wt%, and to allow only 1–2 wt% resin loss during molding.

Figure 5.9 shows the schematic of a bag-molding process. The mold surface is covered with a Teflon-coated glass fabric separator (used for preventing sticking in the mold) on which the prepreg plies are laid up in the desired fiber orientation angle as well as in the desired sequence. Plies are trimmed from the prepreg roll into the desired shape, size, and orientation by means of a cutting device, which may simply be a mat knife. Laser beams, high-speed water jets, or trimming dies are also used. The layer-by-layer stacking operation can be performed either manually (by hand) or by numerically controlled automatic tape-laying machines. Before laying up the prepreg, the backup release film is peeled off from each ply. Slight compaction pressure is applied to adhere the prepreg to the Teflon-coated glass fabric or to the preceding ply in the layup.

After the layup operation is complete, a porous release cloth and a few layers of bleeder papers are placed on top of the prepreg stack. The bleeder papers are used to absorb the excess resin in the prepreg as it flows out during the molding process. The complete layup is covered with another sheet of

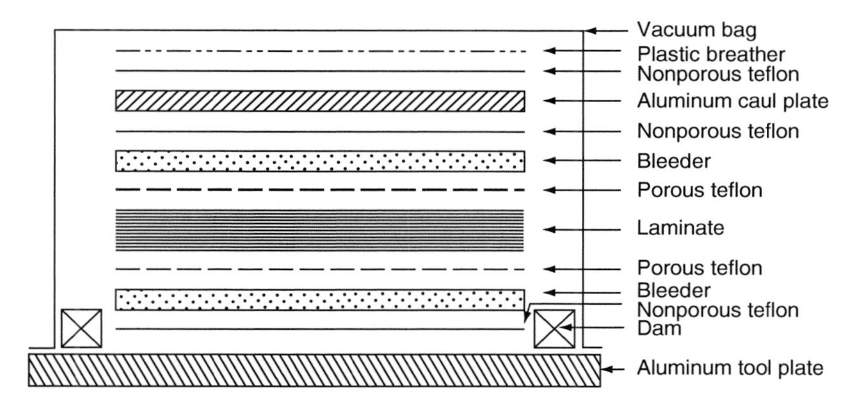

FIGURE 5.9 Schematic of a bag-molding process.

Teflon-coated glass fabric separator, a caul plate, and then a thin heat-resistant vacuum bag, which is closed around its periphery by a sealant. The entire assembly is placed inside an autoclave where a combination of external pressure, vacuum, and heat is applied to consolidate and densify separate plies into a solid laminate. The vacuum is applied to remove air and volatiles, while the pressure is required to consolidate individual layers into a laminate.

As the prepreg is heated in the autoclave, the resin viscosity in the B-staged prepreg plies first decreases, attains a minimum, and then increases rapidly (gels) as the curing (cross-linking) reaction begins and proceeds toward completion. Figure 5.10 shows a typical two-stage cure cycle for a carbon fiber–epoxy prepreg. The first stage in this cure cycle consists of increasing the temperature at a controlled rate (say, 2°C/min) up to 130°C and dwelling at this temperature for nearly 60 min when the minimum resin viscosity is reached.

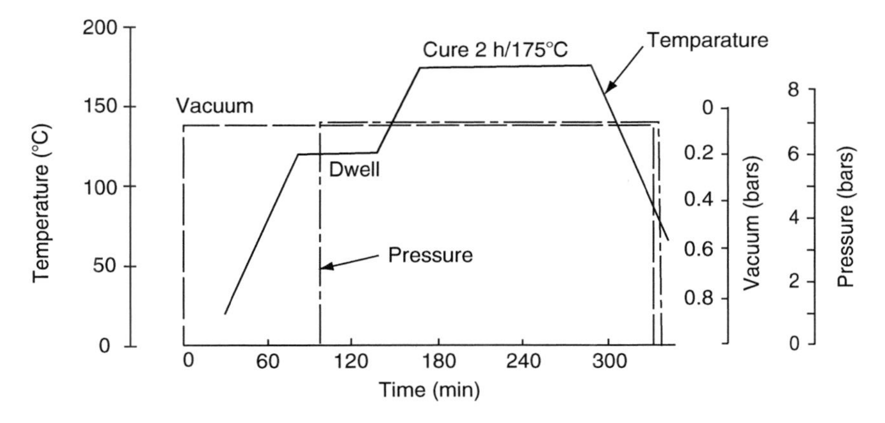

FIGURE 5.10 Typical two-stage cure cycle for a carbon fiber–epoxy prepreg.

During this period of temperature dwell, an external pressure is applied on the prepreg stack that causes the excess resin to flow out into the bleeder papers. The resin flow is critical since it allows the removal of entrapped air and volatiles from the prepreg and thus reduces the void content in the cured laminate. At the end of the temperature dwell, the autoclave temperature is increased to the actual curing temperature for the resin. The cure temperature and pressure are maintained for 2 h or more until a predetermined level of cure has occurred. At the end of the cure cycle, the temperature is slowly reduced while the laminate is still under pressure. The laminate is removed from the vacuum bag and, if needed, postcured at an elevated temperature in an air-circulating oven.

The flow of excess resin from the prepregs is extremely important in reducing the void content in the cured laminate. In a bag-molding process for producing thin shell or plate structures, resin flow by face bleeding (normal to the top laminate face) is preferred over edge bleeding. Face bleeding is more effective since the resin-flow path before gelation is shorter in the thickness direction than in the edge directions. Since the resin-flow path is relatively long in the edge directions, it is difficult to remove entrapped air and volatiles from the central areas of the laminate by the edge bleeding process.

The resin flow from the prepregs reduces significantly and may even stop after the gel time, which can be increased by reducing the heat-up rate as well as the dwell temperature (Figure 5.11). Dwelling at a temperature lower than the curing temperature is important for two reasons: (1) it allows the layup to achieve a uniform temperature throughout the thickness and (2) it provides

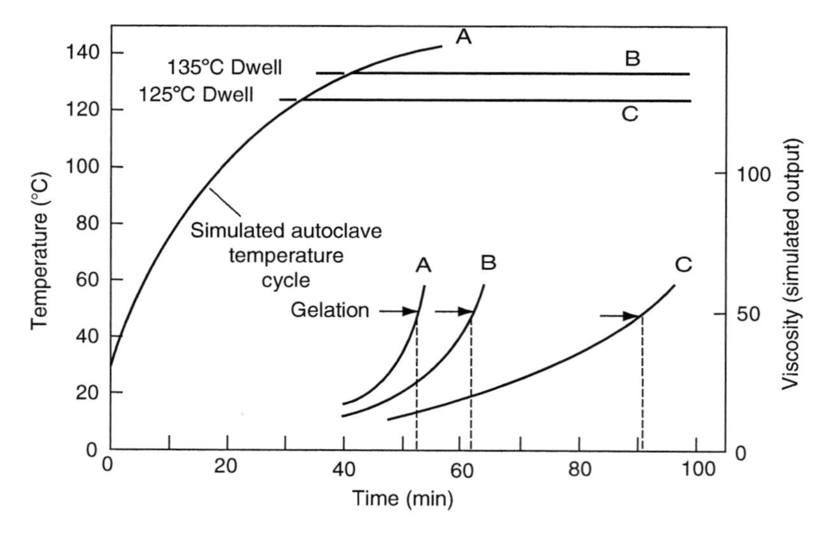

FIGURE 5.11 Effect of dwelling on gel time. (Adapted from Purslaw, D. and Childs, R., *Composites*, 17, 757, 1986.)

time for the resin to achieve a low viscosity. A small batch-to-batch variation in dwell temperature may cause a large variation in gel time, as evidenced in Figure 5.11.

The cure temperature and pressure are selected to meet the following requirements:

1. The resin is cured uniformly and attains a specified degree of cure in the shortest possible time.
2. The temperature at any position inside the prepreg does not exceed a prescribed limit during the cure.
3. The cure pressure is sufficiently high to squeeze out all of the excess resin from every ply before the resin gels (increases in viscosity) at any location inside the prepreg.

Loos and Springer [9] developed a theoretical model for the complex thermo-mechanical phenomenon that takes place in a vacuum bag-molding process. Based on their model and experimental works, the following observations can be made regarding the various molding parameters.

The maximum temperature inside the layup depends on (1) the maximum cure temperature, (2) the heating rate, and (3) the initial layup thickness. The maximum cure temperature is usually prescribed by the prepreg manufacturer for the particular resin–catalyst system used in the prepreg and is determined from the time–temperature–viscosity characteristics of the resin–catalyst system. At low heating rates, the temperature distribution remains uniform within the layup. At high heating rates and increased layup thickness, the heat generated by the curing reaction is faster than the heat transferred to the mold surfaces and a temperature "overshoot" occurs.

Resin flow in the layup depends on the maximum pressure, layup thickness, and heating rate, as well as the pressure application rate. A cure pressure sufficient to squeeze out all excess resin from 16 to 32 layups was found to be inadequate for squeezing out resin from the layers closer to the bottom surface in a 64-ply layup. Similarly, if the heating rate is very high, the resin may start to gel before the excess resin is squeezed out from every ply in the layup.

Loos and Springer [9] have pointed out that the cure cycle recommended by prepreg manufactures may not be adequate to squeeze out excess resin from thick layups. Since the compaction and resin flow progress inward from the top, the plies adjacent to the bottom mold surface may remain uncompacted and rich in resin, thereby creating weak interlaminar layers in the laminate.

Excess resin must be squeezed out of every ply before the gel point is reached at any location in the prepreg. Therefore, the maximum cure pressure should be applied just before the resin viscosity in the top ply becomes sufficiently low for the resin flow to occur. If the cure pressure is applied too early, excess resin loss would occur owing to very low viscosity in the pregel period.

TABLE 5.2
Cure Time for 90% Degree of Cure in a 32-Ply Carbon Fiber–Epoxy Laminate[a]

Cure Temperature, °C (°F)	Heating Rate, °C/min (°F/min)	Cure Time (min)
135 (275)	2.8 (5)	236
163 (325)	2.8 (5)	110
177 (351)	2.8 (5)	89
177 (351)	5.6 (10)	65
177 (351)	11.1 (20)	52

[a] Based on a theoretical model developed by Loos and Springer — Loos, A.C. and Springer, G.S., *J. Compos. Mater.*, 17, 135, 1983.

If on the other hand the cure pressure is applied after the gel time, the resin may not be able to flow into the bleeder cloth because of the high viscosity it quickly attains in the postgel period. Thus the pressure application time is an important molding parameter in a bag-molding process. In general, it decreases with increasing cure pressure as well as increasing heating rate.

The uniformity of cure in the laminate requires a uniform temperature distribution in the laminate. The time needed for completing the desired degree of cure is reduced by increasing the cure temperature as well as increasing the heating rate (Table 5.2).

Besides voids and improper cure, defects in bag-molded laminates relate to the ply layup and trimming operations. Close control must be maintained over the fiber orientation in each ply, the stacking sequence, and the total number of plies in the stack. Since prepreg tapes are not as wide as the part itself, each layer may contain a number of identical plies laid side by side to cover the entire mold surface. A filament gap in a single layer should not exceed 0.76 mm (0.03 in.), and the distance between any two gaps should not be <38 mm (1.5 in.) [10]. Care must also be taken to avoid filament crossovers. Broken filaments, foreign matter, and debris should not be permitted. To prevent moisture pickup, the prepreg roll on removal from the cold storage should be warmed to room temperature before use.

5.3 COMPRESSION MOLDING

Compression molding is used for transforming sheet-molding compounds (SMC) into finished products in matched molds. The principal advantage of compression molding is its ability to produce parts of complex geometry in short periods of time. Nonuniform thickness, ribs, bosses, flanges, holes, and shoulders, for example, can be incorporated during the compression-molding

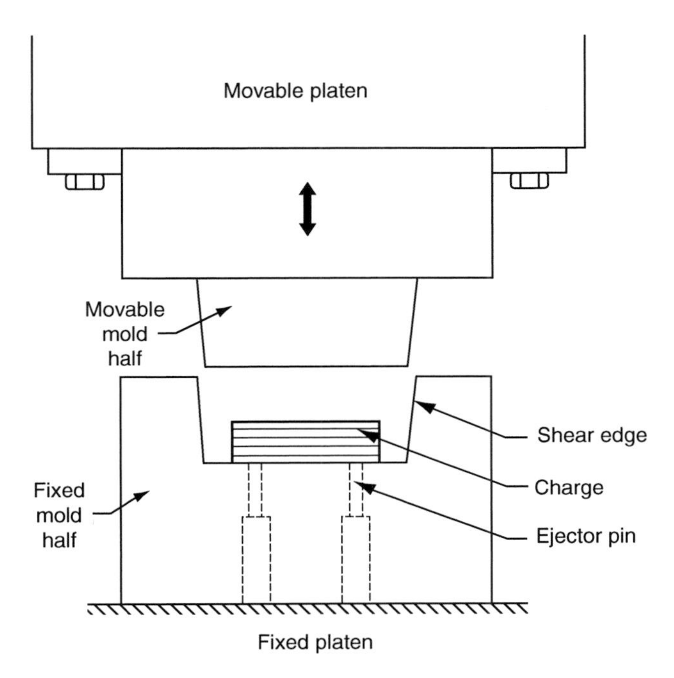

FIGURE 5.12 Schematic of a compression-molding process.

process. Thus, it allows the possibility of eliminating a number of secondary finishing operations, such as drilling, forming, and welding. The entire molding process, including mold preparation and placement of SMC in the mold, as well as part removal from the mold, can be automated. Thus, the compression-molding process is suitable for the high-volume production of composite parts. It is considered the primary method of manufacturing for many structural automotive components, including road wheels, bumpers, and leaf springs.

The compression-molding operation begins with the placement of a precut and weighed amount of SMC, usually a stack of several rectangular plies called the charge, onto the bottom half of a preheated mold cavity (Figure 5.12). The ply dimensions are selected to cover 60%–70% of the mold surface area. The mold is closed quickly after the charge placement, and the top half of the mold is lowered at a constant rate until the pressure on the charge increases to a preset level. With increasing pressure, the SMC material in the mold starts to flow and fill the cavity. Flow of the material is required to expel air entrapped in the mold as well as in the charge. Depending on the part complexity, length of flow, and fiber content (which controls the viscosity of SMC), the molding pressure may vary from 1.4 to 34.5 MPa (200–5000 psi). Usually, high pressures are required for molding parts that contain deep ribs and bosses. The mold temperature is usually in the range of 130°C–160°C (270°F–320°F). After a

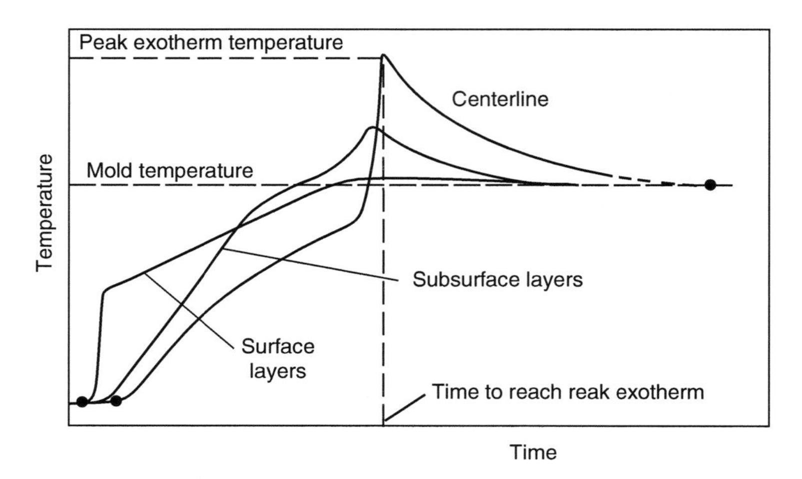

FIGURE 5.13 Temperature distribution at various locations across the thickness of an SMC during the compression-molding operation. (After Mallick, P.K. and Raghupathi, N., *Polym. Eng. Sci.*, 19, 774, 1979.)

reasonable degree of cure is achieved under pressure, the mold is opened and the part is removed, often with the aid of ejector pins.

During molding, a complex heat transfer and a viscous flow phenomenon take place in the cavity. A review of the current understanding of the flow and cure characteristics of compression-molded SMC is given in Ref. [11]. Temperature–time curves measured at the outer surface, subsurface, and centerline of thick E-glass fiber-SMC moldings (Figure 5.13) show that the charge surface temperature quickly attains the mold temperature and remains relatively uniform compared with the centerline temperature. However, owing to the low thermal conductivity of E-glass fiber-SMC, the centerline temperature increases slowly until the curing reaction is initiated at the mid-thickness of the part. Since the SMC material has a relatively low thermal conductivity, the heat generated by the exothermic curing reaction in the interior of the SMC charge is not efficiently conducted to the mold surface and the centerline temperature increases rapidly to a peak value. As the curing reaction nears completion, the centerline temperature decreases gradually to the mold surface temperature. For thin parts, the temperature rise is nearly uniform across the thickness and the maximum temperature in the material seldom exceeds the mold temperature.

Since the surface temperature first attains the resin gel temperature, curing begins first at the surface and progresses inward. Curing occurs more rapidly at higher mold temperatures (Figure 5.14); however, the peak exotherm temperature may also increase. Since peak exotherm temperature of 200°C or higher may cause burning and chemical degradation in the resin, high molding temperatures in thick parts should be avoided.

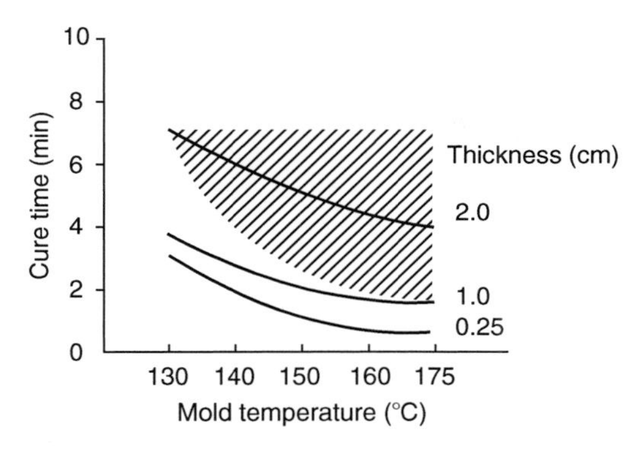

FIGURE 5.14 Curing time vs. mold temperature of SMC sheets. Note that the shaded area represents undesirable molding conditions due to the exotherm temperature exceeding 200°C. (After Panter, M.R., The effect to processing variables on curing time and thermal degradation of compression molded SMC, *Proceedings 36th Annual Conference*, Society of the Plastics Industry, February 1981.)

Increasing the filler content in SMC formulations decreases the peak exotherm temperature since it replaces part of the resin and thereby decreases the total amount of heat liberated. It also acts as a heat sink within the material. The time to reach peak exotherm, which increases almost linearly with the part thickness, is also reduced with increasing filler content. Thus fillers can play a significant role in reducing the cure cycle of a part. Another efficient way of reducing the cure time in the mold is to preheat the charge to pregel temperatures outside the mold and finish curing with high mold-closing speeds inside the mold. Preheating can be accomplished by dielectric heaters that increase the temperature rapidly and uniformly throughout the charge volume. During molding, the thermal gradient remains nearly constant across the thickness of a preheated charge, which allows uniform curing in the thickness direction. As a result, residual curing stresses in the molded part are also reduced.

As the temperature of SMC charge increases in the mold, the network structure created by the thickening reaction with MgO (see Chapter 2) breaks down and the resin viscosity is reduced (Figure 5.15). If the material does not attain a low viscosity before gelling, its flow in the mold is severely restricted. If premature gelation occurs before the mold is filled, the molded part will be incomplete and may contain voids and interlaminar cracks. A number of investigators have studied the basic flow behavior of random fiber SMC with multicolored layers in flat plaque mold cavities [12–14]. At fast mold-closing speeds, the layers flow with uniform extension (plug flow), with slip occurring at the mold surface (Figure 5.16a). The charge thickness does not influence this flow pattern at fast mold-closing speeds. At slow mold-closing speeds, on the other hand, SMC flow pattern depends very much on the charge thickness.

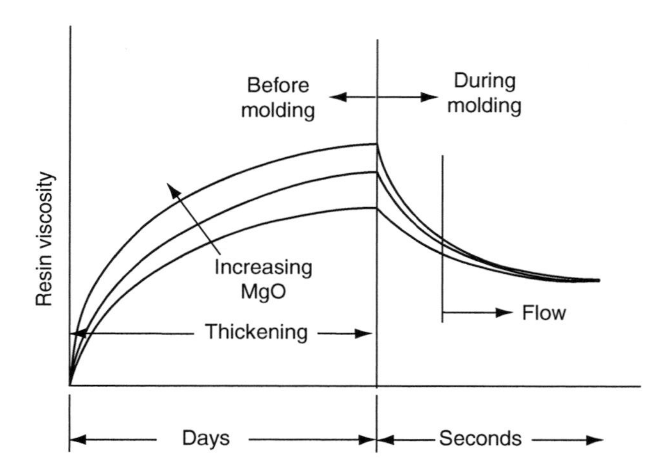

FIGURE 5.15 Viscosity variation of an SMC before and during the compression-molding operation.

For thick charges, the viscosity of SMC in layers adjacent to the hot mold surfaces decreases rapidly while the viscosity in the interior layers is still quite high. As a result, the outer layers begin to flow before the interior layers and may even squirt into the uncovered areas of the mold (Figure 5.16b). Thus the outer layers in this case undergo greater extensional deformation than

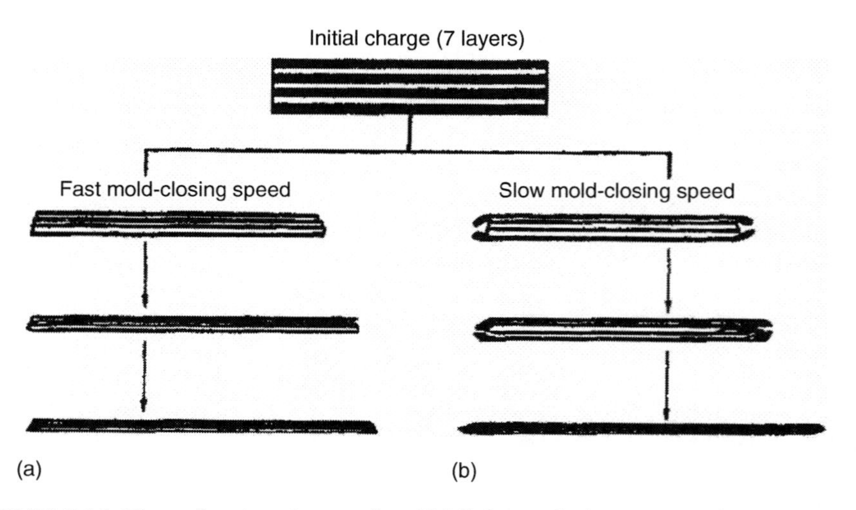

FIGURE 5.16 Flow of various layers of an SMC charge during compression molding at (a) fast mold-closing speeds and (b) slow mold-closing speeds. (After Barone, M.R. and Caulk, D.A., *Polym. Compos.*, 6, 105, 1985.)

TABLE 5.3
Common Surface Defects in Compression-Molded SMC

Defect	Possible Contributing Factors
Pinhole	Coarse filler particles, filler particle agglomeration
Long-range waviness or ripple	Resin shrinkage, glass fiber distribution
Craters	Poor dispersion of the lubricant (zinc stearate)
Sink marks	Resin shrinkage, fiber distribution, fiber length, fiber orientation
Surface roughness	Resin shrinkage, fiber bundle integrity, strand dimensions, fiber distribution
Dark areas	Styrene loss from the surface
Pop-up blisters in painted parts	Subsurface voids due to trapped air and volatiles

the interior layers, with slip occurring between the layers as well as at the mold surface. As the charge thickness is reduced, the extensional deformation becomes more uniform and approaches the same flow pattern observed at fast mold-closing speeds. For a good molded part, a rapid mold-closing speed is desirable since it avoids the possibility of premature gelation and produces the most uniform flow pattern regardless of the charge thickness [14].

Compression-molded SMC parts may contain a wide variety of surface and internal defects (Table 5.3). The surface defects usually create a poor surface appearance or unacceptable surface finish, and the internal defects may affect performance of the molded part. The origin of some of these defects is discussed as follows.

Porosity is the result of small internal voids or surface pits (Figure 5.17a) caused by the entrapment of air or other gases in the molded part. Air is introduced into the SMC at a number of stages, namely, (1) in the resin paste during mechanical blending of liquid resin, styrene monomer, and fillers, (2) at the fiber–resin interface owing to inefficient wetting, (3) in the SMC sheet during compaction between carrier films, (4) between layers of SMC sheets in the charge, and (5) in the closed mold. Air entrapped in the SMC before mold closure is squeezed into small volumes by the pressure exerted during molding. A substantial amount of these air volumes can be carried away by the material flowing toward the vents and shear edges. However, if proper venting is not provided in the mold or the material viscosity is high during its flow, these air volumes may remain entrapped as voids in the molded part.

Blisters are interlaminar cracks (Figure 5.17b) formed at the end of molding due to excessive gas pressure in the interior region of the molded part. The internal gas pressure is generated during molding from unreacted styrene monomer in undercured parts or from large pockets of entrapped air between the stacked layers. If this internal pressure is high, interlaminar cracks may form at the time of mold opening. The delaminated area near the surface may bulge

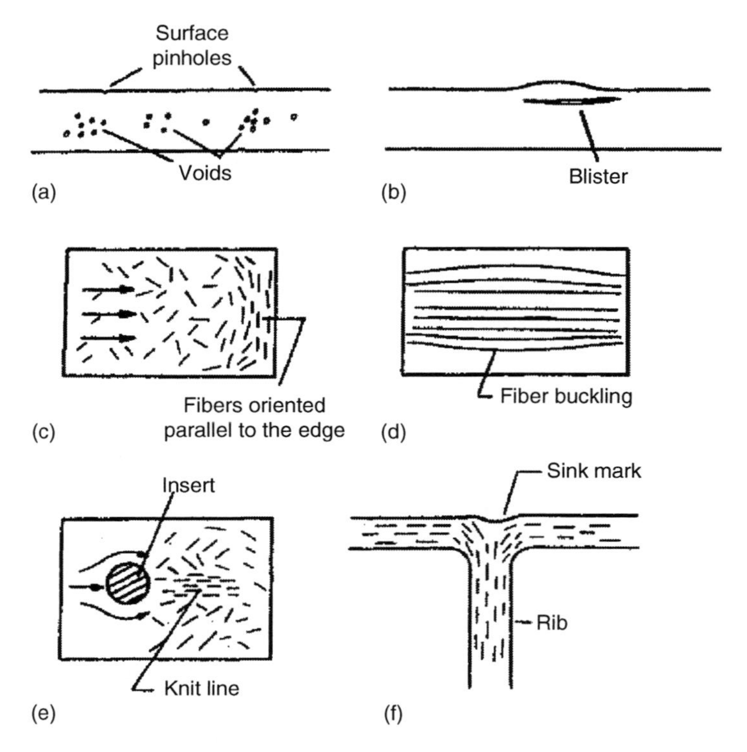

FIGURE 5.17 Various defects in a compression-molded SMC part.

into a dome-shaped blister by the entrapped gas pressure. Blisters may also appear during some postmolding operations, such as high-temperature baking in a paint oven, which causes expansion of entrapped air or gases.

Griffith and Shanoski [15] suggested two possible ways of reducing blisters:

1. Minimize the entrapped air. The most effective method for minimizing the entrapped air is vacuum molding, in which air from the mold is evacuated just as the mold is closed. A second method of reducing the entrapped air is to allow more radial flow by stacking more plies over a smaller area instead of stacking fewer plies over a larger area.
2. Increase the interlaminar shear strength by changing the resin type, using coupling agents, reducing contamination between layers, decreasing the molding temperature, and assuring proper cure before the mold pressure is released.

In any molding operation involving long flow paths, it is extremely difficult to control the preferential orientation of fibers. With compression molding of SMC-R, abrupt changes in thickness, any obstruction in the flow path, or the presence of high shear zones can create fiber orientations that deviate from the ideal random orientation. As a result, the molded part may become locally anisotropic with its strength and modulus higher in the direction of flow than in the transverse direction. In the compression molding of SMC-R, it is common practice to cover only 60%–70% of the mold surface with the charge, and then use high normal pressure to spread it over the entire mold cavity. When the flow front contacts the cavity edges, discontinuous fibers in the SMC-R tend to rotate normal to the flow direction (Figure 5.17c). This results in strength reduction normal to the flow direction and makes the edges prone to early cracking.

Compression molding of SMC-CR or XMC containing continuous fibers is normally performed with 90%–95% initial mold surface coverage. For these materials, flow is possible only in the transverse direction of fibers. If excessive transverse flow is allowed, continuous fibers in the surface and subsurface layers of both SMC-CR and XMC may buckle (bow out) near the end of the flow path (Figure 5.17d). In addition, the included angle between the X-patterned fibers in XMC may also increase. As a result, the longitudinal tensile strengths of SMC-CR and XMC are reduced in areas with fiber misorientation [16]. However, since severe fiber misorientations are generally restricted to the outer layers, increasing the number of plies improves the longitudinal strength to the level observed with no misorientation.

Knit lines are linear domains of aligned fiber orientation and are formed at the joining of two divided flow fronts (Figure 5.17e), such as behind a metal insert or pin or where two or more separate flow fronts arising from multiple charge pieces meet. Multiple charge pieces are used for compression molding of large and complex parts. Since fibers tend to align themselves along the knit line, the strength of the part in a direction normal to the knit line is reduced.

The formation of knit lines can be reduced by proper charge placement in the mold. A common location of knit lines in behind a core pin used in forming molded-in holes. Thus if the holes are in a high-stress area, it is better to drill them instead of using core pins, since knit lines formed behind such core pins may extend to the edge and initiate premature cracking.

Warpage is critical in thin-section moldings and is caused by variations in cooling rate between sections of different thicknesses or different fiber orientations. Differential cooling rates may also lead to complex residual stresses, which may ultimately reduce the strength of a molded part.

Nonuniform cure is critical in thick-section moldings and can create a gradient of properties in the thickness direction. Since the curing reaction is initiated at the surfaces and progresses inward, it is likely that insufficient molding time will leave the interior undercured. As a result, the interlaminar shear strength of the molded part is reduced.

The effect of various molding times on the development of through-thickness properties of a thick-section molding is demonstrated in Figure 5.18. This figure was developed by sectioning 12 mm (0.5 in.) thick compression-molded specimens along the center plane and testing each half in flexure, one with the

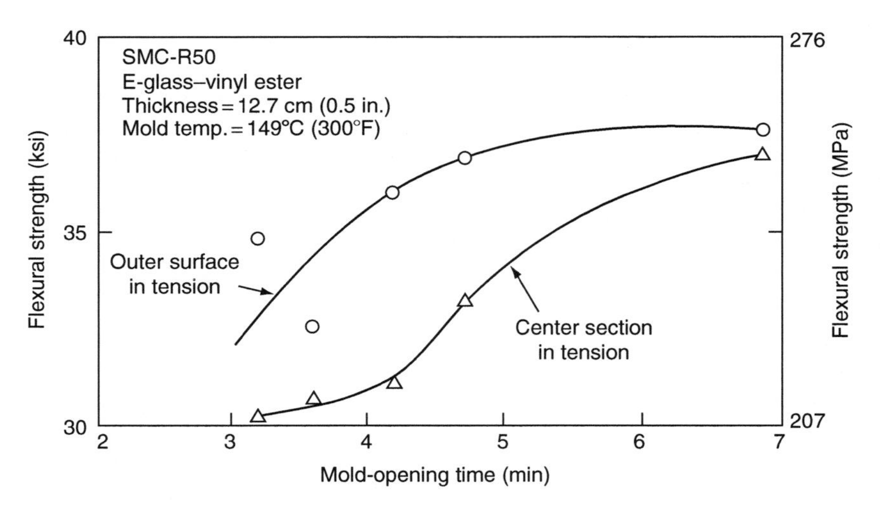

FIGURE 5.18 Effect of cure time on the development of flexural strength in a compression-molded SMC-R laminate. (After Mallick, P.K. and Raghupathi, N., *Polym. Eng. Sci.*, 19, 774, 1979.)

outer skin in tension and the other with the exposed center in tension. For short mold-opening times the center has a much lower strength than the outer skin, indicating that the part was removed before completion of cure at the center. The difference in strength is reduced at higher molding times.

Sink marks are small surface depressions normally observed above the ribs in compression-molded SMC parts (Figure 5.17f). Jutte [17] has shown that the flow of material into a rib creates a fiber-rich zone near its base and resin-rich zone near the opposite surface. Since the resin-rich zone has a higher coefficient of thermal contraction, it shrinks more than the surrounding material, which contains uniform fiber distribution. As a result, the surface opposite to a rib will depress and a sink mark will appear.

A nonuniform flow pattern of material is generally considered the reason for the separation of resin from fibers at or near the base of a rib. Smith and Suh [18] have shown that protruding rib corners (Figure 5.19) create less sink

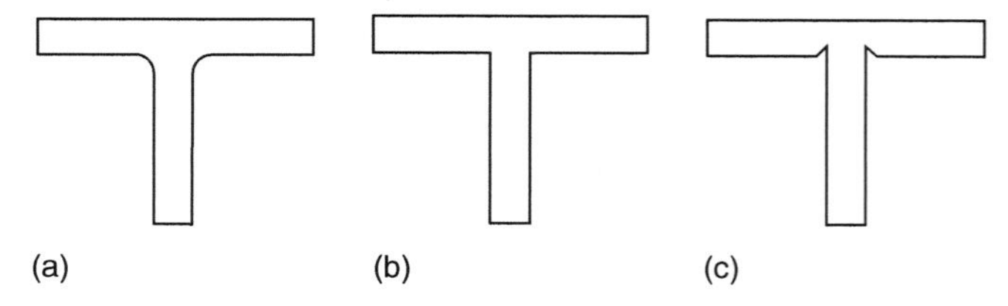

FIGURE 5.19 Rib design in SMC parts: (a) rounded corner, (b) sharp corner, and (c) protruded corner.

depths than either sharp or rounded rib corners. Their experiments also show that sink depths in 12 mm (0.5 in.) long fiber-reinforced SMC are lower than those in 25 mm (1 in.) long fiber-reinforced SMC. Uneven part thickness on the two sides of a rib tends to reduce the sink depth as well as shift the sink mark toward the thicker section.

A poor surface finish caused by sink marks is undesirable in highly visible exterior automotive body panels, such as a hood or a door panel, made of compression-molded SMC. Short ribs are commonly used on the back surface of these panels to improve their flexural stiffness. However, sink marks formed on the top surface reduce the surface finish to lower than the Class A (mirror) finish. Although sink depths can be controlled by using longer ribs, a combination of long and short fibers in the SMC-R sheets, or a low-profile resin, they are not completely eliminated. The current approach is to mask these and other surface imperfections by coating the outer surface with a flexible paint. Just after the completion of the cure cycle, the top mold is retracted by a small amount and the liquid paint is injected over the top surface. This process is known as *in-mold coating*.

5.4 PULTRUSION

Pultrusion is a continuous molding process for producing long, straight structural members of constant cross-sectional area. Among the common pultruded products are solid rods, hollow tubes, flat sheets, and beams of a variety of cross sections, including angles, channels, hat sections, and wide-flanged sections. Pultrusion processes for producing variable cross sections along the length as well as curved members have also been developed.

The major constituent in a pultruded product is longitudinally oriented continuous strand rovings. Several layers of mats or woven rovings are added at or near the outer surface (Figure 5.20) to improve its transverse strength. The total fiber content in a pultruded member may be as high as 70% by weight; however, owing to the presence of mats or woven rovings, its longitudinal

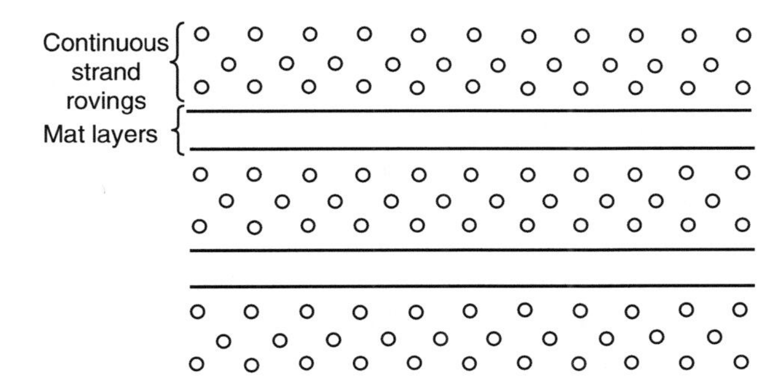

FIGURE 5.20 Typical construction of a pultruded sheet.

TABLE 5.4
Mechanical Properties of Pultruded E-Glass–Polyester Sheets

Total Fiber Content (wt%)	70	60	50	40	30
Continuous roving content (wt%)	38.8	28.8	18.8	18.8	16.1
Mat content (wt%)	31.2	31.2	31.2	20.8	13.9
Roving–mat ratio	1.24	0.92	0.60	0.90	1.16
Roving end count	79	58	38	29	33
No. mat layers	2	3	3	3	2
Mat weight (oz)	1.5	1.5	1.5	1.5	1
Tensile strength, MPa (ksi)					
Longitudinal	373.1 (54.1)	332.4 (48.2)	282.1 (40.9)	265.5 (38.5)	217.2 (31.5)
Transverse	86.9 (12.6)	93.1 (13.5)	94.5 (13.7)	84.8 (12.3)	67.6 (9.8)
Tensile modulus, GPa (Msi)					
Longitudinal	28.8 (4.17)	23.6 (3.42)	18.4 (2.67)	17.1 (2.48)	15.4 (2.24)
Transverse	8.34 (1.21)	9.31 (1.35)	8.55 (1.24)	7.1 (1.03)	5.24 (0.76)
Flexural strength, MPa (ksi)					
Longitudinal	412.4 (59.8)	375.9 (54.5)	325.5 (47.2)	338.6 (49.1)	180.7 (26.2)
Transverse	204.1 (29.6)	199.3 (28.9)	220.0 (31.9)	181.4 (26.3)	169 (24.5)

Source: Adapted from Evans, D.J., Classifying pultruded products by glass loading, *Proceedings 41st Annual Conference*, Society of the Plastics Industry, January 1986.

strength and modulus are lower than those obtained with all unidirectional 0° fiber strands. The ratio of continuous strand rovings to mats or woven rovings determines its mechanical properties (Table 5.4).

In commercial applications, polyester and vinyl ester resins are used as the matrix material. Epoxies have also been used; however, they require longer cure times and do not release easily from the pultrusion die. Pultrusion process has also been used with thermoplastic polymers, such as PEEK and polysulfone.

Figure 5.21 is a schematic of a typical pultrusion line. Continuous strand rovings and mats are pulled from one end of the line into a resin bath that contains liquid resin, curing agent (initiator), and other ingredients, such as colorant, ultraviolet (UV) stabilizer, and fire retardant. The viscosity of the liquid resin, residence time, and mechanical action on the fibers (such as looping of fibers) in the resin bath are adjusted to ensure a complete wet-out of fibers with the resin. Thermoplastic polyester surfacing veils are added to the fiber–resin stream just outside the resin bath to improve the surface smoothness of the molded product. The fiber–resin stream is pulled first through a series of preformers and then through a long preheated die. The preformers distribute

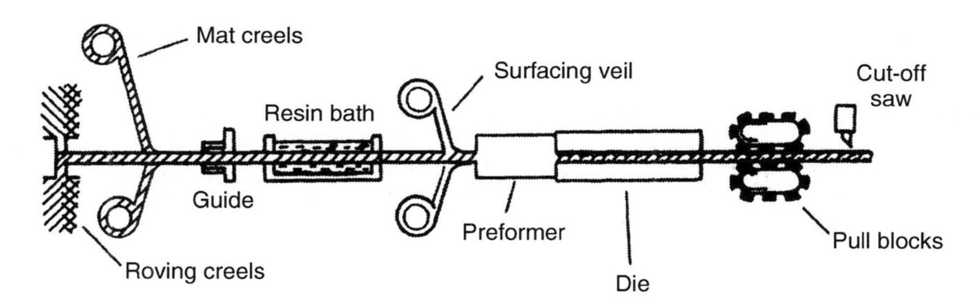

FIGURE 5.21 Schematic of a pultrusion process.

the fiber bundles evenly, squeeze out the excess resin, and bring the material into its final configuration. Final shaping, compaction, and curing take place in the die, which has a gradually tapering section along its length. The entrance section of the die is usually water cooled to prevent premature gelling, and the rest of the die is heated in a controlled manner either by oil heaters or by electric heaters. Infrared heating has also been used to speed up the curing process. A number of pulling rolls or blocks pull the cured pultruded member out of the die. The die temperature, die length, and pulling speed are controlled to ensure that the resin has cured completely before the pultruded member exits from the die. After cooling with air or water, it is cut into desired lengths by a diamond-impregnated saw at the end of the line.

The most important factor controlling the mechanical performance of a pultruded member is the fiber wet-out. The ability to wet out the fibers with the resin depends on the initial resin viscosity, residence time in the resin bath, resin bath temperature, and mechanical action applied to fibers in the resin bath. For a given resin viscosity, the degree of wet-out is improved as (1) the residence time is prolonged by using slower line speeds or longer baths, (2) the resin bath temperature is increased (which reduces the resin viscosity), or (3) the degree of mechanical working on fibers is increased. Since each roving pulled through the resin bath contains a large number of fiber bundles, it is extremely important that the resin penetrates inside the roving and coats each bundle uniformly. Resin penetration takes place through capillary action as well as lateral squeezing between the bundles. Lateral pressure at the resin squeeze-out bushings (located at the resin bath exit), preformers, and die entrance also improves the resin penetration in the bundles. Generally, slower line speed and lower resin viscosity favor resin penetration by capillary action, and faster line speed and higher resin viscosity improve the amount of resin pickup owing to increased drag force [19]. The fiber and resin surface energies are also important parameters in improving the amount of resin coating on fiber rovings. Thus, Kevlar 49 fibers, by virtue of their high surface energies, pick up more resin in the resin bath than either E-glass or carbon fibers under similar process conditions.

The resin viscosity in commercial pultrusion lines may range from 0.4 to 5 Pa s (400–5000 cP). Resin viscosities >5 Pa s may result in poor fiber wet-out, slower line speed, and frequent fiber breakage at the resin squeeze-out bushings. On the other hand, very low resin viscosities may cause excessive resin draining from the fiber–resin stream after it leaves the resin bath. Resin viscosity can be lowered by increasing the bath temperature; however, if it reduces to 0.2 Pa s or lower, the fiber–resin stream must be cooled at the resin bath exit to increase the resin viscosity and prevent excessive draining.

As the fiber–resin stream enters the heated die, the resin viscosity first decreases, which aids in the continued wet-out of uncoated fibers. However, the curing reaction begins a short distance from the die entrance, and soon after the resin viscosity increases rapidly, as shown in Figure 5.22. If the die temperature is not gradually increased in the die entrance zone, a cured resin skin may quickly form on the die walls. The separation of uncured material from the skin results in poor surface quality for the pultruded product. This problem can be alleviated by preheating the fiber–resin stream just outside the die, which reduces the temperature gradient at the die entrance zone.

The curing reaction continues at an increasing rate as the fiber–resin stream moves toward the exit end of the die. Heat generated by the exothermic curing reaction raises the temperature in the fiber–resin stream. The location of the exothermic peak depends on the speed of pulling the fiber–resin stream through

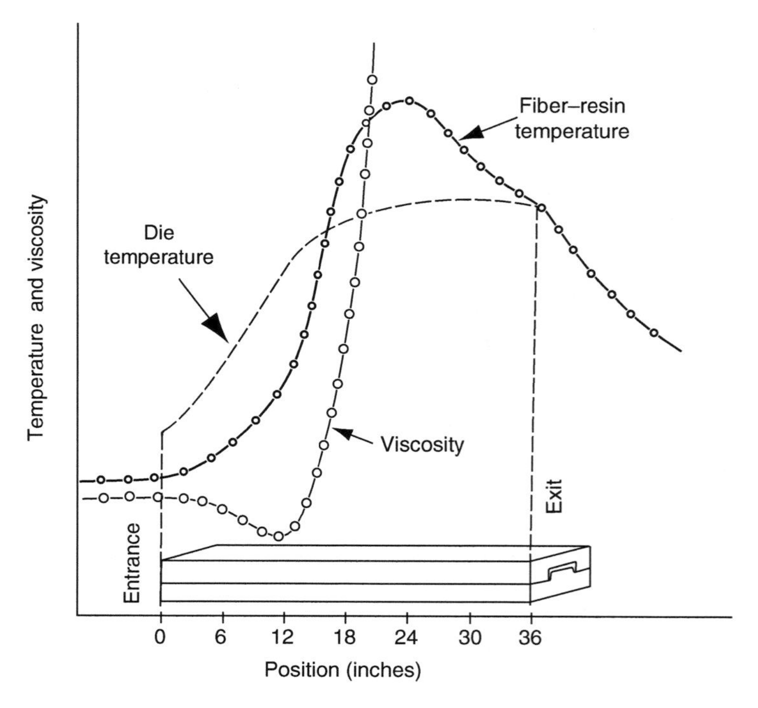

FIGURE 5.22 Viscosity change of a thermosetting resin in a pultrusion die.

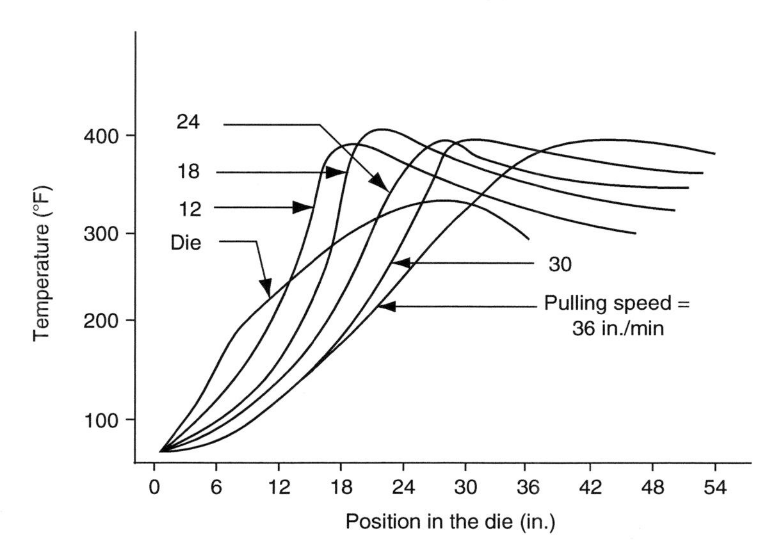

FIGURE 5.23 Temperature distribution along the length of a pultrusion die. (After Sumerak, J.E. and Martin, J.D., Applying internal temperature measurement data to pultrusion process control, *Proceedings 41st Annual Conference*, Society of the Plastics Industry, January 1986.)

the die (Figure 5.23). As the curing reaction nears completion, the exotherm temperature decreases and a cooling period begins. The rate of heat transfer from the cured material into the die walls is increased owing to a lower die temperature near the exit zone. This allows rapid cooling of the entire cured section while it is still under the confinement of the die walls. If the temperature in the interior of the cured section remains high at the time of exit from the die, interlaminar cracks may form within the pultruded member.

Unlike many other molding processes, no external pressure is applied in a pultrusion process. However, experiments performed by Sumerak [20] have demonstrated that the pressure in the die entrance zone is in the range of 1.7–8.6 MPa (250–1250 psi). The principal source for such high internal pressure is the volumetric expansion of the resin as it is heated in the die entrance zone. However, as the curing reaction begins, the polymerization shrinkage reduces the pressure to near-zero values at approximately the mid-length of the die. In general, the internal pressure can be increased by controlling the resin chemistry and the fiber volume fraction. Although increasing the internal pressure may also result in a higher pulling force, it will improve fiber–resin consolidation in the pultruded section.

Depending on the part complexity, resin viscosity, and cure schedule, the line speed in a commercial pultrusion process may range from 50–75 mm/min (2–3 in./min) to 3–4.5 m/min (10–15 ft./min). High line speeds usually shift the location of the peak exotherm temperature toward the exit end of the die and increase the pulling force. Although the production rate is increased at high line speeds, the product quality may deteriorate owing to poor fiber wet-out, unfinished curing, and roving migration within the cross section. If high output is desired, it is often better to use multiple dies instead of high line speeds.

The pulling force applied to the pulling mechanism at the end of the pultrusion line must overcome the combined effects of (1) the frictional force of fibers sliding against the die wall, (2) the shear viscous force between a very thin layer of resin and the die wall, and (3) the backflow force or drag resistance between the fibers and the backflowing resin at the die entrance. The contribution from each of these forces to the total pulling force varies along the length of the die. For example, the shear viscous force and the backflow force have larger contributions near the die entrance. As curing progresses along the die length and the liquid resin transforms into a solid mass, the frictional force becomes more predominant. The theoretical model developed by Bibbo and Gutowski [21] shows that the pulling force increases with increasing fiber volume fraction, resin viscosity, line speed, and compaction ratio (i.e., the ratio of the die entrance opening to the die exit opening). However, contributions from each of these parameters on the frictional force, shear viscous force, and backflow force are not uniform. Control of the pulling force and the design of the fiber guidance system are extremely important since they influence the fiber alignment as well as fiber wet-out. Some of the defects found in pultruded products, such as fiber bunching, fiber shifting, wrinkles, and folding of mats or woven rovings, are related to these factors.

5.5 FILAMENT WINDING

In a filament-winding process, a band of continuous resin-impregnated rovings or monofilaments is wrapped around a rotating mandrel and cured to produce axisymmetric hollow parts. Among the applications of filament winding are automotive drive shafts, helicopter blades, oxygen tanks, pipelines, spherical pressure vessels, conical rocket motor cases, and large underground gasoline storage tanks. The filament-winding process is also used to manufacture prepreg sheets or continuous fiber-reinforced sheet-molding compounds, such as XMC. The sheet is formed by slitting the wound shape parallel to the mandrel axis.

Figure 5.24 shows the schematic of a basic filament-winding process. A large number of fiber rovings are pulled from a series of creels into a liquid resin bath containing liquid resin, catalyst, and other ingredients, such as pigments and UV absorbers. Fiber tension is controlled using the fiber guides or scissor bars located between each creel and the resin bath. Just before entering the resin bath, the rovings are usually gathered into a band by passing them through a textile thread board or a stainless steel comb.

At the end of the resin tank, the resin-impregnated rovings are pulled through a wiping device that removes the excess resin from the rovings and

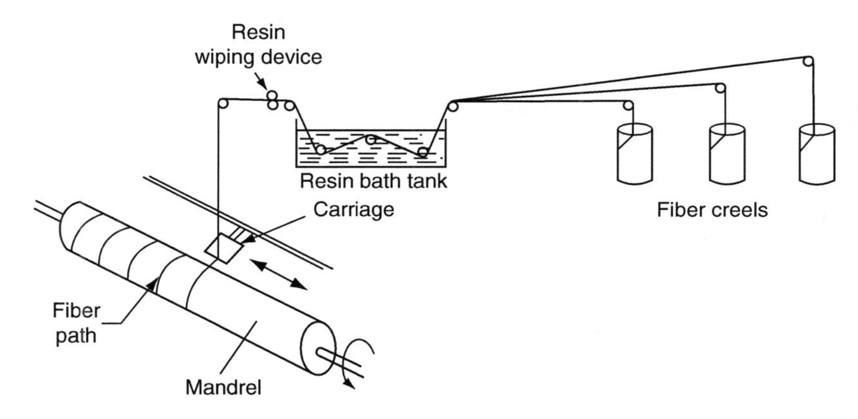

FIGURE 5.24 Schematic of a filament-winding process.

controls the resin coating thickness around each roving. The most commonly used wiping device is a set of squeeze rollers in which the position of the top roller is adjusted to control the resin content as well as the tension in fiber rovings. Another technique for wiping the resin-impregnated rovings is to pull each roving separately through an orifice, very much like the procedure in a wire drawing process. This latter technique provides better control of resin content. However, in the case of fiber breakage during a filament-winding operation, it becomes difficult to rethread the broken roving line through its orifice.

Once the rovings have been thoroughly impregnated and wiped, they are gathered together in a flat band and positioned on the mandrel. Band formation can be achieved by using a straight bar, a ring, or a comb. The band former is usually located on a carriage, which traverses back and forth parallel to the mandrel, like a tool stock in a lathe machine. The traversing speed of the carriage and the winding speed of the mandrel are controlled to create the desired winding angle patterns. Typical winding speeds range from 90 to 110 linear m/min (300–360 linear ft./min). However, for more precise winding, slower speeds are recommended.

The basic filament-winding process described earlier creates a helical winding pattern (Figure 5.25) and is called the helical winding process. The angle of the roving band with respect to the mandrel axis is called the wind angle. By adjusting the carriage feed rate and the mandrel's rotational speed, any wind angle between near 0° (i.e., longitudinal winding) to near 90° (i.e., hoop winding) can be obtained. Since the feed carriage moves backward and forward, fiber bands crisscross at plus and minus the wind angle and create a weaving or interlocking effect. It is also possible to produce a helical winding by keeping the feed carriage stationary and traversing the rotating mandrel back and forth. The mechanical properties of the helically wound part depend strongly on the

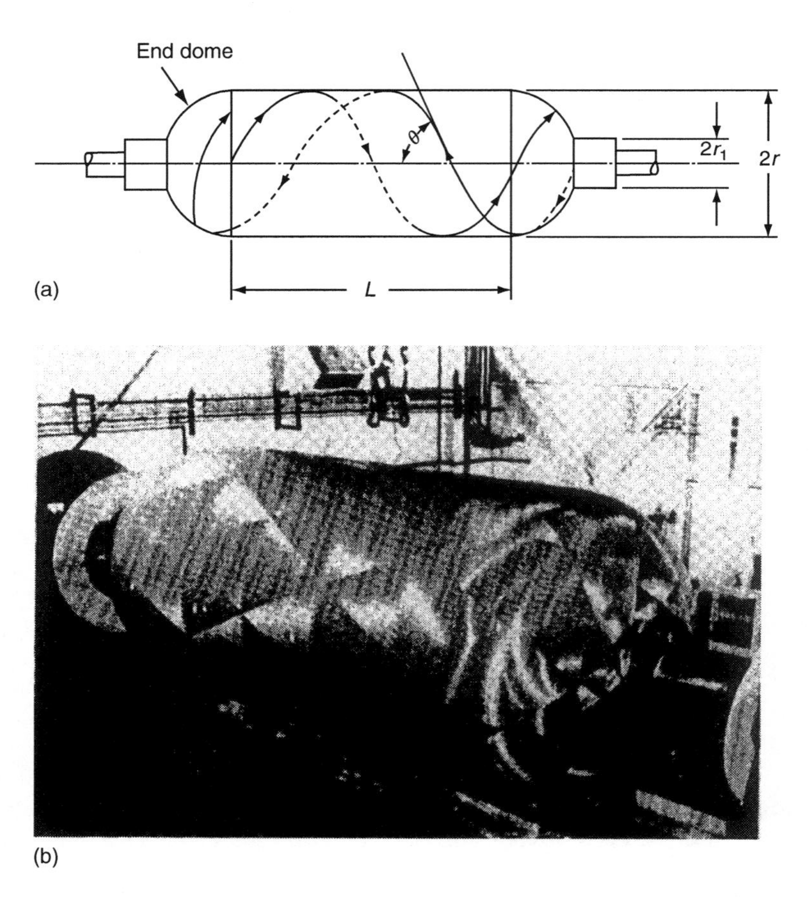

FIGURE 5.25 (a) Schematic of helical winding pattern and (b) a helically wound cylindrical tank. (Courtesy of En-Tec Technology, Inc. With permission.)

wind angle, as shown in Figure 5.26. In another type of filament-winding process, called polar winding, the carriage rotates about the longitudinal axis of a stationary (but indexable) mandrel. After each rotation of the carriage, the mandrel is indexed to advance one fiber bandwidth. Thus, the fiber bands lie adjacent to each other and there are no fiber crossovers. A complete wrap consists of two plies oriented at plus and minus the wind angle on two sides of the mandrel (Figure 5.27).

After winding a number of layers to generate the desired thickness, the filament-wound part is generally cured on the mandrel. The mandrel is then extracted from the cured part. To facilitate mandrel extraction, collapsible mandrels, either segmented or inflatable, are used for products in which the end closures are integrally wound, as in pressure vessels. For prototyping or for

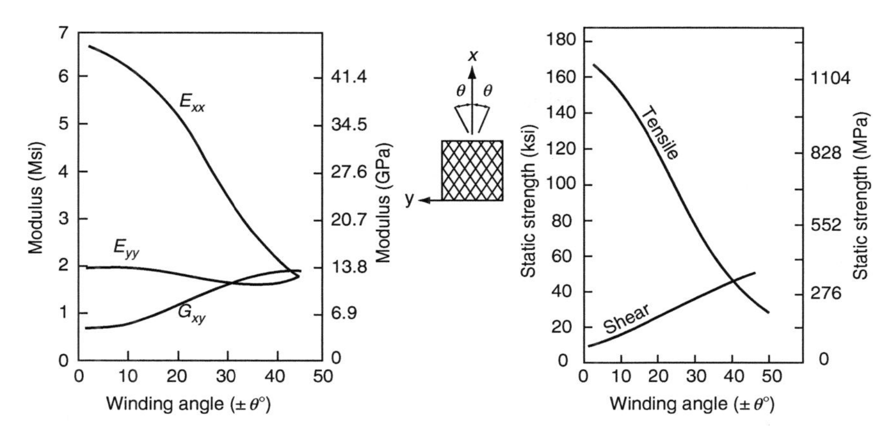

FIGURE 5.26 Mechanical property variation in a filament-wound part as a function of wind angle.

low-volume productions, soluble plasters, eutectic salts, or low-melting alloys are also used. However, a proper mandrel material must be able to resist sagging due to its own weight, withstand the applied winding tension, and keep its form during curing at elevated temperatures.

Both helical and polar winding processes require winding the fiber band around the mandrel ends. Hemispherical domes with central openings are commonly used at the mandrel ends for manufacturing pressure vessels. The central openings are necessary to extract the mandrel from the cured pressure vessel. Pins or rounded edges at the mandrel ends are used for manufacturing open-ended products, such as a pipe or a drive shaft.

Conventional filament-winding machines use a driving motor to rotate the mandrel and a chain and sprocket to move the carriage back and forth parallel to the mandrel (Figure 5.28). The main sprocket is connected to the mandrel shaft through a set of gears so that the carriage feed can be controlled in relation to the mandrel rotation by changing the gear ratios or the sprocket size. For a

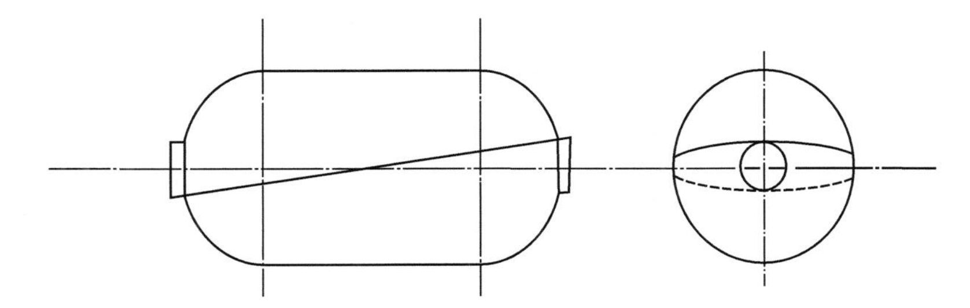

FIGURE 5.27 Polar winding pattern.

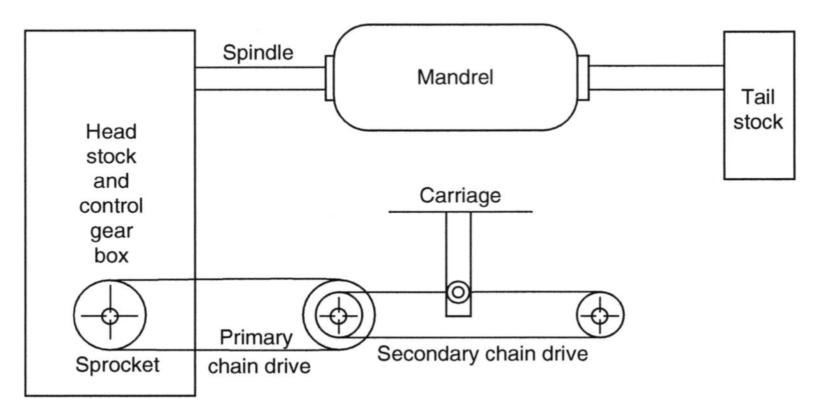

FIGURE 5.28 Schematic of a conventional filament-winding machine.

circular mandrel rotating with a constant rotational speed of N revolutions per minute and a constant carriage feed of V, the wind angle is given by

$$\theta = \frac{2\pi N r}{V}, \tag{5.12}$$

where r is the radius of the mandrel. Equation 5.12 shows that a constant wind angle can be maintained in a thick part only if the ratio N/V is adjusted from layer to layer. Winding with the conventional machine also requires the carriage to travel extra lengths on both sides of the mandrel.

The versatility of filament-winding process is improved tremendously if numerical controls are added to the filament-winding machine [22,23]. In numerically controlled machines, independent drives are used for the mandrel as well as for the carriage. In addition, a cross (transverse) feed mechanism and a rotating pay-out eye (Figure 5.29) allow an unequal fiber placement on the mandrel. The cross feed mechanism is mounted on the carriage and can move in and out radially; the pay-out eye can be controlled to rotate about a horizontal axis. The combination of these two motions prevents fiber slippage as well as fiber bunching on mandrels of irregular shape. Although each mechanism is driven by its own hydraulic motor, their movements are related to the mandrel rotation by numerical controls. Since no mechanical connections are involved, wind angles can be varied without much manual operation. With conventional filament-winding machines, the shapes that can be created are limited to surfaces of revolution, such as cylinders of various cross sections, cones, box beams, or spheroids (Figure 5.30). The computer-controlled multiaxis machines can wind irregular and complex shapes with no axis of symmetry, such as the aerodynamic shape of a helicopter blade.

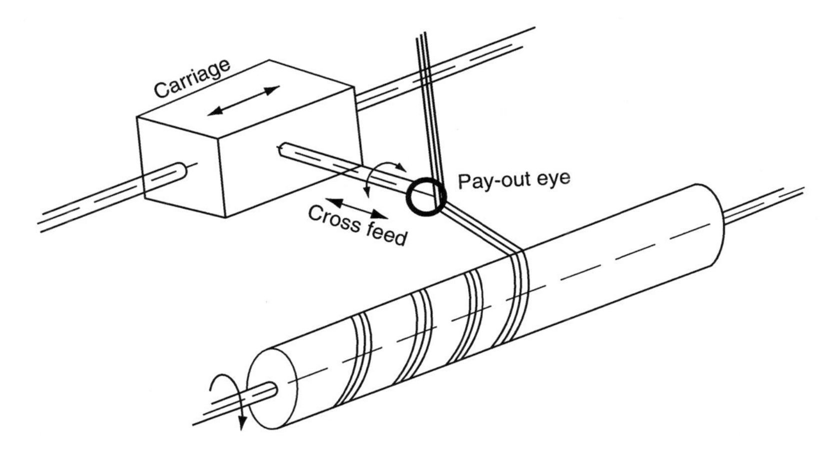

FIGURE 5.29 Schematic of a numerically controlled filament-winding machine.

The important process parameters in a filament-winding operation are fiber tension, fiber wet-out, and resin content. Adequate fiber tension is required to maintain fiber alignment on the mandrel as well as to control the resin content in the filament-wound part. Excessive fiber tension can cause differences in resin content in the inner and outer layers, undesirable residual stresses in the finished product, and large mandrel deflections. Typical tension values range from 1.1 to 4.4 N (0.25–1 lb) per end.

Fiber tension is created by pulling the rovings through a number of fiber guides placed between the creels and the resin bath. Figure 5.31 illustrates three

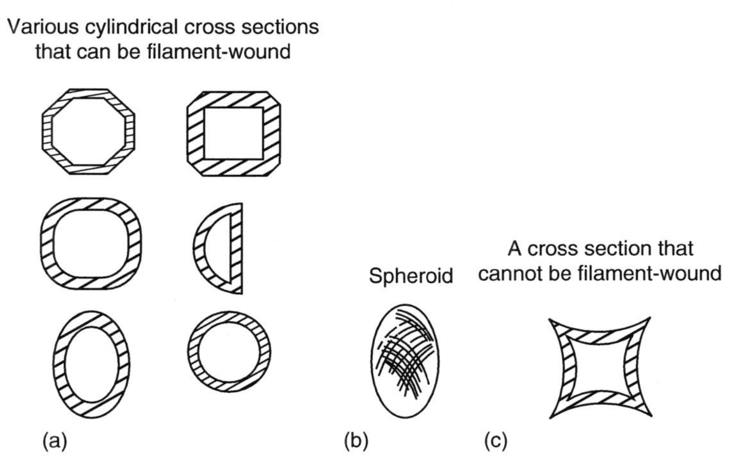

FIGURE 5.30 Cross sections of possible filament-wound parts (a and b); (c) demonstrates a cross section that cannot be filament wound.

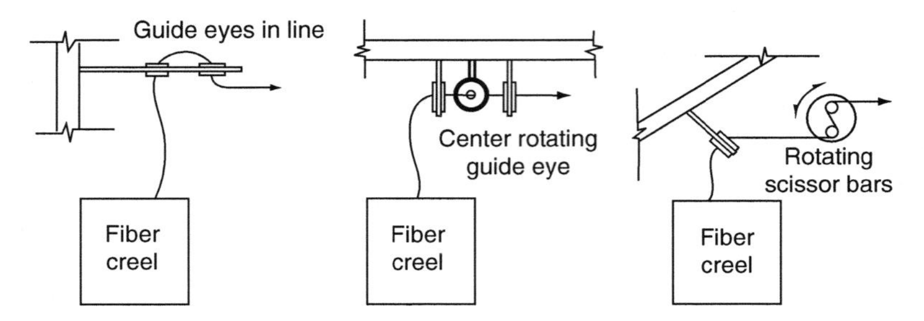

FIGURE 5.31 Typical fiber guides used in controlling fiber tension in a filament-winding line.

common types of fiber guides. Mechanical action on the fibers in the resin bath, such as looping, generates additional fiber tension.

Good fiber wet-out is essential for reducing voids in a filament-wound part. The following material and process parameters control the fiber wet-out:

1. Viscosity of the catalyzed resin in the resin bath, which depends on the resin type and resin bath temperature, as well as cure advancement in the resin bath
2. Number of strands (or ends) in a roving, which determines the accessibility of resin to each strand
3. Fiber tension, which controls the pressure on various layers already wound around the mandrel
4. Speed of winding and length of the resin bath

There are two essential requirements for the resin used in filament winding:

1. The viscosity of the mixed resin system (which may include a solvent) should be low enough for impregnating the moving fiber strands in the resin bath, yet not so low that the resin drips and runs out easily from the impregnated fiber strands. Usually, a viscosity level of 1–2 Pa s (1000–2000 cP) is preferred.
2. The resin must have a relatively long pot life so that large structures can be filament-wound without premature gelation or excessive exotherm. Furthermore, the resin bath is usually heated to lower the viscosity level of the mixed resin system. Since increased temperature of the resin bath may reduce the pot life, a resin with short pot life at room temperature has a limited usefulness in filament winding.

As a rule of thumb, each roving should be under the resin surface level for 1/3–1/2 s. In a line moving at 60 m/min (200 ft./min), this means that the length of roving under the resin surface level be ~30 cm (1 ft.). For good wetting, the minimum roving length under the resin surface level is 15 cm (6 in.).

Proper resin content and uniform resin distribution are required for good mechanical properties as well as for weight and thickness control. Resin content is controlled by proper wiping action at the squeegee bars or stripper die, fiber tension, and resin viscosity. Dry winding in which prepregs are wound around a mandrel often provides a better uniform resin distribution than the wet winding process.

Fiber collimation in a multiple-strand roving is also an important consideration to create uniform tension in each strand as well as to coat each strand evenly with the resin. For good fiber collimation, single-strand rovings are often preferred over conventional multiple-strand rovings. Differences in strand lengths in conventional multiple-strand rovings can cause sagging (catenary) in the filament-winding line.

In a helical winding operation, the fiber bands crisscross at several points along the length of the mandrel, and one complete layer consists of a balanced helical pattern with fiber oriented in the $+\theta°/-\theta°$ directions. The thickness of a layer depends on the band density (i.e., the number of rovings per unit length of a band), the roving count (i.e., the number of strands or ends per roving), and the resin content. For the same band density, a high roving count results in larger amounts of fibers and, therefore, thicker layers. Increasing the resin content also results in thicker layers. An example is shown in Figure 5.32.

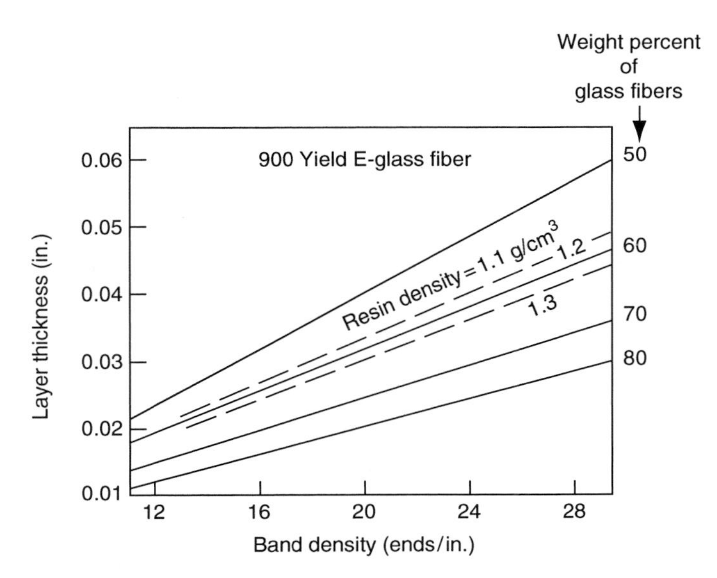

FIGURE 5.32 Effects of roving count, fiber weight fraction, and matrix density (in g/cm^3) of the thickness of a filament-wound E-glass fiber-reinforced epoxy part. (Adapted from Filament Winding, Publication No. 5-CR-6516, Owens-Corning Fiberglas, 1974.)

The common defects in filament-wound parts are voids, delaminations, and fiber wrinkles. Voids may appear because of poor fiber wet-out, the presence of air bubbles in the resin bath, an improper bandwidth resulting in gapping or overlapping, or excessive resin squeeze-out from the interior layers caused by high winding tension. In large filament-wound parts, an excessive time lapse between two consecutive layers of windings can result in delaminations, especially if the resin has a limited pot life. Reducing the time lapse and brushing the wound layer with fresh resin just before starting the next winding are recommended for reducing delaminations. Wrinkles result from improper winding tension and misaligned rovings. Unstable fiber paths that cause fibers to slip on the mandrel may cause fibers to bunch, bridge, and improperly orient in the wound part.

5.6 LIQUID COMPOSITE MOLDING PROCESSES

In liquid composite molding (LCM) processes, a premixed liquid thermoset resin is injected into a dry fiber perform in a closed mold. As the liquid spreads through the preform, it coats the fibers, fills the space between the fibers, expels air, and finally as it cures, it transforms into the matrix. This section describes two LCM processes, namely RTM and structural reaction injection molding (SRIM).

5.6.1 Resin Transfer Molding

In RTM, several layers of dry continuous strand mat, woven roving, or cloth are placed in the bottom half of a two-part mold, the mold is closed, and a catalyzed liquid resin is injected into the mold via a centrally located sprue. The resin injection point is usually at the lowest point of the mold cavity. The injection pressure is in the range of 69–690 kPa (10–100 psi). As the resin flows and spreads throughout the mold, it fills the space between the fiber yarns in the dry fiber preform, displaces the entrapped air through the air vents in the mold, and coats the fibers. Depending on the type of the resin–catalyst system used, curing is performed either at room temperature or at an elevated temperature in an air-circulating oven. After the cured part is pulled out of the mold, it is often necessary to trim the part at the outer edges to conform to the exact dimensions.

Instead of using flat-reinforcing layers, such as a continuous strand mat, the starting material in an RTM process can be a preform that already has the shape of the desired product (Figure 5.33). The advantages of using a preform are good moldability with complicated shapes (particularly with deep draws) and the elimination of the trimming operation, which is often the most labor-intensive step in an RTM process.

There are various methods of producing preforms. One of them is the spray-up process in which 12.7–76.2 mm (0.5–3 in.) long fibers, chopped from

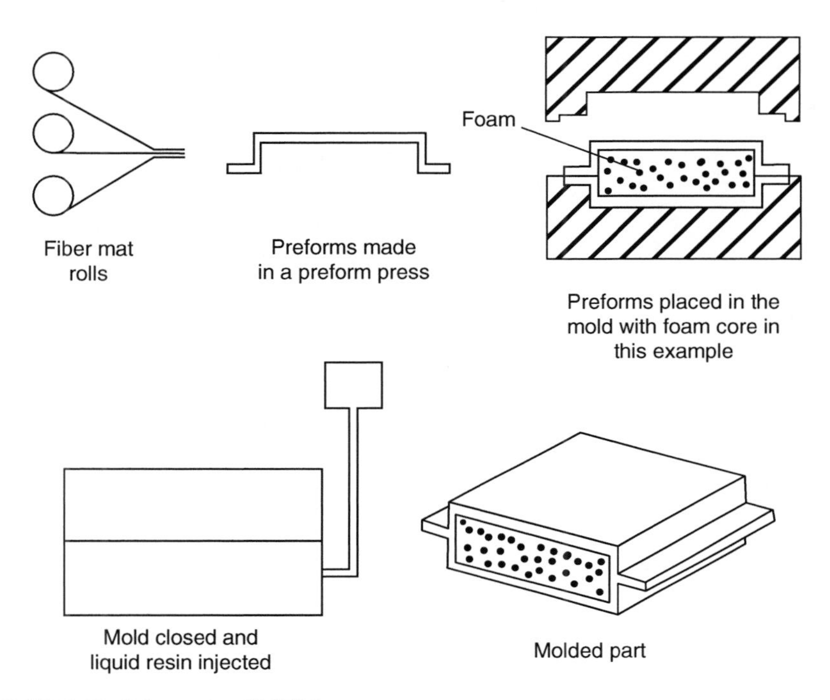

FIGURE 5.33 Schematic of RTM process.

continuous fiber rovings, are sprayed onto a preshaped screen. Vacuum applied to the rear side of the screen holds the fibers securely on the screen. A binder sprayed with the fibers keeps them in place and maintains the preformed shape. Continuous strand mats containing random fibers can be preformed by a stamping operation using a simple press and a preshaped die (Figure 5.34). Both thermoplastic and thermoset binders are available for retaining the formed shape after stamping. With woven fabric containing bidirectional fibers, a "cut and sew" method is used in which various patterns are first cut from the fabric and then stitched together by polyester, glass, or Kevlar sewing threads into the shape of the part that is being produced. Braiding and textile weaving processes have also been used to produce two- or three-dimensional preforms [24]. Braiding is particularly suitable for producing tubular preforms.

Compared with the compression-molding process, RTM has a very low tooling cost and simple mold clamping requirements. In some cases, a ratchet clamp or a series of nuts and bolts can be used to hold the two mold halves together. RTM is a low-pressure process, and therefore parts can be resin transfer molded in low-tonnage presses. A second advantage of the RTM process is its ability to encapsulate metal inserts, stiffeners, washers, and so on within the molded laminate. It is also possible to encapsulate a foam core between the top and bottom preforms of a hollow part, which adds stiffness to

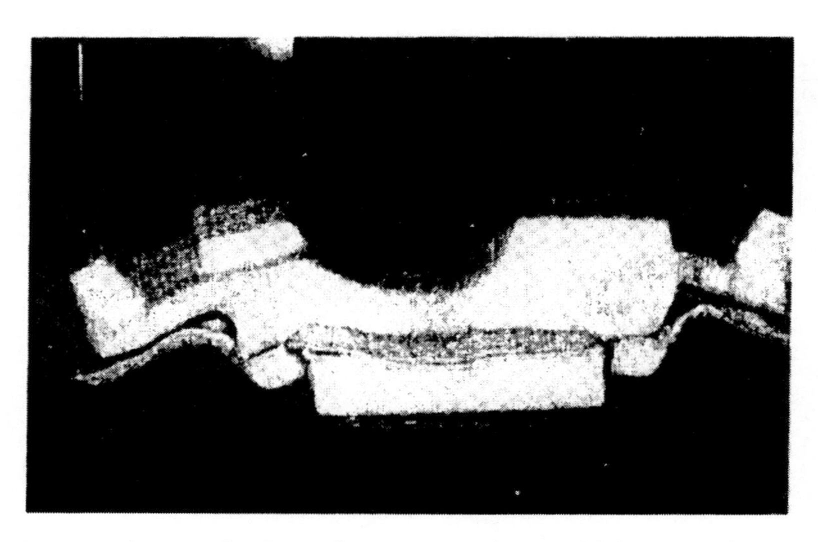

FIGURE 5.34 Photograph of a preform stamped from multiple layers of continuous strand random glass fiber mat. (Courtesy of Ford Motor Co. With permission.)

the structure and allows molding of complex three-dimensional shapes in one piece. The RTM process has been successfully used in molding such parts as cabinet walls, chair or bench seats, hoppers, water tanks, bathtubs, and boat hulls. It also offers a cost-saving alternative to the labor-intensive bag-molding process or the capital-intensive compression-molding process. It is particularly suitable for producing low- to midvolume parts, say 5,000–50,000 parts a year.

There are several variations of the basic RTM process described earlier. In one of these variations, known as vacuum-assisted RTM (VARTM), vacuum is used in addition to the resin injection system to pull the liquid resin into the preform. Another variation of the RTM process is SCRIMP, which stands for Seemann's Composite Resin Infusion Molding Process, a patented process named after its inventor William Seemann. Vacuum is also used in SCRIMP to pull the liquid resin into the dry fiber preform, but in this process, a porous layer is placed on the preform to distribute the resin uniformly throughout the preform. The porous layer is selected such that it has a very low resistance to flow and it provides the liquid resin an easy flow path to the preform. In both VARTM and SCRIMP, a single-sided hard mold is used. The preform is placed on the hard mold surface (Figure 5.35) and covered with a vacuum bag, much like the bag-molding process described in Section 5.2.

5.6.2 Structural Reaction Injection Molding

Another manufacturing process very similar to RTM is called structural reaction injection molding (SRIM). It also uses dry fiber preform that is

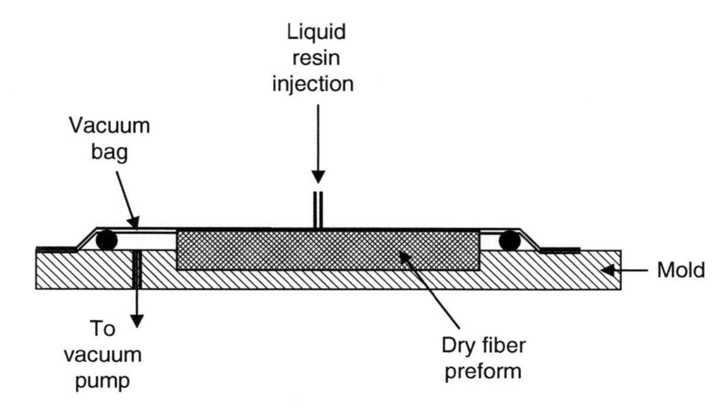

FIGURE 5.35 Vacuum-assisted resin transfer molding (VARTM).

placed in the mold before resin injection. The difference in RTM and SRIM is mainly in the resin reactivity, which is much higher for SRIM resins than for RTM resins. SRIM is based on the reaction injection molding (RIM) technology [25] in which two highly reactive, low-viscosity liquid streams are impinged on each other at high speeds in a mixing chamber immediately before injecting the liquid mix into a closed mold cavity (Figure 5.36).

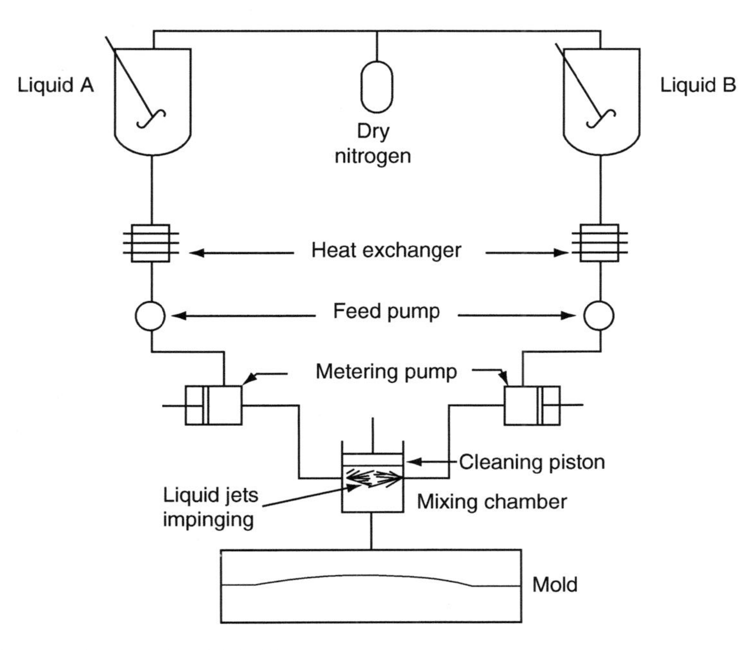

FIGURE 5.36 Schematic of SRIM process.

Commercial RIM resins are mostly based on polyurethane chemistry, although epoxies have also been used. The curing temperature for polyurethane resins is between 60°C and 120°C.

The reaction rate for the resins used in RIM or SRIM is much faster than epoxy, polyester, or vinyl ester resins that are commonly used for the RTM process. The curing time for the SRIM resins is in the range of a few seconds compared with a few minutes for the RTM resins. The molding pressure for both processes is in the range of 0.5–1.5 MPa (70–210 psi).

In both RTM and SRIM processes, the liquid resin flows through layers of dry fiber preform while the curing reaction continues. For producing good quality parts, it is imperative that the resin fills the mold completely and wets out the reinforcement before arriving at the gel point. Therefore, the resin viscosity in both processes must be low. However, since the curing reaction is much faster in SRIM, the initial viscosity of SRIM resins must be lower than that for RTM resins. Preferred room temperature viscosity range for SRIM resins is 0.01–0.1 Pa s (10–100 cP) compared with 0.1–1 Pa s (100–1000 cP) for RTM resins. Since the reaction rate of the liquid resin mix in SRIM is very high, its viscosity increases rapidly, and therefore, the mold must be filled very quickly. For this reason, preforms in SRIM are, in general, lower in fiber content and simpler in shape than in RTM.

The quality of liquid composite-molded parts depends on resin flow through the dry fiber preform, since it determines mold filling, fiber surface wetting, and void formation. The principal molding problems observed are incomplete filling, dry spots, nonuniform resin distribution, void formation, nonuniform cure, and low degree of cure. The main source of void formation is the air entrapped in the complex fiber network in the preform. Good resin flow and mold venting are essential in reducing the void content in the composite. There may also be fiber displacement and preform distortion as the liquid resin moves through the fiber preform, especially if the viscosity increases rapidly before the mold filling is complete.

5.7 OTHER MANUFACTURING PROCESSES

5.7.1 Resin Film Infusion

In the resin film infusion (RFI) process, a precatalyzed resin film placed under the dry fiber preform provides the liquid resin that flows through the preform and on curing, becomes the matrix. The process starts by covering the mold surface with the resin film and then placing the dry fiber preform on top of the resin film (Figure 5.37). The thickness of the resin film depends on the quantity of resin needed to completely infiltrate the preform.

RFI can be carried out using the bag-molding technique described in Section 5.2. In that case, the assembly of resin film and dry fiber preform is covered with a vacuum bag and placed inside an autoclave. The full vacuum

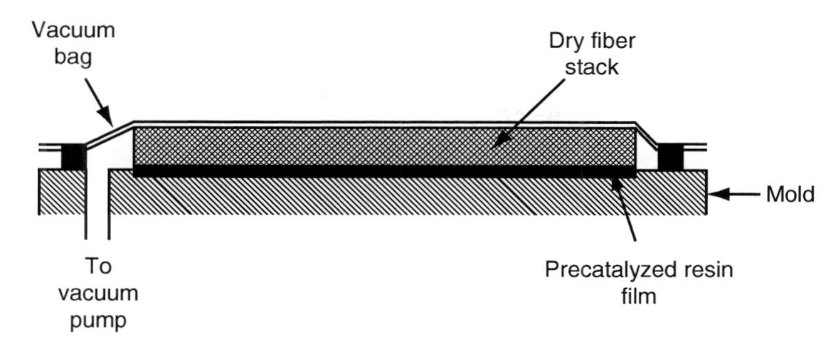

FIGURE 5.37 Schematic of resin film infusion process.

is applied at the beginning to remove the trapped air from the preform. As the temperature is increased in the autoclave, the resin viscosity decreases and the resin starts to flow through the dry fiber preform. Pressure is applied to force the liquid resin to infiltrate the preform and wet out the fibers. With the temperature now raised to the prescribed curing temperature, the curing reaction begins and the liquid resin starts to gel. If an epoxy film is used, the curing cycle is similar to that shown in Figure 5.10 and may take several minutes to several hours depending on the resin type and the curing conditions used.

5.7.2 Elastic Reservoir Molding

In elastic reservoir molding (ERM), a sandwich of liquid resin-impregnated open-celled foam and face layers of dry continuous strand mat, woven roving, or cloth placed in a heated mold (Figure 5.38) are pressed with a molding pressure of 520–1030 kPa (75–150 psi). The foam in the center of the sandwich

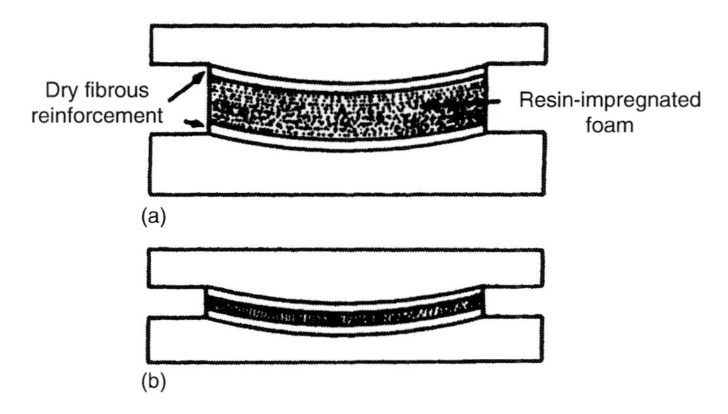

FIGURE 5.38 Schematic of the elastic reservoir molding process.

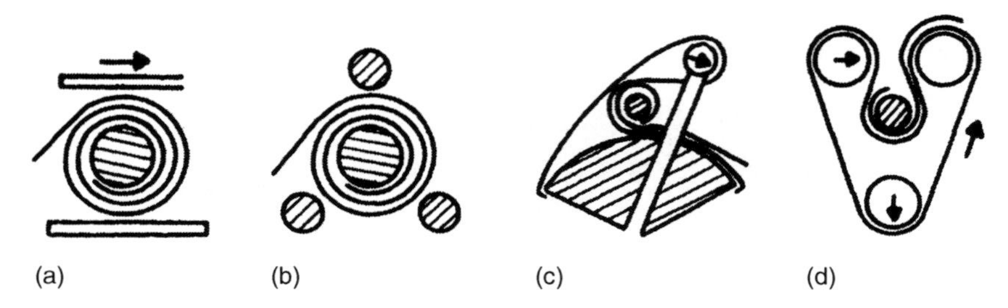

FIGURE 5.39 Schematic of various tube-rolling methods.

is usually a flexible polyurethane that acts as an elastic reservoir for the catalyzed liquid resin. As the foam is compressed, resin flows out vertically and wets the face layers. On curing, a sandwich of low-density core and fiber-reinforced skins is formed.

Among the advantages of an ERM process are low-cost tooling, better control of properties (since there is no horizontal flow), and a better stiffness–weight ratio (due to the sandwich construction). It is generally restricted to molding thin panels of simple geometry. Examples of ERM applications are bus roof panels, radar reflecting surfaces, automotive body panels, and luggage carriers.

5.7.3 Tube Rolling

Circular tubes for space truss or bicycle frame, for example, are often fabricated from prepregs using the tube-rolling technique. In this process, precut lengths of a prepreg are rolled onto a removable mandrel, as illustrated in Figure 5.39. The uncured tube is wrapped with a heat-shrinkable film or sleeve and cured at elevated temperatures in an air-circulating oven. As the outer wrap shrinks tightly on the rolled prepreg, air entrapped between the layers is squeezed out through the ends. For a better surface finish, the curing operation can be performed in a close-fitting steel tube or a split steel mold. The outer steel mold also prevents the mandrel from sagging at the high temperatures used for curing. After curing, the mandrel is removed and a hollow tube is formed. The advantages of the tube-rolling process over the filament-winding process are low tooling cost, simple operation, better control of resin content and resin distribution, and faster production rate. However, this process is generally more suitable for simple layups containing 0° and 90° layers.

5.8 MANUFACTURING PROCESSES FOR THERMOPLASTIC MATRIX COMPOSITES

Many of the manufacturing processes described earlier for thermoset matrix composites are also used for thermoplastic matrix composites. However, there are some critical differences that arise due to the differences in physical and thermal characteristics of these two types of polymers. For example, thermoplastic prepregs are not tacky (sticky). They are also not very flexible, which poses problems in draping them into contoured mold surfaces. To overcome the problem associated with the lack of stickiness, thermoplastic prepreg layups are spot-welded together along the outside edges. One method of spot welding is to use a hot soldering iron and light pressure, which causes the matrix to melt and fuse at the edges. Laser beams have also been used to melt and fuse thermoplastic prepregs in continuous tape-laying processes.

The processing temperatures required for thermoplastic matrix composites are much higher than the curing temperatures required for thermoset matrix composites. Therefore, if a bag-molding process is used, the bagging material and sealant tapes must be of high-temperature type.

Unlike thermoset matrix composites, no chemical reaction occurs during the processing of thermoplastic matrix composites. However, individual plies in the stack must still be consolidated to form a laminate, which requires both high temperature and pressure. The stacked layup can be heated rapidly by means of quartz lamps or infrared heaters. The consolidation time may range from a few seconds to several minutes, depending on the laminate thickness and geometry. After consolidation, the laminate must be cooled at a controlled rate to solidify the matrix without causing residual stresses, warpage, and so on. For a semicrystalline thermoplastic matrix, such as PEEK, the crystallinity in the solidified matrix depends very much on the cooling rate (Figure 5.40). Slower cooling rate produces a higher crystallinity in the matrix, which in turn may influence the matrix properties, particularly its fracture toughness.

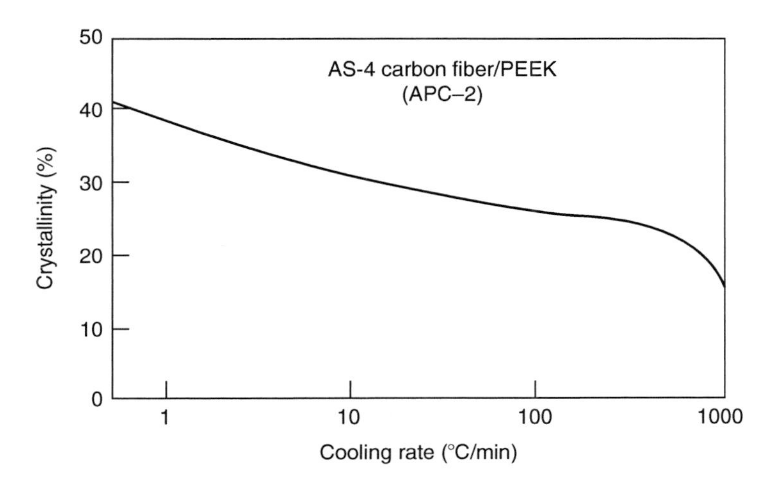

FIGURE 5.40 Crystallinity in PEEK thermoplastic matrix composite as a function of cooling rate.

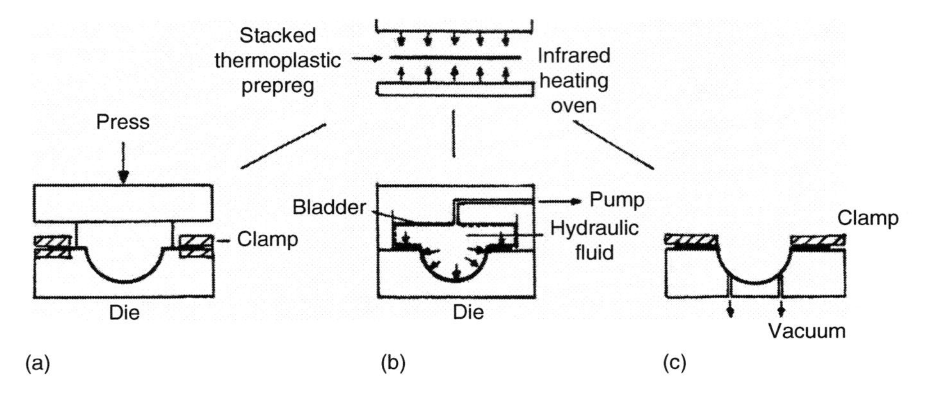

FIGURE 5.41 Forming methods for thermoplastic matrix composites: (a) matched die forming, (b) hydroforming, and (c) thermoforming.

Since thermoplastic matrix composites can be shaped and formed repeatedly by the application of heat and pressure, they can be processed using some metal-working as well as thermoplastic forming techniques, such as matched die forming, hydroforming, and thermoforming [26]. These are highly efficient processes for converting flat sheets into three-dimensional objects at relatively high production rates.

Matched die forming is a widely used forming technique for sheet metals. It uses two matching metal dies mounted in a hydraulic press (Figure 5.41a). The deformation produced in the sheet during forming is a combination of stretching, bending, and drawing. It is suitable for forming constant thickness parts, since the dies are generally designed to a fixed gap with close tolerances. The forming pressure for thermoplastic matrix composites is considerably lower than that for metals. However, if a thickness variation exists in the part, the pressure distribution will be nonuniform, which may result in nonuniform ply consolidation.

Hydroforming uses a hydraulic fluid inside an elastomeric diaphragm to generate pressure required for deforming and consolidating the layup (Figure 5.41b). Only one metal die is required for the hydroforming operation. Although this process is limited due to temperature limitation of the elastomeric diaphragm, it allows a better control on the forming pressure distribution in parts with nonuniform thickness.

Thermoforming is a common manufacturing technique in the plastics industry for forming unreinforced thermoplastic sheets into trays, cups, packages, bathtubs, small boats, and so on. In this process, the sheet is preheated to the forming temperature, placed in the mold, clamped around the edges, and formed into the mold cavity by the application of either vacuum, pressure, or both (Figure 5.41c). The forming temperature is usually higher than the glass transition temperature T_g of the polymer. After forming, the part is cooled

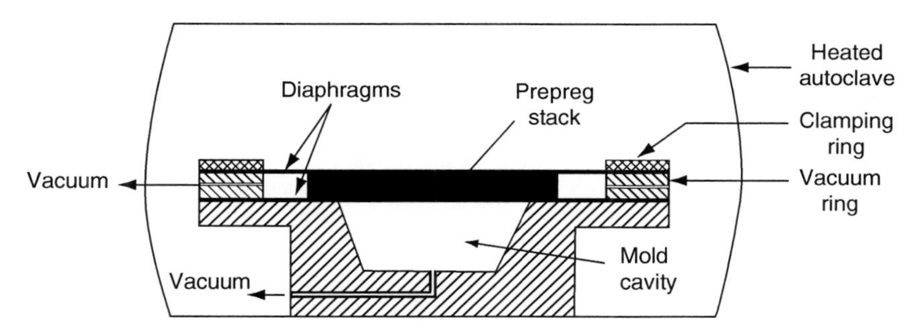

FIGURE 5.42 Schematic of diaphragm forming.

under pressure to below the glass transition temperature before it is removed from the mold. If this process is applied to thermoplastic matrix composites, various layers in the layup must be consolidated into a laminated sheet before thermoforming. If consolidation takes place during the thermoforming process, the forming temperature will be close to the melt temperature of the polymer.

In the forming operation, the sheet is both stretched and drawn into the final shape. However, for thermoplastic matrix composites containing continuous fibers that are inextensible, it is not possible to stretch the individual plies in the fiber direction without breaking the fibers. Therefore, the thermoplastic matrix composite layup cannot be clamped and fixed around its edges. On the other hand, if the plies are not in tension during forming, some of them may wrinkle. To overcome this problem, the layup is placed between two thin, highly deformable diaphragms, which are clamped around their edges (Figure 5.42). As the forming pressure is applied, the deformation of diaphragms creates a biaxial tension in the layup, which prevents the plies from wrinkling. Common diaphragm materials are superplastic aluminum alloys and polyimide films.

A typical process cycle for diaphragm forming of carbon fiber-reinforced PEEK in an autoclave is shown in Figure 5.43. The temperature in the autoclave is slowly increased to 380°C, which is ~40°C above the melting point of PEEK. The autoclave pressure is increased to ~0.4 MPa (60 psi) to force the layup into the mold cavity and is maintained at that pressure to consolidate the laminate. The thickness of the layup decreases during consolidation as the free spaces between the plies are eliminated and interlaminar adhesion builds up. There may also be some rearrangement of fibers and matrix squeeze-out during consolidation.

Figure 5.44 depicts the four basic mechanisms that may occur during consolidation and forming of thermoplastic matrix composites [27]:

1. Percolation or flow of molten polymer through fiber layers
2. Transverse flow (movement in the thickness direction) of fibers or fiber networks

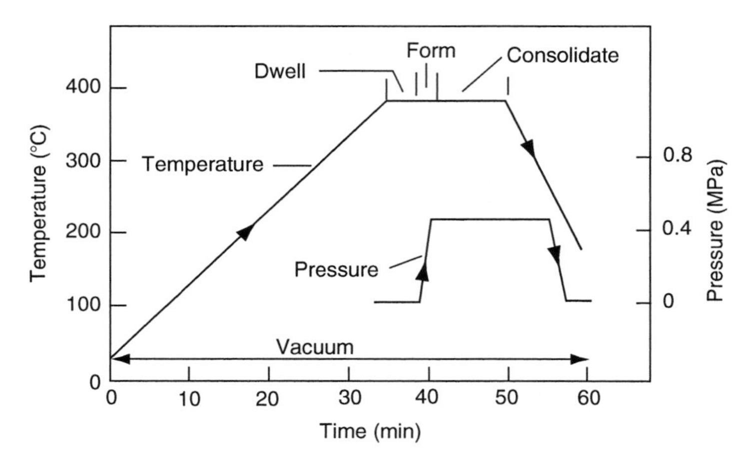

FIGURE 5.43 Typical process cycle in diaphragm forming. (Adapted from Mallon, P.J., O'Bradaigh, C.M., and Pipes, R.B., *Composites*, 20, 48, 1989.)

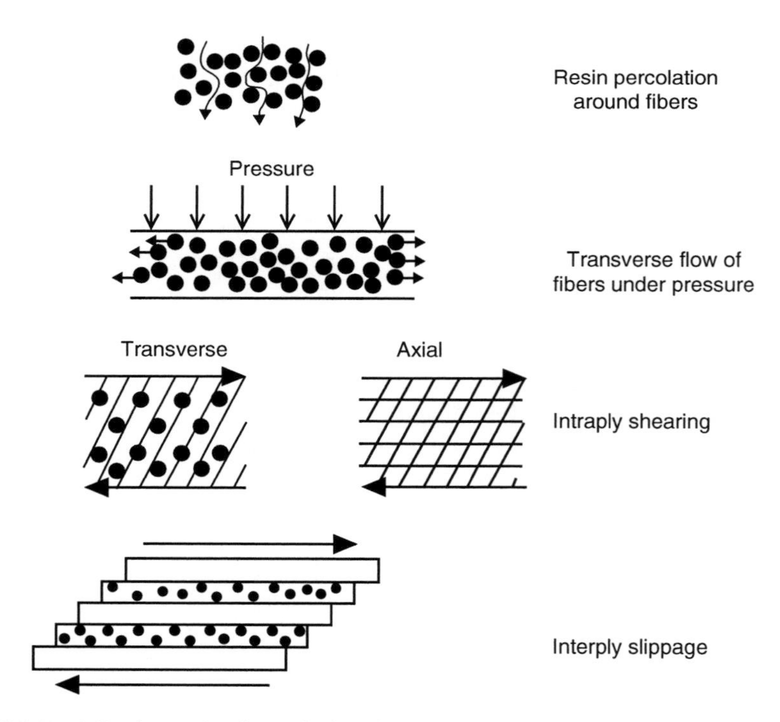

FIGURE 5.44 Basic mechanisms during forming of thermoplastic matrix composites. (Adapted from Cogswell, F.N., *Int. Polym. Proc.*, 1, 157, 1987.)

3. Intraply shearing which allows fibers within each ply to move relative to each other in the axial as well as in the transverse direction
4. Interply slip, by which plies slide over each other while conforming to the shape of the die

The first two mechanisms are essential for good consolidation, since they help reduce the gaps and voids that exist between the plies as well as within each ply and create good interlaminar adhesion. The other two mechanisms help in forming contoured shapes without fiber wrinkling, splitting, or local thinning.

5.9 QUALITY INSPECTION METHODS

The mechanical properties of a molded part and their variations from part to part depend on many factors, including the quality of raw materials, the degree of cure, and the presence of molding defects in the part. Many of the molding defects can be either eliminated or reduced by proper part and mold designs as well as by controlling the various process parameters described in the earlier sections. Since complete elimination of all molding defects is not possible, it is important that the quality of the molded parts be inspected regularly for critical defects that can cause premature failure in a part during its service. The criticality of a defect or a group of defects is established through extensive testing at the prototype development stage. The quality inspection techniques and the part acceptance criteria are also established at the prototype development stage.

5.9.1 Raw Materials

Among the raw materials, the two most important candidates for close quality inspection are fibers and resin. Measurements of tensile strength and modulus are the primary screening methods for continuous fibers. For the resin, curing agents, or diluents, the standard quality inspection items are the density, viscosity, color, and moisture content. Additional tests recommended for the incoming resin and other components [28] are (1) wet chemical analysis to determine the amount of functional groups (such as epoxide groups in epoxies or acid numbers in polyesters) in the material, (2) infrared (IR) spectroscopy or nuclear magnetic resonance (NMR) spectroscopy to fingerprint the chemical structure and impurities, and (3) liquid chromatography or gel permeation chromatography (GPC) to determine the weight-average molecular weight and molecular weight distribution of the resin molecules.

The average molecular weight and molecular weight distribution are two very important characteristics of a resin that control its viscosity and mechanical properties. Burns et al. [29] have shown that high fractions of low-molecular-weight molecules in a polyester resin reduce the rate of thickening (viscosity increase) in SMC sheets. The viscosity of a polyester resin with low-average molecular weight may not attain the level required for proper handling (0.5–1.3 $\times$ 10^6 P) even after a thickening period of 7 days or more. Another factor in controlling the thickening rate of polyester resins was found to be the moisture content.

Raw materials are often purchased by the part manufacturer in the form of prepreg rolls. The prepreg characteristics that influence its moldability as well as its mechanical properties are resin content, volatile content, filament count, filament diameter, gel time, and resin flow. The gel-time and resin-flow tests can indicate cure advancement in the B-staged resin, which in turn is related to the tackiness and drapability of the prepreg as well as to the fluidity of the resin during the molding process. Proper tack is required to (1) adhere the prepreg to the mold surface as well as to the preceding ply and (2) release the backup film without separating the resin from the prepreg. The prepreg should also be sufficiently drapable to conform to the contour of the mold surface.

The gel-time and resin-flow tests are inadequate to detect variations in resin formulation. To improve the quality assurance of the B-staged resin in the prepreg, chemical and rheological tests, such as liquid chromatography, differential scanning calorimetry (DSC), and dynamic mechanical analysis (DMA), should be adopted [30].

5.9.2 Cure Cycle Monitoring

Cure cycle monitoring during production runs is important to ensure that each molded part has the same degree of cure as has been established earlier in the prototype development stage. Accurate control over the cure cycle helps achieve greater consistency between the parts and reduce the amount of scrap or waste.

Many research tools are available for monitoring the curing process of a fiber-reinforced thermoset polymer. They include wet chemical analysis, infrared spectroscopy, NMR, DSC, DMA, and dielectrometry. Among these, dielectrometry is considered the most promising technique to monitor cure cycle during a production molding operation [31]. In this technique, the dielectric loss factor of the resin is measured as a function of cure time. The instrumentation includes two conductor probes (electrodes), which are usually embedded in the top and bottom mold surfaces and are connected to an alternating electric field. Since polymers are dielectric materials, the combination of these probes and fiber–resin system forms a parallel-plate capacitor. The charge accumulated in this capacitor depends on the ability of the dipoles and ions present in polymer molecules to follow the alternating electric field at different stages of curing. The loss factor of a resin represents the energy expended in aligning its dipoles and moving its ions in accordance with the direction of the alternating electric field.

A typical dielectric loss factor plot obtained during the molding of a fiber-reinforced thermoset polymer is shown in Figure 5.45. At the beginning of the

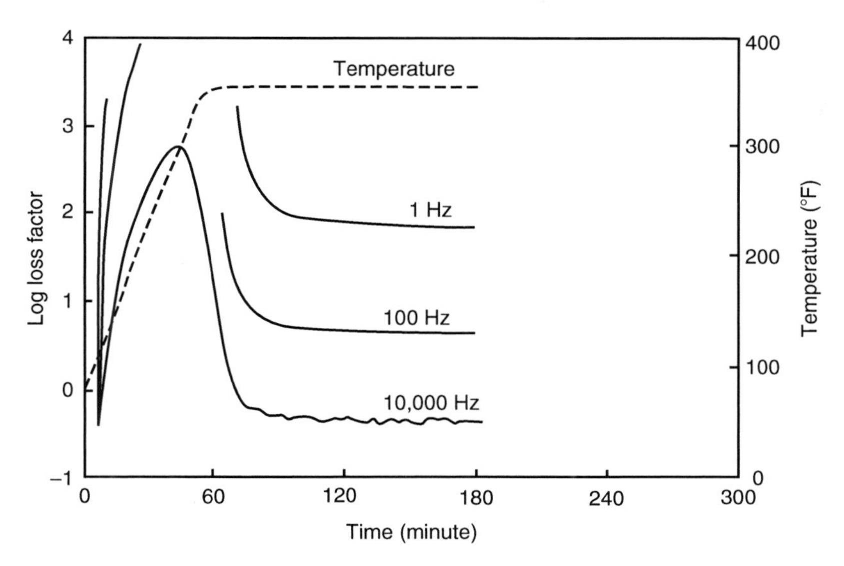

FIGURE 5.45 Dielectric loss factor as a function of cure time and frequency of the oscillating electric field.

cure cycle, the resin viscosity in the uncured prepreg is relatively high so that the dipole and ion mobilities are restricted. This results in low loss factor just after the uncured prepreg is placed in the mold. As the temperature of the prepreg increases in the mold, the resin viscosity is reduced and the loss factor increases owing to greater dipole and ion mobilities. As soon as the gel point is reached, the resin viscosity increases rapidly and the loss factor decreases. At a full degree of cure, the loss factor levels off to a constant value.

5.9.3 Cured Composite Part

Quality inspection of cured composite parts includes both destructive and nondestructive tests (NDT). Examples of routine destructive tests are burn-off tests for checking the fiber weight fraction and tension tests on specimens cut from the finished parts for checking their strength and modulus values. Performance tests on randomly selected parts are also recommended. Simple NDT include thickness measurements, visual inspection for surface defects, and proof tests. In proof tests, each part is loaded to predetermined stress levels, which are usually lower than the design stress level.

A cured composite part may contain a multitude of internal defects, such as voids, delaminations, fiber misorientations, and nonuniform fiber distribution. Some of these internal defects may act as or grow into critical flaws during the service operation of a part and severely affect its performance. During a production process, these defects are detected by NDT, and parts are either accepted or rejected on the basis of defect quality standards developed earlier at the prototype development stage. In the event of service failure, the NDT records can also serve a useful purpose in analyzing the cause of failure. At the present time, both ultrasonic and radiographic tests are performed on structural composites used in aircraft or aerospace applications. Other NDT methods, such as the acoustic emission test, thermography, and the acousto-ultrasonic test, are used mostly as research tools to monitor damage development during mechanical tests of composite specimens. A common problem with all these tests, including ultrasonic and radiography, is the lack of standards that can be used to distinguish between critical and noncritical defects.

5.9.3.1 Radiography

In radiographic techniques, the internal structure of a molded part is examined by impinging a beam of radiation on one of its surfaces and recording the intensity of the beam as it emerges from the opposite surface. Conventional radiography uses x-rays (in the range of 7–30 keV) as the source of radiation and records the internal defects as shadow images on a photographic film. Gamma rays are more useful in thicker parts, since they possess shorter wavelengths and, hence, greater penetrating power than x-rays. Other radiation beams, such as beta irradiation and neutron radiation, are also used. Imaging techniques, such as displaying the image on a fluorescent screen (fluoroscopy) or cross-sectional scanning [32] by computed tomography (CT), are also available. The former is more useful for on-line inspection of production parts than the photographic technique.

In fiber-reinforced composite structures, radiography is capable of detecting large voids, foreign inclusions, translaminar cracks, and nonuniform fiber distribution, as well as fiber misorientation (such as weld lines or fiber wrinkles). These defects change the intensity of the radiation beam by varying amounts and create images of different shades and contrasts on the photographic film. Thus, for example, large voids appear as dark spots and fiber-rich areas appear as light streaks on an x-ray film. The detection of microvoids and delaminations is possible by using radiopaque penetrants, such as sulfur, trichloroethylene, or carbon tetrachloride. The detection of fiber misorientation may require the use of lead glass tracers in the prepreg or SMC.

5.9.3.2 Ultrasonic

Ultrasonic inspection uses the energy levels of high-frequency (1–25 MHz) sound waves to detect internal defects in a test material. The ultrasonic sound energy is generated by electrically exciting a piezoelectric transducer and is introduced into the surface of a molded part by means of a coupling medium. As the ultrasonic waves propagate through the material, their energy levels are attenuated by the presence of defects. Although some of the attenuated

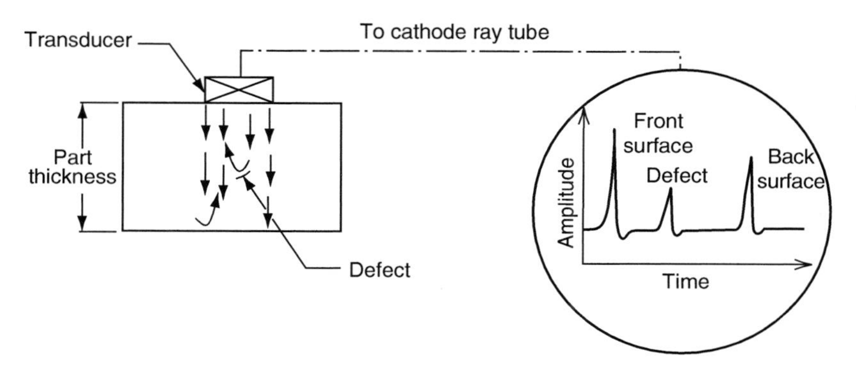

FIGURE 5.46 Pulse-echo method of ultrasonic testing.

ultrasonic waves are transmitted through the part thickness, others are reflected back to the input surface. The energy levels of these transmitted and reflected ultrasonic waves are converted into electrical signals by a receiving transducer and are then compared with a preset threshold and displayed on a cathode ray tube (CRT) screen or a computer screen.

The following methods are commonly used for ultrasonic inspection of defects in a fiber-reinforced composite material.

Pulse-echo method: In this method, echos reflecting from the front surface, back surface, and internal defects are picked up either by the transmitting transducer or by a separate receiving transducer. All reflected pulses are displayed as distinct peaks on the monitor (Figure 5.46). Pulse-echo depths are determined by measuring the time intervals between the front surface reflection peak and other significant peaks. Knowing the ultrasonic wave velocity in the material, these time intervals can be converted into defect location (depth) or part thickness measurements.

Through-transmission method: In this method, ultrasonic waves transmitted through the part thickness are picked up by a receiving transducer on the other side of the part (Figure 5.47). Since the transmitted wave interacting with a

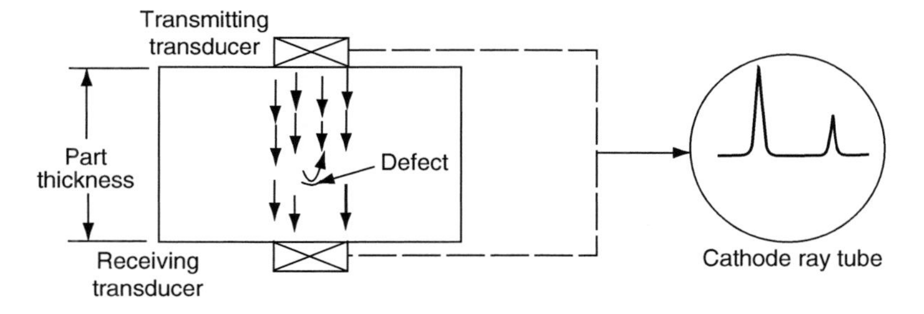

FIGURE 5.47 Through-transmission method of ultrasonic testing.

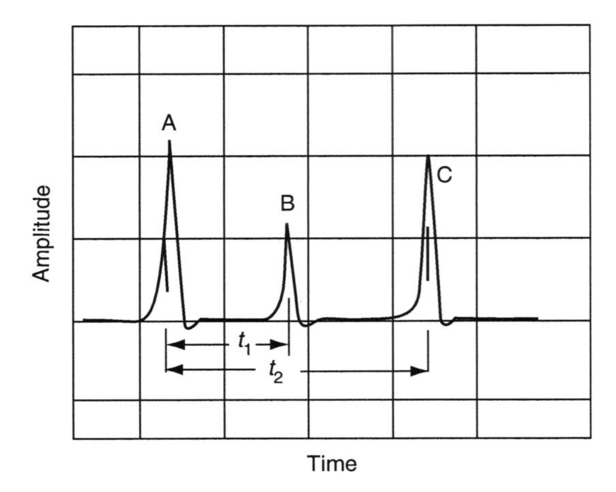

FIGURE 5.48 A-scan representation of internal defects: peak A for front surface reflection, peak B for reflection from a defect, and peak C for back surface reflection.

defect has a lower energy level than an uninterrupted wave, it is displayed as a smaller peak. In contrast to the pulse-echo method, the through-transmission method requires access to both surfaces of the part.

In general, part surfaces are scanned at regular intervals by piezoelectric transducers and an ultrasonic map of the entire part is generated. The three different procedures used for data presentation are A scan, B scan, and C scan. In the A-scan procedure, output signal amplitudes are displayed against a time scale (Figure 5.48) and the depths of various defect locations are judged from the positions of the signal peaks on the time sweep. The B-scan procedure profiles the top and bottom surfaces of a defect (Figure 5.49). The C-scan procedure, on the other hand, displays the plan view of the defect boundaries in the material (Figure 5.50).

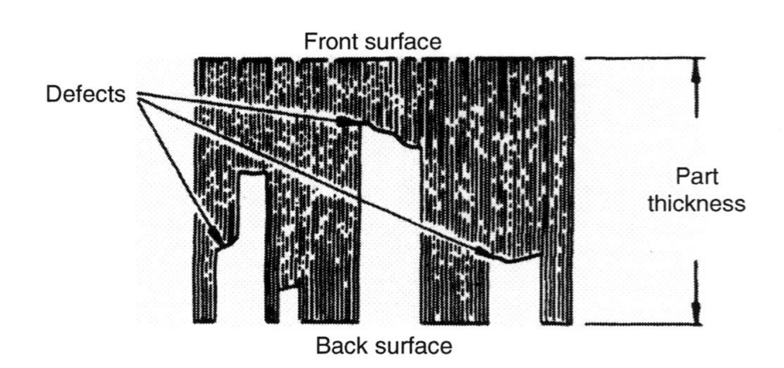

FIGURE 5.49 B-scan representation of internal defects.

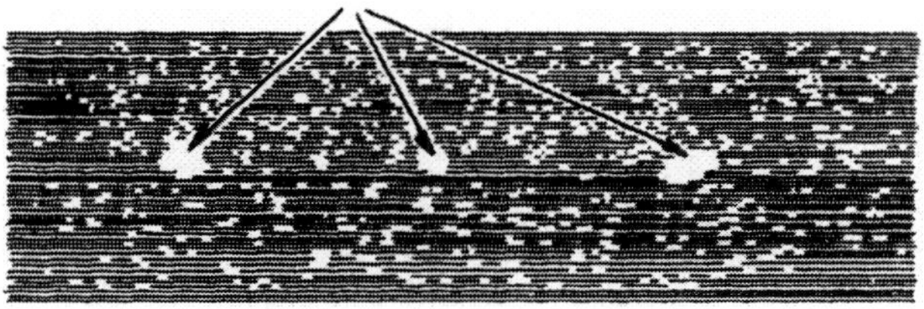

FIGURE 5.50 C-scan representation of internal defects.

C-scan images of through-transmission waves are commonly used for on-line inspection of large molded parts. In gray-level C scans, weaker transmitted signals are either dark gray or black. Thus, defects are identified as dark patches in a light gray background. The through-the-thickness location of any defect observed in a C scan can be obtained by using the pulse-echo method and by recording the A-scan image of the reflected pulse.

The ultrasonic inspection has been successfully used to detect large voids, delaminations, clusters of microvoids, and foreign materials. Reynolds [33] has reported the widespread application of ultrasonic C scanning in the aircraft industry. Water is the most commonly used coupling medium for ultrasonic scanning. The composite part may be squirted with water on its surface or may be completely immersed in a water tank for more uniform coupling.

5.9.3.3 Acoustic Emission

Acoustic emission (AE) refers to the transient elastic stress waves generated by the release of sound energy at stress-induced microscopic damage sites in the material. In metals, localized plastic deformation, crack initiation, and crack propagation are the primary sources of acoustic emission. In fiber-reinforced composites, acoustic emission is generated by the development of matrix microcracking, fiber–matrix interfacial debonding, localized delamination, fiber pullout, and fiber breakage. If the applied stress level is relatively low, these failure modes are likely to occur at the sites of inherent defects. Thus by monitoring the acoustic emission at low proof stress levels, it may be possible to locate and map critical defects in a molded part.

Acoustic emissions in a stressed part are detected by an array of highly sensitive piezoelectric transducers attached at various locations on its surface. These transducers measure the surface displacements caused by stress waves originating at the defect sites. Through proper instrumentation and a multiple-channel recorder, the electrical signal output from each transducer is amplified, stored, and recorded for further analysis.

Figure 5.51 illustrates the electrical signal output associated with an acoustic emission event. The most important AE parameter obtained from such output records is the cumulative event count. This is defined as the total

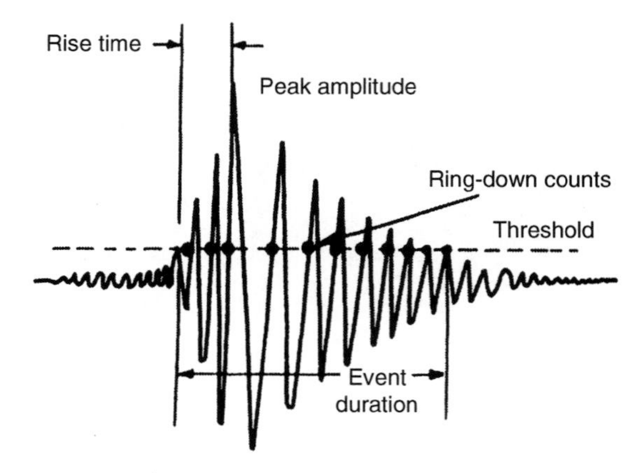

FIGURE 5.51 An acoustic emission signal (event) and associated nomenclatures.

number of events that contain signal amplitudes higher than a preset threshold. In general, the cumulative event count increases almost exponentially with increasing stress level. If the stress is held constant, the cumulative event count levels off, unless the part continues to deform owing to creep.

Figure 5.52 shows the cumulative AE event count recorded during a quasi-static loading–unloading cycle of a notched carbon–epoxy tensile specimen.

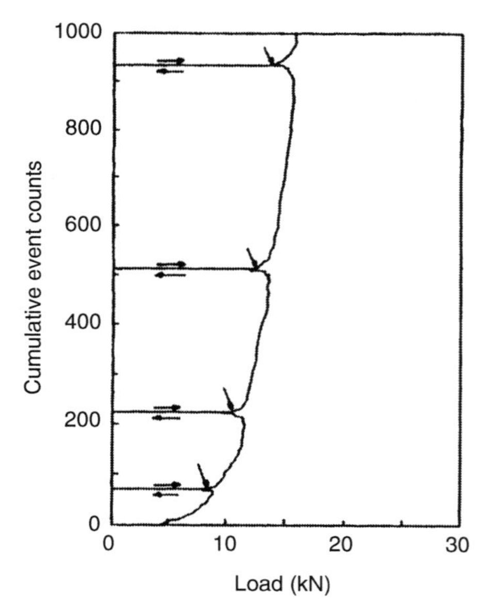

FIGURE 5.52 Typical AE event count as a function of loading on a fiber-reinforced laminate.

The acoustic emission begins only after the applied load in the first cycle exceeds a minimum value, and then the cumulative event count increases rapidly with increasing load. If the specimen is unloaded, the acoustic emission stops. If the load is reapplied, the acoustic emission occurs even before the previous maximum load is reached. This unique AE characteristic of a fiber-reinforced composite is not observed in metals and is called the felicity effect [34]. The load at which the AE begins on the $(n + 1)$th load cycle divided by the maximum load reached on the nth load cycle is termed the felicity ratio. The value of the felicity ratio indicates the extent of damage in the material and can be used as a potential acceptance or rejection criterion [35].

The difference in arrival times of a given event at various transducers can be used to locate the source of an acoustic emission. However, since a fiber-reinforced composite structure emits numerous bursts of emission in successive order, the transducers may not always be able to isolate a specific event. Furthermore, signal attenuation in fiber-reinforced polymers is high. Since low-frequency waves (typically 20–111 kHz) attenuate less and therefore travel longer distances than high-frequency waves, transducers designed to detect this frequency range are preferred for fiber-reinforced composite structures. Usually, high-frequency transducers are used in local areas of probable emission in conjunction with low-frequency transducers.

Acoustic emission has been used by a number of investigators to study the development and progression of damage with increasing time or stress. Some of these studies have also included the effects of various process and material parameters, such as cure temperature, cooling rate, matrix flexibility, and fiber orientation, on the acoustic emission activities of fiber-reinforced polymers. Standard practices have been developed for acoustic emission testing of glass fiber-reinforced plastic tanks and pressure vessels [36]. However, as Hamstad [37] has noted, acoustic emission in fiber-reinforced composites is a complex phenomenon and may lead to erroneous source identification if proper attention is not paid to the nature of wave attenuation and extraneous noise generation.

5.9.3.4 Acousto-Ultrasonic

Acousto-ultrasonic testing combines aspects of ultrasonic testing with acoustic emission signal analysis techniques to measure the severity of subtle internal flaws and inhomogeneity in a material. Examples of such flaws in a composite material are porosity, fiber breaks, fiber bunching, resin-rich zones, improper curing, fiber misorientations, and fiber–matrix debonding. While these flaws are not always considered as critical as delaminations or large voids, collectively they can reduce the mechanical properties or structural response of a material [38].

In acousto-ultrasonic method, a broad-band piezoelectric transducer is used to introduce a series of repeating ultrasonic pulses in the material. When these pulses interact with various flaws in the material, spontaneous stress

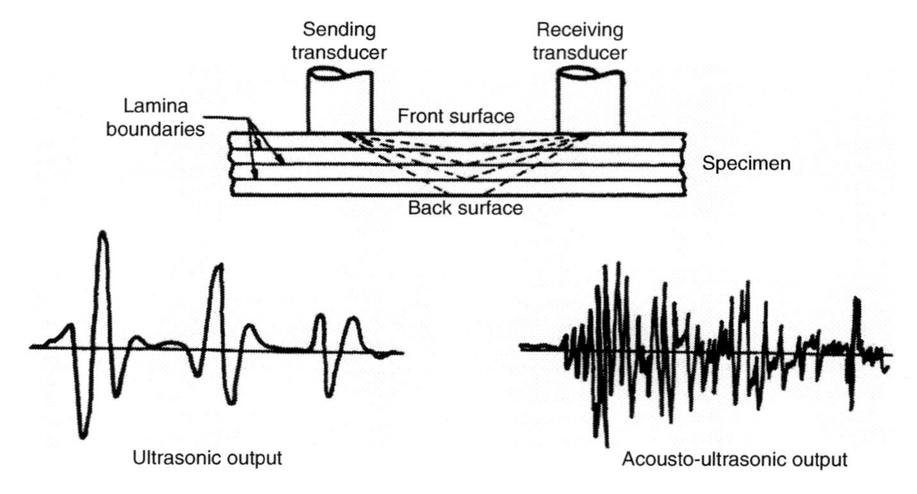

FIGURE 5.53 Acousto-ultrasonic measurements.

waves are produced that simulate acoustic emission events. The attenuated stress waves are detected by another piezoelectric transducer, which is commonly located on the same surface, but at some distance away from the sender transducer (Figure 5.53).

Acousto-ultrasonic signals received at the second transducer resemble acoustic emission burst-type waves that decay exponentially. They are processed using acoustic emission methodology, for example, ring-down count or voltage peak. The degree of attenuation of the transmitted waves is converted into a numerical value, called the stress wave factor (SWF). It is used as an indirect measure for the criticality of total flaw population present in the material. A low SWF that occurs due to high attenuation of stress waves indicates a severe flaw population. High values of SWF indicate an efficient transmission of stress waves through the material and, therefore, a relatively defect-free material.

The SWF has been correlated with mechanical properties, such as tensile strength or interlaminar shear strength. The SWF is also a sensitive indicator of accumulated damage in composite laminates subjected to fatigue cycling or impact loads. However, it is not suited for detecting isolated gross flaws, such as delaminations or voids, for which ultrasonic testing is recommended.

5.9.3.5 Thermography

Thermography or thermal imaging is based on the principle that the thermal conductivity of a material is altered locally due to the presence of defects, such as delaminations or inclusions. If the material containing such defects is

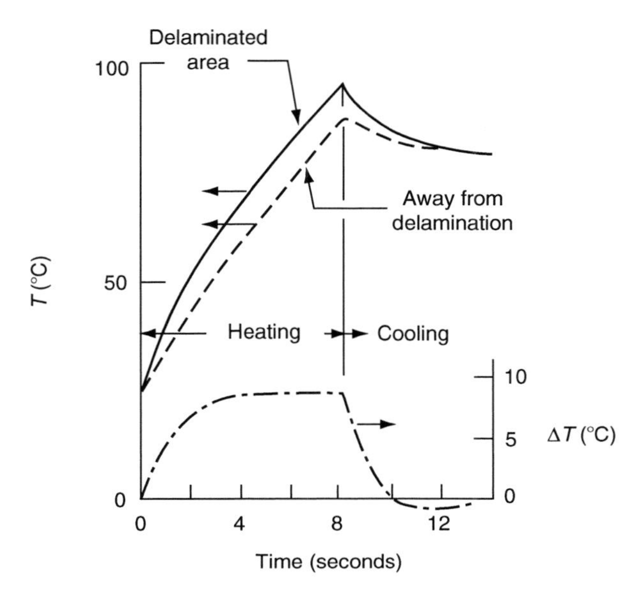

FIGURE 5.54 Surface temperature difference over delaminated and away from delaminated areas in a eight-ply carbon fiber–epoxy laminate. The delamination size is 5 mm between plies 1 and 2. (Adapted from Mirchandani, M.G. and McLaughlin, P.V., Jr., *Quantitative Nondestructive Evaluation*, Vol. 5B, Plenum Press, New York, 1986.)

subjected to a uniform heat source on one surface, the heat flow through its thickness will be uneven as a result of local variation of thermal conductivity. This will result in a nonuniform surface temperature distribution, which is detected and related to the presence of defects in the interior of the material [39]. For example, heat flow obstructed by a delaminated area increases the surface temperature above the delaminated area compared with that in the surrounding area (Figure 5.54).

The most common technique in thermography is to apply a uniform heat source on one surface of the material and record the transient temperature contours on the other surface by means of either liquid crystals or an infrared heat detection camera. The heat source can be as simple as hot water bags, hot air-dryers, a bank of light bulbs, or a photographic flash gun. Planar location and planar size of the defect can be easily detected from the thermogram.

A modified form of thermography is called vibrothermography [40], in which low-amplitude mechanical vibrations are used to induce localized heating in the material. It has been observed that localized heating occurs preferentially around internal defects, such as delaminations and large matrix cracks. The resulting temperature contours are recorded as before.

5.10 COST ISSUES

One major obstacle to greater use of fiber-reinforced composites in structural applications is their cost. Although carbon fiber-reinforced composites are used in large quantities in aerospace applications, their high cost is still a concern. The automotive industry is significantly more cost-conscious and has used carbon fiber-reinforced composites in a very limited amount. Most of the structural composite parts used in automobiles today are manufactured from glass fiber-reinforced polyesters or vinyl esters using compression molding or glass fiber-reinforced polyurethane using SRIM. Even with their lower cost, they cannot always compete with low carbon steel, which is less than half as expensive. It has been shown that carbon fiber-reinforced polymers can greatly reduce the weight of automotive body and chassis components due to their high modulus–weight ratio and fatigue resistance, but the high cost of carbon fibers has prevented their use in the automotive industry. The same argument can be made for many other nonaerospace industries as well.

The high cost of fibers is not the only reason for the high cost of fiber-reinforced polymers. The cost of the matrix can also be very high, especially if it is a high-temperature polymer. For the total material cost, one must also include the cost of preparing the material, for example, for making prepreg or for making sheet-molding compound sheet. The standard unidirectional carbon fiber–epoxy prepreg is typically about 1.5–2 times the cost of carbon fibers [41]. The cost of making toughened epoxy prepreg or high temperature thermoplastic matrix prepreg can be as much as three times higher than that of making standard epoxy prepregs.

The second item in the cost consideration is the fabrication cost of fiber-reinforced polymer parts, which depends on part design and the manufacturing process selected for fabrication. In the early days of composite applications in the aerospace industry, hand layup of prepreg stacks followed by vacuum bagging and autoclave curing was the most common method of producing carbon fiber-reinforced epoxy parts. The processing steps involved in autoclave molding are

1. Cutting the prepreg material (template layout, actual cutting and marking of the cut plies)
2. Laying up the plies (orienting the plies, building up the stack, and tacking or debulking)
3. Bagging and autoclave preparation
4. Autoclave cycle
5. Demolding
6. Trimming the part to final dimensions
7. Inspection, including nondestructive testing

Most of the operations involved in autoclave molding are done manually, may take several minutes to several hours, and require the use of highly skilled workers. It is a labor-intensive process and the direct labor cost for such a process may be 3–3.5 times the direct material cost. While vacuum bagging is still the most common manufacturing process in the aerospace industry, the use of computer-controlled prepreg cutting and tape-laying machines has reduced the labor cost; albeit at a higher equipment cost. RTM is also increasingly used in manufacturing many aircraft composite parts. Filament winding and pultrusion are less labor-intensive, can be highly automated, and can produce higher number of parts per hour; however, as discussed earlier in this chapter, they are suitable for making relatively simple structures.

Another aspect of the cost issue is the tooling cost, which includes the mold material cost and mold making cost. The selection of the mold material depends on part design (both size and complexity), production rate, and volume of production. For thermoset matrix prepregs, alloy steel tools are commonly used for high-volume production runs. For thermoplastic matrix prepregs, the processing temperature is higher, typically in the range of 260°C–300°C compared with 175°C or lower for thermoset matrix prepregs. At high processing temperatures, the difference in thermal expansions of the tool material and the prepreg must be minimized; otherwise, there may be fiber motion during cooling, which can cause fiber buckling and fiber breakage. Many low CTE mold materials are available—for example, monolithic graphite, castable ceramic, and chemically bonded ceramic. However, these materials are very expensive and they are easily damaged during transport, demolding, and temperature ramp-up.

Next, the cost of assembly must also be considered. Even though reduction in the number of parts achieved through parts integration reduces the number of assembly operations, there may still be a few needed to construct a large or complex composite structure. Assembly operations are also needed to attach composite parts with metal parts. Mechanical assembly using rivets and bolts is the primary assembly technique used in the aerospace industry. Cocuring and cobonding are also becoming assembly techniques. Mechanical assembly is often preferred over adhesive bonding, since it does not require surface preparation and long cycle time as is needed for the adhesive curing process and it allows disassembly for inspection and repair. However, the close tolerance required for mechanical assembly or secondary adhesive bonding is often not achieved with fiber-reinforced polymers, and shimming is used to assure a good fit and avoid damage to composite part during assembly. Thus, even with mechanical assembly, the assembly cost may be close to the part fabrication cost.

Finally, the cost of quality inspection must also be added. For high-performance parts requiring safe operation during their life time, 100% of the parts may need to be inspected for manufacturing defects, such as voids, delaminations, ply gap, resin-rich or resin-starved areas, misoriented plies, and so on. In the aerospace industry, the most common quality inspection technique for many of these defects is the through-transmission ultrasonic testing. Depending on the part complexity and the level as well as frequency of inspection needed, the cost of quality inspection can be in the range of 25%–100% of the fabrication cost [41].

REFERENCES

1. W.I. Lee, A.C. Loos, and G.S. Springer, Heat of reaction, degree of cure, and viscosity of Hercules 3501-6 resin, *J. Compos. Mater., 16*:510 (1982).
2. K.W. Lem and C.D. Han, Thermokinetics of unsaturated polyester and vinyl ester resins, *Polym. Eng. Sci., 24*:175 (1984).
3. D.S. Lee and C.D. Han, Chemorheology and curing behavior of low-profile polyester resin, *Proceedings 40th Annual Technical Conference*, Society of the Plastics Industry, January (1985).
4. M.R. Kamal and S. Sourour, Kinetics and thermal characterization of thermoset cure, *Polym. Eng, Sci., 13*:59 (1973).
5. M.R. Kamal, Thermoset characterization for moldability analysis, *Polym. Eng. Sci., 14*:231 (1974).
6. B.R. Gebart, Permeability of unidirectional reinforcements in RTM, *J. Compos. Mater., 26*:1100 (1992).
7. P. Hubert and A. Poursartip, A review of flow and compaction modelling relevant to thermoset matrix laminate processing, *J. Reinf. Plast. Compos., 17*:286 (1998).
8. T.G. Gutowski, T. Morgaki, and Z. Cai, The consolidation of laminate composites, *J. Compos. Mater., 21*:172 (1987).
9. A.C. Loos and G.S. Springer, Curing of epoxy matrix composites, *J. Compos. Mater., 17*:135 (1983).
10. *Structural Composites Fabrication Guide*, Vol. 1, Manufacturing Technology Division, U.S. Air Force Materials Laboratory (1982).
11. P.K. Mallick, Compression molding, *Composite Materials Technology* (P.K. Mallick and S. Newman, eds.), Hanser Publishers, New York (1990).
12. L.F. Marker and B. Ford, Flow and curing behavior of SMC during molding, *Mod. Plast., 54*:64 (1977).
13. P.J. Costigan, B.C. Fisher, and M. Kanagendra, The rheology of SMC during compression molding and resultant material properties, *Proceedings 40th Annual Conference*, Society of the Plastics Industry, January (1985).
14. M.R. Barone and D.A. Caulk, Kinematics of flow in sheet molding compounds, *Polym. Compos., 6*:105 (1985).
15. R.M. Griffith and H. Shanoski, Reducing blistering in SMC molding, *Plast. Des. Process., 17*:10 (1977).
16. P.K. Mallick, Effect of fiber misorientation on the tensile strength of compression molded continuous fiber composites, *Polym. Compos., 7*:14 (1986).
17. R.B. Jutte, SMC-sink mechanisms and techniques of minimizing sing, International Automotive Engineering Congress, Detroit, Paper No. 730171, Society of Automotive Engineers (1973).
18. K.L. Smith and N.P. Suh, An approach towards the reduction of sink marks in sheet molding compound, *Polym. Eng. Sci., 19*:829 (1979).

19. A.N. Dharia and N.R. Schott, Resin pick-up and fiber wet-out associated with coating and pultrusion processes, *Proceedings 44th Annual Technical Conference*, Society of Plastics Engineers, p. 1321, April (1986).
20. J.E. Sumerak, Understanding pultrusion process variables, *Mod. Plast., 62*:58 (1985).
21. M.A. Bibbo and T.G. Gutowski, An analysis of the pulling force in pultrusion, *Proceedings 44th Annual Technical Conference*, Society of Plastics Engineers, p. 1430, April (1986).
22. C.D. Hermansen and R.R. Roser, Filament winding machine: which type is best for your application? *Proceedings 36th Annual Conference*, Society of the Plastics Industry, February (1981).
23. J.F. Kober, Microprocessor-controlled filament winding, *Plastics Machinery Equipment*, June (1979).
24. D. Brosius and S. Clarke, Textile preforming techniques for low cost structural composites, *Advanced Composite Materials: New Developments and Applications*, ASM International (1991).
25. C.W. Macosko, *Fundamentals of Reaction Injection Molding*, Hanser Publishers, New York (1989).
26. R.K. Okine, Analysis of forming parts from advanced thermoplastic composite sheet materials, *SAMPE J., 25*:9 (1989).
27. F.N. Cogswell, The processing science of thermoplastic structural composites, *Int. Polym. Proc., 1*:157 (1987).
28. L.S. Penn, Physiochemical characterization of composites and quality control of raw materials, *Composite Materials: Testing and Design* (*Fifth Conference*), *ASTM STP, 674*:519 (1979).
29. R. Burns, K.S. Gandhi, A.G. Hankin, and B.M. Lynskey, Variability in sheet molding compound (SMC). Part I. The thickening reaction and effect of raw materials, *Plast. Polym., 43*:228 (1975).
30. C.M. Tung and P.J. Dynes, Chemorheological characterization of B-stage printed wiring board resins, *Composite Materials: Quality Assurance and Processing, ASTM STP, 797*:38 (1983).
31. M.L. Bromberg, Application of dielectric measurements to quality control, *Proceedings Annual Technical Conference*, Society of Plastics Engineers, p. 403, April (1986).
32. W.H. Pfeifer, Computed tomography of advanced composite materials, *Proceedings Advanced Composites Conference*, American Society for Metals, December (1985).
33. W.N. Reynolds, Nondestructive testing (NDT) of fibre-reinforced composite materials, *Mater. Des., 5*:256 (1985).
34. T.J. Fowler, Acoustic emission testing of fiber reinforced plastics, American Society of Civil Engineers Fall Convention, Preprint 3092, October (1977).
35. J. Awerbuch, M.R. Gorman, and M. Madhukar, Monitoring acoustic emission during quasi-static loading/unloading cycles of filament-wound graphite/epoxy laminate coupons, *First International Symposium on Acoustic Emission from Reinforced Composites*, Society of the Plastics Industry, July (1983).
36. C.H. Adams, Recommended practice for acoustic emission testing of fiberglass reinforced plastic tanks/vessels, *Proceedings 37th Annual Conference*, Society of the Plastics Industry, January (1982).
37. M.A. Hamstad, A review: acoustic emission, a tool for composite materials studies, *Exp. Mech., 26*:7 (1986).

38. A. Vary, Acousto-ultrasonics, *Non-Destructive Testing of Fibre-Reinforced Plastics Composites*, Vol. 2 (J. Summerscales, ed.), Elsevier Applied Science, London (1990).
39. D.J. Hillman and R.L. Hillman, Thermographic inspection of carbon epoxy structures, *Delamination and Debonding of Materials, ASTM STP, 873*:481 (1985).
40. E.G. Henneke, II and T.S. Jones, Detection of damage in composite materials by vibrothermography, *Nondestructive Evaluation and Flaw Criticality of Composite Materials, ASTM STP, 696*:83 (1979).
41. T.G. Gutowski, Cost, automation, and design, *Advanced Composites Processing* (T.G. Gutowski, ed.), John Wiley & Sons, New York, NY (1997).

PROBLEMS

P5.1. The following isothermal cure rate equation was found to fit the DSC data for a thermoset resin:

$$\frac{d\alpha_c}{dt} = k(1 - \alpha_c)^2,$$

where k is a temperature-dependent constant. Solve this differential equation to show that time required to achieve a cure level α_c is given by

$$t = \frac{\alpha_c}{k(1 - \alpha_c)}.$$

P5.2. In Problem 5.1, the constant k (per minute) is given by the following Arrhenius equation:

$$k = 2.3 \times 10^6 \exp\left(-\frac{14.8}{RT}\right),$$

where
R is the universal gas constant (kcal/mol K)
T is the absolute temperature (K)

Calculate the cure times for 50%, 80%, and 99.9% cure levels in the thermoset resin at 100°C and 150°C. State the observations you will make from these calculations.

P5.3. The degree of cure of a vinyl ester resin used for RTM as a function of cure time is given by the following equation.

$$\alpha_c = \frac{kt}{1 + kt},$$

where

$$k = 1.25 \times 10^6 \exp\left(-\frac{5000}{T}\right)$$

t is the cure time in minutes
T is the temperature in °K

(a) Assuming that the presence of fibers does not influence the cure kinetics, determine the temperature that should be used to achieve 80% cure in 1 min cure time.
(b) What is the cure rate at 80% cure in (a)?
(c) What temperature should be used if the cure rate needs to be doubled at 80% cure? What will be the cure time now?

P5.4. The isothermal rate of cure for a thermoset polyester system is given by the following equation:

$$\frac{d\alpha_c}{dt} = k\alpha_c^m(1 - \alpha_c)^n.$$

Show that the maximum rate of cure for this polyester system is obtained at a cure level

$$\alpha_c = \frac{m}{m+n}.$$

P5.5. The isothermal cure rate of a DGEBA epoxy cured with *m*-phenylene diamine is given by Equation 5.4. The kinetic parameters, determined from DSC data, are

$$m = 0.45$$
$$n = 1.55$$
$$k_1 = 2.17 \times 10^6 \exp(-E_1/RT) \text{ per minute}$$
$$k_2 = 1.21 \times 10^6 \exp(-E_2/RT) \text{ per minute}$$
$$E_1 = 13{,}600 \text{ cal/mol}$$
$$E_2 = 13{,}700 \text{ cal/mol}$$

1. Graphically compare the cure rates of this epoxy resin at 100°C, 125°C, and 150°C as a function of the degree of cure, α_c. State the observations you will make from the graphs.
2. Determine the maximum cure rate at each temperature.

P5.6. The following isothermal viscosity data were obtained in a parallel-plate viscometer for an epoxy system cured at 125°C:

Time (min)	Viscosity (cP)
0	79.5
20	100
40	148
60	162
80	263
100	512
120	1,480
140	6,160
160	44,600
180	513,000
200	14,781,000

Plot the data on a semilog scale and schematically show how the nature of this curve may change if (a) the curing temperature is 150°C, (b) the curing temperature is 100°C, and (c) the epoxy system is B-staged before curing at 125°C.

P5.7. Lee et al. [1] fitted the following equation to the viscosity (η) data (in Pa s) for an epoxy system:

$$\eta = 7.9 \times 10^{-14} \exp\left(\frac{E}{RT} + 14\alpha_c\right) \quad \text{for } \alpha_c < 0.5,$$

where

E = activation energy for viscosity = 9×10^4 J/mol
R = universal gas constant (J/mol K)
T = absolute temperature (K)
α_c = degree of cure

Using this equation, estimate the viscosity of the epoxy system at 100°C and 150°C (a) at the beginning of the curing reaction, (b) at 20% cure, and (c) at 40% cure.

P5.8. Roller* has represented the isothermal viscosity data for a B-staged epoxy resin, by the following four-parameter equation:

* M.B. Roller, Characterization of the time–temperature–viscosity behavior of curing B-staged epoxy resin, *Polym. Eng. Sci., 15*:406 (1975).

$$\ln \eta = \ln \eta_\infty + \frac{E_1}{RT} + tk \exp\left(-\frac{E_2}{RT}\right),$$

where

E_1 and E_2 are activation energies
k is a kinetic parameter (which depends on the resin type, curing agent concentration, and other factors)
η_∞ is the calculated viscosity at $T = \infty$
T is the absolute temperature
t is the time at temperature T
R is the universal gas constant

For a particular B-staged epoxy resin, the four parameters in the isothermal viscosity equation are: $E_1 = 27{,}000$ cal/mol, $E_2 = 19{,}000$ cal/mol, $k = 6.4 \times 10^7$ per second, and $\eta_\infty = 2 \times 10^{-11}$ cP. Determine the gel-time viscosity of the resin at 170°C if the gel time is 200 s at this temperature.

P5.9. A gel-time test on an epoxy resulted in the following data:

T (°C)	Time (min)
140	4
130	7.5
120	14
110	25
105	34.3

1. Plot ln(gel time) vs. $1/T$ is (where T is the temperature in K).
2. Assume that gel time can be represented by Arrhenius equation:

$$t_{gel} = A \exp\left(\frac{E}{RT}\right),$$

where

E = activation energy
R = universal gas constant
A = constant

Determine the values of A and E.

3. Determine the gel time at 100°C and 150°C using the Arrhenius equation.

P5.10. The following figure shows two cure cycles and the corresponding viscosity–time curves for an epoxy-based prepreg. Which of these two cure cycles is expected to produce better and more uniform mechanical properties, and why?

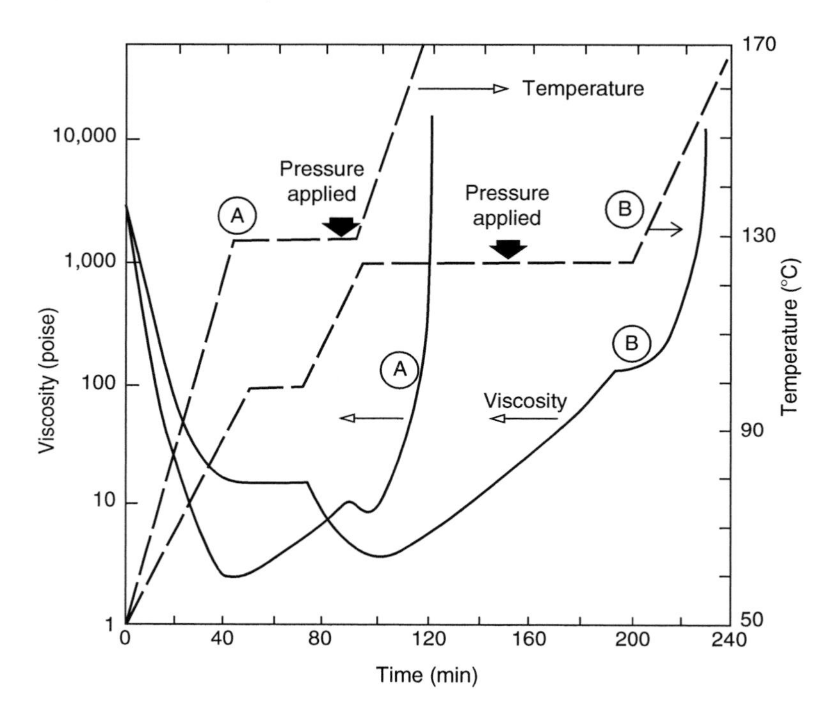

P5.11. The following figure shows the viscosity and gel-time curves for two epoxy resin systems. Which of the two resin systems will be more suitable for a bag-molding process, and why?

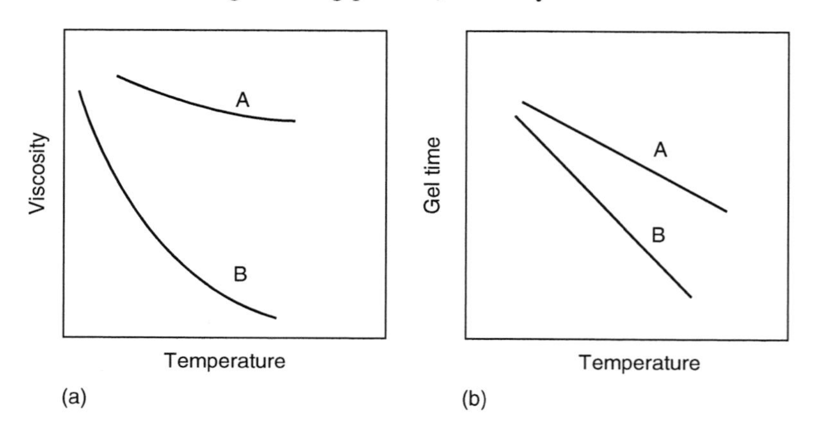

P5.12. The following figure shows a typical isothermal DSC curve for a polyester resin system. It contains *t*-butyl perbenzoate (TBPB) as the initiator for the curing reaction. The three important time parameters relevant for the compression-molding cycle of this material are also indicated in the figure. Explain how these time parameters may be affected by (a) an increase in cure temperature, (b) an increase in the initiator concentration, (c) the presence of an inhibitor, (d) the use of a lower temperature initiator, and (e) a combination of TBPB and a lower temperature initiator. From the standpoint of flow in the mold, which one of these changes would be more desirable?

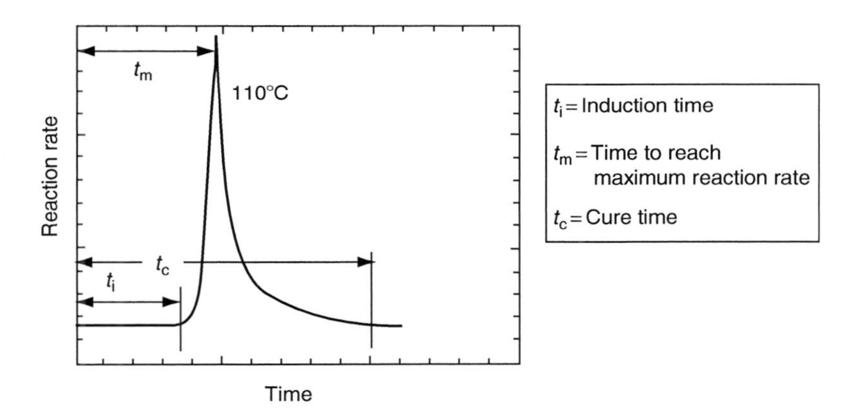

P5.13. As a part of the manufacturing study of an SMC-R material, you are asked to investigate the effects of SMC machine variables as well as molding variables on the mechanical properties of compression-molded flat plaques. The study was instigated by the wide variation in tensile strengths and failure locations observed in a prototype molded part. Select the experimental variables that might be of primary importance in this investigation, and design an experimental program (including test methods, specimen location, and so on) for this study.

P5.14. A thin-walled cylinder with closed ends contains oxygen at a high internal pressure, which creates a hoop tensile stress twice the axial tensile stress in its wall. Assume that the fibers carry all load and there is no interaction between the layers. Show that the optimum helix angle in the filament-wound cylinder is 54.74°.

P5.15. The price of carbon fibers is lower when there are 25,000–50,000 filaments per tow instead of 1,000–5,000 filaments per tow. Discuss the advantages and disadvantages of using such large tows in filament winding.

P5.16. One common specimen used in testing the quality of materials in filament winding is the Naval Ordnance Laboratory (NOL) ring, which contains only circumferential fibers. The specimen dimensions are shown in the figure.

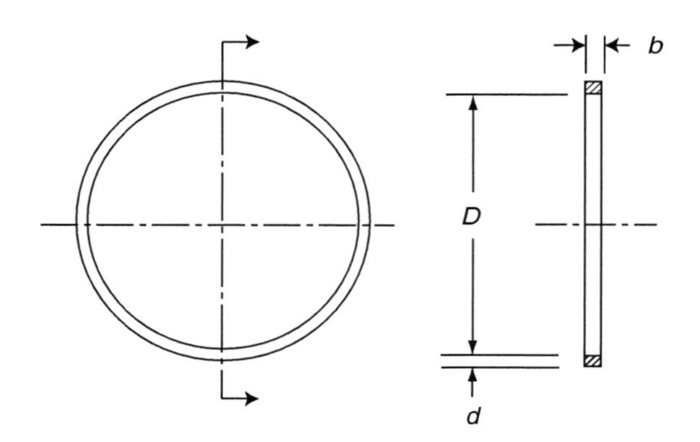

$D = 146.05$ mm (5.75 in.)
$b = 6.35$ mm (0.25 in.)
$d = 1.52$ mm (0.06 in.)

Develop a method of producing NOL ring specimens and the test techniques that can be used to measure the tensile, compressive, interlaminar shear, and in-plane shear strengths using such specimens. Discuss the advantages and disadvantages of NOL ring tests compared with the test methods presented in Chapter 4.

P5.17. List material (both resin and preform), process, and tool design parameters that may influence the quality of a resin transfer molded part.

P5.18. In an LCM process, liquid resin is injected under isothermal condition through a dry-fiber bed at a constant pressure P_o. Assuming that the flow is one-dimensional and is in the x direction, show that the fill time t is proportional to x^2, where x is the fill distance in time t.

P5.19. RTM is used to mold 500 mm long × 50 mm wide × 2.5 mm thick E-glass fiber-reinforced epoxy plates. The following details are given.

$$v_f = 32\%,\ \rho_f = 2.54\ \text{g/cm}^3,\ \rho_m = 1.2\ \text{g/cm}^3,$$
$$\text{Permeability } (P) = 4 \times 10^{-4}\ \text{mm}^2.$$

The resin viscosity during mold filling $= 1.15 \times 10^{-6}$ MPa s.

The gate is located at one of the edges normal to 500 mm length of the flat plate mold.

1. How much resin (in kilograms) must be pumped to fill the cavity?
2. How much time will be needed to fill the cavity at a constant flow rate of 130 mm^3/s?
3. What will be the maximum injection pressure needed to fill the cavity at this constant flow rate?

P5.20. Suppose in Problem 5.19, it is decided to fill the cavity at constant pressure instead of constant fill rate. If the constant pressure is 1.5 MPa, how much time will be needed to fill the cavity?

P5.21. Ultrasonic testing of an impact-tested quasi-isotropic composite panel containing carbon fibers in PEEK matrix shows the presence of local delaminations, even though there are no apparent damages on the outside. Propose, in as much detail as possible, a plan to repair the internal damage and recommend tests to validate the repair.

P5.22. Eddy current technique has been used for nondestructively detecting cracks and inclusions in metals. Investigate how this technique can be used for fiber-reinforced composites and discuss its limitations.

6 Design

In the preceding chapters, we have discussed various aspects of fiber-reinforced polymers, including the constituent materials, mechanics, performance, and manufacturing methods. A number of unique characteristics of fiber-reinforced polymers that have emerged in these chapters are listed in Table 6.1. Many of these characteristics are due to the orthotropic nature of fiber-reinforced composites, which has also necessitated the development of new design approaches that are different from the design approaches traditionally used for isotropic materials, such as steel or aluminum alloys. This chapter describes some of the design methods and practices currently used for fiber-reinforced polymers including the failure prediction methods, the laminate design procedures, and the joint design considerations. A number of design examples are also included.

6.1 FAILURE PREDICTION

Design analysis of a structure or a component is performed by comparing stresses (or strains) due to applied loads with the allowable strength (or strain capacity) of the material. In the case of biaxial or multiaxial stress fields, a suitable failure theory is used for this comparison. For an isotropic material that exhibits yielding, such as a mild steel or an aluminum alloy, either the maximum shear stress theory or the distortional energy theory (von Mises yield criterion) is commonly used for designing against yielding. Fiber-reinforced polymers are not isotropic, nor do they exhibit gross yielding. Thus, failure theories developed for metals or other isotropic materials are not applicable to composite materials. Instead, many new failure theories have been proposed for fiber-reinforced composites, some of which are discussed in this section.

6.1.1 FAILURE PREDICTION IN A UNIDIRECTIONAL LAMINA

We consider the plane stress condition of a general orthotropic lamina containing unidirectional fibers at a fiber orientation angle of θ with respect to the x axis (Figure 6.1). In Chapter 3, we saw that four independent elastic constants, namely, E_{11}, E_{22}, G_{12}, and ν_{12}, are required to define its elastic characteristics. Its strength properties are characterized by five independent strength values:

S_{Lt} = longitudinal tensile strength
S_{Tt} = transverse tensile strength

TABLE 6.1
Unique Characteristics of Fiber-Reinforced Polymer Composites

- Nonisotropic
  - Orthotropic
  - Directional properties
  - Four independent elastic constants instead of two
  - Principal stresses and principal strains not in the same direction
  - Coupling between extensional and shear deformations
- Nonhomogeneous
  - More than one macroscopic constituent
  - Local variation in properties due to resin-rich areas, voids, fiber misorientation, etc.
  - Laminated structure
- Laminated structure
  - Extensional–bending coupling
  - Planes of weakness between layers
  - Interlaminar stresses
  - Properties depend on the laminate type
  - Properties may depend on stacking sequence
  - Properties can be tailored according to requirements
  - Poisson's ratio can be greater than 0.5
- Nonductile behavior
  - Lack of plastic yielding
  - Nearly elastic or slightly nonelastic stress–strain behavior
  - Stresses are not locally redistributed around bolted or riveted holes by yielding
  - Low strains-to-failure in tension
- Noncatastrophic failure modes
  - Delamination
  - Localized damage (fiber breakage, matrix cracking, debonding, fiber pullout, etc.)
  - Less notch sensitivity
  - Progressive loss in stiffness during cyclic loading
  - Interlaminar shear failure in bending
- Low coefficient of thermal expansion
  - Dimensional stability
  - Zero coefficient of thermal expansion possible
  - Attachment problem with metals due to thermal mismatch
- High internal damping: High attenuation of vibration and noise
- Noncorroding

S_{Lc} = longitudinal compressive strength
S_{Tc} = transverse compressive strength
S_{LTs} = in-plane shear strength

Experimental techniques for determining these strength properties have been presented in Chapter 4. Note that the in-plane shear strength S_{LTs} in the principal material directions does not depend on the direction of the shear stress although both the longitudinal and transverse strengths may depend on the direction of the normal stress, namely, tensile or compressive.

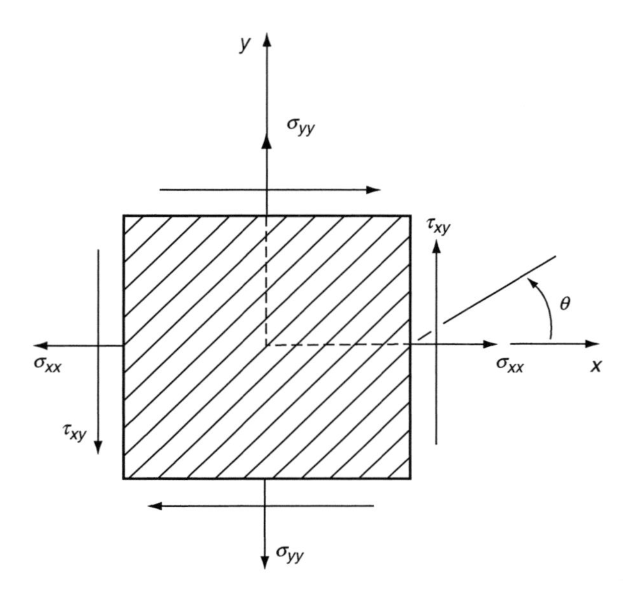

FIGURE 6.1 Two-dimensional stress state in a thin orthotropic lamina.

Many phenomenological theories have been proposed to predict failure in a unidirectional lamina under plane stress conditions. Among these, the simplest theory is known as the maximum stress theory; however, the more commonly used failure theories are the maximum strain theory and the Azzi–Tsai–Hill failure theory. We discuss these three theories as well as a more generalized theory, known as the Tsai–Wu theory. To use them, applied stresses (or strains) are first transformed into principal material directions using Equation 3.30. The transformed stresses are denoted σ_{11}, σ_{22}, and τ_{12}, and the applied stresses are denoted σ_{xx}, σ_{yy}, and τ_{xy}.

6.1.1.1 Maximum Stress Theory

According to the maximum stress theory, failure occurs when any stress in the principal material directions is equal to or greater than the corresponding ultimate strength. Thus to avoid failure,

$$\begin{aligned} -S_{\mathrm{Lc}} &< \sigma_{11} < S_{\mathrm{Lt}}, \\ -S_{\mathrm{Tc}} &< \sigma_{22} < S_{\mathrm{Tt}}, \\ -S_{\mathrm{LTs}} &< \tau_{12} < S_{\mathrm{LTs}}. \end{aligned} \tag{6.1}$$

For the simple case of uniaxial tensile loading in the x direction, only σ_{xx} is present and $\sigma_{\mathrm{yy}} = \tau_{xy} = 0$. Using Equation 3.30, the transformed stresses are

$$\sigma_{11} = \sigma_{xx} \cos^2 \theta,$$
$$\sigma_{22} = \sigma_{xx} \sin^2 \theta,$$
$$\tau_{12} = -\sigma_{xx} \sin \theta \ \cos \theta.$$

Thus, using the maximum stress theory, failure of the lamina is predicted if the applied stress σ_{xx} exceeds the smallest of $(S_{Lt}/\cos^2\theta)$, $(S_{Tt}/\sin^2\theta)$, and $(S_{LTs}/\sin\theta \cos\theta)$. Thus the safe value of σ_{xx} depends on the fiber orientation angle θ, as illustrated in Figure 6.2. At small values of θ, longitudinal tensile failure is expected, and the lamina strength is calculated from $(S_{Lt}/\cos^2\theta)$. At high values of θ, transverse tensile failure is expected, and the lamina strength is calculated from $(S_{Tt}/\sin^2\theta)$. At intermediate values of θ, in-plane shear failure of the lamina is expected and the lamina strength is calculated from $(S_{LTs}/\sin\theta \cos\theta)$. The change from longitudinal tensile failure to in-plane shear failure occurs at $\theta = \theta_1 = \tan^{-1} S_{LTs}/S_{Lt}$ and the change from in-plane shear failure to

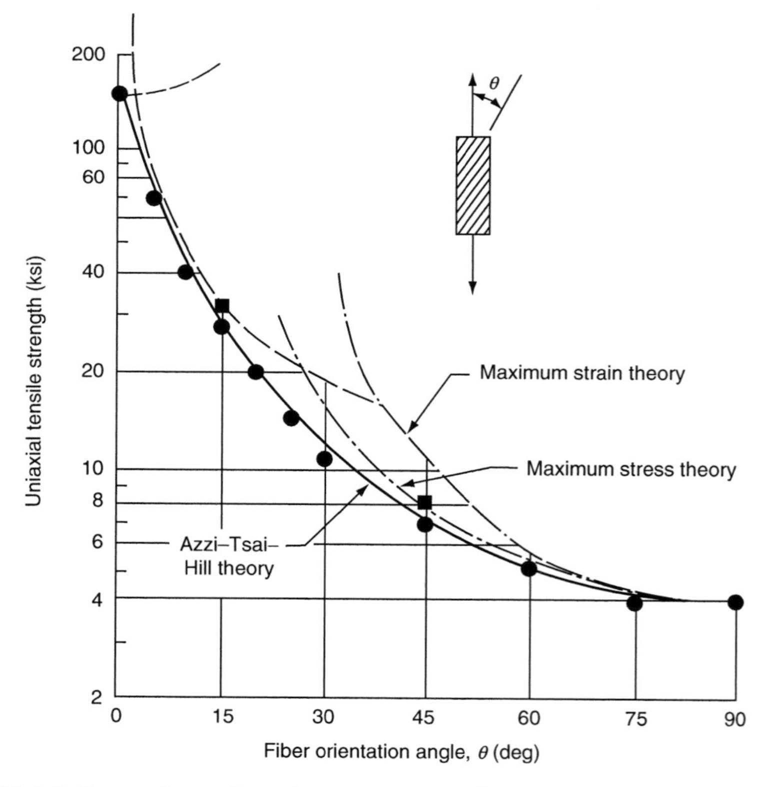

FIGURE 6.2 Comparison of maximum stress, maximum strain, and Azzi–Tsai–Hill theories with uniaxial strength data of a glass fiber-reinforced epoxy composite. (After Azzi, V.D. and Tsai, S.W., *Exp. Mech.*, 5, 283, 1965.)

transverse tensile failure occurs at $\theta = \theta_2 = \tan^{-1} S_{\text{Tt}}/S_{\text{LTs}}$. For example, for an E-glass fiber–epoxy composite with $S_{\text{Lt}} = 1100$ MPa, $S_{\text{Tt}} = 96.5$ MPa, and $S_{\text{LTs}} = 83$ MPa, $\theta_1 = 4.3°$ and $\theta_2 = 49.3°$. Thus, according to the maximum stress theory, longitudinal tensile failure of this composite lamina will occur for $0° \leq \theta < 4.3°$, in-plane shear failure will occur for $4.3° \leq \theta \leq 49.3°$ and transverse tensile failure will occur for $49.3° < \theta \leq 90°$.

EXAMPLE 6.1

A unidirectional continuous T-300 carbon fiber-reinforced epoxy laminate is subjected to a uniaxial tensile load P in the x direction. The laminate width and thickness are 50 and 2 mm, respectively. The following strength properties are known:

$$S_{\text{Lt}} = S_{\text{Lc}} = 1447.5 \text{ MPa}, \ S_{\text{Tt}} = 44.8 \text{ MPa, and } S_{\text{LTs}} = 62 \text{ MPa}.$$

Determine the maximum value of P for each of the following cases: (a) $\theta = 0°$, (b) $\theta = 30°$, and (c) $\theta = 60°$.

Solution

The laminate is subjected to a uniaxial tensile stress σ_{xx} due to the tensile load applied in the x direction. In all three cases, $\sigma_{xx} = \frac{P}{A}$, where A is the cross-sectional area of the laminate.

1. Since $\theta = 0°$, $\sigma_{11} = \sigma_{xx}$, $\sigma_{22} = 0$, and $\tau_{12} = 0$.
   Therefore, in this case the laminate failure occurs when $\sigma_{11} = \sigma_{xx} = S_{\text{Lt}} = 1447.5$ MPa.
   Since $\sigma_{xx} = \frac{P}{A} = \frac{P}{(0.05 \text{ m})(0.002 \text{ m})}$, the tensile load P at which failure occurs is 144.75 kN. The mode of failure is the longitudinal tensile failure of the lamina.
2. Since $\theta = 30°$, using Equation 3.30,

$$\sigma_{11} = \sigma_{xx} \cos^2 30° = 0.75 \, \sigma_{xx},$$
$$\sigma_{22} = \sigma_{xx} \sin^2 30° = 0.25 \, \sigma_{xx},$$
$$\tau_{12} = \sigma_{xx} \sin 30° \cos 30° = 0.433 \, \sigma_{xx}.$$

According to Equation 6.1, the maximum values of σ_{11}, σ_{22}, and τ_{12} are

(1) $\sigma_{11} = 0.75\sigma_{xx} = S_{\text{Lt}} = 1447.5$ MPa, which gives $\sigma_{xx} = 1930$ MPa
(2) $\sigma_{22} = 0.25\sigma_{xx} = S_{\text{Tt}} = 44.8$ MPa, which gives $\sigma_{xx} = 179.2$ MPa
(3) $\tau_{12} = 0.433\sigma_{xx} = S_{\text{LTs}} = 62$ MPa, which gives $\sigma_{xx} = 143.2$ MPa

Laminate failure occurs at the lowest value of σ_{xx}. In this case, the lowest value is 143.2 MPa. Using $\sigma_{xx} = \frac{P}{A} = 143.2$ MPa, $P = 14.32$ kN. The mode of failure is the in-plane shear failure of the lamina.

3. Since $\theta = 60°$, using Equation 3.30,

$$\sigma_{11} = \sigma_{xx}\cos^2 60° = 0.25\,\sigma_{xx},$$
$$\sigma_{22} = \sigma_{xx}\sin^2 60° = 0.75\,\sigma_{xx},$$
$$\tau_{12} = \sigma_{xx}\sin 60°\cos 60° = 0.433\,\sigma_{xx}.$$

According to Equation 6.1, the maximum values of σ_{11}, σ_{22}, and τ_{12} are

(1) $\sigma_{11} = 0.25\sigma_{xx} = S_{\mathrm{Lt}} = 1447.5$ MPa, which gives $\sigma_{xx} = 5790$ MPa
(2) $\sigma_{22} = 0.75\sigma_{xx} = S_{\mathrm{Tt}} = 44.8$ MPa, which gives $\sigma_{xx} = 59.7$ MPa
(3) $\tau_{12} = 0.433\sigma_{xx} = S_{\mathrm{LTs}} = 62$ MPa, which gives $\sigma_{xx} = 143.2$ MPa

Laminate failure occurs at the lowest value of σ_{xx}. In this case, the lowest value is 59.7 MPa. Using $\sigma_{xx} = \frac{P}{A} = 59.7$ MPa, $P = 5.97$ kN. The mode of failure is transverse tensile failure of the lamina.

6.1.1.2 Maximum Strain Theory

According to the maximum strain theory, failure occurs when any strain in the principal material directions is equal to or greater than the corresponding ultimate strain. Thus to avoid failure,

$$-\varepsilon_{\mathrm{Lc}} < \varepsilon_{11} < \varepsilon_{\mathrm{Lt}},$$
$$-\varepsilon_{\mathrm{Tc}} < \varepsilon_{22} < \varepsilon_{\mathrm{Tt}},$$
$$-\gamma_{\mathrm{LTs}} < \gamma_{12} < \gamma_{\mathrm{LTs}}. \tag{6.2}$$

Returning to the simple case of uniaxial tensile loading in which a stress σ_{xx} is applied to the lamina, the safe value of this stress is calculated in the following way.

1. Using the strain–stress relationship, Equation 3.72, and the transformed stresses, the strains in the principal material directions are

$$\varepsilon_{11} = S_{11}\sigma_{11} + S_{12}\sigma_{22} = (S_{11}\cos^2\theta + S_{12}\sin^2\theta)\,\sigma_{xx},$$
$$\varepsilon_{22} = S_{12}\sigma_{11} + S_{22}\sigma_{22} = (S_{12}\cos^2\theta + S_{22}\sin^2\theta)\,\sigma_{xx},$$
$$\gamma_{12} = S_{66}\tau_{22} = -S_{66}\sin\theta\cos\theta\,\sigma_{xx},$$

where

$$S_{11} = \frac{1}{E_{11}}$$
$$S_{12} = -\frac{\nu_{12}}{E_{11}} = -\frac{\nu_{21}}{E_{22}}$$
$$S_{22} = \frac{1}{E_{22}}$$
$$S_{66} = \frac{1}{G_{12}}$$

2. Using the maximum strain theory, failure of the lamina is predicted if the applied stress σ_{xx} exceeds the smallest of

(1) $$\frac{\varepsilon_{\mathrm{Lt}}}{S_{11}\cos^2\theta + S_{12}\sin^2\theta} = \frac{E_{11}\varepsilon_{\mathrm{Lt}}}{\cos^2\theta - \nu_{12}\sin^2\theta} = \frac{S_{\mathrm{Lt}}}{\cos^2\theta - \nu_{12}\sin^2\theta}$$

(2) $$\frac{\varepsilon_{\mathrm{Tt}}}{S_{12}\cos^2\theta + S_{22}\sin^2\theta} = \frac{E_{22}\varepsilon_{\mathrm{Tt}}}{\sin^2\theta - \nu_{21}\cos^2\theta} = \frac{S_{\mathrm{Tt}}}{\sin^2\theta - \nu_{21}\cos^2\theta}$$

(3) $$\frac{\gamma_{\mathrm{LTs}}}{S_{66}\sin\theta\cos\theta} = \frac{G_{12}\gamma_{\mathrm{LTs}}}{\sin\theta\cos\theta} = \frac{S_{\mathrm{LTs}}}{\sin\theta\cos\theta}$$

The safe value of σ_{xx} for various fiber orientation angles is also shown in Figure 6.2. It can be seen that the maximum strain theory is similar to the maximum stress theory for θ approaching 0°. Both theories are operationally simple; however, no interaction between strengths in different directions is accounted for in either theory.

EXAMPLE 6.2

A T-300 carbon fiber-reinforced epoxy lamina containing fibers at a + 10° angle is subjected to the biaxial stress condition shown in the figure. The following material properties are known:

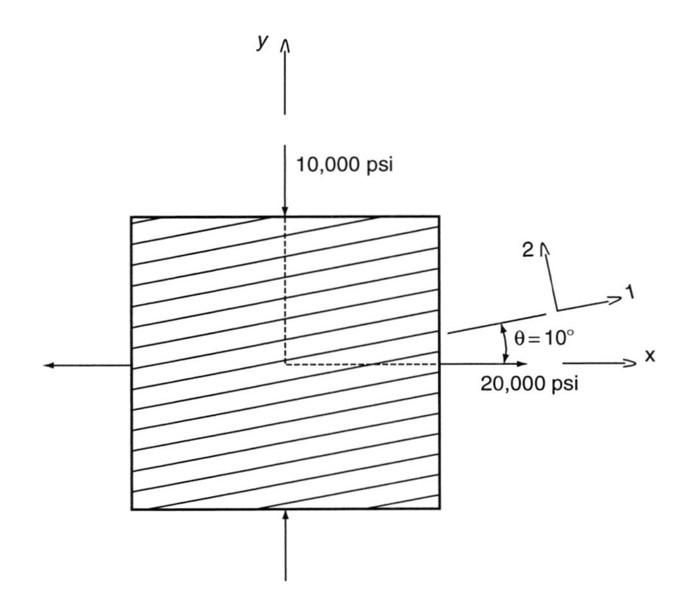

$$E_{11t} = E_{11c} = 21 \times 10^6 \text{ psi}$$
$$E_{22t} = 1.4 \times 10^6 \text{ psi}$$
$$E_{22c} = 2 \times 10^6 \text{ psi}$$
$$G_{12} = 0.85 \times 10^6 \text{ psi}$$
$$\nu_{12t} = 0.25$$
$$\nu_{12c} = 0.52$$
$$\varepsilon_{Lt} = 9{,}500 \ \mu\text{in./in.}$$
$$\varepsilon_{Tt} = 5{,}100 \ \mu\text{in./in.}$$
$$\varepsilon_{Lc} = 11{,}000 \ \mu\text{in./in.}$$
$$\varepsilon_{Tc} = 14{,}000 \ \mu\text{in./in.}$$
$$\gamma_{LTs} = 22{,}000 \ \mu\text{in./in.}$$

Using the maximum strain theory, determine whether the lamina would fail.

Solution

Step 1: Transform σ_{xx} and σ_{yy} into σ_{11}, σ_{22}, and τ_{12}.

$$\sigma_{11} = 20{,}000 \cos^2 10° + (-10{,}000) \sin^2 10° = 19{,}095.9 \text{ psi},$$
$$\sigma_{22} = 20{,}000 \sin^2 10° + (-10{,}000) \cos^2 10° = -9{,}095.9 \text{ psi},$$
$$\tau_{12} = (-20{,}000 - 10{,}000) \sin 10° \cos 10° = -5{,}130 \text{ psi}.$$

Step 2: Calculate ε_{11}, ε_{22}, and γ_{12}.

$$\varepsilon_{11} = \frac{\sigma_{11}}{E_{11t}} - \nu_{21c} \frac{\sigma_{22}}{E_{22c}} = 1134.3 \times 10^{-6} \text{ in./in.},$$
$$\varepsilon_{22} = -\nu_{12t} \frac{\sigma_{11}}{E_{11t}} + \frac{\sigma_{22}}{E_{22c}} = -4774.75 \times 10^{-6} \text{ in./in.},$$
$$\gamma_{12} = \frac{\tau_{12}}{G_{12}} = -6035.3 \times 10^{-6} \text{ in./in.}$$

Step 3: Compare ε_{11}, ε_{22}, and γ_{12} with the respective ultimate strains to determine whether the lamina has failed. For the given stress system in this example problem,

$$\varepsilon_{11} < \varepsilon_{Lt},$$
$$-\varepsilon_{Tc} < \varepsilon_{22},$$
$$-\gamma_{LTs} < \gamma_{12}.$$

Thus, the lamina has not failed.

6.1.1.3 Azzi–Tsai–Hill Theory

Following Hill's anisotropic yield criterion for metals, Azzi and Tsai [1] proposed that failure occurs in an orthotropic lamina if and when the following equality is satisfied:

$$\frac{\sigma_{11}^2}{S_{\mathrm{Lt}}^2} - \frac{\sigma_{11}\sigma_{22}}{S_{\mathrm{Lt}}^2} + \frac{\sigma_{22}^2}{S_{\mathrm{Tt}}^2} + \frac{\tau_{12}^2}{S_{\mathrm{LTs}}^2} = 1, \tag{6.3}$$

where σ_{11} and σ_{22} are both tensile (positive) stresses. When σ_{11} and σ_{22} are compressive, the corresponding compressive strengths are used in Equation 6.3.

For the uniaxial tensile stress situation considered earlier, failure is predicted if

$$\sigma_{xx} \geq \frac{1}{\left(\frac{\cos^4\theta}{S_{\mathrm{Lt}}^2} - \frac{\sin^2\theta\cos^2\theta}{S_{\mathrm{Lt}}^2} + \frac{\sin^4\theta}{S_{\mathrm{Tt}}^2} + \frac{\sin^2\theta\cos^2\theta}{S_{\mathrm{Lts}}^2} \right)^{1/2}}.$$

This equation, plotted in Figure 6.2, indicates a better match with experimental data than the maximum stress or the maximum strain theories.

EXAMPLE 6.3

Determine and draw the failure envelope for a general orthotropic lamina using Azzi–Tsai–Hill theory.

SOLUTION

A failure envelope is a graphic representation of failure theory in the stress coordinate system and forms a boundary between the safe and unsafe design spaces. Selecting σ_{11} and σ_{22} as the coordinate axes and rearranging Equation 6.3, we can represent the Azzi–Tsai–Hill failure theory by the following equations.

1. In the $+\sigma_{11}/+\sigma_{22}$ quadrant, both σ_{11} and σ_{22} are tensile stresses. The corresponding strengths to consider are S_{Lt} and S_{Tt}.

$$\frac{\sigma_{11}^2}{S_{\mathrm{Lt}}^2} - \frac{\sigma_{11}\sigma_{22}}{S_{\mathrm{Lt}}^2} + \frac{\sigma_{22}^2}{S_{\mathrm{Tt}}^2} = 1 - \frac{\tau_{12}^2}{S_{\mathrm{LTs}}^2}$$

2. In the $+\sigma_{11}/-\sigma_{22}$ quadrant, σ_{11} is tensile and σ_{22} is compressive. The corresponding strengths to consider are S_{Lt} and S_{Tc}.

$$\frac{\sigma_{11}^2}{S_{\mathrm{Lt}}^2} + \frac{\sigma_{11}\sigma_{22}}{S_{\mathrm{Lt}}^2} + \frac{\sigma_{22}^2}{S_{\mathrm{Tc}}^2} = 1 - \frac{\tau_{12}^2}{S_{\mathrm{LTs}}^2}$$

3. In the $-\sigma_{11}/+\sigma_{22}$ quadrant, σ_{11} is compressive and σ_{22} is tensile. The corresponding strengths to consider are S_{Lc} and S_{Tt}.

$$\frac{\sigma_{11}^2}{S_{\mathrm{Lc}}^2} + \frac{\sigma_{11}\sigma_{22}}{S_{\mathrm{Lc}}^2} + \frac{\sigma_{22}^2}{S_{\mathrm{Tt}}^2} = 1 - \frac{\tau_{12}^2}{S_{\mathrm{LTs}}^2}$$

4. In the $-\sigma_{11}/-\sigma_{22}$ quadrant, both σ_{11} and σ_{22} are compressive stresses. The corresponding strengths to consider are S_{Lc} and S_{Tc}.

$$\frac{\sigma_{11}^2}{S_{\mathrm{Lc}}^2} - \frac{\sigma_{11}\sigma_{22}}{S_{\mathrm{Lc}}^2} + \frac{\sigma_{22}^2}{S_{\mathrm{Tc}}^2} = 1 - \frac{\tau_{12}^2}{S_{\mathrm{LTs}}^2}$$

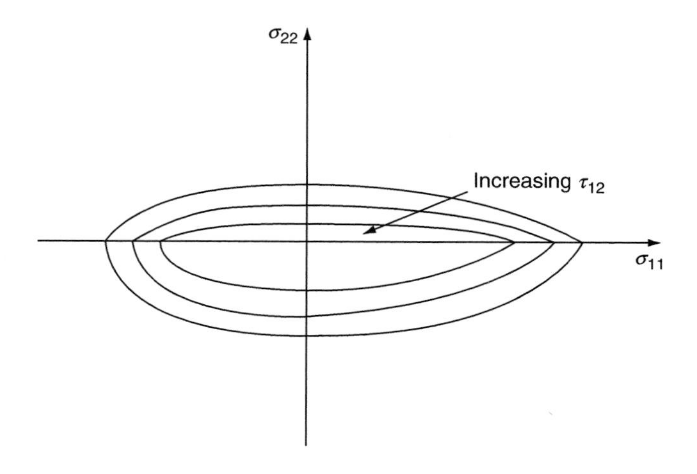

A failure envelope based on these equations is drawn in the figure for various values of the $\tau_{12}/S_{\mathrm{LTs}}$ ratio. Note that, owing to the anisotropic strength characteristics of a fiber-reinforced composite lamina, the Azzi–Tsai–Hill failure envelope is not continuous in the stress space.

6.1.1.4 Tsai–Wu Failure Theory

Under plane stress conditions, the Tsai–Wu failure theory [2] predicts failure in an orthotropic lamina if and when the following equality is satisfied:

$$F_1\sigma_{11} + F_2\sigma_{22} + F_6\tau_{12} + F_{11}\sigma_{11}^2 + F_{22}\sigma_{22}^2 + F_{66}\tau_{12}^2 + 2F_{12}\sigma_{11}\sigma_{22} = 1, \quad (6.4)$$

where F_1, F_2, and so on are called the strength coefficients and are given by

$$F_1 = \frac{1}{S_{\mathrm{Lt}}} - \frac{1}{S_{\mathrm{Lc}}}$$
$$F_2 = \frac{1}{S_{\mathrm{Tt}}} - \frac{1}{S_{\mathrm{Tc}}}$$
$$F_6 = 0$$
$$F_{11} = \frac{1}{S_{\mathrm{Lt}} S_{\mathrm{Lc}}}$$
$$F_{22} = \frac{1}{S_{\mathrm{Tt}} S_{\mathrm{Tc}}}$$
$$F_{66} = \frac{1}{S_{\mathrm{LTs}}^2}$$

and F_{12} is a strength interaction term between σ_{11} and σ_{22}. Note that F_1, F_2, F_{11}, F_{22}, and F_{66} can be calculated using the tensile, compressive, and shear strength properties in the principal material directions. Determination of F_{12} requires a suitable biaxial test [3]. For a simple example, consider an equal biaxial tension test in which $\sigma_{11} = \sigma_{12} = \sigma$ at failure. Using Equation 6.4, we can write

$$(F_1 + F_2)\,\sigma + (F_{11} + F_{22} + 2F_{12})\,\sigma^2 = 1,$$

from which

$$F_{12} = \frac{1}{2\sigma^2}\left[1 - \left(\frac{1}{S_{\mathrm{Lt}}} - \frac{1}{S_{\mathrm{Lc}}} + \frac{1}{S_{\mathrm{Tt}}} - \frac{1}{S_{\mathrm{Tc}}}\right)\sigma - \left(\frac{1}{S_{\mathrm{Lt}} S_{\mathrm{Lc}}} + \frac{1}{S_{\mathrm{Tt}} S_{\mathrm{Tc}}}\right)\sigma^2\right].$$

Since reliable biaxial tests are not always easy to perform, an approximate range of values for F_{12} has been recommended [4]:

$$-\frac{1}{2}(F_{11}F_{22})^{1/2} \leq F_{12} \leq 0. \tag{6.5}$$

In the absence of experimental data, the lower limit of Equation 6.5 is frequently used for F_{12}.

Figure 6.3 shows a comparison of the maximum strain theory, the Azzi–Tsai–Hill Theory, and the Tsai–Wu theory with a set of experimental data for a carbon fiber–epoxy lamina. The Tsai–Wu theory appears to fit the data best, which can be attributed to the presence of the strength interaction terms in Equation 6.4. Note that, for a given value of τ_{12}, the failure envelope defined by the Tsai–Wu failure theory is a continuous ellipse in the (σ_{11}, σ_{22}) plane. The inclination of the ellipse in the σ_{11}, σ_{22} plane and the lengths of its semi-axes are controlled by the value of F_{12}. The ellipse intercepts the σ_{11} axis at S_{Lt} and $-S_{\mathrm{Lc}}$, and the σ_{22} axis at S_{Tt} and $-S_{\mathrm{Tc}}$.

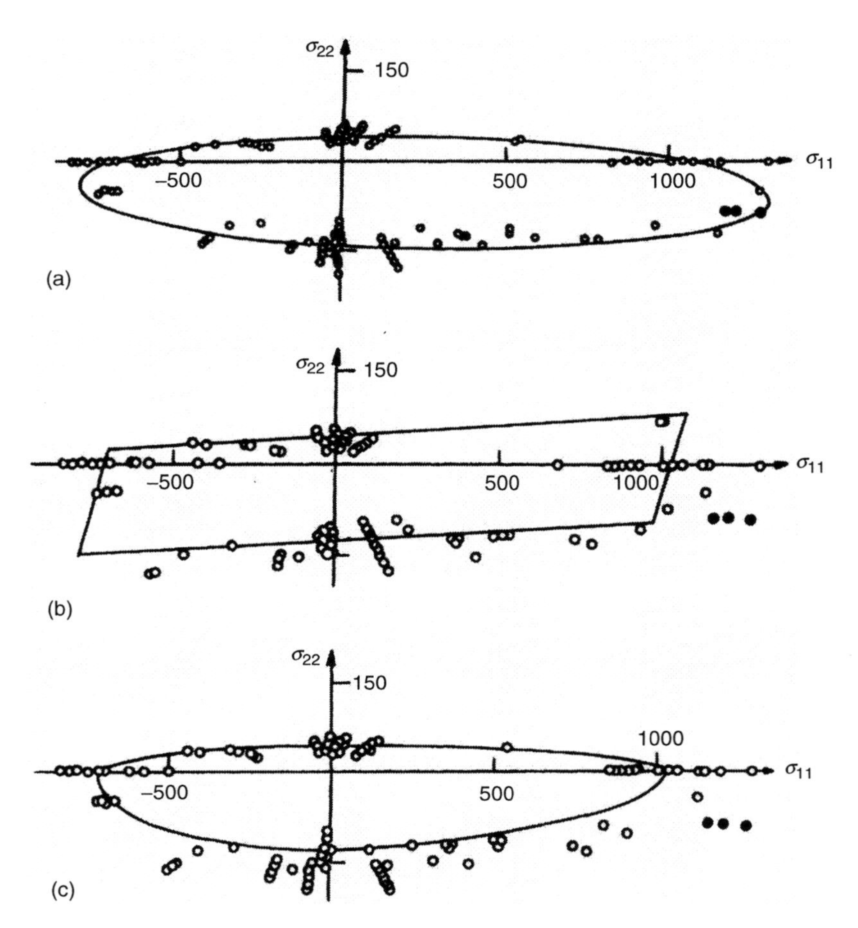

FIGURE 6.3 Comparison of (a) Tsai–Wu, (b) maximum strain, and (c) Azzi–Tsai–Hill failure theories with biaxial strength data of a carbon fiber-reinforced epoxy composite (note that the stresses are in MPa). (After Tsai, S.W. and Hahn, H.T., *Inelastic Behavior of Composite Materials*, C.T. Herakovich, ed., American Society of Mechanical Engineers, New York, 1975.)

EXAMPLE 6.4

Estimate the failure strength of a unidirectional lamina in an off-axis tension test using the Tsai–Wu theory. Assume that all strength coefficients for the lamina are known.

Solution

An off-axis tension test on a unidirectional lamina is performed at a fiber orientation angle θ with the loading axis. The stress state in the gage section of the lamina is shown in the figure. The stress σ_{xx} in the loading direction creates the following stresses in the principal material directions:

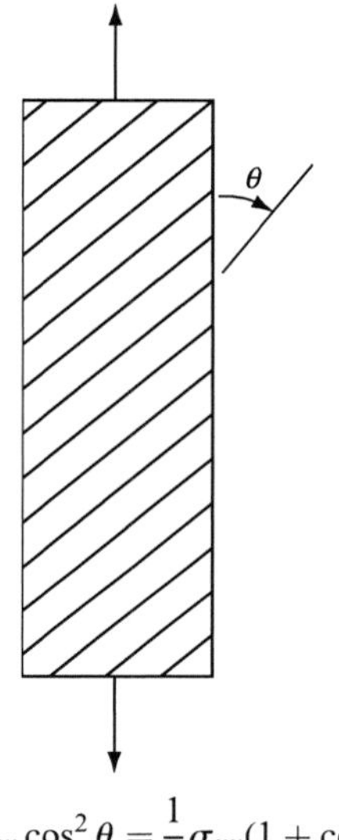

$$
\begin{aligned}
\sigma_{11} &= \sigma_{xx}\cos^2\theta = \frac{1}{2}\sigma_{xx}(1+\cos 2\theta),\\
\sigma_{22} &= \sigma_{xx}\sin^2\theta = \frac{1}{2}\sigma_{xx}(1-\cos 2\theta),\\
\tau_{12} &= -\sigma_{xx}\sin\theta\cos\theta = -\frac{1}{2}\sigma_{xx}\sin 2\theta.
\end{aligned}
$$

At failure, $\sigma_{xx} = S_\theta$, where S_θ denotes the failure strength in the off-axis tension test. Substituting for σ_{11}, σ_{22}, and τ_{12} in Equation 6.4 gives

$$
\begin{aligned}
&S_\theta^2[(3F_{11}+3F_{22}+2F_{12}+F_{66})+4(F_{11}-F_{22})\cos 2\theta+(F_{11}+F_{22}-2F_{12}-F_{66})\\
&\quad\cos 4\theta]+4S_\theta[(F_1+F_2)+(F_1-F_2)\cos 2\theta]-8=0.
\end{aligned}
$$

This represents a quadratic equation of the form

$$
AS_\theta^2 + BS_\theta + C = 0,
$$

which can be solved to calculate the failure strength S_θ.

6.1.2 Failure Prediction for Unnotched Laminates

Failure prediction for a laminate requires knowledge of the stresses and strains in each lamina, which are calculated using the lamination theory described in Chapter 3. The individual lamina stresses or strains in the loading directions are transformed into stresses or strains in the principal material directions for each lamina, which are then used in an appropriate failure theory to check whether the lamina has failed. After a lamina fails, the stresses and strains in the remaining laminas increase and the laminate stiffness is reduced; however, the laminate may not fail immediately. Furthermore, the failed lamina may not cease to carry its share of load in all directions.

6.1.2.1 Consequence of Lamina Failure

Several methods have been proposed to account for the failed lamina and the subsequent behavior of the laminate [5]. Among them are the following:

Total discount method: In this method, zero stiffness and strength are assigned to the failed lamina in all directions.

Limited discount method: In this method, zero stiffness and strength are assigned to the failed lamina for the transverse and shear modes if the lamina failure is in the matrix material. If the lamina fails by fiber rupture, the total discount method is adopted.

Residual property method: In this method, residual strength and stiffness are assigned to the failed lamina.

EXAMPLE 6.5

A quasi-isotropic $[0/\pm45/90]_S$ laminate made from T-300 carbon–epoxy is subjected to an in-plane normal load N_{xx} per unit width. With increasing values of N_{xx}, failure occurs first in the 90° layers owing to transverse cracks.

Determine the stiffness matrices before and after the first ply failure (FPF). Assume that each ply has a thickness t_0. Use the same material properties as in Example 3.6.

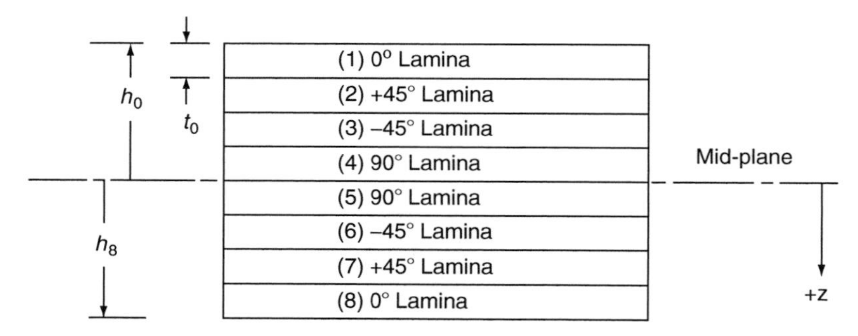

SOLUTIONS

Referring to the figure, we observe $h_8 = -h_0 = 4t_0$, $h_7 = -h_1 = 3t_0$, $h_6 = -h_2 = 2t_0$, $h_5 = h_3 = t_0$, and $h_4 = 0$. In addition, note that

$$\begin{aligned}
(\bar{Q}_{mn})_1 &= (\bar{Q}_{mn})_8 = (\bar{Q}_{mn})_{0^\circ}, \\
(\bar{Q}_{mn})_2 &= (\bar{Q}_{mn})_7 = (\bar{Q}_{mn})_{+45^\circ}, \\
(\bar{Q}_{mn})_3 &= (\bar{Q}_{mn})_6 = (\bar{Q}_{mn})_{-45^\circ}, \\
(\bar{Q}_{mn})_4 &= (\bar{Q}_{mn})_5 = (\bar{Q}_{mn})_{90^\circ}.
\end{aligned}$$

Since the given laminate is symmetric about the midplane, $[B] = [0]$. For in-plane loads, we need to determine the elements in the $[A]$ matrix.

$$\begin{aligned} A_{mn} &= \sum_{j=1}^{8} (\bar{Q}_{mn})_j (h_j - h_{j-1}) \\ &= 2t_0[(\bar{Q}_{mn})_{0^\circ} + (\bar{Q}_{mn})_{+45^\circ} + (\bar{Q}_{mn})_{-45^\circ} + (Q_{mn})_{90^\circ}]. \end{aligned}$$

Note that A_{mn} does not depend on the stacking sequence, since $(h_j - h_{j-1}) = t_0$ regardless of where the jth lamina is located.

Before the 90° layers fail: From Example 3.6, we tabulate the values of various $\bar{Q}_{mn}$ as follows. The unit of $\bar{Q}_{mn}$ is GPa.

	0°	**+45°**	**−45°**	**90°**
$\bar{Q}_{11}$	134.03	40.11	40.11	8.82
$\bar{Q}_{12}$	2.29	33.61	33.61	2.29
$\bar{Q}_{16}$	0	31.30	−31.30	0
$\bar{Q}_{22}$	8.82	40.11	40.11	134.03
$\bar{Q}_{26}$	0	31.30	−31.30	0
$\bar{Q}_{66}$	3.254	34.57	34.57	3.254

Therefore,

$$[A]_{\text{before}} = \begin{bmatrix} 446.14t_0 & 143.60t_0 & 0 \\ 143.60t_0 & 446.14t_0 & 0 \\ 0 & 0 & 151.30t_0 \end{bmatrix}.$$

After the 90° layers fail:

1. Total discount method: For the failed 90° layers, we assume $\bar{Q}_{11} = \bar{Q}_{12} = \bar{Q}_{16} = \bar{Q}_{22} = \bar{Q}_{26} = \bar{Q}_{66} = 0$.
   Therefore,

$$[A]_{\text{after}} = \begin{bmatrix} 428.50t_0 & 139.02t_0 & 0 \\ 139.02t_0 & 178.08t_0 & 0 \\ 0 & 0 & 144.79t_0 \end{bmatrix}$$

2. Limited discount method: Since the 90° layers failed by transverse cracking, we assume $\bar{Q}_{11} = \bar{Q}_{12} = \bar{Q}_{16} = \bar{Q}_{26} = \bar{Q}_{66} = 0$.
   However, $\bar{Q}_{22} = 134.03$ GPa. Therefore,

$$[A]_{\text{after}} = \begin{bmatrix} 428.50t_0 & 139.02t_0 & 0 \\ 139.02t_0 & 446.14t_0 & 0 \\ 0 & 0 & 144.79t_0 \end{bmatrix}$$

6.1.2.2 Ultimate Failure of a Laminate

Steps for the ultimate failure prediction of a laminate are as follows.

1. Calculate stresses and strains in each lamina using the lamination theory
2. Apply an appropriate failure theory to predict which lamina failed first
3. Assign reduced stiffness and strength to the failed lamina
4. Recalculate stresses and strains in each of the remaining laminas using the lamination theory
5. Follow through steps 2 and 3 to predict the next lamina failure
6. Repeat steps 2–4 until ultimate failure of the laminate occurs

Following the procedure outlined earlier, it is possible to generate failure envelopes describing the FPF as well as the ultimate failure of the laminate. In practice, a series of failure envelopes is drawn in a two-dimensional normal stress space in which the coordinate axes represent the average laminate stresses N_{xx}/h and N_{yy}/h. The area bounded by each failure envelope represents the safe design space for a constant average laminate shear stress N_{xy}/h (Figure 6.4).

Experimental verification for the laminate failure prediction methods requires the use of biaxial tests in which both normal stresses and shear stresses are present. Thin-walled large-diameter tubes subjected to various combinations of internal and external pressures, longitudinal loads, and torsional loads are the most suitable specimens for this purpose [6]. From the limited number of experimental results reported in the literature, it can be concluded

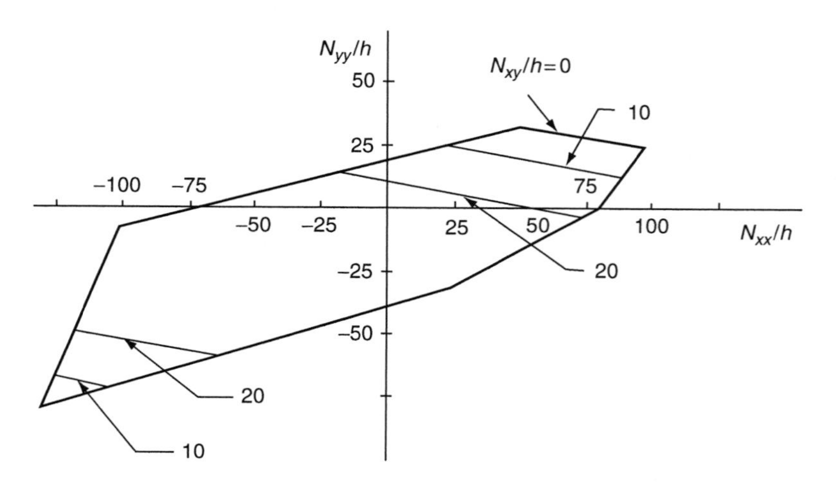

FIGURE 6.4 Theoretical failure envelopes for a carbon fiber–epoxy $[0/90\pm45]_S$ laminate (note that the in-plane loads per unit laminate thickness are in ksi).

that no single failure theory represents all laminates equally well. Among the various deficiencies in the theoretical prediction methods are the absence of interlaminar stresses and nonlinear effects. The assumption regarding the load transfer between the failed laminas and the active laminas can also introduce errors in the theoretical analyses.

The failure load prediction for a laminate depends strongly on the lamina failure theory selected [7]. In the composite material industry, there is little agreement on which lamina failure theory works best, although the maximum strain theory is more commonly used than the others [8]. Recently, the Tsai–Wu failure theory is finding more applications in the academic field.

6.1.3 Failure Prediction in Random Fiber Laminates

There are two different approaches for predicting failure in laminates containing randomly oriented discontinuous fibers.

In the Hahn's approach [9], which is a simple approach, failure is predicted when the maximum tensile stress in the laminate equals the following strength averaged over all possible fiber orientation angles:

$$S_r = \frac{4}{\pi}\sqrt{S_{Lt}S_{Tt}}, \tag{6.6}$$

where

S_r = strength of the random fiber laminate
S_{Lt} = longitudinal strength of a 0° laminate
S_{Tt} = transverse strength of a 0° laminate

In the Halpin–Kardos approach [10], the random fiber laminate is modeled as a quasi-isotropic $[0/\pm 45/90]_S$ laminate containing discontinuous fibers in the 0°, ±45°, and 90° orientations. The Halpin–Tsai equations, Equations 3.49 through 3.53, are used to calculate the basic elastic properties, namely, E_{11}, E_{22}, ν_{12}, and G_{12}, of the 0° discontinuous fiber laminas. The ultimate strain allowables for the 0° and 90° laminas are estimated from the continuous fiber allowables using the Halpin–Kardos equations:

$$\varepsilon_{Lt(d)} = \varepsilon_{Lt}\left[\left(\frac{E_f}{E_m}\right)^{-0.87} + 0.50\right] \quad \text{for } l_f > l_c \tag{6.7}$$

and

$$\varepsilon_{Td(d)} = \varepsilon_{Tt}\left(1 - 1.21 v_f^{2/3}\right).$$

The procedure followed by Halpin and Kardos [10] for estimating the ultimate strength of random fiber laminates is the same as the ply-by-ply analysis used for continuous fiber quasi-isotropic $[0/\pm 45/90]_S$ laminates.

6.1.4 Failure Prediction in Notched Laminates

6.1.4.1 Stress Concentration Factor

It is well known that the presence of a notch in a stressed member creates highly localized stresses at the root of the notch. The ratio of the maximum stress at the notch root to the nominal stress is called the stress concentration factor. Consider, for example, the presence of a small circular hole in an infinitely wide plate (i.e., $w \gg R$, Figure 6.5). The plate is subjected to a uniaxial tensile stress σ far from the hole. The tangential stress σ_{yy} at the two ends of the horizontal diameter of the hole is much higher than the nominal stress σ. In this case, the hole stress concentration factor K_T is defined as

$$K_{\mathrm{T}} = \frac{\sigma_{yy}(R,0)}{\sigma}.$$

For an infinitely wide isotropic plate, the hole stress concentration factor is 3. For a symmetric laminated plate with orthotropic in-plane stiffness properties, the hole stress concentration factor is given by

$$K_{\mathrm{T}} = 1 + \sqrt{\frac{2}{A_{22}}\left(\sqrt{A_{11}A_{22}} - A_{12} + \frac{A_{11}A_{22} - A_{12}^2}{2A_{66}}\right)}, \tag{6.8}$$

where A_{11}, A_{12},, A_{22}, and A_{66} are the in-plane stiffnesses defined in Chapter 3.

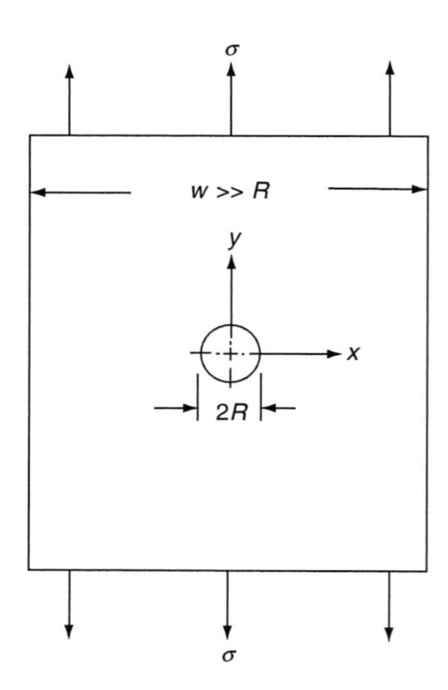

FIGURE 6.5 A uniaxially loaded infinite plate with a circular hole.

TABLE 6.2
Circular Hole Stress Concentration Factors

Material	Laminate	Circular Hole Stress Concentration Factor (K_T)
Isotropic material	—	3
S-glass–epoxy	0	4
	$[0_2/\pm45]_S$	3.313
	$[0/90/\pm45]_S$	3
	$[\pm45]_S$	2.382
T-300 carbon–epoxy	0	6.863
	$[0_4/\pm45]_S$	4.126
	$[0/90/\pm45]_S$	3
	$[0/\pm45]_S$	2.979
	$[\pm45]_S$	1.984

Note that, for an infinitely wide plate, the hole stress concentration factor K_T is independent of the hole size. However, for a finite width plate, K_T increases with increasing ratio of hole diameter to plate width. No closed-form solutions are available for the hole stress concentration factors in finite width orthotropic plates. They are determined either by finite element methods [11,12] or by experimental techniques, such as strain gaging, moire interferometry, and birefringent coating, among other techniques [13]. Appendix A.7 gives the finite width correction factor for isotropic plates, which can be used for approximate calculation of hole stress concentration factors for orthotropic plates of finite width.

Table 6.2 lists values of hole stress concentration factors for a number of symmetric laminates. For each material, the highest value of K_T is observed with 0° fiber orientation. However, K_T decreases with increasing proportions of ±45° layers in the laminate. It is interesting to note that a $[0/90/\pm45]_S$ laminate has the same K_T value as an isotropic material and a $[\pm45°]_S$ laminate has a much lower K_T than an isotropic material.

6.1.4.2 Hole Size Effect on Strength

The hole stress concentration factor in wide plates containing very small holes ($R \leq w$) is constant, yet experimental results show that the tensile strength of many laminates is influenced by the hole diameter instead of remaining constant. This hole size effect has been explained by Waddoups et al. [14] on the basis of intense energy regions on each side of the hole. These energy regions were modeled as incipient cracks extending symmetrically from the hole boundary perpendicular to the loading direction. Later, Whitney and Nuismer [15,16] proposed two stress criteria to predict the strength of notched composites. These two failure criteria are discussed next.

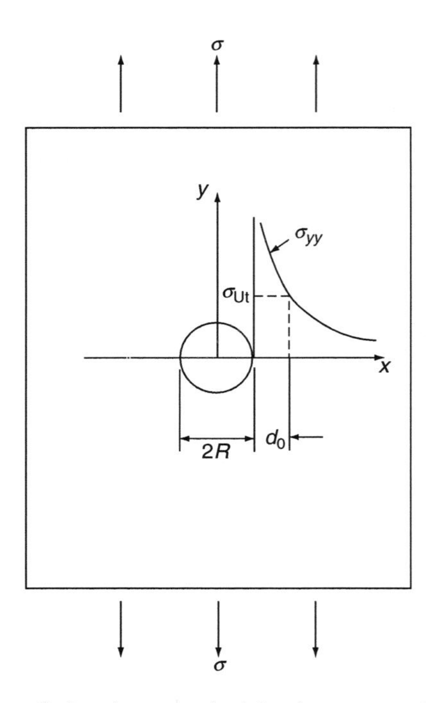

FIGURE 6.6 Failure prediction in a notched laminate according to the point stress criterion.

Point Stress Criterion: According to the point stress criterion, failure occurs when the stress over a distance d_0 away from the notch (Figure 6.6) is equal to or greater than the strength of the unnotched laminate. This characteristic distance d_0 is assumed to be a material property, independent of the laminate geometry as well as the stress distribution. It represents the distance over which the material must be critically stressed to find a flaw of sufficient length to initiate failure.

To apply the point stress criterion, the stress field ahead of the notch root must be known. For an infinitely wide plate containing a circular hole of radius R and subjected to a uniform tensile stress σ away from the hole, the most significant stress is σ_{yy} acting along the x axis on both sides of the hole edges. For an orthotropic plate, this normal stress component is approximated as [17]:

$$\sigma_{yy}(x,0) = \frac{\sigma}{2}\left\{2+\left(\frac{R}{x}\right)^2+3\left(\frac{R}{x}\right)^4-(K_\mathrm{T}-3)\left[5\left(\frac{R}{x}\right)^6-7\left(\frac{R}{x}\right)^8\right]\right\}, \qquad (6.9)$$

which is valid for $x \geq R$. In this equation, K_T is the hole stress concentration factor given in Equation 6.8.

According to the point stress criterion, failure occurs at $\sigma = \sigma_N$ for which

$$\sigma_{yy}(R + d_0, 0) = \sigma_{Ut},$$

where σ_N is the notched tensile strength and σ_{Ut} is the unnotched tensile strength for the laminate.

Thus from Equation 6.9, the ratio of notched to unnotched tensile strength is

$$\frac{\sigma_N}{\sigma_{Ut}} = \frac{2}{2 + \lambda_1^2 + 3\lambda_1^4 - (K_T - 3)(5\lambda_1^6 - 7\lambda_1^8)}, \tag{6.10}$$

where

$$\lambda_1 = \frac{R}{R + d_0}.$$

Average Stress Criterion: According to the average stress criterion, failure of the laminate occurs when the average stress over a distance a_0 ahead of the notch reaches the unnotched laminate strength. The characteristic distance a_0 is assumed to be a material property. It represents the distance over which incipient failure has taken place in the laminate owing to highly localized stresses.

In a plate containing a circular hole of radius R, failure by the average stress criterion occurs when

$$\frac{1}{a_0}\int_R^{R+a_0} \sigma_{yy}(x, 0)\mathrm{d}x = \sigma_{Ut}.$$

If the plate is made of a symmetric laminate with orthotropic properties, substitution of Equation 6.9 gives

$$\frac{\sigma_N}{\sigma_{Ut}} = \frac{2(1 - \lambda_2)}{2 - \lambda_2^2 - \lambda_2^4 + (K_T - 3)(\lambda_2^6 - \lambda_2^8)}, \tag{6.11}$$

where

$$\lambda_2 = \frac{R}{R + a_0}.$$

Both Equations 6.10 and 6.11 show that the notched tensile strength σ_N decreases with increasing hole radius. At very small hole radius, that is, as $R \to 0$, $\sigma_N \to \sigma_{Ut}$. At very large hole radius, as λ_1 or $\lambda_2 \to 1$, $\sigma_N \to \frac{\sigma_{Ut}}{K_T}$.

The following points should be noted in applying the point stress and the average stress criteria for notched laminates.

1. The application of both the criteria requires the knowledge of the overall stress field surrounding the notch tip. Since closed-form solutions are seldom available for notch geometries other than circular holes, this stress field must be determined by either numerical or experimental methods.
2. As a first approximation, the characteristic lengths d_0 and a_0 appearing in Equations 6.10 and 6.11 are considered independent of the notch geometry and the laminate configuration. Thus, the values d_0 and a_0 determined from a single hole test on one laminate configuration can be used for predicting the notched laminate strength of any laminate of the same material system. Nuismer and Whitney [16] have observed that $d_0 = 1.02$ mm (0.04 in.) and $a_0 = 3.81$ mm (0.15 in.) are applicable for a variety of laminate configurations of both E-glass fiber–epoxy and T-300 carbon fiber–epoxy composites.
3. Both the failure criteria make adequate failure predictions for notched laminates under uniaxial loading conditions only. The point stress criterion is simpler to apply than the average stress criterion. However, the errors resulting from the approximate analysis of the notch tip stresses tend to have less effect on the average stress criterion because of the averaging process itself.

It is important to note that, for many laminates, the unnotched tensile strength is strongly affected by the stacking sequence and the notched tensile strength is relatively insensitive to the stacking sequence. An example of this behavior is given in Table 6.3. In uniaxial tensile loading, unnotched $[\pm 45/90/0]_S$

TABLE 6.3
Tensile Strengths of Unnotched and Notched Laminates[a]

	Average Tensile Strength, MPa (ksi)	
Test Condition	$[\pm 45/0/90]_S$	$[90/0/\pm 45]_S$
Unnotched	451 (65.4)	499.3 (72.4)
Notched		
2.5 mm (0.1 in.) hole	331.7 (48.1)	322.8 (46.8)
7.5 mm (0.3 in.) hole	273.1 (39.6)	273.1 (39.6)
15.0 mm (0.6 in.) hole	235.2 (34.1)	233.1 (33.8)
2.5 mm (0.1 in.) crack	324.2 (47.0)	325.5 (47.2)
7.5 mm (0.3 in.) crack	268.3 (38.9)	255.9 (37.1)
15.0 mm (0.6 in.) crack	222.1 (32.2)	214.5 (31.1)

Source: Adapted from Whitney, J.M. and Kim, R.Y., *Composite Materials: Testing and Design* (*Fourth Conference*), *ASTM STP*, 617, 229, 1977.

[a] Material: T-300 carbon–epoxy.

laminates fail by gross edge delaminations due to the presence of tensile interlaminar normal stress σ_{zz} throughout the thickness. In contrast, the interlaminar normal stress at the free edges of $[90/0/\pm45]_S$ laminates under similar loading conditions is compressive in nature and no free-edge delaminations are found in these laminates. When notched, both laminates fail by the initiation and propagation of tensile cracks from the hole boundary, regardless of the interlaminar stress distributions at the free edges of the hole or the straight boundaries.

EXAMPLE 6.6

Failure Prediction in a Centrally Cracked Plate. Using the point stress criterion, estimate the strength of an infinitely wide symmetric laminated plate containing a central straight crack of length $2c$ subjected to a uniform tensile stress applied normal to the crack plane at infinity.

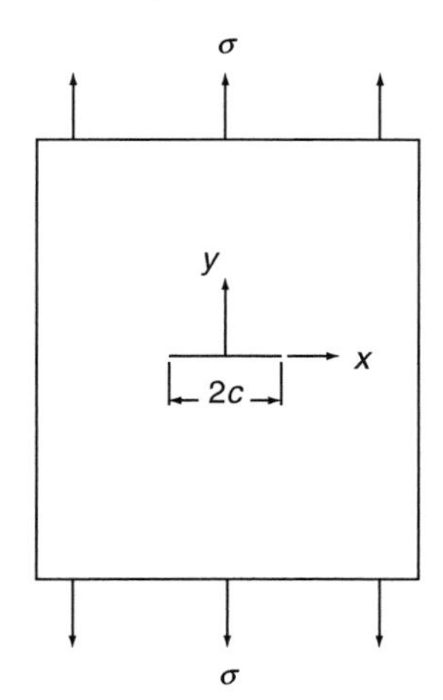

SOLUTION

The expression for the normal stress σ_{yy} ahead of the crack is

$$\sigma_{yy}(x, 0) = \frac{x}{\sqrt{x^2 - c^2}}\sigma,$$

which is valid for $x > c$.

According to the point stress criterion, failure occurs at $\sigma = \sigma_N$ for which $\sigma_{yy}(c + d_0, 0) = \sigma_{Ut}$.

Thus,

$$\frac{(c + d_0)\sigma_N}{\sqrt{(c + d_0)^2 - c^2}} = \sigma_{Ut}$$

or

$$\sigma_N = \sigma_{Ut}\sqrt{1-\lambda_3^2},$$

where

$$\lambda_3 = \frac{c}{c+d_0}.$$

The mode I stress intensity factor for this condition can be written as

$$K_I = \sigma_N\sqrt{\pi c} = \sigma_{Ut}\sqrt{\pi c(1-\lambda_3^2)}.$$

6.1.5 Failure Prediction for Delamination Initiation

Delamination or ply separation due to interlaminar stresses is a critical failure mode in many composite laminates. It can reduce the failure strength of some laminates well below that predicted by the in-plane failure theories discussed in Section 6.1.1.

Brewer and Lagace [18] as well as Zhou and Sun [19] proposed the following quadratic failure criterion to predict the initiation of delamination at the free edges:

$$\frac{\bar{\sigma}_{zz}^2}{S_{zt}^2} + \frac{\bar{\tau}_{xz}^2}{S_{xz}^2} + \frac{\bar{\tau}_{yz}^2}{S_{yz}^2} = 1, \tag{6.12}$$

where

S_{zt} = tensile strength in the thickness direction
S_{xz}, S_{yz} = interlaminar shear strengths

$\bar{\sigma}_{zz}$, $\bar{\tau}_{xz}$, and $\bar{\tau}_{yz}$ are the average interlaminar stresses defined by

$$(\bar{\sigma}_{zz}, \bar{\tau}_{xz}, \bar{\tau}_{yz}) = \frac{1}{x_c}\int_0^{x_c} (\sigma_{zz}, \tau_{xz}, \tau_{yz})\mathrm{d}x,$$

where x_c is the critical distance over which the interlaminar stresses are averaged.

Since the interlaminar strength data are not usually available, Zhou and Sun [19] have suggested using $S_{xz} = S_{yz} = S_{LTs}$ and $S_{zt} = S_{Tt}$.. They also recommend using x_c equal to twice the ply thickness.

EXAMPLE 6.7

The average interlaminar shear stresses in a $[\pm 45]_{2S}$ laminate under an in-plane tensile force $N_{xx} = 410$ kN/m are given as:

Interface	$\bar{\sigma}_{zz}$ (MPa)	$\bar{\tau}_{xz}$ (MPa)	$\bar{\tau}_{yz}$ (MPa)
1	−0.9	−5.67	68.71
2	−1.45	−2.71	26.60
3	0.73	9.91	67.47

Using $S_{zt} = 42.75$ MPa and $S_{xz} = S_{yz} = 68.95$ MPa, investigate whether any of the interfaces will fail by delamination at this load.

SOLUTION

Interface 1: Since $\bar{\sigma}_{zz}$ is compressive, we will not consider it in the failure prediction by Equation 6.12. Thus, the left-hand side (LHS) of Equation 6.12 is

$$\left(\frac{-5.67}{68.95}\right)^2 + \left(\frac{68.71}{68.95}\right)^2 \approx 1.$$

Interface 2: The term $\bar{\sigma}_{zz}$ is also compressive at this interface. Therefore, we will not consider σ_{zz} in the failure prediction by Equation 6.12. Thus the LHS of Equation 6.12 is

$$\left(\frac{-2.71}{68.95}\right)^2 + \left(\frac{26.6}{68.95}\right)^2 = 0.1504.$$

Interface 3: We will consider all three interlaminar stresses at this interface and compute the LHS of Equation 6.12 as:

$$\left(\frac{0.73}{42.75}\right)^2 + \left(\frac{9.91}{68.95}\right)^2 + \left(\frac{67.47}{68.95}\right)^2 = 0.9784.$$

Thus, according to Equation 6.12, only interface 1 is expected to fail by delamination at $N_{xx} = 410$ kN/m.

6.2 LAMINATE DESIGN CONSIDERATIONS

6.2.1 DESIGN PHILOSOPHY

The design of a structure or a component is in general based on the philosophy of avoiding failure during a predetermined service life. However, what constitutes failure depends principally on the type of application involved. For example, the most common failure mode in a statically loaded structure made of a ductile metal is yielding beyond which a permanent deformation may occur in the structure. On the other hand, the design of the same structure in a fatigue load application must take into account the possibility of a brittle failure accompanied by negligible yielding and rapid crack propagation.

Unlike ductile metals, composite laminates containing fiber-reinforced thermoset polymers do not exhibit gross yielding, yet they are also not classic brittle materials. Under a static tensile load, many of these laminates show nonlinear characteristics attributed to sequential ply failures. The question that often arises in designing with such composite laminates is, "Should the design be based on the ultimate failure or the first ply failure?" The current design practice in the aircraft or aerospace industry uses the FPF approach, primarily since cracks appearing in the failed ply may make the neighboring plies susceptible to mechanical and environmental damage. In many laminated constructions, the ultimate failure occurs soon after the FPF (Table 6.4), and therefore with these laminates an FPF design approach is justified. For many other laminates, the difference between the FPF stress level and the ultimate strength level is quite high. An FPF design approach with these laminates may be considered somewhat conservative.

The behavior of a fiber-reinforced composite laminate in a fatigue load application is also quite different from that of metals. In a metal, nearly 80%–90% of its fatigue life is spent in the formation of a critical crack. Generally, the fatigue crack in a metal is not detectable by the present-day NDT techniques until it reaches the critical length. However, once the fatigue crack attains the critical length, it propagates rapidly through the structure, failing it in a catastrophic manner (Figure 6.7). In many polymer matrix composites, fatigue damage may appear at multiple locations in the first few hundred to a thousand cycles. Some of these damages, such as surface craze marks, fiber splitting, and edge delaminations, may also be visible in the early stages of the fatigue life. Unlike metals, the propagation or further accumulation of damage in a fiber-reinforced composite takes place in a progressive manner resulting in a gradual loss in the stiffness of the structure. Thus, the laminated composite structure continues to carry the load without failing catastrophically; however, the loss of its stiffness may create gradually increasing deflections or vibrations. In these situations, a fatigue design approach based on the appearance of the first mechanical damage may again be considered conservative.

Since the history of their development is new, very few long-term field performance experiences exist. Design data in the areas of combined stresses, cumulative fatigue, repeated impact, environmental damage, and so on are not available. There is very little agreement among designers about what constitutes a structural failure and how to predict it. Industry-wide standards for material specifications, quality control, test methods, and failure analysis have not yet been developed. For all these reasons, the development of fiber-reinforced composite parts often relies on empirical approaches and requires extensive prototyping and testing.

6.2.2 Design Criteria

In general, the current design practice for fiber-reinforced composite structures uses the same design criteria as those used for metals. For example, the primary

TABLE 6.4
Predicted Tensile Properties of $[0_2/\pm45]_S$ and $[0/90]_S$ Laminates

		First Ply Failure (FPF)			Ultimate Failure		
Material	Laminate	Stress, MPa (ksi)	Strain (%)	Modulus, GPa (Msi)	Stress, MPa (ksi)	Strain (%)	Modulus, GPa (Msi)
S-glass–epoxy	$[0_2/\pm45]_S$	345.5 (50.1)	1.34	25.5 (3.7)	618.0 (89.6)	2.75	19.3 (2.8)
	$[0/90]_S$	89.7 (13.0)	0.38	23.4 (3.4)	547.6 (79.4)	2.75	19.3 (2.8)
HTS carbon–epoxy	$[0_2/\pm45]_S$	591.1 (85.7)	0.72	82.1 (11.9)	600.0 (87.0)	0.83	82.8 (12.0)
	$[0/90]_S$	353.1 (51.2)	0.45	78.6 (11.4)	549.0 (79.6)	0.72	72.4 (10.5)

Source: Adapted from Halpin, J.C., *J. Compos. Mater.*, 6, 208, 1972.

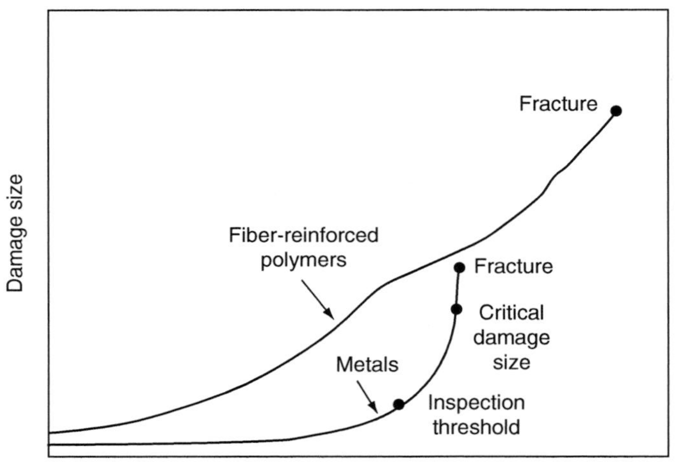

FIGURE 6.7 Schematic representation of damage development in metals and fiber-reinforced polymers. (Adapted from Salkind, M.J., *Composite Materials: Testing and Design* (*2nd Conference*), *ASTM STP*, 497, 143, 1972.)

structural components in an aircraft, whether made from an aluminum alloy or a carbon fiber-reinforced epoxy, are designed on the basis of the following criteria.

1. They must sustain the design ultimate load (DUL) in static testing.
2. The fatigue life must equal or exceed the projected vehicle life.
3. Deformations resulting from the applications of repeated loads and limit design load must not interfere with the mechanical operation of the aircraft, adversely affect its aerodynamic characteristics, or require repair or replacement of parts.

The DUL consists of the design limit load (DLL) multiplied by a specified ultimate factor of safety. The DLL is the maximum load that the structure (or any of its parts) is likely to experience during its design life. At the DLL, the structure should not undergo any permanent deformation. For metallic components, the most commonly used factor of safety is 1.5, although in fatigue-critical components it may be raised to 1.95. A higher factor of safety, such as 2 or more, is often used with fiber-reinforced composite materials, principally owing to the lack of design and field experience with these materials.

6.2.3 Design Allowables

Design allowable properties of a composite laminate are established by two different methods, namely, either by testing the laminate itself or by using the lamination theory along with a ply-level material property database. Considering the wide variety of lamination possibilities with a given fiber–resin combination, the second method is preferred by many designers since it has the flexibility of creating new design allowables without recourse to extensive testing.

The ply-level database is generated by testing unnotched unidirectional specimens in the tension, compression, and shear modes. The basic characteristics of the lamina, such as its longitudinal, transverse, and shear moduli, Poisson's ratios, and longitudinal, transverse, and shear strengths, as well as strains-to-failure, are determined by the various static test procedures described in Chapter 4. These tests are usually performed at room temperature; however, the actual application environment should also be included in the test program. The lamination theory combined with a failure criterion and a definition of failure is then used to predict the design allowables for the selected lamination configuration. This approach is particularly advantageous in generating design charts (also called carpet plots) for a family of laminates with the same basic ply orientations. Two of these charts for the $[0/\pm45/90]_S$ family of a carbon fiber–epoxy composite are shown in Figures 6.8 and 6.9. Such charts are very useful in selecting the proportions of the various ply orientations required to meet the particular design criteria involved [20].

Owing to the statistical nature of the ultimate properties, the design allowable strengths and strains are usually presented on one of the following bases:

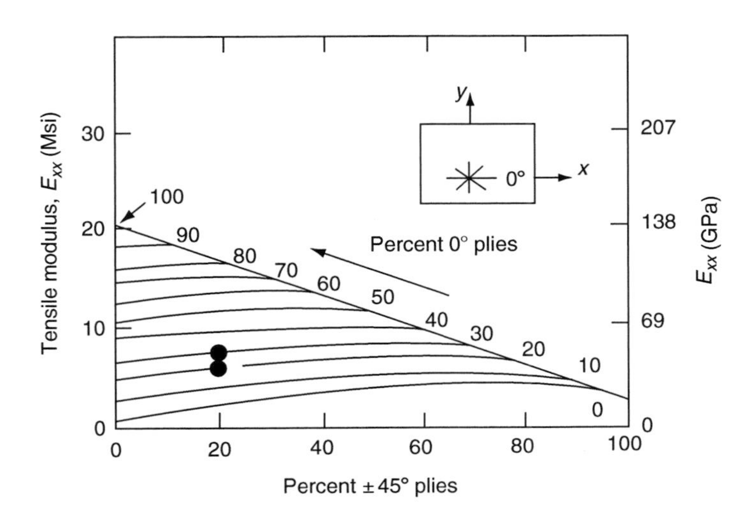

FIGURE 6.8 Carpet plot for tensile modulus E_{xx} of a carbon fiber–epoxy $[0/90/\pm45]_S$ laminate family. Note that the percentage of 90° plies is equal to 100 – (percentage of 0° plies) – (percentage of ±45° plies).

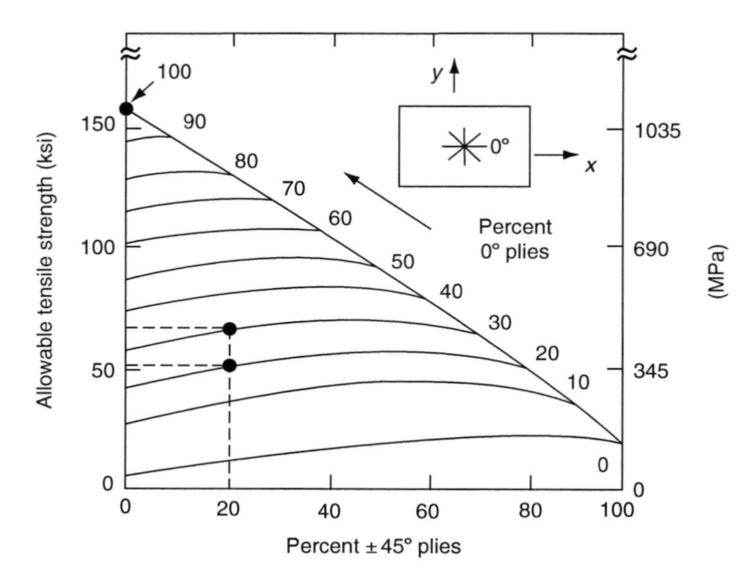

FIGURE 6.9 Carpet plot for tensile strength σ_{xx} of a carbon fiber–epoxy $[0/90/\pm45]_S$ laminate family.

1. *A basis*: Designed on the A-basis strength or strain, a component has at least a 99% probability of survival with a confidence level of 95%.
2. *B basis*: Designed on the B-basis strength or strain, a component has at least a 90% probability of survival with a confidence level of 95%.

Statistical methods to generate A- and B-basis design allowables are briefly described in Appendix A.8. The B-basis design allowables are commonly used for fiber-reinforced composite laminates, since the failure of one or more plies in these materials does not always result in the loss of structural integrity. Ekvall and Griffin [21] have described a step-by-step procedure of formulating the B-basis design allowables for T-300 carbon fiber–epoxy unidirectional and multidirectional laminates. Their approach also takes into account the effects of a 4.76 mm (0.1875 in.) diameter hole on the design allowable static strengths of these laminates.

In establishing the fatigue strength allowables for composite helicopter structures, Rich and Maass [22] used the mean fatigue strength minus 3 standard deviations. They observed that extrapolating the tension–tension fatigue data to the tension–compression or compression–compression mode may not be applicable for composite materials since significant fatigue strength reductions are possible in these two modes. Ply-level static and fatigue tests in their program were conducted under a room-temperature dry (RTD) as well as at an elevated-temperature wet (ETW) condition. The ETW condition selected was more severe than the actual design environment for helicopters. However, this

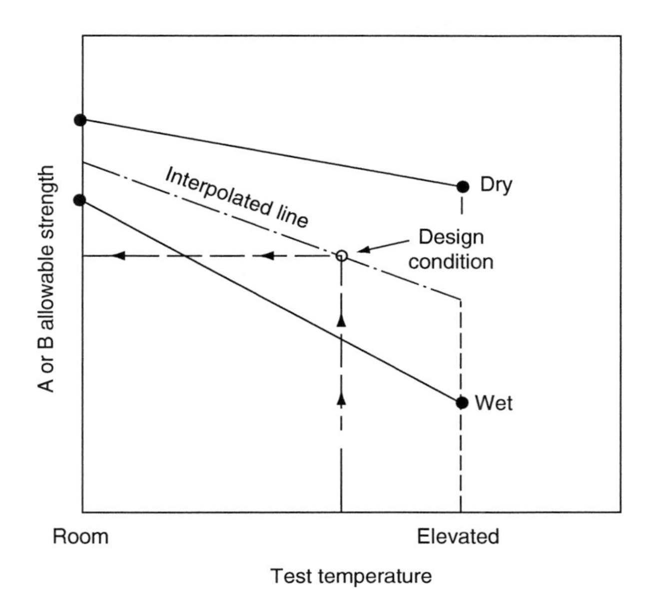

FIGURE 6.10 Linear interpolation method for determining design allowables.

procedure of testing at an ETW exceeding the design condition allows the designer to determine the allowable values for a variety of environmental conditions within the range investigated by interpolation rather than by extrapolation (Figure 6.10).

6.2.4 General Design Guidelines

The principal steps in designing a composite laminate are

1. Selection of fiber, resin, and fiber volume fraction
2. Selection of the optimum fiber orientation in each ply and the lamina stacking sequence
3. Selection of the number of plies needed in each orientation, which also determines the final thickness of the part

Considering these variables, it is obvious that a large variety of laminates may be created even if the ply orientations are restricted to a single family, such as a $[0/\pm45/90]_S$ family. Thus in most cases there is no straightforward method of designing a composite laminate unless the problem involves a simple structure, such as a rod or a column, and the loading is uniaxial.

From the standpoint of design as well as analytic simplicity, symmetric laminates are commonly preferred over unsymmetric laminates. This eliminates the extension–bending coupling represented by the $[B]$ matrix. The presence of extension–bending coupling is also undesirable from the stiffness standpoint, since it reduces the effective stiffness of the laminate and thereby increases its deflection, reduces the critical buckling loads, and decreases the natural frequency of vibration. Similar but lesser effects are observed if the laminate has bending–twisting coupling due to the presence of D_{16} and D_{26} terms. However, unless the fibers are at 0, 90, or 0/90 combinations, a symmetric laminate cannot be designed with $D_{16} = D_{26} = 0$.

The deleterious free-edge effects in a laminate can be reduced through proper selection of lamina stacking sequence. If angle-ply laminates are used, the layers with $+\theta$ and $-\theta$ orientations should be alternated instead of in a clustered configuration [23]. Thus, for example, an eight-layer laminate with four layers of $+\theta$ orientations and four layers of $-\theta$ orientations should be designed as $[+\theta/-\theta/+\theta/-\theta]_S$ instead of $[\theta/\theta/-\theta/-\theta]_S$ or $[-\theta/-\theta/+\theta/+\theta]_S$. However, if a laminate contains 0, 90, and $\pm\theta$ layers, adjacent $+\theta$ and $-\theta$ layers should be avoided. For example, in the quasi-isotropic laminate family containing 0, 90, and ± 45 layers, a $[45/0/90/-45]_S$ configuration is preferred over a $[90/+45/-45/0]_S$ or a $[0/+45/-45/90]_S$ configuration.

6.2.4.1 Laminate Design for Strength

When the state of stress in a structure is known and does not change during the course of its service operation, the lamina orientations may be selected in the following way [24].

Using the standard Mohr's circle technique, determine the principal normal loads and the principal directions. Analytically, the principal normal loads are

$$N_1 = \frac{1}{2}(N_{xx} + N_{yy}) + \left[\left(\frac{N_{xx} - N_{yy}}{2}\right)^2 + N_{xy}^2\right]^{1/2},$$

$$N_2 = \frac{1}{2}(N_{xx} + N_{yy}) - \left[\left(\frac{N_{xx} - N_{yy}}{2}\right)^2 + N_{xy}^2\right]^{1/2}, \tag{6.13}$$

and the principal direction with respect to the x axis is

$$\tan 2\theta = \frac{2N_{xy}}{N_{xx} - N_{yy}}. \tag{6.14}$$

Select a $[0_i/90_j]_S$ cross-ply configuration with the 0° layers aligned in the direction of the maximum principal load N_1 and the 90° layers aligned in the direction of the minimum principal load N_2. Thus with respect to the x axis, the laminate configuration is $[\theta_i/(90+\theta)_j]_S$. The ratio of 0° to 90° plies, i/j, is equal to the principal load ratio N_1/N_2.

When the stress state in a structure varies in direction or is unknown, a common approach in laminate design is to make it quasi-isotropic, for example,

$[0_i/\pm 45_j/90_k]_S$. The design procedure is then reduced to the selection of ply ratios ($i/j/k$) and the total thickness of the laminate. Design charts or carpet plots of the types shown in Figures 6.8 and 6.9 can be used to select the initial ply ratios; however, the final design must include a ply-by-ply analysis of the entire laminate.

Massard [25] has used an iterative ply-by-ply approach for designing symmetric laminates under in-plane and bending loads. In this approach, an initial laminate configuration is assumed and additional plies are added in a stepwise fashion to achieve the most efficient laminate that can sustain the given loading condition. At each step, strains in each lamina and the margin of safety (ratio of lamina strength to effective lamina stress) for each lamina are calculated. A margin of safety greater than unity indicates a safe ply in the laminate. The process is repeated until the margin of safety in each lamina is greater than unity.

Park [26] has used a simple optimization procedure to determine the fiber orientation angle θ for maximum FPF stress in symmetric laminates, such as $[\pm\theta]_S$, $[-\theta/0/\theta]_S$, and so on. The objective function F was expressed as

$$F = \varepsilon_{xx}^{\circ 2} + \varepsilon_{yy}^{\circ 2} + \frac{1}{2}\gamma_{xy}^{\circ 2}. \tag{6.15}$$

For a given laminate configuration, the midplane strain components are functions of the applied in-plane loads (N_{xx}, N_{yy}, and N_{xy}) as well as the fiber orientation angle θ. If the in-plane loads are specified, the design optimization procedure reduces to finding θ for which the objective function is minimum.

EXAMPLE 6.8

Using the carpet plots in Figures 6.8 and 6.9, determine the number of layers of 0, 90, and $\pm 45°$ orientations in a quasi-isotropic $[0/90/\pm 45]_S$ laminate that meets the following criteria:

1. Minimum modulus in the axial (0°) direction $= 6 \times 10^6$ psi
2. Minimum B-allowable strength in the axial (0°) direction $=$ 65 ksi

Solution

Step 1: Referring to Figure 6.8, determine the ply ratio that gives $E_{xx} = 6 \times 10^6$ psi.

$$0{:}90{:}\pm 45 = 20\%{:}\,60\%{:}\,20\%$$

Step 2: Referring to Figure 6.9, check the B-allowable strength for the ply ratio determined in Step 1, which in our case is 51 ksi. Since this value is less than the minimum required, we select a new ply ratio that will give a minimum B-allowable strength of 65 ksi. This new ply ratio is 0:90:±45 $=$ 30:50:20. Referring back to Figure 6.8, we find that this ply ratio gives an axial modulus of 7×10^6 psi, which is higher than the minimum required in the present design. Thus, the laminate configuration selected is $[0_3/90_5/\pm 45_2]_S$.

Step 3: Assuming that the ply thickness is 0.005 in., we determine the ply thickness for each fiber orientation as

$$0^\circ\text{: } 6 \times 0.005 \text{ in.} = 0.03 \text{ in.}$$

$$90^\circ\text{: } 10 \times 0.005 \text{ in.} = 0.05 \text{ in.}$$

$$+45^\circ\text{: } 4 \times 0.005 \text{ in.} = 0.02 \text{ in.}$$

$$-45^\circ\text{: } 4 \times 0.005 \text{ in.} = 0.02 \text{ in.}$$

Thus, the total laminate thickness is 0.12 in.

6.2.4.2 Laminate Design for Stiffness

The stiffness of a member is a measure of its resistance to deformation or deflection owing to applied loads. If the member is made of an isotropic material, its stiffnesses are given by

$$\begin{aligned} \text{Axial stiffness} &= EA_0, \\ \text{Bending stiffness} &= EI_c, \\ \text{Torsional stiffness} &= GJ_c, \end{aligned} \tag{6.16}$$

where

E = modulus of elasticity
G = shear modulus
A_0 = cross-sectional area
I_c = moment of inertia of the cross section about the neutral axis
J_c = polar moment of inertia of the cross section

The stiffness equations for composite members are in general more involved than those given in Equation 6.16. If the composite member is made of a symmetric laminate, its stiffness against the in-plane loads is related to the elements in the [A] matrix, whereas its stiffness against bending, buckling, and torsional loads is related to the elements in the [D] matrix. The elements in both [A] and [D] matrices are functions of the fiber type, fiber volume fraction, fiber orientation angles, lamina thicknesses, and the number of layers of each orientation. In addition, the elements in the [D] matrix depend strongly on the lamina stacking sequence.

Except for 0, 90, and 0/90 combinations, the [D] matrix for all symmetric laminates contains nonzero D_{16} and D_{26} terms. Closed-form solutions for bending deflections, buckling loads, and vibrational frequencies of general symmetric laminates are not available. The following closed-form solutions [27,28] are valid for the special class of laminates for which $D_{16} = D_{26} = 0$ and the elements in the [B] matrix are negligible:

1. Center deflection of a simply supported rectangular plate carrying a uniformly distributed load p_0:

$$w \cong \frac{16p_0R^4b^4}{\pi^6}\left[\frac{1}{D_{11}+2(D_{12}+2D_{66})R^2+D_{22}R^4}\right] \quad (6.17)$$

2. Critical buckling load for a rectangular plate with pinned edges at the ends of its long dimension:

$$N_{cr} \cong \frac{\pi^2[D_{11}+2(D_{12}+2D_{66})R^2+D_{22}R^4]}{b^2R^2} \quad (6.18)$$

3. Fundamental frequency of vibration for a simply supported rectangular plate:

$$f^2 \cong \frac{\pi^4}{\rho R^4b^4}[D_{11}+2(D_{12}+2D_{66})R^2+D_{22}R^4] \quad (6.19)$$

where
a = plate length
b = plate width
R = plate aspect ratio = a/b
ρ = density of the plate material

A few closed-form solutions are also available in the literature for unsymmetric laminates with a nonzero [*B*] matrix [27,28]. However, it has been shown that the coupling effect of the [*B*] matrix becomes small when the laminate contains more than six to eight layers [29]. Therefore, for most practical laminates, Equations 6.17 through 6.19 can be used for initial design purposes.

6.2.5 Finite Element Analysis

Design analysis of a laminated composite structure almost invariably requires the use of computers to calculate stresses and strains in each ply and to investigate whether the structure is "safe." For simple structures, such as a plate or a beam, the design analysis can be performed relatively easily. If the structure and the loading are complex, it may be necessary to perform the design analysis using finite element analysis. Commercially available finite element softwares, such as MSC-NASTRAN, ANSYS, ABAQUS, and LS-DYNA, have the capability of combining the lamination theory with the finite element codes. Many of these packages are capable of calculating in-plane as well as interlaminar stresses, incorporate more than one failure criterion, and contain a library of plate, shell, or solid elements with orthotropic material properties [30].

Although finite element analyses for both isotropic materials and laminated composite materials follow the same procedure, the problem of preparing the input data and interpreting the output data for composite structures is much more complex than in the case of metallic structures. Typical input information for an isotropic element includes its modulus, Poisson's ratio, and thickness. Its properties are assumed invariant in the thickness direction. An element for a composite structure may contain the entire stack of laminas. Consequently, in this case, the element specification must include the fiber orientation angle in each lamina, lamina thicknesses, and the location of each lamina with respect to the element midplane. Furthermore, the basic material property data for plane stress analysis of thin fiber-reinforced composite structure include four elastic constants, namely, the longitudinal modulus, transverse modulus, major Poisson's ratio, and shear modulus. Thus, the amount of input information even for a static load analysis of a composite structure is quite large compared with a similar analysis of a metallic structure.

The stress output from the finite element analysis of an isotropic material includes only three stress components for each element. In contrast, the stress output for a composite structure can be very large since it contains three in-plane stresses in each individual lamina as well as the interlaminar stresses between various laminas for every element. The lamina in-plane stresses are usually computed in the material principal directions, which vary from layer to layer within the same element. To examine the occurrence of failure in an element, a preselected failure criterion is applied to each lamina. In many finite element packages, the stress output may be reduced by calculating stress resultants, which are integrals of the lamina stresses through the thickness. However, interpretation of these stress resultants is difficult since they do not provide information regarding the adequacy of a design.

6.3 JOINT DESIGN

The purpose of a joint is to transfer loads from one member to another in a structure. The design of joints has a special significance in fiber-reinforced composite structures for two reasons: (1) the joints are often the weakest areas in a composite structure and (2) the composite materials do not possess the forgiving characteristics of ductile metals, namely, their capacity to redistribute local high stresses by yielding.

For composite laminates, the basic joints are either mechanical or bonded. Mechanical joints are created by fastening the substrates with bolts or rivets; bonded joints use an adhesive interlayer between the substrates (commonly called the adherends). The advantages and disadvantages of these two types of joints are listed as follows.

Mechanical Joints:

1. Permit quick and repeated disassembly for repairs or replacements without destroying the substrates
2. Require little or no surface preparation

3. Are easy to inspect for joint quality
4. Require machining of holes that interrupt the fiber continuity and may reduce the strength of the substrate laminates
5. Create highly localized stress concentrations around the joints that may induce failure in the substrates
6. Add weight to the structure
7. May create a potential corrosion problem, for example, in an aluminum fastener if used for joining carbon fiber–epoxy laminates

Bonded Joints:

1. Distribute the load over a larger area than mechanical joints
2. Require no holes, but may need surface preparation (cleaning, pretreatment, etc.)
3. Add very little weight to the structure
4. Are difficult to disassemble without either destroying or damaging substrates
5. May be affected by service temperature, humidity, and other environmental conditions
6. Are difficult to inspect for joint quality

We will now discuss the general design considerations with these two types of joints.

6.3.1 Mechanical Joints

The strength of mechanical joints depends on the following.

1. Geometric parameters, such as the ratios of edge distance to bolt hole diameter (e/d), width to bolt hole diameter (w/d), and laminate thickness to bolt hole diameter (h/d). In multibolt joints, spacing between holes and their arrangements are also important.
2. Material parameters, such as fiber orientation and laminate stacking sequence. Some of these material parameters are discussed in Chapter 4 (see Section 4.4.1).

In applications involving mechanical joints, three basic failure modes are observed in the substrates, namely, shear-out, net tension, and bearing failure (Figure 6.11). If the laminate contains nearly all 0° fibers, cleavage failure is also possible. From a safe design standpoint, a bearing failure is more desirable than either a shear-out or a net tension failure. However, unless the e/d and w/d ratios are very large, the full bearing strength is seldom achieved. In general, shear-out and net tension failures are avoided if $e/d > 3$ and $w/d > 6$. The actual geometric parameters are usually determined by conducting pin-bearing tests on the specific laminates involved. However, a bolted joint

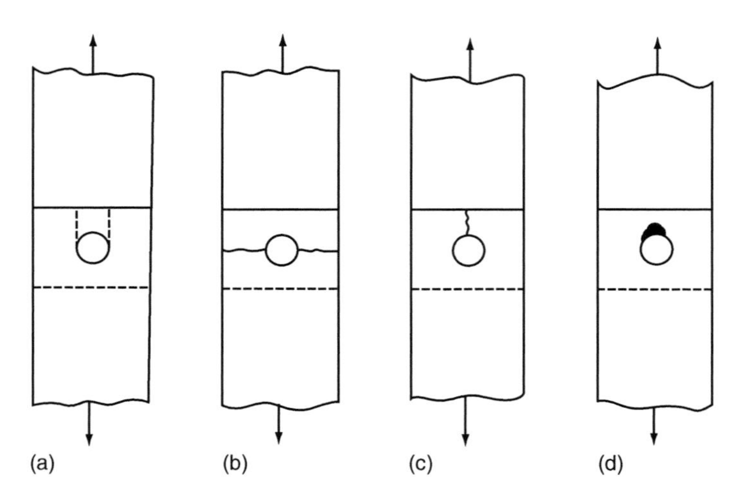

FIGURE 6.11 Basic failure modes in bolted laminates: (a) shear-out, (b) net tension failure, (c) cleavage, and (d) bearing failure. Combinations of these failure modes are possible.

differs from a pinned joint, since in the former the clamping torque is an added factor contributing to the joint strength. The edge distance needed to reduce shear-out failure can be reduced by increasing the laminate thickness at the edge or by inserting metal shims between various composite layers near the bolted area.

The strength of a mechanical joint can be improved significantly by relieving the stress concentrations surrounding the joint. The following are some of the methods used for relieving stress concentrations.

Softening strips of lower modulus material are used in the bolt bearing area. For example, strips made of E-glass fiber plies can be used to replace some of the carbon fiber plies aligned with the loading direction in a carbon fiber-reinforced laminate.

Laminate tailoring method [31] divides the bolt hole area of structure into two regions, namely, a primary region and a bearing region. This is demonstrated in Figure 6.12 for a structure made of a $[0_2/\pm45]_{2S}$ laminate. In the bearing region surrounding the bolt hole, the 0° plies in the laminate are replaced by ±45° plies. Thus, the lower modulus bearing region is bounded on both sides by the primary region of $[0_2/\pm45]$ laminate containing 20%–60% 0° plies. The majority of the axial load in the joint is carried by the high-modulus primary region, which is free of fastener holes. The combination of low axial stress in the bearing region and the relatively low notch sensitivity of the [±45] laminate delays the onset of the net tension failure commonly observed in non-tailored $[0_2/\pm45]_S$ laminates.

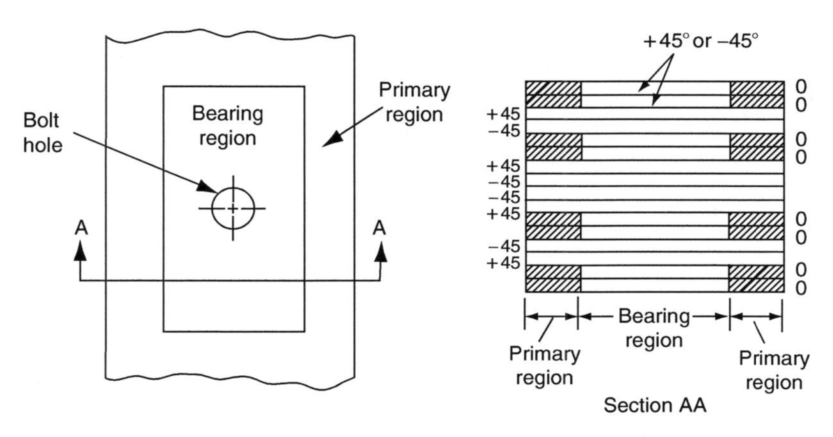

FIGURE 6.12 Laminate tailoring method for improving bolted joint strength.

Interference fit fasteners increase the possibility of localized delamination instead of fiber failure in the joint area (Figure 6.13). This leads to a redistribution of high stresses surrounding the joint. However, care must be taken not to damage the fibers while installing the interference fit fasteners.

Holes for mechanical joints can either be machined in a postmolding operation or formed during the molding of the part. Machining is preferred, since molded

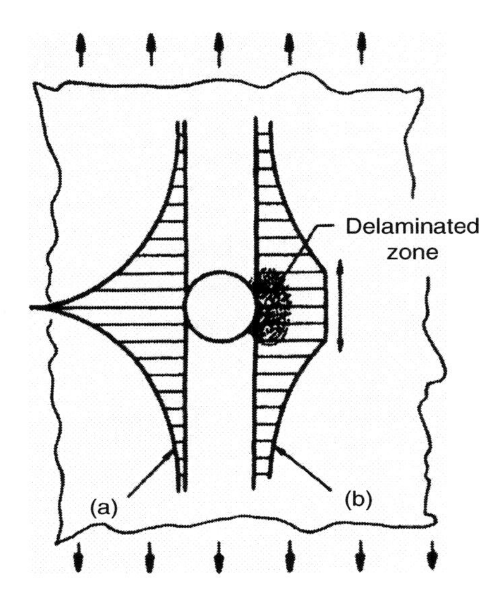

FIGURE 6.13 Stress distributions in areas adjacent to a bolt hole with (a) no delaminated zone and (b) a delaminated zone. (After Jones, R.M., Morgan, H.S., and Whitney, J.M., *J. Appl. Mech.*, 40, 1143, 1973.)

holes may be surrounded by misoriented fibers, resin-rich areas, or knit lines. Drilling is the most common method of machining holes in a cured laminate; however, unless proper cutting speed, sharp tools, and fixturing are used, the material around the drilled hole (particularly at the exit side of the drill) may be damaged. High-speed water jets or lasers produce cleaner holes and little or no damage compared with the common drilling process.

6.3.2 Bonded Joints

The simplest and most widely used bonded joint is a single-lap joint (Figure 6.14a) in which the load transfer between the substrates takes place through a distribution of shear stresses in the adhesive. However, since the loads applied at the substrates are off-centered, the bending action sets up a normal (peel) stress in the thickness direction of the adhesive. Both shear and normal stress distributions exhibit high values at the lap ends of the adhesive layer, which tends to reduce the joint strength. The double-lap joint, shown in Figure 6.14b, eliminates much of the bending and normal stresses present in the single-lap joint. Since the average shear stress in the adhesive is also reduced by nearly one-half, a double-lap joint has a higher joint strength than a single-lap joint (Figure 6.15). The use of a long-bonded strap on either side or on each side of the substrates (Figure 6.14c) also improves the joint strength over that of single-lap joints.

Stepped lap (Figure 6.14d) and scarf joints (Figure 6.14e) can potentially achieve very high joint strengths, however in practice, the difficulty in machining the steps or steep scarf angles often overshadows their advantages. If a

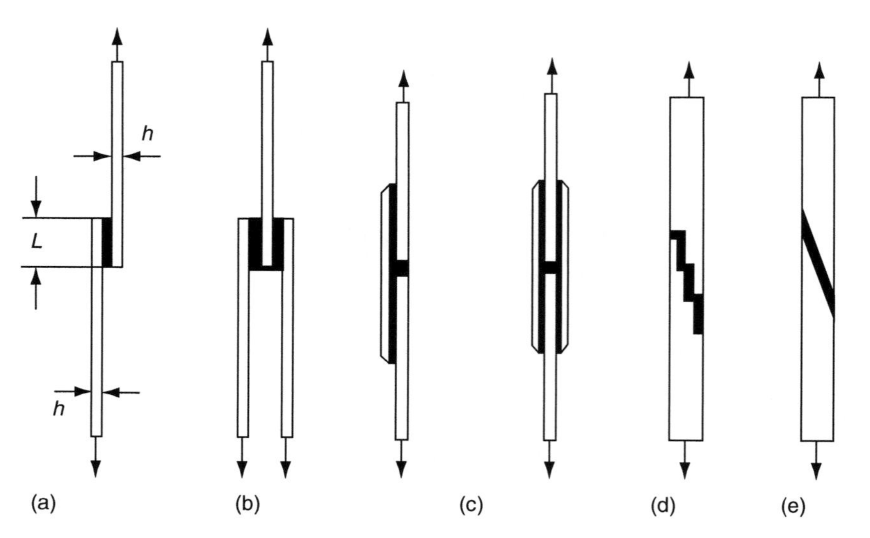

FIGURE 6.14 Basic bonded joint configurations: (a) single-lap joint, (b) double-lap joint, (c) single- and double-strap joints, (d) stepped lap joint, and (e) scarf joint.

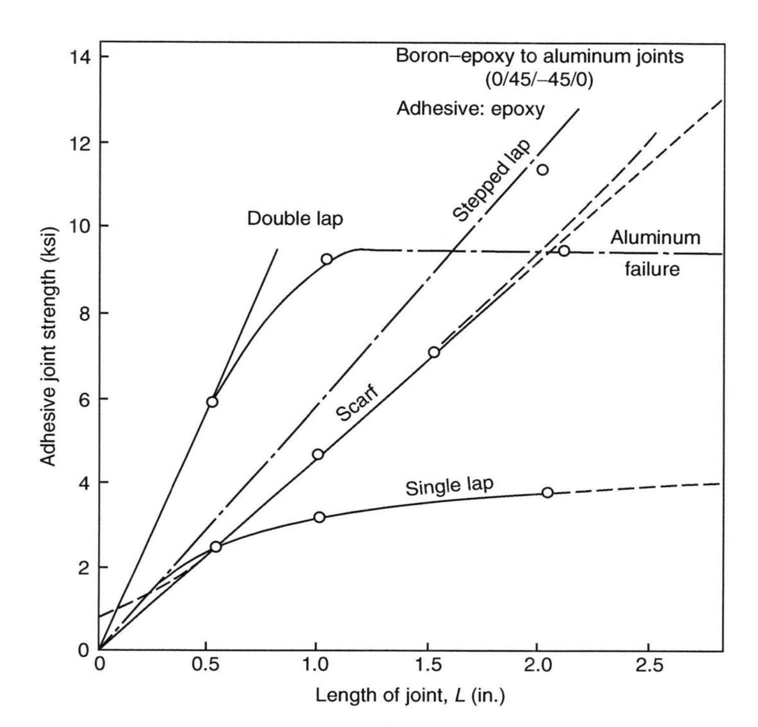

FIGURE 6.15 Increase in the joint strength of single-lap bonded joints with lap length. (After Griffin, O.H., *Compos. Technol. Rev.*, 4, 136, 1982.)

stepped lap joint is used, it may often be easier and less expensive to lay up the steps before cure. This eliminates the machining operation and prevents damage to the fibers.

The following points are important in designing a bonded joint and selecting an appropriate adhesive for the joint.

1. Increasing the ratio of lap length to substrate thickness h improves the joint strength significantly at small L/h ratios. At high L/h ratios, the improvement is marginal (Figure 6.16).
2. Tapering the substrate ends at the ends of the overlap reduces the high normal stresses at these locations [32].
3. Equal axial stiffnesses for the substrates are highly desirable for achieving the maximum joint strength (Figure 6.16). Since stiffness is a product of modulus E and thickness h, it is important to select the proper thickness of each substrate so that $E_1h_1 = E_2h_2$. If the two substrates are of the same material, their thicknesses must be equal.
4. The important characteristics of a good adhesive are high shear and tensile strengths but low shear and tensile moduli. An efficient way of

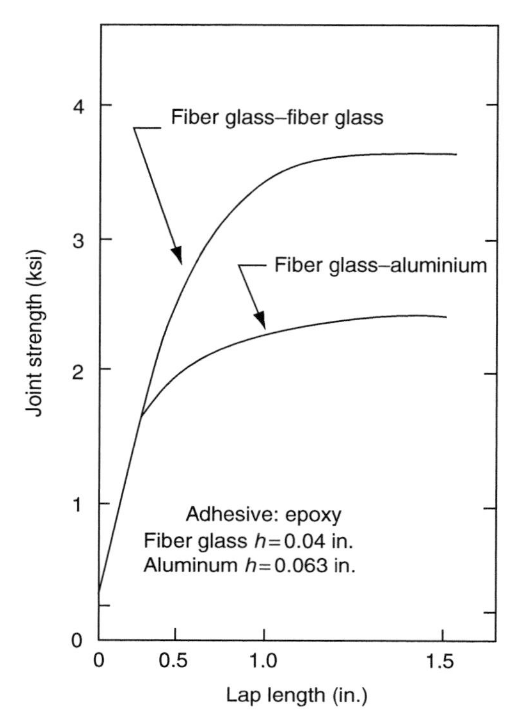

FIGURE 6.16 Comparison of joint strengths of various bonded joint configurations. (After Griffin, O.H., *Compos. Technol. Rev.*, 4, 136, 1982.)

increasing the joint strength is to use a low-modulus adhesive only near the ends of the overlap, which reduces stress concentrations, and a higher modulus adhesive in the central region, which carries a large share of the load. High ductility for the adhesive becomes an important selection criterion if the substrates are of dissimilar stiffnesses or if the joint is subjected to impact loads.

5. Fiber orientation in the laminate surface layers adjacent to the lap joints should be parallel to the loading direction (Figure 6.17). Otherwise, a scarf joint should be considered even though machining is required to produce this configuration.

6.4 DESIGN EXAMPLES

6.4.1 Design of a Tension Member

The simplest and the most efficient structure to design with a fiber-reinforced composite material is a two-force tension member, such as a slender rod or a slender bar subjected to tensile forces along its axis. Since the fibers have

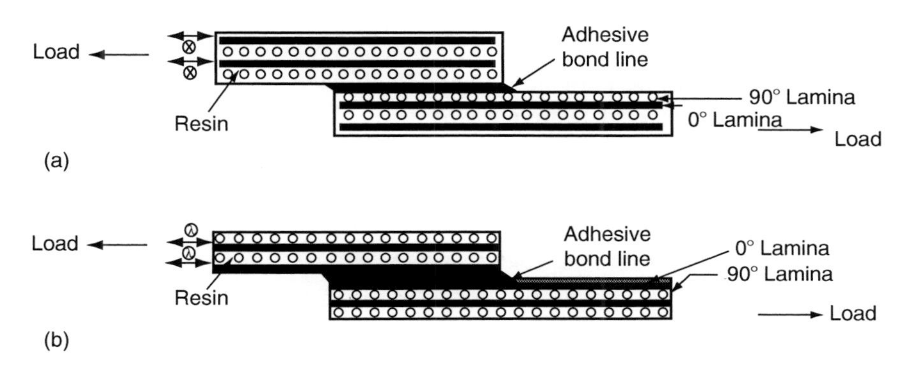

FIGURE 6.17 Examples of (a) poor and (b) good laminate designs for bonded joints. (After Ref. Griffin, O.H., *Compos. Technol. Rev.*, 4, 136, 1982.)

exceptionally high tensile strength–weight ratios, the load-carrying capacity of a tension member with fibers oriented parallel to its axis can be very high. The static tension load that can be supported by a tension member containing longitudinal continuous fibers is

$$P = S_{\mathrm{Lt}} A_0 \cong \sigma_{\mathrm{fu}} \mathrm{v}_{\mathrm{f}} A_0, \tag{6.20}$$

where

σ_{fu} = ultimate tensile strength of the fibers
v_{f} = fiber volume fraction
A_0 = cross-sectional area

The axial stiffness of the tension member is

$$EA_0 \cong E_{\mathrm{f}} \mathrm{v}_{\mathrm{f}} A_0, \tag{6.21}$$

so that its axial elongation can be written as

$$\Delta = \frac{PL_0}{EA_0} = \frac{\sigma_{\mathrm{fu}}}{E_{\mathrm{f}}} L_0. \tag{6.22}$$

Equations 6.20 and 6.21 indicate the importance of selecting the proper fiber type as well as the fiber volume fraction for maximum load-carrying capacity and stiffness of a tension member. For good fiber wet-out, the practical limit for the maximum fiber volume fraction is about 0.6. Thus, for a specified design load, the minimum cross-sectional area is obtained by selecting the strongest fiber. In many applications, however, the maximum elongation may also be specified. In that case, the ratio of fiber strength to fiber modulus should also be checked. Although selection of the matrix has little influence on the load-carrying capacity or the elongation, it can influence the manufacturing and environmental considerations for the member.

TABLE 6.5
Tensile Fatigue Strength Coefficient *b* for Various 0° Laminates

Fiber Type	*b*
Ultrahigh-modulus carbon	0.021
High-modulus carbon	0.035
High-strength carbon	0.05
Kevlar 49	0.035
S-glass	0.088
E-glass	0.093

For most 0° continuous fiber composites, the fatigue strength in tension–tension cycling can be approximated as

$$S = S_{\mathrm{Lt}}(1 - b \log N), \tag{6.23}$$

where the value of the constant b depends primarily on the fiber type (Table 6.5). If the tension member is exposed to tension–tension fatigue, its design should be based on the fatigue strength of the material at the desired number of cycles. It should be noted that ranking of the fibers based on the fatigue strength can be different from that based on the fiber tensile strength.

The most critical design issue for a tension member involves the joints or connections at its ends. A few joint design ideas other than the simple bolted or bonded joints are shown in Figure 6.18.

6.4.2 Design of a Compression Member

Tubular compression members made from fiber-reinforced polymeric materials are finding applications in many aerospace structures, such as satellite trusses, support struts, and flight control rods. Since these compression members are mostly slender tubes, their design is usually based on preventing overall column buckling as well as local buckling [33].

The compressive stress on a thin tube of radius r and wall thickness t is

$$\sigma = \frac{P}{A_0} = \frac{P}{2\pi r t}, \tag{6.24}$$

where P is the axial compressive load.

For a pin-ended column, the overall buckling stress is given by

$$\sigma_{\mathrm{col}} = \left(\frac{2L^2}{\pi^2 E_{xx} r^2} + \frac{2}{G_{xy}} \right)^{-1}, \tag{6.25}$$

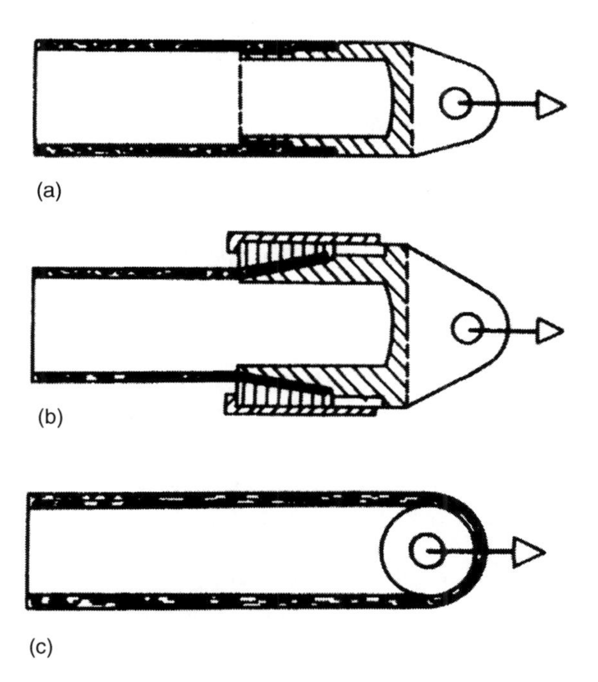

FIGURE 6.18 Joints in tension members: (a) tube with bonded shear fitting, (b) tube with wedge fitting, and (c) wrapping around a bushing. (Adapted from Taig, I.C., *Composites—Standards, Testing and Design*, National Physical Laboratory, London, 1974.)

where

L = length of the compression member
E_{xx} = modulus of elasticity in the axial direction
G_{xy} = shear modulus

Note that the second term in Equation 6.25 represents the shear effect on the critical buckling stress.

The local buckling stress of a thin-walled tube is given by

$$\sigma_{\text{local}} = \beta_0 \frac{t}{r}, \tag{6.26}$$

where

$$\beta_0 = \frac{\gamma\Phi\sqrt{E_{xx}E_{yy}}}{\sqrt{3(1-\nu_{xy}\nu_{yx})}}$$

γ = a correlation coefficient

$$\Phi = \left[\frac{2G_{xy}}{\sqrt{E_{xx}E_{yy}}}(1+\nu_{xy}\nu_{yx})\right]^{1/2} \quad \text{or 1, whichever is smaller}$$

Using Equations 6.24 through 6.26, Maass [33] developed the following optimum stress equation for a compression tube design:

$$\sigma_{\mathrm{opt}}^{3}\left[\frac{4}{\pi E_{xx}\beta_0(P/L^2)}\right]+\sigma_{\mathrm{opt}}\left(\frac{2}{G_{xy}}\right)=1. \tag{6.27}$$

Neglecting the shear term, an approximate σ_{opt} can be calculated as

$$\sigma_{\mathrm{opt}}=\left[\frac{\pi E_{xx}\beta_0(P/L^2)}{4}\right]^{1/3}, \tag{6.28}$$

where (P/L^2) is called the loading index.

Knowing σ_{opt} from either Equation 6.27 or 6.28, the actual tube dimensions can be determined from the following equations:

$$r_{\mathrm{opt}}=\left(\frac{P\beta_0}{2\pi\sigma_{\mathrm{opt}}^{2}}\right)^{1/2},$$

$$t_{\mathrm{opt}}=\frac{\sigma_{\mathrm{opt}}}{\beta_0}r_{\mathrm{opt}}. \tag{6.29}$$

Equations 6.27 through 6.29 show that the optimum design of a compression tube depends very much on the laminate configuration. Maass [33] used these equations to determine the optimum stress and the corresponding ply ratio for various $[0/\pm\theta]_S$ tubes made of high-strength carbon–epoxy laminates. For a unit loading index, the highest optimum stress occurs for a $[0_3/\pm45]_S$ tube, although a $[\pm15]_S$ tube with no 0° fibers exhibits approximately the same optimum stress. The optimum design with increasing off-axis angle θ is obtained with increasing percentages of 0° layers in the tube.

6.4.3 Design of a Beam

Beams are slender structural members designed principally to resist transverse loads. In general, the stress state at any point in the beam consists of an axial normal stress σ_{xx} and a transverse shear stress τ_{xz}. Both these stresses are nonuniformly distributed across the thickness (depth) of the beam. In an isotropic homogeneous beam, these stress distributions are continuous with the maximum and minimum normal stresses occurring at the outermost surfaces and the maximum shear stress occurring at the neutral axis. In laminated beams, the normal stress and shear stress distributions are not only nonuniform, but also they are discontinuous at the interfaces of dissimilar laminas. Depending on the lamination configuration, it is possible to create maximum normal stresses in the interior of the beam thickness and the maximum shear stress away from the midplane of the beam. For this reason, the actual stress distribution in a laminated composite beam should always be calculated using the lamination theory instead of the homogeneous beam theory.

Except all 0, all 90, and 0/90 combinations for which $D_{16} = D_{26} = 0$, there will be a bending–twisting coupling in all symmetric beams. This means that a bending moment will create not only bending deformations, but also tend to twist the beam. Whitney et al. [34] have shown that the deflection equation for a symmetric beam has the same form as that for a homogeneous beam, namely,

$$\frac{d^2 w}{dx^2} = \frac{bM_{xx}}{E_b I}, \tag{6.30}$$

where

w = beam deflection
b = beam width
M_{xx} = bending moment per unit width
I = moment of inertia of the cross section about the midplane
E_b = effective bending modulus of the beam

The effective bending modulus E_b is defined as

$$E_b = \frac{12}{h^3 D_{11}^*}, \tag{6.31}$$

where

h is the beam thickness
D_{11}^* is the first element in the inverse $[D]$ matrix (see Example 3.13)

Equation 6.31 neglects the effect due to transverse shear, which can be significant for beams with small span-to-thickness ratio. For long beams, for which the effect of the transverse shear is negligible, the maximum deflection can be calculated by replacing the isotropic modulus E with the effective bending modulus E_b in the deflection formulas for homogeneous beams. From Equation 6.31, the effective bending stiffness for a laminated beam can be written as

$$E_b I = \frac{b}{D_{11}^*}. \tag{6.32}$$

For a symmetric beam containing isotropic layers or specially orthotropic layers (such as all 0°, all 90°, or combinations of 0° and 90° layers), the effective bending stiffness becomes

$$(EI)_b = \Sigma (E_{11})_j I_j, \tag{6.33}$$

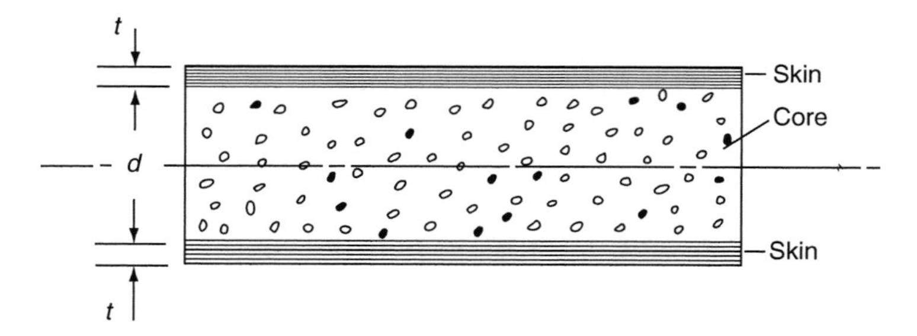

FIGURE 6.19 Construction of a sandwich beam.

where $(E_{11})_j$ is the longitudinal modulus of the jth layer and I_j is the moment of inertia of the jth layer with respect to the midplane

The most effective method of reducing the weight of a beam (or a panel) without sacrificing its bending stiffness is to use a sandwich construction (Figure 6.19). This consists of a lightweight, low-modulus foam or honeycomb core adhesively bonded to high-modulus fiber-reinforced laminate skins (face-sheets). The bending stiffness of the sandwich beam is

$$(EI)_b = E_s \frac{bt^3}{6} + 2bE_s t \left(\frac{d+t}{2} \right)^2 + E_c \frac{bd^3}{12}, \tag{6.34}$$

where

E_s = modulus of the skin material
E_c = modulus of the core material ($E_c << E_s$)
b = beam width
t = skin thickness
d = core thickness

Equation 6.34 shows that the bending stiffness of a sandwich beam can be increased significantly by increasing the value of d, that is, by using a thicker core. Since the core material has a relatively low density, increasing its thickness (within practical limits) does not add much weight to the beam. However, it should be noted that the core material also has a low shear modulus. Thus, unless the ratio of span to skin thickness of the sandwich beam is high, its deflection will be increased owing to the transverse shear effect.

Commonly used core materials are honeycombs with hexagonal cells made of either aluminum alloys or aramid fiber-reinforced phenolics.* The strength and stiffness of such cores depend on the cell size, cell wall thickness, and the material used in the honeycomb. High core strength is desirable to resist

* Trade name: Nomex, manufactured by Du Pont.

transverse shear stresses as well as to prevent crushing of the core under the applied load. While the facings in a sandwich beam or panel resist tensile and compressive stresses induced due to bending, the core is required to withstand transverse shear stresses, which are high near the center of the beam cross section. The core must also have high stiffness to resist not only the overall buckling of the sandwich structure but also local wrinkling of the facing material under high compressive loads.

Proper fiber selection is important in any beam design. Although beams containing ultrahigh-modulus carbon fibers offer the highest flexural stiffness, they are brittle and exhibit a catastrophic failure mode under impact conditions. The impact energy absorption of these beams can be increased significantly by using an interply hybrid system of ultrahigh-modulus carbon fibers in the skin and glass or Kevlar 49 fibers in the core. Even with lower modulus carbon fibers, hybridization is recommended since the cost of a hybrid beam is lower than an all-carbon beam. Beams containing only Kevlar 49 fibers are seldom used, since composites containing Kevlar 49 fibers have low compressive strengths. In some beam applications, as in the case of a spring, the capacity of the beam to store elastic strain energy is important. In selecting fibers for such applications, the elastic strain energy storage capacity of the fibers should be compared (Table 6.6).

EXAMPLE 6.9

Design of a Hybrid Beam. Determine the thickness of 25.4 mm (1 in.) wide hybrid beam containing three layers of HMS carbon–epoxy and two layers of S-glass–epoxy to replace a steel beam of bending stiffness 26.2 kN m^2 (150 lb in.^{2}). Fibers in the composite beam are parallel to the beam axis. Assuming that each carbon fiber ply is 0.15 mm (0.006 in.) thick and each glass fiber ply is 0.13 mm (0.005 in.) thick, determine the number of plies required for each fiber type.

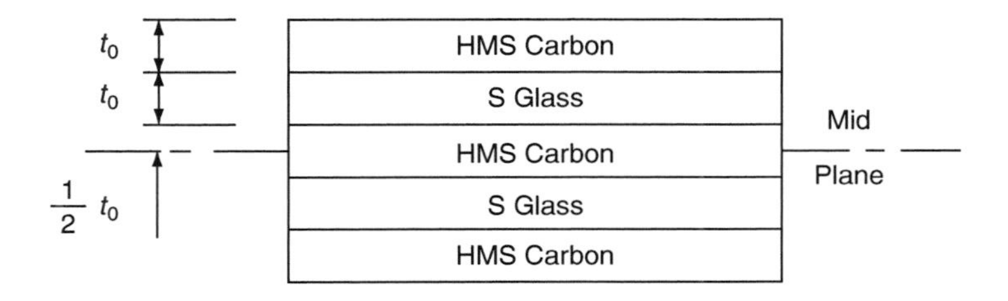

SOLUTION

For maximum stiffness, we place two carbon fiber layers on the outside surfaces. For symmetry, the layers just below the outside carbon layers will be the S-glass layers, which will leave the remaining carbon fiber layer at the center of the cross section. We assume that each layer has a thickness t_0. Since fibers in each layer are

TABLE 6.6
Strain Energy Storage Capacity of E-Glass and Carbon Fiber Laminates

Material	Density, g/cm^3 (lb/in.^{3})	Strength, MPa (ksi)		Modulus, GPa (Msi)	Strain Energy Capacity[a]		
		Static	Fatigue		Per Unit Volume, kN m/m^3 (lb in./in.^{3})	Per Unit Weight, kN m/kg (lb in./lb)	Per Unit Cost,[b] kN m/$ (lb in./$)
Spring steel	7.84 (0.283)	1448 (210)	724 (105)	200 (29)	1310 (190)	167 (672)	253 (2240)
E-glass–epoxy	1.77 (0.064)	690 (100)	241 (35)	38 (5.5)	765 (111)	432 (1734)	245 (2167)
High-strength carbon–epoxy	1.50 (0.054)	1035 (150)	672 (97.5)	145 (21)	1558 (226)	1041 (4185)	23.6 (209)

[a] Strain energy per unit volume = strength^2/(2 × modulus). In this table, strain energy is calculated on the basis of fatigue strength.
[b] Cost: Steel = $0.30/lb, E-glass–epoxy = $0.80/lb, high-strength carbon–epoxy = $20/lb.

at a 0° orientation with the beam axis, we can apply Equation 6.33 to calculate the bending stiffness of the hybrid beam. Thus,

$$
\begin{aligned}
(EI)_{\text{hybrid}} &= 2E_c I_1 + 2E_g I_2 + E_c I_3 \\
&= 2E_c\left[\frac{1}{12}bt_0^3 + bt_0\left(t_0 + \frac{1}{2}t_0 + \frac{1}{2}t_0\right)^2\right] \\
&\quad + 2E_g\left[\frac{1}{12}bt_0^3 + bt_0\left(\frac{1}{2}t_0 + \frac{1}{2}t_0\right)^2\right] + E_c\left[\frac{1}{12}bt_0^3\right] \\
&= \frac{1}{12}bt_0^3(99E_c + 26E_g).
\end{aligned}
$$

Substituting $E_c = 207$ GPa, $E_g = 43$ GPa (from Appendix A.5), and $b = 0.0254$ m into the equation for bending stiffness and equating it to 26.2 kN m^2, we calculate $t_0 = 8.3$ mm (0.33 in.). Since there are five layers, the total thickness of the beam is 41.5 mm (1.65 in.). In comparison, the thickness of the steel beam of the same width is 39 mm (1.535 in.).

Now we calculate the number of plies for each layer by dividing the layer thickness by the ply thickness. For HMS carbon fibers, the number of plies in each layer is 55.3, or 56, and that for the S-glass layers is 63.8, or 64.

6.4.4 Design of a Torsional Member

The shear modulus of many fiber-reinforced composites is lower than that for steel. Thus for an equivalent torsional stiffness, a fiber-reinforced composite tube must have either a larger diameter or a greater thickness than a steel tube. Among the various laminate configurations, $[\pm 45]_S$ laminates possess the highest shear modulus and are the primary laminate type used in purely torsional applications.

In general, the shear modulus of a laminate increases with increasing fiber modulus. Thus, for example, the shear modulus of a GY-70 carbon–epoxy $[\pm 45]_S$ laminate is 79.3 GPa (11.5 Msi), which is equivalent to that of steel. The shear modulus of an AS carbon–epoxy $[\pm 45]_S$ laminate is 31 GPa (4.49 Msi), which is slightly better than that of aluminum alloys. Glass fiber laminates have even lower shear modulus, and Kevlar 49 fiber laminates are not generally used in torsional applications because of their low shear strengths. The shear strengths of both GY-70 and AS carbon–epoxy $[\pm 45]_S$ laminates are comparable with or even slightly better than those for mild steel and aluminum alloys.

The maximum torsional shear stress in a thin-walled tube of balanced symmetric laminate constructions [35] is

$$\tau_{xy} = \frac{T}{2\pi r^2 t} \tag{6.35}$$

and the angle of twist per unit length of the tube is given by

$$\phi = \frac{T}{2\pi G_{xy} r^3 t}, \tag{6.36}$$

where

T = applied torque
r = mean radius
t = wall thickness

For very thin-walled tubes, the possibility of torsional buckling exists. For symmetrically laminated tubes of moderate lengths, the critical buckling torque [36] is

$$T_{cr} = 24.4CD_{22}^{5/8}A_{11}^{3/8}r^{5/4}L^{-1/2}, \tag{6.37}$$

where C is end-fixity coefficient, which is equal to 0.925 for simply supported ends and 1.03 for clamped ends.

EXAMPLE 6.10

Design of an Automotive Drive Shaft. Select a laminate configuration for an automotive drive shaft that meets the following design requirements:

1. Outer diameter = 95.25 mm (3.75 in.)
2. Length = 1.905 m (75 in.)
3. Minimum resonance frequency = 90 Hz
4. Operating torque = 2,822 N m (25,000 in. lb)
5. Overload torque = 3,386 N m (30,000 in. lb)

Use a carbon–epoxy laminate containing 60% by volume of T-300 carbon fibers. The ply thickness is 0.1524 mm (0.006 in.). The elastic properties of the material are given in Example 3.6. The static shear strength for a $[\pm 45]_S$ laminate of this material is 455 MPa (66,000 psi).

SOLUTION

Step 1: Select an initial laminate configuration, and determine the minimum wall thickness for the drive shaft.

The primary load on the drive shaft is a torsional moment for which we select a $[\pm 45]_{kS}$ laminate, where k stands for the number of ±45° layers in the laminate. The minimum wall thickness for the laminate is determined from the following equation:

$$\tau_{all} = \frac{S_{xys}}{n} = \frac{T_{max}}{2\pi r^2 t},$$

where

T_{max} = maximum torque = 3386 N m
r = mean radius = 0.048 m
t = wall thickness
S_{xys} = static shear strength = 455×10^6 N/m^2
n = factor of safety = 2.2 (assumed)

Using this equation, we calculate $t = 0.001131$ m = 1.131 mm. Since each ply is 0.1524 mm thick, the minimum number of 45° plies is (1.131/0.1524) = 7.42. Assume eight plies, so that the initial laminate configuration is $[\pm 45]_{2S}$.

Step 2: Check for the minimum resonance frequency. The fundamental resonance frequency corresponding to the critical speed of a rotating shaft is

$$f_{cr} = \frac{1}{2\pi}\left[\frac{\pi^2}{L^2}\sqrt{\frac{E_{xx} I_c}{\rho A_0}}\right],$$

where

E_{xx} = axial modulus
A_0 = $2\pi r t$
I_c = $\pi r^3 t$
ρ = density

Substituting for A_0 and I_c, we obtain

$$f_{cr} = \frac{\pi}{2}\frac{r}{L^2}\sqrt{\frac{E_{xx}}{2\rho}}.$$

To meet the minimum resonance frequency, the shaft must have an adequate axial modulus. Since the axial modulus of a $[\pm 45]_{2S}$ laminate is rather low, we add four plies of 0° layers to the previous ±45° layers so that the new laminate configuration is $[\pm 45/0_2/\pm 45]_S$. The two 45° layers are placed in the outer diameter instead of the 0° layers to resist the maximum shear stress due to the torsional moment. Using the lamination theory, we calculate E_{xx} as 53.39 GPa. Since the ply material contains 60% by volume of T-300 carbon fibers, its density is calculated as 1556 kg/m^3. Using these values, we calculate the resonance frequency as 86 Hz, which is less than the minimum value required.

To improve the resonance frequency, we add two more layers of 0° plies, which brings the laminate configuration to $[\pm 45/0_3/\pm 45]_S$. Using the lamination theory, we recalculate E_{xx} as 65.10 GPa. The resonance frequency of this new shaft is 95 Hz, which exceeds the minimum value required.

Step 3: Check for the maximum torsional shear stress. We need to check for the maximum torsional shear stress since the laminate configuration is different from that assumed in Step 1. The $[\pm 45/0_3/\pm 45]_S$ laminate contains 14 plies with a wall thickness of 14×0.1524 mm = 2.13 mm. The maximum torsional shear stress is calculated as 109.8 MPa.

The maximum static shear strength of the $[\pm 45/0_3/\pm 45]_S$ laminate is not known. Since 57% of this laminate is ±45° layers, we estimate its shear strength as 57% of the static shear strength of the $[\pm 45]_S$ laminate, or 0.57 × 455 MPa = 260 MPa. Comparing the maximum torsional shear stress with this estimated shear strength, we find the factor of safety as $n = 260/109.8 = 2.37$, which is adequate for the torsional shear stress.

Step 4: Check the critical buckling torque. Using the lamination theory, we calculate $D_{22} = 71.48$ N m and $A_{11} = 1714.59 \times 10^5$ N/m. Substitution of these values into Equation 6.37 gives $T_{cr} = 6420$ N m, which is nearly twice the maximum application torque. Thus the $[\pm 45/0_3/\pm 45]_S$ laminate is safe against torsional buckling. If this were not the case, the easiest way to increase the critical buckling torque would be to increase D_{22}, which is achieved by adding one or more 90° plies on both sides of the laminate midplane.

Although this example does not address the problem of the end fitting attachments, it is a critical design issue for an automotive drive shaft. The common methods of attaching the metal end fittings are bonded or interference joints for low applied torques and bonded or bolted joints for high applied torques. If bolting is used, it is recommended that the joint area be locally reinforced either by using a tubular metal insert or by using additional layers in the laminate.

6.5 APPLICATION EXAMPLES

6.5.1 INBOARD AILERONS ON LOCKHEED L-1011 AIRCRAFT [37]

Ailerons are adjustable control surfaces hinged to the wing trailing edges of an aircraft for controlling its roll (rotation about the longitudinal axis). Their angular positions are manipulated by hydraulic actuators. Each aileron has a wedge-shaped one-cell box configuration consisting of a front spar, a rear spar, upper and lower covers, and a number of reinforcing ribs. Other parts in the aileron assembly are leading edge shrouds, end fairings, trailing edge wedge, shroud supports, feedback fittings, and hinge and actuator fittings. The primary load on the aileron surfaces is the air pressure. Ailerons are not considered primary structural components in an aircraft. Like other secondary components, their design is governed by stiffness instead of strength.

In Lockheed L-1011 aircraft, inboard ailerons are located between the outboard and inboard flaps on each wing (Figure 6.20). At the front spar, each aileron is 2.34 m (92 in.) in length and ~250 mm (10 in.) deep. Its width is 1.27 m (50 in.). The composite ailerons in L-1011 aircraft are designed with the following goals.

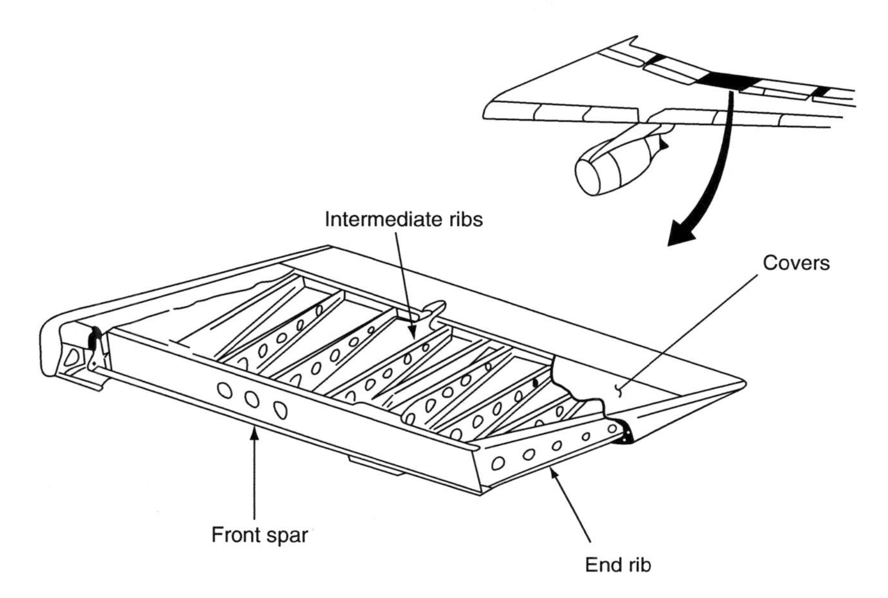

FIGURE 6.20 Construction of a Lockheed L-1011 composite aileron. (*Note*: The hinge and actuator fittings are not shown. They are located on the front spar.)

1. They must directly replace production aluminum ailerons in fit, form, function, and stiffness, and result in weight reduction.
2. As in the case of aluminum ailerons, the composite ailerons must meet the fail–safe design criteria for limit flight loads in accordance with the U.S. Federal Aviation Administration (FAA) requirements.
3. The material selected for the aileron structure must not severely degrade at temperatures ranging from −54°C to 82°C (−65°F to 180°F) or at high humidity conditions.

After careful evaluation of a large number of material as well as design alternatives, the following material and laminate constructions have been selected for the principal structural components of composite ailerons.

Upper and lower covers: The cover panels have a sandwich construction consisting of three layers of T-300 carbon fiber–epoxy tape on each side of a 0.95 mm (0.0375 in.) thick syntactic epoxy core. The laminate configuration is [45/0/−45/syntactic core/−45/0/45], with the 0° plies oriented in the spanwise direction. The syntactic epoxy core is a film epoxy adhesive filled with hollow glass microspheres. Near the main rib as well as at the ends of each cover, the syntactic core is replaced by five plies of T-300 carbon fiber–epoxy tape oriented in the chordwise direction.

Front spar: The front spar is a constant-thickness channel section constructed of a $[45/0/-45/90/0]_S$ T-300 carbon fiber–epoxy tape laminate. The

0° plies in the front spar laminate are in the spanwise direction. Holes are machined in the spar web for access and inspection purposes. The flange width of the spar caps is increased locally to facilitate mountings of main ribs and rib backup fittings.

Main ribs: Main ribs are used at three hinge or actuator fitting locations for transferring loads from the fittings to the aileron covers and spars. They are constant-thickness channel sections constructed with four plies of T-300 carbon fiber–epoxy bidirectional fabric oriented at $[45/90_2/45]$, where 0° represents the lengthwise direction for the rib. Five plies of unidirectional 0° T-300 carbon fiber–epoxy tape are added to the rib cap to increase the stiffness and strength at the rib ends.

Other ribs: In addition to the three main ribs, the aileron assembly has five intermediate ribs and two end closeout ribs that support the covers and share the air pressure load. These ribs are constant-thickness channel sections consisting of five plies of T-300 carbon fiber–epoxy bidirectional fabric oriented at $[45/90/-45/90/45]$, where the 0° direction represents the lengthwise direction for each rib. Five holes are machined in each rib to reduce its weight.

Rear spar: No material substitution is made for the rear spar, since the usage of composites is considered too expensive for the small amount of weight saved over the existing constant-thickness channel section of 7075-T6 clad aluminum.

In the aileron assembly, the upper cover, all ribs, and two spars are permanently fastened with titanium screws and stainless steel collars. The removable lower cover, trailing edge wedge, leading edge shroud, and fairings are fastened with the same type of screws, but with stainless steel nut plates attached to these substructures with stainless steel rivets. To prevent galvanic corrosion, all aluminum parts are anodized, primed with epoxy, and then painted with a urethane top coat. All carbon fiber–epoxy parts are also painted with a urethane coat.

The composite aileron is 23.2% lighter than the metal aileron. It also contains 50% fewer parts and fasteners (Table 6.7). A summary of the ground

TABLE 6.7
Comparison of Composite and Metal Ailerons

	Composite	Aluminum
Weight (lb)	100.1	140.4
Number of ribs	10	18
Number of parts	205	398
Number of fasteners	2574	5253

Source: Adapted from Griffin, C.F., Design development of an advanced composite aileron, Paper No. 79-1807, AIAA Aircraft Systems and Technology Meeting, August 1979.

TABLE 6.8
Ground Tests on Lockheed L-1011 Composite Ailerons

Vibration in the flapping mode	Resonance frequencies comparable with those of metal ailerons
Vibration in the torsional mode	Resonance frequencies comparable with those of metal ailerons
Chordwise static bending stiffness	Composite ailerons 27% less stiff than metal ailerons
Static torsional stiffness	Comparable with metal ailerons
Static loading	124% Design ultimate load without failure at 12° down-aileron positions 139% Design ultimate load at 20° up-aileron positions with postbuckling of the hinge and backup rib webs
Impact loading to cause visible damage at four locations followed by one lifetime flight-by-flight fatigue loading	Slight growth of damage (caused by impact loading) during the fatigue cycling
Simulated lightning followed by static loading	Burn-through and delamination over a small area; however, no evidence of growth of this damage during static testing

Source: Adapted from Griffin, C.F., Design development of an advanced composite aileron, Paper No. 79–1807, AIAA Aircraft Systems and Technology Meeting, August 1979.

tests performed on the aileron assemblies is given in Table 6.8. Additionally, a number of composite aileron prototypes have also been tested on the aircraft during engine run-up, level flights, and high-speed descends. The performance of composite aileron prototypes has been judged equal to or better than the performance of metal ailerons in these tests. As part of the maintenance evaluation program, five sets of composite ailerons were installed on commercial aircrafts and placed in service in September 1981.

6.5.2 Composite Pressure Vessels [38]

Composite pressure vessels with S-glass or Kevlar 49 fiber-reinforced epoxy wrapped around a metal liner are used in many space, military, and commercial applications. The liner is used to prevent leakage of the high-pressure fluid through the matrix microcracks that often form in the walls of filament-wound fiber-reinforced epoxy pressure vessels. The winding is done on the liner, which also serves as a mandrel. The winding tension and the subsequent curing action create compressive stresses in the liner and tensile stresses in the fiber-reinforced epoxy overwrap. After fabrication, each vessel is pressurized with an internal proof pressure (also called the ‘‘sizing’’ pressure) to create tensile yielding in the metal liner and additional tensile stresses in the overwrap. When the proof pressure is released, the metal liner attains a compressive residual

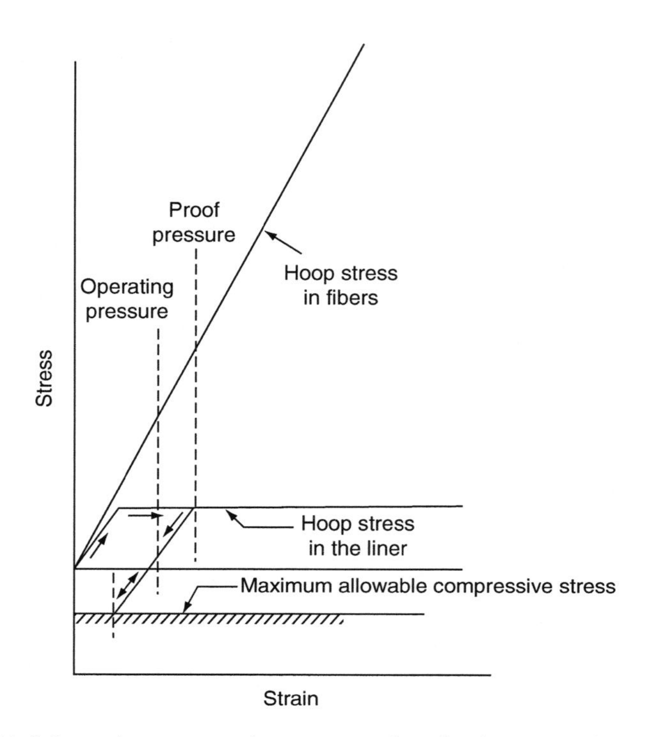

FIGURE 6.21 Schematic stress–strain representations in the composite overwrap and metal liner in a pressure vessel.

stress and the overwrap remains in tension. In service, the metal liner operates elastically from compression to tension and the composite overwrap operates in tension mode (Figure 6.21).

A commercial application of the metal liner–composite overwrap concept is the air-breathing tank that firefighters carry on their backs during a firefighting operation. It is a thin-walled pressure vessel with closed ends containing air or oxygen at pressures as high as 27.6 MPa (4000 psi). The internal pressure generates tensile normal stresses in the tank wall in both the hoop (circumferential) and axial directions. The hoop stress for the most part is twice the axial stress. The fiber orientation pattern in the composite overwrap is shown in Figure 6.22. The metal liner is usually a seamless 6061-T6 aluminum tube with a closed dome at one end and a dome with a threaded port at the other end. The tanks are designed to withstand a maximum (burst) pressure three times the operating pressure. Selected numbers of tanks are tested up to the burst pressure after subjecting them to 10,000 cycles of zero to operating pressure and 30 cycles of zero to proof pressure. Leakage before catastrophic rupture is considered the desirable failure mode during this pressure cycling. Other major

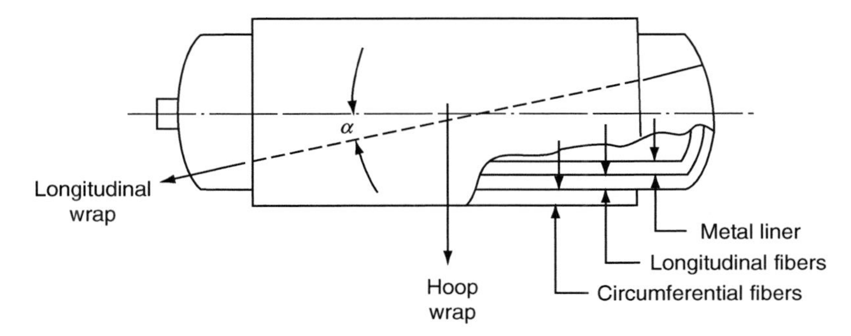

FIGURE 6.22 Fiber orientation in the composite overwrap of a pressure vessel.

qualification tests for the air-breathing tanks are drop impacts, exposure to high temperatures in the pressurized condition, and exposure to direct fire.

6.5.3 Corvette Leaf Springs [39]

The first production application of fiber-reinforced polymers in an automotive structural component is the 1981 Corvette leaf spring manufactured by the General Motors Corporation. It is a single-leaf transverse spring weighing about 35.3 N (7.95 lb) that directly replaces a 10-leaf spring weighing 182.5 N (41 lb).

The material in the 1981 Corvette composite spring is an E-glass fiber-reinforced epoxy with fibers oriented parallel to the length direction of the spring. Although the cross-sectional area of the spring is uniform, its width and thickness are varied to achieve a constant stress level along its length. This design concept can be easily understood by modeling the spring as a simply supported straight beam with a central vertical load P (Figure 6.23). If the

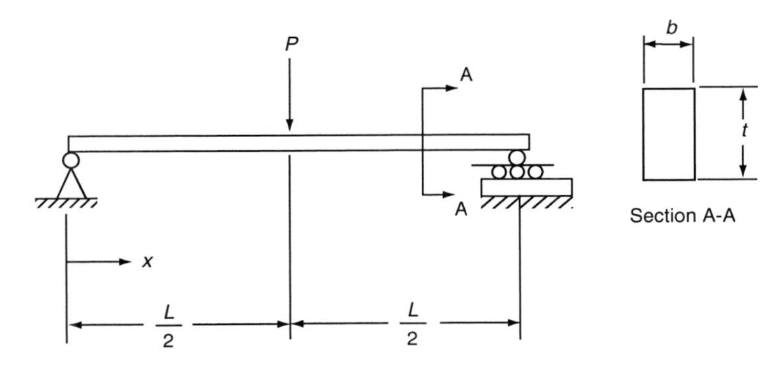

FIGURE 6.23 Simplified model of a leaf spring.

beam has a rectangular cross section, the maximum normal stress at any location x in the beam is given by

$$\sigma_{xx} = \frac{3Px}{bt^2}, \tag{6.38}$$

where b and t are the width and thickness of the beam, respectively.

For a uniform cross-sectional area beam, $bt = \text{constant} = A_0$. Furthermore, the beam is designed for a constant maximum stress, $\sigma_{xx} = \text{constant} = \sigma_0$. Thus, using Equation 6.38, we can write the thickness variation for each half length of the beam as

$$t = \frac{3P}{A_0\sigma_0}x \tag{6.39}$$

and correspondingly, its width variation as

$$b = \frac{A_0^2\sigma_0}{3P}\frac{1}{x}. \tag{6.40}$$

Equations 6.39 and 6.40 show that an ideal spring of uniform cross-sectional area and constant stress level has zero thickness and infinite width at each end. The production Corvette composite spring is ~15 mm (0.6 in.) thick by 86 mm (3.375 in.) wide at each end and 25 mm thick (1 in.) by 53 mm (2.125 in.) wide at the center. Two of these springs are filament-wound in the mold cavities, which are machined on two sides of an elliptic mandrel. After winding to the proper thickness, the mold cavities are closed and the springs are compression-molded on the mandrel at elevated temperature and pressure. The pressure applied during the molding stage spreads the filament-wound material in the mold cavities and creates the desired cross-sectional shapes. Each cured spring has a semi-elliptic configuration in the unloaded condition. When the spring is installed under the axle of a Corvette and the curb load is applied, it assumes a nearly flat configuration.

Prototype Corvette composite springs are tested in the laboratory to determine their static spring rates as well as their lives in jounce-to-rebound stroke-controlled fatigue tests. The test springs are required to survive a minimum of 500,000 jounce-to-rebound cycles with a load loss not exceeding 5% of the initially applied load at both high (above 100°C) and low (below 0°C) temperatures. Stress relaxation tests are performed for 24 h at elevated temperatures and high-humidity conditions. Other laboratory tests include torsional fatigue and gravelometer test (to evaluate the effect of gravel impingement on the surface coating). Prototype composite springs are also vehicle-tested to determine their ride and durability characteristics.

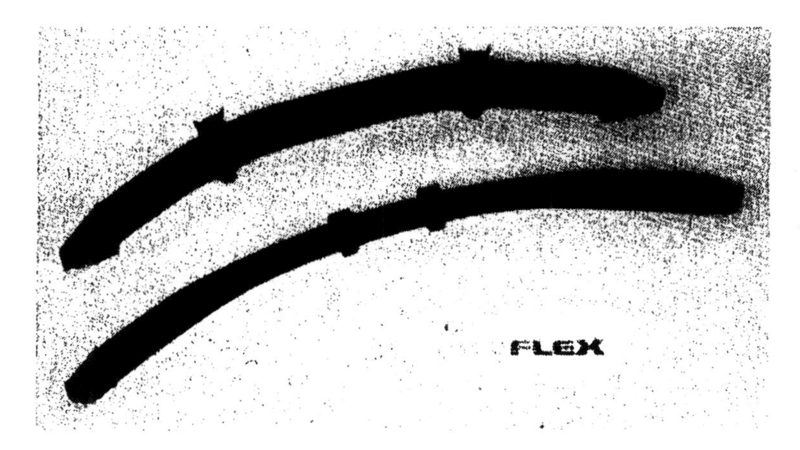

FIGURE 6.24 E glass–epoxy front (*top*) and rear (*bottom*) springs for 1984 Corvette. (Courtesy of General Motors Corporation.)

Figure 6.24 shows photographs of 1984 Corvette front and rear springs made of E-glass–epoxy composite material. The front spring has a constant width, but the rear spring has a variable width. Both springs are transversely mounted in the car. At maximum wheel travel, the front and rear springs support 13,000 N (2,925 lb.) and 12,000 N (2,700 lb.), respectively.

6.5.4 Tubes for Space Station Truss Structure [40]

The truss structure in low earth orbiting (LEO) space stations is made of tubular members with a nominal diameter of 50 mm (2 in.). Lengths of these tubes are 7 m (23 ft.) for the diagonal members and 5 m (16.4 ft.) for other members. The function of the truss structure is to support the crew and lab modules as well as the solar arrays.

The important design criteria for the tubes are

1. Maximum axial load = ± 5.33 kN (± 1200 lb.)
2. Coefficient of thermal expansion = $0 \pm 0.9 \times 10^{-6}/°C$ (CTE in the axial direction including the end fittings)
3. Low outgassing
4. Joints that allow easy tube replacements while in operation
5. 30 year service life

In addition, there are several environmental concerns in using polymer matrix composites for the space station applications:

1. *Atomic oxygen (AO) degradation*: Atomic oxygen is the major component in the LEO atmosphere. On prolonged exposure, it can substantially reduce the thickness of carbon fiber-reinforced epoxy tubes and reduce their properties. The AO degradation of such tubes can be controlled by wrapping them with a thin aluminum foil or by cladding them with an aluminum layer.
2. *Damage due to thermal cycling*: It is estimated that the space station, orbiting at 250 nautical miles with an orbital period of 90 min, will experience 175,000 thermal cyclings during a 30 year service life. Unless protected by reflective aluminum coatings, the transient temperature variation may range from −62°C to 77°C. In the "worst case" situation, for example, when the tube is always shadowed, the lowest steady-state temperature may reach −101°C.
3. *Damage due to low-velocity impact during assembly or due to extravehicular activities*. This type of damage may occur when two tubes accidentally strike one another or when a piece of equipment strikes the tube. These incidents can cause internal damages in the tube material and reduce its structural properties. They can also damage the AO protective coating and expose the tube material to atomic oxygen.

Since the tubes have large slenderness ratios (length-to-diameter ratios) and are subjected to axial loading, column buckling is considered to be the primary failure mode. Using Euler's buckling formula for pin-ended columns, the critical axial force is written as

$$P_{cr} = \frac{\pi^2 EI}{L^2}, \tag{6.41}$$

where
E = axial modulus for the tube material
I = moment of inertia of the tube cross section
L = tube length

Setting P_{cr} = 5.33 kN, the minimum allowable flexural stiffness (EI) is calculated as 26.49 kN m^2 for the 7 m long diagonal tubes and 13.51 kN m^2 for the 5 m long nondiagonal tubes.

The CTE requirement for the entire tube including its end fittings is $0 \pm 0.9 \times 10^{-6}$/°C. Assuming that the end fittings are made of aluminum and are 5% of the total length, the CTE requirement for the tube is $-0.635 \pm 0.5 \times 10^{-6}$/°C.

Bowles and Tenney [40] used the lamination theory to calculate the axial modulus (E) and CTE for several carbon fiber-reinforced composites. The first three composites are 177°C (350°F) cure carbon fiber–epoxies containing either T-300, T-50, or P-75 carbon fibers (having E_f = 207, 344.5, and 517 GPa, respectively). Two different ply orientations were examined for each of these material systems:

1. $[15/0/\pm10/0/-15]_S$, containing only small-angle off-axis plies to provide high axial modulus and low CTE
2. $[60/0/\pm10/0/-60]_S$, containing 60° and −60° plies to provide higher hoop modulus and strength than (1); but lower axial modulus and higher CTE than (1).

A hybrid construction with ply orientations as just described but containing T-50 carbon fiber–epoxy in the ±15° and ±60° plies and P-75 carbon fiber–epoxy in the 0° and ±10° plies was also investigated. For AO protection, thin aluminum foils (0.05 mm thick) were used on both inside and outside of the tubes made of these materials. A 0.075 mm thick adhesive layer is used between the aluminum foil and the composite tube. The fourth material was a sandwich construction with unidirectional P-75 carbon fiber-reinforced epoxy in the core and 0.125–0.25 mm thick aluminum claddings in the skins.

Figure 6.25 shows the axial modulus vs. CTE values for all composite laminates investigated by Bowles and Tenney. It appears that the CTE requirement is met by the following materials/constructions:

1. $[15/0/\pm10/0/-15]_S$ T-50 carbon fiber–epoxy
2. Both P-75 carbon fiber–epoxy laminates
3. Hybrid construction

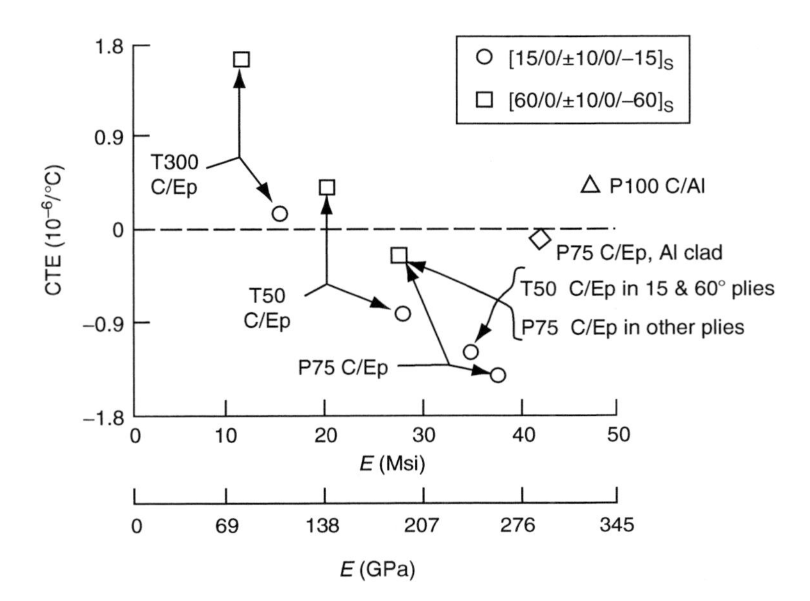

FIGURE 6.25 Axial modulus vs. CTE values for various laminates considered for space station truss structure tubes. (Adapted from Bowles, D.E. and Tenney, D.R., *SAMPE J.*, 23, 49, 1987.)

For comparison, the modulus and CTE values of a P-100 carbon fiber/6061 aluminum alloy composite, also shown in Figure 6.25, are 330 GPa and $0.36 \times 10^{-6}/°C$, respectively. The 6061 aluminum alloy has a modulus of 70 GPa and a CTE of $22.9 \times 10^{-6}/°C$.

After selecting the laminate type based on the CTE requirement, the next step is to examine which laminate provides the required flexural stiffness and has the minimum weight per unit length. The flexural stiffness EI is a function of the cross-sectional dimensions of the tube. Using an inner radius of 25.4 mm, EI values are plotted as a function of the tube wall thickness in Figure 6.26, along with the range of EI values required for this application. A comparison of tube weight per unit length is made in Figure 6.27, which shows P-75 carbon fiber–epoxy to be the lightest of all candidate materials considered.

Although both $[15/0/\pm10/0/-15]_S$ and $[60/0/\pm10/0/-60]_S$ laminates meet the structural requirements, it is necessary to compare the residual thermal stresses that may be induced in these laminates due to cooling from the curing temperature to the use temperature. These residual stresses can be high enough to cause matrix microcracking, and change the mechanical and environmental characteristics of the laminate.

Figure 6.28 shows the residual thermal stresses in the principal material directions (1–2 directions) through the thickness of a $[15/0/\pm10/0/-15]_S$

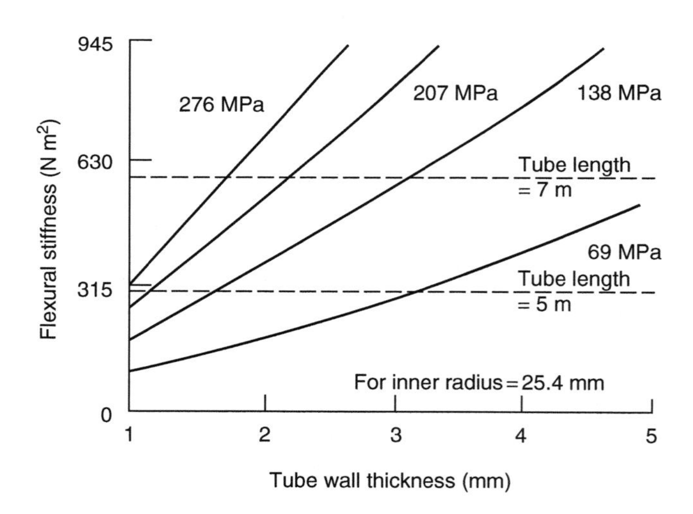

FIGURE 6.26 Flexural stiffness (EI) as a function of the tube wall thickness for different axial modulus values. (Adapted from Bowles, D.E. and Tenney, D.R., *SAMPE J.*, 23, 49, 1987.)

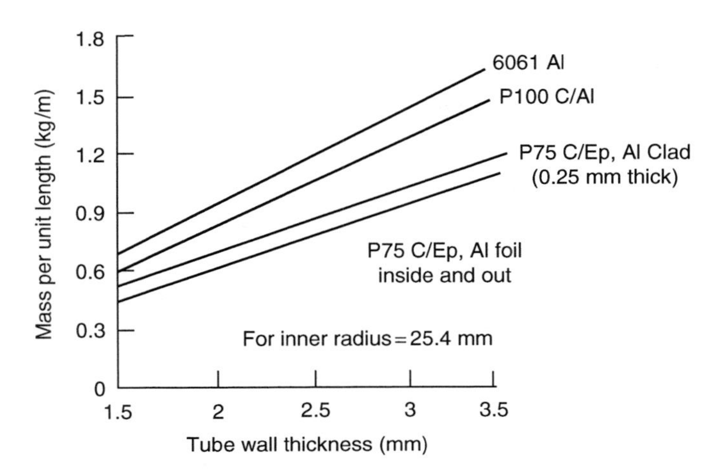

FIGURE 6.27 Mass per unit length of various materials as a function of tube wall thickness. (Adapted from Bowles, D.E. and Tenney, D.R., *SAMPE J.*, 23, 49, 1987.)

laminate. The normal stress σ_{22} (which is transverse to the fiber direction and controls the matrix microcracking) is tensile in all the plies and has the largest magnitude in the 15° plies. A comparison of maximum transverse normal stresses (σ_{22}) in $[15/0/\pm10/0/-15]_S$ and $[60/0/\pm10/0/-60]_S$ laminates indicates that the latter is more prone to matrix microcracking.

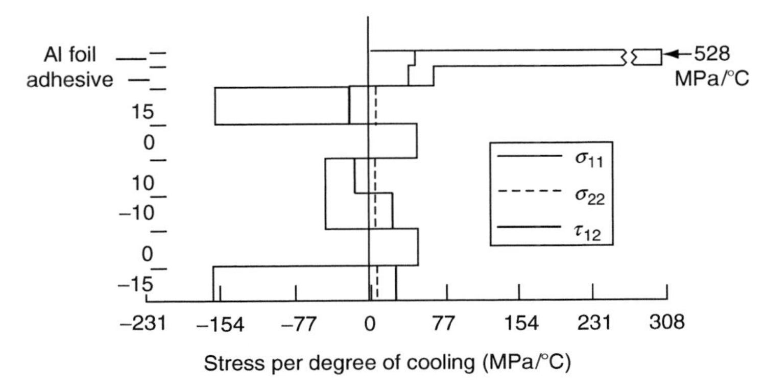

FIGURE 6.28 Thermally induced lamina stresses in a $[15/0/\pm10/0/-15]_S$ carbon fiber–epoxy laminate. (Adapted from Bowles, D.E. and Tenney, D.R., *SAMPE J.*, 23, 49, 1987.)

REFERENCES

1. V.D. Azzi and S.W. Tsai, Anisotropic strength of composites, *Exp. Mech., 5*:283 (1965).
2. S.W. Tsai and E.M. Wu, A general theory of strength for anisotropic materials, *J. Compos. Mater., 5*:58 (1971).
3. E.M. Wu, Optimal experimental measurements of anisotropic failure tensors, *J. Compos. Mater., 6*:472 (1972).
4. S.W. Tsai and H.T. Hahn, Failure analysis of composite materials, *Inelastic Behavior of Composite Materials* (C.T. Herakovich, ed.), American Society of Mechanical Engineers, New York, p. 73 (1975).
5. P.H. Petit and M.E. Waddoups, A method of predicting the non-linear behavior of laminated composites, *J. Compos. Mater., 3*:2 (1969).
6. R.E. Rowlands, Flow and failure of biaxially loaded composites: experimental–theoretical correlation, *Inelastic Behavior of Composite Materials* (C.T. Herakovich, ed.), American Society of Mechanical Engineers, New York, p. 97 (1975).
7. S.R. Soni, A comparative study of failure envelopes in composite laminates, *J. Reinforced Plast. Compos., 2*:34 (1983).
8. R.C. Burk, Standard failure criteria needed for advanced composites, *Astronaut. Aeronaut., 21*:6 (1983).
9. H.T. Hahn, On approximations for strength of random fiber composites, *J. Compos. Mater., 9*:316 (1975).
10. J.C. Halpin and J.L. Kardos, Strength of discontinuous reinforced composites: I. Fiber reinforced composites, *Polym. Eng. Sci., 18*:6 (1978).
11. C.S. Hong and J.H. Crews, Jr., Stress concentration factors for finite orthotropic laminates with a circular hole and uniaxial loading, NASA Technical Paper No. 1469 (1979).
12. B. Pradhan, Effect of width and axial to transverse elastic stiffness ratio on SCF in uniaxially loaded FRP composite plates containing circular holes, *Fibre Sci. Technol., 17*:245 (1982).
13. J.M. Whitney, I.M. Daniel, and R.B. Pipes, *Experimental Mechanics of Fiber Reinforced Composite Materials*, Society for Experimental Mechanics, Brookfield Center, CT (1984).
14. M.E. Waddoups, J.R. Eisenmann, and B.E. Kaminski, Macroscopic fracture mechanics of advanced composite materials, *J. Compos. Mater., 5*:446 (1971).
15. J.M. Whitney and R.J. Nuismer, Stress fracture criteria for laminated composites containing stress concentrations, *J. Compos. Mater., 8*:253 (1974).
16. R.J. Nuismer and J.M. Whitney, Uniaxial failure of composite laminates containing stress concentrations, *Fracture Mechanics of Composites, ASTM STP, 593*:117 (1975).
17. H.J. Konish and J.M. Whitney, Approximate stresses in an orthotropic plate containing a circular hole, *J. Compos. Mater., 9*:157 (1975).
18. J.C. Brewer and P.A. Lagace, Quadratic stress criterion for initiation of delamination, *J. Compos. Mater., 22*:1141 (1988).
19. S.G. Zhou and C.T. Sun, Failure analysis of composite laminates with free edge, *J. Compos. Tech. Res., 12*:91 (1990).

20. H. Kulkarni and P. Beardmore, Design methodology for automotive components using continuous fibre-reinforced composite materials, *Composites*, *11*:225 (1980).
21. J.C. Ekvall and C.F. Griffin, Design allowables for T300/5208 graphite/epoxy composite materials, *J. Aircraft, 19*:661 (1982).
22. M.J. Rich and D.P. Maass, Developing design allowables for composite helicopter structures, *Test Methods and Design Allowables for Fibrous Composites, ASTM STP, 734*:181 (1981).
23. C.T. Herakovich, Influence of layer thickness on the strength of angle-ply laminates, *J. Compos. Mater., 16*:216 (1982).
24. S.L. Donaldson, Simplified weight saving techniques for composite panels, *J. Reinforced Plast. Compos., 2*:140 (1983).
25. T.N. Massard, Computer sizing of composite laminates for strength, *J. Reinforced Plast. Compos., 3*:300 (1984).
26. W.J. Park, An optimal design of simple symmetric laminates under first ply failure criterion, *J. Compos. Mater., 16*:341 (1982).
27. J.M. Whitney and A.W. Leissa, Analysis of heterogeneous anisotropic plates, *J. Appl. Mech., 36*:261 (1969).
28. J.M. Whitney, Bending–extensional coupling in laminated plates under transverse loading, *J. Compos. Mater., 3*:20 (1969).
29. R.M. Jones, H.S. Morgan, and J.M. Whitney, Buckling and vibration of antisymmetrically laminated angle-ply rectangular plates, *J. Appl. Mech., 40*:1143 (1973).
30. O.H. Griffin, Evaluation of finite-element software packages for stress analysis of laminated composites, *Compos. Technol. Rev., 4*:136 (1982).
31. J.R. Eisenmann and J.L. Leonhardt, Improving composite bolted joint efficiency by laminate tailoring, *Joining of Composite Materials, ASTM STP, 749*:117 (1979).
32. L.J. Hart-Smith, Designing to minimize peel stresses in adhesive-bonded joints, *Delamination and Debonding of Materials, ASTM STP, 876*:238 (1985).
33. D.P. Maass, Tubular composite compression members—design considerations and practices, *Proceedings 41st Annual Conference*, Society of the Plastics Industry, January (1986).
34. J.M. Whitney, C.E. Browning, and A. Mair, Analysis of the flexural test for laminated composite materials, *Composite Materials: Testing and Design (Third Conference), ASTM STP, 546*:30 (1974).
35. J.M. Whitney and J.C. Halpin, Analysis of laminated anisotropic tubes under combined loading, *J. Compos. Mater., 2*:360 (1968).
36. G.J. Simitses, Instability of orthotropic cylindrical shells under combined torsion and hydrostatic pressure, *AIAA J., 5*:1463 (1967).
37. C.F. Griffin, Design development of an advanced composite aileron, Paper No. 79-1807, AIAA Aircraft Systems and Technology Meeting, August (1979).
38. E.E. Morris, W.P. Patterson, R.E. Landes, and R. Gordon, Composite pressure vessels for aerospace and commercial applications, *Composites in Pressure Vessels and Piping* (S.V. Kulkarni and C.H. Zweben, eds.), American Society of Mechanical Engineers, New York, PVP-PB-021 (1977).
39. B.E. Kirkham, L.S. Sullivan, and R.E. Bauerie, Development of the Liteflex suspension leaf spring, Society of Automotive Engineers, Paper No. 820160 (1982).
40. D.E. Bowles and D.R. Tenney, Composite tubes for the space station truss structure, *SAMPE J., 23*:49 (1987).

PROBLEMS

P6.1. A Kevlar 49–epoxy composite has the following material properties: $E_{11} = 11 \times 10^6$ psi, $E_{22} = 0.8 \times 10^6$ psi, $G_{12} = 0.33 \times 10^6$ psi, $\nu_{12} = 0.34$, $S_{Lt} = 203$ ksi, $S_{Tt} = 1.74$ ksi, $S_{Lc} = 34$ ksi, $S_{Tc} = 7.7$ ksi, and $S_{LTs} = 4.93$ ksi. A unidirectional laminate of this material is subjected to uniaxial tensile loading in the x direction. Determine the failure stress of the laminate using (a) the maximum stress theory, (b) the maximum strain theory, and (c) the Azzi–Tsai–Hill theory for $\theta = 0°, 30°, 45°, 60°$, and $90°$.

P6.2. The Kevlar 49–epoxy composite in Problem P6.1 has a fiber orientation angle of 45° and is subjected to a biaxial normal stress field ($\tau_{xy} = 0$). Determine the failure stress of the laminate using (a) the maximum stress theory, (b) the maximum strain theory, and (c) the Azzi–Tsai–Hill theory for the normal stress ratios of 0, 1, and 2.

P6.3. Biaxial tension–compression tests on closed-ended 90° tubes (with fibers oriented in the hoop direction of the tube) are performed to determine the normal stress interaction parameter F_{12}, which appears in the Tsai–Wu failure criterion.

The desired stress state is created by a combination of the internal pressure and axial compressive load. In one particular experiment with carbon fiber–epoxy composites, the biaxial stress ratio σ_{11}/σ_{22} was -9. The internal tube diameter was 2 in. and the tube wall thickness was 0.05 in. If the burst pressure was recorded as 2700 psi, determine (a) the axial compressive load at the time of failure and (b) the value of F_{12} for this carbon fiber–epoxy composite.

The following strength properties for the material are known: $S_{Lt} = 185$ ksi, $S_{Tt} = 7.5$ ksi, $S_{Lc} = 127$ ksi, $S_{Tc} = 34$ ksi, and $S_{LTs} = 11$ ksi.

P6.4. Average tensile strengths of 15°, 45°, and 60° boron–epoxy off-axis tensile specimens are 33.55, 12.31, and 9.28 ksi, respectively. Determine F_{12} for these three cases using the Tsai–Wu failure theory. Which of the three F_{12} values is in the permissible range? What conclusion can be made about the use of an off-axis tensile test for determining the F_{12} value? The following properties are known for the boron–epoxy system: $S_{Lt} = 188$ ksi, $S_{Lc} = 361$ ksi, $S_{Tt} = 9$ ksi, $S_{Tc} = 45$ ksi, and $S_{LTs} = 10$ ksi.

P6.5. A $[0/45]_{8S}$ T-300 carbon fiber–epoxy laminate is subjected to a uniaxial tensile force F_{xx}.

Each ply in this laminate is 0.1 mm thick. The laminate is 100 mm wide. The ply-level elastic properties of the material are given in Example 3.6. The basic strength properties of the material are as follows:

$S_{Lt} = S_{Lc} = 1447.5$ MPa, $S_{Tt} = S_{Tc} = 44.8$ MPa, and $S_{LTs} = 62$ MPa. Assuming that the maximum stress failure theory applies to this material, determine F_{xx} at (a) FPF and (b) ultimate failure.

P6.6. A $[0/90/\pm 45]_S$ T-300 carbon fiber–epoxy laminate is subjected to the following in-plane loads: $N_{xx} = 1000$ lb/in., $N_{yy} = 200$ lb/in., and $N_{xy} = -500$ lb/in. Each ply in the cured laminate is 0.006 in. thick. The basic elastic and ultimate properties of the material are as follows: $E_{11} = 20 \times 10^6$ psi, $E_{22} = 1.3 \times 10^6$ psi, $G_{12} = 1.03 \times 10^6$ psi, $\nu_{12} = 0.3$, $\varepsilon_{Lt} = 0.0085$, $\varepsilon_{Lc} = 0.0098$, $\varepsilon_{Tt} = 0.0045$, $\varepsilon_{Tc} = 0.0090$, and $\gamma_{LTs} = 0.015$. Using the maximum strain theory, determine whether any of the laminas in this laminate would fail at the specified load.

P6.7. If the laminate in Problem P6.6 is subjected to an increasing unaxial load in the x direction, determine the minimum load at which the FPF would occur.

P6.8. Show that, for an isotropic material, Equation 6.8 gives a hole stress concentration factor of 3.

P6.9. Show that the hole stress concentration factor for a 0° laminate is

$$K_T = 1 + \sqrt{2\left(\sqrt{\frac{E_{11}}{E_{22}}} - \nu_{12}\right) + \frac{E_{11}}{G_{12}}}.$$

P6.10. Compare the hole stress concentration factors of $[0/90]_{4S}$, $[0/90/\pm 45]_{2S}$, and $[0/90/\pm 60]_{2S}$ T-300 carbon fiber–epoxy laminates. The basic lamina properties are: $E_{11} = 21 \times 10^6$ psi, $E_{22} = 1.35 \times 10^6$ psi, $\nu_{12} = 0.25$, and $G_{12} = 0.83 \times 10^6$ psi.

P6.11. A 10 mm diameter hole is drilled at the center of the 100 mm wide $[0/45]_{8S}$ laminate in Problem P6.5. Calculate the hole stress concentration factor of the laminate, and state how it may change if (a) some of the 45° layers are replaced with −45° layers, (b) some of the 45° layers are replaced with 90° layers, and (c) some of the 45° layers are replaced with 0° layers.

P6.12. Using the point stress criterion, estimate the notched tensile strength of a $[0/\pm 30/90]_{8S}$ T-300 carbon fiber–epoxy laminate containing a central hole of (a) 0.25 in. diameter and (b) 1 in. diameter. Assume that the characteristic distance d_0 for the material is 0.04 in. The basic elastic properties for the material are given in Problem P6.10. Assume $\sigma_{Ut} = 61$ ksi.

P6.13. Rework Problem P6.12 using the average stress criterion. Assume that the characteristic distance a_0 is 0.15 in.

P6.14. A 300 mm wide SMC-R65 panel contains a 12 mm diameter hole at its center. The unnotched tensile strength of the material is 220 MPa. During the service operation, the panel may be subjected to an axial force of 25 kN. Using a characteristic distance d_0 of 0.8 mm in the point stress criterion, estimate the notched tensile strength of the material and determine the minimum safe thickness of the panel.

P6.15. A T-300 carbon fiber–epoxy panel is made of alternate layers of fibers at right angles to each other. For the various loading conditions shown in the figure, determine the proportion of the two types of layers and their orientations with the x axis. The total laminate thickness may not exceed 0.100 in.

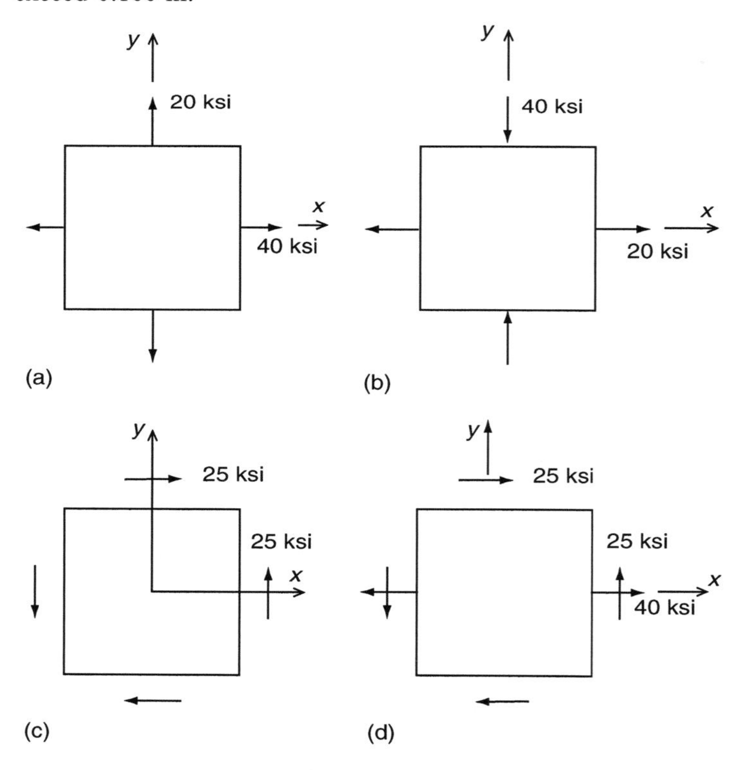

P6.16. The primary load on a rectangular plate, 1 m long × 0.25 m wide, is a 1000 N load acting parallel to its length. The plate is to be made of a symmetric cross-plied T-300 carbon fiber–epoxy laminate with 0°

outside layers. Assuming that the plate is pinned along its width, determine the minimum number of 0° and 90° plies required to avoid failure due to buckling. Each cured layer in the laminate is 0.125 mm thick. Basic elastic properties of the material are given in Appendix A.5.

P6.17. A 20 in. long E-glass–polyester pultruded rod ($v_f = 50\%$) with a solid round cross section is designed to carry a static tensile load of 1000 lb. The longitudinal extension of the rod may not exceed 0.05 in. Determine the minimum diameter of the rod. Laboratory tests have shown that the tensile strength and tensile modulus for the material is 100,000 psi and 5.5×10^6 psi, respectively. Assume a factor of safety of 2.0.

P6.18. A 1 m tension bar of solid round cross section is to be designed using unidirectional GY-70 carbon fiber-reinforced epoxy with 60% fiber volume fraction. The maximum load on the rod is expected to be 445 kN. The rod may be subjected to tension–tension fatigue cycling at an average cycling rate of 10 cycles/s for a total time period of 10 years in an environment where the temperature may fluctuate between −20°C and 100°C. The elongation of the rod should not exceed 0.2 mm.

1. Determine the diameter of the rod using a factor of safety of 3.
2. Assume the rod will be pin-connected at each end to another structure, propose two conceptual designs for the end fittings for the rod and discuss their applicability.

P6.19. A $[0/\pm45/90]_{4S}$ T-300 carbon fiber–epoxy laminate is used in a beam application. Each layer in the cured laminate is 0.005 in. thick. The beam is 0.5 in. wide. Using the basic elastic properties in Problem P6.10, calculate the effective bending stiffness of the laminated beam.

P6.20. Determine the effective bending stiffness and the failure load of a $[(0/90)_8/0]_S$ E-glass–epoxy beam having a rectangular cross section, 12.7 mm wide × 4.83 mm thick. Assume that each layer in the beam has the same cured thickness. Use Appendix A.5 for the basic material properties.

P6.21. A cantilever beam, 0.1 m long × 50 mm wide × 10 mm thick, has a sandwich construction with $[0/\pm45/90]_S$ carbon fiber–epoxy facings and an aluminum honeycomb core. Each layer in the cured laminate is 0.125 mm thick. Assuming that the core has a negligible bending stiffness, determine the end deflection of the beam if it is subjected to a 2000 N load at its free end. Basic ply-level elastic properties of the material are the same as in Example 3.6.

P6.22. A 2 m long, 100 mm wide, simply supported rectangular beam is subjected to a central load of 5 kN. The beam material is pultruded E-glass–polyester containing 60 wt% continuous fibers and 20 wt% mat. Determine the thickness of the beam so that its central deflection does not exceed 70 mm. Laboratory tests have shown that the tensile modulus of the material is 35.2 GPa. Assume its flexural modulus to be 20% less than the tensile modulus.

P6.23. A 30 in. long automotive transmission member has a hat section with uniform thickness. It is connected to the frame by means of two bolts at each end. The maximum load acting at the center of the member is estimated not to exceed 600 lb during its service life. The material considered for its construction is SMC-C20R30.

Modeling the transmission member as a simply supported beam, determine its thickness and the maximum deflection at its center. What special attention must be given at the ends of the transmission member where it is bolted to the frame? The fatigue strength of the SMC material at 10^6 cycles is 45% of its static tensile strength.

P6.24. Using the same material and design requirements as in Example 6.10, design the wall thickness of an automotive drive shaft with a $[\pm 15]_{nS}$ T-300 carbon fiber–epoxy laminate.

P6.25. Design a constant stress cantilever leaf spring of uniform width (70 mm) using (a) E-glass–epoxy and (b) AS carbon–epoxy. Free length of the spring is 500 mm. It is subjected to a reversed fatigue load of ± 10 kN. What will be a suitable manufacturing method for this spring?

7 Metal, Ceramic, and Carbon Matrix Composites

In the earlier chapters of this book, we considered the performance, manufacturing, and design issues pertaining to polymer matrix composites. In this chapter, we review the thermomechanical properties of metal, ceramic, and carbon matrix composites and a few important manufacturing methods used in producing such composites.

The history of development of metal, ceramic, and carbon matrix composites is much more recent than that of the polymer matrix composites. Initial research on the metal and ceramic matrix composites was based on continuous carbon or boron fibers, but there were difficulties in producing good quality composites due to adverse chemical reaction between these fibers and the matrix. With the development of newer fibers, such as silicon carbide or aluminum oxide, in the early 1980s, there has been a renewed interest and an accelerated research activity in developing the technology of both metal and ceramic matrix composites. The initial impetus for this development has come from the military and aerospace industries, where there is a great need for materials with high strength-to-weight ratios or high modulus-to-weight ratios that can also withstand severe high temperature or corrosive environments. Presently, these materials are very expensive and their use is limited to applications that can use their special characteristics, such as high temperature resistance or high wear resistance. With developments of lower cost fibers and more cost-effective manufacturing techniques, it is conceivable that both metal and ceramic matrix composites will find commercial applications in automobiles, electronic packages, sporting goods, and others.

The carbon matrix composites are more commonly known as carbon–carbon composites, since they use carbon fibers as the reinforcement for carbon matrix. The resulting composite has a lower density, higher modulus and strength, lower coefficient of thermal expansion, and higher thermal shock resistance than conventional graphite. The carbon matrix composites have been used as thermal protection materials in the nose cap and the leading edges of the wing of space shuttles. They are also used in rocket nozzles, exit cones, and aircraft brakes, and their potential applications include pistons in internal combustion engines, gas turbine components, heat exchangers, and biomedical implants.

7.1 METAL MATRIX COMPOSITES

The metal matrix composites (MMC) can be divided into four general categories:

1. Fiber-reinforced MMC containing either continuous or discontinuous fiber reinforcements; the latter are in the form of whiskers with approximately 0.1–0.5 μm in diameter and have a length-to-diameter ratio up to 200.
2. Particulate-reinforced MMC containing either particles or platelets that range in size from 0.5 to 100 μm. The particulates can be incorporated into the metal matrix to higher volume fractions than the whiskers.
3. Dispersion-strengthened MMC containing particles that are <0.1 μm in diameter.
4. In situ MMC, such as directionally solidified eutectic alloys.

In this chapter, we focus our attention on the first two categories, more specifically on whisker- and particulate-reinforced MMCs. More detailed informations on MMC can be found in Refs. [1–4].

Continuous carbon or boron fiber-reinforced MMCs have been under development for >20 years; however, they have found limited use due to problems in controlling the chemical reaction between the fibers and the molten metal at the high processing temperatures used for such composites. The result of this chemical reaction is a brittle interphase that reduces the mechanical properties of the composite. Fiber surface treatments developed to reduce this problem increase the cost of the fiber. Additionally, the manufacturing cost of continuous carbon or boron fiber-reinforced MMC is also high, which makes them less attractive for many applications. Much of the recent work on MMC is based on silicon carbide whiskers (SiC_w) or silicon carbide particulates (SiC_p). SiC is less prone to oxidative reactions at the processing temperatures used. Furthermore, not only they are less expensive than carbon or boron fibers, but also they can be incorporated into metal matrices using common manufacturing techniques, such as powder metallurgy and casting.

7.1.1 MECHANICAL PROPERTIES

In Chapter 2, we discussed simple micromechanical models in relation to polymer matrix composites in which fibers carry the major portion of the composite load by virtue of their high modulus compared with the polymer matrix, such as epoxy. The same micromechanical models can be applied to MMC with some modifications. The modulus of metals is an order of magnitude higher than that of polymers (Table 7.1). Many metals are capable of

TABLE 7.1
Properties of Some Metal Alloys Used in Metal Matrix Composites

Material	Density, g/cm^3	Tensile Modulus, GPa (Gsi)	YS, MPa (ksi)	UTS, MPa (ksi)	Failure Strain, %	CTE 10^{-6} per °C	Melting Point, °C
Aluminum alloy							
2024-T6	2.78	70 (10.1)	468.9 (68)	579.3 (84)	11	23.2	
6061-T6	2.70	70 (10.1)	275.9 (40)	310.3 (45)	17	23.6	
7075-T6	2.80	70 (10.1)	503.5 (73)	572.4 (83)	11	23.6	
8009	2.92	88 (12.7)	407 (59)	448 (64.9)	17	23.5	
380 (As cast)	2.71	70 (10.1)	165.5 (24)	331 (48)	4	—	540
Titanium alloy							
Ti-6A1-4V (Solution-treated and aged)	4.43	110 (16)	1068 (155)	1171 (170)	8	9.5	1650
Magnesium alloy							
AZ91A	1.81	45 (6.5)	158.6 (23)	234.5 (34)	3	26	650
Zinc–aluminum alloy							
ZA-27 (Pressure die-cast)	5	78 (11.3)	370 (53.6)	425 (61.6)	3	26	375

undergoing large plastic deformations and strain hardening after yielding. In general, they exhibit higher strain-to-failure and fracture toughness than polymers. Furthermore, since the processing temperature for MMCs is very high, the difference in thermal contraction between the fibers and the matrix during cooling can lead to relatively high residual stresses. In some cases, the matrix may yield under the influence of these residual stresses, which can affect the stress–strain characteristics as well as the strength of the composite.

7.1.1.1 Continuous-Fiber MMC

Consider an MMC containing unidirectional continuous fibers subjected to a tensile load in the fiber direction. Assume that the matrix yield strain is lower than the fiber failure strain. Initially, both fibers and matrix deform elastically. The longitudinal elastic modulus of the composite is given by the rule of mixtures:

$$E_L = E_f v_f + E_m v_m. \tag{7.1}$$

After the matrix reaches its yield strain, it begins to deform plastically, but the fiber remains elastic. At this point, the stress–strain diagram begins to deviate from its initial slope (Figure 7.1) and exhibits a new longitudinal modulus, which is given by:

$$E_L = E_f v_f + \left(\frac{d\sigma}{d\varepsilon}\right)_m v_m, \tag{7.2}$$

where $\left(\frac{d\sigma}{d\varepsilon}\right)_m$ is slope of the stress–strain curve of the matrix at the composite strain ε_c. The stress–strain diagram of the composite in this region is not elastic. In addition, it may not be linear if the matrix has a nonuniform strain-hardening rate.

For brittle fiber MMCs, such as SiC fiber-reinforced aluminum alloys, the composite strength is limited by fiber fracture, and the MMCs fail as the composite strain becomes equal to the fiber failure strain. For ductile fiber MMCs, such as tungsten fiber-reinforced copper alloys [5] and beryllium fiber-reinforced aluminum alloys [6], the fiber also yields and plastically deforms along with the matrix. In addition, the composite strength is limited by the fiber failure strain, unless the fibers fail by necking. If the fibers exhibit necking before failure and its failure strain is lower than that of the matrix, the strain at the ultimate tensile stress of the composite will be greater than that at the ultimate tensile stress of the fiber alone.

If the composite failure is controlled by the fiber failure strain, the longitudinal composite strength is given by

$$\sigma_{Ltu} = \sigma_{fu} v_f + \sigma'_m (1 - v_f), \tag{7.3}$$

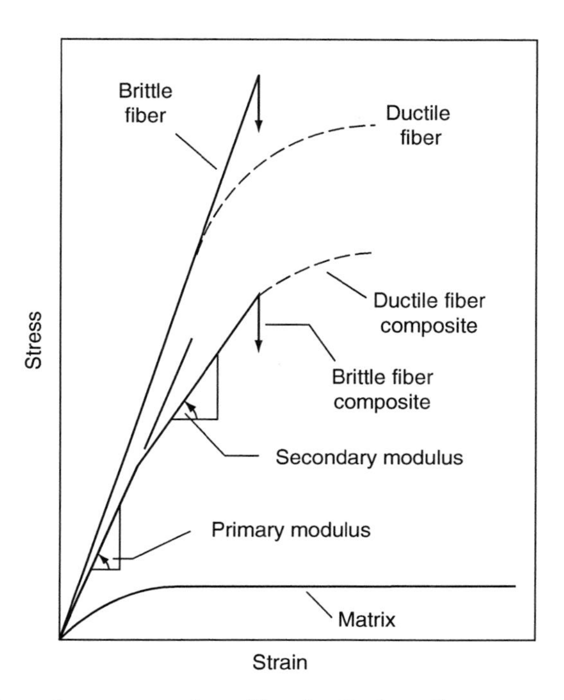

FIGURE 7.1 Schematic representation of longitudinal tensile stress–strain diagram of a unidirectional continuous fiber-reinforced MMC.

where $\sigma'_{\rm m}$ is the matrix flow stress at the ultimate fiber strain that is determined from the matrix stress–strain diagram.

Equation 7.3 appears to fit the experimental strength values for a number of MMCs, such as copper matrix composites (Figure 7.2) containing either brittle or ductile tungsten fibers [5]. In general, they are valid for MMCs in which (1) there is no adverse interfacial reaction between the fibers and the matrix that produces a brittle interphase, (2) there is a good bond between the fibers and the matrix, and (3) the thermal residual stresses at or near the interface are low.

The longitudinal tensile strength predicted by Equation 7.3 is higher than the experimental values for carbon fiber-reinforced aluminum alloys. In these systems, unless the carbon fibers are coated with protective surface coating, a brittle Al_4C_3 interphase is formed. Cracks initiated in this interphase cause the fibers to fail at strains that are lower than their ultimate strains. In some cases, the interfacial reaction is so severe that it weakens the fibers, which fail at very low strains compared with the unreacted fibers [7]. If the matrix continues to carry the load, the longitudinal tensile strength of the composite will be

$$\sigma_{\rm Ltu} = \sigma_{\rm mu}(1 - {\rm v_f}), \tag{7.4}$$

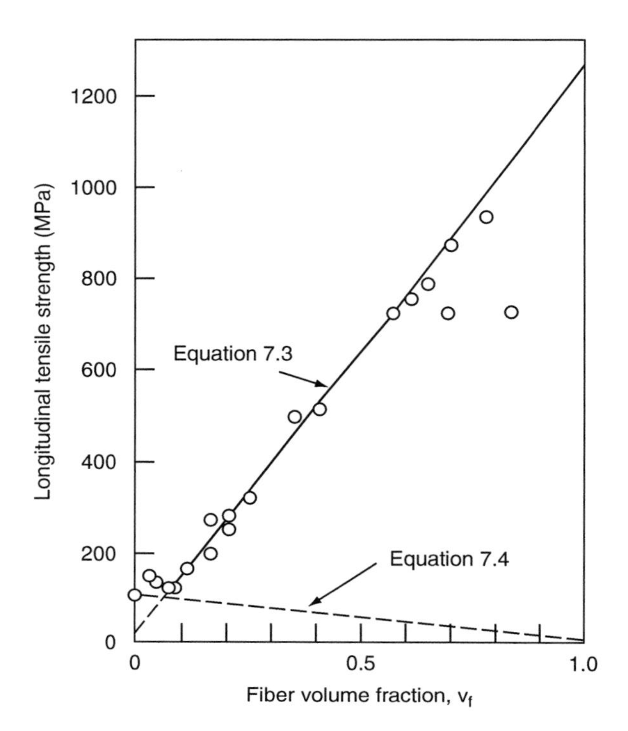

FIGURE 7.2 Longitudinal tensile strength variation of a unidirectional continuous tungsten fiber/copper matrix composite at various fiber volume fractions. (Adapted from Kelly, A. and Davies, G.J., *Metall. Rev.*, 10, 1, 1965.)

which is less than the matrix tensile strength σ_{mu}. Thus, in this case, the matrix is weakened in the presence of fibers instead of getting strengthened (Figure 7.3).

7.1.1.2 Discontinuously Reinforced MMC

In recent years, the majority of the research effort has been on SiC_w- and SiC_p-reinforced aluminum alloys [8]. Titanium, magnesium, and zinc alloys have also been used; however, they are not discussed in this chapter. Reinforcements other than SiC, such as Al_2O_3, have also been investigated. Tensile properties of some of these composites are given in Appendix A.9.

McDanels [9] has reported the mechanical properties of both SiC_w- and SiC_p-reinforced aluminum alloys, such as 6061, 2024/2124, 7075, and 5083. Reinforcement content is in the range of 10–40 vol%. These composites were produced by powder metallurgy, followed by extrusion and hot rolling. His observations are summarized as follows:

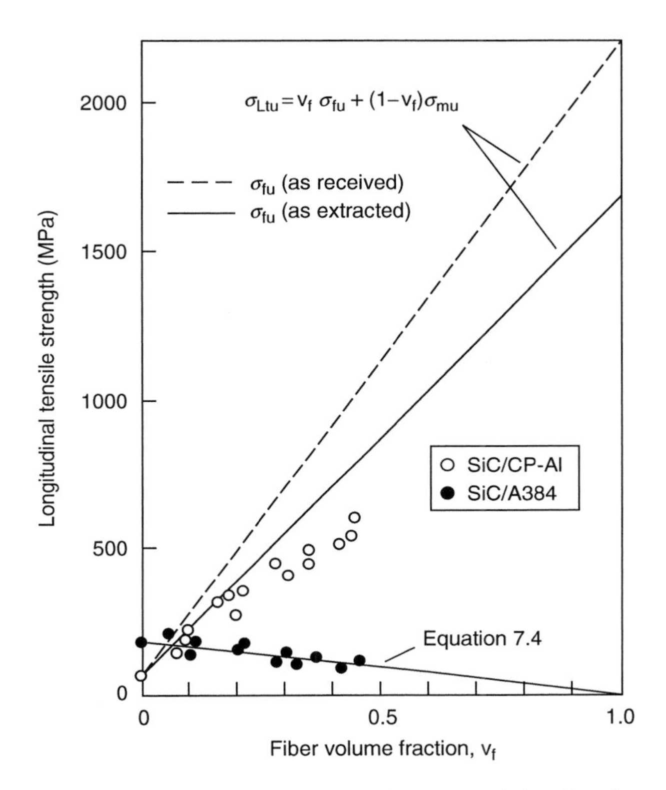

FIGURE 7.3 Longitudinal tensile strength variation of a unidirectional continuous SiC fiber-reinforced high-purity aluminum and A384 aluminum alloy composites at various fiber volume fractions. (Adapted from Everett, R.K. and Arsenault, R.J., eds., *Metal Matrix Composites: Processing and Interfaces*, Academic Press, San Diego, 1991.)

1. The tensile modulus of the composite increases with increasing reinforcement volume fraction; however, the increase is not linear. The modulus values are much lower than the longitudinal modulus predicted by Equation 7.1 for continuous-fiber composites. Furthermore, the reinforcement type has no influence on the modulus.
2. Both yield strength and tensile strength of the composite increase with increasing reinforcement volume fractions; however, the amount of increase depends more on the alloy type than on the reinforcement type. The higher the strength of the matrix alloy, the higher the strength of the composite.
3. The strain-to-failure decreases with increasing reinforcement volume fraction (Figure 7.4). The fracture mode changes from ductile at low volume fractions (below 15%) to brittle (flat and granular) at 30–40 vol%.

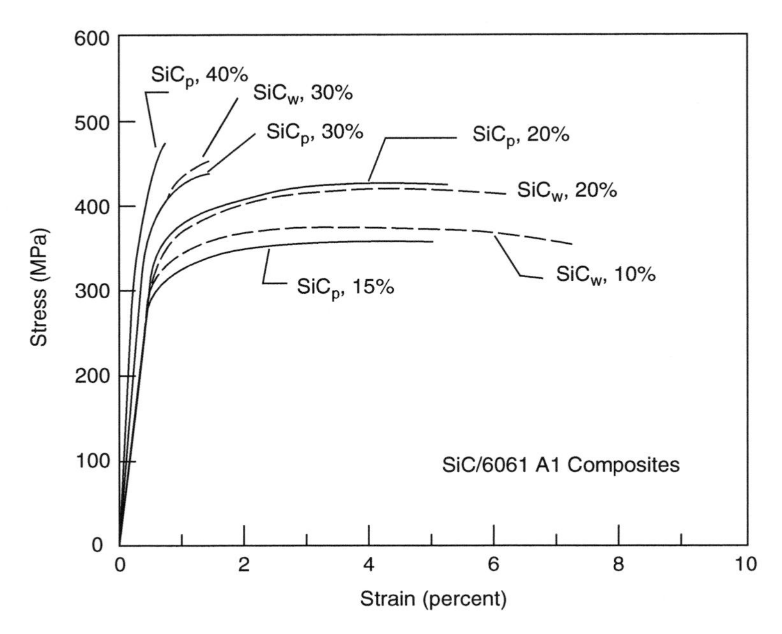

FIGURE 7.4 Tensile stress–strain diagrams of SiC_p- and SiC_w-reinforced 6061-T6 aluminum alloy composites. (Adapted from McDanels, D.L., *Metall. Trans.*, 16A, 1105, 1985.)

McDanels [9] did not observe much directionality in SiC-reinforced aluminum alloys. Since MMCs manufactured by powder metallurgy are transformed into bars and sheets by hot rolling, it is possible to introduce differences in orientation in SiC_w-reinforced alloys with more whiskers oriented in the rolling direction. Repeated rolling through small roll gaps can break whiskers and particulates into smaller sizes, thereby reducing the average particle size or the average length-to-diameter ratio of the whiskers. Both whisker orientation and size reduction may affect the tensile properties of rolled MMCs.

Johnson and Birt [10] found that the tensile modulus of both SiC_p- and SiC_w-reinforced MMCs can be predicted reasonably well using Halpin–Tsai equations (Equations 3.49 through 3.53). However, the strength and ductility of MMCs with discontinuous reinforcements are difficult to model in terms of reinforcement and matrix properties alone, since the matrix microstructure in the composite may be different from the reinforcement-free matrix due to complex interaction between the two. The particle size has a significant influence on yield strength, tensile strength, and ductility of SiC_p-reinforced MMCs [11]. Both yield and tensile strengths increase with decreasing particle size. Such behavior is attributed to the generation of thermal residual stresses, increase in dislocation density, and constraints to dislocation motion, all due to the presence of particles. The ductility of the composite also increases with decreasing particle size; however, after attaining a maximum value at particle diameters between 2 and 4 μm, it decreases rapidly to low values. The failure of the composite is initiated by cavity formation at the interface or by particle fracture.

Other observations on the thermomechanical properties of SiC_p- or SiC_w-reinforced aluminum alloys are

1. Both CTE and thermal conductivity of aluminum alloys are reduced by the addition of SiC_p [11,12].
2. The fracture toughness of aluminum alloys is reduced by the addition of SiC_p. Investigation by Hunt and his coworkers [13] indicates that fracture toughness is also related to the particle size. They have also observed that overaging, a heat treatment process commonly used for 7000-series aluminum alloys to enhance their fracture toughness, may produce lower fracture toughness in particle-reinforced aluminum alloys.
3. The long-life fatigue strength of SiC_w-reinforced aluminum alloys is higher than that of the unreinforced matrix, whereas that of SiC_p-reinforced aluminum alloys is at least equal to that of the unreinforced matrix (Figure 7.5).
4. The high-temperature yield strength and ultimate tensile strength of SiC-reinforced aluminum alloys are higher than the corresponding values of unreinforced alloys. The composite strength values follow similar functional dependence on temperature as the matrix strength values (Figure 7.6).

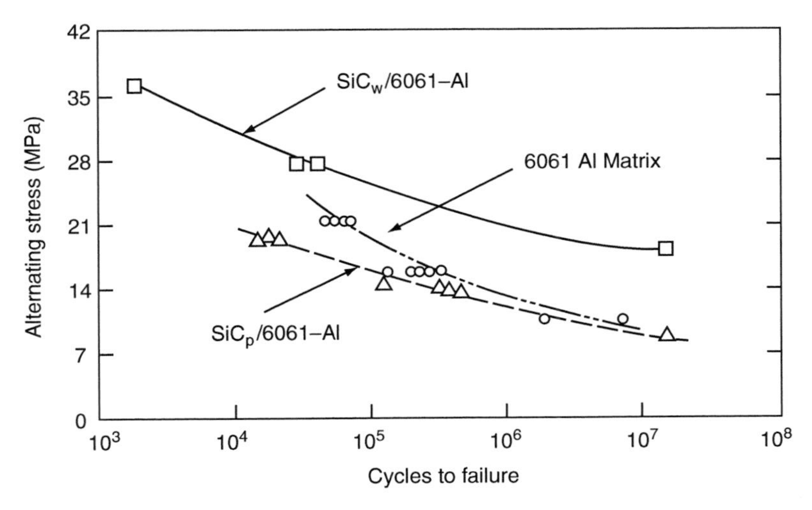

FIGURE 7.5 *S–N* curves for SiC_w- and SiC_p-reinforced 6061 aluminum alloy. (Adapted from Rack, H.J. and Ratnaparkhi, P., *J. Metals*, 40, 55, 1988.)

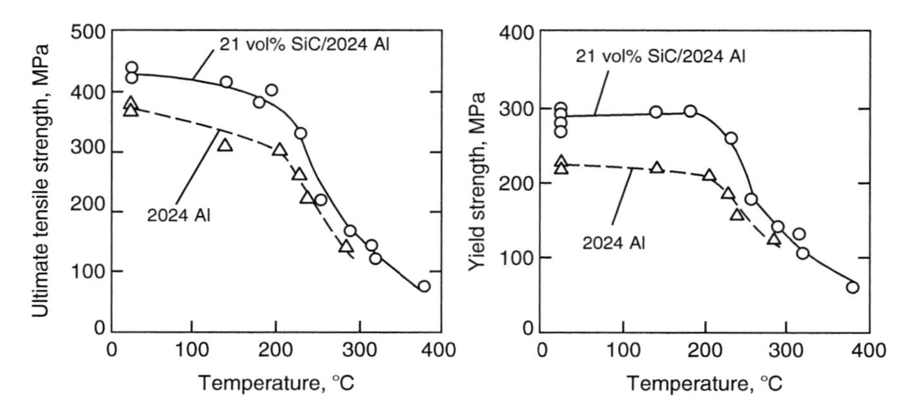

FIGURE 7.6 Effects of increasing temperature on the ultimate tensile strength and yield strength of unreinforced and 21 vol% SiC-reinforced 2024 aluminum alloy. (Adapted from Nair, S.V., Tien, J.K., and Bates, R.C., *Int. Metals Rev.*, 30, 275, 1985.)

5. Karayaka and Sehitoglu [14] conducted strain-controlled fatigue tests on 20 vol% SiC_p-reinforced 2xxx-T4 aluminum alloys at 200°C and 300°C. Based on stress range, the reinforced alloys have a superior fatigue performance than the unreinforced alloys.
6. Creep resistance of aluminum alloys is improved by the addition of either SiC_w or SiC_p. For example, Morimoto et al. [15] have shown that the second-stage creep rate of 15 vol% SiC_w-reinforced 6061 aluminum alloy is nearly two orders of magnitude lower than that of the unreinforced alloy (Figure 7.7).

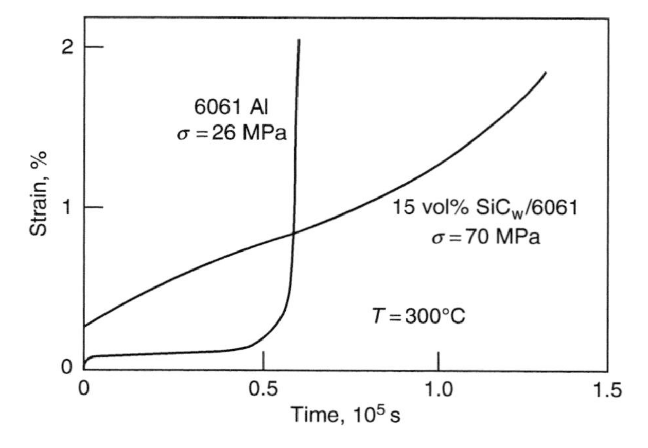

FIGURE 7.7 Comparison of creep strains of unreinforced and 15 vol% SiC_w-reinforced 6061 aluminum alloy. (Adapted from Morimoto, T., Yamaoka, T., Lilholt, H., and Taya, M., *J. Eng. Mater. Technol.*, 110, 70, 1988.)

7.1.2 Manufacturing Processes

7.1.2.1 Continuously Reinforced MMC

Vacuum hot pressing (VHP) is the most common method of manufacturing continuous fiber-reinforced MMC. The starting material (precursor) is made either by drum winding a set of parallel fibers on a thin matrix foil with the fibers held in place by a polymeric binder or by plasma spraying the matrix material onto a layer of parallel fibers that have been previously wound around a drum (Figure 7.8). Sheets are then cut from these preformed tapes by shears or dies and stacked into a layered structure. The fiber orientation in each layer can be controlled as desired.

The layup is inserted into a vacuum bag, which is then placed between the preheated platens of a hydraulic press. After the vacuum level and the temperature inside the vacuum bag reach the preset values, the platen pressure

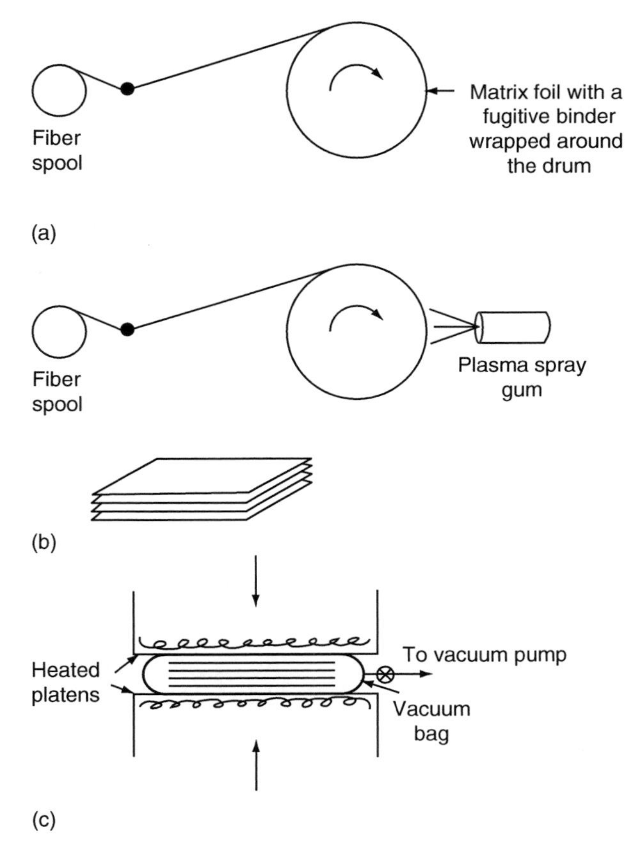

FIGURE 7.8 Schematic of drum winding and vacuum hot pressing for manufacturing continuous-fiber MMC (a) drum winding, (b) cut and stack, and (c) vacuum hot pressing.

is raised and held for a length of time to complete the consolidation of layers into the final thickness. Typically, the temperature is 50%–90% of the solidus temperature of the matrix, and the pressure ranges from 10 to 120 MPa (1.45–17.4 ksi). Afterward, the cooling is carried out at a controlled rate and under pressure to reduce residual stresses as well as to prevent warping.

The consolidation of layers in VHP takes place through diffusion bonding, plastic deformation, and creep of the metal matrix [16]. The basic process parameters that control the quality of consolidation are temperature, pressure, and holding time. Another important parameter is the cleanliness of the layers to be joined. Polymeric binder residues or oxide scales can cause poor bonding between the fibers and the matrix as well as between the matrices in various layers. Therefore, careful surface preparation before stacking the layers can be very beneficial.

7.1.2.2 Discontinuously Reinforced MMC

7.1.2.2.1 Powder Metallurgy

In this process, atomized metal powders (typically with a mean diameter of 15 μm or −325 mesh size) are blended with deagglomerated whiskers or particulates. Deagglomeration may involve agitation of a liquid suspension of the reinforcement in an ultrasonic bath. The blended mixture is cold-compacted in a graphite die, outgassed, and vacuum hot-pressed to form a cylindrical billet. The pressure is applied in the hot-pressing stage only after the temperature is raised above the solidus temperature of the matrix alloy.

Currently available SiC_p-reinforced aluminum billets range from 15 cm (6 in.) to 44.5 cm (17.5 in.) in diameter. Most billets are extruded to rods or rectangular bars using lubricated conical or streamline dies. Various structural shapes can be fabricated from these rods and bars using hot extrusion, while sheets and plates can be produced by hot rolling on conventional rolling mills. Other metalworking processes, such as forging and shear spinning, can also be used for shaping them into many other useable forms.

Typical tensile properties of SiC_w- and SiC_p-reinforced aluminum alloys are given in Table 7.2. All of these composites were manufactured by powder metallurgy and then extruded to rods or bars. Sheets were formed by hot rolling the bars. Following observations can be made from the table:

1. The secondary processing of extrusion and rolling can create significant difference in tensile properties in the longitudinal and transverse directions, especially in whisker-reinforced composites.
2. Ultimate tensile strength, yield strength, and modulus of discontinuously reinforced aluminum alloys increase with increasing reinforcement content, but elongation to failure (which is related to ductility) decreases.
3. For the same reinforcement content, whiskers produce stronger and stiffer composites than particulates.

TABLE 7.2
Tensile Properties of Discontinuously Reinforced Aluminum Alloy Sheets and Extrusions

Material (Alloy/$SiC_{w\ or\ p}$/ vol. frac.-temper)	Orientation	UTS, MPa (ksi)	YS, MPa (ksi)	Elongation, %	Modulus, GPa (Msi)
2009/SiC_w/15%-T8 (Sheet)	L	634 (91.9)	483 (70)	6.4	106 (15.37)
	T	552 (80)	400 (58)	8.4	98 (14.21)
2009/SiC_p/20%-T8 (Sheet)	L	593 (86)	462 (67)	5.2	109 (15.8)
	T	572 (82.9)	421 (61)	5.3	109 (15.8)
6013/SiC_p/15%-T6 (Extrusion)	L	517 (75)	434 (62.9)	6.3	101 (14.64)
6013/SiC_p/20%-T6 (Extrusion)	L	538 (78)	448 (65)	5.6	110 (15.95)
6013/SiC_p/25%-T6 (Extrusion)	L	565 (81.9)	469 (68)	4.3	121 (17.54)
6013/SiC_w/15%-T6 (Extrusion)	L	655 (95)	469 (68)	3.2	119 (17.25)
6090/SiC_p/25%-T6 (Extrusion)	A	483 (70)	393 (57)	5.5	117 (17)
6090/SiC_p/40%-T6 (Extrusion)	A	538 (78)	427 (61.9)	2.0	138 (20)
7475/SiC_p/25%-T6 (Extrusion)	A	655 (95)	593 (86)	2.5	117 (17)

Sources: From Geiger, A.L. and Andrew Walker, J., *J. Metals*, 43, 8, 1991; Harrigan, W.C., Jr., *J. Metals*, 43, 22, 1991.

Notes: L, longitudinal; T, transverse; A, average.

7.1.2.2.2 Casting/Liquid Metal Infiltration

The conventional casting methods can be adopted for producing MMC components for low costs and high production rates. In casting MMC, the liquid metal is poured over a fiber preform and forced to infiltrate the preform either by gravity or by the application of moderate to high pressure. After the infiltration is complete, the liquid metal is allowed to cool slowly in the mold. The pressure is usually applied by means of a hydraulic ram as in die-casting. Vacuum may also be applied to remove air and gaseous by-products of any chemical reaction that may take place between the fibers and matrix as the liquid metal flows into the fiber preform.

Two most important requirements for liquid metal infiltration are (1) the flowing liquid must have a low viscosity and (2) the liquid metal must wet the

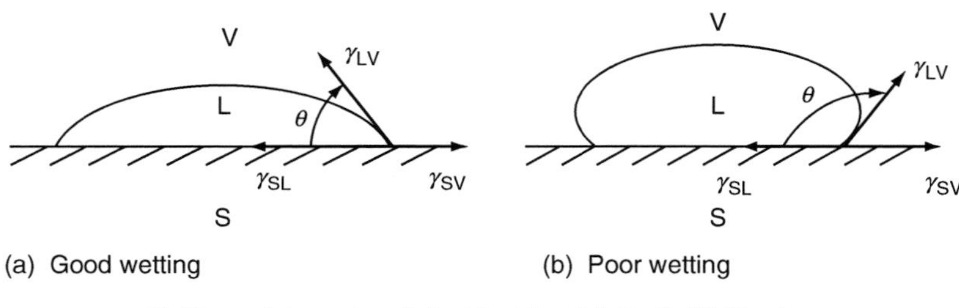

FIGURE 7.9 Surface energy requirement for good fiber surface wetting.

fiber surface. The viscosity of molten aluminum or other alloys is very low (almost two orders of magnitude lower than that of liquid epoxy and five to seven orders of magnitude lower than that of liquid thermoplastics). However, there is, in general, poor wettability between the fibers and the liquid metal, which may affect the infiltration process. The wettability is improved by increasing fiber/atmosphere surface tension or by reducing the liquid metal/fiber surface tension [17]. Both approaches reduce the fiber wetting angle (Figure 7.9) and, therefore, improve wetting between the fibers and the matrix. Common methods of improving wettability are

1. Control the chemical environment in which the infiltration is conducted. For example, controlling the oxygen content of the environment can improve the wettability of aluminum alloys with carbon fibers. This is because aluminum has a strong affinity toward atmospheric oxygen. As the hot liquid metal comes in contact with air, a thin, adherent aluminum oxide layer is formed on its surface, which interferes with wetting.
2. Add an alloying element that modifies or disrupts the surface oxide layer. For example, magnesium added to aluminum alloys disrupts the surface oxide and improves wettability with most reinforcements.
3. Use a coating on the reinforcement surface that promotes wetting. Suitable coatings may also prevent unwanted chemical reaction between the fibers and the matrix. For example, aluminum reacts with carbon fibers above 550°C to form Al_4C_3 platelets, which cause pitting on the carbon fiber surface and reduce the fiber tensile strength.

In general, for high cooling rates and low reinforcement levels (i.e., large interfiber spacings), the matrix microstructure in the solidified MMC is similar to that observed in a reinforcement-free alloy. However, for slower cooling rates or high reinforcement levels, the matrix microstructure in the MMC can be significantly different from that of the reinforcement-free alloy [18,19]. Some of the observed microstructural differences are

1. Matrix microsegregation may be reduced.
2. The crystal morphology may change from cellular dendritic to completely featureless.
3. Certain primary phases may nucleate preferentially at the reinforcement surface (e.g., silicon in a hypereutectic Al–Si alloy nucleating on the surfaces of carbon fibers or SiC particles).
4. If the reinforcement is relatively mobile, it can be pushed by growing dendrites into the last freezing zone, thereby creating an uneven distribution of the reinforcement in the composite.
5. If the fibers are held below the liquidus temperature of the matrix alloy, the liquid melt is very rapidly cooled as it comes in contact with the fibers and solidification begins at the fiber surface. This leads to finer grain size, which is of the order of the fiber diameter.

7.1.2.2.3 Compocasting

When a liquid metal is vigorously stirred during solidification by slow cooling, it forms a slurry of fine spheroidal solids floating in the liquid. Stirring at high speeds creates a high shear rate, which tends to reduce the viscosity of the slurry even at solid fractions that are as high as 50%–60% by volume. The process of casting such a slurry is called rheocasting. The slurry can also be mixed with particulates, whiskers, or short fibers before casting. This modified form of rheocasting to produce near net-shape MMC parts is called compocasting.

The melt-reinforcement slurry can be cast by gravity casting, die-casting, centrifugal casting, or squeeze casting. The reinforcements have a tendency to either float to the top or segregate near the bottom of the melt due to the differences in their density from that of the melt. Therefore, a careful choice of the casting technique as well as the mold configuration is of great importance in obtaining uniform distribution of reinforcements in a compocast MMC [20].

Compocasting allows a uniform distribution of reinforcement in the matrix as well as a good wet-out between the reinforcement and the matrix. Continuous stirring of the slurry creates an intimate contact between them. Good bonding is achieved by reducing the slurry viscosity as well as increasing the mixing time. The slurry viscosity is reduced by increasing the shear rate as well as increasing the slurry temperature. Increasing mixing time provides longer interaction between the reinforcement and the matrix.

7.1.2.2.4 Squeeze Casting

Squeeze casting is a net-shape metal casting process and involves solidification of liquid metal under pressure. It differs from the more familiar process of pressure die-casting in which pressure is used only to inject the liquid metal into a die cavity and the solidification takes place under little or no pressure. In squeeze casting, high pressure is maintained throughout solidification. This leads to a fine equiaxed grain structure and very little porosity in the cast component. In general, squeeze-cast metals have higher tensile strengths as well as greater strains-to-failure than the gravity-cast or die-cast metals. In addition, since the use of pressure increases castability, a wide variety of alloy compositions can be squeeze cast, including the wrought alloys that are usually considered unsuitable for casting because of their poor fluidity.

Squeeze casting, as it applies to MMC, starts by placing a preheated fiber preform in an open die cavity, which is mounted in the bottom platen of a hydraulic press. A measured quantity of liquid metal is poured over the preform, the die cavity is closed, and pressure up to 100 MPa is applied to force the liquid metal into the preform. The pressure is released only after the solidification is complete.

Squeeze casting has been used to produce a variety of MMCs, including those with wrought aluminum alloys, such as 2024, 6061, and 7075. However, several investigators [21,22] have suggested that improved properties are obtained by alloy modification of commercial alloys. Other variables that may influence the quality of a squeeze-cast MMC are the temperatures of the molten metal as well as the fiber preform, infiltration speed, and the final squeeze pressure [22,23].

7.2 CERAMIC MATRIX COMPOSITES

Structural ceramics such as silicon carbide (SiC), silicon nitride (Si_3N_4), and aluminum oxide (Al_2O_3) are considered candidate materials for applications in internal combustion engines, gas turbines, electronics, and surgical implants. They are high-modulus materials with high temperature resistance (Table 7.3); however, they have low failure strains and very low fracture toughness. Poor structural reliability resulting from their brittleness and high notch sensitivity is the principal drawback for widespread applications of these materials.

The impetus for developing ceramic matrix composites comes from the possibility of improving the fracture toughness and reducing the notch sensitivity of structural ceramics [24]. Early work on ceramic matrix composites used carbon fibers in reinforcing low- to intermediate-modulus ceramics, such as glass and glass-ceramics. Recent work on ceramic matrix composites has focused on incorporating SiC fibers or whiskers into high-temperature-resistant polycrystalline ceramics, such as SiC and Si_3N_4. Since the modulus values of these materials are close to those of fibers, there is very little reinforcement effect from the fibers. Instead, the heterogeneous nature of the composite contributes to the increase in matrix toughness through several microfailure mechanisms as described in the following section.

TABLE 7.3
Properties of Ceramics and Fibers Used in Ceramic Matrix Composites

Material	Density, g/cm^3	Modulus, GPa (Msi)	Strength, GPa (ksi)	CTE, 10^{-6} per °C	Melting Point, °C
Ceramic matrix					
α-Al_2O_3	3.95	380 (55)	200–310 (29–45)	8.5	2050
SiC	3.17	414 (60)	—	4.8	2300–2500
Si_3N_4	3.19	304 (44)	350–580 (51–84)	2.87	1750–1900
ZrO_2	5.80	138 (20)	—	7.6	2500–2600
Borosilicate glass	2.3	60 (8.7)	100 (14.5)	3.5	—
Lithium aluminosilicate glass-ceramic	2.0	100 (14.5)	100–150 (14.5–21.8)	1.5	—
Fibers					
SiC (Nicalon)	2.55	182–210 (26.4–30.4)	2520–3290 (365.4–477)	—	—
SiC (SCS-6)	3	406 (58.9)	3920 (568.3)	—	—
Al_2O_2 (Nextel-440)	3.05	189 (27.4)	2100 (304.5)	—	—

7.2.1 MICROMECHANICS

Consider a ceramic matrix of modulus E_m containing unidirectional continuous fibers of modulus E_f. Under a tensile load applied in the fiber direction, the initial response of the composite is linear and the composite modulus is given by

$$E_L = E_f v_f + E_m v_m. \tag{7.5}$$

Unlike polymer or metal matrices, the ceramic matrix has a failure strain that is lower than that of the fiber. As a result, failure in ceramic matrix composites initiates by matrix cracking, which originates from preexisting flaws in the matrix. Depending on the fiber strength as well as the fiber–matrix interfacial bond strength, two different failure modes are observed:

1. If the fibers are weak and the interfacial bond is strong, the matrix crack runs through the fibers, resulting in a catastrophic failure of the material. The stress–strain behavior of the composite, in this case, is linear up to failure ((a) in Figure 7.10).
2. If the fibers are strong and the interfacial bond is relatively weak, the matrix crack grows around the fibers and the fiber–matrix interface is debonded (both in front as well as the wake of the matrix crack tip) before the fiber failure. Thus, the matrix crack, in this case, does not result in a catastrophic failure. The ultimate failure of the composite occurs at strains that can be considerably higher than the matrix failure strain.

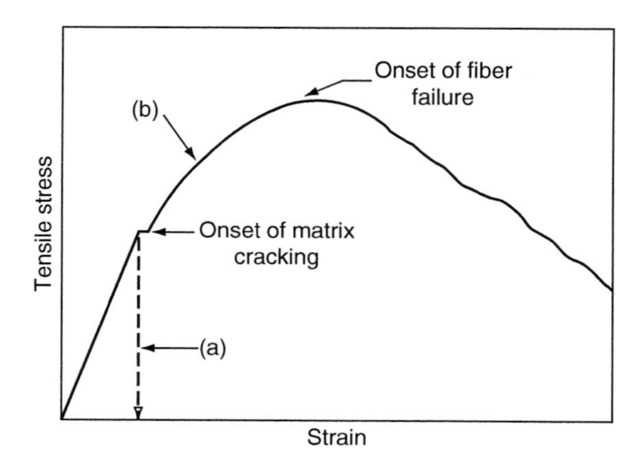

FIGURE 7.10 Schematic tensile stress–strain diagrams of ceramic matrix composites.

Assuming that the failure mode is of the second type, increasing load leads to the formation of multiple matrix cracks [25], which divide the composite into blocks of matrix held together by intact fibers (Figure 7.11). Fibers also form bridges between the crack faces and stretch with the opening of cracks. The stress–strain behavior of the composite becomes nonlinear at the onset of matrix cracking ((b) in Figure 7.10). The ultimate strength of the composite is determined by fiber failure. However, even after the fibers fail, there may be substantial energy absorption as the broken fibers pull out from the matrix. This is evidenced by a long tail in the stress–strain diagram.

The onset of matrix cracking occurs at a critical strain, $\varepsilon_{\mathrm{m}}^{*}$, which may be different from the matrix failure strain, $\varepsilon_{\mathrm{mu}}$. This strain is given by

$$\varepsilon_{\mathrm{m}}^{*} = \left[\frac{24\gamma_{\mathrm{m}}\tau_{\mathrm{i}}\mathrm{v}_{\mathrm{f}}^{2}E_{\mathrm{f}}}{d_{\mathrm{f}}(1-\mathrm{v}_{\mathrm{f}})E_{\mathrm{m}}^{2}E_{\mathrm{L}}}\right]^{1/3}, \tag{7.6}$$

where

γ_{m} = fracture energy of the unreinforced matrix
τ_{i} = fiber–matrix interfacial shear stress (or the resistance to sliding of fibers relative to the matrix)
d_{f} = fiber diameter
v_{f} = fiber volume fraction

Once the matrix crack has formed, the stress in the matrix at the crack plane reduces to zero and builds up slowly over a stress transfer length l_{m}, which is inversely proportional to the fiber–matrix interfacial shear stress. Fiber stress, on the other hand, is maximum at the crack plane and reduces to a lower value

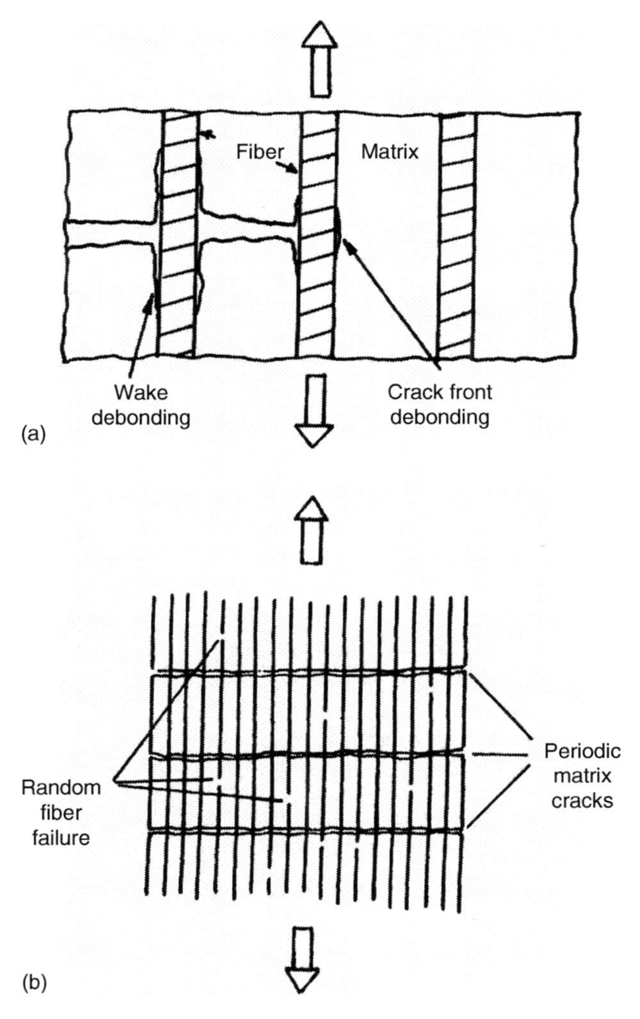

FIGURE 7.11 (a) Debonding and (b) matrix cracking in CMC.

over the same stress transfer length. If the sliding resistance is high, the fiber stress decays rapidly over a small stress transfer length, and the possibility of fiber failure close to the crack plane increases. If the fiber fails close to the crack plane so that the pull-out length is small, the pull-out energy is reduced, and so is the fracture toughness of the composite.

Debonding instead of fiber failure, and subsequently fiber pull-out are essential in achieving high fracture toughness of a ceramic matrix composite.

Evans and Marshall [26] have made the following suggestions for promoting debonding and fiber pull-out:

1. The fracture energy required for debonding, G_{ic}, should be sufficiently small compared with that for fiber fracture, G_{fc}:

$$\frac{G_{ic}}{G_{fc}} \leq \frac{1}{4}.$$

2. Residual radial strain at the interface due to matrix cooling from the processing temperature should be tensile instead of compressive.
3. Sliding resistance between the fibers and the matrix should be small, which means the friction coefficient along the debonded interface should also be small (~0.1).
4. Long pull-out lengths are expected with fibers having a large variability in their strength distribution; however, for high composite strength, the median value of the fiber strength distribution should be high.

As suggested in the preceding discussion and also verified experimentally, the fracture toughness of a ceramic matrix composite is improved if there is poor bonding between the fibers and the matrix. This can be achieved by means of a fiber coating or by segregation at the fiber–matrix interface. The most common method is to use a dual coating in which the inner coating meets the requirements for debonding and sliding, and the outer coating protects the fiber surface from degradation at high processing temperatures.

7.2.2 Mechanical Properties

7.2.2.1 Glass Matrix Composites

Development of ceramic matrix composites began in the early 1970s with carbon fiber-reinforced glass and glass-ceramic composites. Research by Sambell et al. [27,28] on carbon fiber-reinforced borosilicate (Pyrex) glass showed the following results:

1. Increasing fiber volume fraction increases both flexural strength and flexural modulus of the composite containing continuous fibers.
2. The fracture energy of the glass matrix, as measured by the area under the load–deflection diagrams of notched flexural specimens, increases significantly by the addition of continuous carbon fibers. For example, the fracture energy of 40 vol% carbon fiber-borosilicate glass is 3000 J/m^2 compared with only 3 J/m^2 for borosilicate glass. The flexural strength for the same composite is 680 MPa compared with 100 MPa for the matrix alone. The load–deflection diagram for the matrix is linear up

to failure, whereas that for the composite is linear up to 340 MPa. The load–deflection diagram becomes nonlinear at this stress due to matrix cracking on the tension side of the flexural specimen.

3. Random fiber orientation of short carbon fibers produces flexural strengths that are lower than that of borosilicates; however, the fracture energy (work of fracture) increases with fiber volume fraction, as shown in Figure 7.12.

Prewo and Brennan [29] have reported that carbon fiber-reinforced borosilicate glass retains its flexural strength up to 600°C. Above this temperature, the composite strength is reduced due to matrix softening. Thus, although glass matrix is attractive for its lower processing temperature, its application temperature is limited to 600°C. An alternative is to replace glass with a glass-ceramic, such as lithium aluminosilicate (LAS), which increases the use temperature to 1000°C or higher. Since carbon fibers are prone to oxidation above 300°C, they

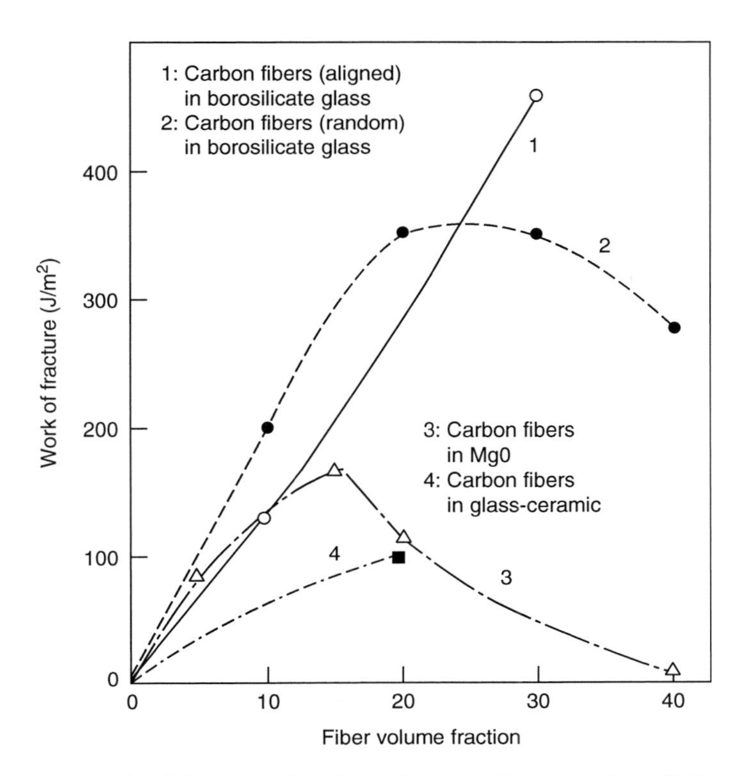

FIGURE 7.12 Work of fracture of various glass matrix composites. (Adapted from Sambell, R.A.J., Bowen, D.H., and Phillips, D.C., *J. Mater. Sci.*, 7, 663, 1972.)

are replaced by SiC fibers, which show no appreciable reduction in either strength or modulus up to 800°C.

Thermomechanical properties of SiC-reinforced glass-ceramics have been studied by a number of investigators [30,31]. The results of investigations on Nicalon (SiC) fiber-reinforced LAS are summarized as follows:

1. Nicalon/LAS composites retain their flexural strength up to 800°C when tested in air, and up to 1100°C when tested in an inert atmosphere, such as argon. The fracture mode of these composites depends strongly on the test environment.
2. Room-temperature tension test in air of unidirectional Nicalon/LAC composites shows multiple matrix cracking and fiber pull-out. The composite fails gradually after the maximum load is reached. When the same composite is tested at 900°C and above, fiber failure occurs after one single matrix crack has formed, resulting in a catastrophic failure with sudden load drop. This change in failure mode is attributed to fiber strength degradation as well as increased fiber–matrix bonding in the oxidative atmosphere. The initial carbon-rich layer between the SiC fiber and the LAS matrix is replaced by an amorphous silicate due to oxidation at 650°C–1000°C. This creates a strong interfacial bond and transforms the composite from a relatively tough material to a brittle material [31].
3. The presence of a carbon-rich interface is very important in obtaining a strong, tough SiC fiber-reinforced LAS composite, since this interface is strong enough to transfer the load from the matrix to the fibers, yet weak enough to debond before fiber failure. This enables the composite to accumulate a significant amount of local damage without failing in a brittle mode.

7.2.2.2 Polycrystalline Ceramic Matrix

Ceramic matrix composites using polycrystalline ceramic matrix, such as SiC, Si_3N_4, and ZrO_2–TiO_2, are also developed. These matrices offer higher temperature capabilities than glasses or glass-ceramics. Research done by Rice and Lewis [32] on unidirectional Nicalon (SiC) fiber-reinforced ZrO_2–TiO_2 and ZrO_2–SiO_2 composites shows the importance of proper fiber–matrix bonding in achieving high flexural strength and fracture toughness in ceramic matrix composites. The BN coating used on Nicalon fibers in their work creates a weaker bond than if the fibers are uncoated. The composite with coated fibers has a flexural strength of 400–900 MPa (60–125 ksi), compared with 70–220 MPa (10–32 ksi) for uncoated fibers. The difference in flexural load–deflection diagrams of coated and uncoated fiber composites is shown in Figure 7.13. The coated fibers not only increased the strength but also created a noncatastrophic failure mode with significant fiber pull-out. Fracture toughness of coated fiber

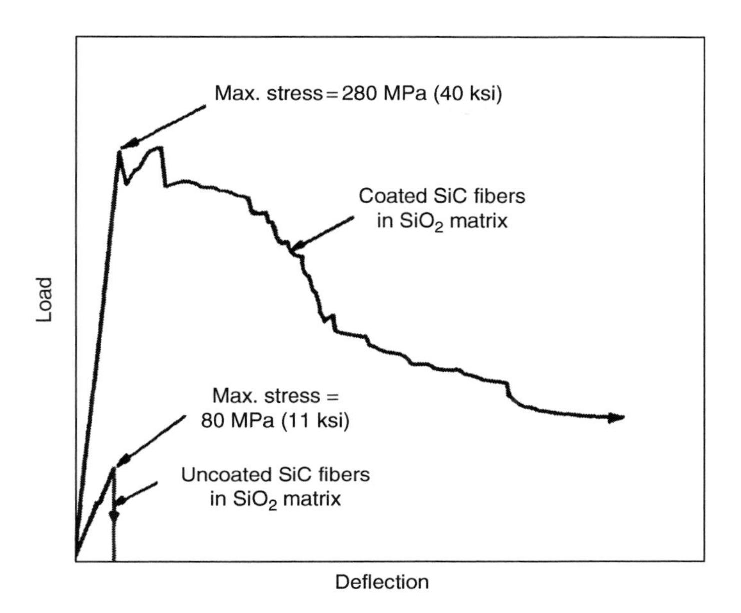

FIGURE 7.13 Effect of fiber surface coating on the room temperature flexural load-deflection diagrams of SiC (Nicalon)-reinforced SiO_2. (Adapted from Rice, R.W. and Lewis, D., III, *Reference Book for Composites Technology*, Vol. 1, S.M. Lee, ed., Technomic Pub. Co., Lancaster, PA, 1989.)

composites is >20 MPa $m^{1/2}$, compared with about 1 MPa $m^{1/2}$ for uncoated fiber composites.

Thomson and LeCostaonec [33] reported the tensile and flexural properties of Si_3N_4 matrix reinforced with unidirectional and bidirectional SiC monofilaments. The fibers are designated as SCS-6 (filament diameter = 140 μm) and SCS-9 (filament diameter = 75 μm), both having a double carbon-rich layer on the outside surface. The purpose of the double layer is to enhance the fiber strength as well as to promote debonding. As demonstrated in Figure 7.14, SCS-reinforced Si_3N_4 composites exhibit considerable load-carrying capability and toughness beyond the initial matrix microcracking at both 23°C and 1350°C.

7.2.3 Manufacturing Processes

The manufacturing processes for ceramic matrix composites can be divided into two groups.

7.2.3.1 Powder Consolidation Process

This is a two-step process of first making a "green" compact, followed by hot pressing the compact into the final shape (Figure 7.15). Hot pressing at

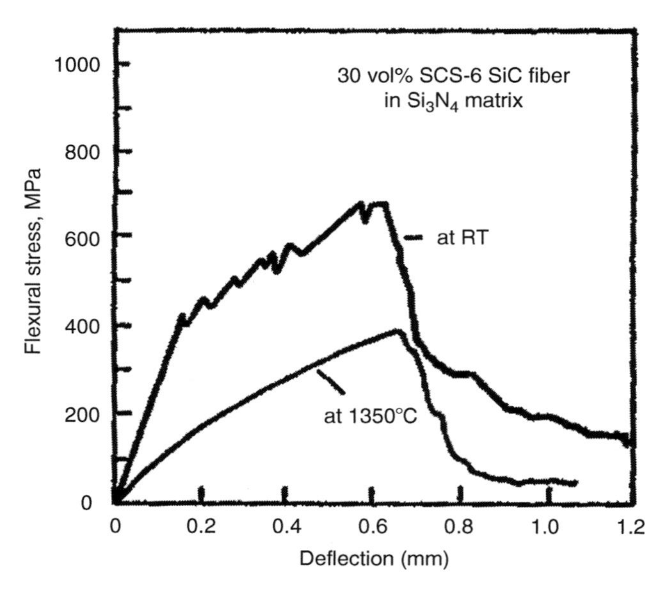

FIGURE 7.14 Flexural stress–deflection diagrams of unidirectional 30 vol% SiC (SCS-6)-reinforced hot-pressed Si_3N_4. (Adapted from Thomson, B. and LeCostaonec, J.F., *SAMPE Q.*, 22, 46, 1991.)

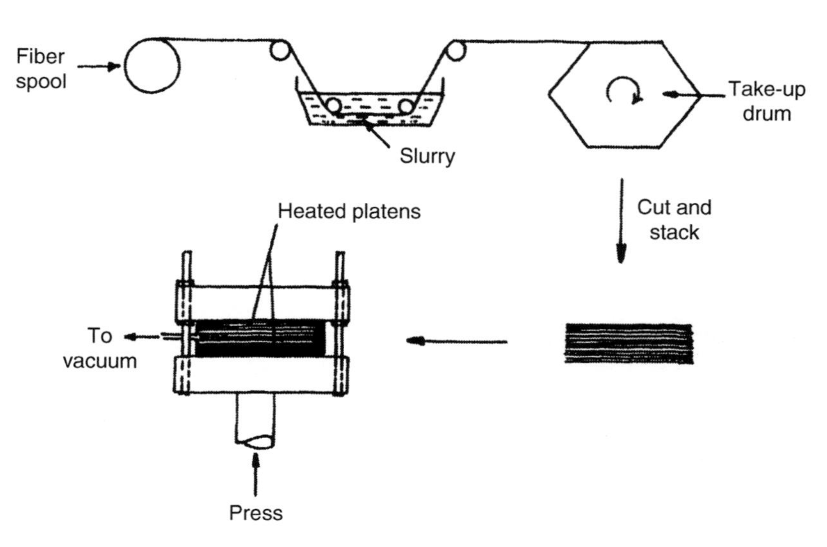

FIGURE 7.15 Powder consolidation process for manufacturing CMC.

temperatures ranging from 1200°C to 1600°C is used to consolidate the compact into a dense material with as little porosity as possible.

There are various ways of making the green compact. For continuous fibers, the compact is made by stacking a number of tapes, which are cut from a sheet of fiber yarns infiltrated with the matrix. The sheet is produced by pulling a row of parallel fiber yarns through a tank containing a colloidal slurry of the matrix and winding it around a drum. The compact can also be made by infiltrating a fiber preform with either a matrix slurry or a colloidal sol prepared from alkoxides and other organic precursors.

The green compact is consolidated and transformed into a dense composite either by VHP in a press or by hot isotactic pressing (HIP) inside an autoclave. VHP is commonly used; however, it is limited to producing only simple shapes, such as plates, rods, or blocks. To reduce porosity in the composite, the hot-pressing temperature is selected in the range of 100°C–200°C higher than that for hot pressing the matrix alone. In some cases, such high temperatures may cause fiber degradation as well as undesirable reactions at the fiber–matrix interface. Rice and Lewis [32] have shown that the strength and toughness of hot-pressed ceramic matrix composites depend strongly on the processing conditions, which include temperature, pressure, and time. Fiber degradation and adverse fiber–matrix reactions can be reduced with a fiber coating as well as by using an inert atmosphere (such as argon) during hot pressing.

The processing temperature is usually lower in HIP. Since pressure is applied equally in all directions instead of uniaxially as in VHP, it is also capable of producing more intricate net-shape structures with higher density and greater uniformity. However, in HIP, the compact must be enclosed in a gas-permeable envelope (called the "can") which is removed after processing. Finding a suitable canning material frequently poses a problem.

7.2.3.2 Chemical Processes

Two common chemical processes are chemical vapor infiltration (CVI) and polymer pyrolysis. They are briefly described as follows.

In CVI, the ceramic matrix is deposited on fibers from a gaseous medium by passing it over a preheated fiber preform in a controlled environment. The gas is selected such that it either reacts or decomposes when it comes in contact with the fibers. The temperature range for CVI is 1000°C–1200°C, which is lower than the hot-pressing temperature. For example, SiC matrix can be deposited on SiC fibers by passing methyltrichlorosilane over the SiC fiber preform in the presence of hydrogen at 1200°C [34].

The advantages of CVI are that (1) it can be used to produce a variety of shapes, (2) there is a uniform distribution of matrix in the composite, and (3) fibers undergo less mechanical damage, although the possibility of chemical degradation due to contact with the gas exists. Controlling the porosity in CVI can be a problem, since pores can seal off between the matrix deposits as they grow from adjacent fibers and impinge on each other.

Polymer pyrolysis is a two-step process of first making a polymer-impregnated preform via standard polymer impregnation techniques, followed by pyrolysis of the polymer at high temperatures to yield the ceramic matrix. For example, the polymer precursor for producing SiC matrix is an organosilicon polymer, such as polycarbosilane or polyborosiloxane.

The pyrolysis temperature is in the range of 700°C–1000°C, which is low compared with that used in hot pressing. Therefore, provided a suitable polymer precursor is available, it is suitable for a wide range of fibers that cannot ordinarily be used with hot pressing. The principal disadvantage is that it results in a highly porous matrix, primarily due to shrinkage cracks originating from the polymerization process during pyrolysis. One method of reducing the polymer shrinkage is to mix it with fine ceramic fillers.

7.3 CARBON MATRIX COMPOSITES

The carbon matrix composites consist of carbon fibers in carbon matrix and are commonly referred to as carbon–carbon (C–C) composites. They are thermally stable up to 3000°C in a nonoxidative environment, but unless protected by a surface coating or chemically modified to provide protection, they oxidize and degrade in presence of oxygen, even at 400°C–500°C. The protection of C–C composites against oxidation at high temperatures requires the use of either an external coating or internal modification. Both oxide coatings, such as SiO_2 and Al_2O_3, and nonoxide coatings, such as SiC, Si_3N_4, and HFB_2, have been used. Low oxygen permeability and thermal expansion matching are the two most critical requirements for the external coating. Thermal expansion mismatch between the coating and the C–C composite may cause cracking in the coating, which in the most severe case may result in spalling of the coating from the C–C composite surface. For applications in oxidizing environments, current coatings limit the maximum use temperature of C–C composites to 1700°C.

Fibers in the C–C composites can be either continuous or discontinuous. Continuous fibers are selected for structural applications. Table 7.4 shows the mechanical and thermal properties of continuous fiber-reinforced C–C composites containing carbon fibers that have strengths up to 2.5 GPa and elastic modulus in the range of 350–450 GPa. For comparison, properties of graphite are also listed in Table 7.4, since graphite is also considered for high temperature applications in which C–C composites are used. As with other fiber-reinforced composites, the tensile and compressive properties of C–C composites depend on the fiber properties and fiber architecture. While they are generally very high, the shear strength and modulus are very low. This is one of the limitations of C–C composites. Another point to note is that the failure strain of the matrix in C–C composites is lower than that of the fibers. Thus if there is a strong bond between the fibers and the matrix in a C–C

TABLE 7.4
Properties of Carbon–Carbon Composites

		Mechanical Properties at 23°C					Thermal Properties	
Material	v_f (%)	**Tensile Strength (MPa)**	**Tensile Modulus (GPa)**	**Compressive Strength (MPa)**	**Shear Strength (MPa)**	**Shear Modulus (GPa)**	**CTE[a] (10^{-6} per °C)**	**Thermal Conductivity[b] (W/m°C)**
C–C with 1D Unidirectional continuous	65	650–1000 (*x*) 2 (*z*)	240–280 (*x*) 3.4 (*z*)	620 (*x*)	7–14 (*xy*)	4–7 (*xy*)	1.1(*x*) 10.1(*z*)	125 (*x*) 10 (*z*)
C–C with 2D Fabric	31 (*x*) 30 (*y*)	300–350 (*x*) 2.8–5 (*z*)	110–125 (*x*) 4.1 (*z*)	150 (*x*)	7–14 (*xy*)	4–7 (*xy*)	1.3 (*x*) 6.1 (*z*)	95 (*x*) 4 (*z*)
C–C with 3D Woven orthogonal fibers	13 (*x*) 13 (*y*) 21 (*z*)	170 (*x*) 300 (*z*)	55 (*x*) 96 (*z*)	140 (*z*)	21–27 (*xy*)	1.4–2.1 (*xy*)	1.3 (*x*) 1.3 (*z*)	57 (*x*) 80 (*z*)
Graphite		20–30	7.5–11	83			2.8	50

Source: Adapted from Sheehan, J.E., Buesking, K.W., and Sullivan, B.J., *Annu. Rev. Mater. Sci.*, 24, 19, 1994.

[a] Coefficient of thermal expansion between 23°C and 1650°C.
[b] At 800°C.

composite, fibers fail immediately after the matrix fails and the composite fails in a brittle manner with low strength. On the other hand, if the fiber–matrix bond is not very strong, matrix cracking at low strain will not produce immediate fiber failure; instead, there will be energy absorption due to debonding, fiber bridging, fiber fracture, and fiber pull-out, and such a composite can possess high fracture toughness. Indeed, the work of fracture test of fully densified unidirectional C–C composites shows fracture energy value of about 2×10^4 J/m^2 [35]. In comparison, the fracture energy of engineering ceramics and premium-grade graphite is $<10^2$ J/m^2.

Two basic fabrication methods are used for making C–C composites [36]: (1) liquid infiltration and (2) chemical vapor deposition (CVD). The starting material in either of these methods is a carbon fiber preform, which may contain unidirectional fibers, bi- or multidirectional fabrics, or a three-dimensional structure of carbon fibers. In the liquid infiltration process, the preform is infiltrated with a liquid, which on heating, carbonizes to yield at least 50 wt% carbon. In the CVD process, a hydrocarbon gas is infiltrated into the preform at high temperature and pressure. The chemical breakdown of the hydrocarbon gas produces the carbon matrix. In general, CVD is used with thin sections and liquid infiltration is used with thick sections.

Two types of liquids are used in the liquid infiltration method: (1) pitch, which is made from coal tar or petroleum and (2) a thermoset resin, such as a phenolic or an epoxy. In the pitch infiltration method, either an isotropic pitch or a mesophase pitch is infiltrated into the dry carbon fiber preform at 1000°C or higher and at pressure ranging from atmospheric to 207 MPa (30,000 psi). At this temperature, carbonization occurs simultaneously with infiltration. If the carbonization is performed at the atmospheric pressure, the carbon yield is 50–60 wt% with isotropic pitch and >80 wt% with mesophase pitch. The carbon yield is higher if the pressure is increased.

In the thermoset resin infiltration method, the starting material is a prepreg containing carbon fibers in a partially cured thermoset resin. A laminate is first constructed using the prepreg layers and the vacuum bag molding process described in Chapter 5. The resin in the laminate is then carbonized at 800°C–1000°C in an inert atmosphere. The volatiles emitted during the carbonization reaction cause shrinkage and a reduction in density. The carbonization process is performed slowly to prevent rapid evolution of volatiles, which may cause high porosity and delamination between the layers. The infiltration/carbonization process is repeated several times to reduce the porosity and increase the density of the composite. The carbon yield in this process is between 50% and 70% depending on the resin and processing conditions used.

The CVD process starts with the fabrication of a dry fiber preform in the shape of the desired part. The preform is heated in a furnace under pressure (typically around 1 psi or 7 kPa) and in the presence of a hydrocarbon gas, such as methane, propane, or benzene. As the gas is thermally decomposed, a layer of pyrolitic carbon is slowly formed and deposited on the hot preform surface. In the isothermal CVD process, the temperature of the furnace is maintained constant at 1100°C. Since the pyrolitic carbon deposited on the surface tends to close the pores located at and very close to the surface, it becomes difficult to fill the internal pores with carbon. Thus it may be necessary to lightly machine the preform surfaces to remove the surface carbon layer and then repeat the infiltration process. The infiltration is better if either a thermal gradient or a pressure gradient is created across the thickness of the preform. In the thermal gradient technique, one surface is maintained at a higher temperature than the other, and the carbon infiltration progresses from the hotter surface to the colder surface. In the pressure gradient technique, a pressure differential is created across the thickness, which forces the hydrocarbon gas to flow through the pores in the preform and deposit carbon throughout the thickness. With both these techniques, very little, if any, crust is formed on the surfaces.

The C–C composite obtained by either liquid infiltration or by CVD can be heated further to transform the carbon matrix from amorphous form to graphitic form. The graphitization temperature is between 2100°C and 3000°C. Graphitization increases both strength and modulus of the C–C composite, while the fracture toughness depends on the graphitization temperature. For example, for a pitch-based matrix, the optimum graphitization temperature is 2700°C, above which fracture toughness starts to decrease [37].

Depending on the starting material and the process condition, the microstructure of carbon matrix in C–C composites may vary from small, randomly oriented crystallites of turbostratic carbon to large, highly oriented, and graphitized crystallites. The carbon matrix may also contain large amount of porosity, which causes the density to be lower than the maximum achievable density of approximately 1.8–2 g/cm^3. There may also be microcracks originating from thermal stresses as the C–C composite cools down from the processing temperature to room temperature. Since porosity affects not only the density, but also the properties of the composite, repeated impregnation and carbonization cycles are used to reduce the porosity. This process is often referred to as densification.

REFERENCES

1. R.K. Everett and R.J. Arsenault (eds.), *Metal Matrix Composites: Mechanisms and Properties*, Academic Press, San Diego (1991).
2. R.K. Everett and R.J. Arsenault (eds.), *Metal Matrix Composites: Processing and Interfaces*, Academic Press, San Diego (1991).
3. T.W. Clyne and P.J. Withers, *An Introduction to Metal Matrix Composites*, Cambridge University Press, Cambridge, UK (1993).
4. N. Chawla and K.K. Chawla, *Metal Matrix Composites*, Springer, New York (2006).
5. A. Kelly and G.J. Davies, The principles of the fibre reinforcement of metals, *Metall. Rev., 10*:1 (1965).

6. A. Toy, Mechanical properties of beryllium filament-reinforced aluminum composites, *J. Mater., 3*:43 (1968).
7. M. Taya and R.J. Arsenault, *Metal Matrix Composites: Thermomechanical Behavior*, Pergamon Press, Oxford (1989).
8. S.V. Nair, J.K. Tien, and R.C. Bates, SiC-reinforced aluminum metal matrix composites, *Int. Metals Rev., 30*:275 (1985).
9. D.L. McDanels, Analysis of stress–strain, fracture, and ductility behavior of aluminum matrix composites containing discontinuous silicon carbide reinforcement, *Metall. Trans., 16A*:1105 (1985).
10. W.S. Johnson and M.J. Birt, Comparison of some micromechanics models for discontinuously reinforced metal matrix composites, *J. Compos. Technol. Res., 13*:168 (1991).
11. A.L. Geiger and J. Andrew Walker, The processing and properties of discontinuously reinforced aluminum composites, *J. Metals, 43*:8 (1991).
12. K. Schmidt, C. Zweben, and R. Arsenault, Mechanical and thermal properties of silicon-carbide particle-reinforced aluminum, *Thermal and Mechanical Behavior of Metal Matrix and Ceramic Matrix Composites, ASTM STP, 1080*:155 (1990).
13. J.J. Lewandowski, C. Liu, and W.H. Hunt, Jr., Microstructural effects on the fracture mechanisms in 7xxx Al P/M-SiC particulate metal matrix composites, *Processing and Properties for Powder Metallurgy Composites* (P. Kumar, K. Vedula, and A. Ritter, eds.), TMS, Warrendale, PA (1988).
14. M. Karayaka and H. Sehitoglu, Thermomechanical fatigue of particulate-reinforced aluminum 2xxx-T4, *Metall. Trans. A, 22A*:697 (1991).
15. T. Morimoto, T. Yamaoka, H. Lilholt, and M. Taya, Second stage creep of whisker/6061 aluminum composites at 573K, *J. Eng. Mater. Technol., 110*:70 (1988).
16. R.K. Everett, Diffusion bonding, *Metal Matrix Composites: Processing and Interfaces*, Chapter 2 (R.K. Everett and R.J. Arsenault, eds.), Academic Press, San Diego (1991).
17. A. Mortensen, J.A. Cornie, and M.C. Flemings, Solidification processing of metal-matrix composites, *J. Metals, 40*:12 (1988).
18. A. Mortensen, M.N. Gungor, J.A. Cornie, and M.C. Flemings, Alloy microstructures in cast metal matrix composites, *J. Metals, 38*:30 (1986).
19. P. Rohatgi and R. Asthana, The solidification of metal-matrix particulate composites, *J. Metals, 43*:35 (1991).
20. P. Rohatgi, Cast aluminum-matrix composites for automotive applications, *J. Metals, 43*:10 (1991).
21. C.R. Cook, D.I. Yun, and W.H. Hunt, Jr., System optimization for squeeze cast composites, *Cast Reinforced Metal Composites* (S.G. Fishman and A.K. Dhingra, eds.), ASM International, Metals Park, OH (1988).
22. T. Kobayashi, M. Yosino, H. Iwanari, M. Niinomi, and K. Yamamoto, Mechanical properties of SiC whisker reinforced aluminum alloys fabricated by pressure casting method, *Cast Reinforced Metal Composites* (S.G. Fishman and A.K. Dhingra, eds.), ASM International, Metals Park, OH (1988).
23. H. Fukunaga, Squeeze casting processes for fiber reinforced metals and their mechanical properties, *Cast Reinforced Metal Composites* (S.G. Fishaman and A.K. Dingra, eds.), ASM International, Metal Park, OH (1988).
24. K.K. Chawla, *Ceramic Matrix Composites*, Chapman & Hall, London (1993).

25. J. Aveston, G.A. Cooper, and A. Kelly, Single and multiple fracture, *The Properties of Fiber/Composites,* IPC Science and Technology Press, Surrey, England (1971).
26. A.G. Evans and D.B. Marshall, The mechanical behavior of ceramic matrix composites, *Acta Metall., 37*:2567 (1989).
27. R.A.J. Sambell, D.H. Bowen, and D.C. Phillips, Carbon fibre composites with ceramic and glass matrices, Part 1: Discontinuous fibres, *J. Mater. Sci., 7*:663 (1972).
28. R.A.J. Sambell, A. Briggs, D.C. Phillips, and D.H. Bowen, Carbon fibre composites with ceramic and glass matrices, Part 2: Continuous fibres, *J. Mater. Sci., 7*:676 (1972).
29. K.M. Prewo and J.J. Brennan, Fiber reinforced glasses and glass ceramics for high performance applications, *Reference Book for Composites Technology*, Vol. 1 (S.M. Lee, ed.), Technomic Pub. Co., Lancaster, PA (1989).
30. J.J. Brennan and K.M. Prewo, Silicon carbide fibre reinforced glass-ceramic matrix composites exhibiting high strength and toughness, *J. Mater. Sci., 17*:2371 (1982).
31. T. Mah, M.G. Mendiratta, A.P. Katz, and K.S. Mazdiyasni, Recent developments in fiber-reinforced high temperature ceramic composites, *Ceramic Bull., 66*:304 (1987).
32. R.W. Rice and D. Lewis III, Ceramic fiber composites based upon refractory polycrystalline ceramic matrices, *Reference Book for Composites Technology*, Vol. 1 (S.M. Lee, ed.), Technomic Pub. Co., Lancaster, PA (1989).
33. B. Thomson and J.F. LeCostaonec, Recent developments in SiC monofilament reinforced Si_3N_4 composites, *SAMPE Q., 22*:46 (1991).
34. R.L. Lehman, Ceramic matrix fiber composites, *Structural Ceramics* (J.B. Wachtman, Jr., ed.), Academic Press, San Diego (1989).
35. J.E. Sheehan, K.W. Buesking, and B.J. Sullivan, Carbon–carbon composites, *Annu. Rev. Mater. Sci., 24*:19 (1994).
36. J.D. Buckley and D.D. Edie, *Carbon–Carbon Materials and Composites*, William Andrews Publishing, Norwich, NY (1993).
37. D.L. Chung, *Carbon Fiber Composites*, Butterworth-Heinemann, Newton, MA (1994).

PROBLEMS

P7.1. Experimentally determined elastic properties of a unidirectional continuous P-100 carbon fiber-reinforced 6061-T6 aluminum alloy are $E_{11} = 403$ GPa, $E_{22} = 24$ GPa, $\nu_{12} = 0.291$, and $G_{12} = 18.4$ GPa. Fiber volume fraction in the composite is 0.5.

1. Compare these values with theoretical predictions and explain the differences, if any.
2. Using the experimental values, determine the off-axis elastic modulus, E_{xx}, at $\theta = 15°$ and 45° and compare them with the experimental values of 192 and 41 GPa, respectively.

P7.2. Suppose both fibers and matrix in a unidirectional continuous fiber MMC are ductile, and their tensile stress–strain equations are given by the general form:

$$\sigma = K\,\varepsilon^m,$$

where K and m are material constants obtained from experimentally determined stress–strain data. K and m for the fibers are different from K and m for the matrix.

Assuming that the longitudinal stress–strain relationship for the composite can also be described by a similar equation, derive the material constants K and m of the composite in terms of fiber and matrix parameters.

P7.3. Referring to Problem P7.2, determine the stress–strain equation for a unidirectional beryllium fiber-reinforced aluminum alloy. Assume that the fiber volume fraction is 0.4. Material constants for the fiber and the matrix are given as follows:

	K (MPa)	**m**
Beryllium fiber	830	0.027
Aluminum matrix	250	0.127

P7.4. Referring to Appendix A.2, determine the thermal residual stresses in a unidirectional SCS-6-reinforced titanium alloy ($v_f = 40\%$) and comment on their effects on the failure mode expected in this composite under longitudinal tensile loading. The fabrication temperature is 940°C. Use the following fiber and matrix characteristics in your calculations:

Fiber: $E_f = 430$ GPa, $\nu_f = 0.25$, $\alpha_f = 4.3 \times 10^{-6}$ per°C,
$\sigma_{fu} = 3100$ MPa, $r_f = 102$ μm.

Matrix: $E_m = 110$ GPa, $\nu_m = 0.34$, $\alpha_m = 9.5 \times 10^{-6}$ per°C,
$\sigma_{my} = 800$ MPa, $\sigma_{mu} = 850$ MPa, $\varepsilon_{mu} = 15\%$.

P7.5. Using Halpin–Tsai equations, determine the tensile modulus of 20 vol% SiC whisker-reinforced 2024-T6 aluminum alloy for (a) $l_f/d_f = 1$ and (b) $l_f/d_f = 10$. Assume a random orientation for the whiskers.

P7.6. SiC_w-reinforced 6061-T6 aluminum alloy ($v_f = 0.25$) is formed by powder metallurgy and then extruded into a sheet. Microscopic examination of the cross section shows about 90% of the whiskers are aligned in the extrusion direction. Assuming $l_f/d_f = 10$, estimate the tensile modulus and strength of the composite. Make reasonable assumptions for your calculations.

P7.7. Coefficient of thermal expansion of spherical particle-reinforced composites is estimated using the Turner equation:

$$\alpha_c = \frac{v_r K_r \alpha_r + v_m K_m \alpha_m}{v_r K_r + v_m K_m},$$

where

K = Bulk modulus = $\frac{E}{3(1 - 2\nu)}$

v = volume fraction

α = CTE

E = modulus

ν = Poisson's ratio

Subscripts r and m represent reinforcement and matrix, respectively

Using the Turner equation and assuming SiC particulates to be spherical, plot the coefficient of thermal expansion of SiC_P-reinforced 6061 aluminum alloy as a function of reinforcement volume fraction. Assume $\alpha_r = 3.8 \times 10^{-6}$ per °C, $E_r = 450$ GPa, $\nu_r = 0.17$.

For comparison, experimental values of α_c are reported as 16.25×10^{-6} and 10.3×10^{-6} per °C at $v_r = 25\%$ and 50%, respectively.

P7.8. The steady-state (secondary) creep rate of both unreinforced and reinforced metallic alloys has been modeled by the following power-law equation:

$$\dot{\varepsilon} = A\sigma^n \exp\left(-\frac{E}{RT}\right),$$

where

$\dot{\varepsilon}$ = steady-state creep rate (per s)

A = constant

σ = stress (MPa)

n = stress exponent

E = activation energy

R = universal gas constant

T = temperature (°K)

1. Plot the following constant temperature creep data (Ref. [15]) obtained at 300°C on an appropriate graph and determine the values of A and n. Assume the activation energies for 6061 aluminum alloy and SiC_w/6061 as 140 and 77 kJ/mol, respectively.
2. Compare the creep rates of the above two materials at 70 MPa and 250°C.

Material	Stress (MPa)	$\dot{\varepsilon}$ (per s)
6061	26	9.96×10^{-10}
	34	3.68×10^{-9}
	38	6.34×10^{-9}
SiC_w/6061	70	7.36×10^{-9}
(v_f = 15%)	86	3.60×10^{-7}
	90	6.40×10^{-7}
	95	2.04×10^{-6}
	100	6.40×10^{-6}

P7.9. A unidirectional Nicalon fiber-reinforced LAS glass matrix composite ($v_f = 0.45$) is tested in tension parallel to the fiber direction. The first microcrack in the matrix is observed at 0.3% strain. Following fiber and matrix parameters are given:
$d_f = 15$ μm, $E_f = 190$ GPa, $\sigma_{fu} = 2.6$ GPa, $E_m = 100$ GPa, $\sigma_{mu} = 125$ MPa, and $\gamma_m = 4 \times 10^{-5}$ MPa m.

Determine (a) the sliding resistance between the fibers and matrix, (b) the stress transfer length, and (c) expected range of spacing between matrix cracks.

P7.10. A continuous fiber-reinforced ceramic matrix composite is tested in three-point flexure. Describe the failure modes you may expect as the load is increased. What material parameters you will recommend to obtain a strong as well as tough ceramic matrix composite in a flexural application.

P7.11. The longitudinal tensile modulus of a unidirectional continuous carbon fiber-reinforced carbon matrix composite is reported as 310 GPa. It is made by liquid infiltration process using epoxy resin. The carbon fiber content is 50% by volume. The carbon fiber modulus is 400 GPa.

1. Calculate the modulus of the carbon matrix and the percentage of the tensile load shared by the carbon matrix. Comment on how this is different from a carbon fiber-reinforced epoxy matrix composite.
2. Assume that the failure strains of the carbon fibers and the carbon matrix are 0.8% and 0.3%, respectively. Estimate the longitudinal tensile strength of the C–C composite. Comment on how this is different from a carbon fiber-reinforced epoxy matrix composite.
3. Using the rule of mixtures, calculate the longitudinal thermal conductivity of the C–C composite. Assume that the thermal conductivity of the carbon fibers is 100 W/m°K and the thermal conductivity of the carbon matrix is 50 W/m°K.

8 Polymer Nanocomposites

Polymer nanocomposites are polymer matrix composites in which the reinforcement has at least one of its dimensions in the nanometer range (1 nanometer (nm) = 10^{-3} μm (micron) = 10^{-9} m). These composites show great promise not only in terms of superior mechanical properties, but also in terms of superior thermal, electrical, optical, and other properties, and, in general, at relatively low-reinforcement volume fractions. The principal reasons for such highly improved properties are (1) the properties of nano-reinforcements are considerably higher than the reinforcing fibers in use and (2) the ratio of their surface area to volume is very high, which provides a greater interfacial interaction with the matrix.

In this chapter, we discuss three types of nanoreinforcements, namely nanoclay, carbon nanofibers, and carbon nanotubes. The emphasis here will be on the improvement in the mechanical properties of the polymer matrix. The improvement in other properties is not discussed in this chapter and can be found in the references listed at the end of this chapter.

8.1 NANOCLAY

The reinforcement used in nanoclay composites is a layered silicate clay mineral, such as smectite clay, that belongs to a family of silicates known as 2:1 phyllosilicates [1]. In the natural form, the layered smectite clay particles are 6–10 μm thick and contain >3000 planar layers. Unlike the common clay minerals, such as talc and mica, smectite clay can be exfoliated or delaminated and dispersed as individual layers, each ~1 nm thick. In the exfoliated form, the surface area of each nanoclay particle is ~750 m^2/g and the aspect ratio is >50.

The crystal structure of each layer of smectite clays contains two outer tetrahedral sheets, filled mainly with Si, and a central octahedral sheet of alumina or magnesia (Figure 8.1). The thickness of each layer is ~1 nm, but the lateral dimensions of these layers may range from 200 to 2000 nm. The layers are separated by a very small gap, called the interlayer or the gallery. The negative charge, generated by isomorphic substitution of Al^{3+} with Mg^{2+} or Mg^{2+} with Li^{+} within the layers, is counterbalanced by the presence of hydrated alkaline cations, such as Na or Ca, in the interlayer. Since the forces that hold the layers together are relatively weak, it is possible to intercalate small organic molecules between the layers.

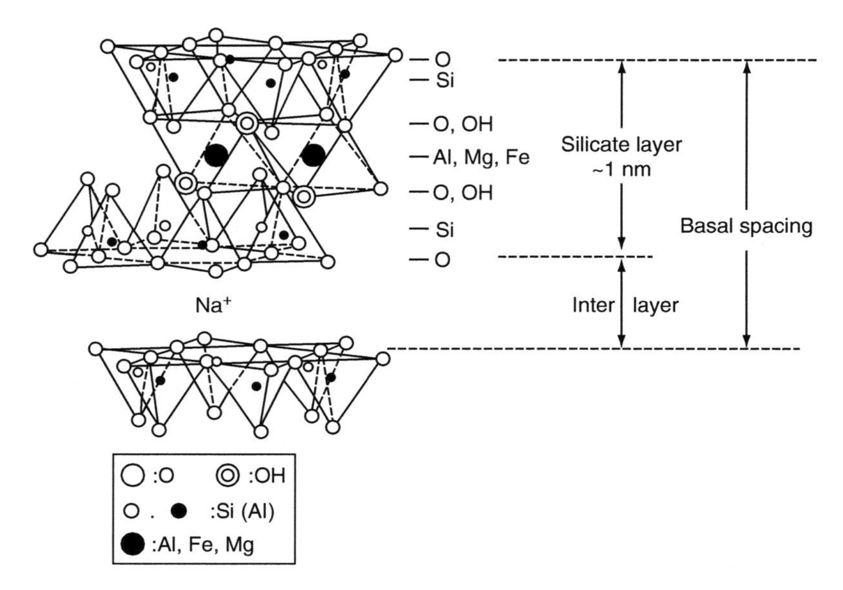

FIGURE 8.1 Crystal structure of smectite clay. (From Kato, M. and Usuki, A., *Polymer–Clay Nanocomposites,* T.J. Pinnavai and Beall, eds., John Wiley & Sons, Chichester, U.K., 2000. With permission.)

One of the common smectite clays used for nanocomposite applications is called montmorillonite that has the following chemical formula

$$M_x(Al_{4-x}Mg_x)Si_8O_{20}(OH)_4,$$

where M represents a monovalent cation, such as a sodium ion, and x is the degree of isomorphic substitution (between 0.5 and 1.3). Montmorillonite is hydrophilic which makes its exfoliation in conventional polymers difficult. For exfoliation, montmorillonite is chemically modified to exchange the cations with alkyl ammonium ions. Since the majority of the cations are located inside the galleries and the alkyl ammonium ions are bulkier than the cations, the exchange increases the interlayer spacing and makes it easier for intercalation of polymer molecules between the layers.

When modified smectite clay is mixed with a polymer, three different types of dispersion are possible. They are shown schematically in Figure 8.2. The type of dispersion depends on the polymer, layered silicate, organic cation, and the method of preparation of the nanocomposite.

1. Intercalated dispersion, in which one or more polymer molecules are intercalated between the silicate layers. The resulting material has a well-ordered multilayered morphology of alternating polymer and silicate layers. The spacing between the silicate layers is between 2 and 3 nm.

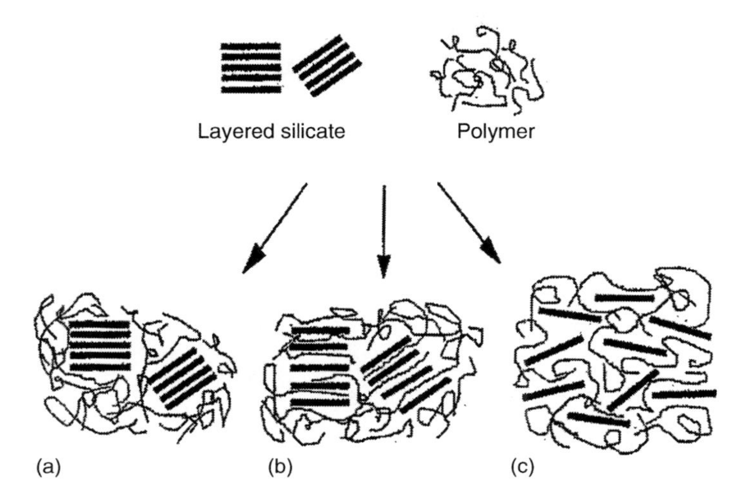

FIGURE 8.2 Three possible dispersions of smectite clay in polymer matrix. (a) phase-separated (microcomposite); (b) intercalated (nanocomposite); and (c) exfoliated (nanocomposite). (From Alexandre, M. and Dubois, P., *Mater. Sci. Eng.*, 28, 1, 2000. With permission.)

2. Exfoliated dispersion, in which the silicate layers are completely delaminated and are uniformly dispersed in the polymer matrix. The spacing between the silicate layers is between 8 and 10 nm. This is the most desirable dispersion for improved properties.
3. Phase-separated dispersion, in which the polymer is unable to intercalate the silicate sheets and the silicate particles are dispersed as phase-separated domains, called tactoids.

Following are the most common techniques used for dispersing layered silicates in polymers to make nanoclay–polymer composites.

1. *Solution method*: In this method, the layered silicate is first exfoliated into single layers using a solvent in which the polymer is soluble. When the polymer is added later, it is adsorbed into the exfoliated sheets, and when the solvent is evaporated, a multilayered structure of exfoliated sheets and polymer molecules sandwiched between them is created. The solution method has been widely used with water-soluble polymers, such as polyvinyl alcohol (PVA) and polyethylene oxide.
2. *In situ polymerization method*: In this method, the layered silicate is swollen within the liquid monomer, which is later polymerized either by heat or by radiation. Thus, in this method, the polymer molecules are formed in situ between the intercalated sheets.

The in situ method is commonly used with thermoset polymers, such as epoxy. It has also been used with thermoplastics, such as polystyrene and polyamide-6 (PA-6), and elastomers, such as polyurethane and thermoplastic polyolefins (TPOs). The first important commercial application of nanoclay composite was based on polyamide-6, and as disclosed by its developer, Toyota Motor Corp., it was prepared by the in situ method [2]. In this case, the montmorillonite clay was mixed with an α,ω-amino acid in aqueous hydrochloric acid to attach carboxyl groups to the clay particles. The modified clay was then mixed with the caprolactam monomer at 100°C, where it was swollen by the monomer. The carboxyl groups initiated the ring-opening polymerization reaction of caprolactam to form polyamide-6 molecules and ionically bonded them to the clay particles. The growth of the molecules caused the exfoliation of the clay particles.

3. *Melt processing method*: The layered silicate particles are mixed with the polymer in the liquid state. Depending on the processing condition and the compatibility between the polymer and the clay surface, the polymer molecules can enter into the interlayer space of the clay particles and can form either an intercalated or an exfoliated structure.

The melt processing method has been used with a variety of thermoplastics, such as polypropylene and polyamide-6, using conventional melt processing techniques, such as extrusion and injection molding. The high melt viscosity of thermoplastics and the mechanical action of the rotating screw in an extruder or an injection-molding machine create high shear stresses which tend to delaminate the original clay stack into thinner stacks. Diffusion of polymer molecules between the layers in the stacks then tends to peel the layers away into intercalated or exfoliated form [3].

The ability of smectite clay to greatly improve mechanical properties of polymers was first demonstrated in the research conducted by Toyota Motor Corp. in 1987. The properties of the nanoclay–polyamide-6 composite prepared by the in situ polymerization method at Toyota Research are given in Table 8.1. With the addition of only 4.2 wt% of exfoliated montmorillonite nanoclay, the tensile strength increased by 55% and the tensile modulus increased by 91% compared with the base polymer, which in this case was a polyamide-6. The other significant increase was in the heat deflection temperature (HDT). Table 8.1 also shows the benefit of exfoliation as the properties with exfoliation are compared with those without exfoliation. The nonexfoliated clay–PA-6 composite was prepared by simply melt blending montmorillonite clay with PA-6 in a twin-screw extruder.

Since the publication of the Toyota research results, the development of nanoclay-reinforced thermoplastics and thermosets has rapidly progressed.

TABLE 8.1
Properties of Nanoclay-Reinforced Polyamide-6

	Wt% of Clay	Tensile Strength (MPa)	Tensile Modulus (GPa)	Charpy Impact Strength (kJ/m^2)	HDT (°C) at 145 MPa
Polyamide-6 (PA-6)	0	69	1.1	2.3	65
PA-6 with exfoliated nanoclay	4.2	107	2.1	2.8	145
PA-6 with nonexfoliated clay	5.0	61	1.0	2.2	89

Source: Adapted from Kato, M. and Usuki, A., in *Polymer–Clay Nanocomposites*, T.J. Pinnavai and G.W. Beall, eds., John Wiley & Sons, Chichester, UK, 2000.

The most attractive attribute of adding nanoclay to polymers has been the improvement of modulus that can be attained with only 1–5 wt% of nanoclay. There are many other advantages such as reduction in gas permeability and increase in thermal stability and fire retardancy [1,4]. The key to achieving improved properties is the exfoliation. Uniform dispersion of nanoclay and interaction between nanoclay and the polymer matrix are also important factors, especially in controlling the tensile strength, elongation at break, and impact resistance.

8.2 CARBON NANOFIBERS

Carbon nanofibers are produced either in vapor-grown form [5] or by electrospinning [6]. Vapor-grown carbon nanofibers (VGCNF) have so far received the most attention for commercial applications and are discussed in this section. They are typically 20–200 nm in diameter and 30–100 μm in length. In comparison, the conventional PAN or pitch-based carbon fibers are 5–10 μm in diameter and are produced in continuous length. Carbon fibers are also made in vapor-grown form, but their diameter is in the range of 3–20 μm.

VGCNF are produced in vapor phase by decomposing carbon-containing gases, such as methane (CH_4), ethane (C_2H_6), acetylene (C_2H_2), carbon monoxide (CO), benzene, or coal gas in presence of floating metal catalyst particles inside a high-temperature reactor. Ultrafine particles of the catalyst are either carried by the flowing gas into the reactor or produced directly in the reactor by the decomposition of a catalyst precursor. The most common catalyst is iron, which is produced by the decomposition of ferrocene, $Fe(CO)_5$. A variety of other catalysts, containing nickel, cobalt, nickel–iron, and nickel–cobalt compounds, have also been used. Depending on the carbon-containing gas, the decomposition temperature can range up to 1200°C. The reaction is conducted in presence of other gases, such as hydrogen sulfide and ammonia, which act as growth promoters. Cylindrical carbon nanofibers grow on the catalyst particles and are collected at the bottom of the reactor. Impurities on their surface, such as tar and other aromatic hydrocarbons, are removed by a subsequent process called pyrolitic stripping, which involves heating them to about 1000°C in a reducing atmosphere. Heat treatment at temperatures up to 3000°C is used to graphitize their surface and achieve higher tensile strength and tensile modulus. However, the optimum heat treatment temperature for maximum mechanical properties is found to be close to 1500°C [5].

The diameter of carbon nanofibers and the orientation of graphite layers in carbon nanofibers with respect to their axis depend on the carbon-containing gas, the catalyst type, and the processing conditions, such as gas flow rate and temperature [7,8]. The catalyst particle size also influences the diameter.

Several different morphologies of carbon nanofibers have been observed [8,9]: platelet, in which the graphite layers are stacked normal to the fiber axis; hollow tubular construction, in which the graphite layers are parallel to the fiber axis, and fishbone or herringbone (with or without a hollow core), in which graphite layers are at an angle between 10° and 45° with the fiber axis (Figure 8.3). Single-wall and double-wall morphologies have been observed in heat-treated carbon nanofibers [10]. Some of the graphite layers in both single-wall and double-wall morphologies are folded, the diameter of the folds remaining close to 1 nm.

Table 8.2 lists the properties of a commercial carbon nanofiber (Pyrograf III) (Figure 8.4) as reported by its manufacturer (Applied Sciences, Inc.). The tensile modulus value listed in Table 8.2 is 600 GPa; however it should be noted that owing to the variety of morphologies observed in carbon nanofibers, they exhibit a range of modulus values, from as low as 110 GPa to as high as 700 GPa. Studies on vapor-grown carbon fibers (VGCF) [11], which are an order of

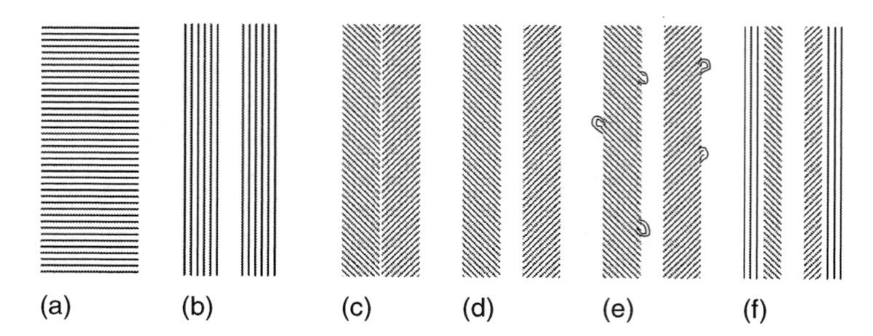

FIGURE 8.3 Different morphologies of carbon nanofibers. (a) Graphite layers stacked normal to the fiber axis; (b) Hollow tubular construction with graphite layers parallel to the fiber axis; (c) and (d) Fishbone or herringbone morphology with graphite layers at an angle with the fiber axis; (e) Fishbone morphology with end loops; and (f) Double-walled morphology.

TABLE 8.2
Properties of Vapor-Grown Carbon Nanofibers

Properties	Carbon Nanofibers[a] Pyrotically Stripped
Diameter (nm)	60–200
Density (g/cm^3)	1.8
Tensile Modulus (GPa)	600
Tensile Strength (GPa)	7
Coefficient of thermal expansion (10^{-6}/°C)	−1.0
Electrical resistivity (μΩ cm)	55

[a] Pyrograf III, produced by Applied Sciences, Inc.

magnitude larger in diameter than the VGCNF, have shown that tensile modulus decreases with increasing diameter, whereas tensile strength decreases with both increasing diameter and increasing length.

Carbon nanofibers have been incorporated into several different thermoplastic and thermoset polymers. The results of carbon nanofiber addition on the mechanical properties of the resulting composite have been mixed.

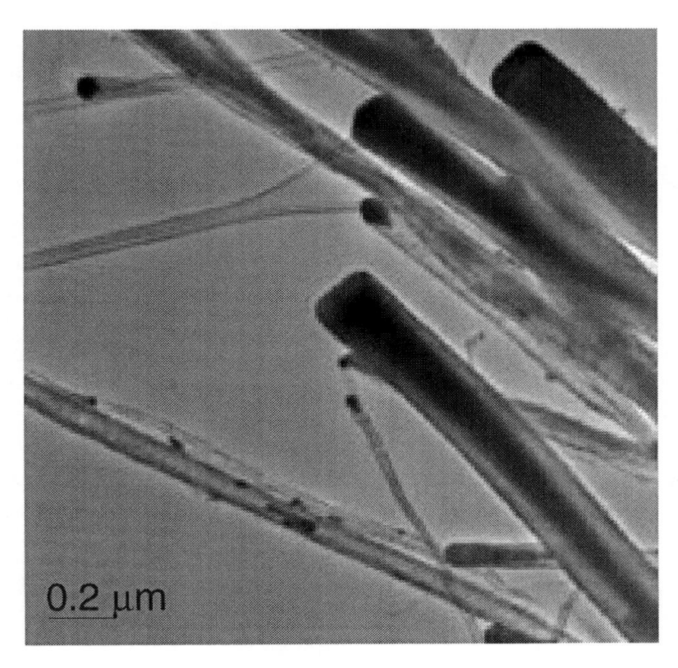

FIGURE 8.4 Photograph of carbon nanofibers. (Courtesy of Applied Sciences, Inc. With permission.)

In general, incorporation of carbon nanofibers in thermoplastics has shown modest to high improvement in modulus and strength, whereas their incorporation in thermosets has shown relatively smaller improvements. An example of each is given as follows.

Finegan et al. [12] conducted a study on the tensile properties of carbon nanofiber-reinforced polypropylene. The nanofibers were produced with a variety of processing conditions (different carbon-containing gases, different gas flow rates, with and without graphitization). A variety of surface treatments were applied on the nanofibers. The composite tensile specimens with 15 vol% nanofibers were prepared using melt processing (injection molding). In all cases, they observed an increase in both tensile modulus and strength compared with polypropylene itself. However, the amount of increase was influenced by the nanofiber production condition and the surface treatment. When the surface treatment involved surface oxidation in a CO_2 atmosphere at 850°C, the tensile modulus and strength of the composite were 4 GPa and 70 MPa, respectively, both of which were greater than three times the corresponding values for polypropylene.

Patton et al. [13] reported the effect of carbon nanofiber addition to epoxy. The epoxy resin was diluted using acetone as the solvent. The diluted epoxy was then infused into the carbon nanofiber mat. After removing the solvent, the epoxy-soaked mat was cured at 120°C and then postcured. Various nanofiber surface treatments were tried. The highest improvement in flexural modulus and strength was observed with carbon nanofibers that were heated in air at 400°C for 30 min. With ~18 vol% of carbon nanofibers, the flexural modulus of the composite was nearly twice that of epoxy, but the increase in flexural strength was only about 36%.

8.3 CARBON NANOTUBES

Carbon nanotubes were discovered in 1991, and within a short period of time, have attracted a great deal of research and commercial interest due to their potential applications in a variety of fields, such as structural composites, energy storage devices, electronic systems, biosensors, and drug delivery systems [14]. Their unique structure gives them exceptional mechanical, thermal, electrical, and optical properties. Their elastic modulus is reported to be >1 TPa, which is close to that of diamond and 3–4 times higher than that of carbon fibers. They are thermally stable up to 2800°C in vacuum; their thermal conductivity is about twice that of diamond and their electric conductivity is 1000 times higher than that of copper.

8.3.1 STRUCTURE

Carbon nanotubes are produced in two forms, single-walled nanotubes (SWNT) and multiwalled nanotubes (MWNT). SWNT is a seamless hollow cylinder and can be visualized as formed by rolling a sheet of graphite layer, whereas MWNT consists of a number of concentric SWNT. Both SWNT and MWNT are closed at the ends by dome-shaped caps. The concentric SWNTs inside an MWNT are also end-capped. The diameter of an SWNT is typically between 1 and 1.4 nm and its length is between 50 and 100 μm. The specific surface area of an SWNT is 1315 m^2/g, and is independent of its diameter [15]. The outer diameter of an MWNT is between 1.4 and 100 nm. The separation between the concentric SWNT cylinders in an MWNT is about 3.45 A°, which is slightly greater than the distance between the graphite layers in a graphite crystal. The specific surface area of an MWNT depends on the number of walls. For example, the specific surface area of a double-walled nanotube is between 700 and 800 m^2/g and that of a 10-walled nanotube is about 200 m^2/g [15].

The structure of an SWNT depends on how the graphite sheets is rolled up and is characterized by its chirality or helicity, which is defined by the chiral angle and the chiral vector (Figure 8.5). The chiral vector is written as

$$\mathbf{C_h} = n\mathbf{a_1} + m\mathbf{a_2}, \tag{8.1}$$

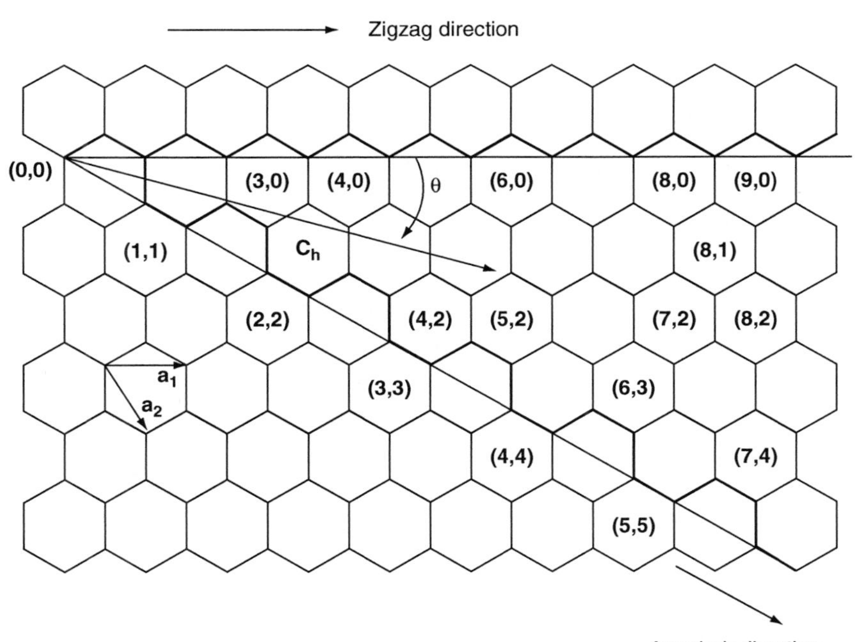

FIGURE 8.5 Chiral vector and chiral angle. (From Govindaraj, A. and Rao, C.N.R., *The Chemistry of Nanomaterials*, Vol. 1, C.N.R. Rao, A. Müller, and A.K. Cheetham, eds., Wiley-VCH, KGaA, Germany, 2004. With permission.)

where $\mathbf{a_1}$ and $\mathbf{a_2}$ are unit vectors in a two-dimensional graphite sheet and (n, m) are called chirality numbers. Both n and m are integers and they define the way the graphite sheet is rolled to form a nanotube.

Nanotubes with $n \neq 0$, $m = 0$ are called the zigzag tubes (Figure 8.6a) and nanotubes with $n = m \neq 0$ are called armchair tubes (Figure 8.6b). In zigzag tubes, two opposite C–C bonds of each hexagon are parallel to the tube's axis, whereas in the armchair tubes, the C–C bonds of each hexagon are perpendicular to the tube's axis. If the C–C bonds are at an angle with the tube's axis, the tube is called a chiral tube (Figure 8.6c). The chiral angle θ is defined as the angle between the zigzag direction and the chiral vector, and is given by

$$\theta = \tan^{-1}\left[\frac{3^{1/2}m}{2n + m}\right] \tag{8.2}$$

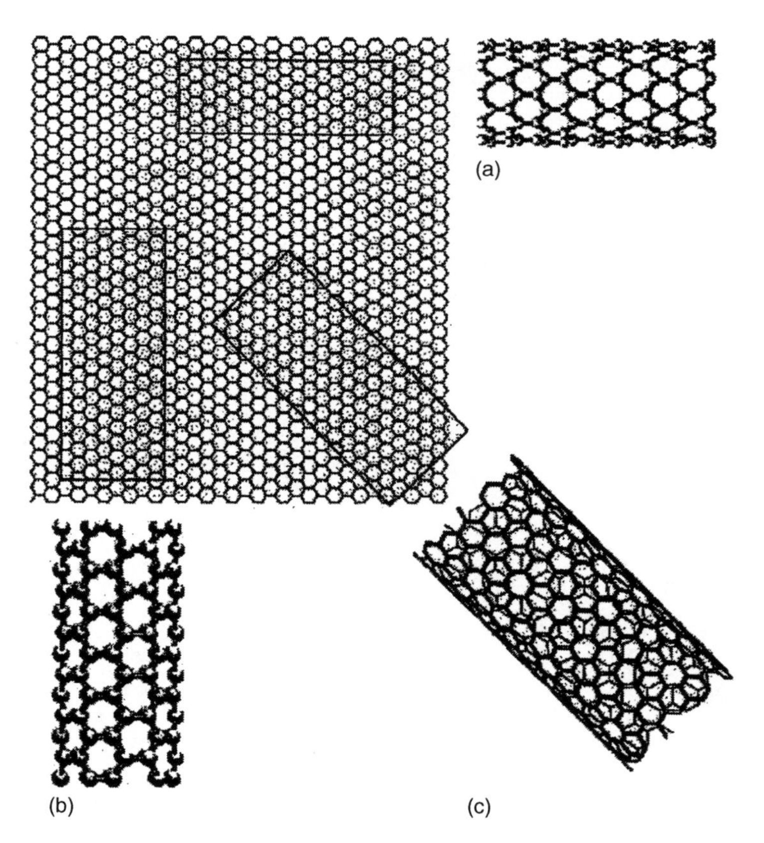

FIGURE 8.6 (a) Zigzag, (b) armchair, and (c) chiral nanotubes. (From Rakov, E.G., *Nanomaterials Handbook*, Y. Gogotsi, ed., CRC Press, Boca Raton, USA, 2006. With permission.)

and $0^\circ \leq \theta \leq 30^\circ$. The diameter of the nanotube is given by

$$d = \frac{a_o\sqrt{3}}{\pi}\sqrt{m^2 + mn + n^2}, \quad (8.3)$$

where a_o = C–C bond length, which is equal to 1.42 Å.

The chirality of a carbon nanotube has a significant influence on its electrical and mechanical properties. Depending on the chirality, a carbon nanotube can behave either as a metallic material or a semiconducting material. For example, armchair nanotubes are metallic; nanotubes with $(n-m) = 2k$, where k is a nonzero integer, are semiconductors with a tiny band gap, and all other nanotubes are semiconductors with a band gap that inversely varies with the nanotube diameter. Chirality also controls the deformation characteristics of carbon nanotubes subjected to tensile stresses and therefore, determines whether they will fracture like a brittle material or deform like a ductile material [16].

The current processing methods used for making carbon nanotubes introduce two types of defect in their structures: topological defects and structural defects. The examples of topological defects are pentagonal and heptagonal arrangements of carbon atoms, which may be mixed with the hexagonal arrangements. The structural defects include nontubular configurations, such as cone-shaped end caps, bent shapes, branched construction with two or more tubes connected together, and bamboo-like structure in which several nanotubes are joined in the lengthwise direction. In general, MWNTs contain more defects than SWNTs.

Carbon nanotubes can form secondary structures. One of these secondary structures is the SWNT rope or bundle, which is a self-assembled close-packed array of many (often thousands or more) SWNTs. The self-assembly occurs at the time of forming the SWNTs, due to the attractive forces between them arising from binding energy of 500–900 eV/nm. If the arrangement of SWNTs in the array is well ordered (e.g., with hexagonal lattice structure), it is called a rope. If the arrangement of SWNTs is not ordered, it is called a bundle [17].

8.3.2 Production of Carbon Nanotubes

There are three main methods for producing carbon nanotubes: electric arc discharge, laser ablation, and chemical vapor deposition (CVD) [18]. The quantity of production of carbon nanotubes by the first two methods is relatively small. Since CVD can produce larger quantities of carbon nanotubes and is more versatile, it has become the focus of attention for industrial production and several different variations of the basic CVD process (e.g., plasma-enhanced CVD or PECVD process and high-pressure carbon monoxide or HiPco process) have been developed. However, the structure of nanotubes produced by CVD is usually different from that of nanotubes produced by the other two methods. For example, the CVD-produced MWNTs are less crystalline and contain more defects than the arc-discharged MWNTs. The CVD-produced MWNTs are longer and less straight than the arc-discharged MWNTs.

The arc discharge method uses two graphite rods; one serving as the cathode and the other serving as the anode, in either helium, argon, or a mixture of helium and argon atmosphere. The graphite rods are placed side by side with a very small gap, typically about 1 mm in size, between them. The pressure inside the reaction chamber is maintained between 100 and 1000 torr. When a stable arc is produced in the gap by passing 50–120 A electric current (at 12–25 V) between the graphite rods, the material eroded from the anode is deposited on the cathode in the form of MWNTs, amorphous carbon, and other carbon particles. To produce SWNTs, the cathode is doped with a small amount of metallic catalyst (Fe, Co, Ni, Y, or Mo). However, the yield of SWNTs is only between 20% and 40% by weight.

The laser ablation method uses either pulsed or continuous-wave laser to vaporize a graphite target held at 1200°C in a controlled atmosphere of argon or helium inside a tube furnace. The vaporized material is collected on a water-cooled copper collector in the form of carbon nanotubes, amorphous carbon, and other carbon particles. To produce SWNTs, the graphite target is doped with metal catalysts such as nickel and cobalt catalysts. The yield of SWNTs is between 20% and 80% by weight.

In the basic CVD method, carbon nanotubes are produced by the decomposition of a carbon-containing gas, such as carbon monoxide and hydrocarbon gases, or by the pyrolysis of carbon-containing solids, such as polymers, at a high pressure inside a furnace. The temperature inside the furnace is typically between 300°C and 800°C for making MWNTs, and 600°C and 1200°C for making SWNTs. MWNTs are produced in an inert gas atmosphere, whereas a mixture of hydrogen and an inert gas is used for SWNTs. A high-temperature substrate, such as alumina, coated with catalyst particles, such as Fe, Ni, and Co, is placed in the furnace. The decomposition of the carbon-containing gas flowing over the substrate causes the growth of carbon nanotubes on the substrate. They are collected after cooling the system down to room temperature. Depending on the carbon-containing gas, catalyst, furnace temperature, pressure, flow rate, residence time for thermal decomposition, and so on, the yield can be between 30% and 99% by weight.

All three production methods produce carbon nanotubes that are contaminated with impurities such as amorphous carbon, carbon soot, other carbon particles, and metal catalysts. Several purification processes have been developed to produce cleaner carbon nanotubes. Two of the processes are gas phase oxidation and liquid phase purification. Purification of MWNTs by gas phase oxidation involves heating them in an oxygen or air atmosphere at temperatures $>$700°C. In the case of SWNT, it involves heating in a mixed atmosphere of hydrochloric acid, chlorine, and water vapor at 500°C.

Liquid phase purification involves refluxing in an acid such as nitric acid at an elevated temperature. In general, carbon nanotubes produced by the arc discharge process requires more extensive purification than the other two processes.

Carbon nanotubes are often formed as long, entangled bundles. The "cutting" process is used to shorten their lengths, disentangle them, open up the ends, and provide active sites for functionalization. The cutting process can be either mechanical (e.g., by ball-milling) or chemical (e.g., by treating them in a 3:1 mixture of concentrated sulfuric acid and nitric acid).

Carbon nanotubes are available in a variety of forms. One of these forms is called the bucky paper, which is a thin film of randomly oriented SWNTs. It is made by filtering SWNTs dispersed in an aqueous or organic solution and then peeling off the nanotube film from filter paper. Carbon nanotube fibers and yarns containing aligned SWNTs have also been produced [19].

8.3.3 Functionalization of Carbon Nanotubes

Carbon nanotubes are functionalized for a variety of purposes. Among them are (1) improve their dispersion in the polymer matrix, (2) create better bonding with the polymer matrix, and (3) increase their solubility in solvents. Functionalization has also been used for joining of nanotubes to form a network structure.

Functionalization can occur either at the defect sites on the nanotube wall or at the nanotube ends. The functional groups are covalently bonded to the nanotubes using either oxidation, fluorination, amidation, or other chemical reactions. Functionalization can also be achieved using noncovalent interactions, for example, by wrapping the nanotubes with polymer molecules or adsorption of polymer molecules in the nanotubes. The covalent functionalization is generally considered to provide better load transfer between the nanotubes and the surrounding polymer matrix, and therefore, improved mechanical properties. On the other hand, noncovalent functionalization may be preferred if it is required that the electronic characteristics of carbon nanotubes remain unchanged.

The covalent bonds can be produced in two different ways: (1) by direct attachment of functional groups to the nanotubes and (2) by a two-step functionalization process in which the nanotubes are chemically treated first to attach simple chemical groups (e.g., –COOH and –OH) at the defect sites or at their ends (Figure 8.7), which are later substituted with more active organic groups. Carbon nanotubes have also been functionalized using silane-coupling agents [20]. Silane-coupling agents are described in Chapter 2.

A variety of chemical and electrochemical functionalization processes have been developed [15]. One of these processes is a two-step functionalization process in which acidic groups, such as the carboxylic (–COOH) groups or the hydroxyl (–OH) groups, are first attached by refluxing carbon nanotubes in concentrated HNO_3 or a mixture of H_2SO_4 and HNO_3. One problem in acid

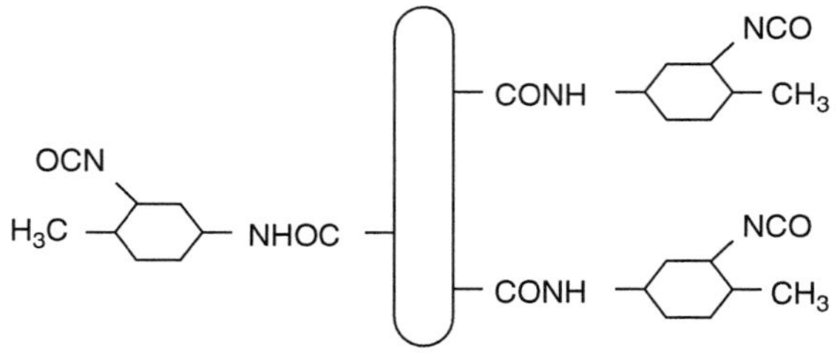

FIGURE 8.7 Schematic of a functionalized SWNT.

refluxing is that it also tends to cut the nanotubes to shorter lengths, thus reducing the fiber aspect ratio. Other milder treatments have also been developed; one of them is the ozone treatment. The ozonized surface can be reacted with several types of reagents, such as hydrogen peroxide, to create the acidic group attachments. In the second step of this process, the carbon nanotubes containing the acidic groups are subjected to amidation reaction either in a mixture of thionyl chloride ($SOCl_2$) and dimethyl formamide or directly in presence of an amine. The second step produces amide functionality on carbon nanotubes that are more reactive than the acid groups.

8.3.4 Mechanical Properties of Carbon Nanotubes

Theoretical calculations for defect-free carbon nanotubes show that the Young's modulus of 1–2 nm diameter SWNT is ~1 TPa and that of MWNT is between 1.1 and 1.3 TPa (Table 8.3). The Young's modulus of SWNT is independent of tube chirality, but it decreases with increasing diameter. The Young's

TABLE 8.3
Properties of Carbon Nanotubes

	Young's Modulus (GPa)	Tensile Strength (GPa)	Density (g/cm^3)
SWNT	1054[a]	75[a]	1.3
SWNT Bundle	563		1.3
MWNT	1200[a]		2.6
Graphite (in-plane)	350	2.5	2.6
P-100 Carbon fiber	758	2.41	2.15

Note: 1 TPa = 10^3 GPa.

[a] Theoretical values.

modulus of MWNT is higher than that of SWNT due to contributions from the van der Waals forces between the concentric SWNTs in MWNT [14].

Experimental determination of the mechanical properties of carbon nanotubes is extremely difficult and has produced a variety of results. Several investigators have used atomic force microscope (AFM) to determine the Young's modulus and strength. The Young's modulus of MWNT determined on AFM has ranged from 0.27 to 1.8 TPa and that of SWNT ranges from 0.32 to 1.47 TPa. Similarly, the strength of MWNT ranges from 11 to 63 GPa and that of SWNT from 10 to 52 GPa. In the TPM experiments, the carbon nanotubes have also shown high tensile strain (up to 15%) before fracture. It has also been observed that carbon nanotubes exhibit nonlinear elastic deformation under tensile, bending, as well as twisting loads. At high strains, they tend to buckle of the wall.

The large variation in Young's modulus and strength is attributed to the fact that nanotubes produced by the current production methods may vary in length, diameter, number of walls, chirality, and even atomic structure [21]. Furthermore, nanotubes produced by different methods contain different levels of defects and impurities, which influence their mechanical properties. For example, the Young's modulus of CVD-produced MWNTs is found to be an order of magnitude lower than that of the arc-discharged MWNTs, which is due to the presence of higher amount of defects in the CVD-produced MWNTs.

8.3.5 Carbon Nanotube–Polymer Composites

The three principal processing methods for combining carbon nanotubes with polymer matrix [22,23] are (1) in situ polymerization, (2) solution processing, and (3) melt processing.

1. *In Situ Polymerization*: In this process, the nanotubes are first dispersed in the monomer and then the polymerization reaction is initiated to transform the monomer to polymer. Depending on the polymer formed and the surface functionality of the nanotubes, the polymer molecules are either covalently bonded to the nanotubes or wrapped around the nanotubes at the completion of the polymerization reaction.
2. *Solution Processing*: In this process, the nanotubes are mixed with a polymer solution, which is prepared by dissolving the polymer in a suitable solvent. The mixing is done using magnetic stirring, high shear mixing, or sonication. The dispersion of the nanotubes in the solution can be improved by treating them with a surfactant, such as derivatives of sodium dodecylsulfate. The solution is poured in a casting mold and the solvent is allowed to evaporate. The resulting material is a cast film or sheet of carbon nanotube-reinforced polymer.
3. *Melt Processing*: Melt processing is the preferred method for incorporating carbon nanotubes in thermoplastics, particularly for high volume

applications. It has been used with a variety of thermoplastics, such as high-density polyethylene, polypropylene, polystyrene, polycarbonate, and polyamide-6. In this process, the nanotubes are blended with the liquid polymer in a high shear mixer or in an extruder. The blend is then processed to produce the final product using injection molding, extrusion, or compression molding. It is important to note that the addition of carbon nanotubes increases the viscosity of the liquid polymer, and therefore, proper adjustments need to be made in the process parameters to mold a good product.

Carbon nanotubes have also been used with thermoset polymers such as epoxy and vinyl ester. They are dispersed in the liquid thermoset prepolymer using sonication. After mixing the blend with a hardener or curing initiator, it is poured into a casting mold, which is then heated to the curing temperature. Curing can be conducted in vacuum to reduce the void content in the composite. To improve the dispersion of the nanotubes, the viscosity of the thermoset prepolymer can be reduced with a solvent that can be evaporated later.

8.3.6 Properties of Carbon Nanotube–Polymer Composites

Based on the mechanical properties of carbon nanotubes described in Section 8.3.4, it is expected that the incorporation of carbon nanotubes in a polymer matrix will create composites with very high modulus and strength. Indeed, many studies have shown that with proper dispersion of nanotubes in the polymer matrix, significant improvement in mechanical properties can be achieved compared with the neat polymer. Three examples are given as follows.

In the first example, CVD-produced MWNTs were dispersed in a toluene solution of polystyrene using an ultrasonic bath [24]. The mean external diameter of the MWNTs was 30 nm and their length was between 50 and 55 μm. No functionalization treatments were used. MWNT-reinforced polystyrene film was produced by solution casting. Tensile properties of the solution cast films given in Table 8.4 show that both elastic modulus and tensile strength increased with increasing MWNT weight fraction. With the addition of 5 wt% MWNT, the elastic modulus of the composite was 120% higher and the tensile strength was 57% higher than the corresponding properties of polystyrene.

The second example involves melt-processed MWNT-reinforced polyamide-6 [25]. The MWNTs, in this example, was also prepared by CVD and functionalized by treating them in nitric acid. The MWNTs and polyamide-6 were melt-compounded in a twin-screw mixer and film specimens were prepared by compression molding. The MWNT content in the composite was 1 wt%. As shown in Table 8.5, both tensile modulus and strength of the composite were significantly higher compared with the tensile modulus and strength of polyamide-6. However, the elongation at break was decreased.

TABLE 8.4
Tensile Properties of Cast MWNT–Polystyrene

	Elastic Modulus (GPa)	Tensile Strength (MPa)
Polystyrene	1.53	19.5
Polystyrene + 1 wt% MWNT	2.1	24.5
Polystyrene + 2 wt% MWNT	2.73	25.7
Polystyrene + 5 wt% MWNT	3.4	30.6

Source: From Safadi, B., Andrews, R., and Grulke, E.A., *J. Appl. Polym. Sci.*, 84, 2660, 2002.

Note: Average values are shown in the table.

The final example considers the tensile properties of 1 wt% MWNT-reinforced epoxy [26]. The MWNT, in this case, was produced by the CVD process, and had an average diameter of 13 nm and an average length of 10 μm. They were acid-treated in a 3:1 mixture of 65% H_2SO_4 and HNO_3 for 30 min at 100°C to remove the impurities. The acid-treated nanotubes were then subjected to two different functionalization treatments: amine treatment and plasma oxidation. The acid-treated and amine-treated MWNTs were first mixed with ethanol using a sonicator and then dispersed in epoxy. The plasma-oxidized MWNTs were dispersed directly in epoxy. The composite was prepared by film casting. As shown in Table 8.6, the addition of MWNT caused very little improvement in modulus, but tensile strength and elongation at break were significantly improved.

TABLE 8.5
Tensile Properties of 1 wt% MWNT-Reinforced Polyamide-6

	Polyamide-6	1 wt% MWNT +Polyamide-6
Tensile modulus (GPa)	0.396	0.852
Yield strength (MPa)	18	40.3
Elongation at break	>150	125

Source: From Zhang, W.D., Shen, L., Phang, I.Y., and Liu, T., *Macromolecules*, 37, 256, 2004.

Note: Average values are shown in the table.

TABLE 8.6
Tensile Properties of 1 wt% MWNT–Epoxy Composites

	Young's Modulus (GPa)	Tensile Strength (MPa)	Elongation at Break (%)
Epoxy	1.21	26	2.33
Untreated MWNT–epoxy	1.38	42	3.83
Acid-treated MWNT–epoxy	1.22	44	4.94
Amine-treated MWNT–epoxy	1.23	47	4.72
Plasma-treated MWNT–epoxy	1.61	58	5.22

Source: From Kim, J.A., Seong, D.G., Kang, T.J., and Youn, J.R., *Carbon*, 44, 1898, 2006.

Note: Average values are shown in the table.

The three examples cited earlier are among many that have been published in the literature. They all show that the addition of carbon nanotubes, even in small concentrations, is capable of improving one or more mechanical properties of polymers. Thermal properties, such as thermal conductivity, glass transition temperature, and thermal decomposition temperature, and electrical properties, such as electrical conductivity, are also increased. However, several problems have been mentioned about combining nanotubes with polymers and achieving the properties that carbon nanotubes are capable of imparting to the polymer matrix. The most important of them are nanotube dispersion and surface interaction with the polymer matrix [22,23]. For efficient load transfer between the nanotubes and the polymer, they must be dispersed uniformly without forming agglomeration. Relatively good dispersion can be achieved with proper mixing techniques (such as ultrasonication) at very small volume fractions, typically less than 1%–2%. At larger volume fractions the nanotubes tend to form agglomeration, which is accompanied by decrease in modulus and strength.

One contributing factor to poor dispersion is that the carbon nanotubes (and other nano-reinforcements) have very high surface area and also very high length-to-diameter ratio [27,28]. While both provide opportunity for greater interface interaction and load transfer, the distance between the individually dispersed nanotubes becomes so small even at 10% volume fraction that the polymer molecules cannot infiltrate between them. Strong attractive forces between the nanotubes also contribute to poor dispersion. This is particularly true with SWNTs, which tend to form bundles (nanoropes) that are difficult to separate. MWNTs, in general, exhibit better dispersion than SWNTs.

The surface interaction between the carbon nanotubes and the polymer matrix requires good bonding between them, which is achieved by functionalization of nanotubes. Functionalization also helps improve the dispersion of nanotubes in the polymer. Table 8.6 shows the effect of functionalization

TABLE 8.7
Tensile Properties of Functionalized SWNT-Reinforced Polyvinyl Alcohol (PVA) Film

Material	Young's Modulus (GPa)	Yield Strength (MPa)	Strain-to-Failure (mm/mm)
PVA	4.0	83	>0.60
PVA + 2.5% purified SWNT	5.4	79	0.09
PVA + 2.5% functionalized SWNT	5.6	97	0.05
PVA + 5% functionalized SWNT	6.2	128	0.06

Source: From Paiva, M.C., Zhou, B., Fernando, K.A.S., Lin, Y., Kennedy, J.M., and Sun, Y.-P., *Carbon*, 42, 2849, 2004.

Note: Average values are shown in the table.

of MWNTs. The effect of functionalization of SWNT is shown in Table 8.7, which gives the tensile properties of functionalized SWNT-reinforced PVA. In this case, the SWNTs were prepared by the arc discharge method and purified by refluxing in an aqueous nitric acid solution. The nanotubes were then functionalized with low-molecular-weight PVA [29] using *N*,*N*′-dicyclohexyl carbodiimide-activated esterification reaction. The functionalized SWNTs were mixed with a water solution of PVA and solution cast into 50–100 μm thick films. Both Young's modulus and yield strength were significantly higher with functionalized SWNTs. The higher yield strength with 2.5 wt% functionalized SWNTs compared with 2.5 wt% purified SWNTs (without functionalization) was attributed to deagglomeration of SWNT ropes into separate nanotubes, better dispersion of them in the PVA matrix, and wetting by PVA.

Another approach to improving the mechanical properties of carbon nanotube-reinforced polymers is to align the nanotubes in the direction of stress. This is a relatively difficult task if the common processing methods are used. Flow-induced alignment of nanotubes is created by solution spinning, melt spinning, or other similar methods to produce carbon nanotube-reinforced polymer fiber or film. Mechanical stretching of carbon nanotube-reinforced thin polymer films also tend to align the nanotubes in the direction of stretching. Other methods of aligning carbon nanotubes in polymer matrix, including the application of magnetic field, are described in Ref. [23].

The effect of alignment of nanotubes on the properties of carbon nanotube-reinforced polymer fibers is demonstrated in Table 8.8. In this case, the polymer was polyacrylonitrile (PAN) and the fibers were prepared by solution spinning [30]. The carbon nanotube content was 5% by weight. All of the mechanical properties, including strain-at-failure and toughness, increased with the addition of SWNT, MWNT, as well as VGCNF. There was a significant decrease

TABLE 8.8
Properties of SWNT, MWNT, and VGCNF-Reinforced PAN Fibers

	PAN Fiber	(PAN + SWNT) Fiber	(PAN + MWNT) Fiber	(PAN + VGCNF) Fiber
Modulus (GPa)	7.8	13.6	10.8	10.6
Strength at break (MPa)	244	335	412	335
Strain-at-failure (%)	5.5	9.4	11.4	6.7
Toughness (MN m/m^3)	8.5	20.4	28.3	14
Shrinkage (%) at 160°C	13.5	6.5	8.0	11.0
T_g (°C)	100	109	103	103

Source: From Chae, H.G., Sreekumar, T.V., Uchida, T., and Kumar, S., *Polymer*, 46, 10925, 2005.

Note: Average values are shown in the table.

in thermal shrinkage and increase in the glass transition temperature (T_g). The improvements in the properties were not only due to the alignment of nano-reinforcements in the fiber, but also due to higher orientation of PAN molecules in the fiber caused by the presence of nano-reinforcements.

REFERENCES

1. M. Alexandre and P. Dubois, Polymer-layered silicate nanocomposites: preparation, properties and uses of a new class of materials, *Mater. Sci. Eng., 28*:1 (2000).
2. M. Kato and A. Usuki, Polymer–clay nanocomposites, in *Polymer–Clay Nanocomposites* (T.J. Pinnavai and G.W. Beall, eds.), John Wiley & Sons, Chichester, UK (2000).
3. H.R. Dennis, D.L. Hunter, D. Chang, S. Kim, J.L. White, J.W. Cho, and D.R. Paul, Effect of melt processing conditions on the extent of exfoliation in organoclay-based nanocomposites, *Polymer, 42*:9513 (2001).
4. S.J. Ahmadi, Y.D. Huang, and W. Li, Synthetic routes, properties and future applications of polymer-layered silicate nanocomposites, *J. Mater. Sci., 39*:1919 (2004).
5. G.G. Tibbetts, M.L. Lake, K.L. Strong, and B.P. Price, A review of the fabrication and properties of vapor-grown carbon nanofiber/polymer composites, *Composites Sci. Tech. 67*:1709 (2007).
6. I.S. Chronakis, Novel nanocomposites and nanoceramics based on polymer nanofibers using electrospinning technology, *J. Mater. Process. Tech., 167*:283 (2005).
7. Y.-Y. Fan, H.-M. Cheng, Y.-L. Wei, G. Su, and Z.-H. Shen, Tailoring the diameters of vapor-grown carbon nanofibers, *Carbon, 38*:921 (2000).
8. J.-H. Zhou, Z.-J. Sui, P. Li, D. Chen, Y.-C. Dai, and W.K. Yuan, Structural characterization of carbon nanofibers formed from different carbon-containing gases, *Carbon, 44*:3255 (2006).
9. O.C. Carnerio, N.M. Rodriguez, and R.T.K. Baker, Growth of carbon nanofibers from the iron–copper catalyzed decomposition of $CO/C_2H_4/H_2$ mixtures, *Carbon, 43*:2389 (2005).

10. T. Uchida, D.P. Anderson, M. Minus, and S. Kumar, Morphology and modulus of vapor grown carbon nano fibers, *J. Mater. Sci., 41*:5851 (2006).
11. F.W.J. van Hattum, J.M. Benito-Romero, A. Madronero, and C.A. Bernardo, Morphological, mechanical and interfacial analysis of vapour-grown carbon fibres, *Carbon, 35*:1175 (1997).
12. I.C. Finegan, G.G. Tibbetts, D.G. Glasgow, J.-M. Ting, and M.L. Lake, Surface treatments for improving the mechanical properties of carbon nanofiber/thermoplastic composites, *J. Mater. Sci., 38*:3485 (2003).
13. R.D. Patton, C.U. Pittman, Jr., L. Wang, and J.R. Hill, Vapor grown carbon fiber composites with epoxy and poly(phenylene sulfide) matrices, *Composites: Part A, 30*:1081 (1999).
14. J. Han, Structure and properties of carbon nanotubes, *Carbon Nanotubes: Science and Applications* (M. Meyappan, ed.), CRC Press, Boca Raton, USA (2005).
15. E.G. Rakov, Chemistry of carbon nanotubes, *Nanomaterials Handbook* (Y. Gogotsi, ed.), CRC Press, Boca Raton, USA, pp. 105–175 (2006).
16. E.T. Thostenson, Z. Ren, and T.-W. Chou, Advances in the science and technology of carbon nanotubes and their composites, *Compos. Sci. Tech., 61*:1899 (2001).
17. J.E. Fischer, Carbon nanotubes: structure and properties, *Nanomaterials Handbook* (Y. Gogotsi, ed.), CRC Press, Boca Raton, USA, pp. 69–103 (2006).
18. D. Mann, Synthesis of carbon nanotubes, *Carbon Nanotubes: Properties and Applications* (M.J. O'Connell, ed.), CRC Press, Boca Raton, USA, pp. 19–49 (2006).
19. H.G. Chae, J. Liu, and S. Kumar, Carbon nanotube-enabled materials, *Carbon Nanotubes: Properties and Applications* (M.J. O'Connell, ed.), CRC Press, Boca Raton, USA, pp. 19–49 (2006).
20. P.C. Ma, J.-K. Kim, and B.Z. Tang, Functionalization of carbon nanotubes using a silane coupling agent, *Carbon, 44*:3232 (2006).
21. N. Grobert, Carbon nanotubes—becoming clean, *Mater. Today, 10*:28 (2007).
22. J.N. Coleman, U. Khan, W.J. Blau, and Y.K. Gun'ko, Small but strong: a review of the mechanical properties of carbon nanotube–polymer composites, *Carbon, 44*:1624 (2006).
23. X.-L. Xie, Y.-W. Mai, and X.-P. Zhou, Dispersion and alignment of carbon nanotubes in polymer matrix: a review, *Mater. Sci. Eng., R49*:89 (2005).
24. B. Safadi, R. Andrews, and E.A. Grulke, Multiwalled carbon nanotube polymer composites: synthesis and characterization of thin films, *J. Appl. Polym. Sci., 84*:2660 (2002).
25. W.D. Zhang, L. Shen, I.Y. Phang, and T. Liu, Carbon nanotubes reinforced nylon-6 composite prepared by simple melt compounding, *Macromolecules, 37*:256 (2004).
26. J.A. Kim, D.G. Seong, T.J. Kang, and J.R. Youn, Effects of surface modification on rheological and mechanical properties of CNT/epoxy composites, *Carbon, 44*:1898 (2006).
27. F.H. Gojny, M.H.G. Wichmann, B. Fiedler, and K. Schulte, Influence of different carbon nanotubes on the mechanical properties of epoxy matrix composites—a comparative study, *Compos. Sci. Tech., 65*:2300 (2005).
28. B. Fiedler, F.H. Gojny, M.H.G. Wichmann, M.C.M. Nolte, and K. Schulte, Fundamental aspects of nano-reinforced composites, *Compos. Sci. Tech., 66*:3115 (2006).

29. M.C. Paiva, B. Zhou, K.A.S. Fernando, Y. Lin, J.M. Kennedy, and Y.-P. Sun, Mechanical and morphological characterization of polymer–carbon nanocomposites from functionalized carbon nanotubes, *Carbon, 42*:2849 (2004).
30. H.G. Chae, T.V. Sreekumar, T. Uchida, and S. Kumar, A comparison of reinforcement efficiency of various types of carbon nanotubes in polyacrylonitrile fiber, *Polymer, 46*:10925 (2005).

PROBLEMS

P8.1. Calculate the specific surface area (unit: m^2/g) of a defect-free single-walled carbon nanotube (SWNT) assuming that the length of the C–C bonds in the curved graphite sheet is the same as that in a planar sheet, which is 0.1421 nm. The atomic mass of carbon is 12 g/mol and the Avogadro's number is 6.023×10^{23} per mole. Assume that the surface area of the end caps is negligible compared with the surface area of the cylindrical side wall.

P8.2. Using the information given in Problem P8.1, calculate the specific surface area of a defect-free double-walled carbon nanotube (DWNT).

P8.3. The specific surface area of SWNT is independent of the nanotube diameter, whereas the specific surface area of MWNT decreases with the nanotube diameter. Why?

P8.4. Assume that SWNTs in a nanorope are arranged in a regular hexagonal array as shown in the following figure. Using the information given in Problem P8.1, calculate the specific surface area of the nanorope.

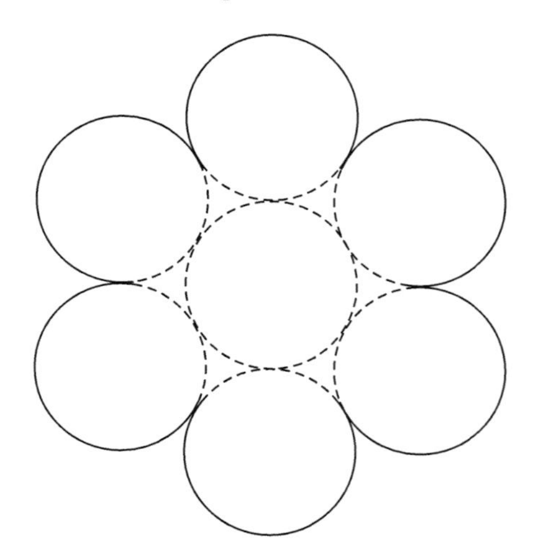

P8.5. Assume that the carbon nanotubes in a nanotube–polymer composite are arranged in a regular face-centered square array as shown in the following figure. Calculate the distance between the nanotubes as a function of nanotube volume fraction. Knowing that the polymer molecules are 0.8 nm in diameter, discuss the problem of incorporating high nanotube concentration in a polymer matrix.

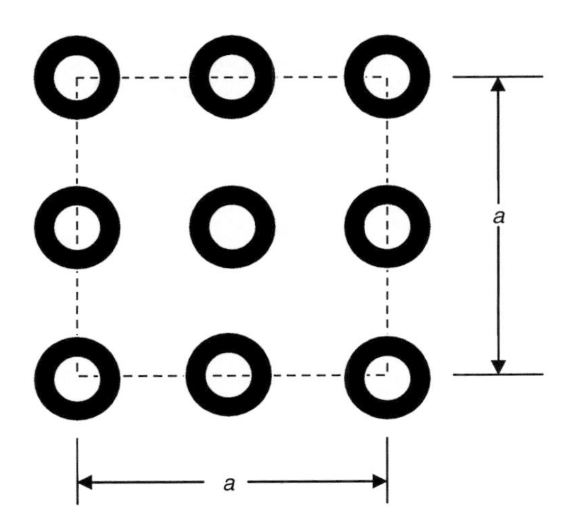

P8.6. In Table 8.8, the tensile modulus of SWNT-reinforced polyacrylonitrile (PAN) fibers is reported as 13.6 GPa. The SWNTs were dispersed in these fibers as bundles and the average bundle diameter was 10 nm. The average length of the SWNT bundles was 300 nm and the SWNT volume fraction was 4.6%.

1. Using the Halpin–Tsai equations (Appendix A.4) and assuming that the SWNT bundles were aligned along the fiber length, calculate the theoretical modulus of the SWNT-reinforced polyacrylonitrile fiber. The modulus of PAN fibers, $E_{PAN} = 7.8$ GPa and the modulus of SWNT, $E_{SWNT} = 640$ GPa.
2. What may be the principal reasons for the difference between the theoretical modulus and the experimentally determined modulus of 13.6 GPa?
3. How will the theoretical modulus change if the SWNT bundles were completely exfoliated into individual SWNTs of 1.2 nm diameter?

Appendixes

A.1 WOVEN FABRIC TERMINOLOGY

Basic woven fabrics consist of two sets of yarns interlaced at right angles to create a single layer. Such biaxial or 0/90 fabrics are characterized by the following nomenclature:

1. *Yarn construction*: May include the strand count as well as the number of strands twisted and plied together to make up the yarn. In case of glass fibers, the strand count is given by the yield expressed in yards per pound or in TEX, which is the mass in grams per 1000 m. For example, if the yarn is designated as 150 2/3, its yield is 150×100 or 15,000 yd/lb. The 2/ after 150 indicates that the strands are first twisted in groups of two, and the /3 indicates that three of these groups are plied together to make up the final yarn. The yarns for carbon-fiber fabrics are called tows. They have little or no twist and are designated by the number of filaments in thousands in the tow. Denier (abbreviated as de) is used for designating Kevlar yarns, where 1 denier is equivalent to 1 g/9000 m of yarn.
2. *Count*: Number of yarns (ends) per unit width in the warp (lengthwise) and fill (crosswise) directions (Figure A.1.1). For example, a fabric count of 60×52 means 60 ends per inch in the warp direction and 52 ends per inch in the fill direction.
3. *Weight*: Areal weight of the fabric in ounces per square yard or grams per square meter.
4. *Thickness*: Measured in thousandths of an inch (mil) or in millimeters.
5. *Weave style*: Specifies the repetitive manner in which the warp and fill yarns are interlaced in the fabric. Common weave styles are shown in Figure A.1.2.
   (a) Plain weave, in which warp and fill yarns are interlaced over and under each other in an alternating fashion.
   (b) Basket weave, in which a group of two or more warp yarns are interlaced with a group of two or more fill yarns in an alternating fashion.
   (c) Satin weave, in which each warp yarn weaves over several fill yarns and under one fill yarn. Common satin weaves are crowfoot satin or four-harness satin, in which each warp yarn weaves over three and under one fill yarn, five-harness satin (over four, under one), and eight-harness satin (over seven, under one).

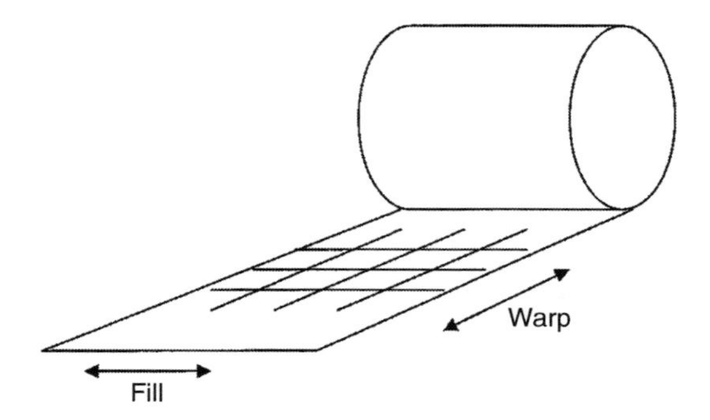

FIGURE A.1.1 Warp and fill directions of fabrics.

Plain weave fabrics are very popular in wet layup applications due to their fast wet-out and ease of handling. They also provide the least yarn slippage for a given yarn count. Satin weave fabrics are more pliable than plain weave fabrics and conform more easily to contoured mold surfaces.

In addition to the biaxial weave described earlier, triaxial (0/60/−60 or 0/45/90) and quadraxial (0/45/90/−45) fabrics are also commercially available. In these fabrics, the yarns at different angles are held in place by tying them with stitch yarns.

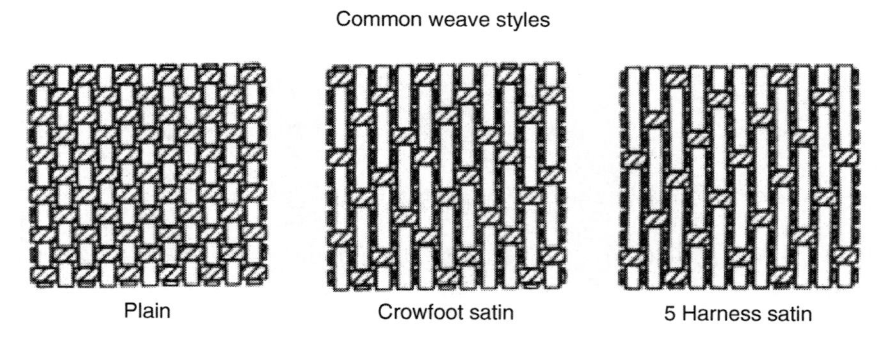

FIGURE A.1.2 Common weave styles. (Courtesy of Hexcel Corporation. With permission.)

A.2 RESIDUAL STRESSES IN FIBERS AND MATRIX IN A LAMINA DUE TO COOLING [1]

The following equations, derived on the basis of a composite cylinder model (Figure A.2.1), can be used to calculate the residual stresses in fibers and matrix in a unidirectional composite lamina developed due to differential thermal shrinkage as it cools down from the high processing temperature to the ambient temperature:

$$\sigma_{r\mathrm{m}} = A_1\left(1 - \frac{r_\mathrm{m}^2}{r^2}\right),$$

$$\sigma_{\theta\mathrm{m}} = A_1\left(1 + \frac{r_\mathrm{m}^2}{r^2}\right),$$

$$\sigma_{z\mathrm{m}} = A_2,$$

$$\sigma_{r\mathrm{f}} = \sigma_{\theta\mathrm{f}} = A_1\left(1 - \frac{r_\mathrm{m}^2}{r_\mathrm{f}^2}\right),$$

$$\sigma_{z\mathrm{f}} = A_2\left(1 - \frac{r_\mathrm{m}^2}{r_\mathrm{f}^2}\right),$$

where

r = radial distance from the center of the fiber

r_f = fiber radius

r_m = matrix radius in the composite cylinder model, which is equal to $(r_\mathrm{f}/\mathrm{v}_\mathrm{f}^{\frac{1}{2}})$

m,f = subscripts for matrix and fiber, respectively

r,θ,z = subscripts for radial, tangential (hoop), and longitudinal directions, respectively.

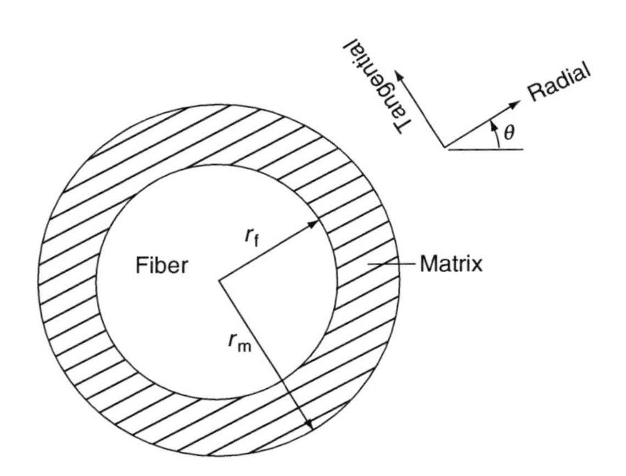

FIGURE A.2.1 Cross section of a composite cylinder model.

The constants A_1 and A_2 are given by the following expressions:

$$A_1 = \left[\frac{(\alpha_m - \alpha_{fl})\Delta_{22} - (\alpha_m - \alpha_{fr})\Delta_{12}}{\Delta_{11}\Delta_{22} - \Delta_{21}\Delta_{12}}\right]\Delta T$$

$$A_2 = \left[\frac{(\alpha_m - \alpha_{fr})\Delta_{11} - (\alpha_m - \alpha_{fl})\Delta_{21}}{\Delta_{11}\Delta_{22} - \Delta_{21}\Delta_{12}}\right]\Delta T$$

where

$$\Delta_{11} = 2\left(\frac{\nu_m}{E_m} + \frac{\nu_{fl}}{E_{fl}}\frac{v_m}{v_f}\right)$$

$$\Delta_{12} = -\left(\frac{v_m}{E_{fl}v_f} + \frac{1}{E_m}\right)$$

$$\Delta_{21} = -\left[\frac{(1-\nu_{fr})v_m}{E_{fr}v_f} + \frac{(1-\nu_m)}{E_m} + \frac{(1+\nu_m)}{E_m v_f}\right]$$

$$\Delta_{22} = \frac{1}{2}\Delta_{11}$$

ΔT = temperature change, which is negative for cooling
E = modulus
ν = Poisson's ratio
α = coefficient of linear thermal expansion
v = volume fraction
fl, fr, m = subscripts indicating fiber (longitudinal and radial) and matrix, respectively

Figure A.2.2 shows the variation of residual stresses for a carbon fiber–epoxy lamina with $v_f = 0.5$. The largest stress in the matrix is the longitudinal stress,

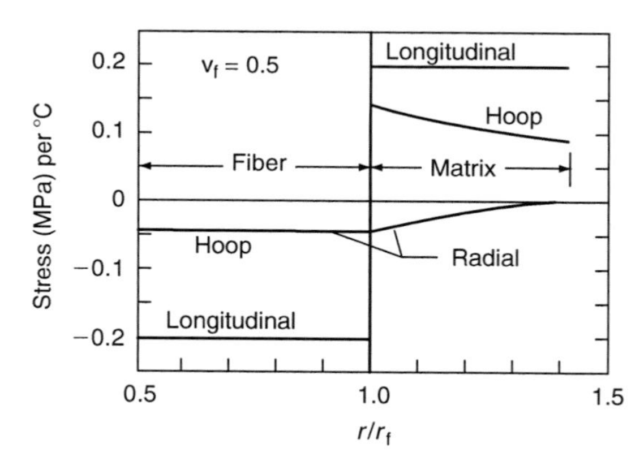

FIGURE A.2.2 Thermal stresses in the fiber and the matrix as a function of radial distance in a 50 vol% AS carbon fiber-reinforced epoxy matrix. (Adapted from Nairn, J.A., *Polym. Compos.*, 6, 123, 1985.)

which is tensile. If it is assumed that the lamina is cured from 177°C to 25°C, the magnitude of this stress will be 29.3 MPa, which is ~25% of the ultimate strength of the matrix. The hoop stress in the matrix is also tensile, while the radial stress is compressive.

REFERENCE

1. J.A. Nairn, Thermoelastic analysis of residual stresses in unidirectional high-performance composites, *Polym. Compos., 6*:123 (1985).

A.3 ALTERNATIVE EQUATIONS FOR THE ELASTIC AND THERMAL PROPERTIES OF A LAMINA

Property	Chamis [1]	Tsai–Hahn [2]
E_{11}	Same as Equation 3.36	Same as Equation 3.36
E_{22}	$\dfrac{E_f E_m}{E_f - \sqrt{v_f}(E_f - E_m)}$	$\dfrac{(v_f + \eta_{22} v_m) E_f E_m}{(E_m v_f + \eta_{22} v_m E_f)}$
G_{12}	$\dfrac{G_f G_m}{G_f - \sqrt{v_f}(G_f - G_m)}$	$\dfrac{(v_f + \eta_{12} v_m) G_f G_m}{(G_m v_f + \eta_{12} v_m G_f)}$
ν_{12}	Same as Equation 3.37	Same as Equation 3.37
α_{11}	Same as Equation 3.58	
α_{22}	$\alpha_{fT}\sqrt{v_f} + (1 - \sqrt{v_f})\left(1 + v_f \nu_m \dfrac{E_f}{E_{11}}\right)\alpha_m$	
K_{11}^*	$K_f v_f + K_m(1 - v_f)$	
K_{22}	$(1 - \sqrt{v_f})K_m + \dfrac{\sqrt{v_f} K_f K_m}{K_f - \sqrt{v_f}(K_f - K_m)}$	

Note: In Tsai–Hahn equations for E_{22} and G_{12}, η_{22} and η_{12} are called stress-partitioning parameters. They can be determined by fitting these equations to respective experimental data. Typical values of η_{22} and η_{12} for epoxy matrix composites are:

	Fiber Type		
	Carbon	Glass	Kevlar-49
η_{22}	0.5	0.516	0.516
η_{12}	0.4	0.316	0.4

* Thermal conductivity.

REFERENCES

1. C.C. Chamis, Simplified composite micromechanics equations for hygral, thermal and mechanical properties, *SAMPE Quarterly, 15*:14 (1984).
2. S.M. Tsai and H.T. Hahn, *Introduction to Composite Materials*, Technomic Publishing Co., Lancaster, PA (1980).

A.4 HALPIN–TSAI EQUATIONS

The Halpin–Tsai equations are simple approximate forms of the generalized self-consistent micromechanics solutions developed by Hill. The modulus values based on these equations agree reasonably well with the experimental values for a variety of reinforcement geometries, including fibers, flakes, and ribbons. A review of their developments is given in Ref. [1].

In the general form, the Halpin–Tsai equations for oriented reinforcements are expressed as

$$\frac{p}{p_m} = \frac{1 + \zeta \eta v_r}{1 - \eta v_r}$$

with

$$\eta = \frac{(p_r/p_m) - 1}{(p_r/p_m) + \zeta}$$

where

p = composite property, such as E_{11}, E_{22}, G_{12}, G_{23}, and ν_{23}
p_r = reinforcement property, such as E_r, G_r, and ν_r
p_m = matrix property, such as E_m, G_m, and ν_m
ζ = a measure of reinforcement geometry, packing geometry, and loading conditions
v_r = reinforcement volume fraction

Reliable estimates for the ζ factor are obtained by comparing the Halpin–Tsai equations with the numerical solutions of the micromechanics equations [2–4]. For example,

$$\zeta = 2\frac{l}{t} + 40v_r^{10} \quad \text{for } E_{11},$$

$$\zeta = 2\frac{w}{t} + 40v_r^{10} \quad \text{for } E_{22},$$

$$\zeta = \left(\frac{w}{t}\right)^{1.732} + 40v_r^{10} \quad \text{for } G_{12},$$

where l, w, and t are the reinforcement length, width, and thickness, respectively. For a circular fiber, $l = l_f$ and $t = w = d_f$, and for a spherical reinforcement, $l = t = w$. The term containing v_r in the expressions for ζ is relatively small up to $v_r = 0.7$ and therefore can be neglected. Note that for oriented continuous fiber-reinforced composites, $\zeta \to \infty$, and substitution of η into the Halpin–Tsai equation for E_{11} gives the same result as obtained by the rule of mixture.

Nielsen [5] proposed the following modification for the Halpin–Tsai equation to include the maximum packing fraction, v_r^*:

$$\frac{p}{p_m} = \frac{1 + \zeta\eta v_r}{1 - \eta\Phi v_r},$$

where $\Phi = 1 + \left(\frac{1 - v_r^*}{v_r^{*2}}\right) v_r$.

Note that the maximum packing fraction, v_r^*, depends on the reinforcement type as well as the arrangement of reinforcements in the composite. In the case of fibrous reinforcements,

1. $v_r^* = 0.785$ if they are arranged in the square array
2. $v_r^* = 0.9065$ if they are arranged in a hexagonal array
3. $v_r^* = 0.82$ if they are arranged in random close packing

REFERENCES

1. J.C. Halpin and J.L. Kardos, The Halpin–Tsai equations: a review, *Polym. Eng. Sci.*, *16*:344 (1976).
2. J.C. Halpin and S.W. Tsai, Environmental factors in composite materials design, U.S. Air Force Materials Laboratory Report, AFML-TR 67-423, June (1969).
3. J.C. Halpin and R.L. Thomas, Ribbon reinforcement of composites, *J. Compos. Mater.*, *2*:488 (1968).
4. J.C. Halpin, *Primer on Composite Materials*: *Analysis*, Chapter 6, Technomic Publishing Co., Lancaster, PA (1984).
5. L.E. Nielsen, *Mechanical Properties of Polymers and Composites*, Vol. 2, Marcel Dekker, New York (1974).

A.5 TYPICAL MECHANICAL PROPERTIES OF UNIDIRECTIONAL CONTINUOUS FIBER COMPOSITES

Property	Boron–Epoxy	AS Carbon–Epoxy	T-300–Epoxy	HMS Carbon–Epoxy	GY-70–Epoxy	Kevlar 49–Epoxy	E-Glass–Epoxy	S-Glass–Epoxy
Density, g/cm^3	1.99	1.54	1.55	1.63	1.69	1.38	1.80	1.82
Tensile properties								
Strength, MPa (ksi) 0°	1585 (230)	1447.5 (210)	1447.5 (210)	827 (120)	586 (85)	1379 (200)	1103 (160)	1214 (176)
90°	62.7 (9.1)	62.0 (9)	44.8 (6.5)	86.2 (12.5)	41.3 (6.0)	28.3 (4.1)	96.5 (14)	—
Modulus GPa (Msi) 0°	207 (30)	127.5 (18.5)	138 (20)	207 (30)	276 (40)	76 (11)	39 (5.7)	43 (6.3)
90°	19 (2.7)	9 (1.3)	10 (1.5)	13.8 (2.0)	8.3 (1.2)	5.5 (0.8)	4.8 (0.7)	—
Major Poisson's ratio	0.21	0.25	0.21	0.20	0.25	0.34	0.30	—
Compressive properties								
Strength, MPa (ksi), 0°	2481.5 (360)	1172 (170)	1447.5 (210)	620 (90)	517 (75)	276 (40)	620 (90)	758 (110)
Modulus, GPa (Msi), 0°	221 (32)	110 (16)	138 (20)	171 (25)	262 (38)	76 (11)	32 (4.6)	41 (6)
Flexural properties								
Strength, MPa (ksi), 0°	—	1551 (225)	1792 (260)	1034 (150)	930 (135)	621 (90)	1137 (165)	1172 (170)
Modulus, GPa (Msi), 0°	—	117 (17)	138 (20)	193 (28)	262 (38)	76 (11)	36.5 (5.3)	41.4 (6)
In-plane shear properties								
Strength, MPa (ksi)	131 (19)	60 (8.7)	62 (9)	72 (10.4)	96.5 (14)	60 (8.7)	83 (12)	83 (12)
Modulus, GPa (Msi)	6.4 (0.93)	5.7 (0.83)	6.5 (0.95)	5.9 (0.85)	4.1 (0.60)	2.1 (0.30)	4.8 (0.70)	—
Interlaminar shear strength, MPa (ksi) 0°	110 (16)	96.5 (14)	96.5 (14)	72 (10.5)	52 (7.5)	48 (7)	69 (10)	72 (10.5)

Source: From Chamis, C.C., *Hybrid and Metal Matrix Composites*, American Institute of Aeronautics and Astronautics, New York, 1977. With permission.

A.6 PROPERTIES OF VARIOUS SMC COMPOSITES

Property	SMC-R25	SMC-R50	SMC-R65	SMC-C20R30	XMC-3
Density, g/cm^3	1.83	1.87	1.82	1.81	1.97
Tensile strength, MPa (ksi)	82.4 (12)	164 (23.8)	227 (32.9)	289 (L) (41.9) 84 (T) (12.2)	561 (L) (81.4) 69.9 (T) (10.1)
Tensile modulus, GPa (Msi)	13.2 (1.9)	15.8 (2.3)	14.8 (2.15)	21.4 (L) (3.1) 12.4 (T) (1.8)	35.7 (L) (5.2) 12.4 (T) (1.8)
Strain-to-failure (%)	1.34	1.73	1.67	1.73 (L) 1.58 (L)	1.66 (T) 1.54 (T)
Poisson's ratio	0.25	0.31	0.26	0.30 (LT) 0.18 (TL)	0.31 (LT) 0.12 (TL)
Compressive strength, MPa (ksi)	183 (26.5)	225 (32.6)	241 (35)	306 (L) (44.4) 166 (T) (24.1)	480 (LT) (69.6) 160 (T) (23.2)
Shear strength, MPa (ksi)	79 (11.5)	62 (9.0)	128 (18.6)	85.4 (12.4)	91.2 (13.2)
Shear modulus, GPa (Msi)	4.48 (0.65)	5.94 (0.86)	5.38 (0.78)	4.09 (0.59)	4.47 (0.65)
Flexural strength, MPa (ksi)	220 (31.9)	314 (45.6)	403 (58.5)	645 (L) (93.6) 165 (T) (23.9)	973 (L) (141.1) 139 (T) (20.2)
Flexural modulus, GPa (Msi)	14.8 (2.15)	14 (2.03)	15.7 (2.28)	25.7 (L) (3.73) 5.9 (T) (0.86)	34.1 (L) (4.95) 6.8 (T) (1.0)
ILSS, MPa (ksi)	30 (4.3)	25 (3.63)	45 (6.53)	41 (5.95)	55 (7.98)
Coefficient of thermal expansion, 10^{-6}/°C	23.2	14.8	13.7	11.3 (L) 24.6 (T)	8.7 (L) 28.7 (T)

Source: From Riegner, D.A. and Sanders, B.A., A characterization study of automotive continuous and random glass fiber composites, *Proceedings of the National Technical Conference*, Society of Plastics Engineers, 1979. With permission.

Note: All SMC composites in this table contain E-glass fibers in a thermosetting polyester resin. XMC-3 contains 50% by weight of continuous strands at ±7.5° to the longitudinal direction and 25% by weight of 25.4 mm (1 in.) long-chopped strands.

A.7 FINITE WIDTH CORRECTION FACTOR FOR ISOTROPIC PLATES

The hole stress concentration factor for an infinitely wide ($w \gg 2R$) isotropic plate is 3. If the plate has a finite width, the hole stress concentration factor (based on gross area) will increase with increasing hole radius. For $w > 8R$, the following equation is used to calculate the hole stress concentration factor of an isotropic plate.

$$\frac{K_T(w)}{K_T(\infty)} = \frac{2 + \left[1 - \left(\frac{2R}{w}\right)\right]^3}{3\left[1 - \left(\frac{2R}{w}\right)\right]},$$

where

$K_T(w)$ = stress concentration factor of a plate of width w
$K_T(\infty)$ = stress concentration factor of an infinitely wide plate
R = hole radius

A.8 DETERMINATION OF DESIGN ALLOWABLES

The statistical analysis for the determination of design allowables depends on the type of distribution used in fitting the experimental data.

A.8.1 NORMAL DISTRIBUTION

If the experimental data are represented by a normal distribution, the A-basis and B-basis design allowables are calculated from the following equations:

$$\sigma_A = \bar{\sigma} - K_A s$$
$$\sigma_B = \bar{\sigma} - K_B s,$$

where

σ_A = A-basis design allowable
σ_B = B-basis design allowable
$\bar{\sigma}$ = mean strength
s = standard deviation
K_A = one-sided tolerance limit factor corresponding to a proportion at least 0.99 of a normal distribution and a confidence coefficient of 0.95
K_B = one-sided tolerance limit factor corresponding to a proportion at least 0.90 of a normal distribution and a confidence coefficient of 0.95

It should be noted that, for a given sample size, K_A is greater than K_B and both K_A and K_B decrease with increasing sample size. Tables of K_A and K_B are given in Ref. [1].

A.8.2 WEIBULL DISTRIBUTION

If the experimental data are represented by a two-parameter Weibull distribution, the A-basis and B-basis design allowables are calculated from the following equation:

$$\sigma_{A,B} = \hat{\sigma}_0 \left[-2n \frac{\ln R}{\chi^2_{(2n,\gamma)}} \right]^{1/\alpha},$$

where

n = sample size
α = Weibull shape parameter
R = 0.99 for the A-basis design allowable and 0.90 for the B-basis design allowable
${\chi_{(2n,\gamma)}}^2$ = value from the χ^2 distribution table corresponding to $2n$ and a confidence limit of 0.95

$$\hat{\sigma}_0 = \left(\frac{1}{n} \sum_{i=1}^{n} \sigma_i^{\alpha} \right)^{1/\alpha}$$

REFERENCE

1. Metallic Materials and Elements for Aerospace Vehicles Structures, MIL-HDBK-5C, U.S. Department of Defense, Washington, D.C., September (1976).

A.9 TYPICAL MECHANICAL PROPERTIES OF METAL MATRIX COMPOSITES

Material	Tensile Strength, MPa (ksi)	Tensile Modulus, GPa (Msi)
6061-T6 aluminum alloy	306 (44.4)	70 (10)
T-300 carbon-6061 Al alloy (v_f = 35%–40%)	1034–1276 (L) (150–185)	110–138 (L) (15.9–20)
Boron-6061 Al alloy (v_f = 60%)	1490 (L) (216)	214 (L) (31)
	138 (T) (20)	138 (T) (20)
Particulate SiC-6061-T6 Al alloy (v_f = 20%)	552 (80)	119.3 (17.3)
GY-70 Carbon-201 Al alloy (v_f = 37.5%)	793 (L) (115)	207 (L) (30)
Al_2O_3-Al alloy (v_f = 60%)	690 (L) (100)	262 (L) (38)
	172–207 (T) (25–30)	152 (T) (22)
Ti-6Al-4V titanium alloy	890 (129)	120 (17.4)
SiC-Ti alloy (v_f = 35%–40%)	820 (L) (119)	225 (L) (32.6)
	380 (T) (55)	—
SCS-6-Ti alloy (v_f = 35%–40%)[a]	1455 (L) (211)	240 (L) (34.8)
	340 (T) (49)	

[a] SCS-6 is a coated SiC fiber.

A.10 USEFUL REFERENCES

A.10.1 Text and Reference Books

1. R.M. Jones, *Mechanics of Composite Materials*, 2nd Ed., Taylor & Francis, Philadelphia, PA (1999).
2. C.T. Herakovich, *Mechanics of Fibrous Composites*, John Wiley & Sons, New York, NY (1998).
3. M.W. Hyer, *Stress Analysis of Fiber-Reinforced Composite Materials*, McGraw-Hill, New York, NY (1998).
4. K.K. Chawla, *Composite Materials*, 2nd Ed., Springer-Verlag, New York, NY (1998).
5. B.D. Agarwal, L.J. Broutman, and K. Chandrasekharan, *Analysis and Performance of Fiber Composites*, 3rd Ed., John Wiley & Sons, New York, NY (2006).
6. I.M. Daniel and O. Ishai, *Engineering Mechanics of Composite Materials*, 2nd Ed., Oxford University Press, London (2005).
7. A.K. Kaw, *Mechanics of Composite Materials*, 2nd Ed., CRC Press, Boca Raton, FL (2006).
8. S.W. Tsai and H.T. Hahn, *Introduction to Composite Materials*, Technomic Publishing Co., Lancaster, PA (1980).
9. J.C. Halpin, *Primer on Composite Materials: Analysis*, Technomic Publishing Co., Lancaster, PA (1984).
10. D. Hull and T.W. Clyne, *An Introduction to Composite Materials*, 2nd Ed., Cambridge University Press, London (1997).
11. M.R. Piggott, *Load Bearing Fibre Composites*, 2nd Ed., Kluwer Academic Press, Dordrecht (2002).
12. K.H.G. Ashbee, *Fundamental Principles of Fiber Reinforced Composites*, Technomic Publishing Co., Lancaster, PA (1989).
13. S.W. Tsai, *Composites Design-1986*, Think Composites, Dayton, OH (1986).
14. J.M. Whitney, I.M. Daniel, and R.B. Pipes, *Experimental Mechanics of Fiber Reinforced Composite Materials*, Society for Experimental Mechanics, Brookfield Center, CT (1984).
15. D.F. Adams, L.A. Carlsson, and R.B. Pipes, *Experimental Characterization of Advanced Composite Materials*, 3rd Ed., CRC Press, Boca Raton, FL (2003).
16. J.R. Vinson and R.L. Sierakowski, *The Behavior of Structures Composed of Composite Materials*, 2nd Ed., Kluwer Academic Publishers, Dordrecht (2002).
17. J.M. Whitney, *Structural Analysis of Laminated Anisotropic Plates*, Technomic Publishing Co., Lancaster, PA (1987).
18. P.K. Mallick and S. Newman (eds.), *Composite Materials Technology: Processes and Properties*, Hanser Publishers, Munich (1990).
19. T.G. Gutowski (ed.), *Advanced Composites Manufacturing*, John Wiley & Sons, New York (1997).
20. S.G. Advani and E.M. Suzer, *Process Modeling in Composites Manufacturing*, Marcel Dekker, New York (2003).
21. G. Lubin (ed.), *Handbook of Composites*, Van Nostrand-Reinhold Co., New York (1982).
22. P.K. Mallick (ed.), *Composites Engineering Handbook*, Marcel Dekker, New York (1997).

A.10.2 Leading Journals on Composite Materials

1. *Journal of Composite Materials*, Sage Publications (www.sagepub.com)
2. *Journal of Reinforced Plastics and Composites*, Sage Publications (www.sagepub.com)
3. *Journal of Thermoplastic Composites*, Sage Publications (www.sagepub.com)
4. *Composites Part A: Applied Science and Manufacturing,* Elsevier (www.elsevier.com)
5. *Composites Part B: Engineering*, Elsevier (www.elsevier.com)
6. *Composites Science and Technology*, Elsevier (www.elsevier.com)
7. *Composite Structures*, Elsevier (www.elsevier.com)
8. *Polymer Composites*, Wiley Interscience (www.interscience.com)
9. *SAMPE Journal*, Society for the Advancement of Material and Process Engineering, Covina, CA

A.10.3 Professional Societies Associated with Conferences and Publications on Composite Materials

1. American Society for Composites (ASC) (www.asc-composites.org)
2. Society for the Advancement of Material and Process Engineering (SAMPE) (www.sampe.org)
3. Society of Plastics Engineers (SPE) (www.4spe.org)
4. American Society for Testing and Materials (ASTM International) (www.astm.org)
5. American Society for Aeronautics and Astronautics (AIAA) (www.aiaa.org)

A.11 LIST OF SELECTED COMPUTER PROGRAMS

Program Name	Program Description	Source
LAMINATOR	Analyzes laminated plates according to classical laminated plate theory. Calculates apparent laminate material properties, ply stiffness and compliance matrices, laminate "ABD" matrices, laminate loads and midplane strains, ply stresses and strains in global and material axes, and load factors for ply failure based on maximum stress, maximum strain, Tsai–Hill, Hoffman, and Tsai–Wu failure theories	1
ESAComp 3.4	Analysis capabilities include fiber–matrix micromechanics, classical laminate theory-based constitutive and thermal analysis of solid and sandwich laminates, first ply failure and laminate failure prediction, notched laminate analysis, probabilistic analysis, load response and failure of plates, stiffened panels, beams and columns, bonded and mechanical joint analysis in laminates. Has import–export interfaces to common finite element packages	2
SYSPLY	Capable of doing stress analysis, buckling analysis, thermomechanical analysis, large displacement and contact analysis of shell structures, and dynamic analysis of composite structures	3
LUSAS Composite	A finite element software capable of performing linear and nonlinear analysis, impact and contact analysis, and dynamic analysis of composite structures	4
GENOA	An integrated stand-alone structural analysis–design software, which uses micro- and macromechanics analysis of composite structures, finite element analysis, and damage evaluation methods. Capable of performing progressive fracture analysis under static, fatigue (including random fatigue), creep, and impact loads	5
MSC-NASTRAN	General purpose finite element analysis package; also performs static, dynamic, and buckling analysis of laminated composite structures; in addition to the in-plane stresses, computes the interlaminar-shear stresses between various laminas	6
ABAQUS	General purpose finite element package with capability of performing both linear and nonlinear analysis	7
LS-DYNA	General purpose finite element package with the capability of performing nonlinear analysis	8
ANSYS	General purpose finite element package with the capability of performing both linear and nonlinear analysis	9

Source:

1. www.thelaminator.net
2. www.componeering.com
3. www.esi-group.com
4. www.lusas.com
5. www.alphastarcorp.com
6. www.mscnastran.com
7. www.simulia.com
8. www.lsdyna.com
9. www.ansys.com

Index

D

E

F

H

I

J

K

M

N

O

P

Q

R

S

T

U

V

W

Y

Z